Functional Analysis

Related Pergamon Titles of Interest

Books

CONSTANTINESCU:
Distributions and their Applications in Physics

COOPER & COOPER:
Introduction to Dynamic Programming

KURATOWSKI:
A Half Century of Polish Mathematics

LAKSHMIKANTHAM & LEELA:
An Introduction to Nonlinear Differential Equations in Space

SINAI:
Rigorous Results in the Theory of Phase Transitions

Journals

Analysis Mathematica

Problems of Control and Information Theory

Functional Analysis

by

L. V. KANTOROVICH

Nobel Prize Winner
Member of the Academy of Sciences
of the USSR

and

G. P. AKILOV

Translated by
HOWARD L. SILCOCK

SECOND EDITION

PERGAMON PRESS

OXFORD · NEW YORK · TORONTO · SYDNEY · PARIS · FRANKFURT

U.K.	Pergamon Press Ltd., Headington Hill Hall, Oxford OX3 0BW, England
U.S.A.	Pergamon Press Inc., Maxwell House, Fairview Park, Elmsford, New York 10523, U.S.A.
CANADA	Pergamon Press Canada Ltd., Suite 104, 150 Consumers Rd., Willowdale, Ontario M2J 1P9, Canada
AUSTRALIA	Pergamon Press (Aust.) Pty. Ltd., P.O. Box 544, Potts Point, N.S.W. 2011, Australia
FRANCE	Pergamon Press SARL, 24 rue des Ecoles, 75240 Paris, Cedex 05, France
FEDERAL REPUBLIC OF GERMANY	Pergamon Press GmbH, 6242 Kronberg-Taunus, Hammerweg 6, Federal Republic of Germany

First edition published by "Nauka" Publishers, 1959

Second edition 1982

Library of Congress Cataloging in Publication Data

Kantorovich, Leonid Vital'evich.
Functional analysis.
Translation of Funktsional'nyi analiz.
Bibliography: p.
Includes index.
1. Functional analysis. I. Akilov, Gleb Pavlovich, joint author. II. Title.
QA320.K283 1982 515.7 80-21734

British Library Cataloguing in Publication Data

Kantorovich, Leonid Vitalevich
Functional analysis.–2nd ed.
1. Functional analysis
I. Title II. Akilov, G P
515.7 QA320 80–41734

ISBN 0–08–023036–9 Hardcover
ISBN 0–08–026486–7 Flexicover

Printed in Great Britain by A. Wheaton & Co. Ltd., Exeter

Dedicated to the memory of our dear teachers,
Grigorii Mikhailovich Fikhtengol'ts and
Vladimir Ivanovich Smirnov

CONTENTS

PREFACE TO THE SECOND EDITION

TWENTY-ODD years have passed since the appearance of the first edition of this book under the title *Functional Analysis in Normed Spaces*. In that time radical changes have taken place, both within mathematics itself and to its status in the system of modern scientific ideas. One important aspect of these changes concerns the place of functional analysis within the mathematical disciplines. Whereas functional analysis was seen, at the appearance of the first edition, as a comparatively new and promising part of analysis, nowadays the term "functional analysis" is used almost interchangeably with "mathematical analysis". What is more, functional analysis now provides a common language for all areas of mathematics involving the concept of continuity. No serious investigation in the theory of functions, differential equations or mathematical physics, in numerical methods, mathematical economics or control theory, or in numerous other fields, takes place—or could take place—without extensive use of the language and results of functional analysis. It is precisely this fact that explains, on the one hand, the rapid development of functional analysis as a mathematical discipline, and, on the other hand, the ever-increasing role played by its techniques in applications.

The authors note these changes with both pride and anxiety. The pride is a natural manifestation of a sense of participation in significant historical events. The anxiety, however, is provoked by imagining the fate of the reader that we envisage for this book, for it is, in fact, now no longer possible to produce a comprehensive textbook of functional analysis (even at an introductory level). Consequently, although we have made significant revisions in preparing the present edition, we thought it expedient to retain the overall plan and, to a large extent, the selection and arrangement of topics adopted for the first edition. However, there are a number of topics for which the account has been substantially changed, particularly in the theory of topological vector spaces and the theory of integral operators. Whereas formerly the account was based on the theory of normed spaces, and topological vector spaces were covered separately (although fairly fully) as optional material, in this edition we have taken the theory of topological vector spaces as the basis of our exposition, in conformity with the logical development of functional analysis: hence the change in the book's title. We have added a chapter dealing with the elements of the theory of partially ordered spaces. Our development of the theory of integral operators and their representations is based on ideal spaces of measurable functions.

As before, most of the book is devoted to the applications of functional analysis to applied analysis, which were a distinctive feature of the first edition. The presence of these sections in the book stimulated the development of the relevant topics, both in the USSR and abroad. In the present edition the account of these has been somewhat extended and modernized. Another essential feature of this edition is the inclusion of some topics of functional analysis connected with applications to mathematical economics and control theory, although we have been unable to give these the space they deserve. Some less topical material has been excluded. The bibliography has been substantially changed.

Chapter I is introductory. In it we present the elements of the theory of topological spaces, the theory of metric spaces, and the theory of abstract measure spaces. Here many results are stated without proofs. The reader is assumed to be familiar with the theory of functions of a real variable and the topology of n-dimensional Euclidean space, roughly to the level of a general university course in mathematical analysis. Subtler and more specialized material, in both the theory and the applications, are marked off by being set in small type, and may be omitted at a first reading. The reader particularly interested in applications of functional analysis may also omit a number of other more abstract sections dealing with topological spaces and topological vector spaces, or—if he is already familiar with the basic ideas of the theory of normed spaces—he may turn immediately to the relevant chapters on applications.

A number of people were of great help in the work of preparing the present edition. First of all we must mention A. V. Bukhvalov, to whom credit is due, not only for editing the entire text of the book, but also for rewriting Chapter X on ordered normed spaces and the related material in § 3 of Chapter IV, § 1 of Chapter VI, § 1 of Chapter XI, and certain other sections. He also substantially revised the exposition in Chapters I–IV; thus in parts of the book he acted, to all intents and purposes, as a co-author.

V. F. Dem'yanov and A. M. Rubinov made some corrections in the exposition in Chapter XV of the method of steepest descent and added the new §§4 and 5. I. K. Daugavet made substantial additions and corrections in Chapters XIV, XV and XVIII. V. P. Il'in made some significant improvements in the exposition of §§3 and 4 of Chapter XI, particularly in the proofs of Lemma 2 and Theorem 1 in § 3 of Chapter XI. G. Sh. Rubinshtein rewrote § 8 of Chapter IX, basing it on work of L.V. Kantorovich and G. Sh. Rubinshtein; some additions to this were made by V. L. Levin. The reviewer of the book, Professor B. Z. Vulikh, made some valuable remarks, as did Yu. A. Abramovich, A. M. Vershik, S. V. Kislyakov, S. S. Kutateladze, G. Ya. Lozanovskii, A. A. Mekler, B. T. Polyak and V. P. Khavin.

The authors express sincere thanks to all those mentioned above, and also to those who helped in reading the manuscript and the proofs.

The bibliography consists of two parts: a list of monographs on functional analysis and related topics, and a list of literature cited, comprising mostly journal articles. A reference of the form Vulikh-II indicates the monograph by B. Z. Vulikh occurring under the author's surname with the number II in the list of monographs, while a reference such as Levin [3] indicates a paper by V. L. Levin in the list of literature cited. The bibliography makes no pretence at completeness. In cases where a result has already appeared in a published monograph we have, as a rule, preferred to give a reference to the book rather than to the original paper.

The book consists of eighteen chapters, each of which is divided into sections and subsections. A reference such as IX.4.2 indicates § 4, subsection 2 of Chapter IX—however, in cross-references within a chapter, the chapter number is omitted. Theorems are numbered consecutively throughout each section, Theorem XIV.3.2 being the second theorem in § 3 of Chapter XIV—though within that section it would be referred to simply as Theorem 2.

L. V. Kantorovich, G. P. Akilov

FROM THE PREFACE TO THE FIRST EDITION

FUNCTIONAL analysis is a scientific discipline of comparatively recent origin. It has taken shape as an independent branch of mathematical analysis only in the last 20–30 years—though this has not prevented it from occupying one of the central positions in contemporary mathematics.

Functional analysis is a most brilliant manifestation of the radical change that is taking place in mathematics at present—a change comparable in importance to the one that occurred when (in the seventeenth century) variables were introduced into mathematics, leading to the development of the differential and integral calculus.

This change has been expressed above all in a new approach to the study of various problems of mathematical analysis. The investigation of individual functions and the relations and equations connecting them has been replaced by a collective investigation—that is, an investigation of spaces of functions and their transformations (functional operators). Thus a differential operator or an integral transform is regarded as applying not to a single function but to a whole class of functions—and one studies the effect of a transformation on a class of functions, and the continuity, in one sense or another, of the operator, and so on.

Another important feature of functional analysis is the general abstract approach to the study of problems of analysis, which makes it possible to combine and subject to a single investigation problems that at first sight appear quite diverse. For example, a study of the functional equation $F(x) = y$, where x and y are objects of a more or less arbitrary domain, makes it possible to bring together such diverse problems as the solution of differential equations, integral equations, boundary-value problems and infinite systems of algebraic equations, and the problem of moments. The transition from individual functions to spaces of functions, though it may at times even be hard to perceive, is nevertheless as important in principle as was the transition from algebraic equations and relations to variables and functional dependence.

This new point of view did not arise from a simple striving to generalize. New problems arising in the development of analysis naturally called for a transition to a new level of abstraction. These included the problem of completeness for a system of functions, the solubility of boundary value problems within a given class of functions, and simultaneous investigations of whole classes of problems, as for instance in the study of the dependence of the solution of a boundary-value problem on the right-hand side of the equation or on the boundary conditions. It was precisely in the formulation and investigation of such problems that the methods of functional analysis proved to be particularly fruitful. Moreover, in many cases it was, remarkably, the generality of the approach that allowed more general, and at the same time deeper and more concrete, regularities and connections to be revealed, because insignificant details of individual problems were brushed aside and no longer obscured the essence of the matter. It is in this way that the relationship between problems of different forms and origins becomes clearer.

The way was paved for the creation of functional analysis by investigations in several areas of classical mathematical analysis—the calculus of variations, integral equations, the theory of orthogonal functions, the approximation theory of Chebyshev, the problem of moments—all of which naturally called for a new approach. In fact the individual problems of functional analysis arose from within these areas—as, for example, with the concept of the functional in the calculus of variations. On the other hand, the development of the set-theoretical disciplines—the theory of functions of a real variable, topology, abstract algebra—prepared techniques for a systematic development, in abstract form, of the new trend. In particular, the theory of abstract spaces was extremely significant for functional analysis.

Functional analysis can be reckoned to have first come into independent existence with the systematic construction (by D. Hilbert and others) of the theory of operators on infinite-dimensional unitary spaces, and the development (in 1918–1923) of a general theory of normed linear spaces in the work of the Hungarian mathematician F. Riesz and, more especially, of the Polish mathematician S. Banach.

Interest in functional analysis intensified further when it turned out that its techniques (the theory of operators in Hilbert space, etc.) had important applications in quantum mechanics. The last 20 years have seen the appearance of new trends in functional analysis, particularly in the work of Soviet mathematicians; its methods and results have had most important applications in theoretical physics, mathematical physics, applied analysis, and other areas of mathematics.

This book does not claim to embrace all the developing trends and applications of functional analysis. It is primarily devoted to the theory of normed spaces, and includes the most important facts of this theory, as originally developed by Riesz and Banach, while also taking into account some later work. Its subject matter is the theory of normed spaces, the theory of operators and the theory of functional equations. Considerable attention is paid to non-linear as well as linear operators and equations, and specific function spaces and operators are discussed at length. In particular, there is a detailed treatment of the spaces of differentiable functions of several variables introduced by S. L. Sobolev. These topics are related to a general investigation of integral operators.

The book is based on a course of lectures given at Leningrad University for students specializing in mathematical analysis and computational mathematics.

PART I

LINEAR OPERATORS AND FUNCTIONALS

I

TOPOLOGICAL AND METRIC SPACES

IN MATHEMATICS an important role is accorded to the concept of a space—that is, a set between whose elements certain relations are prescribed by means of axioms. The set is said, in this situation, to have been endowed with the structure of the relevant space. In this chapter the basic subjects of investigation will be topological and metric spaces: these are sets for whose elements a concept of closeness has been postulated. Topological spaces were introduced in 1910 by Hausdorff (see Hausdorff) and metric spaces a little earlier by Fréchet [1].

§1. General information on sets. Ordered sets

1.1. In this subsection we recall some elementary concepts and notation from the general theory of sets.* In this context we maintain an informal point of view, taking the concept of *set*, or *collection*, to be intuitively clear and not in need of precise definition. By the elements of a set we mean the objects of which it is composed.

We denote the set of all natural numbers by $\mathbb{N}$, the set of all real numbers by $\mathbb{R}$, and the set of all complex numbers by $\mathbb{C}$.

Let A and B be sets. The notation $a \in A$ indicates that the object a belongs to the set A; the notation $a \notin A$ that a does not belong to A. We say that A is a *subset* of a set B, and write $A \subset B$ (or $B \supset A$), if every element of A also belongs to B. If $A \subset B$ and $B \subset A$, then we say that A and B are *equal*, and write $A = B$. The *empty* set (that is, the set containing no elements) is denoted by the symbol $\varnothing$. If (P) is any statement pertaining to elements of a set A, then the subset consisting of all $a \in A$ for which (P) is satisfied is denoted by $\{a \in A : (P)a\}$, or briefly $\{a : (P)a\}$.

Let A and B be sets. If with each element $a \in A$ there is associated, by some definite rule, a unique element $f(a) \in B$, then a *mapping* f from A to B is said to be defined, and we write $f: A \to B$. For each $X \subset A$, the *image* $f(X)$ of X is defined to be the set $\{b \in B : \text{there exists } a \in X \text{ with } b = f(a)\}$. For each $Y \subset B$, the *inverse image* (or *preimage*) $f^{-1}(Y)$ of Y is defined to be the set $\{a \in A : f(a) \in Y\}$. A mapping $f: A \to B$ is said to be *one-to-one* if $f(a_1) = f(a_2)$ implies $a_1 = a_2$. It is said to be a mapping *onto* B if $f(A) = B$. A mapping $f: A \to B$ is said to be a *bijection* if it is both one-to-one and "onto". If f is a bijection, then the mapping g defined by $g(f(a)) = a$, $a \in A$, is called the *inverse mapping* to f, and is denoted by f^{-1}.

We now give the definitions of the basic set-theoretical operations. If with each element α of some non-empty index set A, there is associated a set X_α, then we say that a *family* of

* The foundations of the general theory of sets were laid by the German mathematician G. Cantor in the latter half of the nineteenth century.

sets $\{X_\alpha\}$ $(\alpha \in \mathrm{A})$ has been defined. The *union* of the family of sets $\{X_\alpha\}$ is the set $\bigcup_{\alpha \in \mathrm{A}} X_\alpha$ consisting of all objects x such that $x \in X_\alpha$ for at least one $\alpha \in \mathrm{A}$. The *intersection* of the family of sets $\{X_\alpha\}$ is the set $\bigcap_{\alpha \in \mathrm{A}} X_\alpha$ consisting of all objects x such that $x \in X_\alpha$ for all $\alpha \in \mathrm{A}$. The *direct* (or *Cartesian*) *product* of the family of sets $\{X_\alpha\}$ is the set $\prod_{\alpha \in \mathrm{A}} X_\alpha$ consisting of all mappings f: $\mathrm{A} \to \bigcup_{\alpha \in \mathrm{A}} X_\alpha$ with $f(\alpha) \in X_\alpha$ for all $\alpha \in \mathrm{A}$. If A consists of the numbers 1, 2, . . . , n, then the following notations are used for the union, intersection and direct product, respectively:

$$\bigcup_{k=1}^{n} X_k = X_1 \cup X_2 \cup \ldots \cup X_n, \qquad \bigcap_{k=1}^{n} X_k = X_1 \cap X_2 \cap \ldots \cap X_n,$$

$$\prod_{k=1}^{n} X_k = X_1 \times X_2 \times \ldots \times X_n.$$

Note that a finite product $\prod_{k=1}^{n} X_k$ may be identified with the set of all ordered n-tuples $(x_1, x_2, \ldots, x_n)$, where $x_k \in X_k$.

If A is the set of all natural numbers $\mathbb{N}$, then the following notations are used:

$$\bigcup_{k=1}^{\infty} X_k, \qquad \bigcap_{k=1}^{\infty} X_k, \qquad \prod_{k=1}^{\infty} X_k.$$

Two sets A and B are said to be *disjoint* (or non-intersecting) if $A \cap B = \varnothing$. The sets X_α $(\alpha \in \mathrm{A})$ are *pairwise disjoint* if $X_{\alpha_1} \cap X_{\alpha_2} = \varnothing$ whenever $\alpha_1 \neq \alpha_2$. A *partition* of a set T is a family of pairwise disjoint sets $\{X_\alpha\}$ $(\alpha \in \mathrm{A})$ such that $T = \bigcup_{\alpha \in \mathrm{A}} X_\alpha$.

The *difference set* $A \setminus B$ is the subset of A consisting of all $a \in A$ such that $a \notin B$. Here, in defining $A \setminus B$, we do not assume that $B \subset A$. If we do have $B \subset A$, then $A \setminus B$ is also called the *complement* of B (with respect to A). The *symmetric difference* of sets A and B is the set

$$A \Delta B = (A \setminus B) \cup (B \setminus A) = (A \cup B) \setminus (A \cap B).$$

If $A \subset X$, then the *characteristic function* χ_A of A is defined by the formula

$$\chi_A(x) = \begin{cases} 1 & \text{if} \quad x \in A, \\ 0 & \text{if} \quad x \in X \setminus A. \end{cases}$$

1.2. A set X is said to be *ordered** if an *ordering relation* $x \geqslant y$ (x is greater than or equal to y) is defined for certain pairs of elements of X, such that the following conditions are satisfied:

1) $x \geqslant x$ for all $x \in X$;
2) if $x \geqslant y$ and $y \geqslant z$, then $x \geqslant z$;
3) if $x \geqslant y$ and $y \geqslant x$, then $x = y$.

The notation $x \leqslant y$ means that $y \geqslant x$; also $x > y$ means that $x \geqslant y$ and $x \neq y$; and $x \geqslant y, z$ means that $x \geqslant y$ and $x \geqslant z$. The set of all real numbers $\mathbb{R}$ provides an

* The term *partially ordered* set is often used to emphasize that the relation "greater than or equal to" need not connect every pair of elements.

example of an ordered set in which every two elements are connected by the relation $\leqslant$ (i.e. every two elements are comparable). For an example of an ordered set containing incomparable elements, we mention the set of all subsets of the set of natural numbers $\mathbb{N}$, where the ordering is inclusion of sets (i.e. $A \geqslant B \Leftrightarrow A \supset B$).*

Let X be an ordered set. A subset $A \subset X$ is said to be *bounded above* if there exists an $x \in X$ such that $a \leqslant x$ for all $a \in A$; then x is called an *upper bound* for A. The terms *bounded below* and *lower bound* are defined analogously. A set A is said to be *bounded* if it is both bounded above and bounded below. If $A \subset X$ then an element $x \in A$ is called: (1) a *greatest* element (of A) if $x \geqslant a$ for all $a \in A$; (2) *maximal* (in A) if the statement $x \leqslant a$, for $a \in A$, implies $x = a$. Note that any greatest element is also maximal, though the converse is, in general, false. The definitions of *least* element and *minimal* are analogous. If $A \subset X$ is bounded above, then the least upper bound of A (if one exists) is called the *supremum* of A and is written sup A. If $A \subset X$ is bounded below, then the greatest lower bound of A (if one exists) is called the *infimum* of A and is written inf A. If the elements of A are given by means of an index set, say $A = \{x_\beta\}$ $(\beta \in \mathrm{B})$, then instead of sup A and inf A one writes, respectively:

$$\sup_{\beta \in \mathrm{B}} x_\beta \text{ or } \sup x_\beta \quad \text{and} \quad \inf_{\beta \in \mathrm{B}} x_\beta \text{ or } \inf x_\beta .$$

If A consists of a finite number of elements $x_1, x_2, \ldots, x_n$, then instead of sup A and inf A one writes $\sup\limits_{k=1}^{n} x_k$ or $x_1 \vee x_2 \vee \ldots \vee x_n$ and $\inf\limits_{k=1}^{n} x_k$ or $x_1 \wedge x_2 \wedge \ldots \wedge x_n$, respectively.

Notice that in the second example at the beginning of this subsection a supremum is clearly a set-theoretical union and an infimum is an intersection.

An ordered set X is said to be *totally ordered* (or *linearly ordered* or a *chain*) if for any elements x and y in X we have $x \geqslant y$ or $x \leqslant y$; that is, if all its elements are mutually comparable.

The following proposition, which is equivalent to the axiom of choice and to the principle of transfinite induction, is useful for various problems; a proof may be found in Dunford and Schwartz-I (Theorem I.2.7) or Kelley.

ZORN'S LEMMA. *If every totally ordered subset of an ordered set X is bounded above, then X has a maximal element.*

In our study of topological spaces, and later of topological vector spaces, the concept of a net will be important. Let X be an arbitrary set and let A be an ordered set which is *directed*: that is, for any $\alpha_1, \alpha_2 \in \mathrm{A}$ there exists an $\alpha \in \mathrm{A}$ such that $\alpha \geqslant \alpha_1$ and $\alpha \geqslant \alpha_2$. A mapping $\alpha \to x_\alpha$ from A into X is called a *net* (or sometimes a *generalized sequence*) and is written $\{x_\alpha\}$ $(\alpha \in \mathrm{A})$, or simply $\{x_\alpha\}$. If $\mathrm{A} = \mathbb{N}$, where $\mathbb{N}$ is the set of natural numbers with the usual ordering, then $\{x_n\}$ $(n \in \mathbb{N})$ is an ordinary sequence. In future, when we wish to indicate an arbitrary net, we shall use Greek letters for the suffices—for instance, $\{x_\alpha\}$, $\{x_\beta\}$, $\{x_\gamma\}$—and, when we wish to indicate a sequence, we shall use Roman letters—$\{x_n\}$, $\{x_m\}$, $\{x_k\}$.

A non-trivial example of a net is the set of all finite subsets of $\mathbb{N}$, ordered by inclusion.

---

* The symbol $\Leftrightarrow$ means "is equivalent to", while the symbol $\Rightarrow$ means "implies".

For nets we have the following generalization of the concept of a subsequence. A net $\{y_\beta\}$ $(\beta \in \mathrm{B})$ is said to be a *subnet* of a net $\{x_\alpha\}$ $(\alpha \in \mathrm{A})$ if for any $\alpha \in \mathrm{A}$ there exists a $\beta(\alpha) \in \mathrm{B}$ such that, whenever $\beta' \in \mathrm{B}$ and $\beta' \geqslant \beta(\alpha)$, we have $x_{\alpha'} = y_{\beta'}$ for some $\alpha' \geqslant \alpha$. Notice that a sequence always has subnets that are not themselves sequences.

We have already come across various examples of ordered sets in this chapter. The material covered in this section may be studied in more detail in the monographs by Bourbaki-I, and Kelley; elementary introductions may be found in the textbooks by Vulikh-III, Kolmogorov and Fomin, Natanson-II. For a detailed study of various classes of ordered sets see the monograph by Birkhoff.

§ 2. Topological spaces

2.1. There are various ways of turning a set into a topological space. One of the most convenient and widely used ways is to specify the collection of open sets of the given space.

A set $\mathbf{X}$ is called a *topological space* if a system $\mathfrak{G}$ of subsets, called the *open* sets, is singled out in $\mathbf{X}$, subject to the following three conditions (the axioms for a topological space):

1) the empty set $\varnothing$ and the whole set $\mathbf{X}$ belong to $\mathfrak{G}$;

2) if $G_\xi \in \mathfrak{G}$ $(\xi \in \Xi)$, then $\bigcup_{\xi \in \Xi} G_\xi \in \mathfrak{G}$; i.e., the union of any number of open sets is open;

3) if $G_1, G_2 \in \mathfrak{G}$, then $G_1 \cap G_2 \in \mathfrak{G}$; i.e., the intersection of a finite number of open sets is open.

If a set $\mathbf{X}$ is turned into a topological space, then we say that $\mathbf{X}$ has been endowed with a *topology*.

Two topological spaces $\mathbf{X}_1$ and $\mathbf{X}_2$ are said to be *homeomorphic* if it is possible to establish a one-to-one correspondence between their elements such that the open sets in $\mathbf{X}_1$ and $\mathbf{X}_2$ correspond to one another. From the point of view of the theory of topological spaces, homeomorphic spaces can clearly be identified with one another.

Suppose that a topology has been introduced on a set $\mathbf{X}$ in two (not necessarily distinct) ways, giving rise to two topological spaces $\mathbf{X}_1$ and $\mathbf{X}_2$ (which coincide in the elements of which they are composed). Let us denote the systems of open sets of the spaces $\mathbf{X}_1$ and $\mathbf{X}_2$ by $\mathfrak{G}_1$ and $\mathfrak{G}_2$, respectively. The topology of $\mathbf{X}_1$ is said to be *stronger* than that of $\mathbf{X}_2$ (or the topology of $\mathbf{X}_2$ *weaker* than that of $\mathbf{X}_1$) if $\mathfrak{G}_1 \supset \mathfrak{G}_2$. In this situation we write $\tau(\mathbf{X}_1) \geqslant \tau(\mathbf{X}_2)$ (or $\tau(\mathbf{X}_2) \leqslant \tau(\mathbf{X}_1)$).

Let $\mathbf{X}$ be a topological space and $\mathfrak{G}$ its system of open sets. Also let $\mathbf{X}_0$ be an arbitrary subset of $\mathbf{X}$. It is easy to verify that the system $\mathfrak{G}_0$ consisting of the sets of the form $G \cap \mathbf{X}_0$ $(G \in \mathfrak{G})$ satisfies the axioms for a topological space (relative to the set $\mathbf{X}_0$), so that $\mathbf{X}_0$ is turned into a topological space. The topology of $\mathbf{X}_0$ is said to be induced by that of $\mathbf{X}$, and $\mathbf{X}_0$ is said to be a *subspace* of $\mathbf{X}$.

2.2. A set F in a topological space $\mathbf{X}$ is said to be *closed* if the set $G = \mathbf{X} \setminus F$ is open. The system $\mathfrak{F}$ of closed sets of the space $\mathbf{X}$ has the following properties:

1) $\varnothing, \mathbf{X} \in \mathfrak{F}$;

2) if $F_\xi \in \mathfrak{F}$ $(\xi \in \Xi)$, then $\bigcap_{\xi \in \Xi} F_\xi \in \mathfrak{F}$; i.e., the intersection of any number of closed sets is closed;

3) if $F_1, F_2 \in \mathfrak{F}$, then $F_1 \cup F_2 \in \mathfrak{F}$; i.e., the union of a finite number of closed sets is closed.

Since the system $\mathfrak{F}$ uniquely determines the system $\mathfrak{G}$ of open sets, it is possible to introduce a topology on $\mathbf{X}$ by specifying the system $\mathfrak{F}$ at the outset, subject to the three conditions above, and declaring the sets in $\mathfrak{F}$ to be closed. The open sets are then defined to be the complements of the closed sets.

If $\mathbf{X}_0 \subset \mathbf{X}$ is a closed set, and we take the topology of $\mathbf{X}_0$ to be that induced by the topology of $\mathbf{X}$, then we see that the system $\mathfrak{F}_0$ of closed sets of the space $\mathbf{X}_0$ consists of those closed sets of $\mathbf{X}$ that lie entirely within $\mathbf{X}_0$.

2.3. Let $\mathbf{X}$ be a topological space. A point $x \in \mathbf{X}$ is called an *interior* point of a set $E \subset \mathbf{X}$ if there exists an open set G in $\mathbf{X}$ such that $x \in G \subset E$. A *neighbourhood* of a point $x \in \mathbf{X}$ is a set $V \subset \mathbf{X}$ of which x is an interior point. A system $\mathfrak{B}_x$ of neighbourhoods of x is called a *fundamental system*, or a *basis of neighbourhoods* of x, if for any neighbourhood V of x there is a neighbourhood $V_x \in \mathfrak{B}_x$ such that $V_x \subset V$. The collection $\mathfrak{B}$ of bases $\mathfrak{B}_x$ for all possible points x of a space is called a *basis for the space.* A basis for a space has the following properties:

1) If $V \in \mathfrak{B}_x$, then $x \in V$.

2) If $V_1, V_2 \in \mathfrak{B}_x$, then there exists $V \in \mathfrak{B}_x$ such that $V \subset V_1 \cap V_2$.

3) For any neighbourhood $V_x \in \mathfrak{B}_x$, there exists $V'_x \in \mathfrak{B}_x$ such that $V'_x \subset V_x$ and such that, for any $y \in V'_x$, there exists $V_y \in \mathfrak{B}_y$ with $V_y \subset V_x$.

Let us explain this last property. Because V_x is a neighbourhood of x, there exists an open set $G \subset V_x$ containing x. The set G is a neighbourhood of x; therefore there exists $V'_x \in \mathfrak{B}_x$ such that $V'_x \subset G$ and so certainly $V'_x \subset V_x$. If $y \in V'_x$, then $y \in G$; and G, being open, is a neighbourhood of y, whence we can find $V_y \in \mathfrak{B}_y$ with $V_y \subset G$. Then, *a fortiori*, we have $V_y \subset V_x$.

Properties 1)–3) characterize a basis for a space. In fact, we have:

THEOREM 1. *Assume that with each point x of a set $\mathbf{X}$ there is associated a system $\mathfrak{B}_x$ of subsets of $\mathbf{X}$ satisfying conditions 1)–3). Call a set $G \subset \mathbf{X}$ open if, for each $x \in G$, there exists $V \subset G$ with $V \in \mathfrak{B}_x$. Then the system of open sets satisfies the axioms for a topology, and each $\mathfrak{B}_x$ is a basis of neighbourhoods of x in the resulting topological space.*

Proof. It follows in an obvious way from conditions 1)–2) that the axioms for a topological space are satisfied. Let us verify that for any $x \in \mathbf{X}$ the system $\mathfrak{B}_x$ is a fundamental system of neighbourhoods of x. Let $V \in \mathfrak{B}_x$. Form the set G consisting of all points $y \in V$ for which there exists $V_y \in \mathfrak{B}_y$ with $V_y \subset V$. We show that G is an open set. Choose any point $z \in G$. There exists a set $V_z \in \mathfrak{B}_z$ such that $V_z \subset V$. By condition 3) we can find $V'_z \in \mathfrak{B}_z$ such that $V'_z \subset V_z$ and such that for any $y \in V'_z$ there exists $V_y \subset V_z$ with $V_y \in \mathfrak{B}_y$. This shows that $y \in G$, and so $V'_z \subset G$; that is, G is an open set, and V is therefore a neighbourhood of x. It remains to verify that $\mathfrak{B}_x$ is a fundamental system of neighbourhoods. Let U be any neighbourhood of x. There exists an open set $G \subset U$ containing x. Consequently, by definition, there exists $V_x \in \mathfrak{B}_x$ such that $V_x \subset G$ and hence $V_x \subset U$.

The theorem we have just proved enables one to introduce a topology by specifying a fundamental system of neighbourhoods at each point of the space. This method of introducing a topology turns out, as a rule, to be very convenient, since it is usually possible to choose sets of a relatively simple structure as neighbourhoods. For example, the topology of Euclidean space is given using the set of all balls (which, it is easy to verify, does yield a basis). Let us now characterize the fundamental concepts that we have introduced for topological spaces in terms of bases.

Suppose that two bases* $\{\mathfrak{B}_x\}$ $(x \in \mathbf{X})$ and $\{\mathfrak{U}_x\}$ $(x \in \mathbf{X})$ have been specified in some set

* By a basis of a set $\mathbf{X}$ we shall understand in what follows a collection of sets satisfying conditions 1)–3).

$\mathbf{X}$. By Theorem 1, each of these determines a topology in $\mathbf{X}$. We write $\mathbf{X}_1$ for the topological space determined by the first basis and $\mathbf{X}_2$ for that determined by the second.

A necessary and sufficient condition for the topology of $\mathbf{X}_1$ to be stronger than that of $\mathbf{X}_2$ is that, for any $x \in \mathbf{X}$ and $V \in \mathfrak{B}_x$, there exists a neighbourhood $U \in \mathfrak{U}_x$ contained in V.

The simple proof of this fact is left to the reader.

It follows from this that the two bases will determine the same topology if and only if they satisfy both the condition just given and the one obtained from it by interchanging $\mathfrak{B}_x$ and $\mathfrak{U}_x$: for any $x \in \mathbf{X}$ and $U \in \mathfrak{U}_x$ there exists $V \in \mathfrak{B}_x$ contained in U. In this situation the bases are said to be *equivalent*.

If $\mathbf{X}_0$ is a set in a topological space $\mathbf{X}$ with a given basis $\{\mathfrak{B}_x\}$ $(x \in \mathbf{X})$, then the induced topology in $\mathbf{X}_0$ may be defined starting from the basis $\{\mathfrak{B}_x^{(0)}\}$ $(x \in \mathbf{X}_0)$, where $\mathfrak{B}_x^{(0)}$ consists of all sets of the form $V \cap \mathbf{X}_0$ $(V \in \mathfrak{B}_x)$. We leave it to the reader to prove that the collection $\{\mathfrak{B}_x^{(0)}\}$ produced in this way satisfies the conditions of Theorem 1, and that the topology determined by this basis is the induced topology.

2.4. Before indicating how the conditions for a set to be closed can be formulated in terms of neighbourhoods, we introduce another important concept. A point x of a topological space $\mathbf{X}$ is called an *adherent point* of a set $E \subset \mathbf{X}$ if, whatever neighbourhood V of x we choose, the intersection $V \cap E$ is non-empty. If, moreover, this intersection never contains only the single point x, then x is called a *limit point* of E (the term *accumulation point* is also used).

In defining an adherent point (respectively, limit point) it is not necessary to take into account all the neighbourhoods of x; one could restrict oneself to neighbourhoods belonging to a fundamental system of neighbourhoods of the point.

The set of all adherent points of a given set E is called the *closure* of E and is denoted by the symbol $\bar{E}$. The following properties of the closure are easily established:

1) $\bar{\varnothing} = \varnothing$;
2) $E \subset \bar{E}$;
3) if $E_1 \subset E_2$, then $\bar{E}_1 \subset \bar{E}_2$;
4) $\overline{E_1 \cup E_2} = \bar{E}_1 \cup \bar{E}_2$;
5) $\bar{\bar{E}} = \bar{E}$.

Let us verify the last two of these properties, which are not so obvious. Let $E = E_1 \cup E_2$. Since $E_i \subset E$, we have $\bar{E}_i \subset \bar{E}$ $(i = 1, 2)$, whence $\bar{E}_1 \cup \bar{E}_2 \subset \bar{E}$. Conversely, if $x \in \bar{E}$, then, assuming $x \notin \bar{E}_1$, we find a neighbourhood V_0 of x such that $V_0 \cap E_1 = \varnothing$. Let V be any neighbourhood of x. We may assume that $V \subset V_0$ (otherwise, consider the intersection $V \cap V_0$). Since $V \cap E \neq \varnothing$, we must have $V \cap E_2 \neq \varnothing$ and therefore $x \in \bar{E}_2 \subset \bar{E}_1 \cup \bar{E}_2$.

To verify the last property it is enough to check that $\bar{\bar{E}} \subset \bar{E}$. Let $x \in \bar{\bar{E}}$. Take any neighbourhood V of x. We may assume that V is an open set, and hence a neighbourhood of each of its points. We have $V \cap \bar{E} \neq \varnothing$; let $y \in V \cap \bar{E}$. Since $y \in \bar{E}$ and V is a neighbourhood of y, we have $V \cap E \neq \varnothing$, and this shows that $x \in \bar{E}$.

A set $F \subset \mathbf{X}$ is closed if and only if $\bar{F} = F$. For if $x \notin F = \bar{F}$, then there exists a neighbourhood V of x that does not intersect with F. This means that $G = \mathbf{X} \setminus F$ is open, and F is therefore closed. The converse is obtained by reversing this argument.

Let us notice another obvious fact: the closure of a set E is obtained by adjoining to E all its limit points, so that a closed set may be characterized as a set that contains all its limit points.

For $E \subset \mathbf{X}$ we write $\mathring{E}$ for the set of all interior points of E, which is called the *interior* of

E. We leave it to the reader to verify that $\mathring{E}$ is the greatest open set contained in E.

Let $\mathbf{X}$ be a topological space. A set $E \subset \mathbf{X}$ is said to be *dense* in a set $\mathbf{X}_0 \subset \mathbf{X}$ if $\overline{E} \supset \mathbf{X}_0$. If $\overline{E} = \mathbf{X}$, then E is said to be *everywhere dense*. A set $E \subset \mathbf{X}$ is said to be *nowhere dense* if the interior of its closure is empty (or, equivalently, if $\mathbf{X} \setminus E$ is everywhere dense). A set $E \subset \mathbf{X}$ is called a *set of the first category* (or a *meagre* set) if it can be expressed as the union of a countable family of nowhere dense sets. A set $E \subset \mathbf{X}$ that is not of the first category is called a *set of the second category* (in $\mathbf{X}$).

A topological space $\mathbf{X}$ is said to be *separable* if it contains a countable everywhere dense set. We shall become familiar with this concept in more detail later.

2.5. Let $\mathbf{X}$ and $\mathbf{Y}$ be two topological spaces. A mapping $f:\mathbf{X} \to \mathbf{Y}$ is said to be *continuous* if the inverse image of every open set is open. A mapping $f:\mathbf{X} \to \mathbf{Y}$ is said to be *continuous at the point* $x \in \mathbf{X}$ if the inverse image of every neighbourhood of $f(x)$ is a neighbourhood of x (clearly this definition will not change if, in place of arbitrary neighbourhoods of $f(x)$, we take neighbourhoods from some basis of neighbourhoods of $f(x)$). In the following theorem we list various conditions equivalent to continuity.

THEOREM 2. *Let $\mathbf{X}$ and $\mathbf{Y}$ be topological spaces and let f be a mapping from $\mathbf{X}$ into $\mathbf{Y}$. Then the following statements are equivalent:*

1) *f is continuous;*
2) *the inverse image of every closed set is closed;*
3) *f is continuous at every point $x \in \mathbf{X}$;*
4) *for every point $x \in \mathbf{X}$ and every neighbourhood U of $f(x)$ there exists a neighbourhood V of x such that $f(V) \subset U$.*

Proof. Statements 1) and 2) are equivalent since $\mathbf{X} \setminus f^{-1}(B) = f^{-1}(\mathbf{Y} \setminus B)$ for any $B \subset \mathbf{Y}$. The equivalence of 3) and 4) is obvious. Since 3) clearly follows from 1), it remains to show that 3) implies 1). Let G be an open set in $\mathbf{Y}$. We prove that $f^{-1}(G)$ is open. If $x \in f^{-1}(G)$ then there exists $y \in G$ with $y = f(x)$. Since G is a neighbourhood of y, the set $f^{-1}(G)$ is a neighbourhood of x, by the definition of continuity at the point x. Hence x is an interior point, and G is open.

Let f be a bijection from $\mathbf{X}$ onto $\mathbf{Y}$. If both f and f^{-1} are continuous, then f is called a *homeomorphism*. It is clear that two spaces $\mathbf{X}$ and $\mathbf{Y}$ are homeomorphic if and only if it is possible to map one onto the other by means of a homeomorphism.

2.6. In the sequel it will often be convenient for us to describe the topology of a space, and other topological concepts, in terms of the convergence of nets (Moore–Smith convergence).

A net $\{x_\alpha\}$ $(\alpha \in \mathrm{A})$ of elements of a topological space $\mathbf{X}$ is said to be *convergent* to $x \in \mathbf{X}$ if for each neighbourhood V of x there exists $\alpha_V \in \mathrm{A}$ such that $x_\alpha \in V$ for $\alpha \geqslant \alpha_V$. This situation is indicated by writing $x_\alpha \underset{\mathrm{A}}{\to} x$ or $x = \lim\limits_{\alpha} x_\alpha$ (in future we shall often omit the index set in this notation). The point x is called a *limit* of the net $\{x_\alpha\}$.

Let us now note some properties of convergent nets.

PROPERTY 1. *If $x_\alpha \underset{\mathrm{A}}{\to} x$ and $\{y_\beta\}$ $(\beta \in \mathrm{B})$ is a subnet of $\{x_\alpha\}$, then $y_\beta \underset{\mathrm{B}}{\to} x$.*

PROPERTY 2. *Let $E \subset \mathbf{X}$. A necessary and sufficient condition for $x \in \overline{E}$ is that there exist a net $\{x_\alpha\}$ such that $x_\alpha \to x$ and $x_\alpha \in E$.*

For if such a net exists, then whatever neighbourhood V of x we choose, $V \cap E \neq \varnothing$, since this intersection contains elements of the net. Therefore $x \in \overline{E}$.

Now assume that $x \in \overline{E}$; and let $\mathfrak{B}_x$ be a fundamental system of neighbourhoods of x. For U, $V \in \mathfrak{B}_x$, let us write $U \leqslant V$ if $U \supset V$. With this ordering, $\mathfrak{B}_x$ is clearly directed.

Now choose a point x_U in the intersection $U \cap E$, which is non-empty because $x \in \bar{E}$. If we do this for every $U \in \mathfrak{B}_x$, we obtain a net $\{x_U\}$ $(U \in \mathfrak{B}_x\}$ which converges to x.

From this follows:

PROPERTY 3. *A necessary and sufficient condition for a set $F \subset \mathbf{X}$ to be closed is that, for any net $\{x_\alpha\}$ $(\alpha \in \mathrm{A})$ such that $x_\alpha \to x$ and $x_\alpha \in F$ $(\alpha \in \mathrm{A})$, we have $x \in F$.*

Topological spaces that satisfy only the three axioms for a topological space may have a highly complex structure; on the other hand, their topological structure may turn out to be so primitive that they cannot be studied by the methods of topology. The latter is the case, for example, if the space has only two open sets: the empty set and the whole set.

For this reason it is usual to introduce additional axioms of one kind or another, singling out more restricted classes of topological spaces—for example, the following separation axiom of Hausdorff, which has the important consequence that the limit of a net is unique.

A topological space $\mathbf{X}$ is called a *Hausdorff space* (or a *separated space*) if for any two distinct points $x, y \in \mathbf{X}$ there exist a neighbourhood U of x and a neighbourhood V of y such that $U \cap V = \varnothing$.

PROPERTY 4. *A necessary and sufficient condition for every convergent net to converge to only one limit is that the space $\mathbf{X}$ be Hausdorff.*

The uniqueness of the limit, under the condition that the space be Hausdorff, follows immediately from the definition of limit.

We now prove the converse. Suppose $\mathbf{X}$ is not a Hausdorff space. Then there exists a pair of distinct points $x, y \in \mathbf{X}$ such that, whatever neighbourhoods U, V of x, y we choose, we have $U \cap V \neq \varnothing$. Denote by A the collection of all pairs (U, V), where U is a neighbourhood of x and V is a neighbourhood of y. If $\alpha' = (U', V')$ and $\alpha'' = (U'', V'')$ are elements of A, then we put $\alpha' \leq \alpha''$ whenever both $U' \supset U''$ and $V' \supset V''$. Let $\alpha = (U, V) \in \mathrm{A}$. Since $U \cap V \neq \varnothing$, we can choose $x_\alpha \in U \cap V$. Then $x_\alpha \underset{\mathrm{A}}{\to} x$ and also $x_\alpha \underset{\mathrm{A}}{\to} y$. Let us verify one of these statements—say the first. For any neighbourhood U_0 of x there exists an $x_{\alpha_0} \in U_0$. We may assume that $\alpha_0 = (U_0, V_0)$, where V_0 is some neighbourhood of y. Since $\alpha = (U, V) \geqslant \alpha_0$ means, in particular, that $U \subset U_0$, we have $x_\alpha \in U \subset U_0$; that is, $x_\alpha \underset{\mathrm{A}}{\to} x$.

One can formulate the property of continuity for a mapping f from a topological space $\mathbf{X}$ into a topological space $\mathbf{Y}$ in terms of the concept of convergence, as follows.

PROPERTY 5. *A necessary and sufficient condition for f to be continuous is that, for every $x \in \mathbf{X}$ and every net $\{x_\alpha\}$ $(\alpha \in \mathrm{A})$ such that $x_\alpha \underset{\mathrm{A}}{\to} x$, we have $f(x_\alpha) \to f(x)$.*

Let us prove the sufficiency. Let $\Phi \subset Y$ be a closed set and let $F = f^{-1}(\Phi)$. Consider a net $\{x_\alpha\}$ $(\alpha \in \mathrm{A})$ of elements of F converging to $x \in \mathbf{X}$. By assumption, $f(x_\alpha) \underset{\mathrm{A}}{\to} f(x)$ and $f(x_\alpha) \in \Phi$ $(\alpha \in \mathrm{A})$, so we have also $f(x) \in \Phi$; consequently, $x \in F$. Thus F is closed, and so f is continuous.

Assume that f is a continuous mapping. Take any $x \in \mathbf{X}$ and a net $\{x_\alpha\}$ converging to x. Let U be a neighbourhood of $y = f(x)$. There exists a neighbourhood V of x such that $f(V) \subset U$. Further, there exists $\alpha_V \in \mathrm{A}$ such that $x_\alpha \in V$ for $\alpha \geq \alpha_V$. For such an α, we have $f(x_\alpha) \in U$, whence it follows that $f(x_\alpha) \underset{\mathrm{A}}{\to} f(x)$.*

PROPERTY 6. *If every point of the space $\mathbf{X}$ has a countable fundamental system of neighbourhoods, then we can use ordinary sequences instead of nets in properties 2–5.*

Let $\mathfrak{B}_x = \{V_n\}$ $(n \in \mathbb{N})$ be a basis of neighbourhoods of x. Write $U_n = V_1$

* This proof shows that, if the condition is satisfied at some given point x, then f is continuous at this point.

$\cap V_2 \cap \ldots \cap V_n$. If $\{x_\alpha\}$ $(\alpha \in A)$ converges to x, then for each $n \in \mathbb{N}$ choose $\alpha_n \in A$ such that $y_n = x_{\alpha_n} \in U_n$. Then the sequence $\{y_n\}$ converges to x. Using this argument, one can easily derive properties of sequences analogous to properties 2–5.

Property 3 of the convergence of nets makes it possible to introduce a topology starting from some preassigned (*a priori*) convergence. In fact, suppose a notion of convergence has been prescribed in a set $\mathbf{X}$: that is, suppose one has distinguished a class of nets, to be called convergent, and suppose that for each of these one has prescribed a limit (for simplicity, we assume it is unique). In addition, suppose that the convergence possesses Property 1. A set F is then called *closed* if it contains the limit of each convergent net of its elements. It is not hard to verify that the system of closed sets thus specified satisfies conditions 1)–3) of 2.2, so we are in fact dealing with a topological space. Because $\mathbf{X}$ has become a topological space, one can introduce convergence in it as in any topological space. We note that the *a priori* convergence and the topological convergence it induces are, in general, distinct.

2.7. One of the most important concepts in general topology is the concept of a compact space, introduced in the early 1920s by P. S. Alexandroff and P. S. Urysohn.

A topological space $\mathbf{X}$ is said to be *compact* (or *bicompact*) if, given any family $\{G_\xi\}$ $(\xi \in \Xi)$ of open sets which is a covering of $\mathbf{X}$, that is, which satisfies

$$\bigcup_{\xi \in \Xi} G_\xi = \mathbf{X},$$

we can choose a finite number of the sets $G_{\xi_1}, G_{\xi_2}, \ldots, G_{\xi_n}$ which also yield a covering of $\mathbf{X}$.

We shall say that a family of sets $\{A_\xi\}$ $(\xi \in \Xi)$ has the *finite intersection property* if the intersection of any finite number of sets in the system is non-empty.

Since the statement that a family of sets $\{G_\xi\}$ $(\xi \in \Xi)$ forms a covering of $\mathbf{X}$ is equivalent to the statement that the intersection of the complements F_ξ of the sets G_ξ is empty, a space $\mathbf{X}$ is compact if and only if every system of closed sets with the finite intersection property has non-empty intersection.

Let $\{x_\alpha\}$ $(\alpha \in A\}$ be an arbitrary net of elements of a topological space $\mathbf{X}$. We say that $\{x_\alpha\}$ is *frequently in* a subset $E \subset \mathbf{X}$ if for each $\alpha \in A$ there is an $\alpha' \in A$ such that $\alpha' \geqslant \alpha$ and $x_{\alpha'} \in E$. A point $x \in \mathbf{X}$ is called a *cluster point* of the net $\{x_\alpha\}$ $(\alpha \in A)$ if $\{x_\alpha\}$ is frequently in every neighbourhood of x (do not confuse these cluster points with the limit points of the set $\{x_\alpha : \alpha \in A\}$!). A net may have more than one cluster point, or exactly one, or none. For example, the sequence $x_n = n$ $(n \in \mathbb{N})$ has no cluster points in $\mathbb{R}$. On the other hand, if we enumerate all the rational numbers in a sequence in any manner, then every real number is a cluster point of this sequence. If a net converges to a point, then this point is a cluster point of the net (in the case of a Hausdorff space, the unique one).

LEMMA 1. *A point x of a topological space $\mathbf{X}$ is a cluster point of a net $\{x_\alpha\}$ $(\alpha \in A)$ if and only if there exists a subnet $\{y_\beta\}$ $(\beta \in B)$ converging to x.*

Proof. Suppose x is a cluster point of $\{x_\alpha\}$, and let $\mathfrak{B}_x$ be the family of all neighbourhoods of x. Consider the set B of all pairs (α, U) such that $\alpha \in A$, $U \in \mathfrak{B}_x$ and $x_\alpha \in U$. If we define an ordering on B by the rule: $(\alpha, U) \geqslant (\alpha_1, U_1)$ if and only if $\alpha \geqslant \alpha_1$ and $U \subset U_1$, then B becomes a directed set. For if $(\alpha, U), (\alpha_1, U_1) \in B$, then there exists $U_2 \in \mathfrak{B}_x$ satisfying $U_2 \subset U \cap U_1$. Since A is directed, there exists α_0 with $\alpha_0 \geqslant \alpha, \alpha_1$. As $\{x_\alpha\}$ is frequently in U_2, there exists an $\alpha_2 \geqslant \alpha_0$ with $x_{\alpha_2} \in U_2$. Then we have $(\alpha_2, U_2) \geqslant (\alpha, U)$, (α_1, U_1).

Now put $y_{(\alpha, U)} = x_\alpha$, where $(\alpha, U) \in \mathrm{B}$. Then $\{y_{(\alpha, U)}\}$ $((\alpha, U) \in \mathrm{B})$ is a subnet of $\{x_\alpha\}$. For, given any $\alpha \in \mathrm{A}$, choose (α, U) so that $U \in \mathfrak{B}_x$ satisfies $x_\alpha \in U$ (for example, $U = \mathbf{X}$). If now $(\alpha_1, U_1) \geqslant (\alpha, U)$, then $\alpha_1 \geqslant \alpha$ and $y_{(\alpha_1, U_1)} = x_{\alpha_1}$. Let us prove that the net $\{y_{(\alpha, U)}\}$ converges to x. Choose $U \in \mathfrak{B}_x$. Since x is a cluster point of $\{x_\alpha\}$, there exists an $\alpha \in \mathrm{A}$ such that $x_\alpha \in U$. If $(\alpha_1, U_1) \geqslant (\alpha, U)$, then $y_{(\alpha_1, U_1)} = x_{\alpha_1} \in U_1 \subset U$, which establishes the convergence.

Conversely, assume that $\{y_\beta\}$ is a subnet converging to x. For any $\alpha \in \mathrm{A}$ there exists $\beta(\alpha) \in \mathrm{B}$ such that, for all $\beta \geqslant \beta(\alpha)$, we have $y_\beta = x_{\alpha'}$ with $\alpha' \geqslant \alpha$. If U is any neighbourhood of x, then there exists $\beta_0 \in \mathrm{B}$ such that $y_\beta \in U$ whenever $\beta \geqslant \beta_0$. Choose $\beta \geqslant \beta(\alpha), \beta_0$. Then there exists $\alpha' \geqslant \alpha$ with $x_{\alpha'} = y_\beta \in U$, which shows that x is a cluster point of $\{x_\alpha\}$ $(\alpha \in \mathrm{A})$.

LEMMA 2. *A topological space* $\mathbf{X}$ *is compact if and only if every net in* $\mathbf{X}$ *has a cluster point.*

Proof. Let $\{x_\alpha\}$ $(\alpha \in \mathrm{A})$ be an arbitrary net in a compact topological space $\mathbf{X}$. Since A is directed, the family of sets $B_\alpha = \{x_{\alpha'} : \alpha' \geqslant \alpha\}$ has the finite intersection property. Hence the family of closures $\overline{B}_\alpha$ will also certainly have the finite intersection property. Hence, as $\mathbf{X}$ is compact, the intersection of all the $\overline{B}_\alpha$ contains a point x. We now show that x is a cluster point of $\{x_\alpha\}$.

Let $\mathfrak{B}_x$ denote the family of all neighbourhoods of x. We must verify that for any $U \in \mathfrak{B}_x$ and any $\alpha \in \mathrm{A}$ there exists $\alpha_1 \in \mathrm{A}$ such that $\alpha_1 \geqslant \alpha$ and $x_{\alpha_1} \in U$. In fact, $x \in \overline{B}_\alpha$ for any $\alpha \in \mathrm{A}$, so for $U \in \mathfrak{B}_x$ there exists $\alpha_1 \geqslant \alpha$ with $x_{\alpha_1} \in U$.

Now let us prove the converse. Let $\mathbf{X}$ be a topological space in which every net has a cluster point. We prove that $\mathbf{X}$ is compact by showing that an arbitrary family $\mathfrak{F}_0$ of closed sets in $\mathbf{X}$ with the finite intersection property has non-empty intersection. Let A be the family of all finite intersections of elements of $\mathfrak{F}_0$.

It is clearly sufficient to prove that the intersection of all the sets in A is non-empty. Since $\mathfrak{F}_0$ has the finite intersection property, if we order A by inclusion ($\alpha_1 \leqslant \alpha_2$ if $\alpha_1, \alpha_2 \in \mathrm{A}$ and $\alpha_1 \supset \alpha_2$), then A is directed. If we choose an arbitrary element $x_\alpha \in \alpha$ for each $\alpha \in \mathrm{A}$, then we obtain a net $\{x_\alpha\}$ $(\alpha \in \mathrm{A})$. By hypothesis, this has a cluster point x. We now take any $\alpha \in \mathrm{A}$ and show that x is an adherent point of α, from which it will follow, since α is closed, that $x \in \alpha$. Then, as α was arbitrary, x will lie in the intersection of all the sets in A, which therefore must be non-empty.

So let U be any neighbourhood of x. Then there exists $\alpha' \in \mathrm{A}$, $\alpha' \geqslant \alpha$, for which $x_{\alpha'} \in U$; hence $x_{\alpha'} \in \alpha' \subset \alpha$, and $x_{\alpha'} \in U \cap \alpha$, showing that x is indeed an adherent point of α.

COROLLARY. *If a net* $\{x_\alpha\}$ *in a compact space* $\mathbf{X}$ *has a unique cluster point* x, *then* $x_\alpha \to x$.

Proof. If $\{x_\alpha\}$ did not converge to x, then there would be a neighbourhood U of x such that $\{x_\alpha : x_\alpha \notin U\}$ is a subnet of $\{x_\alpha\}$. By Lemma 2, this subnet must have a cluster point; but this cluster point cannot be x.

The next theorem now follows from Lemmas 1 and 2.

THEOREM 3. *A topological space* $\mathbf{X}$ *is compact if and only if every net in* $\mathbf{X}$ *has a subnet converging to a point of* $\mathbf{X}$.

A compact Hausdorff space is called a *compactum*.

A set E in a topological space $\mathbf{X}$ is said to be *compact* if it is compact as a topological space (with the topology induced by that of $\mathbf{X}$). Since the open sets in E have the form $G \cap E$, where G is an open set in $\mathbf{X}$, one can formulate the definition of compactness of a set E as follows: given any system $\{G_\xi\}$ of open (in $\mathbf{X}$) sets covering E—that is, satisfying $\bigcup_\xi G_\xi \supset E$—one can find a finite number of these, $G_{\xi_1}, G_{\xi_2}, \ldots, G_{\xi_n}$ also covering E.

THEOREM 4. *If* $\mathbf{X}$ *is a Hausdorff space, then every compact set* $E \subset \mathbf{X}$ *is closed.*

Proof. Assume that E has an adherent point $x_0 \notin E$. For each $x \in E$ there exist non-intersecting open neighbourhoods V_x of x and $V_{x_0}^{(x)}$ of x_0. The system $\{V_x\}$ $(x \in E)$ clearly covers E, and thus there exist a finite number of points $x_1, x_2, \ldots, x_n \in E$ such that

$$G = \bigcup_{k=1}^{n} V_{x_k} \supset E.$$

The intersection

$$V_0 = \bigcap_{k=1}^{n} V_{x_0}^{(x_k)}$$

is evidently a neighbourhood of x_0. However, $V_0 \cap E \subset V_0 \cap G = \varnothing$, which contradicts the hypothesis that x_0 was an adherent point of E.

REMARK. If $\mathbf{X}$ is a compact space and E is a closed set in $\mathbf{X}$, then E is compact.

For the closed sets in E are also closed in $\mathbf{X}$. Therefore, if $\{F_\xi\}$ $(\xi \in \Xi)$ is a system of closed sets in E with the finite intersection property, then this system has non-empty intersection, and this intersection is contained in E.

We call a set E in a topological space $\mathbf{X}$ *relatively compact* if its closure is compact. Obviously any set in a compact space is relatively compact.

Let $\mathbf{X}$ be a compact space and let f be a continuous mapping from $\mathbf{X}$ into a space $\mathbf{Y}$. Then we have

THEOREM 5. *The set* $f(\mathbf{X})$ *is compact in* $\mathbf{Y}$.

Proof. Write $A = f(\mathbf{X})$, and let $\{G_\xi\}$ $(\xi \in \Xi)$ be a system of open sets covering A. Let $G'_\xi = f^{-1}(G_\xi)$. The sets G'_ξ are open and $\bigcup_{\xi \in \Xi} G'_\xi = \mathbf{X}$. Consequently there exist $\xi_1, \xi_2, \ldots, \xi_n$ such that $\bigcup_{k=1}^{n} G_{\xi_k} \supset A$. Thus the theorem is proved.

COROLLARY 1. *If* E *is a compact set in* $\mathbf{X}$, *then* $f(E)$ *is a compact set in* $\mathbf{Y}$.

Using Theorem 4, we deduce

COROLLARY 2. *If* $\mathbf{Y}$ *is a Hausdorff space, then the image* $f(E)$ *of a compact set* $E \subset \mathbf{X}$ *is closed.*

COROLLARY 3. *If* $\mathbf{X}$ *is compact,* $\mathbf{Y}$ *is Hausdorff, and* f *is a one-to-one continuous mapping from* $\mathbf{X}$ *onto* $\mathbf{Y}$, *then the inverse mapping* f^{-1} *is continuous; that is,* f *is a homeomorphism.*

For every closed set $F \subset \mathbf{X}$ is compact. The inverse image of such a set under f^{-1} is $f(F)$, which is closed by Corollary 2. Thus the inverse image of every closed set is closed, and hence f^{-1} is continuous.

Let us note, in addition, the following important property of continuous functions on compacta, which is a generalization of Weierstrass's Theorem for functions of a real variable.

THEOREM 6. *Every continuous real-valued function* f *on a compact space* $\mathbf{X}$ *attains both its supremum and its infimum.*

Proof. We give the proof for the case of the supremum only. Let $M = \sup\{f(x): x \in \mathbf{X}\}$. By Theorem 5, M is finite. For any $n \in \mathbb{N}$, write

$$F_n = \{x \in \mathbf{X}:\ f(x) \geqslant M - 1/n\}.$$

The sets F_n are clearly closed and non-empty. Furthermore, $F_n \subset F_m$ for $n \geqslant m$. Consequently $\{F_n\}$ is a system of closed sets with the finite intersection property, and so

$\bigcap_{n=1}^{\infty} F_n \neq \varnothing$. If $x \in \bigcap_{n=1}^{\infty} F_n$, then $f(x) \geqslant M$, whence $f(x) = M$; that is, f attains its supremum at x.

We shall need two further properties closely connected with the concept of compactness. Let $\mathbf{X}$ be a topological space. A set $E \subset \mathbf{X}$ is said to be *sequentially compact* if every sequence in E has a subsequence that converges to a point of E (cf. Theorem 3). A set $E \subset \mathbf{X}$ is said to be *relatively sequentially compact* if every sequence in E has a subsequence that converges to a point of $\mathbf{X}$. A set $E \subset \mathbf{X}$ is said to be *countably compact* if every sequence in E has a cluster point $x \in E$ (cf. Lemma 2). A set $E \subset \mathbf{X}$ is said to be *relatively countably compact* if every sequence in E has a cluster point $x \in \mathbf{X}$. Note that, in general, compactness does not imply sequential compactness, or conversely. If a set is compact, then it is obviously also countably compact. The relationships between these concepts in various special cases will be considered below.

2.8. Let $\{\mathbf{X}_\alpha\}$ $(\alpha \in \mathrm{A})$ be a family of topological spaces, and consider their direct product $\mathbf{X} = \prod_{\alpha \in \mathrm{A}} \mathbf{X}_\alpha$ (see §1). We introduce a topology on $\mathbf{X}$ by taking a basis of neighbourhoods of a point $f \in \mathbf{X}$ to consist of all sets of the form $\prod_{\alpha \in \mathrm{A}} U_\alpha$, where each U_α is a neighbourhood of the point $f(x)$ in $\mathbf{X}_\alpha$ and where U_α is distinct from $\mathbf{X}_\alpha$ for at most a finite number of suffices $\alpha \in \mathrm{A}$. We leave it to the reader to verify that this family of sets does actually form a basis. It is easy to prove that a net $\{f_\beta\}$ $(\beta \in \mathrm{B})$ in $\mathbf{X}$ converges to a point $f \in \mathbf{X}$ if and only if $f_\beta(\alpha) \underset{1}{\rightarrow} f(\alpha)$ in the space $\mathbf{X}_\alpha$, for each α. Thus the convergence of a net $\{f_\beta\}$ is coordinatewise—that is, each "coordinate" $f_\beta(\alpha)$ converges to the corresponding "coordinate" of the limit f in $\mathbf{X}_\alpha$.

Later we shall need the following theorem of Tychonoff, for which a proof may be found, for example, in Kelley (Chapter 5, Theorem 13).

THEOREM 7. *If $\mathbf{X}_\alpha$ is a compact space, for each $\alpha \in \mathrm{A}$, then so is $\mathbf{X} = \prod_{\alpha \in \mathrm{A}} \mathbf{X}_\alpha$.*

Further details on the material in this section may be found in the monographs by Bourbaki-II and Kelley.

We conclude with a remark on terminology. Since it will be necessary from time to time to consider different topologies on the same set $\mathbf{X}$ simultaneously, we adopt an abbreviated notation to avoid confusion. If τ is a topology on a space $\mathbf{X}$, then the terms τ-closure, τ-compact set, etc., will mean closure in the topology τ, compact set in the topology τ, etc., respectively.

§ 3. Metric spaces

3.1. One important class of topological spaces is the class of metric spaces. A set $\mathbf{X}$ is called a *metric space* if to each pair of elements $x, y \in \mathbf{X}$ there is associated a real number $\rho(x, y)$, the *distance* between x and y, subject to the following conditions:

1) $\rho(x, y) \geqslant 0$; $\rho(x, y) = 0$ if and only if $x = y$;
2) $\rho(x, y) = \rho(y, x)$;
3) $\rho(x, y) \leqslant \rho(x, z) + \rho(z, y)$ for any $z \in \mathbf{X}$ (the *triangle inequality*).

Such a function $\rho: \mathbf{X} \times \mathbf{X} \rightarrow \mathbb{R}$ is called a *metric*. Simple examples of metric spaces are n-

dimensional Euclidean space* $\mathbf{R}^n$, the line interval, and the circumference of a circle, if one defines the distance to be the shortest arc length between the two points.

Let $\mathbf{X}$ be a metric space, with metric ρ. The *open ball* of radius $\varepsilon > 0$ with centre at the point $x_0 \in \mathbf{X}$ is the set

$$K_\varepsilon(x_0) = \{x \in \mathbf{X}:\ \rho(x, x_0) < \varepsilon\}.$$

The *closed ball* (henceforth usually simply the *ball*) of radius $\varepsilon > 0$ with centre at the point $x_0 \in \mathbf{X}$ is the set

$$B_\varepsilon(x_0) = \{x \in \mathbf{X}:\ \rho(x, x_0) \leqslant \varepsilon\}.$$

THEOREM 1. *The family of sets $K_{1/n}(x)$ $(n \in \mathbb{N},\ x \in \mathbf{X})$ satisfies conditions 1)–3) in subsection 2.3: that is, it forms a basis for* $\mathbf{X}$.

Proof. Since $\rho(x, x) = 0$, condition 1) is obviously satisfied. If $n, m \in \mathbb{N}$ and $p = \max(n, m)$, then $K_{1/p}(x) \subset K_{1/n}(x) \cap K_{1/m}(x)$. Hence condition 2) is satisfied. If $V_x = K_{1/n}(x)$, then we set $V'_x = K_{1/2n}(x)$. Then $V'_x \subset V_x$. Furthermore, if $y \in V'_x$ and $V_y = K_{1/2n}(y)$, then $V_y \subset V_x$, since for any $z \in V_x$ the triangle inequality gives $\rho(x, z) \leqslant \rho(x, y) + \rho(y, z) < 1/2n + 1/2n = 1/n$. Thus condition 3) is also satisfied.

The construction described in Theorem 2.1 turns a metric space $\mathbf{X}$, in a canonical way, into a topological space with a basis consisting of open balls. Theorem 1 shows, in addition, that every point of $\mathbf{X}$ has a countable basis. Therefore, to describe the topology of $\mathbf{X}$, we need only consider the convergence of sequences (see 2.6, Property 6). Notice that $x_n \to x$ in the resulting space if and only if $\rho(x_n, x) \to 0$.

A topological space whose topology arises from some metric is said to be *metrizable* (by no means every topological space is metrizable; we shall meet examples to show this later). One should bear in mind that the same topology may arise from different metrics.

THEOREM 2. 1) *$\rho(x, y)$ is a continuous function of its arguments; that is, if $x_n \to x_0$ and $y_n \to y_0$ then $\rho(x_n, y_n) \to \rho(x_0, y_0)$.*

2) *A metric space is Hausdorff (and hence a convergent sequence can have only one limit).*

Proof. 1) By the triangle inequality, we have

$$\rho(x', y') - \rho(x, y) \leqslant \rho(y, y') + \rho(x, x').$$

Interchanging x, y and x', y' yields the inequality with the opposite sign, so we have

$$|\rho(x', y') - \rho(x, y)| \leqslant \rho(x, x') + \rho(y, y'). \tag{1}$$

Using this, we obtain

$$|\rho(x_n, y_n) - \rho(x_0, y_0)| \leqslant \rho(x_n, x_0) + \rho(y_n, y_0) \to 0.$$

2) If $x \neq x_0$, then $\varepsilon = \rho(x, x_0) > 0$, and by the triangle inequality the open balls $K_\varepsilon(x)$ and $K_\varepsilon(x_0)$ do not intersect.

Later we shall need the concept of distance of a point from a set $E \subset \mathbf{X}$. As in the case of Euclidean space, we take this to be the number

$$\rho(x_0, E) = \inf_{x \in E} \rho(x_0, x).$$

It is not hard to see that $\rho(x_0, E) = 0$ is equivalent to $x_0 \in \overline{E}$.

---

* We write $\mathbf{R}^n$ for n-dimensional real space, and $\mathbf{C}^n$ for n-dimensional complex space.

Let $\mathbf{X}_0$ be a set in a metric space $\mathbf{X}$. Since the distance is defined for every pair of points in $\mathbf{X}$, it is also defined in $\mathbf{X}_0$. Moreover, the axioms 1)–3) for a metric space are clearly satisfied, so $\mathbf{X}_0$ turns out to be a metric space, in a natural way. The metric in $\mathbf{X}_0$ is said to be *induced* by that of $\mathbf{X}$, and $\mathbf{X}_0$ is called a *subspace* of $\mathbf{X}$.

Now assume that the elements of one metric space $\mathbf{X}$ can be put into a one-to-one correspondence with those of another metric space $\mathbf{Y}$ in such a way that the distance between any two points of $\mathbf{X}$ is equal to that between the corresponding points of $\mathbf{Y}$. Then the spaces are said to be *isometric*. It is clear that all the metric relations in one of these spaces will also hold in the other; therefore the difference between the spaces is only in the concrete nature of their elements and does not concern the *essential* properties of the spaces (i.e. those connected with distance). This is the justification for *identifying* isometric spaces, as we shall in future.

3.2. Let us give some more complicated examples of metric spaces. In this book, spaces whose elements are functions will play a fundamental role. Here, and in what follows, whenever we introduce a space $\mathbf{X}$ whose elements are numerical functions, we shall, unless we say otherwise, be introducing two spaces at once: a real space $\mathbf{X}$, consisting of all the real-valued functions satisfying the relevant conditions, and a complex space $\mathbf{X}$, consisting of the complex functions satisfying the same conditions. As a rule, we shall make no distinction in notation between these spaces. If, in some statement, we do not specify what space is in question—the real space or the complex one—then this will mean that the statement holds in both cases.

1) Let K be a compactum. The space $\mathbf{C}(K)$ is the set of all continuous functions on K, the distance between functions x and y being defined as follows:

$$\rho(x, y) = \sup_{t \in K} |x(t) - y(t)|.$$

Verification of conditions 1)–3) presents no difficulty, so we omit it. Since a continuous function on a compactum attains its maximum (see Theorem 2.6), we can also write

$$\rho(x, y) = \max_{t \in K} |x(t) - y(t)|.$$

Convergence of a sequence $\{x_n\}$ of elements of $\mathbf{C}(K)$ to a point x_0 amounts to uniform convergence of the sequence of functions $x_n(t)$ to the function $x_0(t)$. For, given $\varepsilon > 0$, if we choose N so that $\rho(x_n, x_0) < \varepsilon$ when $n \geqslant N$, then, for such values of n,

$$\sup_{t \in K} |x_n(t) - x_0(t)| < \varepsilon,$$

so that for all $t \in K$ we have

$$|x_n(t) - x_0(t)| < \varepsilon \quad (n \geqslant N),$$

from which the uniform convergence follows, as required.

The converse is also true: if a sequence of continuous functions converges uniformly to a continuous function, then the corresponding sequence of elements of $\mathbf{C}(K)$ is convergent in $\mathbf{C}(K)$. When $K = [a, b]$, we denote the space by $\mathbf{C}[a, b]$.

2) The space $\mathbf{s}$ is the set of all numerical sequences, the distance between sequences $x = (\xi_1, \xi_2, \dots, \xi_k, \dots)$ and $y = (\eta_1, \eta_2, \dots, \eta_k, \dots)$ being defined by the formula

$$\rho(x, y) = \sum_{k=1}^{\infty} \frac{1}{2^k} \frac{|\xi_k - \eta_k|}{1 + |\xi_k - \eta_k|}.$$

Verification of axioms 1) and 2) for a metric space presents no difficulty. Condition 3) follows from the fact that the function $\phi(\lambda) = \lambda/(1+\lambda)$ is increasing for $\lambda \geqslant 0$, so that we have the numerical inequality

$$\frac{|\alpha+\beta|}{1+|\alpha+\beta|} \leqslant \frac{|\alpha|+|\beta|}{1+|\alpha|+|\beta|} \leqslant \frac{|\alpha|}{1+|\alpha|} + \frac{|\beta|}{1+|\beta|}. \tag{2}$$

If a sequence $\{x_n\}$ $(x_n = (\xi_1^{(n)}, \xi_2^{(n)}, \ldots, \xi_k^{(n)}, \ldots), n = 1, 2, \ldots)$ converges to the element $x_0 = (\xi_1^{(0)}, \xi_2^{(0)}, \ldots, \xi_k^{(0)}, \ldots)$, then this means that

$$\lim_{n\to\infty} \xi_k^{(n)} = \xi_k^{(0)} \qquad (k = 1, 2, \ldots), \tag{3}$$

that is, the convergence of sequences of points in **s** is coordinatewise—in other words, each coordinate of x_n converges to the corresponding coordinate of x_0.

For the inequality

$$\frac{1}{2^k}\frac{|\xi_k^{(n)} - \xi_k^{(0)}|}{1+|\xi_k^{(n)} - \xi_k^{(0)}|} \leqslant \rho(x_n, x_0) \qquad (k = 1, 2, \ldots) \tag{4}$$

shows that, if $x_n \to x_0$, then $\xi_k^{(n)} \to \xi_k^{(0)}$ $(k = 1, 2, \ldots)$.

Conversely, if condition 3) is satisfied, then, since the series

$$\sum_{k=1}^{\infty} \frac{1}{2^k}\frac{|\xi_k^{(n)} - \xi_k^{(0)}|}{1+|\xi_k^{(n)} - \xi_k^{(0)}|} = \rho(x_n, x_0)$$

converges uniformly in n (it is majorized by the series $\sum_{k=1}^{\infty} 1/2^k$), we can take limits term-by-term and, since each term tends to zero, we have $\rho(x_n, x_0) \to 0$.

From what we have proved it follows that **s** is the topological product of a countable family of real lines.

We mention in addition the space $\mathbf{C}^{(l)}[a, b]$ (to distinguish this from the space of complex numbers $\mathbf{C}^n$ we write the superscript here in round brackets), whose elements are functions defined on the interval $[a, b]$ and having continuous derivatives up to the lth order (inclusive) in this interval. The distance between x and y may be defined by

$$\rho(x, y) = \sum_{k=0}^{l} \max_{a \leqslant t \leqslant b} |x^{(k)}(t) - y^{(k)}(t)|$$

$$(x^{(0)}(t) = x(t), \qquad y^{(0)}(t) = y(t)).$$

Convergence in $\mathbf{C}^{(l)}[a, b]$ of a sequence means that the sequence of functions and the sequences of kth derivatives $(1 \leqslant k \leqslant l)$ all converge uniformly.

One can also consider the space $\mathbf{C}^{(l)}(D)$, consisting of continuous functions having continuous partial derivatives up to the lth order in a domain D of multidimensional space (see IV.4.4).

We remark also that $\mathbf{C}(K)$ and **s** become ordered sets (but not totally ordered) if we introduce an ordering on $\mathbf{C}(K)$ by setting $x \geqslant y$ if and only if $x(t) \geqslant y(t)$ for all $t \in K$, and an ordering on **s** by setting $x = (\xi_k)_{k=1}^{\infty} \geqslant y = (\eta_k)_{k=1}^{\infty}$ if and only if $\xi_k \geqslant \eta_k$ for all $k \in \mathbb{N}$.

§ 4. Completeness and separability. Sets of the first and second categories*

From the axioms for a metric space one can derive a number of other properties of distance and limit, analogous to well-known properties of the real numbers: for example, if $x_n^{(k)} \to x_n$ as $k \to \infty$ and $x_n \to x_0$, then there is a sequence $\{x_n^{(k_n)}\}$ converging to x_0. However, certain important propositions about limits which hold for sets of real numbers cannot be deduced from the axioms for a metric space and, generally speaking, do not hold for an arbitrary metric space. Therefore it is natural to single out in the collection of all metric spaces certain classes of spaces whose metrics satisfy additional conditions of one kind or another, generalizing other essential properties of distance and convergence in the space of real numbers.

4.1. The most important of these properties is completeness. For real numbers, the principle of completeness may be expressed in various forms (Dedekind's principle, the existence of suprema and infima for bounded sets, and so on); however, there is only one of these—Cauchy's convergence test—that does not make use of non-metrical concepts. In fact, we shall take the availability of a generalization of Cauchy's test as the definition of completeness for an arbitrary metric space. More precisely, we make the following definition.

A sequence $\{x_n\}$ of points in a metric space $\mathbf{X}$ is said to be a *Cauchy* (or *fundamental*) *sequence* if $\rho(x_m, x_n) \to 0$ as $m, n \to \infty$: that is, if for each $\varepsilon > 0$ there is a number N_ε such that $\rho(x_m, x_n) < \varepsilon$ whenever $m, n \geqslant N_\varepsilon$.

A metric space $\mathbf{X}$ is said to be *complete* if every Cauchy sequence $\{x_n\}$ converges; that is, if there exists $x_0 \in \mathbf{X}$ such that $x_n \to x_0$.

Obviously, every convergent sequence is a Cauchy sequence, in any metric space.

Thus, for a complete space, we have the *Cauchy convergence test*: a sequence $\{x_n\}$ is convergent if and only if it is a Cauchy sequence.

The property of completeness is preserved when we pass from a space $\mathbf{X}$ to a closed subspace $\mathbf{X}_0$. For if $\{x_n\}$ is a Cauchy sequence of points in $\mathbf{X}_0$ then, since $\mathbf{X}$ is complete, there exists a point $x_0 = \lim_{n\to\infty} x_n$ in $\mathbf{X}$. As $\mathbf{X}_0$ is closed, we must have $x_0 \in \mathbf{X}_0$: that is, $\{x_n\}$ is convergent in $\mathbf{X}_0$.

It is easy to see that the above statement will no longer be true if we omit the condition that $\mathbf{X}_0$ be closed. For an example one need only take $\mathbf{X}$ to be the space of all real numbers and $\mathbf{X}_0$ to be the space of rational numbers, which is contained in $\mathbf{X}$.

Checking that a metric space is complete is often facilitated by

LEMMA 1. *If a Cauchy sequence $\{x_n\}$ contains a subsequence $\{x_{n_k}\}$ which converges to a point x, then $x_n \to x$.*

Proof. Since $\{x_n\}$ is a Cauchy sequence, for every $\varepsilon > 0$ there exists an N such that $\rho(x_n, x_m) < \varepsilon$ whenever $n, m \geqslant N$. For $n_k \geqslant N$, we have

$$\rho(x_n, x_{n_k}) < \varepsilon. \tag{1}$$

Now $x_{n_k} \to x$, so if we take limits in (1), we obtain $\rho(x_n, x) \leqslant \varepsilon$, whence it follows that $x_n \to x$, as claimed.

4.2. All the specific spaces introduced in § 3 are complete. Let us prove this.

1) *The space* $\mathbf{C}(K)$. Let $\{x_n\}$ be a Cauchy sequence of elements of $\mathbf{C}(K)$. If $\varepsilon > 0$, then for

* Sets of the first and second categories were introduced by R. Baire.

sufficiently large m and n ($m, n \geqslant N_\varepsilon$),

$$\rho(x_m, x_n) = \max_{t \in K} |x_m(t) - x_n(t)| < \varepsilon.$$

Thus, for any $t \in K$,

$$|x_m(t) - x_n(t)| < \varepsilon. \tag{2}$$

Fixing $t \in K$, we see that the numerical sequence $\{x_n(t)\}$ is a Cauchy sequence, and consequently $\lim_{n \to \infty} x_n(t)$ exists; let us denote this limit by $x_0(t)$. It remains to prove that x_0 belongs to $\mathbf{C}(K)$ and that x_n converges to x_0 in the metric of $\mathbf{C}(K)$. If we let $m \to \infty$ in equation (2), we obtain

$$|x_0(t) - x_n(t)| \leqslant \varepsilon.$$

From this it is clear that the sequence of functions $\{x_n(t)\}$ converges uniformly to $x_0(t)$, which is therefore continuous, so that x_n converges to x_0 in $\mathbf{C}(K)$.

The completeness of the space $\mathbf{C}^{(l)}$ may be verified similarly.

2) *The space* **s**. It is very simple to show that **s** is complete. If $x_n = (\xi_1^{(n)}, \xi_2^{(n)}, \ldots, \xi_k^{(n)}, \ldots)$ is a Cauchy sequence, then it is easy to check, using an inequality like (4) in § 3, that each of the numerical sequences $\xi_k^{(1)}, \xi_k^{(2)}, \ldots, \xi_k^{(n)}, \ldots$ is also a Cauchy sequence, and so converges to a limit $\xi_k^{(0)} = \lim_{n \to \infty} \xi_k^{(n)}$ $(k = 1, 2, \ldots)$. If we now set $x_0 = (\xi_1^{(0)}, \xi_2^{(0)}, \ldots, \xi_k^{(0)}, \ldots)$, then we see that $x_n \to x_0$ in **s** (since convergence in **s** is coordinatewise).

4.3. Just as the set of rational numbers is embedded in the set of real numbers, one can also embed an arbitrary metric space in a complete metric space.

The smallest complete metric space containing a given metric space **X** is called the *completion* of **X**. (The term "smallest" is to be understood in the sense that the completion is contained in every other complete metric space containing **X**. Here isometric spaces are identified, as explained in 3.1.)

THEOREM 1. *Every metric space has a completion.*

*Proof.** Let **X** be a metric space. The Cauchy sequences $\{x_n\}$ and $\{x'_n\}$ are said to be *equivalent* if $\rho(x_n, x'_n) \to 0$. It is clear that if one of a pair of equivalent sequences converges, then so does the other, and to the same point. For if, say, $x_n \to x$, then

$$\rho(x'_n, x) \leqslant \rho(x_n, x'_n) + \rho(x_n, x) \to 0$$

and therefore $x'_n \to x$.

We now partition the set of all Cauchy sequences into classes, assigning all mutually equivalent sequences to the same class. Let Ξ denote the set of all such classes. Clearly two sequences that are both equivalent to a third are equivalent to one another, so a sequence cannot belong to two distinct classes.

Let ξ and η be classes in Ξ. Choose a sequence $\{x_n\}$ in ξ in any way, and a sequence $\{y_n\}$ in η. Using inequality (1) of §3, we obtain

$$|\rho(x_m, y_m) - \rho(x_n, y_n)| \leqslant \rho(x_m, x_n) + \rho(y_m, y_n).$$

Now $\{x_n\}$ and $\{y_n\}$ are Cauchy sequences, so the right-hand side tends to zero, and the

* See Hausdorff. The proof follows, in essence, the Cantor–Méray process for introducing the real numbers.

numerical sequence $\{\rho(x_n, y_n)\}$ is a Cauchy sequence. We set

$$\rho(\xi, \eta) = \lim_{n\to\infty} \rho(x_n, y_n).$$

The classes ξ and η determine $\rho(\xi, \eta)$ uniquely. For if $\{x'_n\}$ is equivalent to $\{x_n\}$ and $\{y'_n\}$ to $\{y_n\}$, then, taking limits in the inequality

$$|\rho(x'_n, y'_n) - \rho(x_n, y_n)| \leqslant \rho(x_n, x'_n) + \rho(y_n, y'_n),$$

we obtain

$$\lim_{n\to\infty} \rho(x_n, y_n) = \lim_{n\to\infty} \rho(x'_n, y'_n).$$

We now verify that $\rho(\xi, \eta)$ satisfies the axioms 1)–3) for a metric space.

To verify condition 1) we need to prove that $\rho(\xi, \eta) = 0$ implies $\xi = \eta$. In the notation used earlier, we see that

$$\lim_{n\to\infty} \rho(x_n, y_n) = 0,$$

that is, the sequences $\{x_n\}$ and $\{y_n\}$ are equivalent; and so the classes ξ and η coincide.

Condition 2) is obvious.

Condition 3) is obtained by taking limits in the inequality

$$\rho(x_n, y_n) \leqslant \rho(x_n, z_n) + \rho(y_n, z_n).$$

Here $\{x_n\}$, $\{y_n\}$, $\{z_n\}$ are sequences belonging to the classes ξ, η, ζ, respectively.

Hence the set Ξ of equivalence classes is a metric space.

Now we show that the original space $\mathbf{X}$ may be regarded as a subspace of Ξ.

For any $x \in \mathbf{X}$ we denote by $\xi_x \in \Xi$ the class of sequences containing the sequence $(x, x, \ldots, x, \ldots)$; in other words, ξ_x is the class of sequences converging to x. Obviously, $\xi_x = \xi_y$ is equivalent to $x = y$. Furthermore, $\rho(\xi_x, \xi_y) = \rho(x, y)$, which is most easily verified by taking the sequences $(x, x, \ldots, x, \ldots)$ and $(y, y, \ldots, y, \ldots)$ as determining the classes ξ_x and ξ_y.

From what has been said it is clear that we embed $\mathbf{X}$ isometrically in Ξ by identifying $x \in \mathbf{X}$ with the class $\xi_x \in \Xi$.

Therefore we may from now on regard $\mathbf{X}$ as a subspace of Ξ; and we shall use the old notation x for the element ξ_x. We also note the following fact. Let $\{x_n\}$ be a sequence determining a class ξ. Then $\xi_{x_n} = x_n \to \xi$ (in Ξ). For $\rho(x_n, x_m) < \varepsilon$ whenever $n, m \geqslant N_\varepsilon$, and thus $\rho(x_n, \xi) = \lim_{m\to\infty} \rho(x_n, x_m) \leqslant \varepsilon$, so it is clear that $x_n \to \xi$ (in Ξ).

By what has been proved, for any $\xi \in \Xi$ we can find an $x \in \mathbf{X}$ such that $\rho(x, \xi) < \varepsilon$ (we can take x to be any x_n with $n \geqslant N_\varepsilon$).

Next we prove that Ξ is complete. Let $\xi^{(1)}, \xi^{(2)}, \ldots, \xi^{(n)}, \ldots$ be a Cauchy sequence of classes. Choose a sequence $\varepsilon_n \to 0$. By what we have proved, for any n, there exists an $x^{(n)} \in \mathbf{X}$ such that $\rho(x^{(n)}, \xi^{(n)}) < \varepsilon_n$. From the inequality

$$\rho(x^{(n)}, x^{(m)}) \leqslant \rho(x^{(n)}, \xi^{(n)}) + \rho(\xi^{(n)}, \xi^{(m)}) + \rho(\xi^{(m)}, x^{(m)}) < \varepsilon_n + \varepsilon_m + \rho(\xi^{(n)}, \xi^{(m)})$$

it is clear that $\{x^{(n)}\}$ is a Cauchy sequence. Consequently it determines a class ξ such that $\rho(x^{(n)}, \xi) \to 0$. But

$$\rho(\xi^{(n)}, \xi) \leqslant \rho(\xi^{(n)}, x^{(n)}) + \rho(x^{(n)}, \xi) < \varepsilon_n + \rho(x^{(n)}, \xi).$$

Hence $\xi^{(n)} \to \xi$ in Ξ, showing that Ξ is complete.

If H is another complete space containing **X** then, by identifying those elements of H which are limits of Cauchy sequences of points in **X** with equivalence classes of sequences, we see, since H is complete, that $\Xi \subset$ H.

Thus we have constructed the completion of **X**.

4.4. Now we consider the properties of denseness and separability, which were introduced for arbitrary topological spaces in 2.4.

Let **X** be a metric space. A set $E \subset \mathbf{X}$ will evidently be dense in a set $\mathbf{X}_0 \subset \mathbf{X}$ if for every $x \in \mathbf{X}_0$ and $\varepsilon > 0$ there exists a point $z \in E$ satisfying $\rho(x, z) < \varepsilon$, or—what is the same—if for each $x \in \mathbf{X}_0$ there exists a sequence $\{x_n\} \subset E$ such that $x_n \to x$.

It is not hard to see that every space **X** is a dense subset of its completion.

For if $x_1, x_2, \ldots, x_n, \ldots$ is a sequence belonging to $\xi \in \Xi$, then, as we showed in the proof of Theorem 1, we have $\lim_{n\to\infty} x_n = \xi$, from which the assertion now follows.

We recall that a metric space **X** is *separable* if it has a countable dense subset.

If a metric space $\mathbf{X}_0$ *is contained in a separable metric space* **X** *then* $\mathbf{X}_0$ *is also separable.*

Let $\mathbf{X}_0$ be a subset of **X** and let $D = \{x_k\}$ be a countable dense subset of **X**. Choose a numerical sequence $\varepsilon_n \to 0$ ($\varepsilon_n > 0$) and, for each $k = 1, 2, \ldots$, let z_{kn} be an element in $\mathbf{X}_0$ such that

$$\rho(x_k, z_{kn}) < \rho(x_k, \mathbf{X}_0) + \varepsilon_n.$$

Let $x \in \mathbf{X}_0$ and $\varepsilon > 0$; there exists $x_k \in D$ such that $\rho(x, x_k) < \varepsilon$; choosing n so large that $\varepsilon_n < \varepsilon$, we have

$$\rho(x, z_{kn}) \leqslant \rho(x, x_k) + \rho(x_k, z_{kn}) < \varepsilon + \rho(x_k, \mathbf{X}_0) + \varepsilon_n \leqslant \varepsilon + \rho(x_k, x) + \varepsilon_n < 3\varepsilon,$$

from which it follows that $D_0 = \{z_{kn}\}$ is dense in $\mathbf{X}_0$.

4.5. *The spaces* $\mathbf{R}^n$, $\mathbf{C}[a, b]$, $\mathbf{C}^{(l)}$, **s** *are separable.** For $\mathbf{R}^n$ this is clear. We can take the set of points with rational coefficients, for example, as the countable dense subset.

In the case of $\mathbf{C}[a, b]$ we can take the set of all algebraic polynomials with rational coefficients as the countable dense subset. For, by a well-known theorem of Weierstrass (see IV.4.1), every continuous function can be uniformly approximated (that is, approximated according to the distance in $\mathbf{C}[a, b]$) by a polynomial, to within any preassigned degree of accuracy. Moreover, by making an arbitrarily small change in the polynomial, we can clearly arrange that its coefficients be rational. Thus for every continuous function $x(t)$ there is a polynomial with rational coefficients arbitrarily close to $x(t)$ in the metric of $\mathbf{C}[a, b]$.

In $\mathbf{C}[a, b]$ we can also take as the countable dense subset the collection of all piecewise linear functions whose graphs are polygonal, with vertices at rational points.

The proof that $\mathbf{C}^{(l)}[a, b]$ is separable is left to the reader.

One countable dense subset of **s** is, for example, the set $\mathbf{r}_0$ of all sequences of rational numbers having a finite number of terms different from zero. For, if we replace all terms from the $(n+1)$st onwards in the sequence $(\xi_1, \xi_2, \ldots, \xi_{n-1}, \xi_n, \ldots)$ determining an element x of **s** by zero, we obtain a sequence $(\xi_1, \xi_2, \ldots, \xi_n, 0, \ldots) = [x]_n$ whose distance, as an element of **s**, from the original element is less than $1/2^n$.

Since the convergence in **s** is coordinatewise, we can clearly approximate the resulting

* We give the proofs only for the real spaces; the complex cases may easily be reduced to the real ones.

sequence—an element of $\mathbf{s}$—by an element of $\mathbf{r}_0$, to within any desired degree of accuracy.

4.6. *Not every metric space is separable.* To construct an example of an *inseparable* space, we consider an arbitrary infinite set T and the set $\ell^\infty(T)$ of all bounded real functions defined on T. If, for $x, y \in \ell^\infty(T)$, we set

$$\rho(x, y) = \sup_{t \in T} |x(t) - y(t)|,$$

then $\ell^\infty(T)$ becomes a complete metric space, in which convergence is uniform convergence of sequences of functions.

These statements are verified exactly as for $\mathbf{C}(K)$. We now prove that $\ell^\infty(T)$ is not separable.

Consider the set $M_0 \subset \ell^\infty(T)$ of all characteristic functions—that is, all functions taking only two values, zero and one. This set is uncountable.* Let D be a dense subset of $\ell^\infty(T)$. With each $x \in M_0$ we associate an arbitrary $z \in D$ such that

$$\rho(x, z) < 1/2.$$

Here different elements x and x' of M_0 correspond to different elements z and z' of D, since if it were the case that

$$\rho(x, z) < 1/2 \quad \text{and} \quad \rho(x', z) < 1/2,$$

then we should have

$$\rho(x, x') \leqslant \rho(x, z) + \rho(x', z) < 1,$$

whereas we also have

$$\rho(x, x') = 1.$$

Thus the uncountable set M_0 is in one-to-one correspondence with a subset of D. Hence D is uncountable.

4.7. It is easy to see that a subset E of a metric space $\mathbf{X}$ is *nowhere dense* if it is not dense in any ball, that is, if for each ball $K_\varepsilon(x)$ there is another ball $K_{\varepsilon_1}(x_1) \subset K_\varepsilon(x)$ which contains no points of E (see 2.4).

An example of a nowhere dense set in two-dimensional Euclidean space (the plane) is any set of points lying on a line. The set of points whose coordinates are both rational is an example of a set of first category: it is the union of a countable set of "one-point" sets. Although this set is of first category, it is nevertheless dense in $\mathbf{R}^2$.

The following theorem answers the question of whether there exist any sets of second category.

THEOREM 2. *A complete metric space* $\mathbf{X}$ *is a set of second category* (*in itself*).

Proof. Assume the contrary; that is, assume that $\mathbf{X}$ is a set of first category, so that $\mathbf{X} = \bigcup_{k=1}^{\infty} E_k$, where E_k is nowhere dense. Since E_1 is nowhere dense, there exists a ball $K_{\varepsilon_1}(x_1)$ containing no points of E_1. Since E_2 is nowhere dense, there exists a ball $K_{\varepsilon_2}(x_2)$, contained in $K_{\varepsilon_1/2}(x_1)$, which contains no points of E_2. Moreover, we may assume that $\varepsilon_2 \leqslant \varepsilon_1/2$.

* Natanson-II.

Proceeding in this way, we construct a sequence of balls $\{K_{\varepsilon_n}(x_n)\}$ such that

$$K_{\varepsilon_n}(x_n) \subset K_{\varepsilon_{n-1}/2}(x_{n-1}), \quad K_{\varepsilon_n}(x_n) \cap E_n = \varnothing,$$
$$\varepsilon_n \leqslant \varepsilon_{n-1}/2 \qquad (n = 2, 3, \ldots).$$

As $K_{\varepsilon_n/2}(x_n)$ contains all the subsequent balls, we have

$$\rho(x_{n+p}, x_n) < \varepsilon_n/2. \tag{3}$$

Obviously, $\varepsilon_n \to 0$, so $\{x_n\}$ is a Cauchy sequence, and hence $\lim_{n\to\infty} x_n = x$ exists, because $\mathbf{X}$ is complete.

Taking limits in (3) (as $p \to \infty$), we obtain

$$\rho(x, x_n) \leqslant \varepsilon_n/2 < \varepsilon_n,$$

so that $x \in K_{\varepsilon_n}(x_n)$, and therefore $x \notin E_n$, for $n = 1, 2, \ldots$. However, $x \in \mathbf{X} = \bigcup_{k=1}^{\infty} E_k$, so x must belong to at least one of the E_n.

REMARK. By slightly changing the proof one can obtain the stronger result that every non-empty open set in a complete space is a set of the second category.

COROLLARY. *In a complete space, a set which is the complement of a set of the first category is always a set of the second category.*

Sets of this type are called *residuals* (or *co-meagre* sets). Notice the following fact: the intersection of a countable set of residuals is a residual. For if A_k are residuals, then $A_k = \mathbf{X} \setminus E_k$, where the E_k are sets of the first category. Since $\bigcap_{k=1}^{\infty} A_k = \mathbf{X} \setminus \bigcup_{k=1}^{\infty} E_k$ and $\bigcup_{k=1}^{\infty} E_k$ is clearly a set of the first category, $\bigcap_{k=1}^{\infty} A_k$ is a residual.

It is also clear that an open dense subset of a complete metric space is a residual, because its complement is obviously nowhere dense.

We leave it to the reader as an exercise to convince himself that the set of real numbers in whose "decimal" expansions in the base k number system some digit does not occur beyond a certain "decimal" place is a set of the first category in the space of real numbers, and to deduce that there exist numbers in all of whose "decimal" expansions (for different bases) each significant digit occurs infinitely many times (the set of such numbers is a residual).

§ 5. Compactness in metric spaces

5.1. In subsection 2.7 we introduced the concepts of compactness and sequential compactness and remarked that these concepts are, in general, distinct. However, it is important to single out classes of spaces for which these concepts coincide. We shall now show that they coincide for the class of metric spaces. Metrizability is, however, by no means a necessary condition for this to be so (see Chapter VIII).

LEMMA 1. *A point x of a metric space $\mathbf{X}$ is a cluster point of a sequence $\{x_n\}$ if and only if there is a subsequence $\{x_{n_k}\}$ converging to x.*

Proof. Let x be a cluster point of the sequence $\{x_n\}$. Write U_n for the ball $K_{1/n}(x)$. Take any point $x_{n_1} \in U_1$. Assume that $x_{n_1}, x_{n_2}, \ldots, x_{n_k}$ have been constructed, where $n_1 < n_2$

$< \ldots < n_k$ and $x_{n_i} \in U_i$ $(i = 1, 2, \ldots, k)$. Since x is a cluster point, there exists $m > n_k$ such that $x_m \in U_{k+1}$. Set $n_{k+1} = m$. Then the resulting sequence $\{x_{n_k}\}$ converges to x. The converse follows from Lemma 2.1.

The requirement of compactness for a space is very strong and distinguishes a comparatively narrow class of spaces: in particular, this class is narrower than the class of complete separable spaces.

THEOREM 1. *A compact metric space is complete.*

Proof. Let $\{x_n\}$ be a Cauchy sequence in a compact metric space $\mathbf{X}$. By Lemma 2.2 this sequence has a cluster point x. Lemma 1 shows that there exists a subsequence $\{x_{n_k}\}$ converging to x. Hence, by Lemma 4.1, $x_n \to x$.

We now introduce an important definition.

Let $\varepsilon > 0$ be a given positive number and M a subset of a metric space $\mathbf{X}$. Then M is called an *ε-net* for a set $E \subset \mathbf{X}$ if for each point $x \in E$ there exists a point z in M such that $\rho(x, z) < \varepsilon$.

THEOREM 2. *Let* $\mathbf{X}$ *be a metric space. The following conditions are equivalent:*

1) $\mathbf{X}$ *is compact;*
2) $\mathbf{X}$ *is sequentially compact;*
3) $\mathbf{X}$ *is countably compact;*
4) $\mathbf{X}$ *is complete and for each* $\varepsilon > 0$ *there exists a finite ε-net (for* $\mathbf{X}$*) in* $\mathbf{X}$.

Proof. 2) $\Rightarrow$ 3) for any topological space; 3) $\Rightarrow$ 2) by Lemma 1; 1) $\Rightarrow$ 3) for any topological space (see Lemma 2 of § 2). We now prove that 2) $\Rightarrow$ 4) $\Rightarrow$ 2) $\Rightarrow$ 1).

2) $\Rightarrow$ 4). The proof that $\mathbf{X}$ is complete is similar to the proof of Theorem 1. Assume that condition 4) does not hold; that is, assume that for some $\varepsilon > 0$ there is no finite ε-net. Take an arbitrary point $x_1 \in \mathbf{X}$. The set consisting of x_1 alone does not form an ε-net for $\mathbf{X}$, so there exists $x_2 \in \mathbf{X}$ such that $\rho(x_2, x_1) \geqslant \varepsilon$. The set $\{x_1, x_2\}$ also does not form an ε-net for $\mathbf{X}$ so there exists $x_3 \in \mathbf{X}$ with $\rho(x_i, x_3) \geqslant \varepsilon$ $(i = 1, 2)$. Proceeding in this way, we obtain a sequence $\{x_n\}$ of points of $\mathbf{X}$ such that $\rho(x_m, x_n) \geqslant \varepsilon$ $(m \neq n;\ m, n = 1, 2, \ldots)$. As one obviously cannot select a convergent subsequence of this, we have a contradiction to the sequential compactness of $\mathbf{X}$.

4) $\Rightarrow$ 2). Assume condition 4) is satisfied. We take an arbitrary sequence $\{x_n\}$ of elements of $\mathbf{X}$ and prove that we can select a convergent subsequence. To this end we choose a numerical sequence $\{\varepsilon_n\}$ $(\varepsilon_n > 0)$ converging to zero, and consider the ε_1-net whose existence is guaranteed by our hypothesis. If we construct balls of radius ε_1 with their centres at the points of this ε_1-net, then every point of $\mathbf{X}$ will occur in at least one of these balls. Since there are a finite number of balls, one of them must contain an infinite number of terms from the sequence $\{x_n\}$. Denote this ball by $K_{\varepsilon_1}(z_1)$. Next take an ε_2-net and consider balls of radius ε_2 with centres at all of its points. As before, one of these must contain an infinite number of terms from our sequence belonging to $K_{\varepsilon_1}(z_1)$: suppose it is the ball $K_{\varepsilon_2}(z_2)$. Continuing in this way, we obtain a sequence of balls $K_{\varepsilon_1}(z_1), K_{\varepsilon_2}(z_2), \ldots, K_{\varepsilon_n}(z_n), \ldots$, such that the intersection of any finite number of these contains an infinite number of points of our sequence. This shows that we can choose

$$x_{n_1} \in K_{\varepsilon_1}(z_1), \qquad x_{n_2} \in K_{\varepsilon_2}(z_2) \cap K_{\varepsilon_1}(z_1) \qquad (n_2 > n_1),$$

and, in general, $x_{n_k} \in \bigcap_{i=1}^{k} K_{\varepsilon_i}(z_i)$ $(n_k > n_{k-1} > \ldots > n_1)$.

Since x_{n_k} and x_{n_l} both belong to $K_{\varepsilon_k}(z_k)$ when $k \leqslant l$, we have

$$\rho(x_{n_k}, x_{n_l}) \leqslant \rho(x_{n_k}, z_k) + \rho(x_{n_l}, z_k) < 2\varepsilon_k,$$

and therefore $\{x_{n_k}\}$ is a Cauchy sequence, which, since $\mathbf{X}$ is complete, must converge to some $x_0 \in \mathbf{X}$. This shows that $\mathbf{X}$ is sequentially compact.

We have now proved that 2)$\Leftrightarrow$4). Next we deduce 1) from 2) and 4), which will conclude the proof of Theorem 2.

First notice that a space possessing property 4) is separable. For if $\varepsilon_k \to 0$ and $M_k (k \in \mathbb{N})$ is an ε_k-net, then $M = \bigcup_{k=1}^{\infty} M_k$ will clearly be a countable dense subset of the space.*

LEMMA 2. *Let $\mathbf{X}$ be any separable metric space. Then any system $\{G_\xi\}(\xi \in \Xi)$ of open sets covering $\mathbf{X}$ has a countable subsystem also covering $\mathbf{X}$.*

Proof. Let D be a countable dense subset of $\mathbf{X}$ and consider all the balls of rational radius with centres at points of D. The set of such balls is obviously countable. Let us enumerate them as follows: $S_1, S_2, \ldots, S_n, \ldots$. For each $x \in \mathbf{X}$ and $\xi \in \Xi$ there is clearly a ball $S_{n(x,\xi)}$ such that

$$x \in S_{n(x,\xi)} \subset G_\xi. \tag{1}$$

As x ranges over the whole of $\mathbf{X}$ and ξ ranges over the whole of Ξ, the suffix $n(x, \xi)$ ranges over some countable set $n_1, n_2, \ldots, n_k, \ldots$. Moreover, the balls S_{n_1}, $S_{n_2}, \ldots, S_{n_k}, \ldots$ cover $\mathbf{X}$. For each $k = 1, 2, \ldots$, choose $\xi_k \in \Xi$ such that $n(x_k, \xi_k) = n_k$, for some $x_k \in \mathbf{X}$. By equation (1) we have

$$\bigcup_{k=1}^{\infty} G_{\xi_k} \supset \bigcup_{k=1}^{\infty} S_{n(x_k, \xi_k)} = \bigcup_{k=1}^{\infty} S_{n_k} = \mathbf{X}.$$

We now complete the proof of Theorem 2. Again suppose that $\mathbf{X}$ is sequentially compact. Consider a cover $\{G_\xi\}$ of $\mathbf{X}$ by open sets. Since, by what we have proved, $\mathbf{X}$ is separable, we may assume, using Lemma 2, that the cover $\{G_\xi\}$ is countable—that is, we may assume that ξ ranges over the set of natural numbers. Assume that the sets

$$F_n = \mathbf{X} \setminus \bigcup_{\xi=1}^{n} G_\xi$$

are non-empty, for all $n \in \mathbb{N}$. Choose a point x_n in F_n. The sequence $\{x_n\}$ has a convergent subsequence $\{x_{n_k}\}$. Suppose that $x_{n_k} \to x$. Since $x_{n_k} \in F_n$ for $n_k \geqslant n$ and F_n is closed, we have also $x \in F_n$ $(n \in \mathbb{N})$. In other words,

$$x \in \bigcap_{n=1}^{\infty} F_n = \mathbf{X} \setminus \bigcup_{\xi=1}^{\infty} G_\xi,$$

which is, however, impossible, as $\bigcup_{\xi=1}^{\infty} G_\xi = \mathbf{X}$.

Hence $\mathbf{X} = \bigcup_{\xi=1}^{n} G_\xi$, for some n, and the space $\mathbf{X}$ is compact. This completes the proof of Theorem 2.

---

* In this connection, we wish to draw attention to a distinction between the notions of compactness and separability. Separability ensures the existence of a countable subset whose elements can be used to approximate arbitrary elements of the space with unbounded accuracy. For a compact space this can also be arranged by taking M to be the union of ε_k-nets. However, in this case we can approximate all the elements of the space uniformly, using a single ε-net with a finite number of elements; and for a separable space this is not, in general, possible.

In the course of the proof we established

COROLLARY 1. *A compact metric space is separable.*

COROLLARY 2. *Let E be a subspace of a metric space* **X**. *The following statements are equivalent*:

1) *E is relatively compact;*
2) *E is relatively sequentially compact;*
3) *E is relatively countably compact.*

Proof. 1) ⇒ 2) by Theorem 2; 2) ⇒ 3) is obvious. Let us show that 3) ⇒ 1). It is enough, by Theorem 2, to show that the closure $\bar{E}$ is countably compact. Let $\{x_n\}$ be a sequence in $\bar{E}$. For each n choose $y_n \in E$ such that $\rho(x_n, y_n) < 1/n$. Since E is relatively countably compact, the sequence $\{y_n\}$ has a cluster point $x \in \bar{E}$. Then x is clearly also a cluster point of $\{x_n\}$.

LEMMA 3. *Let E be a subset of a metric space* **X**. *If there is a finite ε-net for E in* **X**, *then there is a finite 2ε-net in E.*

Proof. Let $M \subset \mathbf{X}$ be a finite ε-net for E. For each $x \in M$, fix $y_x \in E$ such that $\rho(x, y_x) < \varepsilon$ (if such a point exists). Then the set $M_1 = \{y_x : x \in M\} \subset E$ is a finite 2ε-net.

Theorem 2 and Lemma 3 yield an important criterion for compactness, due to Hausdorff.

THEOREM 3 (Hausdorff). *For a subset E of a metric space* **X** *to be relatively compact, it is necessary—and, if* **X** *is complete, also sufficient—that* **X** *have a finite ε-net for E, for each* $\varepsilon > 0$.

REMARK. For a set to be relatively compact, it is also sufficient (in the case of a complete space **X**) for **X** to have a relatively compact ε-net for each $\varepsilon > 0$.

For then there will exist a finite ε-net for this relatively compact ε-net, and this finite ε-net will clearly be a 2ε-net for the original set, which therefore must be relatively compact.

5.2. Let us now turn to a study of compactness in specific spaces. A classical theorem of Bolzano and Weierstrass asserts that a subset of n-dimensional space $\mathbf{R}^n$ (or $\mathbf{C}^n$) is compact if and only if it is closed and bounded.

Let us now record some conditions for a set in the space **s** to be compact. It is not hard to see that a set $E \subset \mathbf{s}$ is relatively compact in **s** if and only if the set of kth coordinates of its points is bounded for each $k \in \mathbb{N}$; that is, if and only if

$$|\xi_k| \leqslant l_k \qquad \text{for} \qquad x = (\xi_1, \xi_2, \ldots, \xi_k, \ldots) \in E.$$

In other words, E must be contained in some parallelepiped in **s**.

The necessity of this condition is clear, for if we take a sequence of elements whose first coordinates, say, increase without bound, then it is obviously impossible to choose a subsequence that converges (coordinatewise).

To establish the sufficiency, we construct an ε-net for E. Choose k so large that $1/2^k < \varepsilon$ and consider the set H of points of the form $[x]_k = (\xi_1, \xi_2, \ldots, \xi_k, 0, \ldots)$, where $x = (\xi_1, \xi_2, \ldots, \xi_k, \xi_{k+1}, \ldots) \in E$. The set H is relatively compact (being essentially a bounded set in k-dimensional space). At the same time it is an ε-net for E, because we obviously have

$$\rho(x, [x]_k) = \sum_{n=k+1}^{\infty} \frac{1}{2^n} \frac{|\xi_n|}{1+|\xi_n|} \leqslant \sum_{n=k+1}^{\infty} \frac{1}{2^n} = \frac{1}{2^k} < \varepsilon.$$

Here, as before, $[x]_k$ denotes the sequence $(\xi_1, \xi_2, \ldots, \xi_k, 0, \ldots)$.

As we remarked above, the existence of a relatively compact ε-net is sufficient for E to be relatively compact.

REMARK. The compactness of parallelepipeds in **s**, which we have just proved, has the following consequence. If we have a collection of sequences, all bounded by the same number, then we can always pick out a sequence that is coordinatewise convergent: that is, a sequence $\{x_n\}$, $x_n = \{\xi_k^{(n)}\}_{k=1}^{\infty}$, such that all the limits

$$\lim_{n\to\infty} \xi_k^{(n)} = \xi_k^{(0)} \qquad (k \in \mathbb{N})$$

exist simultaneously. We shall use this fact repeatedly in what follows.

Next we give a compactness criterion for sets in $\mathbf{C}(K)$, where K is a metric compactum with metric r.

THEOREM 4 (Arzelà–Ascoli). *The following are necessary and sufficient conditions for a set E of continuous functions to be relatively compact in $\mathbf{C}(K)$:*

1) *the functions in E are bounded in aggregate: that is, there exists a constant M such that*

$$|x(t)| \leqslant M \qquad (x \in E, t \in K);$$

2) *the functions in E are equicontinuous: that is, for any $\varepsilon > 0$, there exists $\delta > 0$ such that $|x(t) - x(t')| < \varepsilon$ for all $x \in E$ whenever $r(t, t') < \delta$.*

Proof. Necessity. Assume that E is relatively compact in $\mathbf{C}(K)$. By Theorem 2 there exists a finite ε-net for E. Let $x_1, x_2, \ldots, x_n$ be the continuous functions forming this net. Since each x_i is bounded and for any $x \in E$ there exists an x_k with $\rho(x, x_k) < \varepsilon$, we have

$$|x(t)| \leqslant |x_k(t)| + |x(t) - x_k(t)| \leqslant \max_{t \in K} |x_k(t)| + \rho(x, x_k) < \max_{t \in K} |x_k(t)| + \varepsilon$$

and condition 1) is therefore satisfied if we take M to be a common upper bound for the functions $|x_k(t)|$ $(k = 1, 2, \ldots, n; t \in K)$, augmented by ε.

Moreover, for each x_k there exists δ_k such that

$$|x_k(t') - x_k(t)| < \varepsilon \qquad \text{for} \qquad r(t', t) < \delta_k.$$

Write $\delta = \min(\delta_1, \delta_2, \ldots, \delta_n)$. Take any function x in E. Let x_k be that element for which we have $\rho(x, x_k) < \varepsilon$. Then

$$|x(t') - x(t)| \leqslant |x(t') - x_k(t')| + |x_k(t') - x_k(t)| + |x_k(t) - x(t)| \leqslant$$
$$\leqslant \rho(x, x_k) + |x_k(t') - x_k(t)| + \rho(x, x_k) < 2\varepsilon + |x_k(t') - x_k(t)|.$$

If $r(t, t') < \delta$, then the second summand here is less than ε, and therefore

$$|x(t') - x(t)| < 3\varepsilon.$$

Hence the functions in E are equicontinuous.

Sufficiency. Since $\mathbf{C}(K)$ is a closed subspace of the space $\ell^\infty(K)$ of all bounded functions on K, it is sufficient to prove that E is relatively compact in $\ell^\infty(K)$. Since $\ell^\infty(K)$ is complete, the remark following Theorem 3 shows that it is sufficient to prove that E has a relatively compact ε-net for each $\varepsilon > 0$. Given $\varepsilon > 0$, choose $\delta > 0$ to satisfy the equicontinuity condition for E. Now K is a metric compactum, so by Hausdorff's Theorem there exists a finite $\delta/3$-net $\{t_k\}_{k=1}^{n}$ for K. Set

$$C_1 = K_{\delta/3}(t_1),$$

$$C_2 = K_{\delta/3}(t_2)\backslash C_1, \ldots, C_n = K_{\delta/3}(t_n)\backslash \bigcup_{k=1}^{n-1} C_k.$$

Since $K = \bigcup_{k=1}^{n} K_{\delta/3}(t_k)$, we have $K = \bigcup_{k=1}^{n} C_k$; moreover, $C_k \cap C_m = \varnothing$ for $k \neq m$, and for any k we have $r(t, t') < 2\delta/3$ if $t, t' \in C_k$. Therefore, by the choice of δ, if $t, t' \in C_k$, then $|x(t) - x(t')| < \varepsilon$ for all $x \in E$. Let x_k be the characteristic function of C_k; that is, $x_k(t) = 1$ if $t \in C_k$ and $x_k(t) = 0$ if $t \notin C_k$. Consider the set H consisting of functions of the form $\sum_{k=1}^{n} \lambda_k x_k(t)$, where the numbers λ_k satisfy $|\lambda_k| \leqslant M$ $(k = 1, 2, \ldots, n)$. The set H is compact in $\ell^{\infty}(K)$, since convergence of a sequence of such functions means convergence of the sequence of corresponding numbers λ_k.

Our proof will be complete if we show that H is a 2ε-net for E. Let $x \in E$. Fix an arbitrary point t_k in each C_k. (If any $C_k = \varnothing$, then we ignore it.) The function $y(t) = \sum_{k=1}^{n} x(t_k)x_k(t)$ belongs to H since $|x(t_k)| \leqslant M$ by condition 1). Choose any $t \in K$. As $K = \bigcup_{k=1}^{n} C_k$, this t lies in some C_m. We thus have

$$|x(t) - y(t)| = |x(t) - x(t_m)| < \varepsilon,$$

since $\rho_k(t, t_m) < 2\delta/3 < \delta$ by the construction of C_m and condition 2).

Since t was arbitrary, we obtain

$$\rho(x, y) = \sup_{t \in K} |x(t) - y(t)| \leqslant \varepsilon < 2\varepsilon,$$

showing that H is indeed a 2ε-net for E.

Let us illustrate the above theorem with some examples.

Consider the family $E = \{\sin nt\}$ $(n = 1, 2, \ldots)$ of functions in the interval $[0, \pi]$. The boundedness condition holds for this E. However, the second condition, on equicontinuity, is violated, since $\sin n\,(\pi/2n) = 1$. The set E is not compact in $\mathbf{C}[0, \pi]$.

Now consider the set E of functions satisfying *Hölder's inequality**

$$|x(t') - x(t)| \leqslant M|t' - t|^{\alpha} \qquad (x \in E, t' \in [a, b], \qquad 0 < \alpha \leqslant 1).$$

If the functions in E are bounded in aggregate, then E is compact. For the equicontinuity of the continuous functions in E is ensured by taking $\delta = (\varepsilon/M)^{1/\alpha}$, and hence both conditions of the Arzelà–Ascoli Theorem are satisfied.

More generally: let E be a bounded set of continuous functions and suppose there exist functions $\omega(\delta)$ such that $\omega(\delta) \to 0$ as $\delta \to 0$ and such that the modulus of continuity†

---

* For $\alpha = 1$ we have the *Lipschitz condition*. A function satisfying Hölder's inequality is often said to satisfy a Lipschitz condition of order α. The class of all such functions is denoted by Lip_{α}.

† Recall that the *modulus of continuity* of a function $x \in \mathbf{C}[a, b]$ is the quantity

$$\omega(x; \delta) = \max_{\substack{a \leqslant t, t' \leqslant b \\ |t' - t| \leqslant \delta}} |x(t') - x(t)| \qquad (0 < \delta \leqslant b - a).$$

$\omega(x;\delta)$ of every $x \in E$ satisfies the inequality $\omega(x;\delta) \leqslant \omega(\delta)$; then E is compact in $\mathbf{C}[a,b]$.
We leave it to the reader to formulate compactness criteria for sets in $\mathbf{C}^{(l)}[a,b]$.

§ 6. Measure spaces

6.1. Under Riemann's classical definition, the integral is defined for a class of functions only slightly wider than the class of continuous functions on an interval. The needs of analysis (in the first place, the theory of trigonometric series) led, in 1902, to the introduction by the French mathematician H. Lebesgue of the concept of a measure and of a more general concept of integral than that of Riemann, for functions on an interval. Subsequent generalizations, by J. Radon, M. Fréchet and C. Carathéodory, led to the construction of a theory of measure for an arbitrary set, and of an integral with respect to this measure. Abstract measure spaces proved very fruitful in analysis and the theory of equations, and also in probability theory, where they enabled A. N. Kolmogorov to develop an axiomatic approach to the science (in 1932).

In this section we present the basic facts of the theory of measure and integral without proofs. All that we require of the reader is a thorough working knowledge of at least the Lebesgue integral on an interval (see Natanson-II). The reader who is interested in the basic applications of functional analysis presented in Part II may take the abstract measure space to be a domain in $\mathbf{R}^n$ or an interval on the real line, with Lebesgue measure. A detailed exposition of the material presented in this section may be found in the following books: Akilov, Makarov and Khavin; Bourbaki-IV; Vulikh-III; Dunford and Schwartz-I; Zaanen-II; Kolmogorov and Fomin; Halmos; Shilov and Gurevich. Our presentation is closest to that of Vulikh-III.

6.2. Let T be an arbitrary set. A non-empty collection Σ of subsets of T is called an *algebra* if it satisfies the following two conditions:

1) if $A, B \in \Sigma$, then $A \cup B \in \Sigma$;
2) if $A \in \Sigma$, then $T \setminus A \in \Sigma$.

It is easy to deduce from the definition that an algebra contains, together with any pair of sets A and B, both their intersection $A \cap B$ and their difference $A \setminus B$. By induction, we see that an algebra is closed under unions and intersections of any finite number of sets. Notice also that we always have $\varnothing \in \Sigma$ and $T \in \Sigma$.

A non-empty collection Σ of subsets of a set T is called a *σ-algebra* if it is an algebra and is closed under unions, not only of a finite number of sets, but also of a countable number—in other words, if it satisfies

3) $A_n \in \Sigma$ $(n \in \mathbb{N})$ implies $A = \bigcup_{n=1}^{\infty} A_n \in \Sigma$.

Obviously, a σ-algebra is also closed under countable intersections of sets.

Let $\overline{\mathbb{R}} = [-\infty, +\infty]$, and let Σ be an algebra of subsets of a set T. A function $\phi : \Sigma \to \overline{\mathbb{R}}$ is called an *additive set function* if for any finite collection of pairwise disjoint sets $A_n \in \Sigma$ we have the equation

$$\phi\Big(\bigcup_n A_n\Big) = \sum_n \phi(A_n). \tag{1}$$

In addition, we shall always assume that $\phi(\varnothing) = 0$. We say that ϕ is *non-negative*, and write $\phi \geqslant \mathbf{0}$, if $\phi(A) \geqslant 0$ for every $A \in \Sigma$.

An additive set function $\phi : \Sigma \to \overline{\mathbb{R}}$ defined on a σ-algebra Σ is said to be *countably additive* if equation (1) holds also for every countable collection of pairwise disjoint sets $A_n \in \Sigma$.

A *measure* on a σ-algebra Σ is a non-negative countably additive set function μ defined on Σ.

A measure μ is said to be *finite* if $\mu(T) < \infty$, and *σ-finite* if T is the union of a countable collection of sets $A_n \in \Sigma$ such that $\mu(A_n) < \infty$. A measure is said to be *complete* if $A \subset B \in \Sigma$, $\mu(B) = 0$, implies $A \in \Sigma$ (and so, of course, $\mu(A) = 0$). Any measure μ can be "completed". For let Σ^* denote the collection of all sets of the form $A \cup N$, where $A \in \Sigma$ and $N \subset B \in \Sigma$, $\mu(B) = 0$. Then Σ^* is a σ-algebra; and, if the domain of definition of μ is extended to Σ^* by setting $\mu(A \cup N) = \mu(A)$, then the extended function will be a measure on Σ^*.

This construction shows that we can usually assume that a measure μ is complete.

We shall say that (T, Σ, μ) is a *measure space** if T is a set, Σ is a σ-algebra of subsets of T, μ is a complete measure on Σ, and the following conditions are satisfied:

1) if $A \subset T$ is a set such that $A \cap B \in \Sigma$ for every set $B \in \Sigma$, $\mu(B) < \infty$, then $A \in \Sigma$;

2) the measure μ is locally finite—that is, for any $A \in \Sigma$ with $\mu(A) > 0$, there exists $B \in \Sigma$ such that $B \subset A$ and $0 < \mu(B) < \infty$.

The conditions we have imposed are not too restrictive. In fact, one can arrange for 1) to be satisfied by extending the σ-algebra Σ to include sets A complying with 1). Moreover, if a set A with $\mu(A) = +\infty$ does not satisfy 2), then we can redefine the measure of A by setting $\mu(A) = 0$.

Notice that conditions 1) and 2) are automatically satisfied for a σ-finite measure. We denote the collection of all sets of finite measure in Σ by $\Sigma(\mu)$. From 1) and 2) we deduce the following property of (T, Σ, μ):

If $A \subset T$ is a set such that $A \cap B \in \Sigma$ and $\mu(A \cap B) = 0$, for every $B \in \Sigma(\mu)$, then $A \in \Sigma$ and $\mu(A) = 0$.

6.3. Let Σ be an algebra of subsets of a set T and $\phi : \Sigma \to \overline{\mathbb{R}}$ an additive set function. The functions associated with ϕ by the following formulae,

$$\phi_+(A) = \sup_{B \subset A,\, B \in \Sigma} \phi(B), \qquad \phi_-(A) = \sup_{B \subset A,\, B \in \Sigma} \{-\phi(B)\},$$

$$|\phi|(A) = \phi_+(A) + \phi_-(A) \quad (A \in \Sigma),$$

are called, respectively, the *positive variation*, the *negative variation* and the *total variation* of ϕ. All three of the functions ϕ_+, ϕ_-, $|\phi|$ are non-negative additive set functions. The total variation can also be defined by the formula

$$|\phi|(A) = \sup \sum_{i=1}^{n} |\phi(A_i)|,$$

where the supremum is taken over all possible partitions of A into a finite number of pairwise disjoint sets $A_i \in \Sigma$. If $\phi(A)$ is finite, for $A \in \Sigma$, then

$$|\phi|(A) = \sup_{\substack{B_1, B_2 \subset A \\ B_1, B_2 \in \Sigma}} \{\phi(B_1) - \phi(B_2)\}.$$

---

* Usually in the definition of a measure space there are no restrictions at all on Σ and μ, but this often leads to pathologies (especially in the case of a measure that is not σ-finite).

THEOREM 1. *For any finite additive set function ϕ defined on an algebra Σ we have the Jordan decomposition*

$$\phi = \phi_+ - \phi_-.$$

Notice that if Σ is a σ-algebra and the function ϕ is countably additive, then so are its variations ϕ_+, ϕ_-, and $|\phi|$. A stronger statement can be made for countably additive functions.

THEOREM 2 (The Hahn decomposition). *Let ϕ be a countably additive set function on a σ-algebra Σ. Then there exists $A_0 \in \Sigma$ such that $\phi(A) \geqslant 0$ if $A \in \Sigma$ and $A \subset A_0$, and $\phi(A) \leqslant 0$ if $A \in \Sigma$ and $A \subset T \setminus A_0$.*

Note that the conditions of Theorem 2 mean that for any $A \in \Sigma$ we have $\phi_+(A) = \phi(A \cap A_0)$, $\phi_-(A) = -\phi(A \cap (T \setminus A_0))$.

6.4. Let T be a set and Σ a σ-algebra of subsets of T. Call the sets in Σ *measurable*. A real-valued function $x(t)$ defined on T is said to be *measurable* if its *Lebesgue sets*

$$\begin{array}{ll} \{t \in T: x(t) > a\}, & \{t \in T: x(t) < a\}, \\ \{t \in T: x(t) \geqslant a\}, & \{t \in T: x(t) \leqslant a\} \end{array} \tag{2}$$

are measurable, for each $a \in \mathbb{R}$.

By a *simple measurable function* (*with respect to* Σ) we mean a function of the form

$$y(t) = \sum_{i=1}^{k} \lambda_i \chi_{A_i}(t),$$

where $\lambda_i \in \mathbb{R}$, $A_i \in \Sigma$ $(i = 1, 2, \ldots, k)$, and where the sets A_i are pairwise disjoint $(i = 1, 2, \ldots, k)$.

We say that a sequence of functions $x_n(t)$ on T is *increasing* ($x_n \uparrow$) if $x_n(t) \leqslant x_m(t)$ $(t \in T)$ whenever $m \geqslant n$. We write $x_n \uparrow x$ if $\{x_n\}$ is increasing and $x_n(t) \to x(t)$ for every $t \in T$.

THEOREM 3. *If a bounded function x is measurable with respect to a σ-algebra Σ, then there exists a sequence of simple measurable functions $\{x_n\}$ that converges to x uniformly on T. If $x(t) \geqslant 0$ $(t \in T)$, then we may assume in addition that $x_n \uparrow x$, $x_n(t) \geqslant 0$ $(t \in T, n \in \mathbb{N})$.*

Proof. Assume that $x(t) \geqslant 0$ $(t \in T)$. Let $M = \sup_{t \in T} x(t)$. For $n = 2^k$ $(k \in \mathbb{N})$ and $i = 0, 1, \ldots, n-1$, set

$$A_i^n = \{t \in T: iM/n < x(t) \leqslant (i+1)M/n\}.$$

Obviously, $A_i^n \in \Sigma$. Define a simple function x_n by the formula

$$x_n(t) = \sum_{i=0}^{n-1} \frac{iM}{n} \chi_{A_i^n}(t).$$

It is easy to see that $x_n \uparrow$. For if $m \geqslant n$ and $t \in T$, then either $x(t) = 0$, in which case $x_n(t) = x_m(t) = 0$, or t belongs to some A_i^m, in which case, since m and n are powers of two, A_i^m is contained in some A_j^n and therefore $i \geqslant j$. Hence $x_m(t) = iM/m \geqslant jM/n = x_n(t)$.

Moreover, if $t \in A_i^n$, then

$$|x(t) - x_n(t)| = |x(t) - iM/n| \leqslant M/n,$$

so that for any $t \in T$ we have

$$|x(t) - x_n(t)| \leqslant M/n.$$

Hence the sequence $\{x_n\}$ converges uniformly to x. If x is an arbitrary function, then, setting $x_+(t) = \max(x(t), 0)$ and $x_-(t) = \max(-x(t), 0)$, we see that x_+ and x_- are non-negative bounded measurable functions, and that $x = x_+ - x_-$. The proof of the Theorem is completed by applying the result just proved to x_+ and x_-.

COROLLARY. *For any measurable function x there exists a sequence of simple measurable functions $\{x_n\}$ such that $x_n(t) \to x(t)$ and $|x_n(t)| \leqslant |x(t)|$ for all $t \in T$. If $x(t) \geqslant 0$, then we may assume that $x_n \uparrow x$, $x_n(t) \geqslant 0$ $(t \in T, n \in \mathbb{N})$.*

Proof. Consider the "truncations" $[x]_n$ $(n \in \mathbb{N})$ of the function $x(t)$, defined by

$$[x]_n(t) = \begin{cases} x(t) & \text{if } |x(t)| \leqslant n, \\ 0 & \text{if } |x(t)| > n. \end{cases}$$

The functions $[x]_n$ are measurable and bounded; hence, if $x(t) \geqslant 0$ $(t \in T)$ then by Theorem 3 we can find a simple measurable function $y_n(t)$ for each $n \in \mathbb{N}$ such that $|[x]_n(t) - y_n(t)| < 1/n$, $0 \leqslant y_n(t) \leqslant [x]_n(t)$ for all $t \in T$. Clearly the sequence of functions

$$x_n(t) = \max_{k=1}^{n} y_k(t)$$

is the one we require. In the general case, we write $x = x_+ - x_-$, as in the proof of Theorem 3. If $0 \leqslant x'_n \uparrow x_+$ and $0 \leqslant y'_n \uparrow x_-$, where x'_n and y'_n are simple functions, then the sequence $x_n(t) = x'_n(t) - y'_n(t)$ is the one we require.

A most important concept in the theory of measurable functions is the concept of an *integral*. Suppose we are given a σ-algebra Σ of subsets of a set T and a countably additive simple function ϕ on Σ. Then in the class of all measurable functions we can single out a class of functions—called the *summable* functions (with respect to ϕ)—with each of which there is associated a certain finite number, called its (Radon) integral.* The integral of a function $x(t)$ (over the set T) is written in one of the following ways:

$$\int_T x(t)\, d\phi(t), \qquad \int_T x(t)\, d\phi, \qquad \int_T x\, d\phi.$$

Similarly, one can consider integrals over an arbitrary set $A \in \Sigma$.

Let us note the most important properties of the integral.

1) Linearity†:

$$\int_A (\lambda x + \mu y)\, d\phi = \lambda \int_A x\, d\phi + \mu \int_A y\, d\phi \qquad (\lambda, \mu \in \mathbb{R}).$$

2) $\int_A x\, d\phi = \int_A x\, d\phi_+ - \int_A x\, d\phi_-$, $\quad \left|\int_A x\, d\phi\right| \leqslant \int |x|\, d|\phi|$.

3) If ϕ is a measure and $x(t) \geqslant 0$ $(t \in A)$, then $\int_A x\, d\phi \geqslant 0$. Moreover, if $\int_A x\, d\phi = 0$, then $\phi(\{t \in A\colon x(t) > 0\}) = 0$:

4) If x is a bounded measurable function, then $\int_A x\, d\phi$ exists $(A \in \Sigma)$.

---

* We do not give the definition of the integral, but refer the reader to the literature cited at the beginning of this section.

† By definition, $(\lambda x + \mu y)(t) = \lambda x(t) + \mu y(t)$.

5) The set function $\nu(A) = \int_A x\, d\phi \ (A \in \Sigma)$ is countably additive.

There are also many other important properties, some of which we shall encounter later on.

In the case where ϕ is a measure, we extend the concept of integral by allowing it also to take infinite values.

Let (T, Σ, μ) be a finite measure space. If a measurable function $x(t) \geqslant 0$ $(t \in A)$ is not summable on the set $A \in \Sigma$, then we set $\int_A x(t)\, d\mu = +\infty$.

As before, for an arbitrary measurable function $x(t)$, we define $x_+(t) = \max(x(t), 0)$, $x_-(t) = \max(-x(t), 0)$ and, if at least one of the two integrals $\int_A x_+\, d\mu$, $\int_A x_-\, d\mu$ is finite, we set

$$\int_A x\, d\mu = \int_A x_+\, d\mu - \int_A x_-\, d\mu. \tag{3}$$

If both the above integrals are infinite, then $x(t)$ has no integral.

Let (T, Σ, μ) be an infinite measure space, and let $\Sigma(\mu)$ be the collection of all sets of finite measure. Using condition 1) of 6.2, we see that a function $x(t)$ is measurable if and only if the functions $x(t)\chi_A(t)$ are measurable for all $A \in \Sigma(\mu)$.

For a set A (of infinite measure) and a measurable function $x(t) \geqslant 0$ $(t \in A)$, we set

$$\int_A x(t)\, d\mu = \sup\left\{\int_B x(t)\, d\mu\colon\ B \in \Sigma(\mu),\ B \subset A\right\} \qquad (A \in \Sigma)$$

(this equation holds trivially if $A \in \Sigma(\mu)$).

For an arbitrary measurable function x we define the integral by equation (3), with the same stipulation about existence as before. The resulting integral has properties 1), 3), 5). As before, functions for which the integral $\int_A x\, d\mu$ is finite are said to be *summable* on $A \in \Sigma$, or simply summable if $A = T$.

Now we consider the problem of extending this theory to functions taking complex values. Let $x\colon T \to \mathbb{C}$ be a complex-valued function. We set $y(t) = \operatorname{Re} x(t)$, $z(t) = \operatorname{Im} x(t)$. The function $x(t)$ is said to be *measurable* (*summable*) if both the real-valued functions $y(t)$ and $z(t)$ are measurable (summable). If $y(t)$ and $z(t)$ are summable, then we set

$$\int_A x(t)\, d\phi = \int_A y(t)\, d\phi + i \int_A z(t)\, d\phi.$$

Sometimes it turns out to be convenient to deal with functions that also take infinite values. As before, a function $x\colon T \to [-\infty, \infty]$ is said to be measurable with respect to a σ-algebra Σ if all the Lebesgue sets (2) are measurable.

Let (T, Σ, μ) be a measure space. If, for each measurable function $x\colon T \to [-\infty, \infty]$, we write

$$A_\infty^+ = \{t \in T\colon\ x(t) = +\infty\},$$

$$A_\infty^- = \{t \in T\colon\ x(t) = -\infty\}, \quad A_\infty = A_\infty^+ \cup A_\infty^-,$$

then, in the case where $\mu(A_\infty) = 0$ and the integral $\int_{T\setminus A_\infty} x\,d\mu$ is defined, we set $\int_T x\,d\mu = \int_{T\setminus A_\infty} x\,d\mu$; if $\mu(A_\infty^+) > 0$, $\mu(A_\infty^-) = 0$, we set $\int_T x\,d\mu = +\infty$; and finally if $\mu(A_\infty^-) > 0$, $\mu(A_\infty^+) = 0$, we set $\int_T x\,d\mu = -\infty$. In the remaining cases, the integral is not defined. Once again, a function $x(t)$ is said to be *summable* if $\int_T x\,d\mu$ is finite—that is, if $\mu(A_\infty) = 0$ and the function $x\chi_{T\setminus A_\infty}$ is summable. The integral $\int_A x\,d\mu$ is defined analogously.

If a function $x(t)$ is only defined on the set $T\setminus A$, where $\mu(A) = 0$, then it is said to be *measurable* (on T) if it is measurable on $T\setminus A$. If $\int_{T\setminus A} x\,d\mu$ is defined, then we set $\int_T x\,d\mu = \int_{T\setminus A} x\,d\mu$.

In all subsequent definitions and theorems in this section, we shall assume that the functions take values in the extended real line or the complex field, with the qualification that if a function occurs in an inequality in the hypotheses then it takes values in $[-\infty, \infty]$.

6.5. Let (T, Σ, μ) be a measure space. We say that a property holds *almost everywhere* (abbreviated to a.e.) if it holds everywhere except possibly on a set of measure zero. For example, if x and y are measurable functions, then $x(t) \geqslant y(t)$ a.e. means that $\mu(\{t: x(t) < y(t)\}) = 0$. Measurable functions x and y are said to be *equivalent* if $x(t) = y(t)$ a.e. By condition 2) of 6.2, this is the same as requiring that $x(t) = y(t)$ for almost all $t \in A$, for every $A \in \Sigma(\mu)$. An integral does not "distinguish" between properties that hold almost everywhere and those that hold everywhere. For example, if $x(t) = y(t)$ a.e., then $\int x\,d\mu = \int y\,d\mu$.

Henceforth in this section we shall assume, unless we state otherwise, that all functions are measurable and take finite values a.e.—that is $\mu(\{t \in T: x(t) = +\infty$ or $x(t) = -\infty\}) = 0$.

One can disregard sets of measure zero not only when dealing with functions, but also when dealing with measurable sets in Σ. In this connection, we introduce the following notation. We write $A \subset B \pmod{\mu}$ to mean that $\mu(A\setminus B) = 0$, and we write $A = B \pmod{\mu}$ to mean that $A \subset B \pmod{\mu}$ and $B \subset A \pmod{\mu}$. Sets A and B are said to be (μ)-*disjoint* if $\mu(A \cap B) = 0$. A family of sets $\{A_\xi\}$ $(\xi \in \Xi)$, $A_\xi \in \Sigma$, is called a (μ)-*partition* of T if the A_ξ are pairwise (μ)-disjoint and $T = \bigcup_{\xi\in\Xi} A_\xi \pmod{\mu}$.

A sequence of measurable functions $\{x_n(t)\}$ is said to be *almost everywhere convergent* to a function $x(t)$ if $\mu(\{t: x_n(t) \not\to x(t)\}) = 0$. (Notice that it is easy to prove that the limit function must be measurable, even if we do not assume this at the outset.) Convergence almost everywhere is indicated by the notation $x_n \to x$ a.e.

Let $x_n(t)$ $(n \in \mathbb{N})$ and $x(t)$ be measurable functions and let $A \in \Sigma$ with $\mu(A) < \infty$. The sequence $\{x_n(t)\}$ is said to *converge in measure* on A to $x(t)$ if, for every $\varepsilon > 0$,

$$\mu(\{t \in A: |x_n(t) - x(t)| \geqslant \varepsilon\}) \xrightarrow[n\to\infty]{} 0.$$

In the general case, where $\mu(A)$ is not assumed finite, the sequence $\{x_n(t)\}$ is said to *converge in measure** on A to $x(t)$ if $\{x_n(t)\}$ converges in measure to $x(t)$ on every $B \in \Sigma$ with $B \subset A$, $\mu(B) < \infty$. Convergence in measure (on T) is indicated by the notation $x_n \to x\ (\mu)$. Convergence in measure of a net $\{x_\alpha(t)\}$ of measurable functions to a measurable function $x(t)$ (written $x_\alpha \to x\ (\mu)$) is defined analogously. The connection between convergence almost everywhere and convergence in measure is described in the following theorem.

THEOREM 4. 1) *If $x_n \to x$ a.e., then $x_n \to x\ (\mu)$.*

2) *If (T, Σ, μ) is a σ-finite measure space and $x_n \to x\ (\mu)$, then there exists a subsequence $\{x_{n_k}\}$ such that $x_{n_k} \to x$ a.e.*

It follows from Theorem 4 that addition and multiplication of two functions are continuous operations with respect to convergence in measure.

Let (T, Σ, μ) be a measure space. A set $A \in \Sigma$ is called an *atom* if $\mu(A) \neq 0$ and $B \in \Sigma$, $B \subset A$ imply that either $\mu(A) = \mu(B)$ or $\mu(B) = 0$. We can now formulate condition 2) of 6.2 as: μ has no atoms of infinite measure. One says that (T, Σ, μ) is a *discrete* measure space if $T = \bigcup_{\alpha \in A} T_\alpha \cup N$, where the T_α are atoms and $\mu(N) = 0$, and that (T, Σ, μ) is a *continuous* measure space if it has no atoms. It can be shown that in a continuous measure space there is always a sequence of measurable functions $\{x_n\}$ such that $x_n \to x(\mu)$ but $x_n \not\to x$ a.e.

Various facts about almost everywhere convergence are collected in the following theorem.

THEOREM 5. *Let (T, Σ, μ) be a σ-finite measure space.*

1) (The Stability of Convergence Theorem) *If $x_n \to 0$ a.e., then there exists an increasing sequence of positive numbers $\lambda_n \to +\infty$ such that $\lambda_n x_n(t) \to 0$ a.e.*

2) (The Convergence Regulator Theorem) *If $x_n \to 0$ a.e., then there exist a measurable, a.e. finite, non-negative function $y(t)$ on T and a sequence of positive numbers $\varepsilon_n \to 0$ such that $|x_n(t)| \leqslant \varepsilon_n y(t)$ a.e.*

3) (The Diagonal Sequence Theorem) *If $x_{nk} \to x_k$ a.e. as $n \to \infty$, for every $k \in \mathbb{N}$, and $x_k \to x$ a.e. as $k \to \infty$, then there exists a sequence $n_1 < n_2 < \ldots < n_k < \ldots$ such that the "diagonal" sequence $\{x_{n_k k}\}$ converges to x a.e. as $k \to \infty$.*

4) (Egoroff's Theorem) *If $\mu(T) < \infty$ and $x_n \to x$ a.e., then for every $\varepsilon > 0$ there exists $A \in \Sigma$ such that $\mu(A) < \varepsilon$ and $x_n(t) \to x(t)$ uniformly on $T \setminus A$.*

6.6. In this subsection we collect together various theorems about taking limits under integral signs. Let (T, Σ, μ) be a measure space.

THEOREM 6 (Lebesgue). *Let $\{x_n(t)\}$ be a sequence of summable functions and suppose that $x_n \to x\ (\mu)$. If there exists a non-negative summable function $y(t)$ such that $|x_n(t)| \leq y(t)$ a.e. ($n \in \mathbb{N}$), then $x(t)$ is also summable, and*

$$\lim \int_T x_n \, d\mu = \int_T x \, d\mu. \tag{4}$$

THEOREM 7 (B. Levi). *If $x_n(t) \geqslant 0$ a.e. and $x_n \uparrow x$ a.e., then (4) is true.*

Notice that Theorem 7 is true even in cases where the function x is not assumed almost everywhere finite.

COROLLARY. *If the summable functions $y_k(t)$ satisfy $y_k(t) \geqslant 0$ a.e. and $\sum_{k=1}^{\infty} \int_T y_k \, d\mu < +\infty$, then $x(t) = \sum_{k=1}^{\infty} y_k(t)$ is almost everywhere finite, and*

* In the case where the measure is infinite, this definition differs from the generally accepted one. This is to avoid the pathological properties of the usual convergence in measure in this case.

$$\int_T x\,d\mu = \sum_{k=1}^{\infty} \int_T y_k\,d\mu.$$

For let us set $x_n(t) = \sum_{k=1}^{n} y_k(t)$. Then $x_n \uparrow x$ a.e., where the function x could *a priori* turn out to be infinite on a set of positive measure. By the remark following Theorem 7, we have $\int_T x\,d\mu = \lim \int_T x_n\,d\mu = \lim \sum_{k=1}^{n} \int_T y_k\,d\mu = \sum_{k=1}^{\infty} \int_T y_k\,d\mu < \infty$. Hence x is summable, and therefore also almost everywhere finite.

THEOREM 8 (Fatou). *If $x_n(t) \geqslant 0$ a.e. and $x_n \to x\,(\mu)$, then*

$$\int_T x\,d\mu \leqslant \sup_n \int_T x_n\,d\mu.$$

Note that from Theorems 6 and 8 one can easily obtain the analogous results for a net $\{x_\alpha\}$ with $x_\alpha \to x\,(\mu)$.

6.7. In this subsection we formulate the Radon–Nikodym Theorem, which plays a fundamental role in applications of measure theory to functional analysis.

Let (T, Σ, μ) be a measure space and let ϕ be a countably additive function on Σ. Then ϕ is said to be *absolutely continuous* with respect to μ if $\mu(A) = 0$ implies $\phi(A) = 0$; and ϕ is said to be *singular* with respect to μ if there exists a set $A_0 \in \Sigma$ such that

$$\mu(A_0) = 0, \quad \phi(A) = \phi(A \cap A_0), \quad A \in \Sigma.$$

THEOREM 9. *If ϕ is a finite countably additive function on Σ, then ϕ can be expressed uniquely in the form $\phi(A) = v(A) + \lambda(A)$, $A \in \Sigma$, where v and λ are countably additive functions on Σ, such that v is absolutely continuous and λ is singular, with respect to μ.*

The expression $\phi = v + \lambda$ is called the *Lebesgue decomposition.* It is clear that, if x is summable, then the set function $v(A) = \int_A x\,d\mu\ (A \in \Sigma)$ is absolutely continuous with respect to μ. The Radon–Nikodym Theorem shows that every absolutely continuous function is obtained in this way.

THEOREM 10 (Radon–Nikodym). *Let (T, Σ, μ) be a σ-finite measure space and let v be a finite set function, defined on Σ and absolutely continuous with respect to μ. Then there exists a function $x(t)$ on T (unique to within measure zero) that is summable with respect to μ and satisfies*

$$v(A) = \int_A x\,d\mu \quad (A \in \Sigma).$$

Moreover, v is non-negative if and only if $x(t) \geqslant 0$ a.e.

6.8. The aim in this subsection is to define the product of two measure spaces. This concept plays an important role in the study of integral operators.

Let (S, Σ_S, v) and (T, Σ_T, μ) be σ-finite measure spaces. We denote by Σ_R^0 the smallest σ-algebra of $R = S \times T$ containing all sets of the form $B \times A$, where $B \in \Sigma_S$, $A \in \Sigma_T$. It can be proved that there is a unique measure λ on Σ_R^0 such that

$$\lambda(B \times A) = v(B)\mu(A)$$

for all $B \in \Sigma_S$, $A \in \Sigma_T$. (We reckon $a \cdot (+\infty) = +\infty$, if $a \neq 0$, and $0 \cdot (+\infty) = 0$.) We "complete" the measure λ in the manner described in 6.2, and denote the resulting σ-algebra by Σ_R and the measure again by λ. The resulting complete σ-finite measure space (R, Σ_R, λ) is called the *product of the spaces* (S, Σ_S, ν) and (T, Σ_T, μ), and λ is called the *product of the measures* ν and μ, which we shall sometimes denote by $\lambda = \nu \times \mu$.

The most-well-known example of a product measure is obtained when we take both the factors to be the interval $[0, 1]$ with Lebesgue measure. The product of these measures is Lebesgue measure of the unit square.

For brevity, we shall refer to functions that are measurable with respect to Σ_S, Σ_T, or Σ_R as ν-measurable, μ-measurable or λ-measurable, respectively.

We now record two theorems on evaluating an integral with respect to a product measure by iterating integrals with respect to the factor-measures.

THEOREM 11 (Fubini). *Assume that the function $K(s, t)$ $(s \in S, t \in T)$ is summable with respect to the measure $\lambda = \nu \times \mu$. Then the function $K_s(t) = K(s, t)$ is summable with respect to μ for almost all (relative to ν) points $s \in S$. Furthermore, the function $H(s) = \int_T K(s, t)\, d\mu(t)$ is summable with respect to ν and*

$$\int_R K(s, t)\, d\lambda(s, t) = \int_S \left\{ \int_T K(s, t)\, d\mu(t) \right\} d\nu(s).$$

For non-negative functions Fubini's Theorem can be sharpened.

THEOREM 12 (Tonelli). *Let $K(s, t)$ be a λ-measurable non-negative function on R. Then the function $K_s(t) = K(s, t)$ is μ-measurable for almost all (relative to ν) points $s \in S$. Furthermore, the function $H_s = \int_T K(s, t) d\mu(t)$ (which may possibly take infinite values on a set of positive measure) is ν-measurable, and*

$$\int_R K(s, t)\, d\lambda(s, t) = \int_S \left\{ \int_T K(s, t)\, d\mu(t) \right\} d\nu(s)$$

regardless of whether these integrals take finite or infinite values.

Let us note some further properties of products of measure spaces:

1) If the measures μ and ν are finite, then for each set $C \in \Sigma_R$ and each $\varepsilon > 0$ there exists a set $C_1 = \bigcap_{k=1}^{n} (B_k \times A_k)$, where $B_k \in \Sigma_S$, $A_k \in \Sigma_T$, such that $\int_R |\chi_C(s, t) - \chi_{C_1}(s, t)|\, d\lambda(s, t) < \varepsilon$.

2) If a function $K(s, t)$ is λ-summable and

$$\int_{B \times A} K(s, t)\, d\lambda(s, t) \geqslant 0,$$

for every $A \in \Sigma_T(\mu)$, $B \in \Sigma_S(\nu)$, then $K(s, t) \geq 0$ a.e. (λ).

6.9. Originally, measure theory was closely connected with topology, and for a long time measures were constructed only on certain classes of topological spaces. Then the efforts of a series of mathematicians (first and foremost, C. Carathéodory) were turned to the construction of a theory of abstract measure spaces, some results of which were briefly surveyed in the preceding subsections. In this subsection we record some facts of measure theory that are connected with topology.

Let K be a compactum. The smallest σ-algebra $\mathscr{B}$ containing all the closed sets of K is called the *Borel σ-algebra* of K, and the sets in $\mathscr{B}$ are called the *Borel sets* of K.

A finite countably additive set function ϕ defined on a σ-algebra Σ of subsets of a compactum K containing all the closed sets is said to be *regular* if for each $A \in \Sigma$ and each $\varepsilon > 0$ there exist a closed set F and an open set G such that $F \subset A \subset G$ and $|\phi|(G \setminus F) < \varepsilon$. If ϕ is regular, then so are its variations $\phi_+, \phi_-, |\phi|$.

Suppose μ_0 is a regular measure defined on a Borel σ-algebra $\mathscr{B}$. Form the completion of μ_0 in the manner described in 6.2. We denote the resulting σ-algebra (which, in general, is larger than $\mathscr{B}$) by Σ and the extended measure by μ. In this way one obtains, for instance, the Lebesgue measure on a compactum in $\mathbf{R}^n$ from its Borel measure. We shall denote the Lebesgue measure of a set A in $\mathbf{R}^n$ by mes (A). Now let μ and Σ be as above. Measurability of a function on K will be understood to mean measurability with respect to Σ, and the term a.e. will mean with respect to μ. Obviously, every continuous function is measurable. The following theorem of Luzin shows that every measurable function is continuous on sets that approximate to compacta with respect to μ..

THEOREM 13 (Luzin). *For an a.e. finite function $x(t)$ on a compactum K, the following statements are equivalent:*

1) *$x(t)$ is measurable;*

2) *for each $\varepsilon > 0$ there exists a closed set $F \subset K$ such that $\mu(K \setminus F) < \varepsilon$ and the restriction of $x(t)$ to F is continuous;*

3) *for each $\varepsilon > 0$ there exists a closed set $F \subset K$ and a continuous function $y(t)$ on K such that $\mu(K \setminus F) < \varepsilon$, $x(t) = y(t)$ for $t \in F$, and $\sup_{t \in F} |x(t)| = \sup_{t \in K} |y(t)|$.*

One easy consequence of Luzin's Theorem is a theorem of Fréchet, asserting that every measurable a.e. finite function $x(t)$ on K is a limit, in the sense of a.e. convergence, of a sequence of continuous functions.

In concluding our survey of measure and integral theory, we consider set functions taking complex values. Additivity, countable additivity, total variation and regularity can be defined for these functions exactly as in the real case. If $\phi: \Sigma \to \mathbb{C}$ is countably additive, then we consider the real-valued countably additive functions $\phi_1(A) = \operatorname{Re} \phi(A)$, $\phi_2(A) = \operatorname{Im} \phi(A)$. The integral can be introduced using the formula $\int_A x(t)\,d\phi = \int_A x(t)\,d\phi_1 + i \int_A x(t)\,d\phi_2$. We shall need complex set functions very infrequently, so unless we specify otherwise all set functions will be assumed real-valued.

6.10. In this book various spaces of measurable functions will play a fundamental role. We now consider the properties of the space of all measurable functions.

Let (T, Σ, μ) be a measure space. We denote by $\mathbf{S}(T, \Sigma, \mu)$ the collection of all measurable functions defined on T and finite almost everywhere. Furthermore, we agree to identify equivalent functions—that is, we reckon them as the same element of $\mathbf{S}(T, \Sigma, \mu)$. Thus, in what follows, elements of $\mathbf{S}(T, \Sigma, \mu)$ will be equivalence classes of functions, and if $x \in \mathbf{S}(T, \Sigma, \mu)$ is an equivalence class of functions, then we shall write $x(t)$ for any measurable function in this class (we can, moreover, always assume that $x(t)$ takes only finite values).

When μ is a σ-finite measure, we can turn $\mathbf{S}(T, \Sigma, \mu)$ into a metric space in which convergence with respect to the metric coincides with convergence in measure.

Hence we now assume, unless we say otherwise, that μ is σ-finite. Let us construct the metric in $\mathbf{S}(T, \Sigma, \mu)$. To do this we take a measurable function $f(t)$ on T satisfying the condition:

$$f(t) > 0 \text{ for every } t \in T; \int_T f(t)\, d\mu(t) = 1. \tag{5}$$

Such functions $f(t)$ do exist. For, as μ is σ-finite, we have $T = \bigcup_{n=1}^{\infty} T_n$, where the T_n are pairwise disjoint and $0 < \mu(T_n) < \infty$ $(n \in \mathbb{N})$. Set

$$f(t) = \sum_{n=1}^{\infty} \chi_{T_n}(t)/(2^n \mu(T_n)).$$

Then $f(t)$ satisfies the condition (5). If $\mu(T) < \infty$, then we may set $f(t) = 1/\mu(T)$ $(t \in T)$. We define the distance between any two elements x, y by the formula

$$\rho(x, y) = \int_T \frac{|x(t) - y(t)|}{1 + |x(t) - y(t)|} f(t)\, d\mu(t). \tag{6}$$

The integral in (6) is finite, because f is summable and the first factor takes values between 0 and 1. Obviously, $\rho(x, y)$ does not depend on the choice of the functions $x(t)$ and $y(t)$ in the corresponding equivalence classes. Let us check the axioms for a metric space (see 3.1).

Since the integrand in (6) is non-negative, $\rho(x, y) \geqslant 0$. Also it is clear that $\rho(x, x) = 0$. Suppose that $\rho(x, y) = 0$. Then $\dfrac{|x(t) - y(t)|}{1 + |x(t) - y(t)|} = 0$ almost everywhere, whence $x(t) = y(t)$ a.e., and such functions are identified in $\mathbf{S}(T, \Sigma, \mu)$. Therefore condition 1) in the definition of a metric space is satisfied. Condition 2) is obviously satisfied. From inequality (2) of § 3 we see that for any $x, y, z \in \mathbf{S}(T, \Sigma, \mu)$ and for almost all $t \in T$ we have

$$\frac{|x(t) - y(t)|}{1 + |x(t) - y(t)|} \leqslant \frac{|x(t) - z(t)|}{1 + |x(t) - z(t)|} + \frac{|y(t) - z(t)|}{1 + |y(t) - z(t)|}. \tag{7}$$

To obtain the triangle inequality, multiply (7) through by $f(t)$ and integrate the resulting inequality.

Let $f(t)$ and $g(t)$ be two functions satisfying (5) and let ρ_f and ρ_g be the metrics associated with them by (6). If f and g are not equivalent, then $\rho_f \neq \rho_g$. However, the topologies induced by these metrics coincide, as is shown, for example, by the following theorem.

THEOREM 14. *A sequence* $\{x_n\} \subset \mathbf{S}(T, \Sigma, \mu)$ *converges to* $x \in \mathbf{S}(T, \Sigma, \mu)$ *relative to the metric if and only if* $x_n \to x(\mu)$.

Proof. Suppose $\rho(x_n, x) \to 0$. Given $\varepsilon > 0$, we set $A_n(\varepsilon) = \{t \in T\colon |x_n(t) - x(t)| f(t) \geqslant \varepsilon\}$. Using the fact that $\phi(\lambda) = \lambda/(1 + \lambda)$ is an increasing function, we have

$$\rho(x_n, x) = \int_T \frac{|x_n(t) - x(t)|}{1 + |x_n(t) - x(t)|} f(t)\, d\mu(t) \geqslant \int_{A_n(\varepsilon)} \frac{|x_n(t) - x(t)|}{1 + |x_n(t) - x(t)|} f(t)\, d\mu(t)$$

$$\geqslant \frac{\varepsilon}{1 + \varepsilon} \mu[A_n(\varepsilon)].$$

Hence $\mu[A_n(\varepsilon)] \to 0$, and so $x_n f \to xf(\mu)$. Since $f(t) > 0$ $(t \in T)$ we have $x_n = (x_n f)/f \to (xf)/f = x(\mu)$.

Conversely, if $x_n \to x(\mu)$ then obviously

$$\frac{|x_n(t) - x(t)|}{1 + |x_n(t) - x(t)|} f(t) \to 0 \ (\mu).$$

Since f is summable, Lebesgue's Theorem shows that we can take limits under the integral sign, giving

$$\rho(x_n, x) = \int_T \frac{|x_n(t) - x(t)|}{1 + |x_n(t) - x(t)|} f(t)\, d\mu(t) \underset{n \to \infty}{\longrightarrow} 0.$$

REMARK. The analogue of Theorem 14 for nets is also true.

If $\rho(x_\alpha, x) \to 0$ then the proof that $x_\alpha \to x(\mu)$ proceeds exactly as in the theorem. Assume that $x_\alpha \to x(\mu)$ but $\rho(x_\alpha, x) \not\to 0$. Then we can find a $\delta > 0$ and a subnet $\{y_\beta\}$ such that $\rho(y_\beta, x) \geqslant \delta$ for all β. As f is summable, there exists a set $A \in \Sigma(\mu)$ for which $\int_{T \setminus A} f\, d\mu < \delta/2$. Since $y_\beta \to x(\mu)$, there exists β_n for each $n \in \mathbb{N}$ such that $\mu(\{t \in A : |y_{\beta_n}(t) - x(t)| > 1/n\}) < 1/n$. Then clearly $y_{\beta_n} \to x(\mu)$ on A and, by Theorem 14,

$$\rho(y_{\beta_n}, x) = \int_A \frac{|x_{\beta_n} - x|}{1 + |y_{\beta_n} - x|} f\, d\mu + \int_{T \setminus A} \frac{|y_{\beta_n} - x|}{1 + |y_{\beta_n} - x|} f\, d\mu < \int_A \frac{|y_{\beta_n} - x|}{1 + |y_{\beta_n} - x|} f\, d\mu + \frac{\delta}{2} \underset{n \to \infty}{\longrightarrow} \frac{\delta}{2},$$

contradicting the fact that $\rho(y_{\beta_n}, x) \geqslant \delta > 0 \, (n \in \mathbb{N})$.

THEOREM 15. *$\mathbf{S}(T, \Sigma, \mu)$ is a complete metric space.*

Proof. Let $\{x_n\}$ be a Cauchy sequence of elements of $\mathbf{S}(T, \Sigma, \mu)$.

Choose n_k such that $\rho(x_n, x_{n_k}) < 2^{-k}$ for $n > n_k$. We may assume that $n_1 < n_2 < \ldots < n_k < \ldots$, and hence that $n_k \to \infty$. The series

$$\sum_{k=1}^{\infty} \rho(x_{n_{k+1}}, x_{n_k}) = \sum_{k=1}^{\infty} \int_T \frac{|x_{n_{k+1}}(t) - x_{n_k}(t)|}{1 + |x_{n_{k+1}}(t) - x_{n_k}(t)|} f(t)\, d\mu(t)$$

is obviously convergent. But if a series of integrals of positive functions is convergent, then by the Corollary to Theorem 7 the series of the functions themselves is also convergent: that is, the series $\sum_{k=1}^{\infty} \alpha_k(t)$ is almost everywhere convergent, where

$$\alpha_k(t) = \frac{|x_{n_{k+1}}(t) - x_{n_k}(t)|}{1 + |x_{n_{k+}}(t) - x_{n_k}(t)|}.$$

If this series is convergent for some t, then $|x_{n_{k+1}}(t) - x_{n_k}(t)| \leqslant 1$ for sufficiently large k. Hence the inequality $|x_{n_{k+1}}(t) - x_{n_k}(t)| \leqslant 2\alpha_k(t)$ holds for such t. Thus the series $x_{n_1}(t) + \sum_{k=1}^{\infty} (x_{n_{k+1}}(t) - x_{n_k}(t))$ is almost everywhere convergent. Denote its sum by $x_0(t)$. This is a measurable function. The convergence of the series shows that $x_{n_k}(t) \to x_0(t)$ almost everywhere, and so we certainly have convergence in measure (*a fortiori*), which is convergence in the space $\mathbf{S}(T, \Sigma, \mu)$; and therefore $\rho(x_{n_k}, x_0) \to 0$.

We have now established that the subsequence converges. But $n_k \to \infty$ as $n \to \infty$, so the inequality $\rho(x_k, x_0) \leqslant \rho(x_k, x_{n_k}) + \rho(x_{n_k}, x_0)$ shows that $\rho(x_k, x_0) \to 0$: that is, that $x_k \to x_0$ in $\mathbf{S}(T, \Sigma, \mu)$.

We obtain important special cases by taking (T, Σ, μ) to be the interval $[a, b]$ with Lebesgue measure—in this case we write $\mathbf{S}(a, b)$ for $\mathbf{S}(T, \Sigma, \mu)$—or by taking T to be $\mathbb{N}$, the σ-algebra Σ to be the collection of all subsets of $\mathbb{N}$, and the measure of each point of T to be unity—in which case we obtain the space $\mathbf{s}$ introduced in §3.

Let us now investigate the separability of $\mathbf{S}(T, \Sigma, \mu)$. First of all, we note that, by the Corollary to Theorem 3, the set of simple measurable functions is dense in $\mathbf{S}(T, \Sigma, \mu)$.

Identifying subsets A and B in Σ whenever $A = B \pmod{\mu}$, we introduce a metric ρ_μ on Σ by setting

$$\rho_\mu(A, B) = \rho(\chi_A, \chi_B); \quad A, B \in \Sigma.$$

The measure μ is said to be *separable* if it is σ-finite and the metric space (Σ, ρ_μ) is separable.

THEOREM 16. *Let μ be a σ-finite measure. The metric space $\mathbf{S}(T, \Sigma, \mu)$ is separable if and only if μ is separable.*

Proof. Assume that $\mathbf{S}(T, \Sigma, \mu)$ is separable. Consider the set $H = \{\chi_A : A \in \Sigma\}$, which we can identify with Σ. Since the metrics ρ and ρ_μ coincide on Σ and separability is hereditary in metric spaces (see 4.4), H is separable, and therefore so is Σ.

Conversely, assume that μ is separable. Let Σ_0 be a countable dense subset of Σ. Consider the set $M = \left\{\sum_{k=1}^{m} r_k \chi_{A_k} : A_k \in \Sigma_0, m \in \mathbb{N}\right\}$, where the r_k range over all the rational numbers.* Obviously, M is countable. We show that M is dense in $\mathbf{S}(T, \Sigma, \mu)$. As we have already mentioned, the set of simple functions is dense in $\mathbf{S}(T, \Sigma, \mu)$, and so therefore is the set of simple functions with rational coefficients. Thus it is sufficient to approximate a function of the form $y(t) = \sum_{k=1}^{m} r_k \chi_{A_k}(t)$, where the r_k are rational and $A_k \in \Sigma$, by functions in M. Since Σ_0 is dense in Σ, there exist sequences $\{A_k^n\}_{n=1}^{\infty}$ $(k = 1, 2, \ldots, m)$ in Σ_0 such that $\rho_\mu(A_k^n, A_k) \to 0$ as $n \to \infty$. Thus $\rho(\chi_{A_k^n}, \chi_{A_k}) \to 0$ as $n \to \infty$, and so $M \ni \sum_{k=1}^{m} r_k \chi_{A_k^n}(t) \to y(t)(\mu)$ as $n \to \infty$, which is the result we require.

Notice that Lebesgue measure on any measurable subset of $\mathbf{R}^n$ is separable. For if D is any measurable subset in $\mathbf{R}^n$ with Lebesgue measure, then let $\mathbf{S}(D)$ denote the corresponding space of measurable functions. We show that $\mathbf{S}(D)$ is separable.

Assume to begin with that D is B_m, the ball of radius m in $\mathbf{R}^n$ with centre at the origin. Then B_m is a metric compactum. We shall prove later (see Theorem IV.4.3) that the space of continuous functions $\mathbf{C}(B_m)$ is separable. Since the topology induced on $\mathbf{C}(B_m)$ by $\mathbf{S}(B_m)$ is clearly weaker than the usual topology of $\mathbf{C}(B_m)$, the separability of $\mathbf{C}(B_m)$ is certainly assured. However, by the Remark following Theorem 13, the continuous functions are dense in $\mathbf{S}(B_m)$, and this shows that $\mathbf{S}(B_m)$ is indeed separable. If $x \in \mathbf{S}(\mathbf{R}^n)$, then $x_m(t)$

---

* In the complex case one must take $M = \{\sum_{k=1}^{m} (r_k + is_k)\chi_{A_k} : A_k \in \Sigma_0, m \in \mathbb{N}\}$, where r_k and s_k range over the rational numbers.

$= x(t)\chi_{B_m}(t) \to x(t)$ for any $t \in \mathbf{R}^n$. Consequently, $\bigcup_{m=1}^{\infty} \mathbf{S}(B_m)$ is dense in $\mathbf{S}(\mathbf{R}^n)$, and so the union of the countable everywhere dense subsets of the sets $\mathbf{S}(B_m)$, for $m \in \mathbb{N}$, is dense in $\mathbf{S}(\mathbf{R}^n)$, which proves that $\mathbf{S}(\mathbf{R}^n)$ is separable. If now D is an arbitrary subset, then $\mathbf{S}(D)$ is separable since it may be identified with a subspace of $\mathbf{S}(\mathbf{R}^n)$.

We conclude this section by investigating the order properties of the real space $\mathbf{S}(T, \Sigma, \mu)$.

We introduce an ordering on the real space $\mathbf{S}(T, \Sigma, \mu)$ by setting $x \leqslant y (x, y \in \mathbf{S}(T, \Sigma, \mu))$ if $x(t) \leqslant y(t)$ a.e.

Let M be a subset of $\mathbf{S}(T, \Sigma, \mu)$ that is bounded above—that is, assume there exists a $y \in \mathbf{S}(T, \Sigma, \mu)$ such that $x \leqslant y$ for all $x \in M$. If M is countable, then clearly $x_0 = \sup M$ exists in the ordered space $\mathbf{S}(T, \Sigma, \mu)$, where x_0 is the equivalence class of functions containing the function

$$x_0(t) = \sup\{x(t)\colon\ x \in M\} \quad (t \in T). \tag{8}$$

The function $x_0(t)$ depends on the choice of the functions $x(t)$ in the classes $x \in M$, but since M is countable (8) will yield equivalent functions. If M is uncountable, this definition is not, in general, possible, for in this case the function defined by (8) may turn out to be non-measurable, or we may obtain two measurable but inequivalent functions from (8) for different choices of the representatives $x(t)$, $x \in M$. For example, if A is a Lebesgue non-measurable subset of the interval $[0, 1]$, and M consists of the characteristic functions of all the one-element subsets of A, then $x_0(t) = \sup\ \{x(t)\colon x \in M\}$ is the characteristic function of A, and is accordingly non-measurable; however, $\sup M$ is clearly the class of all functions equal to zero almost everywhere.

By (8), $(x \vee y)(t) = \max(x(t), y(t))$ and $(x \wedge y)(t) = \min(x(t), y(t))$.

THEOREM 17. *Assume that the measure μ is σ-finite, and let M be an arbitrary non-empty subset of $\mathbf{S}(T, \Sigma, \mu)$ that is bounded above. Then*

a) $x_0 = \sup M \in \mathbf{S}(T, \Sigma, \mu)$ *exists;*

b) *there exists a countable subset* $\{x_n\} \subset M$ *such that* $\sup x_n = \sup M$.

Proof. Take any function $x' \in M$ and consider the set M' consisting of all functions of the form $x \vee x'$, where $x \in M$. Clearly $\sup M$ and $\sup M'$ must exist simultaneously and be equal. Next consider the set M'' consisting of all functions of the form $x - x'$, where $x \in M'$. This consists of non-negative functions and is bounded above. If we show that $y_0 = \sup M''$ exists, then $\sup M = x' + y_0$. Thus we may assume at the outset, without losing any generality, that M consists of non-negative functions and contains, together with any functions $x_1, x_2, \ldots, x_p$, also the function $x_1 \vee x_2 \vee \ldots \vee x_p$.

For any $x \in \mathbf{S}(T, \Sigma, \mu)$, we write $m(x) = \rho(x, 0)$. Since the function $\phi(\lambda) = \lambda/(1 + \lambda)$ is increasing on $[0, +\infty)$, we have $m(x_1) \leqslant m(x_2)$ whenever $0 \leqslant x_1 \leqslant x_2$.

Now we put $m_0 = \sup\{m(x)\colon x \in M\}$ and choose a sequence $\{x_n\} \subset M$ such that $m(x_n) \to m_0$. Replacing x_n by $x_1 \vee x_2 \vee \ldots \vee x_n$ where necessary, we may assume that the sequence $\{x_n\}$ is increasing.

Let $x_0(t) = \sup x_n(t) = \lim x_n(t)$. Since $x_0 \leqslant y$, we have $x_0 \in \mathbf{S}(T, \Sigma, \mu)$. By Lebesgue's Theorem, $m(x_n) \to m(x_0)$, and so $m(x_0) = m_0$. If we verify that $x_0 = \sup M$, then Theorem 17 will be proved.

If y is an upper bound for M, then clearly $y \geqslant x_0$. It remains to show that x_0 is an upper bound for M. Take any $x \in M$ and set $x_n' = x_n \vee x$, $x_0' = x_0 \vee x$. Since $x_n \uparrow x$, we have $x_n' \uparrow x_0'$. Therefore $m(x_n') \to m(x_0')$. In view of the properties of M, we have $x_n' \in M$, and so

$m(x'_0) \leqslant m_0$. On the other hand, $x'_0 \geqslant x_0$, and therefore $m(x'_0) \geqslant m(x_0) = m_0$, so that $m(x'_0) = m_0$. Consequently,

$$\int_T \left(\frac{x'_0}{1+x'_0} - \frac{x_0}{1+x_0} \right) d\mu = 0.$$

As the integrand is non-negative, $x'_0(t) = x_0(t)$ a.e., which shows that $x_0 = \sup M = \sup x_n$, as required.

COROLLARY 1. *Assume that the measure μ is σ-finite, and let M be any non-empty subset of $\mathbf{S}(T, \Sigma, \mu)$ that is bounded below. Then*

a′) $x_0 = \inf M \in \mathbf{S}(T, \Sigma, \mu)$ *exists;*

b′) *there exists a countable subset $\{x_n\} \subset M$ such that* $\inf x_n = \inf M$.

To prove the Corollary, simply consider the set $-M = \{-x : x \in M\}$.

From the proof of Theorem 17 it is simple to deduce the following more general statement.

COROLLARY 2. *If M is a subset of $\mathbf{S}(T, \Sigma, \mu)$ such that* $\sup\{x(t) : x \in M\} < \infty$ *a.e., then* $x_0 = \sup M \in \mathbf{S}(T, \Sigma, \mu)$ *exists. Furthermore, if M is directed, then one can choose a sequence $\{x_n\} \subset M$ such that $x_n \uparrow x$ a.e.*

Statement (b) of Theorem 17 holds for every subset M that is bounded above if and only if μ is σ-finite. Statement (a) holds for a significantly wider class of measure spaces.

We shall say that a measure space (T, Σ, μ) has the *direct sum property* if the following condition is satisfied:

There exists a family of pairwise disjoint subsets $T_\xi \in \Sigma$, $0 < \mu(T_\xi) < \infty$ $(\xi \in \Xi)$ such that, for any $A \in \Sigma$, $\mu(A) < +\infty$, there is a countable set of suffices $\Xi_0 \subset \Xi$ and a set N of measure zero such that $A = \bigcup_{\xi \in \Xi_0} (A \cap T_\xi) \cup N$.

Clearly $\bigcup_{\xi \in \Xi} T_\xi \in \Sigma$ and $\mu(T \setminus \bigcup_{\xi \in \Xi} T_\xi) = 0$. It is easy to deduce from Theorem 17 that if a space (T, Σ, μ) has the direct sum property, then condition (a) of Theorem 17 is satisfied in $\mathbf{S}(T, \Sigma, \mu)$.

REMARK. If $\{x_\alpha\}$ is a net such that $x_\alpha \to x\,(\mu)$ on each set T_ξ $(\xi \in \Xi)$, then $x_\alpha \to x\,(\mu)$.

For if $A \in \Sigma(\mu)$, then by the definition of $\{T_\xi\}$ we have $A = \bigcup_{\xi \in \Xi_0} (A \cap T_\xi) \cup N$, where $\mu(N) = 0$ and Ξ_0 is countable. Since

$$\mu(A) = \sum_{\xi \in \Xi_0} \mu(A \cap T_\xi) < \infty,$$

there exists, for each $\delta > 0$, a finite subset $\Xi_1 \subset \Xi_0$ such that

$$\sum_{\xi \in \Xi_0 \setminus \Xi_1} \mu(A \cap T_\xi) < \delta.$$

Choose any $\varepsilon > 0$. If n is the number of elements in Ξ_1, then for $\alpha \geqslant \alpha_0$ we have

$$\mu(\{t \in T_\xi : |x_\alpha(t) - x(t)| \geqslant \varepsilon\}) < \delta/n \quad (\xi \in \Xi_1).$$

Therefore

$$\mu(\{t\in A\colon |x_\alpha(t)-x(t)|\geqslant\varepsilon\})\leqslant\sum_{\xi\in\Xi_1}\mu(\{t\in T_\xi\colon |x_\alpha(t)-x(t)|\geqslant\varepsilon\}+$$

$$+\sum_{\xi\in\Xi_0\setminus\Xi_1}\mu(A\cap T_\xi)<n\delta/n+\delta=2\delta.$$

II

VECTOR SPACES

§ 1. Basic definitions

1.1. A vector space over the field of real or complex numbers is a natural generalization of the familiar three-dimensional Euclidean space. In it two algebraic operations are defined: addition of vectors and multiplication of a vector by a scalar (number), subject to certain conditions.

Let $\mathbb{K}$ be the field of real or complex numbers (the field of scalars). A set $\mathbf{X}$ is called a *vector* (or *linear*) *space* over $\mathbb{K}$ if for every two of its elements x and y there is defined a sum $x+y$—an element of $\mathbf{X}$—and if for every element $x \in \mathbf{X}$ and every number $\lambda \in \mathbb{K}$ there is defined a product λx—also an element of $\mathbf{X}$—such that the following axioms are satisfied:

1) $(x+y)+z = x+(y+z)$ (associativity of addition);
2) $x+y = y+x$ (commutativity of addition);
3) there exists an element $\mathbf{0}$ in $\mathbf{X}$ such that $0 \,.\, x = \mathbf{0}$ for every $x \in \mathbf{X}$;
4) $(\lambda+\mu)x = \lambda x + \mu x$ } (distributivity);
5) $\lambda(x+y) = \lambda x + \lambda y$ }
6) $(\lambda\mu)x = \lambda(\mu x)$ (associativity of multiplication);
7) $1 \,.\, x = x$.

If operations of addition and multiplication by scalars are introduced on a set $\mathbf{X}$ such that $\mathbf{X}$ is turned into a vector space, then $\mathbf{X}$ is said to be endowed with a vector space structure. A vector space over $\mathbb{R}$ is called a *real* vector space, and one over $\mathbb{C}$ is called a *complex* vector space. We use the term vector space whenever (as is usually the case) the field of scalars has no significance for us, or when it is clear from the context.

All the spaces $\mathbf{C}(K)$, $\mathbf{C}^{(l)}[a, b]$, $\mathbf{S}(T, \Sigma, \mu)$, $\mathbf{s}$, $\ell^{\infty}(T)$ introduced in Chapter I are vector spaces, if the sum of two of their elements is defined to be the function equal to the sum of the two corresponding functions and the product of an element and a scalar is defined similarly. In all these sets the role of the zero element is played by the function identically equal to zero.

Another example of a vector space is the set $\mathbf{K}^n$ of all n-dimensional vectors (with coordinates in $\mathbb{K}$), the operations of addition and multiplication by scalars in $\mathbb{K}$ being defined "coordinatewise". This vector space is a basic subject of study in linear algebra.

1.2. Let us indicate the simplest consequences of axioms 1)–7) (here x, y, z denote elements of the same vector space, and $\lambda, \mu \in \mathbb{K}$).

a) $x + \mathbf{0} = x$.

For $x + \mathbf{0} = 1 \,.\, x + 0 \,.\, x = (1+0)x = 1 \,.\, x = x$ (by axioms 3), 4), 7)).

b) To each x there corresponds a unique element x' such that $x + x' = \mathbf{0}$. In fact, $x' = (-1) \,.\, x$. The element x' is usually written as $-x$ and called the *negative* of x.

As explained, we take $x' = (-1).x$. Then, by axioms 3), 4), 7) again, $x + x' = 1.x + (-1).x = (1 + (-1)).x = 0.x = \mathbf{0}$. Moreover, if there were another element $\bar{x}$ for which $x + \bar{x} = \mathbf{0}$, then by the associativity and commutativity of addition we see, using a), that

$$x' = x' + \mathbf{0} = x' + (x + \bar{x}) = (x' + x) + \bar{x} = (x + x') + \bar{x} = \mathbf{0} + \bar{x} = \bar{x} + \mathbf{0} = \bar{x}.$$

c) $-(\alpha x) = (-\alpha) . x = \alpha . (-x)$.
For by 6) and 7) we have

$$-(\alpha x) = (-1) . (\alpha x) = (-\alpha) . x = \alpha((-1) . x) = \alpha . (-x).$$

d) For any pair of elements x and y there exists a unique element z such that $z + y = x$; this element is called the *difference* of the elements x and y, and is denoted by $z = x - y$.
We put $z = x + (-y)$. Using what we have proved above, we see that $z + y = (x + (-y)) + y = x + (y + (-y)) = x + \mathbf{0} = x$. If another element $\bar{z}$ satisfied the same condition, then we should have

$$\bar{z} = \bar{z} + \mathbf{0} = \bar{z} + (y + (-y)) = (\bar{z} + y) + (-y) = x + (-y) = z.$$

e) $x = y$ is equivalent to $x - y = \mathbf{0}$.
For if $x = y$, then obviously

$$x - y = x + (-y) = y + (-y) = \mathbf{0}.$$

If, on the other hand, $x - y = \mathbf{0}$, then $y = y + (x - y) = [y + (-y)] + x = x + \mathbf{0} = x$.
f) $\lambda(x - y) = \lambda x - \lambda y$; $(\lambda - \mu)x = \lambda x - \mu x$.
Since multiplication is distributive, b) and c) show that $\lambda(x - y) = \lambda[x + (-y)] = \lambda x + \lambda(-y) = \lambda x + (-\lambda)y = \lambda x - \lambda y$ and $(\lambda - \mu)x = \lambda x + (-\mu)x = \lambda x + (-\mu x) = \lambda x - \mu x$.
g) $\lambda . \mathbf{0} = \mathbf{0}$.
For $\lambda . \mathbf{0} = \lambda(0 . x) = (\lambda . 0)x = 0 . x = \mathbf{0}$.
h) If $\lambda x = \mathbf{0}$ and $\lambda \neq 0$, then $x = \mathbf{0}$.
For $x = 1 . x = \left(\frac{1}{\lambda} . \lambda\right)x = \frac{1}{\lambda}(\lambda x) = \frac{1}{\lambda}\mathbf{0} = \mathbf{0}$.
i) If $\lambda x = \lambda y$ and $\lambda \neq 0$, then $x = y$.
This is an obvious consequence of h) and e).
j) If $\lambda x = \mathbf{0}$ and $x \neq \mathbf{0}$, then $\lambda = 0$.
For if we had $\lambda \neq 0$, then by h) we should have $x = \mathbf{0}$.
k) If $\lambda x = \mu x$ and $x \neq \mathbf{0}$, then $\lambda = \mu$.
This is obvious.
In conclusion, we note that the associativity of addition enables us to omit brackets and write simply $x + y + z$ in place of $(x + y) + z$ or $x + (y + z)$; and similarly for expressions with a larger number of summands.

1.3. As in the case of metric spaces, we shall not distinguish between vector spaces $\mathbf{X}$ and $\mathbf{Y}$ if it is possible to establish a *linear isomorphism* between their elements: that is, a one-to-one correspondence $x \leftrightarrow y$ such that $x_1 \leftrightarrow y_1$ and $x_2 \leftrightarrow y_2$ imply $\lambda x_1 + \mu x_2 \leftrightarrow \lambda y_1 + \mu y_2$.

From this point of view, the set of real numbers and the set of points on the real line must be considered as one and the same vector space, as we know is customary in analysis.

1.4. Consider a set $\mathbf{X}_0$ contained in a vector space $\mathbf{X}$, such that if $x, y \in \mathbf{X}_0$ then also any linear combination $\lambda x + \mu y$ belongs to $\mathbf{X}_0$. Then the operations of addition and multiplication of an element by a scalar are defined in $\mathbf{X}_0$ and yield elements belonging to $\mathbf{X}_0$. Moreover, axiom 3) for vector spaces ($\mathbf{0} = 0 . x$) is satisfied in $\mathbf{X}_0$. The other axioms hold in $\mathbf{X}_0$ because they hold in $\mathbf{X}$. Thus $\mathbf{X}_0$ turns out to be a vector space in a natural way; it is called a *linear manifold* (or *linear set*) in $\mathbf{X}$.

The intersection of any collection of linear manifolds in $\mathbf{X}$ is clearly also a linear manifold. Therefore, if E is any set in $\mathbf{X}$, there exists a smallest linear manifold $\mathscr{L}(E)$ containing E. This is the intersection of all linear manifolds containing E. The space $\mathscr{L}(E)$ is called the *linear hull* of E.

It is easy to convince oneself that $\mathscr{L}(E)$ coincides with the set $\tilde{L}$ of all elements x of the form $\lambda_1 x_1 + \lambda_2 x_2 + \ldots \lambda_n x_n$, where $x_1, x_2, \ldots, x_n$ is any set of elements and $\lambda_1, \lambda_2, \ldots, \lambda_n$ are arbitrary scalars. Such elements x are called *linear combinations* of the elements $x_1, x_2, \ldots, x_n$.

For $\tilde{L}$ is evidently a linear manifold and contains E. On the other hand, every linear manifold containing E must also contain all linear combinations of elements of E, that is, it must contain $\tilde{L}$. Hence $\mathscr{L}(E) = \tilde{L}$.

Elements $x_1, x_2, \ldots, x_n$ are said to be *linearly independent* if a relation of the form $\sum_{k=1}^{n} \lambda_k x_k = \mathbf{0}$ is possible only with $\lambda_1 = \lambda_2 = \ldots = \lambda_n = 0$. Otherwise $x_1, x_2, \ldots, x_n$ are said to be *linearly dependent*. Thus, for example, the elements x and $-x$ are linearly dependent, since $1 . x + (-1) . x = \mathbf{0}$. If there is an element equal to zero among the $x_1, x_2, \ldots, x_n$, then these elements are linearly dependent.

An infinite system of elements is said to be *linearly independent* if every finite subset of distinct elements of the system is linearly independent.

If the elements $\{x_\xi\}$ form a linearly independent system, then it is clear that an equation

$$\sum_{k=1}^{n} \lambda_k x_{\xi_k} = \sum_{k=1}^{n} \mu_k x_{\xi_k}$$

implies that

$$\lambda_k = \mu_k \quad (k = 1, 2, \ldots, n).$$

An example of a linearly independent system in the space $\mathbf{C}[a, b]$ is the system $\{x_n\}$ $(x_n(t) = t^n)$.

A linearly independent system $\{x_\xi\}$ is called an *algebraic basis* of a vector space $\mathbf{X}$ if $\mathscr{L}(\{x_\xi\}) = \mathbf{X}$. Thus every element $x \in \mathbf{X}$ can be expressed as a linear combination of elements of an algebraic basis, and it follows from what was said above that the expression is unique.

From this point of view the simplest vector spaces are those having a finite algebraic basis. Such spaces are called *finite-dimensional*, and the number of elements forming a basis is called the *dimension* of the space. The dimension of a vector space can be shown to be an invariant of the space: that is, it does not depend on which basis we choose to define it.

Let $\mathbf{X}$ be a finite-dimensional vector space (of dimension n). As we have already noted, each element $x \in \mathbf{X}$ can be expressed uniquely in the form $x = \lambda_1 x_1 + \ldots + \lambda_n x_n$, where $x_1, x_2, \ldots, x_n$ is an algebraic basis. By associating with x the vector $\tilde{x} \in \mathbf{K}^n$ having components $(\lambda_1, \ldots, \lambda_n)$, we define a one-to-one correspondence between $\mathbf{X}$ and $\mathbf{K}^n$, which is a linear isomorphism since $\lambda x+\mu y \leftrightarrow \lambda \tilde{x}+\mu \tilde{y}$ whenever $x \leftrightarrow \tilde{x}$ and $y \leftrightarrow \tilde{y}$. In accordance with what was said earlier, we are now justified in identifying $\mathbf{X}$ and $\mathbf{K}^n$, and thus considering $\mathbf{X}$ as a set of n-dimensional vectors.

For this reason, the elements of an arbitrary vector space are often referred to as *vectors*.

1.5. We now introduce some notation that will be needed in the sequel.

For any $x \in \mathbf{X}$ and $E \subset \mathbf{X}$

$$x+E=\{x+y\colon y\in E\}.$$

For any $E_1 \subset \mathbf{X}$ and $E_2 \subset \mathbf{X}$

$$E_1+E_2=\{x+y\colon x\in E_1,\ y\in E_2\}.$$

For any $\lambda \in \mathbb{K}$ and $E \subset \mathbf{X}$

$$\lambda E=\{\lambda x\colon\ x\in E\}.$$

Notice that, in general, $E+E \neq 2E$, and we can only say that $2E \subset E+E$.

1.6. Let us define some operations which enable one to define new vector spaces from given ones.

A vector space $\mathbf{X}$ is said to be the *algebraic direct sum* of vector spaces $\mathbf{X}_1$ and $\mathbf{X}_2$ if $\mathbf{X}_1$ and $\mathbf{X}_2$ are linear manifolds in $\mathbf{X}$ and each $x \in \mathbf{X}$ can be expressed uniquely in the form $x = x_1+x_2$, where $x_1 \in \mathbf{X}_1$ and $x_2 \in \mathbf{X}_2$.

If $\mathbf{X}_1$ and $\mathbf{X}_2$ are vector spaces (over the field $\mathbb{K}$) then the direct product $\mathbf{X}=\mathbf{X}_1 \times \mathbf{X}_2$ becomes a vector space if the operations are defined in it by the equations

$$(x_1,x_2)+(y_1,y_2)=(x_1+y_1,x_2+y_2),$$
$$\lambda(x_1,x_2)=(\lambda x_1,\lambda x_2).$$

Let $\mathbf{X}$ be a vector space, $\mathbf{X}_0$ a linear manifold in $\mathbf{X}$.

We group the elements of $\mathbf{X}$ into classes, assigning two elements x' and x'' to the same class if $x'-x'' \in \mathbf{X}_0$. It is then clear that distinct classes have no elements in common, and that every $x \in \mathbf{X}$ lies in one (and, for the reason just given, only one) class. Let $\bar{x}$ be one of these classes and suppose $x \in \bar{x}$. It follows from the definition that $\bar{x}=x+\mathbf{X}_0$. Conversely, a set of the form $x+\mathbf{X}_0$ is a class, namely the class containing x.

We can introduce algebraic operations in the set $\mathbf{X}/\mathbf{X}_0$ of all classes by setting

$$\bar{x}+\bar{y}=x+y+\mathbf{X}_0,\quad \lambda\bar{x}=\lambda x+\mathbf{X}_0$$
$$(\bar{x},\bar{y}\in\mathbf{X}/\mathbf{X}_0,\quad x\in\bar{x},\quad y\in\bar{y}).$$

It is easy to check that these definitions do not depend on the choice of elements x, y representing the classes $\bar{x}$, $\bar{y}$. With these definitions, $\mathbf{X}/\mathbf{X}_0$ becomes a vector space, called a *factor space* (or *quotient space*), in which the role of zero element is evidently taken by the class containing the zero element of $\mathbf{X}$—that is, by the subspace $\mathbf{X}_0$.

§ 2. Linear operators and functionals

2.1. Let $\mathbf{X}$, $\mathbf{Y}$ be vector spaces over $\mathbb{K}$. A mapping $U:\mathbf{X}\to\mathbf{Y}$ is called a *linear mapping* or *linear operator* if

$$U(\lambda x+\mu y)=\lambda U(x)+\mu U(y)$$

for all $\lambda, \mu \in \mathbb{K}$ and $x, y \in \mathbf{X}$. The set of all linear mappings from $\mathbf{X}$ into $\mathbf{Y}$, which is denoted by $L(\mathbf{X}, \mathbf{Y})$, is turned into a vector space when algebraic operations are defined as follows.

Let $U_1, U_2 \in L(\mathbf{X}, \mathbf{Y})$. By definition, $U = U_1 + U_2$ is the operator from $\mathbf{X}$ into $\mathbf{Y}$ defined by

$$U(x) = U_1(x) + U_2(x) \quad (x \in \mathbf{X}). \tag{1}$$

Obviously, $U \in L(\mathbf{X}, \mathbf{Y})$. If $U \in L(\mathbf{X}, \mathbf{Y})$ and $\lambda \in K$, then $\tilde{U} = \lambda U$ is defined by

$$\tilde{U}(x) = \lambda U(x) \quad (x \in \mathbf{X}). \tag{2}$$

Obviously, $\tilde{U} \in L(\mathbf{X}, \mathbf{Y})$. We leave it to the reader to verify that, with these definitions, $L(\mathbf{X}, \mathbf{Y})$ is a vector space over $\mathbb{K}$, remarking only that the role of the zero element of $L(\mathbf{X}, \mathbf{Y})$ is taken by the mapping $U_0 = \mathbf{0}$ which is identically equal to zero:

$$U_0(x) = \mathbf{0} \quad (x \in \mathbf{X}).$$

Notice that for any $U \in L(\mathbf{X}, \mathbf{Y})$ we have $U(\mathbf{0}) = \mathbf{0}$ and $U(-x) = -U(x)$ $(x \in \mathbf{X})$. The *kernel* of a mapping $U \in L(\mathbf{X}, \mathbf{Y})$ is the set $\operatorname{Ker} U = U^{-1}(\mathbf{0})$, which is clearly a linear manifold in $\mathbf{X}$. It is easy to see that the mapping U is one-to-one if and only if $\operatorname{Ker} U = \{\mathbf{0}\}$.

A one-to-one linear mapping $U: \mathbf{X} \to \mathbf{Y}$ that maps $\mathbf{X}$ onto $\mathbf{Y}$ is called a *linear isomorphism* from $\mathbf{X}$ onto $\mathbf{Y}$, and in this situation the spaces $\mathbf{X}$ and $\mathbf{Y}$ are said to be *linearly isomorphic* (this definition clearly agrees with that of 1.3).

A linear mapping f from a vector space into the field of scalars $\mathbb{K}$ is called a *linear functional.*

2.2. Everything we have said so far applied equally well to both real and complex vector spaces. In connection with the Hahn–Banach Theorem, which we shall prove below, and also in connection with later investigations in the theory of operators, we shall require certain auxiliary methods for the complex case.

Let $\mathbf{X}$ be any vector space. Formulae (1) and (2) turn the set of linear functionals into a vector space $L(\mathbf{X}, \mathbb{K})$. However, in the case of a complex space $\mathbf{X}$, we define multiplication of a functional f by a complex number λ according to the equation*

$$(\lambda f)(x) = \bar{\lambda} f(x) \quad (x \in \mathbf{X}). \tag{3}$$

If $\lambda \in \mathbb{R}$ then we obtain equation (2). It is easy to see that, with the algebraic operations defined in this way, $L(\mathbf{X}, \mathbf{Y})$ is a vector space, which we call the *algebraic dual* (or *algebraic conjugate*) of $\mathbf{X}$ and denote by $\mathbf{X}^+$.

REMARK. In the case of operators that are not functionals, the usual formula (2) is retained in the complex case too.

Let $\mathbf{X}$ be a complex vector space. If we retain the former definition of addition in $\mathbf{X}$ but take multiplication by a scalar to be defined only in the case where the scalar is real—in which case it is to be as before—then we obtain a real vector space $\mathbf{X}_{\mathbb{R}}$, associated with $\mathbf{X}$. By a *real linear functional* on $\mathbf{X}$ we shall mean a linear functional on $\mathbf{X}_{\mathbb{R}}$.

Let f be a linear functional on $\mathbf{X}$. Consider the functional ϕ defined on $\mathbf{X}_{\mathbb{R}}$ as follows:

$$\phi(x) = \operatorname{Re} f(x) \quad (x \in \mathbf{X}). \tag{4}$$

It is easy to check that ϕ is a real linear functional. For if $x_1, x_2 \in \mathbf{X}$, $\lambda, \mu \in \mathbb{R}$, then

---

* Here, and in what follows, $\bar{\lambda}$ is the complex conjugate of λ.

$$\phi(\lambda x_1+\mu x_2) = \mathrm{Re} f(\lambda x_1+\mu x_2) = \mathrm{Re}[\lambda f(x_1)+\mu f(x_2)] =$$
$$= \lambda\,\mathrm{Re} f(x_1)+\mu\,\mathrm{Re} f(x_2) = \lambda\phi(x_1)+\mu\phi(x_2).$$

We shall show that for any $x\in\mathbf{X}$

$$f(x) = \phi(x)-i\phi(ix), \tag{5}$$

where i is the imaginary unit. In fact, $f(x) = \mathrm{Re} f(x)+i\,\mathrm{Im} f(x) = \mathrm{Re} f(x)-i\,\mathrm{Re}\,(if(x)) = \mathrm{Re} f(x)-i\,\mathrm{Re} f(ix) = \phi(x)-i\phi(ix)$.

Thus we have shown that f satisfies (2) on $\mathbf{X}$. Conversely, if ϕ is any real linear functional on $\mathbf{X}$, then the f defined by (5) is a linear functional on $\mathbf{X}$, and (4) holds (the verification is left to the reader).

2.3. Let $\mathbf{X}$ be a vector space. A linear manifold H in $\mathbf{X}$ is called a *hyperspace* if $H \neq \mathbf{X}$ and $\mathbf{X} = \mathscr{L}(H, x_0)$ for some $x_0\in\mathbf{X}$.

LEMMA 1. *Let $\mathbf{X}$ be a vector space, let H be a hyperspace in $\mathbf{X}$, and suppose $x_1\in\mathbf{X}\setminus H$. Then every $x\in\mathbf{X}$ can be expressed in the form $x = \lambda x_1+h$, where $\lambda\in\mathbb{K}$, $h\in H$; moreover, this expression is unique.*

Proof. By the definition of a hyperspace, there exists $x_0\in\mathbf{X}$ such that $\mathbf{X} = \mathscr{L}(H, x_0)$. Hence $x_1 = \lambda_0 x_0+\lambda_1 h_1+\ldots+\lambda_n h_n$, where $h_i\in H$, $(1\leqslant i\leqslant n)$. If $h_0 = \sum_{i=1}^{n}\lambda_i h_i$, then $h_0\in H$ and $x_1 = \lambda_0 x_0+h_0$; moreover, $\lambda_0\neq 0$ because $x_1\notin H$. For an arbitrary $x\in\mathbf{X}$ there exist $\mu\in\mathbb{K}$ and $h\in H$ such that

$$x = \mu x_0+h = \mu\frac{x_1-h_0}{\lambda_0}+h = \frac{\mu}{\lambda_0}x_1+\left(h-\frac{\mu}{\lambda_0}h_0\right).$$

Since $h-(\mu/\lambda_0)h_0\in H$, this is the required expression. Let us now prove that this expression is unique. Suppose $x = \lambda x_1+h = \mu x_1+g$ $(h, g\in H)$. Then $(\lambda-\mu)x_1 = g-h\in H$, but $x_1\notin H$, so $\lambda = \mu$ and therefore $h = g$.

It follows from Lemma 1 that if a linear manifold $\mathbf{X}_0\subset\mathbf{X}$ contains H, then either $\mathbf{X}_0 = H$ or $\mathbf{X}_0 = \mathbf{X}$.

A *hyperplane* in a vector space $\mathbf{X}$ is a set of the form $x+H$, where $x\in\mathbf{X}$ and H is a hyperspace. Hyperplanes are closely connected with linear functionals, as the following theorems show.

THEOREM 1. *Let f be a linear functional that is not identically equal to zero on a vector space $\mathbf{X}$. Then*

1) *$H = f^{-1}(0)$ is a hyperspace;*
2) *for any $\lambda\in\mathbb{K}$, we have $f^{-1}(\lambda) = x_\lambda+H$, where $f(x_\lambda) = \lambda$.*

Proof. 1) Choose $x_0\in\mathbf{X}$ such that $f(x_0) = \lambda_0\neq 0$. For any $x\in\mathbf{X}$ we have

$$x = \frac{f(x)}{\lambda_0}x_0+\left(x-\frac{f(x)}{\lambda_0}x_0\right).$$

Then $h = x-(f(x)/\lambda_0)x_0\in H$. For we have $f(h) = f(x)-(f(x)/\lambda_0)f(x_0) = 0$, so $h\in H$ and $x = (f(x_0)/\lambda_0)x_0+h$; that is, H is a hyperspace.

2) If $x_\lambda = (\lambda/\lambda_0)x_0$, then $f(x_\lambda) = \lambda$; thus such points x_λ do exist. Let x_λ be any element such that $f(x_\lambda) = \lambda$. If $y\in f^{-1}(\lambda)$, then $f(y) = \lambda$ and $y = x_\lambda+(y-x_\lambda)$. Obviously, $y-x_\lambda\in H$, so $y\in x_\lambda+H$. Conversely, if $z\in x_\lambda+H$, then $z = x_\lambda+h\,(h\in H)$, and so $f(z) = \lambda$ and $z\in f^{-1}(\lambda)$. Therefore $f^{-1}(\lambda) = x_\lambda+H$.

THEOREM 2. *Let H be a hyperspace in a vector space $\mathbf{X}$, let $x_0 \notin H$, $\lambda \neq 0$. Then there exists a unique linear functional f on $\mathbf{X}$* such that

1) $f^{-1}(0) = H$;
2) $f(x_0) = \lambda$.

Proof. Since H is a hyperspace, Lemma 1 shows that each $x \in \mathbf{X}$ has a unique expression of the form $x = \mu x_0 + h$ $(h \in H)$. Write $f(x) = \mu\lambda$. Let us check that f is a linear functional. If $y = \mu' x_0 + h'$ $(h' \in H)$, then $x + y = (\mu + \mu')x_0 + (h + h')$, so $f(x+y) = (\mu + \mu')\lambda = f(x) + f(y)$. If $\alpha \in \mathbb{K}$, then $\alpha x = \alpha\mu x_0 + \alpha h$, so $f(\alpha x) = \alpha\mu\lambda = \alpha f(x)$.

Now $x_0 = 1 \,.\, x_0 + \mathbf{0}$, so $f(x_0) = \lambda$. For $x \in H$, we have $x = 0 \,.\, x_0 + x$, so $f(x) = 0$ and therefore $f^{-1}(0) \supset H$. As $\lambda \neq 0$, $f \not\equiv 0$ on $\mathbf{X}$. Thus $f^{-1}(0)$ is a hyperspace, by Theorem 1. By the remark following the proof of Lemma 1, this implies that $f^{-1}(0) = H$.

We now prove that f is unique. Suppose a linear functional g satisfies 1) and 2). For any $x = \mu x_0 + h$ $(h \in H)$, we have

$$g(x) = g(\mu x_0 + h) = \mu g(x_0) + g(h) = \mu\lambda = f(x).$$

COROLLARY. *If f and g are linear functionals on a vector space $\mathbf{X}$ and $f^{-1}(0) = g^{-1}(0)$, then there exists $\alpha \in \mathbb{K}$ such that $g = \alpha f$.*

Proof. If $f^{-1}(0) = g^{-1}(0) = \mathbf{X}$, then $f = g = 0$. Suppose $f^{-1}(0) = g^{-1}(0) = H$ is a hyperspace. Choose $x_0 \notin H$. Then $f(x_0) = \lambda \neq 0$, $g(x_0) = \mu \neq 0$. If $\alpha = \mu/\lambda$, then $g^{-1}(0) = H$, $(\alpha f)^{-1}(0) = H$, $g(x_0) = \mu$, $(\alpha f)(x_0) = \mu$. In view of the uniqueness proved in Theorem 2, this means that $g = \alpha f$.

Theorems 1 and 2 and the Corollary to Theorem 2 can be collected together as follows.

THEOREM 3. *A subset $M \subset \mathbf{X}$ is a hyperplane if and only if $M = \{x \in \mathbf{X} : f(x) = \lambda\}$ for some $\lambda \in \mathbb{K}$ and some non-zero linear functional f on $\mathbf{X}$. Moreover, f and λ are determined by H to within a common factor μ, $\mu \in \mathbb{K}$, $\mu \neq 0$.*

In conclusion, we look at the connection between the real and complex cases.

Let $\mathbf{X}$ be a complex vector space, $\mathbf{X}_{\mathbb{R}}$ the real vector space associated with $\mathbf{X}$. A *real hyperplane in* $\mathbf{X}$ is a hyperplane in $\mathbf{X}_{\mathbb{R}}$. It is easy to see that a hyperplane M is real if and only if $M = \{x \in \mathbf{X} : f(x) = \lambda\}$ where $\lambda \in \mathbb{R}$ and f is a real-valued linear functional on $\mathbf{X}$.

LEMMA 2. *Let $\mathbf{X}$ be a complex vector space. If M is a real hyperspace in $\mathbf{X}$, then $M \cap iM$ is a hyperspace in M. Every hyperplane in $\mathbf{X}$ is the intersection of two uniquely determined real hyperplanes.*

Proof. If M is a real hyperspace, then $M = \{x : g(x) = 0\}$, where g is a real linear functional. Taking $f(x) = g(x) - ig(ix)$, we have $M \cap iM = \{x : f(x) = 0\}$, and so $M \cap iM$ is a hyperspace.

If H is a hyperplane, then $H = \{x : f(x) = \lambda + i\mu\}$, where λ, $\mu \in \mathbb{R}$, and f is a linear functional. If $g(x) = \operatorname{Re} f(x)$, then $H = \{x : g(x) = \lambda\} \cap \{x : g(ix) = -\mu\}$.

§ 3. Convex sets and seminorms

3.1. Let $\mathbf{X}$ be a vector space. A set $E \subset \mathbf{X}$ is said to be *convex* if, for every pair of points x, $y \in E$, all elements of the form $\lambda x + (1-\lambda)y$ $(0 \leqslant \lambda \leqslant 1)$ also belong to E. The geometric meaning of this concept is that E, like any plane, contains together with any two of its points x, y also the whole *interval* $\{\lambda x + (1-\lambda)y : 0 \leqslant \lambda \leqslant 1\}$ containing x and y. A set $E \subset \mathbf{X}$ is said to be *balanced* if, for any $x \in E$ and $\lambda \in \mathbb{K}$ such that $|\lambda| \leqslant 1$, we have $\lambda x \in E$. A set $E \subset \mathbf{X}$ is said to be *absolutely convex* if, for any pair of points x, $y \in E$ and any λ, $\mu \in \mathbb{K}$ such that $|\lambda| + |\mu| \leqslant 1$, we have $\lambda x + \mu y \in E$.

Let us record some simple consequences of these definitions.

a) A set E is absolutely convex if and only if it is both convex and balanced.

For an absolutely convex set is clearly convex and balanced. Conversely, assume that E is a convex balanced set and let $x, y \in E$ and $|\lambda| + |\mu| \leqslant 1$. If $\lambda = 0$ or $\mu = 0$, then obviously $\lambda x + \mu y \in E$. If, however, $\lambda \neq 0$ and $\mu \neq 0$, then

$$\frac{\lambda}{|\lambda|} x \in E, \; \frac{\mu}{|\mu|} y \in E \text{ and } \frac{|\lambda|}{|\lambda| + |\mu|} + \frac{|\mu|}{|\lambda| + |\mu|} = 1.$$

Hence

$$\lambda x + \mu y = (|\lambda| + |\mu|)\left(\frac{|\lambda|}{|\lambda| + |\mu|}\frac{\lambda x}{|\lambda|} + \frac{|\mu|}{|\lambda| + |\mu|}\frac{\mu y}{|\mu|}\right) \in E.$$

b) If E_1, E_2 are convex subsets of $\mathbf{X}$, $\lambda \in \mathbb{K}$, then the sets $E_1 + E_2$ and λE_1 are convex. The same is true with convexity replaced by absolute convexity.

c) If E is a non-empty absolutely convex set, then $\mathbf{0} \in E$, and if $|\lambda| \leqslant |\mu|$, then $\lambda E \subset \mu E$.

The elementary proofs of b) and c) are left to the reader.

If E is an arbitrary non-empty set in $\mathbf{X}$, then the set of all finite linear combinations $\Sigma \lambda_i x_i$, where $\lambda_i \geqslant 0$, $\Sigma \lambda_i = 1$ and all $x_i \in E$, is called the *convex hull* of E and is denoted by co (E). Obviously co (E) is the smallest convex set containing E. The set of all finite linear combinations $\Sigma\, \lambda_i x_i$, where $\Sigma\, |\lambda_i| \leqslant 1$ and all $x_i \in E$, is called the *absolutely convex hull* of E and is denoted by abs co (E). Obviously abs co (E) is the smallest absolutely convex set containing E.

A subset E of a vector space E is said to be *absorbent* if for any $x \in \mathbf{X}$ there exists $\lambda > 0$ such that $x \in \mu E$ for all μ with $|\mu| \geqslant \lambda$. Geometrically, this means that on any ray through the origin there exists an interval with one end at the origin lying wholly within E. In view of c), an absolutely convex set E is absorbent if and only if for each $x \in \mathbf{X}$ there exists $\lambda > 0$ such that $x \in \lambda E$; that is, if and only if $\mathbf{X} = \bigcup_{\lambda > 0} \lambda E$, or even (again in view of c)) $\mathbf{X} = \bigcup_{n=1}^{\infty} nE$.

3.2. Let $\mathbf{X}$ be a vector space. A real-valued function p defined on $\mathbf{X}$ is said to be *semi-additive* if, for any pair of elements $x_1, x_2 \in \mathbf{X}$,

$$p(x_1 + x_2) \leqslant p(x_1) + p(x_2),$$

to be *positive homogeneous* if, for $\lambda \geqslant 0$,

$$p(\lambda x) = \lambda p(x),$$

and to be *homogeneous* if, for any λ,

$$p(\lambda x) = |\lambda| p(x).$$

A semi-additive positive homogeneous function is called a *gauge function.* A homogeneous gauge function is called a *seminorm.* We note some properties of these functions.

a) $p(\mathbf{0}) = 0$ for any gauge function p.

b) If p is a seminorm, then $p(x) \geqslant 0$ for any $x \in \mathbf{X}$.

This follows from the relation $0 = p(\mathbf{0}) = p(x + (-x)) \leqslant p(x) + p(-x) = 2p(x)$.

c) If p is a seminorm, then

$$|p(x)-p(y)| \leqslant p(x-y).$$

For, by the semi-additivity of p, we have $p(x) = p(x-y+y) \leqslant p(x-y)+p(y)$. Interchanging x and y and using the fact that $p(x-y) = p(y-x)$, we obtain the required result.

c′) If p is a gauge function, then we have similarly $|p(x)-p(y)| \leqslant \max(p(x-y), p(y-x))$.

By a), we have $p(\mathbf{0}) = 0$ for any seminorm; however, it may happen that $p(x) = 0$ for $x \neq \mathbf{0}$. Seminorms for which $p(x) = 0$ implies $x = \mathbf{0}$ are called *norms*.

The connection between the class of functions just introduced and convex sets is indicated in the following lemma.

LEMMA 1. 1) *Let p be a non-negative gauge function. Then for any $\lambda > 0$ the sets $\{x : p(x) < \lambda\}$ and $\{x : p(x) \leqslant \lambda\}$ are convex and absorbent. If p is a seminorm, then these sets are absolutely convex.*

2) *To every convex absorbent set $U \subset \mathbf{X}$ there corresponds a non-negative gauge function p_U, called the Minkowski functional of U, defined by*

$$p_U(x) = \inf\{\lambda \colon \lambda > 0,\ x \in \lambda U\},$$

and we have

$$\{x\colon p_U(x) < 1\} \subset U \subset \{x\colon p_U(x) \leqslant 1\}. \tag{1}$$

If, in addition, U is absolutely convex, then p_U is a seminorm.

Proof. 1) We prove only that the set $E_\lambda = \{x : p(x) < \lambda\}$ is absorbent, leaving the elementary verification of the remaining statements to the reader. If $m = \max(p(x), p(-x))$, then, for $|\mu| \geqslant \lambda$, we have*

$$p(x/\mu) = 1/|\mu|\, p(\operatorname{sign}\mu\,.\,x) \leqslant \lambda/(m+1)\, p(\operatorname{sign}\mu\,.\,x) < \lambda,$$

and so $x \in \mu E_\lambda$. Hence E_λ is absorbent.

2) Since U is absorbent, $p_U(x) < +\infty$. Clearly $p_U(\mathbf{0}) = 0$. Hence in verifying that p_U is positive homogeneous we may assume that $\lambda > 0$. Using the fact that $\lambda x \in \mu U$ if and only if $x \in (\mu/\lambda)U$, we obtain

$$p_U(\lambda x) = \inf\{\mu > 0\colon \lambda x \in \mu U\} = \lambda \inf\{(\mu/\lambda)\colon \mu > 0, \quad x \in (\mu/\lambda)U\} = \lambda p_U(x).$$

If U is absolutely convex, then it is balanced, so $\lambda x \in \mu U$ if and only if $x \in (\mu/|\lambda|)U$, whence

$$p_U(\lambda x) = \inf\{\mu > 0\colon \lambda x \in \mu U\} = |\lambda| \inf\{(\mu/|\lambda|)\colon \mu > 0, \quad x \in (\mu/|\lambda|)U\} = |\lambda|\, p_U(x).$$

It now remains to verify that p_U is semi-additive. Let $x, y \in \mathbf{X}$ and let $\varepsilon > 0$. There exist λ, $\mu > 0$ such that

$$p_U(x) < \lambda < p_U(x)+\varepsilon, \qquad p_U(y) < \mu < p_U(y)+\varepsilon.$$

Thus x/λ, $y/\mu \in U$. Since U is convex,

$$\frac{x+y}{\lambda+\mu} = \frac{\lambda}{\lambda+\mu}\frac{x}{\lambda} + \frac{\mu}{\lambda+\mu}\frac{y}{\mu} \in U,$$

---

* If $\mu \in \mathbb{C}$, then $\operatorname{sign}\mu = \begin{cases} |\mu|/\mu & \text{if } \mu \neq 0, \\ 0 & \text{if } \mu = 0. \end{cases}$

and so

$$p_U(x+y) \leqslant \lambda + \mu < p_U(x) + p_U(y) + 2\varepsilon.$$

Since ε was arbitrary, it follows that $p_U(x+y) \leqslant p_U(x) + p_U(y)$. It is now obvious that (1) holds. This completes the proof of Lemma 1.

§ 4. The Hahn–Banach Theorem

4.1. In this section we present the so-called analytic form of the Hahn–Banach Theorem (see Banach) on the extension of continuous functionals.* This theorem has numerous applications in the theory of topological vector spaces and the theory of normed spaces and their applications.

THEOREM 1 (The analytic form of the Hahn–Banach Theorem). *Let p be a gauge function defined on a real vector space $\mathbf{X}$.*

Suppose f_0 is a linear functional defined on a linear manifold $\mathbf{X}_0 \subset \mathbf{X}$, such that

$$f_0(x) \leqslant p(x) \quad (x \in \mathbf{X}_0). \tag{1}$$

Then there exists a linear functional f, defined on the whole of $\mathbf{X}$, which coincides with f_0 on $\mathbf{X}_0$ and is such that, throughout $\mathbf{X}$, we have

$$f(x) \leqslant p(x) \quad (x \in \mathbf{X}). \tag{2}$$

Proof. The existence of the required functional is obtained with the aid of Zorn's Lemma. To apply this, we consider the set $\mathfrak{M}$ of all pairs (L, g) satisfying the following conditions:

1) L is a linear manifold in $\mathbf{X}$, $L \supset \mathbf{X}_0$;
2) g is a linear functional defined on L and extending f_0;
3) $g(x) \leqslant p(x)$ for every $x \in L$.

The set $\mathfrak{M}$ is non-empty because $(\mathbf{X}_0, f_0) \in \mathfrak{M}$. We introduce an ordering on $\mathfrak{M}$ by setting $(L_1, g_1) \leqslant (L_2, g_2)$ whenever $L_2 \supset L_1$ and g_2 is an extension of g_1.

We shall show that the conditions of Zorn's Lemma are satisfied in the ordered set $\mathfrak{M}$, and hence that $\mathfrak{M}$ has a maximal element. Let $\mathfrak{M}_0$ be a totally ordered subset of $\mathfrak{M}$. Let $L_0 = \bigcup\{L : (L, g) \in \mathfrak{M}_0\}$. We now show that L_0 is a linear manifold. If $x, y \in L_0$, then by the definition of L_0 there exist linear manifolds L_1, L_2 such that $x \in L_1, y \in L_2$ and (L_1, g_1), $(L_2, g_2) \in \mathfrak{M}_0$. As $\mathfrak{M}_0$ is totally ordered, the elements (L_1, g_1) and (L_2, g_2) are comparable. Suppose, for definiteness, that $(L_1, g_1) \geqslant (L_2, g_2)$. Then $L_1 \supset L_2$ and so $x, y \in L_1$, whence $\lambda x + \mu y \in L_1 \subset L_0$ for all $\lambda, \mu \in \mathbb{R}$. Every $x \in L_0$ belongs to some manifold L such that $(L, g) \in \mathfrak{M}_0$. Set $g_0(x) = g(x)$. An argument like that above shows that this yields a well-defined linear functional g_0 on L_0. It is clear that $(L_0, g_0) \in \mathfrak{M}_0$ and that (L_0, g_0) is an upper bound for the set $\mathfrak{M}_0$. By Zorn's Lemma, there exists a maximal element $(L_{\max}, f_{\max}) \in \mathfrak{M}$. If it is shown that $L_{\max} = \mathbf{X}$, then the functional $f_{\max}$ will clearly be the one we are seeking.

Assume the contrary: that is, assume that $L_{\max} \neq \mathbf{X}$. If we show that for any $(L, g) \in \mathfrak{M}$ such that $x_0 \notin L$ there exists $(L_1, g_1) \in \mathfrak{M}$, where L_1 is the linear hull of L and x_0, then we evidently obtain a contradiction to the maximality of $L_{\max}$, and the proof of the theorem

* We recall that if $\mathbf{X}_0 \subset \mathbf{X}$ and f_0 is a function on $\mathbf{X}_0$, then a function f defined on $\mathbf{X}$ is said to be an *extension* (or *prolongation*) of f_0 if $f_0(x) = f(x)$ for every $x \in \mathbf{X}_0$.

will be complete. Hence we may assume that $\mathbf{X}$ is a simple extension of $\mathbf{X}_0$—that is, that every $x \in \mathbf{X}$ is expressible in the form

$$x = \lambda x_0 + x' \quad (x' \in \mathbf{X}_0). \tag{3}$$

If $x', x'' \in \mathbf{X}_0$, then by (1) we find that

$$f_0(x') + f_0(x'') = f_0(x' + x'') \leqslant p((x_0 + x') + (-x_0 + x'')) \leqslant p(x_0 + x') + p(-x_0 + x''),$$

so that

$$f_0(x'') - p(-x_0 + x'') \leqslant -f_0(x') + p(x_0 + x'),$$

and therefore, since x' and x'' here are arbitrary, we have

$$A = \sup_{x'' \in \mathbf{X}_0} [f_0(x'') - p(-x_0 + x'')] \leqslant \inf_{x' \in \mathbf{X}_0} [-f_0(x') + p(x_0 + x')] = B.$$

Assume that $A \leqslant t_0 \leqslant B$. Define a functional f on $\mathbf{X}$ by

$$f(x) = \lambda t_0 + f_0(x') \quad (x = \lambda x_0 + x',\ x' \in \mathbf{X}_0).$$

Clearly f is an additive homogeneous functional; moreover, f is an extension of f_0. Let us show that (2) holds. Assume that $\lambda \neq 0$ in (3). Suppose $\lambda > 0$. Then

$$f(x) = \lambda t_0 + f_0(x') \leqslant \lambda B + f_0(x') \leqslant \lambda\left[-f_0\left(\frac{x'}{\lambda}\right) + p\left(x_0 + \frac{x'}{\lambda}\right)\right] + f_0(x') = -f_0(x')$$
$$+ p(\lambda x_0 + x') + f_0(x') = p(x).$$

In exactly the same way (using the inequality $t_0 \geqslant A$) we deal with the case where $\lambda < 0$. Hence the theorem is proved.

COROLLARY. *If a functional p satisfies the conditions of the theorem, then there exists an additive homogeneous functional f defined on* $\mathbf{X}$ *such that*

$$f(x) \leqslant p(x), \quad x \in \mathbf{X}. \tag{4}$$

To see the truth of this, one need only take $\mathbf{X}_0 = \{\mathbf{0}\}$ and take f_0 to be the functional $f_0(\mathbf{0}) = 0$, and then apply the theorem.

Notice that f satisfies

$$-p(-x) \leqslant f(x) \leqslant p(x). \tag{5}$$

For by (4) we have

$$f(x) = -f(-x) \geqslant -p(-x).$$

The construction of the simple extension used the fact that f was real-valued. The Hahn–Banach Theorem is also valid for complex spaces, although in a rather less general form.

THEOREM 2. *Let p be a seminorm on an arbitrary vector space* $\mathbf{X}$. *Let* f_0 *be a linear functional defined on a linear manifold* $\mathbf{X}_0 \subset \mathbf{X}$, *such that*

$$|f_0(x)| \leqslant p(x) \quad (x \in \mathbf{X}_0). \tag{6}$$

Then there exists a linear functional f, defined on the whole of $\mathbf{X}$, *which coincides with* f_0 *on* $\mathbf{X}_0$, *and is such that, throughout* $\mathbf{X}$, *we have*

$$|f(x)| \leqslant p(x) \quad (x \in \mathbf{X}). \tag{7}$$

Proof. If $\mathbf{X}$ is a real vector space, then the required result follows from the Corollary to Theorem 1 and inequality (5).

Now assume that $\mathbf{X}$ is a complex vector space. If $\mathbf{X}_{\mathbb{R}}$ is the real vector space associated with $\mathbf{X}$, then $(\mathbf{X}_0)_{\mathbb{R}}$ is a linear manifold in $\mathbf{X}_{\mathbb{R}}$. Writing $\phi_0(x) = \operatorname{Re} f_0(x)$ $(x \in \mathbf{X})$, we obtain a real-valued linear functional ϕ_0 on $(\mathbf{X}_0)_{\mathbb{R}}$; also, in view of (2) in § 2, for any $x \in \mathbf{X}$ we have

$$f_0(x) = \phi_0(x) - i\phi_0(ix). \tag{8}$$

By (6) we have $|\phi_0(x)| = |\operatorname{Re} f_0(x)| \leqslant |f_0(x)| \leqslant p(x)$ for $x \in \mathbf{X}_0$. By what has already been proved, there exists a real functional ϕ on $\mathbf{X}$ which is an extension of ϕ_0 and satisfies

$$|\phi(x)| \leqslant p(x) \quad (x \in \mathbf{X}). \tag{9}$$

Write $f(x) = \phi(x) - i\phi(ix)$ for every $x \in \mathbf{X}$. As we observed in § 2, f is a linear functional on $\mathbf{X}$, and $\operatorname{Re} f(x) = \phi(x)$ $(x \in \mathbf{X})$. Hence, by (8), we see that f is an extension of f_0. It remains to verify the inequality (7).

For any $x \in \mathbf{X}$ there exists $\theta \in \mathbb{R}$ such that

$$e^{i\theta} f(x) = |f(x)| \geqslant 0.$$

Therefore $f(e^{i\theta}x) = e^{i\theta} f(x)$ is real, and so $f(e^{i\theta}x) = \phi(e^{i\theta}x)$. From (9) we now see that

$$|f(x)| = e^{i\theta} f(x) = f(e^{i\theta}x) = \phi(e^{i\theta}x) \leqslant p(e^{i\theta}x) = |e^{i\theta}| p(x) = p(x),$$

which proves (7), as required.

4.2. Theorem 1 has an elegant application in the theory of measure and integral.*

We pose the following problem: with each real-valued periodic function $x(t)$ of period 1 we wish to associate an "integral"—that is, a certain real number $\int_0^1 x(t)\,dt$—in such a way that the following conditions are satisfied:

1) $\int_0^1 [\alpha x_1(t) + \beta x_2(t)]\,dt = \alpha \int_0^1 x_1(t)\,dt + \beta \int_0^1 x_2(t)\,dt \quad (\alpha, \beta \in \mathbb{R})$;

2) if $x(t) \geqslant 0$ in $[0, 1]$, then $\int_0^1 x(t)\,dt \geqslant 0$;

3) $\int_0^1 x(t + t_0)\,dt = \int_0^1 x(t)\,dt$ (for any real number t_0);

4) $\int_0^1 x(1 - t)\,dt = \int_0^1 x(t)\,dt$;

5) if $x_0(t) \equiv 1$, then $\int_0^1 x_0(t)\,dt = 1$.

THEOREM 3. *The problem posed above has at least one solution.*

Proof. Denote the set of all bounded periodic functions (of period 1) by $\mathbf{M}$. Clearly $\mathbf{M}$ is a vector space. Let $x \in \mathbf{M}$ and let $\alpha_1, \alpha_2, \ldots, \alpha_n$ be any set of real numbers. Write

$$\pi(x; \alpha_1, \alpha_2, \ldots, \alpha_n) = \sup_{-\infty < t < \infty} \frac{1}{n} \sum_{k=1}^{n} x(t + \alpha_k)$$

and set

$$p(x) = \inf \pi(x; \alpha_1, \alpha_2, \ldots, \alpha_n),$$

* See Banach.

where the infimum is taken over all finite sets of numbers $\alpha_1, \alpha_2, \ldots, \alpha_n$. We shall prove that p satisfies the conditions of Theorem 1. Clearly, we need only prove that p is semi-additive.

Let $\alpha_1, \alpha_2, \ldots, \alpha_m$ and $\beta_1, \beta_2, \ldots, \beta_n$ be sets of numbers such that $\pi(x_1; \alpha_1, \ldots, \alpha_m) < p(x_1)+\varepsilon$ and $\pi(x_2; \beta_1, \ldots, \beta_n) < p(x_2)+\varepsilon$. Write $\gamma_{j,k} = \alpha_j+\beta_k$. Then we have, on the one hand,

$$p(x_1+x_2) \leqslant \pi(x_1+x_2; \gamma_{1,1}, \gamma_{1,2}, \ldots, \gamma_{m,n}). \tag{10}$$

On the other hand,

$$\pi(x_1+x_2; \gamma_{1,1}, \gamma_{1,2}, \ldots, \gamma_{m,n}) = \frac{1}{mn} \sup_{-\infty<t<\infty} \sum_{j,k} [x_1(t+\gamma_{j,k})+x_2(t+\gamma_{j,k})] \leqslant$$

$$\leqslant \frac{1}{mn} \sup_{-\infty<t<\infty} \sum_{j,k} x_1(t+\gamma_{j,k}) + \frac{1}{mn} \sup_{-\infty<t<\infty} \sum_{j,k} x_2(t+\gamma_{j,k}) \leqslant$$

$$\leqslant \frac{1}{n}\sum_{k=1}^{n} \sup \frac{1}{m}\sum_{j=1}^{m} x_1(t+\beta_k+\alpha_j) + \frac{1}{m}\sum_{j=1}^{m} \sup \frac{1}{n}\sum_{k=1}^{n} x_2(t+\alpha_j+\beta_k) =$$

$$= \pi(x_1; \alpha_1, \alpha_2, \ldots, \alpha_m)+\pi(x_2; \beta_1, \beta_2, \ldots, \beta_m) < p(x_1)+p(x_2)+2\varepsilon.$$

Comparing this with (10) and bearing in mind that ε was arbitrary, we obtain

$$p(x_1+x_2) \leqslant p(x_1)+p(x_2).$$

Let f be the functional whose existence is asserted in the Corollary to Theorem 1. If $x(t) \geqslant 0$, then $p(x) \geqslant 0$, $p(-x) \leqslant 0$, and hence, by (5), $f(x) \geqslant 0$. Further, if we set $x'(t) = x(t+t_0)-x(t)$, then taking $\alpha_k = (k-1)t_0 (k = 1, 2, \ldots, n+1)$, we obtain

$$p(x') \leqslant \pi(x'; \alpha_1, \alpha_2, \ldots, \alpha_{n+1}) = \frac{1}{n+1} \sup_{-\infty<t<\infty} [x(t+(n+1)t_0)-x(t)] \underset{n\to\infty}{\rightarrow} 0.$$

Therefore $p(x') \leqslant 0$ and, in exactly the same way, $p(-x') \leqslant 0$. Hence, using (5) again, we have $f(x') = 0$.

Finally, if $x_0(t) \equiv 1$, then clearly $p(x_0) = 1$; $p(-x_0) = -1$. Therefore $f(x_0) = 1$.

To complete the proof we need only set

$$\int_0^1 x(t)\,dt = \frac{1}{2}[f(x)+f(\tilde{x})] \quad (\tilde{x}(t) = x(1-t)). \tag{11}$$

REMARK. It is not hard to prove that the integral we have just constructed coincides with the Riemann integral when the latter exists. In general, one cannot say the same of the Lebesgue integral. However, one can choose the functional f in such a way that the integral (11) does coincide with the Lebesgue integral for all measurable functions.

Using the generalized integral (11), one can construct a generalized measure for subsets of the interval $E_0 = [0, 1]$. In fact, we have the following theorem.

THEOREM 4. *With each set $e \subset E_0 = [0, 1]$ it is possible to associate a number $\mu(e)$—the "measure" of e—such that the following conditions are satisfied*:

1) $\mu(e_1 \cup e_2) = \mu(e_1)+\mu(e_2)$ *if* $e_1 \cap e_2 = \varnothing$;
2) $\mu(e) \geqslant 0$;
3) *if e_1 is congruent to e_2, then* $\mu(e_1) = \mu(e_2)$;
4) $\mu(E_0) = 1$.

Proof. Let χ_e be the characteristic function of the set $e \subset E_0$. If we set

$$\mu(e) = \int_0^1 \chi_e(t)\,dt,$$

then properties 1)–4) can be derived without difficulty from the properties of the generalized integral.

REMARK. A generalized measure satisfying 1)–4) can also be defined for the square $[0, 1; 0, 1]$. However, it should be noted that this problem has still not been solved for the three-dimensional cube.*

4.3. By analogy with the generalized integral, one can define a generalized limit for an arbitrary bounded sequence.

Consider the vector space l^∞ of bounded real sequences. Let $x = (\xi_1, \xi_2, \ldots) \in l^\infty$. We write

$$\pi(x; n_1, n_2, \ldots, n_k) = \overline{\lim_{n\to\infty}} \frac{1}{k} \sum_{j=1}^{k} \xi_{n+n_j},$$

$$p(x) = \inf \pi(x; n_1, n_2, \ldots, n_k),$$

where the infimum is taken over all sets of natural numbers $n_1, n_2, \ldots, n_k$.

As in the proof of Theorem 3, one can show that p is a semi-additive positive homogeneous functional. Hence there exists a linear functional f satisfying condition (5). If we set

$$\operatorname*{Lim}_{n\to\infty} \xi_n = f(x),$$

then, arguing exactly as in the proof of Theorem 3, one can establish the following properties of this functional:

1) $\operatorname*{Lim}_{n\to\infty} [\alpha\xi_n' + \beta\xi_n''] = \alpha \operatorname*{Lim}_{n\to\infty} \xi_n' + \beta \operatorname*{Lim}_{n\to\infty} \xi_n''$;

2) $\operatorname*{Lim}_{n\to\infty} \xi_n \geqslant 0, \quad \text{if} \quad \xi_n \geqslant 0 \quad (n = 1, 2, \ldots)$;

3) $\operatorname*{Lim}_{n\to\infty} \xi_{n+1} = \operatorname*{Lim}_{n\to\infty} \xi_n$;

4) $\operatorname*{Lim}_{n\to\infty} \xi_n^{(0)} = 1, \quad \text{if} \quad \xi_n^{(0)} = 1 \quad (n = 1, 2, \ldots)$;

5) $\underline{\lim}\, \xi_n \leqslant \operatorname*{Lim}_{n\to\infty} \xi_n \leqslant \overline{\lim_{n\to\infty}}\, \xi_n$.

It follows from the last of these that, if $\lim \xi_n$ exists, then we must have $\operatorname*{Lim}_{n\to\infty} \xi_n = \lim_{n\to\infty} \xi_n$. Bearing in mind also the other properties of the number $\operatorname*{Lim}_{n\to\infty} \xi_n$, one naturally calls this number the (Banach) *generalized limit* of the sequence $\{\xi_n\}$.

* For a proof see the book: I. P. Natanson, *Theory of Functions of a Real Variable* (Russian), Gostekhizdat, 1950, p. 277.

III

TOPOLOGICAL VECTOR SPACES

IN THE majority of cases, when we are dealing with a specific vector space **X**, there is already a "natural" convergence that defines a topology in **X**, and this topology is compatible, in a certain sense, with the algebraic operations in **X**. In the case that will be our principal concern in this book, this topology will be definable by means of a norm—that is, **X** will be a normed space. However, we first consider the more general case of a topological vector space. Our motivation for doing this is, first, that many problems on normed spaces have natural solutions in the more general setting, and secondly that, even to study normed spaces in their own right, one needs to use the so-called weak topology, which is not normable in the infinite-dimensional case. The introduction that we present below to the elementary theory of topological vector spaces is directed only towards the goals just referred to, and we therefore make no claim to completeness or conclusiveness (we do not even go into the most important concepts of barrelled, bornological or nuclear spaces). For a detailed account of the theory of topological vector spaces, see Bourbaki-III; Dunford and Schwartz-I; Yosida; Robertson and Robertson; Schaefer; Edwards.

§ 1. General definitions

1.1. Suppose **X** is a vector space which is at the same time a topological space. Then **X** is called a *topological vector space* (or TVS, for short) if the algebraic operations are continuous in the topology of **X**; that is, if:

1) for each pair of elements $x, y \in \mathbf{X}$ and each neighbourhood V_{x+y} of $x+y$, there exist a neighbourhood V_x of x and a neighbourhood V_y of y such that

$$V_x + V_y \subset V_{x+y};$$

2) given any element $x \in \mathbf{X}$, any number λ, and any neighbourhood $V_{\lambda x}$ of λx, we can find a neighbourhood V_x of x and a number $\delta > 0$ such that, for any μ with $|\mu - \lambda| < \delta$, we have

$$\mu V_x \subset V_{\lambda x}.$$

It is not hard to see that a linear manifold in a TVS **X** is itself a TVS under the topology and algebraic operations induced from **X**. We call such a TVS a subspace of **X**.

We note some simple consequences of the definition of a TVS.

I. If $G \subset \mathbf{X}$ is an open set, then $x_0 + G$ is also an open set.

For let $x \in x_0 + G$, so that $x = x_0 + x'$, where $x' \in G$. Let $V_{x'}$ be a neighbourhood of x' contained in G. Since $x' = x + (-x_0)$, condition 1) shows that there exist a neighbourhood

V_x of x and a neighbourhood V_{-x_0} of $-x_0$ such that $V_x + V_{-x_0} \subset V_{x'}$. Since $-x_0 \in V_{-x_0}$, we have $-x_0 + V_x \subset V_{x'} \subset G$ and therefore $V_x \subset x_0 + G$: that is, x is an interior point of $x_0 + G$.

Similarly, one can prove

II. If G is an open set and $\lambda \neq 0$, then λG is also an open set.

Furthermore, the analogous statements are true of closed sets.

As a consequence of I, we have

III. Every neighbourhood of a point $x \in \mathbf{X}$ has the form $x + V$, where V is a neighbourhood of the zero element of $\mathbf{X}$. Moreover, if V ranges over a fundamental system of neighbourhoods of zero, then $x + V$ ranges over a fundamental system of neighbourhoods of x.

This last property means that one can restrict one's attention to neighbourhoods of the zero element.

THEOREM 1. *Any* TVS $\mathbf{X}$ *has a fundamental system* $\mathfrak{B}$ *of neighbourhoods of zero with the following properties*:

1) *for any* $V_1, V_2 \in \mathfrak{B}$, *there exists* $V_3 \in \mathfrak{B}$ *such that* $V_3 \subset V_1 \cap V_2$;
2) *every* $V \in \mathfrak{B}$ *is a balanced set*;
3) *every* $V \in \mathfrak{B}$ *is an absorbent set*;
4) *for any* $V \in \mathfrak{B}$, *there exists* $U \in \mathfrak{B}$ *such that* $U + U \subset V$.

Conversely, if $\mathbf{X}$ *is a linear space in which a family* $\mathfrak{B}$ *of subsets has been singled out, subject to conditions* 1)–4), *then by taking the neighbourhoods of an element* $x \in \mathbf{X}$ *to be the sets of the form* $x + V$ $(V \in \mathfrak{B})$ *we turn* $\mathbf{X}$ *into a* TVS *in which* $\mathfrak{B}$ *is a fundamental system of neighbourhoods of zero.*

Proof. If $\mathfrak{B}$ is a fundamental system of neighbourhoods of zero in a TVS $\mathbf{X}$, then condition 1) is obviously satisfied. Also condition 4) is satisfied since $\mathbf{0} + \mathbf{0} = \mathbf{0}$. We now show that every neighbourhood of zero V is an absorbent set.

Since $0 \,.\, x = \mathbf{0}$, the definition of a TVS implies that there exists a neighbourhood V_x of x and a number $\delta > 0$ such that $\lambda V_x \subset V$ for $|\lambda| \leqslant \delta$. In particular, $x \in (1/\lambda)V$ whenever $|1/\lambda| \geqslant 1/\delta$.

To complete the proof of the first part of the theorem, it is sufficient to establish that the balanced neighbourhoods form a fundamental system of neighbourhoods of zero. Let V be any neighbourhood of zero. Since $0 \,.\, \mathbf{0} = \mathbf{0}$, we can find a neighbourhood V_1 of zero and a number $\delta > 0$ such that $\lambda V_1 \subset V$ whenever $|\lambda| \leqslant \delta$. Write $V_0 = \bigcup_{|\lambda| \leqslant \delta} \lambda V_1$. Since $V_0 \supset \delta V_1$ and since δV_1 is a neighbourhood of zero by Property III, it follows that V_0 is also a neighbourhood of zero. If $|\alpha| \leqslant 1$, then $\alpha V_0 = \bigcup_{|\lambda| \leqslant \delta} \alpha\lambda V_1 \subset V_0$, so V_0 is a balanced neighbourhood. It remains only to note that $V_0 \subset V$.

Now we turn to the proof of the second part of the theorem.

We verify first that, if the neighbourhoods in $\mathbf{X}$ are defined as in the statement of the theorem, then $\mathbf{X}$ becomes a topological space (see Theorem I.2.1).

1) Every neighbourhood of x contains x. In fact, every $V \in \mathfrak{B}$ is a balanced set and so contains zero; hence $x \in x + V$.

2) The intersection of two neighbourhoods of x contains a third. This follows at once from condition 1) of the theorem.

3) For any neighbourhood V_x of x, there exists a neighbourhood V_x' of x such that V_x contains a neighbourhood of every $y \in V_x'$. We confine ourselves to the case where $x = \mathbf{0}$.

Let $V \in \mathfrak{B}$ be an arbitrary neighbourhood of zero and let $U \in \mathfrak{B}$ be the neighbourhood of zero whose existence is stipulated in condition 4) of the theorem. We can take $V_0' = U$. For if $y \in U$ then $y + U$ is a neighbourhood of y and $y + U \subset U + U \subset V$.

Thus **X** is a topological space. Next we verify that the algebraic operations in **X** are continuous. The continuity of addition is easily obtained from condition 4). Before turning to multiplication, we note a consequence of conditions 2) and 3). For every $E \subset \mathbf{X}$ we have $2E \subset E + E$; therefore, by condition 4), for every neighbourhood $V \in \mathfrak{B}$ we can find $\tilde{V} \in \mathfrak{B}$ such that $2\tilde{V} \subset V$, and similarly, for any natural number n, there exists $V^{(n)} \in \mathfrak{B}$ such that $2^n V^{(n)} \subset V$. Let λ be an arbitrary number. Let n be any natural number large enough that $|\lambda| \leqslant 2^n$. Since $V^{(n)}$ is balanced, so is $2^n V^{(n)}$, and thus

$$\lambda V^{(n)} = \frac{\lambda}{2^n}(2^n V^{(n)}) \subset 2^n V^{(n)} \subset V.$$

It is now easy to prove that multiplication is continuous. Let $x \in \mathbf{X}$ and let λ be any scalar. We have

$$\mu y - \lambda x = (\mu - \lambda)(y - x) + (\mu - \lambda)x + \lambda(y - x).$$

Taking this and condition 4) into account, we now need only establish three facts:

1) for any $V \in \mathfrak{B}$, we can find $V_1 \in \mathfrak{B}$ and a number $\delta > 0$ such that $\alpha V_1 \subset V$ ($|\alpha| \leqslant \delta$);
2) for any $V \in \mathfrak{B}$, we can find $\delta > 0$ such that $\alpha x \in V$ whenever $|\alpha| \leqslant \delta$;
3) for any $V \in \mathfrak{B}$, we can find $V_1 \in \mathfrak{B}$ such that $\lambda V_1 \subset V$.

The first of these statements is a consequence of the fact that V is balanced, which shows that we can take $V_1 = V$, $\delta = 1$.

The second statement is also true. For V is balanced and absorbent, so there exists $\lambda' > 0$ such that $x \in \lambda' V$. If we set $\delta = 1/\lambda'$, then for $|\alpha| \leqslant \delta$ we have

$$\alpha(\lambda' V) = (\alpha\lambda')V \subset V,$$

since $|\alpha\lambda'| \leqslant 1$. It follows from this that $\alpha x \in V$ ($|\alpha| \leqslant \delta$).

Finally, we have already noted that the third statement is true. The theorem is therefore proved.

COROLLARY 1. *Every* TVS *has a fundamental system of closed balanced neighbourhoods.*

For it is sufficient to prove that the zero element has such a system of neighbourhoods. The closures of the neighbourhoods belonging to any fundamental system $\mathfrak{B}$ of balanced neighbourhoods of zero form such a system. For if $V \in \mathfrak{B}$ and if $U \in \mathfrak{B}$ is a neighbourhood such that $U + U \subset V$, then $\bar{U} \subset V$, because if $x_0 \notin V$ then the neighbourhood $x_0 + U$ of x_0 does not intersect with U. To complete the proof it is enough to observe that the property of being balanced is inherited by the closure of a set.

COROLLARY 2. *Let* $\mathfrak{B}$ *be a basis of neighbourhoods of zero in a* TVS **X**. *A necessary and sufficient condition for* **X** *to be Hausdorff is that*

$$\bigcap_{V \in \mathfrak{B}} V = \{\mathbf{0}\}. \tag{1}$$

For if **X** is Hausdorff and $x \neq \mathbf{0}$, then there exists $V \in \mathfrak{B}$ not containing x, so that (1) is true. Conversely, if (1) is true and $x \neq y$, then there exists $V \in B$ not containing $x - y$. By Theorem 1, there exists a balanced neighbourhood U of zero such that $U + U \subset V$. Then $x + U$ and $y + U$ are disjoint neighbourhoods of x and y, as $z \in (x + U) \cap (y + U)$ implies $x - y = (z - y) - (z - x) \in U - U = U + U \subset V$. Therefore **X** is Hausdorff.

In future we shall assume that every TVS is Hausdorff.

1.2. Let us formulate a few more statements which hold in a TVS $\mathbf{X}$.

I. The closure of a linear manifold $\mathbf{X}_0$ in $\mathbf{X}$ is a linear manifold.

For let $x, y \in \overline{\mathbf{X}}_0$ and let α, β be arbitrary scalars. Also let V_z be a neighbourhood of $z = \alpha x + \beta y$. There exist neighbourhoods V_x of x and V_y of y such that $\alpha V_x + \beta V_y \subset V_z$. Now V_x contains points of $\mathbf{X}_0$: let x' be one such point. Similarly, let $y' \in V_y \cap \mathbf{X}_0$. Since $z' = \alpha x' + \beta y'$ belongs to $\mathbf{X}_0$ and also $z' \in \alpha V_x + \beta V_y \subset V_z$, the intersection $V_z \cap \mathbf{X}_0$ is non-empty, which implies, as V_z was arbitrary, that $z \in \overline{\mathbf{X}}_0$.

A very similar proof shows that

II. The closure of a convex set is convex. The closure of an absolutely convex set is absolutely convex.

Let E be any non-empty subset of a TVS $\mathbf{X}$. The closure of the linear hull of E is called the *closed linear hull* of E and is denoted by $\overline{\mathscr{L}}(E)$. By property I, $\overline{\mathscr{L}}(E)$ is the smallest closed linear manifold of $\mathbf{X}$ containing E. The closure of the convex (respectively, absolutely convex) hull of E is called the *closed convex* (respectively, *closed absolutely convex*) *hull* of E, and is denoted by $\overline{\mathrm{co}}\,(E)$ (respectively, $\overline{\mathrm{abs\,co}}\,(E)$). By II, $\overline{\mathrm{co}}\,(E)$ (respectively, $\overline{\mathrm{abs\,co}}\,(E)$) is the smallest closed convex (respectively, absolutely convex) set in $\mathbf{X}$ containing E.

From III in the preceding subsection it is easy to deduce

III. A necessary and sufficient condition for a net $\{x_\alpha\}$ $(\alpha \in \mathrm{A})$ to converge to $x \in \mathbf{X}$ is that $x_\alpha - x \underset{\mathrm{A}}{\to} \mathbf{0}$.

The proof of the next property is a little more complicated.

IV. Let K_1 and K_2 be compact subsets in a TVS $\mathbf{X}$. Then $\lambda_1 K_1 + \lambda_2 K_2$ is compact.

For the set $K = K_1 \times K_2$ is compact in the product space $\mathbf{X} \times \mathbf{X}$, by I.2.8, and the mapping ϕ defined by

$$\phi(x, y) = \lambda_1 x + \lambda_2 y \quad (x, y \in \mathbf{X})$$

from $\mathbf{X} \times \mathbf{X}$ into $\mathbf{X}$ is continuous; therefore, by I.2.5, the set $\phi(K) = \lambda_1 K_1 + \lambda_2 K_2$ is compact in $\mathbf{X}$.

Let $\mathbf{X}$ and $\mathbf{Y}$ be TVSs. A mapping $f: \mathbf{X} \to \mathbf{Y}$ that is simultaneously a vector space isomorphism and a homeomorphism is called an *isomorphism* between $\mathbf{X}$ and $\mathbf{Y}$, and in this situation $\mathbf{X}$ and $\mathbf{Y}$ are said to be *isomorphic*. In keeping with our earlier point of view, we shall identify isomorphic TVSs.

A TVS $\mathbf{X}$ is said to be *metrizable* if its topology can be defined by means of a metric.

1.3. We now give some examples of TVSs.

1) The space of measurable functions $\mathbf{S}(T, \Sigma, \mu)$.

For any set $A \in \Sigma$ of finite measure and any number $\varepsilon > 0$, let us write

$$V(A, \varepsilon) = \left\{ x \in \mathbf{S}(T, \Sigma, \mu) : \int_A \frac{|x|}{1 + |x|} \, d\mu < \varepsilon \right\}.$$

It is easy to show that the system of all sets of the form $V(A, \varepsilon)$ is a fundamental system of neighbourhoods of zero for a topology on $\mathbf{S}(T, \Sigma, \mu)$. It follows from Theorem I.6.14 that a net $\{x_\alpha\}$ converges to x in this topological space if and only if $x_\alpha \to x\ (\mu)$. Since the algebraic operations are continuous in the topology of convergence in measure, $\mathbf{S}(T, \Sigma, \mu)$ thus becomes a TVS. As we showed in I.6.10, this TVS is metrizable if μ is σ-finite.

If μ is not σ-finite, then $\mathbf{S}(T, \Sigma, \mu)$ is not metrizable. For the collection of all sets of finite measure is directed by the ordering of inclusion and gives rise to the net $\{x_\alpha\}$ of characteristic functions of such sets. Clearly $x_\alpha \to \mathbf{1}(\mu)$, where $\mathbf{1}$ is the function identically equal to unity on T. If $\mathbf{S}(T, \Sigma, \mu)$ were metrizable, then there would exist a sequence $\{x_{\alpha_n}\}$ with $x_{\alpha_n} \to \mathbf{1}(\mu)$. Suppose that x_{α_n} is the characteristic function of A_n. Write $B = \bigcup_{n=1}^{\infty} A_n$. Since μ is not σ-finite, $\mu(T \setminus B) \neq 0$, and since μ has no atoms of infinite measure, there exists $A \in \Sigma$ such that $A \subset T \setminus B$, $0 < \mu(A) < \infty$. On A we have $x_{\alpha_n} = 0$, contradicting the fact that $x_{\alpha_n} \to \mathbf{1}(\mu)$.

A generalization of the space $\mathbf{s}$ is the space $\mathbf{s}(T)$, which consists of all the real-valued functions defined on an abstract set T. One fundamental system of neighbourhoods of zero in $\mathbf{s}(T)$ is the system $\mathfrak{B}$ of all sets $V_{t_1, t_2, \ldots, t_n; \varepsilon}$, where $t_1, t_2, \ldots, t_n$ are arbitrary elements of T and ε is any positive number, and $x \in V_{t_1, t_2, \ldots, t_n; \varepsilon}$ is defined to mean that

$$|x(t_k)| \leqslant \varepsilon \quad (k = 1, 2, \ldots, n).$$

We leave it to the reader to verify that this system satisfies the conditions of Theorem 1, so that $\mathbf{s}(T)$ is a Hausdorff TVS, coinciding with $\mathbf{s}$ if T is a countable set.

As we remarked in I.6.10, the space $\mathbf{s}$ is a special case of the space $\mathbf{S}(T, \Sigma, \mu)$. In exactly the same way, $\mathbf{s}(T)$ is a special case of $\mathbf{S}(T, \Sigma, \mu)$. We leave it to the reader to verify that the corresponding vector spaces are isomorphic as TVSs. By what we have proved, $\mathbf{s}(T)$ is a metrizable TVS if and only if T is countable—that is, if and only if $\mathbf{s}(T) = \mathbf{s}$.

If $T = \{1, 2, \ldots, n\}$, then we can identify $\mathbf{s}(T)$ with the n-dimensional space $\mathbf{K}^n$, and regard the latter as a TVS. However, we postpone the discussion of finite-dimensional TVSs to the following chapter.

2) For the next example we consider the space $\mathbf{C}(\mathbf{R}^1)$, whose elements are all the continuous functions defined on the whole of the real line. The topology in $\mathbf{C}(\mathbf{R}^1)$ is introduced by means of the fundamental system of neighbourhoods of zero consisting of the sets $V_{n;\varepsilon}$, where n is a natural number and $\varepsilon > 0$. Here $x \in V_{n;\varepsilon}$ means that

$$|x(t)| \leqslant \varepsilon \qquad (|t| \leqslant n).$$

As before, it is left to the reader to verify that the conditions of Theorem 1 are satisfied, and that $\mathbf{C}(\mathbf{R}^1)$ is Hausdorff.

3) In the theory of generalized functions, the space $\mathbf{D}[a, b]$ of all indefinitely differentiable functions vanishing outside the interval $[a, b]$ plays a major role. This space has a fundamental system of neighbourhoods of zero consisting of the sets $V_{n;\varepsilon}$, one for each natural number n and each positive number ε, consisting of all x satisfying the condition

$$|x^{(k)}(t)| \leqslant \varepsilon \quad (k = 0, 1, \ldots, n; \quad t \in [a, b]).$$

4) Let us give one more example of a TVS. Denote by $\mathbf{L}_\omega$ the set of all measurable functions defined on $[0, 1]$ and summable to any power. Each pair of numbers $p > 1$ and $\varepsilon > 0$ determines a neighbourhood $V_{p;\varepsilon}$ where $x \in V_{p;\varepsilon}$ if

$$\left[\int_0^1 |x(t)|^p \, dt\right]^{1/p} \leqslant \varepsilon.$$

The spaces $\mathbf{C}(\mathbf{R}^1)$, $\mathbf{D}[a, b]$, and $\mathbf{L}_\omega$ are metrizable TVSs, as we shall show below.

1.4. A subset E of a TVS $\mathbf{X}$ is said to be *bounded* if for each neighbourhood of zero V in $\mathbf{X}$ there exists a number λ such that $E \subset \lambda V$. In verifying that a set is bounded it is clearly enough to consider neighbourhoods belonging to a fundamental system of neighbourhoods of zero.

A subset G of a TVS $\mathbf{X}$ is said to be *totally bounded* if for any neighbourhood V of zero in $\mathbf{X}$

there exists a finite subset $\{x_k\}_{k=1}^n \subset G$ such that $G \subset \bigcup_{k=1}^{n} (x_k + V)$.

Let us note some simple facts about bounded and totally bounded sets.

I. Let E_1 and E_2 be bounded (respectively, totally bounded) subsets of a TVS $\mathbf{X}$. Then the following subsets of $\mathbf{X}$ are bounded (respectively, totally bounded): $E_1 \cup E_2, E_1 + E_2, \lambda E_1$ (where λ is a scalar).

In the case of $E_1 \cup E_2$ and λE_1, the assertion is obvious. To obtain the statement for $E_1 + E_2$ we note that by 4) of Theorem 1 the sets of the form $V+V$ form a basis of neighbourhoods of zero, where V ranges over a basis of neighbourhoods of zero.

Property I shows that a finite set is (totally) bounded.

II. Every totally bounded set is bounded.

If E is totally bounded, then, given a neighbourhood V, which we take to be balanced, we can find points $x_k \in E$ such that $E \subset \bigcup_{k=1}^{n} (x_k + V)$. As we have already remarked, the finite set $\{x_k\}_{k=1}^n$ is bounded, so there exists λ_1 such that $\{x_k\}_{k=1}^n \subset \lambda_1 V$, and hence $E \subset \lambda_1 V + V \subset \max(|\lambda_1|, 1)\ (V+V)$. By Theorem 1, the sets $V+V$ form a basis of neighbourhoods of zero, as V ranges over a basis of neighbourhoods of zero.

Since one can construct a fundamental system of neighbourhoods of zero consisting of closed sets (Corollary to Theorem 1), we have

III. The closure of a bounded (respectively, totally bounded) set is bounded (respectively, totally bounded).

THEOREM 2. *A necessary and sufficient condition for a subset E of a* TVS $\mathbf{X}$ *to be bounded is that for any sequence $\{x_n\} \subset E$ and any sequence $\{\lambda_n\}$ of real numbers with $\lambda_n \to 0$ we have $\lambda_n x_n \to \mathbf{0}$.*

Proof. Necessity. Let $\{x_n\}$ and $\{\lambda_n\}$ be sequences with the above properties and let V be any balanced neighbourhood of zero. There exists $\lambda > 0$ such that $E \subset \lambda V$. In particular, $x_n \in \lambda V$ $(n = 1, 2, \ldots)$. Thus, if n is chosen large enough that $|\lambda_n| \leqslant 1/\lambda$, then $\lambda_n \lambda V \subset V$; in other words, $\lambda_n x_n \to \mathbf{0}$.

Sufficiency. If, under the conditions of the theorem, the set E were not bounded, then there would be a neighbourhood of zero V such that $E \setminus \lambda V$ is empty for every $\lambda > 0$. By taking $\lambda = 1, 2, \ldots$ in turn, we should then obtain a sequence of elements

$$x_n \in E \setminus nV \quad (n = 1, 2, \ldots).$$

Since then $x_n \in E$ $(n = 1, 2, \ldots)$ on the one hand, and $(1/n)x_n \notin V$ on the other hand, this would contradict the hypotheses.

Later we shall require the following lemma on bounded sets in $\mathbf{S}(T, \Sigma, \mu)$.

LEMMA 1. *If a sequence of real-valued functions $\{x_n\}$ is bounded in the* TVS $\mathbf{S}(T, \Sigma, \mu)$ *and satisfies $0 \leqslant x_1(t) \leqslant x_2(t) \leqslant \ldots \leqslant x_n(t) \leqslant x_{n+1}(t) \leqslant \ldots$ a.e., then there exists a function $x \in \mathbf{S}(T, \Sigma, \mu)$ such that $x_n(t) \to x(t)$ a.e.*

Proof. It is sufficient to verify that $\lim_{n \to \infty} x_n(t)$ is finite for almost all $t \in T$. Assume that there exists a set $A \in \Sigma$ such that $\mu(A) > 0$ and $x_n(t) \to +\infty$ for every $t \in A$. Since $\{x_n\}$ is bounded, it is absorbed by the neighbourhood $U(A, \frac{1}{2}\varepsilon)$, where $\varepsilon = \mu(A)$; that is, for some $\lambda > 0$ we have

$$\int_A \frac{|\lambda x_n(t)|}{1 + |\lambda x_n(t)|} d\mu(t) < \varepsilon/2 \quad (n \in \mathbb{N}). \tag{2}$$

Since $x_n(t) \uparrow +\infty$ for $t \in A$, we have

$$\frac{|\lambda x_n(t)|}{1+|\lambda x_n(t)|} \xrightarrow[n\to\infty]{} 1,$$

and hence, by Lebesgue's Theorem (Theorem I.6.6),

$$\lim_{n\to\infty} \int_A \frac{|\lambda x_n(t)|}{1+|\lambda x_n(t)|} d\mu(t) = \int_A d\mu(t) = \varepsilon,$$

contradicting (2). Thus $x_n(t) \uparrow x(t) < +\infty$ for almost all $t \in T$.

1.5. We conclude with the concept of completeness for a TVS. Let $\{x_\alpha\}$ $(\alpha \in \mathrm{A})$ be a net. Write A^2 for the set of all pairs (α', α''), where $\alpha', \alpha'' \in \mathrm{A}$. We make A^2 into a directed set by introducing the ordering $(\alpha'_1, \alpha''_1) \geqslant (\alpha'_2, \alpha''_2)$ if and only if both $\alpha'_1 \geqslant \alpha'_2$ and $\alpha''_1 \geqslant \alpha''_2$.

A net $\{x_\alpha\}$ in a TVS $\mathbf{X}$ is called a *Cauchy net* if the net $\{x_{(\alpha', \alpha'')}\}$ $((\alpha', \alpha'') \in \mathrm{A}^2)$, where $x_{(\alpha', \alpha'')} = x_{\alpha'} - x_{\alpha''}$ converges to zero in $\mathbf{X}$; that is, if for any neighbourhood of zero U there exists $\alpha_0 \in \mathrm{A}$ such that $x_{\alpha'} - x_{\alpha''} \in U$ whenever $\alpha', \alpha'' > \alpha_0$. A subset E of a TVS $\mathbf{X}$ is said to be *complete* if every Cauchy net of elements of E converges to an element of E. In keeping with this terminology, the TVS $\mathbf{X}$ is said to be *complete* if every Cauchy net in $\mathbf{X}$ is convergent. If we restrict ourselves to bounded nets here, then the TVS is said to be *quasi-complete*; if we restrict ourselves to ordinary sequences, then it is said to be *sequentially complete* (or *semi-complete*). Note that a closed subset of a complete space is complete.

As the set of elements in a Cauchy net is not necessarily bounded, completeness is a stronger requirement than quasi-completeness (that it is genuinely stronger we shall see below by means of examples). On the other hand, it is not hard to see that a Cauchy sequence is bounded, so quasi-completeness implies sequential completeness. In a metrizable TVS, the three types of completeness coincide, as is clear from the following theorem.

THEOREM 3. *A sequentially complete* TVS $\mathbf{X}$ *having a countable fundamental system of neighbourhoods of zero* is complete.*

Proof. Let $\{x_\alpha\}$ $(\alpha \in \mathrm{A})$ be a Cauchy net of elements of $\mathbf{X}$. Let $\{V_n\}$ $(n = 1, 2, \ldots)$ be a countable fundamental system of neighbourhoods of zero. For each $n = 1, 2, \ldots$, there exists a pair (α'_n, α''_n) in A^2 such that

$$x_{\alpha'} - x_{\alpha''} \in V_n \quad ((\alpha', \alpha'') \geqslant (\alpha'_n, \alpha''_n)). \tag{3}$$

Let α_n be an element of A such that $\alpha_n \geqslant \alpha'_n$, $\alpha_n \geqslant \alpha''_n$. We may assume that $\alpha_1 \leqslant \alpha_2 \leqslant \ldots \leqslant \alpha_n \leqslant \ldots$. Consider the sequence $\{x_{\alpha_n}\}$. Since it is clearly a Cauchy sequence, there exists, by hypothesis, an $x \in \mathbf{X}$ such that $x_{\alpha_n} \to x$. We shall prove that $x_\alpha \to x$. Take any neighbourhood V_m and choose V_k so that $V_k + V_k \subset V_m$. We can find $n \geqslant k$ such that

$$x_{\alpha_n} - x \in V_k.$$

On the other hand, when $\alpha \geqslant \alpha_n$, it follows from (3) that

$$x_\alpha - x_{\alpha_n} \in V_k.$$

---

* This condition is in fact equivalent to metrizability (see Schaefer, I.6.1).

Therefore

$$x_\alpha - x = (x_\alpha - x_{\alpha_n}) + (x_{\alpha_n} - x) \in V_k + V_k \subset V_m.$$

This proves the theorem.

In the case of a metrizable TVS $\mathbf{X}$ there are two concepts of completeness: completeness as a TVS and completeness as a metric space. In general, these are distinct. However, if the metric ρ defining the topology in $\mathbf{X}$ satisfies the condition $\rho(x, y) = \rho(x - y, \mathbf{0})$ for all $x, y \in \mathbf{X}$, then the two concepts will clearly coincide. The metric does have this property in all the specific TVSs that we shall be considering in this book. For more details on completeness, see Bourbaki-IV and Kelley.

Every TVS $\mathbf{X}$ can be completed. In other words, there exists a complete TVS $\hat{\mathbf{X}}$ such that $\mathbf{X}$ is a subspace of $\hat{\mathbf{X}}$ and $\mathbf{X}$ is dense in $\hat{\mathbf{X}}$. This theorem is proved on the pattern of Theorem I.4.1, replacing equivalence classes of Cauchy sequences by classes of nets (the verification that $\hat{\mathbf{X}}$, in this situation, is a vector space is analogous to IV.1.4). The space $\hat{\mathbf{X}}$ is called the *completion* of $\mathbf{X}$. Completions of TVSs will not be needed in the sequel.

Using Theorem 3, it is easy to prove that the TVSs $\mathbf{C}(\mathbf{R}^1)$ and $\mathbf{D}[a, b]$ are complete. The metrizable TVS $\mathbf{L}_\omega$ is easily shown to be complete, using the results of Chapter IV on the completeness of $\mathbf{L}^p$ spaces. Theorem I.6.15 shows that the metrizable TVS $\mathbf{S}(T, \Sigma, \mu)$ is complete, in the case of a σ-finite measure. It is simple to deduce from this that the TVS $\mathbf{S}(T, \Sigma, \mu)$ is complete if the space (T, Σ, μ) has the direct sum property (see I.6.10).

Let us now turn to a study of the connections between totally bounded sets and compact sets. The following theorem may be regarded as an analogue of Hausdorff's theorem (see I.5.1) on compact sets in metric spaces. In fact both these results are corollaries of a more general result on so-called uniform spaces (see Bourbaki-II).

THEOREM 4. *A subset of a* TVS $\mathbf{X}$ *is compact if and only if it is totally bounded and complete.*

Proof. Let K be a compact set. We show first that K is totally bounded. Let V be any open neighbourhood of zero. The family of open sets $\{x + V\}$ $(x \in K)$ is clearly a covering of K and so, by the definition of compactness, we see that K is totally bounded.

Next we prove that K is complete. Let $\{x_\alpha\}$ $(\alpha \in \mathrm{A})$ be a Cauchy net of elements of K. By Theorem I.2.3, there exists a subnet $\{y_\beta\}$ $(\beta \in \mathrm{B})$ converging to $x \in K$. We show that $x_\alpha \to x$. Take any neighbourhood of zero V. Then there exists a neighbourhood of zero U such that $U + U \subset V$. Because $\{x_\alpha\}$ is a Cauchy net, there exists $\alpha_0 \in \mathrm{A}$ such that $x_\alpha - x_{\alpha'} \in U$ whenever $\alpha, \alpha' \geqslant \alpha_0$. On the other hand, since $y_\beta \to x$, there exists $\beta_0 \in \mathrm{B}$ such that $x - y_\beta \in U$ whenever $\beta \geqslant \beta_0$. By the definition of subnet, there is a $\beta(\alpha_0) \in \mathrm{B}$ such that if $\beta' \in \mathrm{B}$ and $\beta' \geqslant \beta(\alpha_0)$ then $y_{\beta'} = x_{\alpha'}$ for some $\alpha' \geqslant \alpha_0$. Choose β' with $\beta' \geqslant \beta(\alpha_0)$, β_0 and let α' be the corresponding element of A. Then, for $\alpha \geqslant \alpha_0$, we have

$$x_\alpha - x = (x_\alpha - x_{\alpha'}) + (y_{\beta'} - x) \in U + U \subset V,$$

and so $x_\alpha \to x$.

Conversely, suppose K is totally bounded and complete. We again use Theorem I.2.3 to verify that K is compact. Let $\{x_\alpha\}$ $(\alpha \in \mathrm{A})$ be any net of elements of K. Let $\mathfrak{M}$ denote the set of all families $\mathfrak{F}$ of subsets E of K with the finite intersection property, such that the net $\{x_\alpha\}$ is frequently in (see I.2.7) every $E \in \mathfrak{F}$. Then $\mathfrak{M}$ is non-empty, for the family consisting of K alone belongs to $\mathfrak{M}$. By Zorn's Lemma, $\mathfrak{M}$ has a maximal element $\mathfrak{F}_0$ under the ordering of inclusion. (The application of Zorn's Lemma is similar to that in the proof of Theorem II.4.1.) The maximality of $\mathfrak{F}_0$ implies that

1) $K \in \mathfrak{F}_0$;

2) if $\{x_\alpha\}$ is frequently in the sets $A_1, A_2, \ldots, A_n$ and $\bigcup_{i=1}^{n} A_i \in \mathfrak{F}_0$, then $A_i \in \mathfrak{F}_0$ for some i.

Statement 1) is clear. Assume that 2) is false. If, on adjoining any of the A_i to $\mathfrak{F}_0$, the resulting family has the finite intersection property, then by the maximality of $\mathfrak{F}_0$ this family must coincide with $\mathfrak{F}$ and so we must have $A_i \in \mathfrak{F}_0$. Since we have assumed that 2) is false, it follows that, for each A_i, there must exist a finite number of sets $B_i^{(k)} \in \mathfrak{F}_0$ ($k = 1, 2, \ldots, k_i$) such that

$$A_i \cap B_i^{(1)} \cap B_i^{(2)} \cap \ldots \cap B_i^{(k_i)} = \varnothing.$$

Then we have $C = \bigcup_{i=1}^{n} A_i \cap \bigcap_{i=1}^{n} (B_i^{(1)} \cap B_i^{(2)} \cap \ldots \cap B_i^{(k_i)}) = \varnothing$. However, in view of the condition $\bigcup_{i=1}^{n} A_i \in \mathfrak{F}_0$, all the $B_i^{(k)}$ belong to $\mathfrak{F}_0$ and $\mathfrak{F}_0$ has the finite intersection property, so that C is non-empty. This contradiction proves 2).

We now consider the set B of all pairs (E, α), where $E \in \mathfrak{F}_0$ and $\alpha \in \mathrm{A}$ and $x_\alpha \in E$. Introduce an ordering on B as follows: $(E_1, \alpha_1) \geqslant (E_2, \alpha_2)$ if and only if $E_1 \subset E_2$ and $\alpha_1 \geqslant \alpha_2$ in A. Using the fact that $\mathfrak{F}_0$ has the finite intersection property, and the fact that $\{x_\alpha\}$ is frequently in each member of $\mathfrak{F}_0$, we see that B is a directed set. For each $(E, \alpha) \in \mathrm{B}$ we set $y_{(E,\alpha)} = x_\alpha$. Then $\{y_{(E,\alpha)}\}$ $((E, \alpha) \in \mathrm{B})$ is a subnet of $\{x_\alpha\}$ $(\alpha \in \mathrm{A})$. For, given $\alpha \in \mathrm{A}$, choose $(K, \alpha) \in \mathrm{B}$. If $(E, \alpha') \in \mathrm{B}$, $(E, \alpha') \geqslant (K, \alpha)$, then $\alpha' \geqslant \alpha$ and $x_{\alpha'} = y_{(E,\alpha')}$, as we require.

We next show that $\{y_{(E,\alpha)}\}$ $((E, \alpha) \in \mathrm{B})$ is a Cauchy net. Since K is complete, this will complete the proof of the theorem.

Take any neighbourhood of zero V. Then we can find a balanced neighbourhood of zero U such that $U + U \subset V$. Since K is totally bounded, there exist points $z_1, z_2, \ldots, z_n \in K$ such that $K \subset \bigcup_{i=1}^{n} (z_i + U)$. If we set $A_i = (z_i + U) \cap K$, then $K = \bigcup_{i=1}^{n} A_i \in \mathfrak{F}_0$, and so by 2) we have $A_i \in \mathfrak{F}_0$. Now choose any $\alpha \in \mathrm{A}$ such that $x_\alpha \in A_i$. If $(E_1, \alpha_1), (E_2, \alpha_2) \in \mathrm{B}$ and $(E_1, \alpha_1), (E_2, \alpha_2) \geqslant (A_i, \alpha)$, then since $E_1, E_2 \subset A_i$, we have

$$y_{(E_1,\alpha_1)} - y_{(E_2,\alpha_2)} = x_{\alpha_1} - x_{\alpha_2} \in E_1 - E_2 \subset A_i - A_i \subset (z_i + U) - (z_i + U) \subset U + U \subset V.$$

We have therefore shown that the subnet $\{x_\alpha\}$ constructed above is a Cauchy net, and so the proof is complete.

Theorem 4 accounts for the widespread use of the term *precompact* as an alternative to totally bounded.

§ 2. Locally convex spaces

2.1. A Hausdorff TVS **X** (over $\mathbb{K}$) is called a *locally convex space* (LCS, for short) if it has a fundamental system of convex neighbourhoods of zero. The theory of LCSs is significantly richer in results than the theory of TVSs, chiefly because there are always plenty of continuous linear functionals on an LCS. Moreover, almost all the concrete spaces that occur in functional analysis are locally convex.

Theorem 1.1 becomes significantly simpler when it is applied to an LCS. By this theorem, an LCS has a basis consisting of closed absolutely convex neighbourhoods of zero.

THEOREM 1. *Suppose that a system $\mathfrak{B}_0$ of absorbent convex sets has been specified in a vector space* **X**, *subject to the following condition:*

for each $x \neq 0$, there exist $V \in \mathfrak{B}_0$ and $\lambda > 0$ such that $x \notin \lambda V$. (1)

Let $\mathfrak{B}$ denote the system of all sets of the form

$$\varepsilon \bigcap_{i=1}^{n} V_i \quad (\varepsilon > 0,\ V_i \in \mathfrak{B}_0,\ n \in \mathbb{N}). \tag{2}$$

Then $\mathfrak{B}$ satisfies all the conditions of Theorem 1.1; *hence* **X** *is turned into an* LCS (*having* $\mathfrak{B}$ *as a fundamental system of neighbourhoods of zero*).

Proof. It is obvious that $\mathfrak{B}$ satisfies conditions 1)–3) of Theorem 1.1, while condition 4) is a consequence of the fact that $\frac{1}{2}E + \frac{1}{2}E = E$ for any convex set E. Thus **X** is a TVS. By Corollary 2 to Theorem 1.1, condition (1) implies that **X** is Hausdorff. Since sets of the form (2) are absolutely convex, **X** is an LCS.

LEMMA 1. *Let* **X** *be a* TVS.

1) *A gauge function p is continuous on* **X** *if and only if it is continuous at the origin.*

2) *The Minkowski functional p_U of a convex absorbent set U is continuous if and only if U is a neighbourhood. In this case,* $\mathring{U} = \{x\colon p_U(x) < 1\}$ *and* $\bar{U} = \{x\colon p_U(x) \leqslant 1\}$.

Proof. 1) If p is continuous at the origin, then for every $\varepsilon > 0$ there is a balanced neighbourhood of zero U every point x of which satisfies $p(x) < \varepsilon$. If now y is any point of **X**, then for all $x \in y + U$ we have $x - y, y - x \in U$ and so II.3.2(c′) shows that $|p(x) - p(y)| \leqslant \max(p(x-y), p(y-x)) < \varepsilon$.

2) By Lemma II.3.1, p_U is a gauge function. If U is a neighbourhood, then for every $\varepsilon > 0$ the statement $x \in \varepsilon U$ implies that $p_U(x) \leqslant \varepsilon$, by the definition of the Minkowski functional, so p_U is continuous at the origin and therefore also on all of **X**.

Conversely, if p_U is continuous, then the set $V = \{x\colon p_U(x) < 1\} = p_U^{-1}((-1, 1))$ is open. Since $V \subset U$, it follows that U is a neighbourhood.

Let $x \in \mathring{U}$. Assuming that $p_U(x) = 1$, we shall show that any neighbourhood V of x contains points y with $y \notin U$, yielding a contradiction to the hypothesis that $x \in \mathring{U}$. In fact, since V is a neighbourhood of x, there exists $\varepsilon > 0$ such that $y = (1+\varepsilon)x \in V$. Thus $p'_U(y) = (1+\varepsilon)p_U(x) = 1 + \varepsilon > 1$, and so $y \notin U$.

Now suppose that $p_U(x) \leqslant 1$. Then there exists a sequence $\{\lambda_n\}$ with $\lambda_n \to 1$ such that $x \in \lambda_n U$ for each n. Hence $y_n = x/\lambda_n \in U$ and $y_n \to x \in \bar{U}$.

This completes the proof of the lemma.

COROLLARY. *If E is a convex set in a* TVS **X**, *then its interior* $\mathring{E}$ *is convex. If* $\mathring{E} \neq \varnothing$, *then the closure of* $\mathring{E}$ *is* $\bar{E}$.

Proof. We may assume that $\mathring{E} \neq \varnothing$. Choose $x \in \mathring{E}$. Then the set $U = x - E$ is a convex neighbourhood of zero, and $\mathring{U} = x - \mathring{E}$. By Lemma 1, $\mathring{U} = \{x\colon p_U(x) < 1\}$ is convex, so $\mathring{E}$ is also convex.

Now let us prove the second statement. Since $\overline{\mathring{U}} = x - \overline{\mathring{E}}$, it is sufficient to prove it for $\mathring{U}$, and to do this we need only check that $\overline{\mathring{U}} \supset U$. Choose $x \in U$ and consider $y_n = \left(1 - \frac{1}{n}\right)x$. Since $p_U(y_n) = \left(1 - \frac{1}{n}\right)p_U(x) < 1$, we have $y_n \in \mathring{U}$. On the other hand, $y_n \to x$, and so $x \in \overline{\mathring{U}}$.

An LCS topology may be specified on a vector space **X** by means of any system of seminorms $\{p_\xi\}$ $(\xi \in \Xi)$ satisfying the condition:

$$\text{for each } x \in \mathbf{X} \text{ there exists } \xi \in \Xi \text{ such that } p_\xi(x) \neq 0. \tag{3}$$

Let $\mathfrak{B}_0$ consist of all sets V_ξ of the form

$$V_\xi = \{x \in \mathbf{X}:\ p_\xi(x) \leqslant 1\} \qquad (\xi \in \Xi).$$

By Lemma II.3.1, the sets V_ξ are absolutely convex and absorbent. It follows from (3) that $\mathfrak{B}_0$ satisfies (1). Now specify a topology in $\mathbf{X}$ as in Theorem 1. The system of sets (2) has the form

$$\{x \in \mathbf{X}:\ \max_{1 \leqslant i \leqslant n} p_{\xi_i}(x) \leqslant \varepsilon\} \quad (\varepsilon > 0; \xi_1, \ldots, \xi_n \in \Xi).$$

The resulting topology, which makes $\mathbf{X}$ into an LCS, is called the topology generated by the family $\{p_\xi\}$ $(\xi \in \Xi)$. If the topology of an LCS $\mathbf{X}$ is generated by a family of seminorms $\{p_\xi\}$ $(\xi \in \Xi)$, then this family is said to be a *generating* or *defining* family for the topology of $\mathbf{X}$. By Lemma 1 all the seminorms in a defining family are continuous.

The topology of any LCS $\mathbf{X}$ is generated by some family of seminorms. For choose a system $\mathfrak{B}_0$ of absolutely convex neighbourhoods of zero such that the sets of the form (2) constitute a fundamental system of neighbourhoods of zero in $\mathbf{X}$ (for example, a basis of absolutely convex neighbourhoods of zero). Then condition (1) is automatically satisfied because $\mathbf{X}$ is Hausdorff. Hence the given topology of $\mathbf{X}$ is generated by the family of seminorms $\{p_V\}$ $(V \in \mathfrak{B}_0)$, where p_V is the Minkowski functional of V.

A generating family of seminorms is, of course, not determined uniquely. For example, if a seminorm p belongs to some such family Q, then the seminorm $q(x) = 2p(x)$ may or may not also belong to Q. We can therefore obtain a generating family of seminorms distinct from Q either by omitting q from Q, in the first case, or by adjoining q to Q, in the second case.

THEOREM 2. *A necessary and sufficient condition for an* LCS $\mathbf{X}$ *to be metrizable is that it be Hausdorff and have a countable generating family of seminorms.*

Proof. Necessity. If $\mathbf{X}$ is metrizable, then it has a countable basis of neighbourhoods of zero. The Minkowski functionals of the absolutely convex hulls of the neighbourhoods in this basis clearly form a countable generating family of seminorms.

Sufficiency. Let $\{p_n\}$ $(n \in \mathbb{N})$ be a generating family of seminorms. This implies that the family of sets

$$V_n = \{x \in \mathbf{X}:\ \max_{1 \leqslant i \leqslant n} p_k(x) \leqslant n^{-1}\} \quad (n \in \mathbb{N})$$

forms a countable basis of neighbourhoods of zero in $\mathbf{X}$. For $x, y \in \mathbf{X}$ we write

$$\rho(x, y) = \sum_{k=1}^{\infty} \frac{1}{2^k} \frac{p_k(x-y)}{1 + p_k(x-y)}. \tag{4}$$

It is easy to check that the function $\rho(x, y)$ defined by (4) is a metric. Let us show that the topology induced by this metric coincides with the given topology on $\mathbf{X}$. Since $\mathbf{X}$ has a countable basis of neighbourhoods of zero, to prove this we need only verify that a sequence $\{x_m\}$ converges to zero in $\mathbf{X}$ if and only if $\rho(x_m, \mathbf{0}) \to 0$. We can identify the sequences $\xi_m = \{p_n(x_m)\}_{n=1}^{\infty}$ with elements of the space $\mathbf{s}$. Since the distance from ξ_m to $\mathbf{0}$ in $\mathbf{s}$ is just $\rho(x_m, \mathbf{0})$ and since convergence in $\mathbf{s}$ is coordinatewise (see I.3.2), it follows that $\rho(x_m, \mathbf{0}) \to 0$ if and only if $p_n(x_m) \to 0$ as $m \to \infty$ for every $n \in \mathbb{N}$. In view of the way we defined the sets V_n, this just means that $x_m \to \mathbf{0}$ in $\mathbf{X}$. Thus the theorem is proved.

Many concepts relating to LCSs take on a simple and intuitive meaning when expressed in terms of generating families of seminorms. Let $\{p_\xi\}$ $(\xi \in \Xi)$ be a family of seminorms generating a topology for an LCS **X**. The following statements are immediate consequences of the definitions.

I. A net $\{x_\alpha\}$ $(\alpha \in \mathrm{A})$ converges to an element x in an LCS **X** if and only if $p_\xi(x_\alpha - x) \underset{\mathrm{A}}{\to} 0$ for every $\xi \in \Xi$.

II. A set $E \subset \mathbf{X}$ is bounded if and only if the set of numbers $\{p_\xi(x): x \in E\}$ is bounded.

The spaces $\mathbf{s}(T)$, $\mathbf{C}(\mathbf{R}^1)$, $\mathbf{D}[a, b]$, $\mathbf{L}_\omega$ introduced in 1.3 are LCSs. For the space $\mathbf{S}(T, \Sigma, \mu)$ the situation is more complicated. If (T, Σ, μ) is not a discrete measure space, the TVS $\mathbf{S}(T, \Sigma, \mu)$ is not locally convex (we prove this below in the special case of $\mathbf{S}(0, 1)$); if μ is discrete, then $\mathbf{S}(T, \Sigma, \mu)$ is an LCS. Suppose that μ is discrete and let T^* be its set of atoms. Then the TVS $\mathbf{S}(T, \Sigma, \mu)$ is isomorphic to the TVS $\mathbf{s}(T^*)$; but the latter is obviously an LCS, so $\mathbf{S}(T, \Sigma, \mu)$ is an LCS in this case.

We conclude this subsection with a result that we shall need in the sequel.

THEOREM 3. *In any* LCS *the convex hull and the absolutely convex hull of a totally bounded set are totally bounded.*

Proof. It is clearly sufficient to consider the case of the absolutely convex hull. Let E be a totally bounded set and E_1 its absolutely convex hull. If V is any absolutely convex neighbourhood of zero in **X**, then we can find elements $x_i \in E$ $(1 \leqslant i \leqslant n)$ such that $E \subset \bigcup_{i=1}^{n} (x_i + V)$. We identify the finite-dimensional space $\mathscr{L}(\{x_i\}_{i=1}^n)$ with $\mathbf{K}^m$, $m \leqslant n$ (see II.1.4). The absolutely convex hull A of the finite set $\{x_i\}_{i=1}^n$ is obviously closed and bounded, and hence also compact, in the Euclidean space $\mathbf{K}^m$. As convergence in $\mathbf{K}^m$ is coordinatewise and the algebraic operations in **X** are continuous, the inclusion mapping $\mathbf{K}^m \to \mathbf{X}$ is continuous, so A is compact in **X** and therefore totally bounded (Theorem 1.4).

Since $A + V$ is absolutely convex and contains E, we have $E_1 \subset A + V$. Since A is totally bounded, there exist elements $y_j \in A$ $(1 \leqslant j \leqslant j_0)$ such that $A \subset \bigcup_{j=1}^{j_0} (y_j + V)$. We thus have $E_1 \subset A + V \subset \bigcup_{j=1}^{j_0} (y_j + 2V)$, so that E_1 is totally bounded. This completes the proof of the theorem.

Using Theorem 1.4, we deduce from Theorem 3 the following

COROLLARY. *In a quasi-complete* LCS *the closed convex hull and the closed absolutely convex hull of a precompact set are both compact.*

2.2. In this subsection we consider some applications of the Hahn–Banach Theorem in its analytic form to the theory of LCSs. In particular, we prove a very important property of LCSs—namely, the existence of several continuous linear functionals.

First we observe that a linear functional f on a TVS **X** is continuous if and only if it is continuous at the origin. For suppose f is continuous at the origin and let $\{x_\alpha\}$ be a net converging to x in **X**. Then $f(x_\alpha) - f(x) = f(x - x_\alpha) \to 0$.

LEMMA 2. *Every hyperplane M in a* TVS **X** *is either closed or dense in* **X**. *A hyperplane $M = \{x: f(x) = \lambda\}$ is closed if and only if the functional f is continuous.*

Proof. Let $M = x + H$, where H is a hyperspace. If H is not closed, then it must be dense in **X**, since its closure $\bar{H}$ is a linear manifold containing H (see Lemma II.2.1). As $\bar{M} = x + \bar{H}$, the first statement now follows. To prove the second statement it is sufficient, by Theorem II.2.1, to verify that $f^{-1}(0)$ is closed if and only if f is continuous. If f is continuous, then $f^{-1}(0)$ is closed, because $\{0\}$ is a closed set in $\mathbb{K}$. Conversely, assume that

$H = f^{-1}(0)$ is closed. It is sufficient to prove that f is continuous at the origin. Suppose $f \neq \mathbf{0}$ (otherwise $f = \mathbf{0}$ is continuous), and let $V = \{x : |f(x)| < \varepsilon\}$ ($\varepsilon > 0$). If we show that V is a neighbourhood of zero in $\mathbf{X}$, then it will follow that f is continuous at zero. Since $f \neq \mathbf{0}$, there exists a point $x_0 \in \mathbf{X}$ for which $f(x_0) = \varepsilon$. As H is closed, there is a balanced neighbourhood of zero U such that $(x_0 + U) \cap H = \varnothing$. We show that $U \subset V$, from which it follows that V is a neighbourhood of zero. Assume the contrary: that is, assume we can find an $x \in U$ for which $|f(x)| \geqslant \varepsilon$. Then $y = -\varepsilon x/f(x) \in U$ and $f(x_0 + y) = f(x_0) - \varepsilon = 0$, and so $y \in (x_0 + U) \cap H$, giving a contradiction.

THEOREM 4 (The geometric form of the Hahn–Banach Theorem). *Let* $\mathbf{X}$ *be a* TVS, *let* E *be a linear manifold in* $\mathbf{X}$, *and let* $x_0 \in \mathbf{X}$. *If* U *is a non-empty convex open subset of* $\mathbf{X}$ *not intersecting with* $x_0 + E$, *then there is a closed hyperplane* H *in* $\mathbf{X}$ *that contains* $x_0 + E$ *and does not intersect with* U.

Proof. Assume first that $\mathbf{X}$ is a real space. Using a translation if necessary, we may assume that $\mathbf{0} \in U$, i.e. that U is a neighbourhood of zero. Denote the Minkowski functional of U by p. Consider the linear manifold $F = \mathscr{L}(x_0, E)$ in $\mathbf{X}$. Now E is a hyperplane in F, so by Theorem II.2.2 there exists a linear functional f_0 on F such that $f_0^{-1}(0) = E$ and $f_0(x_0) = 1$. We shall show that for any $x = \lambda x_0 + y$ ($y \in E$) we have $f_0(x) \leqslant p(x)$, or, what is the same, $\lambda \leqslant p(\lambda x_0 + y)$. Since the last inequality is obvious if $\lambda \leqslant 0$, it is sufficient to prove that for any $\lambda > 0$ we have

$$p\left(\frac{\lambda x_0 + y}{\lambda}\right) \geqslant 1.$$

By Lemma 1, the latter inequality means that $x_0 + y/\lambda \notin U$. But this is true since $y/\lambda \in E$.

By the analytic form of the Hahn–Banach Theorem (see Theorem II.4.1), there exists a linear functional f on $\mathbf{X}$ which is an extension of f and satisfies $f(x) \leqslant p(x)$ for every $x \in \mathbf{X}$.

Since $-p(-x) \leqslant f(x) \leqslant p(x)$ and U is a neighbourhood of zero, f is continuous on $\mathbf{X}$. Thus, by Lemma 2, $H = f^{-1}(1)$ is a closed hyperplane in $\mathbf{X}$. Clearly $H \supset x_0 + E$. Let us verify that $H \cap U = \varnothing$. In fact, if $x \in U$ then $p(x) < 1$, while if $x \in H$ then $1 = f(x) \leqslant p(x)$. Thus the theorem is proved when $\mathbf{X}$ is a real vector space.

Now let $\mathbf{X}$ be a complex vector space. Using a translation if necessary, we may assume that $\mathbf{0} \in x_0 + E$—i.e. that $x_0 + E$ is a linear manifold. By what we have proved, there exists a closed real hyperspace H_1 such that $H_1 \supset x_0 + E$, $H_1 \cap U = \varnothing$. By Lemma II.2.2, $H = H_1 \cap (iH_1)$ is a hyperspace; and this H clearly fulfils the requirements of the theorem.

Theorem 4 is valid for any TVS, but its application requires the existence of open convex sets, and it is precisely in LCSs that these are in abundance. We defer applications of Theorem 4 to the next subsection, and now present a number of corollaries to the analytic form of the Hahn–Banach Theorem.

Let $\mathbf{X}$ be an LCS and $\mathbf{X}_0$ a linear manifold in $\mathbf{X}$. Then $\mathbf{X}_0$ is clearly an LCS in the topology induced from $\mathbf{X}$.

COROLLARY 1. *Let* f_0 *be a continuous linear functional on* $\mathbf{X}_0$. *There exists a continuous linear functional on* $\mathbf{X}$ *extending* f_0.

Proof. As f_0 is continuous, there is an absolutely convex neighbourhood of zero U such that $|f_0(x)| \leqslant 1$ on $U \cap \mathbf{X}_0$. Then $|f_0(x)| \leqslant p_U(x)$ on $\mathbf{X}_0$. By Theorem II.4.2, f_0 has an extension f to all of $\mathbf{X}$ which satisfies $|f(x)| \leqslant p_U(x)$ and is therefore continuous.

COROLLARY 2. *For each point* x_0 *of a vector space* $\mathbf{X}$ *and each seminorm* p, *there exists a linear functional* f *on* $\mathbf{X}$ *such that* $|f(x)| \leqslant p(x)$ *and* $f(x_0) = p(x_0)$.

Proof. We set $f_0(\lambda x_0) = \lambda p(x_0)$ on the one-dimensional subspace $\mathbf{X}_0$ spanned by x_0, and extend f_0 to all of $\mathbf{X}$ by Theorem II.4.2.

COROLLARY 3. *Let* $\mathbf{X}$ *be an* LCS. *If* $f(x) = 0$ *for each continuous linear functional* f *on* $\mathbf{X}$, *then* $x = \mathbf{0}$.

Proof. If $x \neq \mathbf{0}$, then by (3) we can find a continuous seminorm p such that $p(x) > 0$. Then by Corollary 2 we can find an f such that $f(x) \neq 0$.

Corollary 3 establishes a most important property of LCS—the existence of a sufficient number of continuous linear functionals. We shall leave the considerations arising from this fact to the next section. The set of all continuous linear functionals on a TVS $\mathbf{X}$ is denoted by $\mathbf{X}^*$ and called the *topological dual* (of $\mathbf{X}$). Clearly $\mathbf{X}^*$ is a linear manifold in the algebraic dual. We have shown that, if $\mathbf{X}$ is an LCS, then $\mathbf{X}^*$ separates points on $\mathbf{X}$. In the case of an arbitrary TVS, it may happen that $\mathbf{X}^* = \{\mathbf{0}\}$ (however, there also exist TVSs $\mathbf{X}$ for which $\mathbf{X}^*$ separates points on $\mathbf{X}$, but which are not LCSs).

We show now that on the TVS $\mathbf{S}(0, 1)$ any continuous linear functional is equal to zero. Suppose $f \in (\mathbf{S}(0, 1))^*, f \neq \mathbf{0}$. The set of linear combinations of characteristic functions of intervals is dense in $\mathbf{S}(0, 1)$, so for any $n \in \mathbb{N}$ we can find an interval Δ_n of length less than $1/n$ such that $f(\chi_{\Delta_n}) = \delta_n \neq 0$. If $x_n = \chi_{\Delta_n}/\delta_n$, then $x_n \to 0$ in measure, and so, by the continuity of f, we have $f(x_n) \to 0$. On the other hand, we have $f(x_n) = 1$, so this is a contradiction.

We conclude this subsection with one further piece of notation. If $\mathbf{X}$ is an LCS, we shall write $\mathbf{X}^*_{\mathbb{R}}$ for the set of all continuous real-valued linear functionals on $\mathbf{X}$. If $\mathbf{X}$ is real, then $\mathbf{X}^*_{\mathbb{R}} = \mathbf{X}^*$.

2.3. Here we record two important theorems on the separation of convex sets of continuous linear functionals, which have numerous applications in convexity and its applications to mathematical economics (see Gol'shtein; Ioffe and Tikhomirov; Nikaido).

Let E and F be subsets of an LCS $\mathbf{X}$. We say that these subsets *can be separated* if there exists a functional $f \in \mathbf{X}^*_{\mathbb{R}}$ such that

$$\sup\{f(x): x \in E\} \leqslant \inf\{f(x): x \in F\}. \tag{5}$$

If we have strict inequality in (5), then we say that E and F *can be strictly separated*. The lack of symmetry in this definition is only apparent—the roles of E and F are interchanged when f is replaced by $-f$.

LEMMA 3. *If* E *is an open set in an* LCS $\mathbf{X}$ *and* $f \in \mathbf{X}^*$, $f \neq \mathbf{0}$, *then* $f(E)$ *is open.*

Proof. If $x \in E$, then $E - x$ is a neighbourhood of zero and therefore an absorbent set. Since $f \neq \mathbf{0}$, we can find an $x_0 \in \mathbf{X}$ such that $f(x_0) = 1$. Hence there exists $\lambda > 0$ such that $\mu x_0 \in E - x$ whenever $|\mu| \leqslant \lambda$. Thus $f(x) + \mu \in f(E)$ whenever $|\mu| \leqslant \lambda$, showing that $f(E)$ is open, as claimed.

THEOREM 5. *Let* E *be a convex subset of an* LCS $\mathbf{X}$ *having non-empty interior* $\mathring{E}$ *and let* F *be a non-empty convex subset of* $\mathbf{X}$ *with* $\mathring{E} \cap F = \varnothing$. *Then* E *and* F *can be separated. If* E *and* F *are open, then they can be strictly separated.*

Proof. By the Corollary to Lemma 1, $\mathring{E}$ is convex. Hence the set $U = \mathring{E} - F$, which is open and does not contain the origin because $\mathring{E} \cap F = \varnothing$, is convex. Hence, by Theorem 4, there is a closed real hyperspace H such that $\mathbf{0} \in H$ and $(\mathring{E} - F) \cap H = \varnothing$. Let $H = \{x : f(x) = 0\}$, where $f \in \mathbf{X}^*_{\mathbb{R}}$. The set $f(\mathring{E} - F)$ is convex and it is therefore an interval in $\mathbb{R}$; and we have $0 \notin f(\mathring{E} - F)$. By a change of sign if necessary, we may assume that $f(\mathring{E} - F) < 0$. Thus $\sup\{f(x) : x \in \mathring{E}\} \leqslant \inf\{f(x) : x \in F\}$. Using the fact that $\mathring{E}$ is dense in E (see the corollary to Lemma 1) and f is continuous, we see that E and F can be separated.

If E and F are open sets, then, by Lemma 3, $f(E)$ and $f(F)$ are open intervals in $\mathbb{R}$, and so the separation is strict.

A closed convex set in an LCS $\mathbf{X}$ having non-empty interior is called a *convex body*. Let E be a subset of an LCS $\mathbf{X}$. A real-valued functional $f \in \mathbf{X}^*_{\mathbb{R}}$ is called a *supporting functional* to E at a point $x_0 \in E$ if there exists $\lambda \in \mathbb{R}$ such that $f(x_0) = \lambda$ and E is contained in $\{x : f(x) \leqslant \lambda\}$ or $\{x : f(x) \geqslant \lambda\}$. In this situation, the real hyperplane $\{x : f(x) = \lambda\}$ is called a *supporting hyperplane* to E at x_0.

COROLLARY. *If C is a convex body in $\mathbf{X}$, then every boundary point of C has a supporting functional.*

Proof. If x_0 is a boundary point of C, then the corollary follows from Theorem 5 when we take $E = C$ and $F = \{x_0\}$.

THEOREM 6. *Let E and F be non-empty, non-intersecting convex subsets of an* LCS $\mathbf{X}$, *where E is closed and F is compact. Then E and F can be strictly separated.*

Proof. We shall show that there is an open convex neighbourhood of zero U such that $E + U$ and $F + U$ do not intersect. Since the latter sets are open and convex, the result will then follow Theorem 15.

It is sufficient to prove the existence of an absolutely convex open neighbourhood of zero V such that $(E + V) \cap F = \varnothing$, since then, taking $U = \frac{1}{2}V$, we have $(E + U) \cap (F + U) = \varnothing$. For if $x = a + \frac{1}{2}v_1 = b + \frac{1}{2}v_2$, where $a \in E$, $b \in F$, v_1, $v_2 \in V$, then $b = x - \frac{1}{2}v_2 \in (E + V) \cap F$.

Let $\mathfrak{B}$ be the basis of all open absolutely convex neighbourhoods of zero in $\mathbf{X}$. Assume that, for all $V \in \mathfrak{B}$, we have $\overline{(E + V)} \cap F \neq \varnothing$. Then $\{\overline{(E + V)} \cap F : V \in \mathfrak{B}\}$ is a system of closed subsets in the compactum F that has the finite intersection property. Hence there is a point $x_0 \in F$ such that $x_0 \in \overline{E + V} \subset E + 2V$, for any $V \in \mathfrak{B}$. Then x_0 is a limit point of E, and since E is closed we must have $x_0 \in E$. But $E \cap F = \varnothing$ by assumption. This contradiction completes the proof.

COROLLARY. *If E is an absolutely convex subset of an* LCS $\mathbf{X}$ *and $x_0 \notin \overline{E}$, then there exists $f \in \mathbf{X}^*$ such that $|f(x)| \leqslant 1$ for all $x \in E$ and* $\operatorname{Re} f(x_0) > 1$.

Proof. Since the one-point set $\{x_0\}$ is compact, it follows from Theorem 6 that there exists a real-valued functional $g \in \mathbf{X}^*_{\mathbb{R}}$ such that $g(x) \leqslant 1$ whenever $x \in E$ and $g(x_0) > 1$. As E is absolutely convex, $|g(x)| \leqslant 1$ for $x \in E$. If $\mathbf{X}$ is a complex space, set $f(x) = g(x) - ig(ix)$. We again apply the method used in the proof of Theorem II.4.2. For each $x \in \mathbf{X}$, there exists $\theta \in \mathbb{R}$ such that $|f(x)| = e^{i\theta} f(x) = f(e^{i\theta}x)$. Hence $f(e^{i\theta}x) = g(e^{i\theta}x)$ and $e^{i\theta}x \in E$, so that $g(e^{i\theta}x) \leqslant 1$.

Using the separation theorem, one can prove the Krein–Milman Theorem.

A point x_0 in a convex subset E of an LCS $\mathbf{X}$ is said to be *extreme* (or *extremal*) if a relation of the form $x_0 = \lambda x + (1 - \lambda)y$, $x, y \in E$, $0 < \lambda < 1$ holds only for $x = y = x_0$. In other words, x_0 is an extreme point of E if it is not an interior point of any interval with endpoints in E. For example, the vertices of a square in the plane are extreme points, but the other boundary points are not extreme.

THEOREM 7 (Krein–Milman). *Every non-empty compact convex subset of an* LCS $\mathbf{X}$ *is the closed convex hull of its extreme points.*

A proof of Theorem 7 may be found, for example, in Schaefer-I. Other applications and refinements of the Krein–Milman Theorem are collected in the book by Phelps. The concept of an extreme point is of great significance in optimal programming and its applications to economic problems.

§ 3. Duality

In this section we consider relationships between an LCS and the space of continuous linear functionals on it.

3.1. Let $\mathbf{X}$ be a vector space and $\mathbf{X}^+$ its algebraic dual. A subset $\mathbf{Y}$ of $\mathbf{X}^+$ is called a *total subset* on $\mathbf{X}$ if $f(x) = 0$ for all $f \in \mathbf{Y}$ implies $x = \mathbf{0}$. Let us fix a total linear manifold $\mathbf{Y}$ in $\mathbf{X}^+$. In this situation we say that $\langle \mathbf{X}, \mathbf{Y} \rangle$ is a *dual pairing*. If $\mathbf{X}$ is an LCS, then $\langle \mathbf{X}, \mathbf{X}^+ \rangle$ and $\langle \mathbf{X}, \mathbf{X}^* \rangle$ are examples of such pairings.

Let $\mathbf{X}$ be an LCS. The topology of $\mathbf{X}$ is said to be *compatible with the dual pairing* $\langle \mathbf{X}, \mathbf{Y} \rangle$ if $\mathbf{X}^* = \mathbf{Y}$. One of the fundamental problems in this section is the problem of describing all possible locally convex topologies compatible with a given dual pairing.

In this section we introduce the so-called weak topology, which will play an important role throughout a large part of the book in the study of normed spaces.

Let $\langle \mathbf{X}, \mathbf{Y} \rangle$ be a dual pairing. The locally convex topology on $\mathbf{X}$ generated by the family of seminorms $p(x) = |f(x)|$, where f ranges over all of $\mathbf{Y}$, is called the *weak topology* in $\mathbf{X}$ determined by $\mathbf{Y}$ and is denoted by $\sigma(\mathbf{X}, \mathbf{Y})$ (since $\mathbf{Y}$ is total, the family of seminorms satisfies (3) in § 2).

The sets

$$\{x: \sup_{1 \leqslant i \leqslant n} |f_i(x)| \leqslant 1\} \quad (f_i \in \mathbf{Y})$$

form a basis of closed neighbourhoods of zero in the topology $\sigma(\mathbf{X}, \mathbf{Y})$. The space $\mathbf{X}$ endowed with the weak topology will be denoted by $(\mathbf{X}, \sigma(\mathbf{X}, \mathbf{Y}))$. Let us show that the topology $\sigma(\mathbf{X}, \mathbf{Y})$ is compatible with the dual pairing $\langle \mathbf{X}, \mathbf{Y} \rangle$.

LEMMA 1. *If $f_1, f_2, \ldots, f_n$ is a linearly independent system of linear functionals on a vector space $\mathbf{X}$, then there exists a system of elements $x_1, x_2, \ldots, x_n \in \mathbf{X}$ satisfying the relations*

$$f_j(x_k) = \begin{cases} 0, & j \neq k, \\ 1, & j = k \end{cases} \quad (j, k = 1, 2, \ldots, n),$$

which is said to be biorthogonal to $\{f_k\}_{k=1}^n$.

Proof. We prove the existence of the required system by induction. If $n = 1$ then, since $f_1 \neq \mathbf{0}$, there exists $x_1 \in \mathbf{X}$ such that $f_1(x_1) = 1$. The element x_1 itself forms the system we seek. Now consider the case where $n > 1$. Assume that a biorthogonal system of elements can be constructed if the number of functionals is less than n. With this assumption, we prove the existence of an element $x_1 \in \mathbf{X}$ such that

$$f_1(x_1) = 1, \quad f_2(x_1) = \ldots = f_n(x_1) = 0. \tag{1}$$

To do this we consider a system of elements $x'_2, \ldots, x'_n \in \mathbf{X}$ biorthogonal to the functionals $f_2, \ldots, f_n$. Each $x \in \mathbf{X}$ has a representation

$$x = \sum_{k=2}^{n} \alpha_k x'_k + x', \tag{2}$$

where x' is an element such that $f_2(x') = \ldots = f_n(x') = 0$. For if we write

$$\alpha_k = f_k(x) \quad (k = 2, \ldots, n), \tag{3}$$

then, with $x' = x - \sum_{k=2}^{n} \alpha_k x'_k$, we have

$$f_j(x') = f_j(x) - \sum_{k=2}^{n} \alpha_k f_j(x'_k) = \alpha_j - \alpha_j = 0 \quad (j = 2, \ldots, n).$$

Note that the coefficients α_n are uniquely determined by the element x; to convince oneself of this, it is sufficient to apply f_j $(j = 2, \ldots, n)$ to both sides of (2).

Now assume that no element $x_1 \in \mathbf{X}$ satisfying (1) exists. This means that if $x' \in \mathbf{X}$ is such that $f_2(x') = \ldots = f_n(x') = 0$, then also $f_1(x') = 0$. Bearing this, and equations (3), in mind, we see from (2) that for any $x \in \mathbf{X}$

$$f_1(x) = \sum_{k=2}^{n} \alpha_k f_1(x'_k) + f_1(x') = \sum_{k=2}^{n} \lambda_k f_k(x),$$

where we have set $\lambda_k = f_1(x'_k)$ $(k = 2, \ldots, n)$. Thus $f_1 = \sum_{k=2}^{n} \lambda_k f_k$, which contradicts the linear independence of the functionals $f_1, f_2, \ldots, f_n$. Hence such an element x_1 exists. Similarly one proves the existence of elements $x_2, \ldots, x_n \in \mathbf{X}$ such that

$$\begin{cases} f_k(x_k) = 1, \qquad f_j(x_k) = 0 \\ (j \neq k; \quad j = 1, 2, \ldots, n; \quad k = 2, \ldots, n), \end{cases}$$

which completes the proof of the lemma.

THEOREM 1. *The weak topology $\sigma(\mathbf{X}, \mathbf{Y})$ is compatible with the dual pairing $\langle \mathbf{X}, \mathbf{Y} \rangle$; that is,* $(\mathbf{X}, \sigma(\mathbf{X}, \mathbf{Y}))^* = \mathbf{Y}$.

Proof. It is clear from the way the weak topology was defined that every functional $f \in \mathbf{Y}$ is a continuous linear functional on $(\mathbf{X}, \sigma(\mathbf{X}, \mathbf{Y}))$.

Conversely, let f be a linear functional on $(\mathbf{X}, \sigma(\mathbf{X}, \mathbf{Y}))$. The set* $V = f^{-1}([-1, 1])$ is a neighbourhood of zero in $(\mathbf{X}, \sigma(\mathbf{X}, \mathbf{Y}))$. This means that there exist functionals $f_1, f_2, \ldots, f_n \in \mathbf{Y}$ such that

$$V \supset \bigcap_{k=1}^{n} V_k \quad (V_k = f_k^{-1}([-1, 1]); \quad k = 1, 2, \ldots, n). \tag{4}$$

Moreover, we may assume that the functionals $f_1, f_2, \ldots, f_n$ are linearly independent. For if one of them, say f_n, is a linear combination of the others,

$$f_n = \sum_{i=1}^{m} \alpha_j f_{k_j} \quad (\alpha_j \neq 0; \, k_j \neq n; \, j = 1, 2, \ldots, m),$$

then

$$V_n \supset \bigcap_{i=1}^{m} \frac{1}{|\alpha_j| m} V_{k_j}$$

and we can therefore omit V_n from the intersection in (4). By Lemma 1, there exist elements

* When the field of scalars is complex, one must replace $[-1, 1]$, here and below, by $\{z \in \mathbb{C} : |z| \leqslant 1\}$.

$x_1, x_2, \ldots, x_n \in \mathbf{X}$ such that

$$f_j(x_k) = \begin{cases} 0, & j \neq k, \\ 1, & j = k \end{cases} \qquad (j, k = 1, 2, \ldots, n).$$

For each $x \in \mathbf{X}$ we have the representation

$$x = \sum_{k=1}^{n} \alpha_k x_k + x', \tag{5}$$

where

$$\alpha_k = f_k(x), \quad f_k(x') = 0 \quad (k = 1, 2, \ldots, n). \tag{6}$$

Applying f to both sides of (5), we have

$$f(x) = \sum_{k=1}^{n} \beta_k f_k(x) + f(x') \quad (\beta_k = f(x_k); k = 1, 2, \ldots, n). \tag{7}$$

We prove that $f(x') = \mathbf{0}$. If we had $f(x') \neq \mathbf{0}$ then for sufficiently large $\alpha > 0$ we would have $|f(\alpha x')| > 1$; that is, $\alpha x' \notin V$, and therefore, by (4), $\alpha x' \notin \lambda_k V_k$ for at least one k, which contradicts (6).

Thus (7) yields $f = \sum_{k=1}^{n} \beta_k f_k$. Since $f_k \in \mathbf{Y}$ $(k = 1, 2, \ldots, n)$, it follows from this that $f \in \mathbf{Y}$, which is what we wanted to prove.

REMARK. It follows from the theorem just proved that if we choose a total linear manifold $\tilde{\mathbf{Y}}$ distinct from the whole space $\mathbf{Y}$, then the topology $\sigma(\mathbf{X}, \tilde{\mathbf{Y}})$ is strictly weaker than the topology $\sigma(\mathbf{X}, \mathbf{Y})$. In this sense the weak topology $\sigma(\mathbf{X}, \mathbf{Y})$ determines the set $\mathbf{Y}$ uniquely.

Let $\mathbf{X}$ be an LCS with topology τ. Clearly the weak topology $\sigma(\mathbf{X}, \mathbf{X}^*)$ is weaker than τ, and each weakly closed set—that is, each closed set in the space $(\mathbf{X}, \sigma(\mathbf{X}, \mathbf{X}^*))$—is therefore also closed in $\mathbf{X}$. The converse is, in general, false. However, we have

THEOREM 2. *In every* LCS $\mathbf{X}$ *the closure of any convex set in the topology* τ *coincides with its weak closure.*

Proof. It is sufficient to prove that any τ-closed convex set E is weakly closed. By Theorem 2.6, for any $x \notin E$ there exists a continuous real-valued functional g_x on $\mathbf{X}$ such that $\sup\{g_x(y): y \in E\} = \alpha_x < g_x(x)$. Thus $E = \bigcap_{x \notin E} \{y \in \mathbf{X}: g_x(y) \leqslant \alpha_x\}$. Each of the sets $\{y \in \mathbf{X}: g_x(y) \leq \alpha_x\}$ is obviously weakly closed, and so E is also weakly closed.

COROLLARY 1. *The closure of a convex set is the same in all locally convex topologies compatible with a given dual pairing.*

COROLLARY 2. *Let* τ_1 *and* τ_2 *be two locally convex topologies compatible with the same dual pairing. If some net* $\{x_\alpha\}$ *converges to* x *in the topology* τ_1 *then there exists a net* $\{y_\beta\}$, *whose elements are convex combinations of the elements* x_α, *such that* $y_\beta \to x$ *in the topology* τ_2.

Proof. The point x evidently belongs to the τ_1-closure of the convex hull E of the set $\{x_\alpha\}$. By Corollary 1, E is the τ_2-closure of the convex hull of $\{x_\alpha\}$. Hence x is a τ_2-limit point of co $(\{x_\alpha\})$.

COROLLARY 3. *Let* $\mathbf{X}$ *be an LCS, and* $\mathbf{X}_0$ *a subspace of* $\mathbf{X}$. *Then:*

1) *the weak topology* $\sigma(\mathbf{X}_0, \mathbf{X}_0^*)$ *coincides with the topology induced on* $\mathbf{X}_0$ *by the space* $(\mathbf{X}, \sigma(\mathbf{X}, \mathbf{X}^*))$;

2) *assume* $\mathbf{X}_0$ *is closed in* $\mathbf{X}$; *if a set* E *is relatively compact in* $(\mathbf{X}, \sigma(\mathbf{X}, \mathbf{X}^*))$, *then* $E \cap \mathbf{X}_0$ *is relatively compact also in* $(\mathbf{X}_0, \sigma(\mathbf{X}_0, \mathbf{X}_0^*))$.

Proof. 1) Since the restriction of a functional $f \in \mathbf{X}^*$ to $\mathbf{X}_0$ is also continuous, we have $\sigma(\mathbf{X}, \mathbf{X}^*) \leqslant \sigma(\mathbf{X}_0, \mathbf{X}_0^*)$ on $\mathbf{X}_0$. The opposite inequality follows from the Hahn–Banach Theorem (see Theorem 2), which shows that each $f_0 \in \mathbf{X}_0^*$ can be extended to a functional $f \in \mathbf{X}^*$.

2) Since the subspace $\mathbf{X}_0$ is $\sigma(\mathbf{X}, \mathbf{X}^*)$-closed, by Corollary 1, the closure of $E \cap \mathbf{X}_0$ in $(\mathbf{X}, \sigma(\mathbf{X}, \mathbf{X}^*))$ is contained in $\mathbf{X}_0$. We see from 1) that this closure is also $\sigma(\mathbf{X}_0, \mathbf{X}_0^*)$-compact.

A set which is bounded in the weak topology is said to be *weakly bounded.*

THEOREM 3. *The bounded sets in an* LCS $\mathbf{X}$ *are the same as those in* $(\mathbf{X}, \sigma(\mathbf{X}, \mathbf{X}^*))$.

We shall not prove Theorem 3 (see Schaefer-I). Later, in Chapter VIII, we shall establish it for the most important case where $\mathbf{X}$ is a normed space.

3.2. Let $\langle \mathbf{X}, \mathbf{Y} \rangle$ be a dual pairing. For $E \subset \mathbf{X}$ we define

$$E^\circ = \{f \in \mathbf{Y} : |f(x)| \leq 1 \text{ for all } x \in E\}.$$

The subset $E^\circ \subset \mathbf{Y}$ is called the *polar* of E.

We now make another definition. For any set $E \subset \mathbf{X}$, the set of all functionals $f \in \mathbf{Y}$ vanishing on E is called the *annihilator* $E^\perp$ of E. The annihilator is obviously a $\sigma(\mathbf{Y}, \mathbf{X})$-closed linear manifold. The simplest properties of polars are collected in the following lemma, whose proof we leave to the reader.

LEMMA 2. 1) *If* $E_1 \subset E_2$, *then* $E_2^\circ \subset E_1^\circ$.

2) *If* $\lambda \neq 0$, *then* $(\lambda E)^\circ = \lambda^{-1} E^\circ$.

3) $\left(\bigcup_{\xi \in \Xi} E_\xi \right)^\circ = \bigcap_{\xi \in \Xi} E_\xi^\circ$.

4) *If* E *is a linear manifold in* $\mathbf{X}$, *then* E° *coincides with the annihilator* $E^\perp$.

Before investigating the deeper properties of polars, we make a remark which will be very important in what follows. Let $\langle \mathbf{X}, \mathbf{Y} \rangle$ be a dual pairing. Each element $x \in \mathbf{X}$ can be viewed as a linear functional F_x, defined on $\mathbf{Y}$ by the formula*

$$F_x(f) = \overline{f(x)} \qquad (f \in \mathbf{Y}).$$

Since $\mathbf{Y}$ is total on $\mathbf{X}$, the mapping $\pi_{\mathbf{Y}} : x \in \mathbf{X} \to F_x$ is a linear isomorphism of $\mathbf{X}$ onto a linear manifold in $\mathbf{Y}^+$. The mapping $\pi_{\mathbf{Y}}$, which is called the *canonical* or *natural embedding* of $\mathbf{X}$ in $\mathbf{Y}^+$, will later play an important role in defining reflexive normed spaces. For the present we observe that what has been said enables us to regard $\mathbf{X}$ as a total vector space of functionals on $\mathbf{Y}$; and, by the same token, $\langle \mathbf{Y}, \mathbf{X} \rangle$ is also a dual pairing, so that one can define on $\mathbf{Y}$ the weak topology $\sigma(\mathbf{Y}, \mathbf{X})$, to which all the results obtained above are applicable.

LEMMA 3. *If* $E \subset \mathbf{X}$, *then* E° *is absolutely convex and* $\sigma(\mathbf{Y}, \mathbf{X})$*-closed.*

Proof. The absolute convexity is clear, and weak closure follows from the formula

$$E^\circ = \bigcap_{x \in E} \{f \in \mathbf{Y} : |f(x)| \leqslant 1\}.$$

Let $\langle \mathbf{X}, \mathbf{Y} \rangle$ and $\langle \mathbf{Y}, \mathbf{Z} \rangle$ be dual pairings, where $\mathbf{X} \subset \mathbf{Z} \subset \mathbf{Y}^+$. If $E \subset \mathbf{X}$, the polar $E^{\circ\circ}$ in $\mathbf{Z}$ of the polar E° is called the *bipolar* of E in $\mathbf{Z}$.

THEOREM 4. *The bipolar* $E^{\circ\circ}$ *in* $\mathbf{Z}$ *of a set* $E \subset \mathbf{X}$ *coincides with its* $\sigma(\mathbf{Z}, \mathbf{Y})$*-closed absolutely convex hull.*

---

* See the definition of multiplication of a functional by a scalar in II.2.2.

Proof. Obviously $E \subset E^{\circ\circ}$. Let G denote the $\sigma(\mathbf{Z}, \mathbf{Y})$-closed absolutely convex hull of E. By Lemma 3, $E^{\circ\circ} \supset G$. If $F_0 \in \mathbf{Z}$, $F_0 \notin G$, then by the corollary to Theorem 2.6 there exists a functional $\phi \in (\mathbf{Z}, \sigma(\mathbf{Z}, \mathbf{Y}))^*$ such that Re $\phi(F_0) > 1$ and $|\phi(F)| \leqslant 1$ for $F \in G$. By Theorem 1, we can find $f \in \mathbf{Y}$ such that $\phi(F) = f(F)$ for every $F \in \mathbf{Z}$. Since $E \subset G$, we have $f \in E^{\circ}$. On the other hand, $|f(F_0)| \geqslant$ Re $f(F_0) > 1$, and so $F_0 \notin E^{\circ\circ}$. Therefore $E^{\circ\circ} \subset G$, and hence $E^{\circ\circ} = G$.

Theorem 4 is called the Bipolar Theorem. Its most important special cases are obtained by taking $\mathbf{Z} = \mathbf{X}$ or $\mathbf{Z} = \mathbf{Y}^+$. By Theorems 2 and 4 we have

COROLLARY 1. *If* $\mathbf{X}$ *is an* LCS, *and we form the polar* E° *of a set* $E \subset \mathbf{X}$ *in* $\mathbf{X}^*$ *and its bipolar* $E^{\circ\circ}$ *in* $\mathbf{X}$, *then* $E^{\circ\circ}$ *coincides with the closed absolutely convex hull of* E.

Let $\mathbf{X}$ be an LCS. A subset E of $\mathbf{X}$ is said to be *fundamental* if $\overline{\mathscr{L}}(E) = \mathbf{X}$. We show now that E is fundamental if and only if $E^{\perp} = \{\mathbf{0}\}$.

COROLLARY 2. *Let E be any subset of an* LCS $\mathbf{X}$. *A necessary and sufficient condition for* $x \in \mathbf{X}$ *to belong to* $\overline{\mathscr{L}}(E)$ *is that* $f(x) = 0$ *for every* $f \in E^{\perp}$, *where* $E^{\perp}$ *is the annihilator in* $\mathbf{X}^*$.

Proof. By Lemma 2, $\mathscr{L}(E)^{\circ} = \mathscr{L}(E)^{\perp}$. By Corollary 1, $\mathscr{L}(E)^{\circ\circ} = \overline{\mathscr{L}}(E)$. Since $\mathscr{L}(E)^{\perp}$ is a linear manifold, $\mathscr{L}(E)^{\circ\circ} = \mathscr{L}(E)^{\perp\circ} = \mathscr{L}(E)^{\perp\perp}$. It is easy to see that $E^{\perp} = \mathscr{L}(E)^{\perp}$. Hence we conclude that $E^{\perp\perp} = \overline{\mathscr{L}}(E)$, which is what we set out to prove.

Once again, let $\langle \mathbf{X}, \mathbf{Y} \rangle$ be a dual pairing. The properties of E in $\mathbf{X}$ are connected with those of its polar E°.

THEOREM 5. *A necessary and sufficient condition for the polar* E° *of a set* $E \subset \mathbf{X}$ *to be absorbent in* $\mathbf{Y}$ *is that E be bounded in the space* $(\mathbf{X}, \sigma(\mathbf{X}, \mathbf{Y}))$.

Proof. Necessity. Consider a neighbourhood of zero V in the space $(\mathbf{X}, \sigma(\mathbf{X}, \mathbf{Y}))$. We may assume that $V = f^{-1}(D)$ $(f \in \mathbf{Y})$, where $D = \{z \in \mathbb{K} : |z| \leqslant 1\}$. The set V is the polar of the set consisting of a single functional f: that is, $V = \{f\}^{\circ}$. Since E is absorbent, we have $\lambda f \in E$ for sufficiently small $\lambda > 0$, and so, passing to the polar, we find that

$$\frac{1}{\lambda}V = \frac{1}{\lambda}\{f\}^{\circ} = \{\lambda f\}^{\circ} \supset E^{\circ\circ}.$$

This shows that $E^{\circ\circ}$ is bounded, so the subset $E \subset E^{\circ\circ}$ is bounded *a fortiori*.

Sufficiency. Let $f \in \mathbf{Y}$. Consider the neighbourhood $V = f^{-1}(D)$. If $E \subset \mathbf{X}$ is bounded, then there exists $\lambda > 0$ such that $\lambda E \subset V$. Taking polars, we have

$$\frac{1}{\lambda}E^{\circ} = (\lambda E)^{\circ} \supset V^{\circ}.$$

But $f \in V^{\circ}$, so $\lambda f \in E^{\circ}$. Thus we have shown that E° is absorbent.

3.3. Let $\langle \mathbf{X}, \mathbf{Y} \rangle$ be a dual pairing. Let us consider the problem of topologizing $\mathbf{X}$. The operation of taking polars associates a set in $\mathbf{Y}$ with each set in $\mathbf{X}$, and therefore presents itself naturally in this connection. In fact, we shall take the set of polars of certain sets in $\mathbf{Y}$ as a system of neighbourhoods generating a locally convex topology. In order that this system satisfy the conditions of Theorem 2.1 it is necessary, as we see from Theorem 5, to demand that these subsets of $\mathbf{Y}$ be bounded.

Thus, let $\mathfrak{A}$ be any set of $\sigma(\mathbf{Y}, \mathbf{X})$-bounded sets in $\mathbf{Y}$ subject to the condition

I) the set $\bigcup_{A \in \mathfrak{A}} A$ is fundamental in $(\mathbf{Y}, \sigma(\mathbf{Y}, \mathbf{X}))$.

Denote by $\mathfrak{B}_0$ the set of polars in $\mathbf{X}$ of sets in $\mathfrak{B}$. We now show that $\mathfrak{B}_0$ satisfies condition (1) of Theorem 2.1. Choose $x \in \mathbf{X}$ and assume that the condition does not hold for this x. Then for every $V \in \mathfrak{B}$ and every $\lambda > 0$ we have $x \in \lambda V$. Since $V = A^{\circ}$, where $A \in \mathfrak{A}$, we have

$|f(nx)| \leqslant 1$ $(n \in \mathbb{N})$ whenever $f \in A$. Thus $f(x) = 0$ if $f \in \bigcup_{A \in \mathfrak{A}} A$. Thus, by Corollary 2 to Theorem 4, $x = \mathbf{0}$.

Hence the system $\mathfrak{B}_0$ satisfies all the conditions of Theorem 2.1, and can therefore be used to turn $\mathbf{X}$ into an LCS. The resulting topology on $\mathbf{X}$ is called the *topology of uniform convergence on the sets of* $\mathfrak{A}$, or the $\mathfrak{A}$-*convergence topology*.

If $\mathfrak{A}$ also satisfies the conditions:

II) if $A, B \in \mathfrak{A}$, then there exists $C \in \mathfrak{A}$ such that $A \cup B \subset C$;

III) if $A \in \mathfrak{A}$, then $\lambda A \in \mathfrak{A}$ for every $\lambda \in \mathbb{K}$,

then it is easy to show that $\mathfrak{B}_0$ is a basis of neighbourhoods of zero for the topology of uniform convergence. This topology can also be described easily in terms of seminorms. For every $A \in \mathfrak{A}$, write

$$p'_A(x) = \sup\{|f(x)|: x \in A\}.$$

Then p'_A is the Minkowski functional of the set A° and the family of seminorms $\{p'_A: A \in \mathfrak{A}\}$ defines the $\mathfrak{A}$-topology.

Among all the conceivable varieties of topology on $\mathbf{X}$ we mention for the present two, which are in a certain sense extremes. In the first place, one can take $\mathfrak{A}$ to consist of all one-element sets. Here the $\mathfrak{A}$-convergence topology is the weak topology $\sigma(\mathbf{X}, \mathbf{Y})$. Secondly, one can take $\mathfrak{A}$ to be the system of all $\sigma(\mathbf{Y}, \mathbf{X})$-bounded sets. The corresponding topology is then called the strong topology and is denoted by $\beta(\mathbf{X}, \mathbf{Y})$. As we already know, the dual of $(\mathbf{X}, \sigma(\mathbf{X}, \mathbf{Y}))$ is $\mathbf{Y}$, so the dual of $(\mathbf{X}, \beta(\mathbf{X}, \mathbf{Y}))$ can be an even larger subspace of $\mathbf{X}^+$. We shall solve the naturally arising problem of describing all $\mathfrak{A}$-convergence topologies in the following subsection. For the present let us make a few more remarks. Obviously we can interchange $\mathbf{X}$ and $\mathbf{Y}$ and endow $\mathbf{Y}$ with the topology of uniform convergence.

Let $\mathbf{X}$ be an LCS. A subset $G \subset \mathbf{X}^*$ is said to be *equicontinuous* if for each $\varepsilon > 0$ there is a neighbourhood of zero U in $\mathbf{X}$ such that $|f(x)| < \varepsilon$ for all $x \in U, f \in G$. It is clear that a set in $\mathbf{X}^*$ is equicontinuous if and only if it is contained in the polar of some neighbourhood of zero in $\mathbf{X}$.

THEOREM 6. *Every locally convex topology is the topology of uniform convergence on the equicontinuous subsets of the dual space.*

Proof. Every LCS $\mathbf{X}$ has a basis $\mathfrak{B}$ of closed absolutely convex neighbourhoods of zero. The sets U° $(U \in \mathfrak{B})$ are equicontinuous and $U^{\circ\circ} = U$ by the Bipolar Theorem (see Corollary 1 to Theorem 4), so the theorem is proved.

Theorem 6 shows the universality of the method of defining locally convex topologies described above.

3.4. Let us, then, turn to a description of all topologies compatible with a given duality.

LEMMA 4. *Let* $\mathbf{X}$ *be a vector space. Then* $\mathbf{X}^+$ *is complete in the topology* $\sigma(\mathbf{X}^+, \mathbf{X})$.

Proof. Let $\{f_\alpha\}$ be a Cauchy net in $\mathbf{X}^+$. Then, for each $x \in \mathbf{X}$, $\{f_\alpha(x)\}$ is a numerical Cauchy net, and so $f(x) = \lim f_\alpha(x)$ exists for each $x \in \mathbf{X}$. Clearly $f \in \mathbf{X}^+$ and $f_\alpha \to f$ in the weak topology.

We now make an observation on finite-dimensional spaces which will prove useful for the proof of the following lemma. If we consider the LCS $\mathbf{K}^n$ (see 1.3), then the collection of sets of the form

$$V_m = \{\{\xi_i\}_{i=1}^n \in \mathbf{K}^n: \max_{1 \leqslant i \leqslant n} |\xi_i| \leqslant m^{-1}\} \qquad (m = 1, 2, \ldots)$$

is a basis of neighbourhoods of zero. It is therefore clear that a set E is bounded in $\mathbf{K}^n$ if and only if there is a constant $c > 0$ such that $|\xi_i| \leqslant c$ for all $\xi = \{\xi_i\}_{i=1}^n \in E$ and for $i = 1, 2, \ldots, n$. Recall that, by the Bolzano–Weierstrass Theorem, bounded sets in $\mathbf{K}^n$ are relatively compact.

LEMMA 5. *Let $\langle \mathbf{X}, \mathbf{Y} \rangle$ be a dual pairing. Then every $\sigma(\mathbf{X}, \mathbf{Y})$-bounded set E is totally bounded in $(\mathbf{X}, \sigma(\mathbf{X}, \mathbf{Y}))$.*

Proof. Let E be a $\sigma(\mathbf{X}, \mathbf{Y})$-bounded set and let $U = \{x \in \mathbf{X} : |f_i(x)| \leqslant 1, f_i \in \mathbf{Y}, f_i \neq \mathbf{0}, i = 1, 2, \ldots, n\}$ be an arbitrary neighbourhood of zero in $(\mathbf{X}, \sigma(\mathbf{X}, \mathbf{Y}))$. With each $x \in \mathbf{X}$ we associate the element $\omega(x) = (f_1(x), \ldots, f_n(x))$ in the n-dimensional space $\mathbf{K}^n$. If $Q = V_1$ then clearly $\omega^{-1}(Q) = U$.

If $z \in \omega(\mathbf{X})$ then $\omega^{-1}(z + Q) = x + U$, where we may take as x any element of $\omega^{-1}(z)$. Since E is weakly bounded, the set $\tilde{E} = \omega(E)$ is bounded in $\mathbf{K}^n$ and is therefore relatively compact. Thus, by Hausdorff's Theorem (or, what is the same, by Theorem 1.4), there exist $z_1, \ldots, z_m \in \tilde{E}$ such that $\tilde{E} \subset \bigcup_{k=1}^{m} (z_k + Q)$. But then, if we denote by x_k elements of E such that $\omega(x_k) = z_k$ $(k = 1, 2, \ldots, m)$, we see, by what was said above, that

$$E \subset \omega^{-1}(\tilde{E}) \subset \omega^{-1}\left(\bigcup_{k=1}^{m} (z_k + Q)\right) = \bigcup_{k=1}^{m} \omega^{-1}(z_k + Q) = \bigcup_{k=1}^{m} (x_k + U),$$

as we set out to prove.

THEOREM 7 (Alaoglu–Bourbaki). *If* $\mathbf{X}$ *is an* LCS, *then the polar U° of each neighbourhood of zero U is $\sigma(\mathbf{X}^*, \mathbf{X})$-compact.*

Proof. Consider the topology $\sigma(\mathbf{X}^+, \mathbf{X})$ on $\mathbf{X}^+$. Since U is an absorbent set in $\mathbf{X}$ we see, by Theorem 5, that U° is bounded in $\mathbf{X}^+$, and therefore also totally bounded (Lemma 5). By Lemma 4, $\mathbf{X}^+$ is complete, and, by Lemma 3, U° is closed. Thus U° is complete, and, by Theorem 1.4, U° is compact. It is clear that $U^\circ \subset \mathbf{X}^*$ and the topologies $\sigma(\mathbf{X}^+, \mathbf{X})$ and $\sigma(\mathbf{X}^*, \mathbf{X})$ on $\mathbf{X}^*$ coincide. Therefore U° is $\sigma(\mathbf{X}^*, \mathbf{X})$-compact.

Let $\langle \mathbf{X}, \mathbf{Y} \rangle$ be a dual pairing. Denote by $\tau(\mathbf{X}, \mathbf{Y})$ the uniform convergence topology on all the absolutely convex $\sigma(\mathbf{Y}, \mathbf{X})$-compact sets in $\mathbf{Y}$. By what was said in 3.3, $\tau(\mathbf{X}, \mathbf{Y})$ is an LCS. The topology $\tau(\mathbf{X}, \mathbf{Y})$ is called the *Mackey topology*.

THEOREM 8 (Mackey–Arens). *A locally convex topology τ on $\mathbf{X}$ is compatible with the duality $\langle \mathbf{X}, \mathbf{Y} \rangle$ (that is, $(\mathbf{X}, \tau)^* = \mathbf{Y}$) if and only if $\sigma(\mathbf{X}, \mathbf{Y}) \leqslant \tau \leqslant \tau(\mathbf{X}, \mathbf{Y})$. In this situation, τ is the topology of uniform convergence on some collection of absolutely convex $\sigma(\mathbf{Y}, \mathbf{X})$-compact sets in $\mathbf{X}$.*

Proof. If $(\mathbf{X}, \tau)^* = \mathbf{Y}$, then τ is the topology of uniform convergence on the collection of sets U°, where the U are all the neighbourhoods of zero in the topology τ. Each U° is absolutely convex and, by Theorem 7, also $\sigma(\mathbf{Y}, \mathbf{X})$-compact. Hence $\tau \leqslant \tau(\mathbf{X}, \mathbf{Y})$. Since $\sigma(\mathbf{X}, \mathbf{Y})$ is the weakest topology of uniform convergence (compatible with the given duality), we have $\sigma(\mathbf{X}, \mathbf{Y}) \leqslant \tau$.

To prove the theorem it remains for us to verify that the Mackey topology $\tau(\mathbf{X}, \mathbf{Y})$ is compatible with the duality $\langle \mathbf{X}, \mathbf{Y} \rangle$. Denote $(\mathbf{X}, \tau(\mathbf{X}, \mathbf{Y}))^*$ by $\mathbf{X}^*$. Clearly $\mathbf{X}^* \supset \mathbf{Y}$. Let us prove the opposite inclusion. Notice that if E is compact in $(\mathbf{Y}, \sigma(\mathbf{Y}, \mathbf{X}))$ then E is compact in the big space $(\mathbf{X}^*, \sigma(\mathbf{X}^*, \mathbf{X}))$. In the following argument polars of sets in $\mathbf{X}$ will be formed in $\mathbf{X}^*$ and polars of sets in $\mathbf{X}^*$ will be formed in $\mathbf{X}$.

Consider a functional $\phi \in \mathbf{X}^*$. The set $V = \phi^{-1}(D)$, where $D = \{z \in \mathbb{K} : |z| \leqslant 1\}$, is a neighbourhood of zero in $(\mathbf{X}, \tau(\mathbf{X}, \mathbf{Y}))$. Thus, by the definition of the Mackey topology, there is an absolutely convex $\sigma(\mathbf{Y}, \mathbf{X})$-compact set A in $\mathbf{Y}$ such that $V \supset A^\circ$, and so $V^\circ \subset A^{\circ\circ}$. By Theorem 4, $A^{\circ\circ}$ is the $\sigma(\mathbf{X}^*, \mathbf{X})$-closure of A. However, A is $\sigma(\mathbf{Y}, \mathbf{X})$-compact and hence, as we have already noted, also $\sigma(\mathbf{X}^*, \mathbf{X})$-compact, so that $A = A^{\circ\circ}$. Since $\phi \in V^\circ$, we have $\phi \in A$ and *a fortiori* $\phi \in \mathbf{Y}$. Thus $\mathbf{X}^* = \mathbf{Y}$, and the proof of the theorem is complete.

It is usual to proceed from the Mackey–Arens Theorem to a more detailed study of the theory of LCSs, dealing first of all with the concepts of bornological and barrelled LCSs and reflexivity. However, we conclude our study of the general theory of TVSs and turn to a detailed investigation of normed spaces—a class of spaces that is extremely important for the applications of functional analysis considered in this book. We should mention, in fairness, that there are many applications of functional analysis involving the study of LCSs that are not normed. These are primarily spaces of generalized functions, or distributions, which have various applications in mathematical physics (see Birman *et al.*; Gel'fand and Vilenkin; Gel'fand and Shilov; Rudin; Hörmander; Schwartz; Shilov; Edwards).

IV

NORMED SPACES

§ 1. Basic definitions and simplest properties of normed spaces*

1.1. The theory of normed spaces and its numerous applications and branches form a very extensive division of functional analysis, and in this book we can only touch upon certain of its aspects, laying special emphasis on applications to the solution of functional equations.

Recall that by a *norm* on a vector space $\mathbf{X}$ we mean a non-negative functional $\|\,.\,\|$ that associates with each element $x \in \mathbf{X}$ a number $\|x\| \geqslant 0$, called the *norm*† of x, satisfying the following conditions:

1) $\|x\| = 0$ if and only if $x = \mathbf{0}$;
2) $\|\lambda x\| = |\lambda|\,\|x\|$, $\lambda \in \mathbb{K}$ (homogeneity of the norm);
3) $\|x+y\| \leqslant \|x\| + \|y\|$ (the triangle inequality).

A vector space $\mathbf{X}$ having a fixed norm on it is called a *normed space*. If we endow it with the topology defined by the norm (that is, by the family of seminorms comprising just this one norm: see III.2.1), then $\mathbf{X}$ becomes an LCS. This topology is called the *norm topology* of $\mathbf{X}$. Let us now see how certain concepts and results in the theory of LCSs are transformed when we consider this very important special case.

1) A normed space $\mathbf{X}$ is a metrizable LCS.

For if we set

$$\rho(x, y) = \|x - y\|,$$

for all $x, y \in \mathbf{X}$, then ρ is a metric on $\mathbf{X}$. For $\rho(x, y) = 0$ implies $\|x - y\| = 0$, which in view of 1) is equivalent to $x - y = \mathbf{0}$, and this in turn is equivalent to $x = y$. The symmetry of the distance function is obvious from its definition. Finally, the triangle inequality for the distance function is a simple consequence of the triangle inequality for norms:

$$\rho(x, y) = \|x - y\| = \|(x-z)+(z-y)\| \leqslant \|x-z\| + \|z-y\| = \rho(x, z) + \rho(z, y).$$

The topology induced on $\mathbf{X}$ by this metric is obviously the same as the topology of $\mathbf{X}$ as an LCS.

A sequence $\{x_n\}$ is said to converge in norm to x if $\|x_n - x\| \to 0$. Convergence in norm clearly coincides with convergence in the norm topology.

2) The closed balls $B_{1/n} = \{x \in \mathbf{X} : \|x\| \leqslant 1/n\}$ $(n \in \mathbb{N})$ form a basis of neighbourhoods of zero in the LCS $\mathbf{X}$.

3) $\big|\|x\| - \|y\|\big| \leqslant \|x - y\|$.

---

* The fundamental concepts in the theory of normed spaces were introduced by Banach [1]; see also Banach.

† If we are considering several normed spaces at the same time, then to distinguish between the corresponding norms we indicate the space as a suffix, writing, for example, $\|x\|_{\mathbf{X}}$.

This property is a consequence of II.3.2(c).

4) $\|x\|$ is a continuous function of x; that is, if $x_n \to x$, then $\|x_n\| \to \|x\|$.

5) A set E is bounded if and only if there is a constant M such that $\|x\| \leqslant M$ for every $x \in E$.

This follows from 2) and the definition of a bounded set (III.1.4).

6) A convergent sequence is bounded.

This follows from 4) and 5).

A TVS whose topology is obtainable from some norm (in the manner described above) is said to be *normable*. Let us now see where normed spaces fit into the class of TVSs.

THEOREM 1 (Kolmogorov). *A necessary and sufficient condition for a Hausdorff* TVS $\mathbf{X}$ *to be normable is that there exists a bounded convex neighbourhood of zero* V_0 *in* $\mathbf{X}$.

Proof. Only the *sufficiency* needs proving. We may assume that the neighbourhood V_0 is absolutely convex and closed (otherwise consider the intersection $(-\overline{V}_0) \cap \overline{V}_0$). For $x \in \mathbf{X}$, write

$$\|x\| = p_{V_0}(x),$$

where p_{V_0} is the Minkowski functional of V_0. Then p_{V_0} has all the properties of a norm except possibly one: $\|x\| = 0$ implies $x = \mathbf{0}$. We show that it has this property too. In fact, if $x \neq \mathbf{0}$, then, as $\mathbf{X}$ is a Hausdorff space, there is a neighbourhood of zero V not containing x. Since V_0 is bounded, we can find $\lambda > 0$ such that $V_0 \subset \lambda V$. Then clearly $\lambda x \notin V_0$, which, by Lemma III.2.1, yields

$$\|x\| = p_{V_0}(x) = \frac{1}{\lambda} p_{V_0}(\lambda x) > \frac{1}{\lambda},$$

so that $\|x\| \neq 0$.

Thus $\mathbf{X}$ is turned into a normed space in which (again by Lemma III.2.1) V_0 is the closed unit ball.

We now prove that the original topology in $\mathbf{X}$ coincides with the topology defined by the norm. For this we need to verify that the set of balls—that is, sets of the form λV_0 $(\lambda > 0)$—is a fundamental system of neighbourhoods of zero in the original topology. This is an immediate consequence of the fact that V_0 is bounded, since for every neighbourhood of zero V there exists $\lambda > 0$ such that $\lambda V_0 \subset V$, which is what we wish to prove.

The paper by Kolmogorov [1] in which Theorem 1 appeared was one of the first in which abstract TVSs were considered.

1.2. Let us consider some examples of normed spaces.

1) The finite dimensional space $\mathbf{K}^n$. A norm on $\mathbf{K}^n$ may be defined in many different ways (which are, in a certain sense, equivalent). For example, if $x = \{\xi_i\}_{i=1}^n$, then set

$$\|x\| = \left[\sum_{i=1}^{n} |\xi_i|^2\right]^{1/2}.$$

This is called the *Euclidean* norm. One can also introduce a norm by the formula $\|x\| = \max_{i=1}^{n} |\xi_i|$ or $\|x\| = \sum_{i=1}^{n} |\xi_i|$.

2) The space of continuous functions $\mathbf{C}(K)$ becomes a normed space, if we define the norm by

$$\|x\| = \max_{t \in K} |x(t)|.$$

3) The space $\ell^{\infty}(T)$ of bounded functions on the set T becomes a normed space if we define the norm by

$$\|x\| = \sup_{t \in T} |x(t)|.$$

4) The space $\mathbf{C}^{(l)}[a, b]$ of l times continuously differentiable functions becomes a normed space if we define a norm by

$$\|x\| = \sum_{k=0}^{l} \max_{t \in [a, b]} |x^{(k)}(t)|.$$

Note that the spaces $\mathbf{S}(0, 1)$ and $\mathbf{s}$ are not normed, which is why we studied them in more detail in the preceding chapters. We shall meet further examples of normed spaces below.

1.3. Let $\mathbf{X}$ and $\mathbf{Y}$ be normed spaces. A linear operator U from $\mathbf{X}$ into $\mathbf{Y}$ is called a *linear isometry* if we have $\|U(x)\| = \|x\|$, for every $x \in \mathbf{X}$. The spaces $\mathbf{X}$ and $\mathbf{Y}$ are said to be *linearly isometric* if there exists a linear isometry mapping $\mathbf{X}$ onto $\mathbf{Y}$. Clearly the relevant isometry U induces both an isometry between $\mathbf{X}$ and $\mathbf{Y}$ as metric spaces and an isomorphism between $\mathbf{X}$ and $\mathbf{Y}$ as vector spaces. To see this, it is enough to convince oneself that an operator U is one-to-one whenever it is an isometry. In fact, if $U(x) = U(y)$ then $\|x - y\| = \|U(x - y)\| = \|U(x) - U(y)\| = 0$.

In the sequel we shall not usually distinguish between linearly isometric spaces. We shall say that normed spaces $\mathbf{X}$ and $\mathbf{Y}$ are *isomorphic* if they are isomorphic as LCSs; that is, if there exists a linear homeomorphism from $\mathbf{X}$ onto $\mathbf{Y}$. Being isomorphic is a substantially weaker property than being isometric.

Two norms $\|.\|_1$ and $\|.\|_2$ on a vector space $\mathbf{X}$ are said to be *equivalent* if there are constants $k_1, k_2 > 0$ such that, for any $x \in \mathbf{X}$, we have

$$k_1 \|x\|_1 \leqslant \|x\|_2 \leqslant k_2 \|x\|_1.$$

It is easy to see that a necessary and sufficient condition for two norms to be equivalent is that the identity mapping of $\mathbf{X}$ onto itself be an isomorphism between the normed spaces $(\mathbf{X}, \|.\|_1)$ and $(\mathbf{X}, \|.\|_2)$.

A linear manifold in a normed space $\mathbf{X}$ is itself a normed space, with respect to the norm induced on it. We shall call such a normed space a *subspace* of the normed space $\mathbf{X}$.

1.4. Among normed spaces, a particularly important role is played by the *complete* spaces, which are called *Banach spaces* or *B-spaces* (after the Polish mathematician S. Banach).

In terms of the norm, a Cauchy sequence $\{x_n\}$ is one such that $\|x_m - x_n\| \to 0$ as m, $n \to \infty$. Hence in a normed space $\mathbf{X}$ the condition for completeness takes the following form: if $\|x_m - x_n\| \to 0$ as $m, n \to \infty$, then there exists an $x_0 \in \mathbf{X}$ such that $x_n \to x_0$.

The spaces $\mathbf{K}^n, \mathbf{C}(\mathbf{K}), \mathbf{C}^{(l)}[a, b], \ell^{\infty}(T)$, being complete, are obviously B-spaces.

If a given normed space $\mathbf{X}$ is not complete, then it can be embedded in a complete *metric* space $\hat{\mathbf{X}}$ (Theorem I.4.1). The algebraic operations and the norm, which are defined only on $\mathbf{X}$, can be uniquely extended to $\hat{\mathbf{X}}$, so that it becomes a B-space.

For the proof of this we shall use the observation, made in I.4.4, that a space $\mathbf{X}$ is dense in its completion $\hat{\mathbf{X}}$.

Let $x, y \in \hat{\mathbf{X}}$. There exist sequences $\{x_n\}$ and $\{y_n\}$ in $\mathbf{X}$ converging to x and y respectively. The operation of addition is defined in $\mathbf{X}$, so the equation $z_n = x_n + y_n (n = 1, 2, \ldots)$

makes sense. Since

$$\|z_n - z_m\| \leqslant \|x_n - x_m\| + \|y_n - y_m\| \xrightarrow[n,\, m \to \infty]{} 0,$$

the sequence $\{z_n\}$ is a Cauchy sequence in $\mathbf{X}$ and so also in $\hat{\mathbf{X}}$. Therefore there exists an element $z = \lim_{n \to \infty} z_n \in \hat{\mathbf{X}}$. Let us write, by definition, $x + y = z$. It is not hard to see that the definition of z does not depend on the choice of the sequences $\{x_n\}$ and $\{y_n\}$. In fact, suppose $x'_n \to x$, $y'_n \to y$ $(x'_n, y'_n \in \mathbf{X};\ n = 1, 2, \ldots)$ and let $z'_n = x'_n + y'_n$. Since

$$\|z'_n - z_n\| \leqslant \|x'_n - x_n\| + \|y'_n - y_n\| \xrightarrow[n \to \infty]{} 0,$$

we have $\lim_{n \to \infty} z'_n = \lim_{n \to \infty} z_n$.

It follows from this, in particular, that, for elements of $\mathbf{X}$, the new definition of the sum reduces to the one already present. For if $x, y \in \mathbf{X}$, then we can set $x_n = x$, $y_n = y$ $(n = 1, 2, \ldots)$, so that $z = \lim_{n \to \infty} (x_n + y_n) = x + y$, where the symbol $+$ is to be understood in its original sense.

The product of an element and a number is defined in $\hat{\mathbf{X}}$ analogously.

If, for $x \in \hat{\mathbf{X}}$, we set

$$\|x\| = \rho(x, \mathbf{0}) = \lim_{n \to \infty} \rho(x_n, \mathbf{0}) = \lim_{n \to \infty} \|x_n\|$$

$$(x_n \to x;\quad x_n \in \mathbf{X};\quad n = 1, 2, \ldots),$$

then it is not hard to see that $\hat{\mathbf{X}}$ satisfies the axioms for a normed space. The uniqueness of the definition of the norm follows from the same equation.

The uniqueness of the definitions of the sum of two elements and the product of an element and a scalar is obtained without difficulty if we use the continuity of these operations. We leave it to the reader to carry out this simple argument for himself.

From now on, when we speak of the *completion* of a normed space we shall assume that it has been linearized and normed in the manner indicated above.

1.5. In a normed space one can consider infinite series

$$\sum_{k=1}^{\infty} x_k = x_1 + x_2 + \ldots + x_n + \ldots. \tag{1}$$

The series (1) is said to be *convergent* if the sequence $\{s_n\}$ of its partial sums $s_n = x_1 + x_2 + \ldots + x_n$ is convergent. By the *sum* of the series is meant the limit of this sequence: $s = \lim_{n \to \infty} s_n$.

In a B-space "absolute" convergence of a series implies ordinary convergence; that is, if the numerical series

$$\sum_{k=1}^{\infty} \|x_k\| = \|x_1\| + \|x_2\| + \ldots + \|x_n\| + \ldots, \tag{2}$$

converges, then so does the series (1); moreover, we have the inequality $\left\| \sum_{k=1}^{\infty} x_k \right\| \leqslant \sum_{k=1}^{\infty} \|x_k\|$.

For if $m > n$ then

$$s_m - s_n = x_{n+1} + x_{n+2} + \ldots + x_m$$

and therefore

$$\|s_m - s_n\| \leqslant \|x_{n+1}\| + \|x_{n+2}\| + \ldots + \|x_m\|.$$

Since the series (2) converges, the right-hand side of this inequality is as small as desired for large enough n. Therefore the sequence $\{s_n\}$ is a Cauchy sequence and so, by the completeness of the space, converges. Passing to the limit as $n \to \infty$ in the inequality

$$\|s_n\| = \left\| \sum_{k=1}^{n} x_k \right\| \leqslant \sum_{k=1}^{n} \|x_k\|$$

and using the continuity of the norm, we obtain the required inequality.

1.6. Let us consider finite-dimensional normed spaces in more detail. We show that all finite-dimensional LCSs having the same algebraic dimension are isomorphic to one another, and thus to the space $\mathbf{K}^n$ with the Euclidean norm.

LEMMA 1. *If* $\mathbf{X}$ *is a finite-dimensional vector space and* $\mathbf{Y}$ *is a total linear manifold of functionals on* $\mathbf{X}$, *then* $\mathbf{Y}$ *coincides with the algebraic dual* $\mathbf{X}^+$.

Proof. Let $\{x_i\}_{i=1}^n$ be an algebraic basis for $\mathbf{X}$ (see II.1.4). Since $\mathbf{X}$ may be considered as a set of functionals on $\mathbf{Y}$, Lemma III.3.1 shows that there is a system of functionals $\{f_i\}_{i=1}^n \subset \mathbf{Y}$ biorthogonal to $\{x_i\}_{i=1}^n$, that is, satisfying

$$f_j(x_k) = \begin{cases} 0, & j \neq k, \\ 1, & j = k \end{cases} \quad (j, k = 1, 2, \ldots, n).$$

Thus $\sum_{i=1}^{n} \lambda_i f_i = 0$ implies $\lambda_k = \left(\sum_{i=1}^{n} \lambda_i f_i \right)(x_k) = 0$ for any k, so the vectors are linearly independent, and hence the space $\mathbf{Y}$ has dimension $\geqslant n$. On the other hand, $\mathbf{Y} \subset \mathbf{X}^+$, and $\mathbf{X}^+$ has dimension equal to n. Thus $\mathbf{Y}$ has dimension n, and so, by well-known property of finite-dimensional spaces, we have $\mathbf{Y} = \mathbf{X}^+$.

THEOREM 2. *All finite-dimensional* (*Hausdorff*) LCSs *with the same algebraic dimension are isomorphic.*

Proof. It is sufficient to prove that any two locally convex topologies on an n-dimensional space $\mathbf{X}$ coincide. To do this, we show that any locally convex topology τ coincides with the weak topology $\sigma(\mathbf{X}, \mathbf{X}^+)$.

By Lemma 1, $(\mathbf{X}, \tau)^* = \mathbf{X}^+$, so that $\tau \geqslant \sigma(\mathbf{X}, \mathbf{X}^+)$. We now prove the reverse inclusion. Fix a basis $\{x_i\}_{i=1}^n$ in $\mathbf{X}$ and consider a biorthogonal system $\{f_i\}_{i=1}^n$ in $\mathbf{X}^+$. Then $\{f_i\}_{i=1}^n$ is a basis in $\mathbf{X}^+$. (See the proof of Lemma 1.) If $\{x_\alpha\}$ is a net such that $x_\alpha \to x$ ($\sigma(\mathbf{X}, \mathbf{X}^+)$) and

$$x = \sum_{i=1}^{n} \xi_i x_i, \qquad x_\alpha = \sum_{i=1}^{n} \xi_i^{(\alpha)} x_i,$$

then $\xi_i^{(\alpha)} \underset{\alpha}{\to} \xi$ for $i = 1, 2, \ldots, n$. Indeed, for any f_i, we have

$$\xi_i^{(\alpha)} = f_i(x_\alpha) \underset{\alpha}{\to} f_i(x) = \xi_i.$$

Now, by the τ-continuity of the algebraic operations, we see that $x_\alpha \to x$ in the topology τ, and so $\sigma(\mathbf{X}, \mathbf{X}^+) \geqslant \tau$. Thus the proof of the theorem is complete.

Since $\mathbf{K}^n$, with the Euclidean norm, is a normed space, we see that every finite-dimensional LCS is normable.

In future we shall identify a finite-dimensional normed space $\mathbf{X}$ with $\mathbf{K}^n$. Let $e_k = (0, \ldots, 1, \ldots, 0)$ (the 1 in the k-th coordinate). Then, if $x = (\xi_1, \xi_2, \ldots, \xi_n)$, we have

$$x = \sum_{k=1}^{n} \xi_k e_k.$$

Applying Theorem 2 to $\mathbf{K}^n$ with the norm $\|x\| = \max_{i=1}^{n} |\xi_i|$, we deduce

COROLLARY 1. 1) *A necessary and sufficient condition for the sequence* $\{x_m\} : x_m = \sum_{k=1}^{n} \xi_k^{(m)} e_k \in \mathbf{X}$ *to converge to* $x = \sum_{k=1}^{n} \xi_k e_k \in \mathbf{X}$ *is that we have coordinatewise convergence:* $\xi_k^{(m)} \to \xi_k$ *as* $m \to \infty$.

2) *A set* $E \subset \mathbf{X}$ *is bounded if and only if there exists a constant* $M > 0$ *such that, for every* $x = \sum_{k=1}^{n} \xi_k e_k \in E$, *we have*

$$|\xi_k| \leqslant M \qquad (k = 1, 2, \ldots, n).$$

From Corollary 1 and the Bolzano–Weierstrass Theorem we deduce

COROLLARY 2. *A necessary and sufficient condition for a set E in a finite-dimensional normed space* $\mathbf{X}$ *to be relatively compact is that it be bounded.*

For if $\{x_m\}$ is a bounded sequence, then, writing $x_m = (\xi_1^{(m)}, \xi_2^{(m)}, \ldots, \xi_n^{(m)})$ $(m = 1, 2, \ldots)$ we infer from Corollary 1 that the numerical sequence $\{\xi_j^{(m)}\}$ is bounded, for each $j = 1, 2, \ldots, n$. Therefore, by the well-known Bolzano–Weierstrass Theorem, there exists a sequence of natural numbers $m_1 < m_2 < \ldots < m_k < \ldots$ such that

$$\xi_j^{(m_k)} \underset{k}{\to} \xi_j^{(0)} \qquad (j = 1, 2, \ldots, n).$$

By Corollary 1, we have $x_{m_k} \to x_0 = (\xi_1^{(0)}, \xi_2^{(0)}, \ldots \xi_n^{(0)})$.

COROLLARY 3. *A finite-dimensional normed space* $\mathbf{X}$ *is complete.*

For if $\{x_m\}$ is a Cauchy sequence, then it is bounded, and by the preceding Corollary we can select a subsequence $\{x_{m_k}\}$ converging to some element $x_0 \in \mathbf{X}$. Then, by Lemma I.4.1, we have $x_m \to x_0$.

COROLLARY 4. *A finite-dimensional linear manifold* $\mathbf{X}_0$ *in a normed space* $\mathbf{X}$ *is closed.*

COROLLARY 5. *Let* $\mathbf{X}$ *be a normed space and* $\mathbf{X}_0$ *a finite-dimensional linear manifold in* $\mathbf{X}$. *For any element* $x \in \mathbf{X}$, *there exists an element* x_0 *in* $\mathbf{X}$ *for which the distance from* x *to* $\mathbf{X}_0$ *is attained: that is,* $\|x - x_0\| = \rho(x, \mathbf{X}_0)$.

Indeed, for each $m = 1, 2, \ldots$, there exists x_m in $\mathbf{X}_0$ such that $\|x - x_m\| < \rho(x, \mathbf{X}_0) + 1/m$. The sequence $\{x_m\}$ is evidently bounded, since

$$\|x_m\| \leqslant \|x\| + \|x - x_m\| < \|x\| + \rho(x, \mathbf{X}_0) + 1 \qquad (m = 1, 2, \ldots),$$

and therefore, by Corollary 2, we can choose a convergent subsequence $\{x_{m_k}\}$ with $x_{m_k} \to x_0$.

Then clearly $\|x - x_0\| \leqslant \rho(x, \mathbf{X}_0)$; and, since $x_0 \in \mathbf{X}_0$, we have the reverse inequality also.

As an application of this last result, we consider the space $\mathbf{C}[a, b]$ and the finite-dimensional subspace $\mathbf{P}^n$ of all algebraic polynomials of degree not exceeding n. By Corollary 5, for any given continuous function x, there is a polynomial x_0 in $\mathbf{P}^n$ such that

$$\max_{a \leqslant t \leqslant b} |x(t) - x_0(t)| = \|x - x_0\| = \rho(x, \mathbf{P}^n),$$

that is, x_0 yields the best approximation to x among all polynomials of degree at most n.

1.7. In conclusion, we show that the result of Corollary 2 to Theorem 2 is reversible: that is, that any normed space in which every bounded set is relatively compact is finite-dimensional.*

As a preliminary, we prove an important lemma.

LEMMA 2. *Let* $\mathbf{X}$ *be a normed space and let* $\mathbf{X}_0 \neq \mathbf{X}$ *be a closed subspace. For every* $\varepsilon > 0$, *there exists a normalized element*† x_0 *such that*

$$\rho(x_0, \mathbf{X}_0) > 1 - \varepsilon.$$

Proof. Since $\mathbf{X}_0$ is a closed set properly contained in $\mathbf{X}$, there exists an element $\bar{x} \in \mathbf{X}$ such that $\rho(\bar{x}, \mathbf{X}_0) = d > 0$. Also there is an element $x' \in \mathbf{X}_0$ such that

$$\|\bar{x} - x'\| < \frac{d}{1 - \varepsilon}.$$

Write

$$x_0 = \frac{\bar{x} - x'}{\|\bar{x} - x'\|} = \alpha(\bar{x} - x') \qquad \left(\alpha = \frac{1}{\|\bar{x} - x'\|} > \frac{1 - \varepsilon}{d}\right).$$

Clearly $\|x_0\| = 1$. Moreover, if $x \in \mathbf{X}_0$, then

$$\|x_0 - x\| = \|\alpha\bar{x} - \alpha x' - x\| = \alpha\left\|\bar{x} - \left(x' + \frac{x}{\alpha}\right)\right\| \geqslant \alpha d > \frac{1 - \varepsilon}{d} d = 1 - \varepsilon.$$

REMARK. A normalized element x_0 having the property that $\rho(x_0, \mathbf{X}_0) = 1$ is in a certain sense perpendicular to $\mathbf{X}_0$ (for in Euclidean space a vector x_0 satisfying this condition is actually orthogonal to $\mathbf{X}_0$). Because of this, we shall call the lemma just proved the Lemma on Almost Perpendicularity.

THEOREM 3. *A necessary and sufficient condition for every bounded set in a normed space* $\mathbf{X}$ *to be relatively compact is that* $\mathbf{X}$ *be finite-dimensional.*

The *sufficiency* of the condition has already been established (Corollary 2 to Theorem 1).

Necessity. Consider any normalized element $x_1 \in \mathbf{X}$ and denote by $\mathbf{X}_1$ the linear hull $\mathscr{L}(\{x_1\})$ of this element; that is, the collection of elements of the form λx_1.

If we assume $\mathbf{X}$ to be infinite-dimensional, then $\mathbf{X}_1 \neq \mathbf{X}$, and by the Lemma on Almost Perpendicularity there is a normalized element x_2 such that $\rho(x_2, \mathbf{X}_1) > \frac{1}{2}$. Form the linear hull of x_1 and x_2, which we denote by $\mathbf{X}_2$. Arguing as before, we obtain a sequence of elements $\{x_n\}$ and a sequence of subspaces $\mathbf{X}_1 \subset \mathbf{X}_2 \subset \ldots \subset \mathbf{X}_n \subset \ldots$ such that

$$\|x_n\| = 1, \quad \mathbf{X}_n = \mathscr{L}(\{x_1, x_2, \ldots, x_n\}),$$
$$\rho(x_{n+1}, \mathbf{X}_n) > 1/2 \quad (n = 1, 2, \ldots). \tag{3}$$

* This fact was established by F. Riesz.

† That is, an element whose norm is equal to 1.

Since the sequence $\{x_n\}$ is bounded, we can choose a convergent subsequence in it. However, by (3),

$$\|x_n - x_m\| > 1/2 \qquad (n > m;\ m, n = 1, 2, \ldots),$$

so neither the sequence $\{x_n\}$ nor any of its subsequences can converge. This proves the theorem.

COROLLARY. *An LCS* $\mathbf{X}$ *is finite-dimensional if and only if it contains a totally bounded neighbourhood of zero.*

Proof. If the LCS $\mathbf{X}$ is finite-dimensional, then by Theorem 2 it is normable, so by Theorem 3 the closure of the unit ball in $\mathbf{X}$ is compact.

Conversely, if V_0 is a totally bounded neighbourhood of zero, then by Theorem 1 the space $\mathbf{X}$ is normable. As there exists $\lambda > 0$ such that the unit ball $B_{\mathbf{X}} = \{x \in \mathbf{X} : \|x\| \leqslant 1\}$ is contained in λV_0, the set $B_{\mathbf{X}}$ is totally bounded, and so any bounded set is also totally bounded. Repeating the proof of Theorem 3 word for word, we easily see that $\mathbf{X}$ is finite-dimensional.

1.8. To conclude this section we introduce some more important concepts from the theory of normed spaces.

Let $\mathbf{X}_1$ and $\mathbf{X}_2$ be normed spaces. We endow their direct product $\mathbf{Z} = \mathbf{X}_1 \times \mathbf{X}_2$ with the following norm: if $z = (x_1, x_2)$, $x_1 \in \mathbf{X}_1$, $x_2 \in \mathbf{X}_2$, then

$$\|z\| = \max(\|x_1\|, \|x_2\|).$$

The normed space $\mathbf{Z}$ is clearly complete if $\mathbf{X}_1$ and $\mathbf{X}_2$ are B-spaces.

Let $\mathbf{X}$ be a normed space and $\mathbf{X}_0$ a closed subspace of $\mathbf{X}$. Consider the factor-space $\mathbf{X}/\mathbf{X}_0$ (II.1.6). Notice that, since $\mathbf{X}_0$ is closed, the class $\bar{x} = x + \mathbf{X}_0$ is a closed set in $\mathbf{X}$.

If we write

$$\|\bar{x}\| = \inf_{x \in \bar{x}} \|x\|,$$

then $\mathbf{X}/\mathbf{X}_0$ becomes a normed space. For if $\bar{x} = \mathbf{0}$, then we may take $x \in \bar{x}$ to be the zero element of $\mathbf{X}$, so that $\|\bar{x}\| = 0$. Conversely, if $\|\bar{x}\| = 0$, then it follows from the definition of $\|x\|$ that there exists a sequence $\{x_n\} \subset \bar{x}$ such that $x_n \to \mathbf{0}$. Since $\bar{x}$ is a closed set, it also contains the limit of this sequence: that is, $\mathbf{0} \in \bar{x}$, and thus $\bar{x}$ is the zero element of $\mathbf{X}/\mathbf{X}_0$.

The homogeneity of the norm is also verified without difficulty. Assuming that $\lambda \neq 0$, we have

$$|\lambda|\,\|\bar{x}\| = |\lambda| \inf_{x \in \bar{x}} \|x\| = \inf_{x \in \bar{x}} \|\lambda x\|.$$

But λx ranges over the class $\lambda\bar{x}$ as x ranges over the class $\bar{x}$, so it follows that

$$\inf_{x \in \bar{x}} \|\lambda x\| = \inf_{z \in \lambda\bar{x}} \|z\| = \|\lambda\bar{x}\|.$$

Finally, let us prove the triangle inequality. For arbitrary elements $x \in \bar{x}$, $y \in \bar{y}$ ($\bar{x}$, $\bar{y} \in \mathbf{X}/\mathbf{X}_0$) we have $x + y \in \bar{x} + \bar{y}$, so

$$\|\bar{x} + \bar{y}\| \leqslant \|x + y\| \leqslant \|x\| + \|y\|.$$

If we pass to the greatest lower bound in the right-hand side, we obtain the triangle inequality.

By associating with each element $x \in \mathbf{X}$ the class $\bar{x} = x + \mathbf{X}_0 = \phi(x)$ in which it lies, we obtain a mapping ϕ, which is called the *natural homomorphism* (or *canonical homomorphism*) of $\mathbf{X}$ onto the factor-space $\mathbf{X}/\mathbf{X}_0$. This operator ϕ is clearly linear.

Since

$$\|\phi(x_0)\| = \inf_{x \in \phi(x_0)} \|x\| \leqslant \|x_0\| \qquad (x_0 \in \mathbf{X}),$$

ϕ is continuous. Moreover, for each $\bar{x} \in \mathbf{X}/\mathbf{X}_0$, there exists $x \in \mathbf{X}$ such that

$$\bar{x} = \phi(x), \quad \|\bar{x}\| \geqslant \frac{1}{2}\|x\|. \tag{4}$$

Using the above property, we show that if the given space $\mathbf{X}$ is complete, then $\mathbf{X}/\mathbf{X}_0$ will also be complete. In fact, let $\{\bar{x}_n\}$ be a Cauchy sequence of elements of $\mathbf{X}/\mathbf{X}_0$. Refining this sequence if necessary, we can arrange that the series $\sum_{n=1}^{\infty} \|\bar{x}_{n+1} - \bar{x}_n\|$ is convergent. By (4), we can find an element $x_n \in \mathbf{X}$ such that

$$\bar{x}_{n+1} - \bar{x}_n = \phi(x_n), \qquad \|\bar{x}_{n+1} - x_n\| \geqslant \frac{1}{2}\|x_n\|$$

$$(n = 0, 1, \ldots, \bar{x}_0 = \mathbf{0}).$$

The series $\sum_{n=1}^{\infty} \|x_n\|$ is clearly convergent. Therefore, as $\mathbf{X}$ is complete, the series $\sum_{n=1}^{\infty} x_n$ is convergent. Denoting its sum by x and writing $\bar{x} = \phi(x)$, we have

$$\bar{x} = \phi(x) = \sum_{n=0}^{\infty} \phi(x_n) = \sum_{n=0}^{\infty} (\bar{x}_{n+1} - \bar{x}_n) = \lim_{n \to \infty} \bar{x}_n,$$

that is, the sequence $\{\bar{x}_n\}$ converges to $\bar{x}$.

1.9. A continuous linear operator P mapping a normed space $\mathbf{X}$ onto a closed subspace $\mathbf{Y}$ of $\mathbf{X}$ is called a *projection* (of $\mathbf{X}$ onto $\mathbf{Y}$) if P fixes the elements of $\mathbf{Y}$: that is, if $P(y) = y$ for every $y \in \mathbf{Y}$.

A closed subspace $\mathbf{Y}$ is said to be *complemented* in the B-space $\mathbf{X}$ if there exists a projection P of $\mathbf{X}$ onto $\mathbf{Y}$. It is easy to see that* $\mathbf{X}/\mathbf{Y} = P^{-1}(\mathbf{0})$. We shall consider the question of the existence of a projection onto a subspace later (see V.3.5).

§ 2. Auxiliary inequalities

In this section we establish certain inequalities which have applications in the study of specific examples of B-spaces (see § 3). For references to the literature on this material, see Hardy, Littlewood and Pólya.

2.1. We first prove a lemma.

LEMMA 1. *Let p and q be positive real numbers connected by the relation*

$$\frac{1}{p} + \frac{1}{q} = 1. \tag{1}$$

For every pair of numbers a and b, we have the inequality

$$|ab| \leqslant \frac{|a|^p}{p} + \frac{|b|^q}{q}. \tag{2}$$

* Here equality means isomorphism of B-spaces.

Proof. We may assume that a and b are positive. Write $m = 1/p$ (so $0 < m < 1$) and consider the function

$$\phi(t) = t^m - mt \qquad (t > 0).$$

Since $\phi'(t) = m(t^{m-1} - 1)$, the function $\phi(t)$ attains a maximum when $t = 1$. Therefore $\phi(t) \leqslant \phi(1)(t > 0)$, so that $t^m - 1 \leqslant m(t-1)$. If we set $t = a^p/b^q$ in the last inequality, we obtain

$$ab^{-q/p} - 1 \leqslant (1/p)(a^p b^{-q} - 1),$$

so that, multiplying each term by b^q and using the fact that $q - q/p = 1$, we obtain (2).

2.2. *Hölder's inequality.* Let $\xi_1, \xi_2, \ldots, \xi_n$ and $\eta_1, \eta_2, \ldots, \eta_n$ be arbitrary numbers. Then we have the following inequality:

$$\sum_{k=1}^{n} |\xi_k \eta_k| \leqslant \left\{ \sum_{k=1}^{n} |\xi_k|^p \right\}^{1/p} \left\{ \sum_{k=1}^{n} |\eta_k|^q \right\}^{1/q}$$

(where p and q are connected by (1)).

Proof. Write $A^p = \sum_{k=1}^{n} |\xi_k|^p$ and $B^q = \sum_{k=1}^{n} |\eta_k|^q$. We may assume that $A, B > 0$. Set $\xi'_k = \xi_k/A$, $\eta'_k = \eta_k/B$. Then, by inequality (2),

$$|\xi'_k \eta'_k| \leqslant \frac{|\xi'_k|^p}{p} + \frac{|\eta'_k|^q}{q}$$

or, summing,

$$\sum_{k=1}^{n} |\xi'_k \eta'_k| \leqslant \frac{\sum_{k=1}^{n} |\xi'_k|^p}{p} + \frac{\sum_{k=1}^{n} |\eta'_k|^q}{q} = \frac{1}{p} + \frac{1}{q} = 1.$$

Hence

$$\sum_{k=1}^{n} |\xi_k \eta_k| \leqslant AB,$$

as we required to prove.

REMARK. Hölder's inequality holds also for a (countably) infinite number of terms; that is, we have the inequality

$$\sum_{k=1}^{\infty} |\xi_k \eta_k| \leqslant \left[\sum_{k=1}^{\infty} |\xi_k|^p \right]^{1/p} \left[\sum_{k=1}^{\infty} |\eta_k|^q \right]^{1/q}, \tag{3}$$

in which convergence of the series on the right-hand side implies convergence of the series on the left-hand side.

In fact, we have already proved the inequality for the partial sums of the series occurring in (3). Passing to the limit, we obtain (3).

2.3. *Hölder's inequality for integrals.* Let $x(t)$ and $y(t)$ be measurable functions on the measure space (T, Σ, μ). We have the inequality

$$\int_T |x(t)y(t)|\, d\mu \leqslant \left[\int_T |x(t)|^p\, d\mu \right]^{1/p} \left[\int_T |y(t)|^q\, d\mu \right]^{1/q}$$

(once again, p and q satisfy (1)).

Proof. In essence, this is a repetition of the proof of Hölder's inequality for sums. We may assume that

$$0 < A^p = \int |x(t)|^p \, d\mu < \infty, \qquad 0 < B^q = \int_T |y(t)|^q du < \infty,$$

since the inequality to be proved is trivial if one of the integrals is equal to zero or infinity.

Write $\bar{x}(t) = x(t)/A$ and $\bar{y}(t) = y(t)/B$. For each $t \in T$, we have, by (1), the inequality

$$|\bar{x}(t)\bar{y}(t)| \leqslant \frac{|\bar{x}(t)|^p}{p} + \frac{|\bar{y}(t)|^q}{q},$$

which, when integrated, yields

$$\int_T |\bar{x}(t)\bar{y}(t)| \, d\mu \leqslant \frac{1}{p} \int_T |\bar{x}(t)|^p \, d\mu + \frac{1}{q} \int_T |\bar{y}(t)|^q \, d\mu = \frac{1}{p} + \frac{1}{q} = 1,$$

so that

$$\int_T |x(t)\, y(t)| \, d\mu \leqslant AB,$$

as we wished to prove.

Notice that, as in the case of the inequality for sums, the integral on the left-hand side is finite whenever those on the right-hand side are.

REMARK. If $p = 2$ (so that $q = 2$ also), then Hölder's inequalities become the well-known Cauchy–Buniakowski inequalities for sums and integrals:

$$\sum_{k=1}^{\infty} |\xi_k \eta_k| \leqslant \left[\sum_{k=1}^{\infty} |\xi_k|^2 \right]^{1/2} \left[\sum_{k=1}^{\infty} |\eta_k|^2 \right]^{1/2},$$

$$\int_T |x(t)y(t)| \, d\mu \leqslant \left[\int_T |x(t)|^2 \, d\mu \right]^{1/2} \left[\int_T |y(t)|^2 \, d\mu \right]^{1/2}.$$

2.4. *The generalized Hölder's inequality.* Suppose positive numbers p, q, r are connected by the relationship

$$\frac{1}{p} + \frac{1}{q} + \frac{1}{r} = 1.$$

Then, for any measurable functions $x(t)$, $y(t)$, $z(t)$ defined on T, we have the inequality

$$\int_T |x(t)y(t)z(t)| \, d\mu \leqslant \left[\int_T |x(t)|^p \, d\mu \right]^{1/p} \left[\int_T |y(t)|^q \, d\mu \right]^{1/q} \left[\int_T |z(t)|^r \, d\mu \right]^{1/r}.$$

Proof. Define p' by the equation $1/p' = 1/p + 1/r$. Then, since $1/p + 1/p' = 1$, Hölder's inequality yields

$$\int_T |x(t)y(t)z(t)| \, d\mu \leqslant \left[\int_T |x(t)|^p d\mu \right]^{1/p} \left[\int_T |y(t)z(t)|^{p'} d\mu \right]^{1/p'}. \tag{4}$$

Now we again apply Hölder's inequality to the second integral, using the exponents q/p' and r/p' (note that $p'/q + p'/r = 1$), to obtain

$$\int_T | y(t)z(t)|^{p'} \, d\mu \leqslant \left\{ \int_T | y(t)|^{p' \frac{q}{p'}} \, d\mu \right\}^{\frac{p'}{q}} \left\{ \int_T |z(t)|^{p' \frac{r}{p'}} \, d\mu \right\}^{\frac{p'}{r}}.$$

Substituting this in (4), we obtain the required inequality.

REMARK. The generalized Hölder's inequality is, of course, also true for sums.

We leave it to the reader to derive the inequality similar to the one just proved involving the integral of a product of n functions on the left-hand side.

2.5. *Minkowski's inequality.* Let $\{\xi_k\}$ and $\{\eta_k\}$ be sequences of numbers. We have the inequality

$$\left[\sum_{k=1}^{\infty} |\xi_k + \eta_k|^p\right]^{1/p} \leqslant \left[\sum_{k=1}^{\infty} |\xi_k|^p\right]^{1/p} + \left[\sum_{k=1}^{\infty} |\eta_k|^p\right]^{1/p} \quad (p \geqslant 1).$$

Proof. We may clearly restrict ourselves to the case where $p > 1$ and all the ξ_k and η_k are non-negative. In addition, we may assume that the sums involved in the inequality have a finite number of terms (the extension to series is carried out by taking limits). With these assumptions, we have

$$\sum_{k=1}^{n} [\xi_k + \eta_k]^p = \sum_{k=1}^{n} \xi_k[\xi_k + \eta_k]^{p-1} + \sum_{k=1}^{n} \eta_k[\xi_k + \eta_k]^{p-1}.$$

Applying Hölder's inequality to each sum on the right-hand side, we obtain (with $1/p + 1/q = 1$)

$$\sum_{k=1}^{n} [\xi_k + \eta_k]^p \leqslant \left[\sum_{k=1}^{n} \xi_k^p\right]^{1/p} \left[\sum_{k=1}^{n} (\xi_k + \eta_k)^{q(p-1)}\right]^{1/q} +$$
$$+ \left[\sum_{k=1}^{n} \eta_k^p\right]^{1/p} \left[\sum_{k=1}^{n} (\xi_k + \eta_k)^{q(p-1)}\right]^{1/q}.$$

But $1/p + 1/q = 1$ implies $q(p-1) = p$; therefore, multiplying both sides of the last inequality by $\left[\sum_{k=1}^{n} (\xi_k + \eta_k)^p\right]^{-1/q}$, we obtain

$$\left[\sum_{k=1}^{n} (\xi_k + \eta_k)^p\right]^{1-1/q} \leqslant \left[\sum_{k=1}^{n} \xi_k^p\right]^{1/p} + \left[\sum_{k=1}^{n} \eta_k^p\right]^{1/p}.$$

Since $1 - 1/q = 1/p$, this is precisely the inequality we wished to prove.

2.6. *Minkowski's inequality for integrals.* Let $x(t)$ and $y(t)$ be measurable functions on a set T. We have the inequality

$$\left[\int_T |x(t) + y(t)|^p \, d\mu\right]^{1/p} \leqslant \left[\int_T |x(t)|^p \, d\mu\right]^{1/p} + \left[\int_T |y(t)|^p \, d\mu\right]^{1/p} \quad (p \geqslant 1).$$

Proof. The inequality is obvious if one of the integrals on the right-hand side is infinite. If the integral on the left-hand side is infinite, then using the estimates

$$\int_T |x(t) + y(t)|^p \, d\mu \leqslant \int_T (|x(t)| + |y(t)|)^p \, d\mu \leqslant$$
$$\leqslant 2^{p/q}\left(\int_T |x(t)|^p \, d\mu + \int_T |y(t)|^p \, d\mu\right),$$

which follow from the numerical inequality

$$(|a|+|b|) \leqslant (|a|^p+|b|^p)^{1/p}(1^q+1^q)^{1/q} = 2^{1/q}(|a|^p+|b|^p)^{1/p}$$

(see subsection 2.2),

we see that at least one of the integrals on the right-hand side is infinite. Therefore we may assume that all the integrals are finite, and the proof is now carried out exactly as in 2.5.

Notice that Hölder's and Minkowski's inequalities for sequences are special cases of the corresponding inequalities for integrals. To convince oneself of this, one need only take (T, Σ, μ) to consist of the natural numbers $\mathbb{N}$ with unit measure at each point.

§ 3. Normed spaces of measurable functions and sequences

3.1. In this section we shall consider spaces whose elements are measurable functions—that is, linear manifolds in $\mathbf{S}(T, \Sigma, \mu)$. Here, as in $\mathbf{S}(T, \Sigma, \mu)$ itself, equivalent functions will be identified. In these spaces, algebraic operations are defined in a natural way. In particular, a function equivalent to zero plays the role of the zero element. We first present a general theory for these spaces, then consider specific examples.

Let (T, Σ, μ) be a σ-finite measure space, and let $\mathbf{S} = \mathbf{S}(T, \Sigma, \mu)$ be the space of all real or complex measurable functions on (T, Σ, μ). For real functions $x, y \in \mathbf{S}$, the notation $x \geqslant y$ means that $x(t) \geqslant y(t)$ a.e. The *support* of an arbitrary element $x \in \mathbf{S}$ is defined by:

$$\operatorname{supp} x = \{t \in T : x(t) \neq 0\}.$$

Clearly, the support of a function is determined to within a set of measure zero. One cannot define the support of an arbitrary subset $E \subset \mathbf{S}$ as the union of the supports of the functions in E, because one could not then guarantee either its uniqueness to within a set of measure zero or even its measurability. We therefore proceed as follows. The support of an arbitrary set $E \subset \mathbf{S}$ is a set $\operatorname{supp} E \in \Sigma$ having the following properties:

1) $\operatorname{supp} x \subset \operatorname{supp} E \pmod{\mu}$, for every $x \in E$;

2) if $A \in \Sigma$ is a set such that $\operatorname{supp} x \subset A \pmod{\mu}$ for every $x \in E$, then $\operatorname{supp} E \subset A \pmod{\mu}$.

Since the uniqueness of the support to within a set of measure zero follows from 2), to justify this definition we need to show that a set $\operatorname{supp} E$ satisfying 1) and 2) does exist. By Theorem I.6.17, $x_0 = \sup\{\chi_A : A = \operatorname{supp} x, x \in E\}$ exists. If we set $\operatorname{supp} E = \operatorname{supp} x_0$, then this is clearly the required set.

For any function $x \in \mathbf{S}$, we write $|x|(t) = |x(t)|$.

If $x, y \in \mathbf{S}$ are real-valued functions, then we define their supremum and infimum, respectively, by the formulae

$$(x \vee y)(t) = \max(x(t), y(t)), \ (x \wedge y)(t) = \min(x(t), y(t)).$$

For a real-valued function x we also write

$$x_+ = x \vee \mathbf{0}, \quad x_- = (-x) \vee \mathbf{0}.$$

Then $|x| = x_+ + x_-$ and $x = x_+ - x_-$. If x is a complex function, then $x(t) = \operatorname{Re} x(t)$

$+ i$ Im $x(t)$. These equations often enable one to restrict one's study to non-negative functions. We write

$$\mathbf{X}_+ = \{x \in \mathbf{X} : x \geqslant \mathbf{0}\}.$$

Elements $x, y \in \mathbf{S}$ are said to be *disjoint* if $|x| \wedge |y| = \mathbf{0}$. The notation $x_n \downarrow$ means that $x_m \geqslant x_n$ when $n \geqslant m$. The notation $x_n \downarrow x$ means that $x_n \downarrow$ and $x_n(t) \to x(t)$ a.e. Analogously we define $x_n \uparrow$ and $x_n \uparrow x$.

An *ideal space* (IS for short) on (T, Σ, μ) is a linear manifold $\mathbf{X}$ in $\mathbf{S}$ such that

$$x \in \mathbf{X}, \quad y \in \mathbf{S}, \quad |y| \leqslant |x| \text{ imply } y \in \mathbf{X}.$$

A *foundation space* (FS for short) on (T, Σ, μ) is an IS $\mathbf{X}$ such that supp $\mathbf{X} = T$. Obviously, every IS $\mathbf{X}$ may be regarded as an FS on supp $\mathbf{X}$.

LEMMA 1. *If* $\mathbf{X}$ *is an* FS *on* (T, Σ, μ), *then for every non-negative function* $x \in \mathbf{S}(T, \Sigma, \mu)$ *there exists a sequence* $x_n \uparrow x$, $\mathbf{0} \leqslant x_n \in \mathbf{X}$.

Proof. Consider the set $E = \{y \in \mathbf{X}_+ : y \leqslant x\}$. By Theorem I.6.17, $\tilde{y} = \sup E$ exists; moreover, there is a countable set $\{y_n\} \subset E$ such that $\tilde{y}(t) = \sup y_n(t)$ a.e. We show that $\tilde{y} = x$. Obviously $\tilde{y} \leqslant x$. Consider the function $z = x - \tilde{y} \geqslant \mathbf{0}$ and the set $A = \{t : z(t) > 0\}$. Assume that $\mu(A) > 0$. Then there exists a non-negative function $y_0 \in \mathbf{X}$ for which $\mu(\text{supp } y_0 \cap A) > 0$. Therefore $\tilde{y}_0 = (y_0 \chi_A) \wedge z > \mathbf{0}$ and $\tilde{y}_0 \in \mathbf{X}$ since $\mathbf{0} \leqslant \tilde{y}_0 \leqslant y_0$. Consequently $\tilde{y} + \tilde{y}_0 > \tilde{y} = \sup E$, while, on the other hand, $\tilde{y} + \tilde{y}_0 \in \mathbf{X}$ and $\tilde{y} + \tilde{y}_0 \leqslant (x - z) + z = x$, giving a contradiction. Hence $x = \tilde{y} = \sup E = \sup y_n$. Write $x_n = y_1 \vee y_2 \vee \ldots \vee y_n$. Clearly $x_n \in \mathbf{X}$ and $\mathbf{0} \leqslant x_n \uparrow x$.

COROLLARY 1. *If* $\mathbf{X}$ *is an* FS *on* (T, Σ, μ), *then for every non-negative function* $x \in \mathbf{S}(T, \Sigma, \mu)$ *there is a non-decreasing sequence of sets* $\{B_n\}_{n=1}^{\infty} \subset \Sigma$ *such that* $x\chi_{B_n} \in \mathbf{X}$ $(n \in \mathbb{N})$ *and* $x\chi_{B_n} \uparrow x$.

Proof. By Lemma 1 there is a sequence $x_n \uparrow x$, $x_n \in \mathbf{X}$. Write

$$B_n = \{t \in T : 2x_n(t) \geqslant x(t)\}.$$

Since $x_n \uparrow x$, we have $x\chi_{B_n} \uparrow x$, and, since $x\chi_{B_n} \leqslant 2x_m \in \mathbf{X}$, we have $x\chi_{B_n} \in \mathbf{X}$.

COROLLARY 2. *If* $\mathbf{X}$ *is an* FS *on* (T, Σ, μ) *then there is a partition* $\{A_n\}_{n=1}^{\infty}$ *of* T *such that* $A_n \in \Sigma(\mu)$ *and* $\chi_{A_n} \in \mathbf{X}$ *for each* $n \in \mathbb{N}$.

Proof. It is sufficient to take x in Corollary 1 to be the function $\mathbf{1}$, which is identically equal to unity on T, and to write $A'_1 = B_1$, $A'_n = B_n \setminus B_{n-1}$ $(n = 2, 3, \ldots)$; and then to decompose each set A'_n into sets of finite measure.

A norm $\|.\|$ on an IS $\mathbf{X}$ is called *monotone* if $x, y \in \mathbf{X}$, $|x| \leqslant |y|$ imply $\|x\| \leqslant \|y\|$.

A *normed ideal space* (NIS for short) on (T, Σ, μ) is an IS equipped with a monotone norm. An NIS which is an FS is called a *normed foundation space* (NFS for short). Finally, an NIS which is complete in its norm is called a *Banach ideal space* (BIS for short), while an NFS which is complete in its norm is called a *Banach foundation space* (BFS for short).

Let us deduce the simplest consequences of monotonicity for a norm on an NIS $\mathbf{X}$. Assume that $x_n \to x$ in the norm of $\mathbf{X}$. Then the statements listed below are true (in these also, convergence means convergence in the norm of $\mathbf{X}$).

1) $|x_n - x| \to \mathbf{0}$.
2) If $y_n \to y$, then $x_n \vee y_n \to x \vee y$ and $x_n \wedge y_n \to x \wedge y$.
3) $(x_n)_+ \to x_+$, $(x_n)_- \to x_-$, $|x_n| \to |x|$.
4) If $x_n \geqslant y_n$ $(n \in \mathbb{N})$ and $y_n \to y$, then $x \geqslant y$.
5) If $A \in \Sigma$, then $x_n \chi_A \to x\chi_A$.

Statement 1) is obvious; 2) follows from the inequality $|(x_n \vee y_n)-(x \vee y)| \leqslant |x_n - x| + |y_n - y|$ and the equation $x \wedge y = -[(-x) \vee (-y)]$; (3) is an immediate consequence of 2); 4) also follows from 2), for, on the one hand, $x_n \vee y_n \to x \vee y$ and, on the other hand, $x_n \vee y_n = x_n \to x$. Therefore $x \vee y = x$, and so $x \geqslant y$. Finally, 5) follows from the inequality $|x_n\chi_A - x\chi_A| \leqslant |x_n - x|$.

We now record a few theorems concerning the classes of spaces just introduced.

THEOREM 1. *Let* $(\mathbf{X}, \|\cdot\|)$ *be an* NIS *on* (T, Σ, μ). *Then*

1) *if* $x_n, x \in \mathbf{X}$ *and* $\|x_n - x\| \to 0$, *then* $x_n \to x\ (\mu)$;

2) *if* $\{x_n\} \subset \mathbf{X}$ *is a Cauchy sequence, then it converges in measure to some* $x \in \mathbf{S}$.

Proof. 1) By Corollary 2 to Lemma 1, there is a partition $\{A_p\}_{p=1}^{\infty}$ of supp $\mathbf{X}$ such that $A_p \in \Sigma(\mu)$, $\chi_{A_p} \in \mathbf{X}$ $(p \in \mathbb{N})$. Assume that $x_n \nrightarrow x(\mu)$. Then, by the remark following the definition of the direct sum property (see I.6.10), $x_n \nrightarrow x(\mu)$ on some A_p. In view of property 5) of convergence in norm in an NIS, we may assume that the function $\mathbf{1}$, identically equal to unity on T, belongs to $\mathbf{X}$, and that $\mu(T) < \infty$.

By passing to a subsequence if necessary, we may also assume that there exist numbers $\varepsilon, \delta > 0$ such that the following conditions are satisfied:

$$\mu(\{t \in T: |x_n(t) - x(t)| \geqslant \varepsilon\}) \geqslant \delta, \tag{1}$$

$$\|x_n - x\| < \varepsilon/2^n. \tag{2}$$

Write

$$B_n = \{t \in T: |x_n(t) - x(t)| \geqslant \varepsilon\}, \quad B = \bigcap_{n=1}^{\infty} \bigcup_{m=n+1}^{\infty} B_m. \tag{3}$$

By (1), we have

$$\mu(B_n) \geqslant \delta \quad (n \in \mathbb{N}), \quad \mu(B) \geqslant \delta. \tag{4}$$

By (2), bearing in mind that $\varepsilon\chi_{B_n} \leqslant |x_n - x|$, we have

$$\|\chi_{B_n}\| < 1/2^n. \tag{5}$$

We now introduce the sets

$$C_{ns} = \bigcup_{m=n+1}^{n+s} (B_m \cap B).$$

Then for every $n \in \mathbb{N}$ the sequence $\{C_{ns}\}_{s=1}^{\infty}$ is non-decreasing and, by (3),

$$B = \bigcup_{s=1}^{\infty} C_{ns}.$$

Hence for each $n \in \mathbb{N}$ there exists a suffix s_n such that

$$\mu(B \setminus C_{ns_n}) < 1/2^{n+1}. \tag{6}$$

Write

$$D_n = \bigcap_{m=n+1}^{\infty} C_{ms_m}.$$

The sequence $\{D_n\}$ is clearly non-decreasing and, since $B \setminus \bigcap_{m=n+1}^{\infty} C_{ms_m} = \bigcup_{m=n+1}^{\infty} (B \setminus C_{ms_m})$, we see from (6) that

$$\mu(B \setminus D_n) = \sum_{m=n+1}^{\infty} \mu(B \setminus C_{ms_m}) < 1/2^n.$$

Therefore $\bigcup_{n=1}^{\infty} D_n = B \pmod{\mu}$. By (5), when $m > n$ we have

$$\|\chi_{D_n}\| \leqslant \|\chi_{C_{ms_m}}\| \leqslant \left\| \sum_{k=m+1}^{m+s_m} \chi_{B_k} \right\| \leqslant \sum_{k=m+1}^{m+s_m} \|\chi_{B_k}\| < 1/2^m,$$

so that $\chi_{D_n} = \mathbf{0}$; that is, $\mu(D_n) = 0$. Therefore $\mu(B) = 0$, contradicting (4).

2) Since $\mathbf{S}(T, \Sigma, \mu)$ is complete (Theorem I.6.15), it is sufficient to prove that, if $\{x_n\}$ is a Cauchy sequence in $\mathbf{X}$, then it is a Cauchy sequence in $\mathbf{S}(T, \Sigma, \mu)$. Assume the contrary. Then, if ρ is the metric in $\mathbf{S}$, there exists $\varepsilon > 0$ such that for each $n \in \mathbb{N}$ we can find m_n and k_n ($m_n, k_n > n$) for which we have $\rho(x_{m_n}, x_{k_n}) \geqslant \varepsilon$. Then $m_n, k_n \longrightarrow \infty$ as $n \to \infty$ and $\|x_{m_n} - x_{k_n}\| \longrightarrow 0$ as $n \to \infty$, so that, by 1), $\rho(x_{m_n}, x_{k_n}) \to 0$, which is the contradiction we require.

COROLLARY. *If a subset E of an* NIS $\mathbf{X}$ *is bounded in norm, then it is also bounded in the* TVS $\mathbf{S}$.

Proof. By Theorem 1, the inclusion mapping of $\mathbf{X}$ in $\mathbf{S}$ is continuous, and every continuous mapping takes bounded sets into bounded sets.

LEMMA 2. *Let* $\mathbf{X}$ *be a* BIS. *If a sequence $\{x_n\}$ converges to x in the norm of* $\mathbf{X}$, *then there exist a subsequence $\{x_{n_k}\}$, a function $r \in \mathbf{X}_+$, and a numerical sequence $\varepsilon_{n_k} \downarrow 0$, such that $|x_{n_k} - x| \leqslant \varepsilon_{n_k} r$.*

Proof. Choose a subsequence $\{x_{n_k}\}$ such that $\|x - x_{n_k}\| < 1/2^k$ ($k \in \mathbb{N}$). Since

$$\sum_{k=1}^{\infty} \|k(x - x_{n_k})\| < \sum_{k=1}^{\infty} k/2^k < \infty, \tag{7}$$

the series $\sum_{k=1}^{\infty} k|x - x_{n_k}|$ converges in the norm of the B-space $\mathbf{X}$ (see 1.5). Write

$$r = \sum_{k=1}^{\infty} k|x - x_{n_k}| \in \mathbf{X}, \qquad \varepsilon_{n_k} = 1/k.$$

By property 4) of convergence in norm in an NIS, we have $k|x - x_{n_k}| < r$ ($k \in \mathbb{N}$), and so $|x_{n_k} - x| \leqslant \varepsilon_{n_k} r$.

Observe that Lemma 1 yields a simpler proof of Theorem 1, in the case where $\mathbf{X}$ is a BIS.

THEOREM 2. *Any two monotone norms $\|\cdot\|_1$ and $\|\cdot\|_2$ turning an* IS $\mathbf{X}$ *into a* BIS *are equivalent.*

Proof. As we remarked in 1.3, it is sufficient to prove that the identity operator $(\mathbf{X}, \|\cdot\|_1) \to (\mathbf{X}, \|\cdot\|_2)$ is an isomorphism. Hence, in view of the symmetry between $(\mathbf{X}, \|\cdot\|_1)$ and $(\mathbf{X}, \|\cdot\|_2)$, it is sufficient to prove that $\|x_n\|_1 \to 0$ implies $\|x_n\|_2 \to 0$. Assume that $\|x_n\|_1 \to 0$ but $\|x_n\|_2 \not\to 0$. Then, passing to a subsequence if necessary, we may assume that $\|x_n\|_2 \geqslant \varepsilon > 0$ ($n \in \mathbb{N}$). By Lemma 2, there exist $\{x_{n_k}\}$, $r \in \mathbf{X}_+$, and $\varepsilon_{n_k} \uparrow 0$ such that $|x_{n_k}| \leqslant \varepsilon_{n_k} r$. Since

$\|\cdot\|_2$ is monotone, it follows from this that $\|x_{n_k}\|_2 \leqslant \varepsilon_{n_k}\|r\|_2 \to 0$. The resulting contradiction proves the theorem.

3.2. We now introduce some important properties that the norm of an NIS **X** may possess.

An NIS **X** is said to have an *order continuous** norm, or to satisfy condition (A), if

$$x_n \downarrow \mathbf{0} \text{ implies } \|x_n\| \to 0.$$

LEMMA 3. *If* **X** *is an* NIS *satisfying condition* (A), *then the set*

$$E = \left\{ y = \sum_{k=1}^{n} \lambda_k \chi_{A_k} : A_k \in \Sigma(\mu), \quad A_k \cap A_{k'} = \varnothing \quad (k \neq k'),\ \chi_{A_k} \in \mathbf{X},\ m \in \mathbb{N} \right\}$$

is dense in **X** *with respect to the norm.*

Proof. It is clearly sufficient to prove that every function $x \in \mathbf{X}_+$ belongs to the closure of E. Since the truncated functions $[x]_n$ satisfy $[x]_n \uparrow x$ (see I.6.4) and **X** satisfies condition (A), $[x]_n \to x$ in norm in **X**. Therefore we need only consider bounded functions x. Since μ is σ-finite, $T = \bigcup_{n=1}^{\infty} T_n$, where $T_n \subset T_{n+1}$ $(n \in \mathbb{N})$ and $\mu(T_n) < \infty$. But, by condition (A) again, $x\chi_{T_n} \to x$ in norm. Therefore it is sufficient to consider the case where x is bounded and μ is finite. By Theorem I.6.3, there exist finite-valued functions x_n such that $\mathbf{0} \leqslant x_n \uparrow x$, and thus $x_n \to x$ in norm. Clearly $x_n \in E$ $(n \in \mathbb{N})$.

In verifying that a specific space is separable, the following theorem often proves useful.

THEOREM 3. *A* BFS **X** *on* (T, Σ, μ) *is separable if and only if* μ *is separable and* **X** *satisfies condition* (A).

Proof. By Lemma 1, the NFS **X** is dense in the metric space **S**. Hence if **X** is separable then so is **S**, and thus by Theorem I.6.16 the measure μ is separable. Condition (A) will be established below in a more general situation (see the Corollary to Theorem X.4.4).

In view of Lemma 3, the converse can be proved by repeating the proof of Theorem I.6.16 word for word.

An NIS **X** is said to have an *order semicontinuous* norm, or to satisfy condition (C), if

$$0 \leqslant x_n \uparrow x \in \mathbf{X} \text{ implies } \|x_n\| \to \|x\|.$$

Condition (A) obviously implies condition (C).

LEMMA 4. *An* NIS **X** *satisfies condition* (C) *if and only if* $x_n \to x\,(\mu), x_n, x \in \mathbf{X}$, *implies* $\|x\| \leqslant \underline{\lim}\, \|x_n\|$.

Proof. Suppose **X** satisfies condition (C). By Theorem I.6.4, we may assume that $x_n \to x$ a.e., and thus $|x_n| \to |x|$ a.e. Write $y_n = \inf\{|x_m| : m \geqslant n\}$ (this infimum exists by the corollary to Theorem I.6.17). Clearly $\mathbf{0} \leqslant y_n \uparrow |x|$. Thus

$$\|x\| = \lim_{n \to \infty} \|y_n\| \leqslant \underline{\lim}\, \|x_n\|,$$

since $\mathbf{0} \leqslant y_n \leqslant |x_n|$.

Let us prove the converse. Let $\mathbf{0} \leqslant x_n \uparrow x \in \mathbf{X}$. Then $x_n \to x\,(\mu)$ and by hypothesis $\|x\| \leqslant \underline{\lim}\, \|x_n\| = \lim \|x_n\|$. Since $\|x\| \geqslant \|x_n\|$ $(n \in \mathbb{N})$, by the monotonicity of the norm, we have $\|x_n\| \to \|x\|$.

* Here, and in what follows, we shall often replace the word "order" by (*o*), for brevity. For example, in this case we should speak of an (*o*)-continuous norm.

An NIS $\mathbf{X}$ is said to have a *monotone complete* norm, or to satisfy condition (B), if

$$\mathbf{0} \leqslant x_n \uparrow, x_n \in \mathbf{X}\ (n \in \mathbb{N}),\ \sup \| x_n \| < \infty$$

imply $x_n \uparrow x$, for some $x \in \mathbf{X}$.

LEMMA 5. *Let* $\mathbf{X}$ *be an* NIS *on* (T, Σ, μ). *The following statements are equivalent:*

1) $\mathbf{X}$ *satisfies conditions* (B) *and* (C);

2) *the unit ball* $B_{\mathbf{X}} = \{x \in \mathbf{X}: \|x\| \leqslant 1\}$ *is closed in* $\mathbf{S}(T, \Sigma, \mu)$; *that is, if* $x_n \in \mathbf{X}$, $x \in \mathbf{S}$, $\|x_n\| \leqslant 1$ $(n \in \mathbb{N})$, $x_n \to x(\mu)$, *then* $x \in \mathbf{X}$ *and* $\|x\| \leqslant 1$.

Proof. 1) $\Rightarrow$ 2). Assume the sequence $\{x_n\}$ satisfies condition 2). As in the proof of Lemma 4, we may assume that $x_n \to x$ a.e. If we set $y_n = \inf \{ |x_m| : m \geqslant n\} \in \mathbf{X}$, then $\mathbf{0} \leqslant y_n \uparrow |x|$, and $\sup \|y_n\| \leqslant \sup \| x_n \| \leqslant 1$, so that, by (B), $x \in \mathbf{X}$. By Lemma 4, we now have $\| x \| \leqslant 1$.

2) $\Rightarrow$ 1). We have to prove that, if $\mathbf{0} \leqslant x_n \uparrow$, $x_n \in \mathbf{X}$ $(n \in \mathbb{N})$, $\sup \| x_n \| < \infty$, then there exists $x \in \mathbf{X}$ such that $x_n \uparrow x$ and $\| x_n \| \to \| x \|$.

We may assume that $\sup \| x_n \| = 1$. Then $x_n \in B_{\mathbf{X}}$. By the corollary to Theorem 1, the set $B_{\mathbf{X}}$ is bounded in the TVS $\mathbf{S}$. Thus, by Lemma III.1.1 (see III.1.4) there exists an element $x \in \mathbf{S}$ such that $x_n \uparrow x$, and so, by condition 2), $x \in \mathbf{X}$ and $\|x\| \leqslant 1$; and since $\|x\| \leqslant \sup \| x_n \| = 1$, we have $\| x \| = 1$, as we set out to prove.

THEOREM 4. *An* NIS $\mathbf{X}$ *that satisfies conditions* (B) *and* (C) *is complete in its norm.*

Proof. Let $\{x_n\}$ be a Cauchy sequence in $\mathbf{X}$. Since every Cauchy sequence is bounded, we may assume that $\| x_n \| \leqslant 1$ $(n \in \mathbb{N})$. By Theorem 1, $\{x_n\}$ is a Cauchy sequence in $\mathbf{S}$ so, since $\mathbf{S}$ is complete (Theorem I.6.15), there exists $x \in \mathbf{S}$ such that $x_n \to x\ (\mu)$. By Lemma 5, $x \in \mathbf{X}$. We show that $x_n \to x$ in norm.

Since $\{x_n\}$ is a Cauchy sequence, for each $\varepsilon > 0$ there is a number N such that, whenever $n, m \geqslant N$, we have

$$\|x_n - x_m\| < \varepsilon. \tag{8}$$

Fix any $n \geqslant N$ and note that $y_n = x_n - x_m \to x_n - x(\mu)$. In view of Lemma 4 and the bound given by (8), we have $\| x_n - x \| \leqslant \underline{\lim}_{m \to \infty} \| x_n - x_m \| \leqslant \varepsilon$.

Therefore $x_n \to x$ in norm, which shows that $\mathbf{X}$ is complete.

As we shall see below (see Chapter X), condition (C) is superfluous in Theorem 4. Let us formulate a very similar completeness criterion for NISs (see Abramovich [1]):

An NIS *is complete in its norm if and only if* $x_n \in \mathbf{X}_+$, $x_n \wedge x_m = \mathbf{0}$ $(n \neq m)$ *and* $\sum_{n=1}^{\infty} \| x_n \| < +\infty$ *imply that* $\sum_{n=1}^{\infty} x_n \in \mathbf{X}$ (*the sum here is to be understood in the sense of almost everywhere convergence*).

The usefulness of this criterion consists in the fact that $\sum_{n=1}^{\infty} x_n$ always exists as an element of $\mathbf{S}$, and it is sufficient to show that this element belongs to $\mathbf{X}$.

The theory of ideal spaces is part of the general theory of vector lattices, which we shall consider in more detail in Chapter X.* We mention that, historically, the theory of vector

* There we shall give also references to the literature; here we refer only to the survey by Zabreiko [1] and the book Zaanen-II, where the study of ISs is not based on the techniques of the theory of vector lattices. There are some interesting results on BFSs in a paper by Lozanovskii [2].

lattices was constructed earlier, and many fundamental facts from the theory of ISs were originally obtained as applications of general theorems concerning vector lattices.

In subsequent chapters we shall also study functionals and operators on ISs. But now we turn to a study of examples of concrete Banach ideal spaces.

3.3. The $\mathbf{L}^p$ spaces, which we shall now consider, play a very important role in functional analysis and in the problems investigated in this book.

As before, we let (T, Σ, μ) be a σ-finite measure space, though in fact for the study of $\mathbf{L}^p$ spaces the requirement of σ-finiteness is generally unnecessary. Fix a number p with $1 \leqslant p < \infty$. Denote by $\mathbf{L}^p(T, \Sigma, \mu)$ the space of all functions $x \in \mathbf{S}(T, \Sigma, \mu)$ for which the expression

$$\|x\| = \left[\int_T |x(t)|^p \, d\mu \right]^{1/p} \tag{9}$$

is finite.

Using Minkowski's inequality for integrals we see that $\mathbf{L}^p(T, \Sigma, \mu)$ is a vector space and that the functional defined by (9) is a norm. Hence it is clear that $\mathbf{L}^p(T, \Sigma, \mu)$ is an NIS.

Convergence in $\mathbf{L}^p(T, \Sigma, \mu)$ is called *convergence in mean of order* p:

$$\int_T |x_n(t) - x(t)|^p \, d\mu \to 0.$$

The theorems of Lebesgue and Beppo Levi (II.6.6) show that conditions (A) and (B) are satisfied in the NIS $\mathbf{L}^p(T, \Sigma, \mu)$, so that, by Theorem 4, $\mathbf{L}^p(T, \Sigma, \mu)$ is a B-space. Since $\chi_A \in \mathbf{L}^p(T, \Sigma, \mu)$ if $A \in \Sigma(\mu)$, the local finiteness of the measure (I.6.2) implies that $\mathbf{L}^p(T, \Sigma, \mu)$ is an FS. All this is consolidated in the following theorem.

THEOREM 5. *For* $1 \leqslant p < \infty$, *the space* $\mathbf{L}^p(T, \Sigma, \mu)$ *is a* BFS *satisfying conditions* (A) *and* (B).

Now we introduce the $\mathbf{L}^p$ space corresponding to $p = \infty$. The space $\mathbf{L}^\infty(T, \Sigma, \mu)$ consists of all functions $x \in \mathbf{S}(T, \Sigma, \mu)$ for which there exists a number α_x such that $|x(t)| \leqslant \alpha_x$ a.e.* For a function $x \in \mathbf{L}^\infty(T, \Sigma, \mu)$ we define the essential supremum $\operatorname{vrai\,sup}_{t \in T} |x(t)|$ of x to be the infimum of the set of numbers $\alpha \in \mathbb{R}$ such that

$$\mu(\{t \in T \colon |x(t)| > \alpha\}) = 0.$$

Clearly $\mathbf{L}^\infty(T, \Sigma, \mu)$ is a linear manifold in $\mathbf{S}$. A norm on $\mathbf{L}^\infty(T, \Sigma, \mu)$ is introduced by the formula

$$\|x\| = \operatorname*{vrai\,sup}_{t \in T} |x(t)|.$$

We leave to the reader the simple verification that $\mathbf{L}^\infty(T, \Sigma, \mu)$ is a BFS satisfying conditions (B) and (C). Condition (A) is almost never satisfied in $\mathbf{L}^\infty(T, \Sigma, \mu)$. The exception is the degenerate case where $\mathbf{L}^\infty(T, \Sigma, \mu)$ is finite-dimensional, which is so if and only if (T, Σ, μ) consists of a finite number of atoms. For suppose $\mathbf{L}^\infty(T, \Sigma, \mu)$ is infinite-dimensional. Then there exists a sequence $\{A_n\} \subset \Sigma(\mu)$ of pairwise disjoint sets of positive

* Such functions are said to be *essentially bounded*.

measure. Consider the sets

$$B_0 = \bigcup_{n=1}^{\infty} A_n, \qquad B_1 = B_0 \setminus A_1 \ldots, \qquad B_k = B_{k-1} \setminus A_k, \ldots .$$

If $x_n = \chi_{B_n}$, then $x_n \in \mathbf{L}^\infty(T, \Sigma, \mu)$ and $x_n \downarrow \mathbf{0}$. However, obviously $\|x_n\| = 1$ $(n \in \mathbb{N})$, so $\mathbf{L}^\infty(T, \Sigma, \mu)$ does not satisfy condition (A).

In future, if it is clear what measure μ we are dealing with, or if, alternatively, this has no significance, we shall write $\mathbf{L}^p$ instead of $\mathbf{L}^p(T, \Sigma, \mu)$. When the measure space is a measurable subset $D \subset \mathbf{R}^n$ with Lebesgue measure (mes), then instead of $\mathbf{L}^p(T, \Sigma, \mu)$ we write $\mathbf{L}^p(D)$; when (T, Σ, μ) is an interval $[a, b]$ with Lebesgue measure we write $\mathbf{L}^p(a, b)$. We shall also keep to similar conventions when studying other concrete spaces.

The results of I.6.10 and Theorems 3 and 5 imply that $\mathbf{L}^p(D)$ is separable for $1 \leqslant p < \infty$. Since $\mathbf{L}^\infty(D)$ does not satisfy condition (A), it is not separable. As a countable dense subset in $\mathbf{L}^p(a, b)$ we may take, for example, the set of polynomials with rational coefficients or the set of step functions taking rational values on intervals with rational endpoints.

To prove this, note first of all that, if $[x]_n$ is the "truncation" of the function x (see I.6.4), then

$$\int_a^b |x(t) - [x]_n(t)|^p \, dt \underset{n \to \infty}{\longrightarrow} 0,$$

since the integrand tends to zero and we may take the limit under the integrand sign because the integral is bounded by the summable function $|x(t)|^p$.

Choose n so that $\|x - [x]_n\| \leqslant \varepsilon$. Then, if we choose a continuous function $y(t)$ which coincides with $[x]_n(t)$ except on a set A with $\operatorname{mes} A < \varepsilon^p/(2n)^p$ and also does not exceed n (see I.6.9), we have

$$\|x_n - y\| = \left\{ \int_T |x_n(t) - y(t)|^p \, dt \right\}^{1/p} \leqslant \left\{ \int_T (2n)^p \, dt \right\}^{1/p} = 2n[\operatorname{mes} A]^{1/p} < \varepsilon.$$

Finally, choose a polynomial $P(t)$ with rational coefficients such that $|y(t) - P(t)| < \dfrac{\varepsilon}{b - a}$ $(t \in [a, b])$. It is then clear that $\|y - P\| < \varepsilon$ and finally $\|x - P\| < 3\varepsilon$.

To obtain a countable dense subset of $\mathbf{L}^p(-\infty, \infty)$ we may take, for instance, the set of step functions on rational intervals taking only rational values only a finite number of which are different from zero.

We now investigate the question of how the $\mathbf{L}^p$ spaces, for various exponents p, are related to one another.

THEOREM 6. *Let μ be a finite measure and suppose $1 \leqslant s < r \leqslant \infty$. Then $\mathbf{L}^r(T, \Sigma, \mu) \subset \mathbf{L}^s(T, \Sigma, \mu)$. Moreover, if*

$$K(r, s) = \begin{cases} \mu(T)^{(r-s)/(rs)} & \text{for } r < \infty, \\ \mu(T)^{1/s} & \text{for } r = \infty, \end{cases}$$

then

$$\|x\|_{\mathbf{L}^s} \leqslant K(r, s) \|x\|_{\mathbf{L}^r}.$$

Proof. We leave it to the reader to investigate the case $r = \infty$. Assume that $r < \infty$. Then, writing $p = r/s$, $q = r/(r - s)$, $(p, q > 1, 1/p + 1/q = 1)$, we deduce from Hölder's inequality for integrals that

$$\|x\|_{\mathbf{L}^s} = \left[\int_T |x(t)|^s \, d\mu\right]^{1/s} \leqslant \left[\int_T |x(t)|^{ps} \, d\mu\right]^{1/(ps)} \left[\int_T 1^q \, d\mu\right]^{1/(qs)} = \|x\|_{\mathbf{L}^r} \mu(T)^{1/(qs)}$$

$$= K(r, s)\|x\|_{\mathbf{L}^r}.$$

Notice that $\|x\|_{\mathbf{L}^\infty} = \lim_{p \to \infty} \|x\|_{\mathbf{L}^p}$ (Hardy, Littlewood and Pólya); however, $\mathbf{L}^\infty \neq \bigcap_{1 \leqslant p < \infty} \mathbf{L}^p$. Theorem 6 is false in the case of an infinite measure.

The $\mathbf{L}^p$ spaces $(1 < p < \infty)$ were introduced by F. Riesz [1]. The space $\mathbf{L}^1$ was introduced by Steinhaus [1].

3.4. Now we consider some important special types of spaces of measurable functions—spaces of sequences. As we have already remarked, if $T = \mathbb{N}$, if Σ is the collection of all subsets of $\mathbb{N}$, and if the measure μ of each point is unity, then $\mathbf{S}(T, \Sigma, \mu)$ is the space of all sequences $\mathbf{s}$. All the definitions and results of the preceding paragraphs are therefore applicable to subspaces of $\mathbf{s}$.

In this case the space $\mathbf{L}^p(T, \Sigma, \mu)$ is denoted by ℓ^p. If $x = \{\xi_k\}_{k=1}^\infty$, then by the definition of the norm in $\mathbf{L}^p(T, \Sigma, \mu)$, we have

$$\|x\|_{\ell^p} = \begin{cases} \left[\sum_{k=1}^\infty |\xi_k|^p\right]^{1/p} & \text{if } 1 \leqslant p < \infty, \\ \sup_{k=1}^{\infty} |\xi_k| & \text{if } p = \infty. \end{cases}$$

The results of 3.3 imply that $\ell^p (1 \leqslant p < \infty)$ is a separable Banach space, while ℓ^∞ is an inseparable Banach space.

We introduce two more spaces of sequences, which are subspaces of ℓ^∞. Denote by $\mathbf{c}$ the space of all convergent sequences, and by $\mathbf{c}_0$ the space of all sequences converging to zero. We equip $\mathbf{c}$ and $\mathbf{c}_0$ with the norms induced from ℓ^∞. Clearly $\mathbf{c}$ and $\mathbf{c}_0$ are linear manifolds in ℓ^∞. We show they are closed, and hence are Banach spaces. We present only the argument for $\mathbf{c}$.

If $\{x_n\}$ is a sequence of elements of $\mathbf{c}$ converging to $x_0 \in \ell^\infty$, then $\|x_n - x_0\| \to 0$. If we write $x_n = \{\xi_k^{(n)}\}_{k=1}^\infty$, then for each $\varepsilon > 0$ we have, for $n \geqslant N_\varepsilon$:

$$\|x_n - x_0\| = \sup_k |\xi_k^{(n)} - \xi_k^{(0)}| < \varepsilon/3,$$

which we can also write as:

$$|\xi_k^{(n)} - \xi_k^{(0)}| < \varepsilon/3 \qquad (k = 1, 2, \ldots; n \geqslant N_\varepsilon).$$

Choose a fixed $n \geqslant N_\varepsilon$. Then $x_n \in \mathbf{c}$, so the sequence $\xi_1^{(n)}, \xi_2^{(n)}, \ldots, \xi_k^{(n)}, \ldots$ converges. Therefore, for sufficiently large k and k', we have $|\xi_k^{(n)} - \xi_{k'}^{(n)}| < \varepsilon/3$. Hence, for such k and k', we have

$$|\xi_k^{(0)} - \xi_{k'}^{(0)}| \leqslant \xi_k^{(0)} - \xi_k^{(n)}| + |\xi_k^{(n)} - \xi_{k'}^{(n)}| + |\xi_{k'}^{(n)} - \xi_{k'}^{(0)}| < \frac{\varepsilon}{3} + \frac{\varepsilon}{3} + \frac{\varepsilon}{3} = \varepsilon.$$

Thus the sequence $\xi_1^{(0)}, \xi_2^{(0)}, \ldots, \xi_k^{(0)}, \ldots$ converges; that is, $x_0 \in \mathbf{c}$.

Notice that $\mathbf{c}_0$ is a BFS satisfying condition (A) (and so a separable B-space) but not satisfying condition (B). Thus condition (B) is not a necessary condition for completeness of an NIS. The space **c** is not an IS.

Notice that for the spaces $\ell^p (1 \leqslant p \leqslant \infty)$ we have the opposite inclusions and inequalities for norms (with $K(r, s) = 1$) to those formulated in Theorem 6.

Finally, we mention the example of the space $\boldsymbol{\phi}$ of terminating sequences. This consists of all sequences $(\xi_k)_{k=1}^{\infty}$ in which not more than a finite number of coordinates are different from zero. The norm in $\boldsymbol{\phi}$ is induced from ℓ^∞. Clearly $\boldsymbol{\phi}$ is an NFS, but it is not a *B*-space since it is dense in $\mathbf{c}_0$.

The spaces ℓ^p were introduced and studied by F. Riesz [3], while ℓ^2 was introduced earlier by D. Hilbert (see Hilbert).

3.5. We now draw attention to the finite-dimensional spaces ℓ_n^p (of dimension n) connected with the spaces of sequences ℓ^p just introduced. Since any two finite-dimensional spaces of the same dimension are isomorphic, the difference between these must consist only in the method of defining the norm. In fact, the norm in ℓ^p is defined as follows: if $x = \{\xi_k\}_{k=1}^n$, then

$$\|x\|_{\ell_n^p} = \begin{cases} \left[\sum_{k=1}^{n} |\xi_k|^p\right]^{1/p} & \text{if } 1 \leqslant p < \infty, \\ \max_{k=1}^{n} |\xi_k| & \text{if } p = \infty. \end{cases}$$

(For $p = 2$, we have the Euclidean norm.)

As Corollary 1 to Theorem 1.2 shows, despite the difference in the norms of these spaces, convergence has exactly the same meaning in all of them, namely coordinatewise convergence. (This can also be seen immediately from the inequality $|\xi_k| \leqslant \|x\|$, which holds in each of these finite-dimensional spaces.)

The space ℓ_n^p may be regarded as a subspace of ℓ^p. For this purpose we identify an element $(\xi_1, \xi_2, \ldots, \xi_n) \in \ell_n^p$ with the element $(\xi_1, \xi_2, \ldots, 0, 0, \ldots) \in \ell^p$.

3.6. In addition to the $\mathbf{L}^p$ spaces, many other spaces of measurable functions are studied in functional analysis.

We begin with definition of the Orlicz spaces, which are generalizations of the $\mathbf{L}^p$ spaces. A continuous, convex, even function $M(u)$, defined on $(-\infty, \infty)$ and positive when $u \neq 0$, is called an *N-function* if

$$\lim_{u \to 0} \frac{M(u)}{u} = 0 \text{ and } \lim_{u \to +\infty} \frac{M(u)}{u} = +\infty.$$

For each N-function, the equation

$$M^*(u) = \sup_{-\infty < v < \infty} (uv - M(v))$$

defines the *complementary* N-function. Obviously, an N-function M is monotonically increasing on $[0, +\infty)$ and $M(0) = 0$. It can be shown that M^* is an N-function and that $M^{**} = M$.

Let (T, Σ, μ) be a finite measure space. Fix an N-function M. The *Orlicz class* $\mathring{\mathbf{L}}_M(T, \Sigma, \mu)$ (or simply $\mathring{\mathbf{L}}_M$) is the collection of all functions $x \in \mathbf{S}(T, \Sigma, \mu)$ such that

$$\int_T M(|x(t)|)\, d\mu < +\infty.$$

An Orlicz class may turn out to be non-linear; more precisely, it can happen that, for $x \in \mathbf{L}_M^\circ$, we have $2x \notin \mathbf{L}_M^\circ$. Obviously $\mathbf{L}^\infty \subset \mathbf{L}_M^\circ$.

The *Orlicz space* $\mathbf{L}_M(T, \Sigma, \mu)$ (or simply $\mathbf{L}_M$) is the collection of all functions $x \in \mathbf{S}(T, \Sigma, \mu)$ for which there exists a number $\lambda = \lambda(x) > 0$ such that

$$\int_T M(|x(t)|/\lambda)\, d\mu < \infty.$$

Clearly, $\mathbf{L}_M^\circ \subset \mathbf{L}_M$. These sets may turn out to be distinct, as we shall see below. The space $\mathbf{L}_M$ is a linear manifold in $\mathbf{S}$. For multiplication by a number clearly does not lead out of $\mathbf{L}_M$. If

$$\int_T M(|x(t)|/\lambda_1)\, d\mu < \infty \qquad \text{and} \qquad \int_T M(|y(t)|/\lambda_2)\, d\mu < \infty$$

then, as M is convex, we have

$$\int_T M\left(\frac{|x(t)+y(t)|}{\lambda_1+\lambda_2}\right) d\mu \leqslant \frac{\lambda_1}{\lambda_1+\lambda_2} \int_T M\left(\frac{|x(t)|}{\lambda_1}\right) d\mu + \frac{\lambda_1}{\lambda_1+\lambda_2} \int_T M\left(\frac{|y(t)|}{\lambda_2}\right) d\mu, \tag{10}$$

so that $x + y \in \mathbf{L}_M$. Since M is monotone, it is clear that $\mathbf{L}_M$ is an FS on (T, Σ, μ). We shall consider the following two norms on an Orlicz space:

$$\|x\|_1 = \sup\left\{ \int_T |xy|\, d\mu: \quad \int_T M^*(y) d\mu \leqslant 1 \right\},$$

$$\|x\|_2 = \inf\left\{ \lambda > 0: \quad \int_T M(|x(t)|/\lambda) d\mu \leqslant 1 \right\}.$$

Let us show that $\|\,.\,\|_1$ and $\|\,.\,\|_2$ are norms. From the definition of M^*, we deduce *Young's inequality*:

$$|uv| \leqslant M(u) + M^*(v).$$

Suppose $\int_T M(|x(t)|/\lambda) d\mu < \infty$. Then, by Young's inequality,

$$\int_T |xy|\, d\mu = \lambda \int_T |x/\lambda|\,|y|\, d\mu \leqslant \lambda \int_T M(|x|/\lambda)\, d\mu + \lambda \int_T M^*(|y|)\, d\mu. \tag{11}$$

It follows from (11) that $\|x\|_1 < \infty$. Since $\mathbf{L}^\infty \subset \mathbf{L}_{M^*}^\circ$, it follows from $\|x\|_1 = 0$ that $x = \mathbf{0}$. The remaining properties of a norm evidently hold for $\|.\|_1$.

It is clear from the definition that $\|x\|_2 < \infty$. The homogeneity is obvious; the triangle inequality is easily obtained from (10). If $\|x\|_2 = 0$, then $x = \mathbf{0}$ because $M(u) > 0$ whenever $u > 0$. Thus $\|\,.\,\|_2$ is a norm. We conclude from (11) that $\|x\| \leqslant 2\|x\|_2$. It can be proved that $\|x\|_2 \leqslant \|x\|_1$. However, we shall not do this but prove only that there exists a constant $k_1 > 0$ such that $\|x\|_1 \geqslant k_1 \|x\|_2 (x \in \mathbf{L}_M)$—that is, that the norms $\|\,.\,\|_1$ and $\|\,.\,\|_2$

are equivalent. We shall deduce the latter result from the fact that $\|.\|_1$ and $\|x\|_2$ are monotone norms turning $\mathbf{L}_M$ into a BFS (see Theorem 2).

THEOREM 7. *The Orlicz space* $\mathbf{L}_M$, *with either norm*, $\|.\|_1$ *or* $\|.\|_2$, *is a* BFS *satisfying conditions* (B) *and* (C).

Proof. Since an NFS satisfying conditions (B) and (C) is complete (Theorem 4), it is sufficient to prove that $(\mathbf{L}_M, \|.\|_i)$ satisfies conditions (B) and (C); that is, to prove that the conditions $\mathbf{0} \leqslant x_n \uparrow$, $x_n \in \mathbf{L}_M$, $\sup \|x_n\|_i < \infty$ imply that there exists $x \in \mathbf{L}_M$ such that $x_n \uparrow x$ and $\sup\limits_{n=1}^{\infty} \|x_n\|_i = \|x\|_i$ $(i = 1, 2)$.

First suppose $i = 1$. Write $x(t) = \sup\limits_n x_n(t)$. *A priori*, x could take infinite values on a set of positive measure; however, by the corollary to Theorem I.6.7, for any y satisfying $\int\limits_T M^*(y)d\mu \leqslant 1$ we have

$$\int\limits_T x|y|\,d\mu = \sup_n \int\limits_T x_n|y|\,d\mu \leqslant \sup \|x_n\|_1. \tag{12}$$

The inequality (12) shows that $(\mathbf{L}_M, \|.\|_1)$ satisfies conditions (B) and (C).

Let us consider the case $i = 2$. As $\|x\|_1 \leqslant 2\|x\|_2$, the space $(\mathbf{L}_M, \|.\|_2)$ satisfies condition (B). It remains to verify that whenever $\mathbf{0} \leqslant x_n \uparrow x \in \mathbf{L}_M$, we have $\|x_n\|_2 \to \|x\|_2$.

First we note the following property of $\|.\|_2$: for every $x \neq \mathbf{0}$ we have

$$\int\limits_T M(|x(t)|/\|x\|_2)\,d\mu \leqslant 1. \tag{13}$$

For choose $\lambda_n \to \|x\|_2$ $(\lambda_n \neq 0)$, where

$$\int\limits_T M(|x(t)|/\lambda_n)\,d\mu \leqslant 1. \tag{14}$$

Passing to the limit in (14), we obtain (13), by Fatou's Theorem.

Write $\lim \|x_n\|_2 = \lambda$. By (13),

$$\int\limits_T M(|x_n(t)|/\|x_n\|_2)\,d\mu \leqslant 1. \tag{15}$$

Passing to the limit in (15), we have, by Fatou's Theorem again,

$$\int\limits_T M(|x(t)|/\lambda)\,d\mu \leqslant 1.$$

Hence $\|x\|_2 \leqslant \lambda = \lim \|x_n\|_2 \leqslant \|x\|_2$, which completes the proof of the theorem.

Let us now investigate the question of when an Orlicz space satisfies condition (A). Since the norms $\|.\|_1$ and $\|.\|_2$ are equivalent, either they both satisfy condition (A) or neither does.

As we have already remarked, we always have $\mathbf{L}^\infty \subset \mathring{\mathbf{L}}_M \subset \mathbf{L}_M$. Denote the closure of $\mathbf{L}^\infty$ in the Orlicz space $\mathbf{L}_M$ by $\mathbf{E}_M$. We shall also regard $\|.\|_1$ and $\|.\|_2$ as norms on $\mathbf{E}_M$.

LEMMA 6. 1) *If* $\{x_n\} \subset \mathbf{L}_M$ *and* $x_n \to x$ *in norm in* $\mathbf{L}_M$, *then the sequence* $\{x_n\}$ *"converges in mean" to* x: *that is,**

$$\int\limits_T M(|x_n(t) - x(t)|)\,d\mu \xrightarrow[n\to\infty]{} 0.$$

* A finite number of these integrals may be infinite.

2) *If, under the hypotheses of* 1), $2x_n \in \mathbf{L}_M^\circ$ $(n \in \mathbb{N})$, *then* $x \in \mathbf{L}_M^\circ$.

3) $\mathbf{E}_M \subset \mathbf{L}_M^\circ$.

Proof. 1) We note first that it follows from the equation $M(0) = 0$ and the convexity of M that

$$M(\alpha u) \leqslant \alpha M(u) \qquad \text{for} \qquad 0 \leqslant \alpha \leqslant 1. \tag{16}$$

Since $\| x_n - x \|_2 \to 0$, we have $\| x_n - x \|_2 \leqslant 1$ when $n \geqslant N$. For such values of n, (16) yields (writing $\alpha = \| x_n - x \|_2$)

$$1 \geqslant \int_T M\left(\frac{x_n - x}{\|x_n - x\|_2}\right) d\mu \geqslant \frac{1}{\|x_n - x\|_2} \int_T M(|x_n - x|)\, d\mu, \tag{17}$$

so that

$$\int_T M(|x_n - x|)\, d\mu \leqslant \|x_n - x\|_2 \to 0.$$

2) This follows from the inequality

$$\int_T M(|x|)\, d\mu \leqslant \frac{1}{2} \int_T M(|2x - 2x_n|)\, d\mu + \frac{1}{2} \int_T M(|2x_n|)\, d\mu.$$

3) If $x \in E_M$, then there exists $\{x_n\} \subset \mathbf{L}^\infty$ such that $x_n \to x$ in norm. Since $2x_n \in \mathbf{L}^\infty \subset \mathbf{L}_M^\circ$, part 2) now yields $x \in \mathbf{L}_M^\circ$.

THEOREM 8. *The space* $\mathbf{E}_M$ *is a* BFS *satisfying condition* (A).

Proof. We need only verify that $\mathbf{E}_M$ satisfies condition (A). Assume that this is not the case.

Then there exists a sequence $\{x_n\} \subset \mathbf{E}_M$ such that $x_n \downarrow \mathbf{0}$, $\| x_n \|_2 > \delta > 0$ $(n \in \mathbb{N})$. By (13) we have

$$\int_T M(|x_n|/\delta)\, d\mu > 1$$

for every $n \in \mathbb{N}$. On the other hand, $x_1/\delta \in \mathbf{E}_M \subset \mathbf{L}_M^\circ$, by Lemma 6. Thus, by Lebesgue's Theorem, we have

$$\int_T M(|x_n|/\delta)\, d\mu \xrightarrow[n \to \infty]{} 0,$$

which yields the required contradiction.

An N-function $M(u)$ is said to satisfy the Δ_2-*condition* if there exist constants $k > 0$, $u_0 \geqslant 0$ such that

$$M(2u) \leqslant kM(u) \qquad (u \geqslant u_0).$$

We assume in the following theorem that the space (T, Σ, μ) is continuous.*

* This assumption is used only in proving that $5) \Rightarrow 6)$.

THEOREM 9. *The following statements are equivalent*:

1) $\mathbf{L}_M$ *satisfies condition* (A);
2) $\mathbf{L}_M = \mathbf{E}_M$;
3) $\mathbf{E}_M$ *satisfies condition* (B);
4) $\mathbf{L}_M = \mathring{\mathbf{L}}_M$;
5) $\mathring{\mathbf{L}}_M$ *is a linear manifold*;
6) *the* N-*function* M *satisfies the* Δ_2-*condition*.

Proof. 1) $\Rightarrow$ 2). If $\mathbf{L}_M$ satisfies condition (A), then by Lemma 3 $\mathbf{L}^\infty$ is dense in $\mathbf{L}_M$, and thus $\mathbf{E}_M = \mathbf{L}_M$.

2) $\Rightarrow$ 3), since $\mathbf{L}_M$ satisfies (B) (Theorem 7).

3) $\Rightarrow$ 2) by Lemma 1.

2) $\Rightarrow$ 4) since $\mathbf{E}_M \subset \mathring{\mathbf{L}}_M$ by Lemma 6.

4) $\Rightarrow$ 5) since $\mathbf{L}_M$ is a linear manifold.

5) $\Rightarrow$ 4). For if $x \in \mathbf{L}_M$, $x \neq \mathbf{0}$, then $x/\|x\|_2 \in \mathring{\mathbf{L}}_M$ by (13). Thus $x = \|x\|_2(x/\|x\|_2) \in \mathring{\mathbf{L}}_M$.

4) $\Rightarrow$ 1) is proved exactly as Theorem 8.

6) $\Rightarrow$ 5) Let $m \in \mathbb{N}$. It is easy to see that the Δ_2-condition implies that, for $u \geqslant u_0$, we have the inequality

$$M(2^m u) \leqslant k^m M(u).$$

Now let $x \in \mathring{\mathbf{L}}_M$ and let λ be a scalar. Then there exists $m \in \mathbb{N}$ such that $|\lambda| \leqslant 2^m$. We deduce that

$$\int_T M(|\lambda x|)\, d\mu \leqslant k^m \int_T M(|x|)\, d\mu + \mu(T) M(2^m u_0) < \infty,$$

and so $\lambda x \in \mathring{\mathbf{L}}_M$.

5) $\Rightarrow$ 6). Assume the N-function M does not satisfy the Δ_2-condition. Then there exists a numerical sequence $\{u_n\}$ such that $0 < u_n \uparrow +\infty$, $M(u_1) > 1$ and

$$M(2u_n) > 2^n M(u_n) \qquad (n = 1, 2, \ldots). \tag{18}$$

By the continuity of (T, Σ, μ), there exist non-intersecting sets $A_n \in \Sigma$ such that

$$\mu(A_n) = \frac{\mu(T)}{2^n M(u_n)} \qquad (n = 1, 2, \ldots). \tag{19}$$

Define a function $x(t)$ by

$$x(t) = \begin{cases} u_n, & \text{if } x \in A_n \quad (n = 1, 2, \ldots), \\ 0, & \text{if } x \notin \bigcup\limits_{n=1}^{\infty} A_n. \end{cases}$$

By (19),

$$\int_T M(x(t))\, d\mu = \sum_{n=1}^{\infty} \int_{A_n} M(x(t))\, d\mu = \sum_{n=1}^{\infty} M(u_n)\mu(A_n) \leqslant \mu(T) < \infty,$$

so that $x \in \mathring{\mathbf{L}}_M$. On the other hand, by (18) and (19) we have

$$\int_{A_n} M(2x(t))\, d\mu = M(2u_n)\mu(A_n) > 2^n M(u_n)\mu(A_n) = \mu(T),$$

so that

$$\int_T M(2x(t))\,d\mu \geqslant \sum_{n=1}^{\infty} \int_{A_n} M(2x(t))\,d\mu = +\infty.$$

Therefore $2x \notin \mathbf{L}^{\circ}_M$. This completes the proof of the theorem.

Let D be a domain in $\mathbf{R}^n$ having finite Lebesgue measure. By Theorems 3 and 9, we have

COROLLARY. *The Orlicz space $\mathbf{L}_M(D)$ is separable if and only if the N-function M satisfies the Δ_2-condition.*

Finally, we note that the $\mathbf{L}^p$ spaces $(1 < p < \infty)$ are contained in the class of Orlicz spaces. For set $M(u) = u^p/p$. Then the norm in the Orlicz space $\mathbf{L}_M$ is equivalent to the norm in $\mathbf{L}^p$. We show that

$$\|x\|_{\mathbf{L}^p} = p^{1/p}\|x\|_2. \tag{20}$$

If $\int_T M(|x|/\|x\|_2)\,d\mu \leqslant 1$, then $\int_T |x(t)|^p/\|x\|_2^p\,d\mu \leqslant p$, so that

$$\left(\int_T |x(t)|^p\,d\mu\right)^{1/p} \leqslant p^{1/p}\|x\|_2.$$

On the other hand,

$$\int_T M\left(\frac{|x(t)|}{p^{-1/p}\left(\int_T |x(t)|^p\,d\mu\right)^{1/p}}\right)d\mu = 1,$$

so that $\|x\|_{\mathbf{L}^p} \geqslant p^{1/p}\|x\|_2$. Comparing this with the inequality just proved, we obtain (20).

A more detailed study of Orlicz spaces and their applications to the solution of non-linear integral equations can be found in the book by Krasnosel'skii and Rutickii, whose exposition we have followed in certain places (see also Zaanen-I).

3.7. Let us briefly touch on two classes of spaces important in the theory of linear integral operators (see Birman *et al.*).

If $x \in \mathbf{S}(0, 1)$, then the function $x^*(\tau)$ on $[0, 1]$ defined by

$$x^*(\tau) = \inf\{\alpha \geqslant 0;\ \mathrm{mes}(\{t: |x(t)| > \alpha\}) \leqslant \tau\}$$

is called the equi-measurable rearrangement of x in non-increasing order.

Let $\psi(t)$ be a continuous concave function on $[0, 1]$, positive for $t \neq 0$, such that $t/\psi(t) \to 0$ as $t \to 0+$.

The *Marcinkiewicz space* $\mathbf{M}(\psi)$ is the Banach space consisting of all $x \in \mathbf{S}(0, 1)$ such that the following norm is finite:

$$\|x\| = \sup\left\{\frac{1}{\psi(\mathrm{mes}(A))}\int_A |x(t)|\,dt:\ A \subset [0, 1] \text{ is measurable and } \mathrm{mes}(A) > 0\right\} = \sup_{0<h<1}\left\{\frac{1}{\psi(h)}\int_0^h x^*(\tau)\,d\tau\right\}.$$

The space $\mathbf{M}(\psi)$ is a BFS satisfying conditions (B) and (C), but not condition (A).

The *Lorentz space* $\Lambda(\psi)$ is the Banach space consisting of all $x \in \mathbf{S}(0, 1)$ such that the norm

$$\|x\| = \int_0^1 x^*(\tau)\, d\psi(\tau)$$

is finite.

The space $\Lambda(\psi)$ is a BFS satisfying conditions (A) and (B).

The spaces $\mathbf{M}(\psi)$ and $\Lambda(\psi)$ are commonly encountered for $\psi(t) = t^\alpha$ $(0 < \alpha < 1)$. In this case one denotes them by $\mathbf{M}(\alpha)$ and $\Lambda(\alpha)$.

There is an important class of BFSs connected with the operation of forming the equimeasurable non-increasing function x^*, namely the class of symmetric spaces (see Birman *et al.*)—that is, those BFSs $\mathbf{X}$ on the interval $[0, 1]$ with Lebesgue measure for which $x \in \mathbf{X}$, $y \in \mathbf{S}(0, 1)$, $x^* = y^*$ imply that $y \in \mathbf{X}$ and $\|x\| = \|y\|$. The spaces $\mathbf{L}^p(0, 1)$, $\mathbf{L}_M(0, 1)$, $\mathbf{E}_M(0, 1)$, $\mathbf{M}(\psi)$ and $\Lambda(\psi)$ are symmetric spaces.

In conclusion, we remark that one can study classes of spaces of measurable functions with values in B-spaces, analogous to BISs. Such spaces have applications in the theory of differential equations, the analytic theory of semigroups, and linear integral operators (see, for example, Bourbaki-V; Dunford and Schwartz-I, II; Dinculeanu; Yosida; Hille and Phillips; Edwards).

§ 4. Other normed spaces of functions

4.1. Let us first look at the by now familiar B-space $\mathbf{C}(K)$ of continuous functions on a compactum K. We now formulate the Stone–Weierstrass Theorem on subsets of $\mathbf{C}(K)$ dense relative to the norm, which is important for applications.

In $\mathbf{C}(K)$ we define an operation of pointwise multiplication: $(xy)(t) = x(t)\,y(t)$. A linear manifold E in $\mathbf{C}(K)$ is called a *subalgebra* if $xy \in E$ whenever $x \in E$, $y \in E$. We write $\mathbf{1}$ for the function identically equal to unity on K.

THEOREM 1 (Stone–Weierstrass). *Let E be a subset of the real B-space* $\mathbf{C}(K)$. *Suppose the following conditions are satisfied:*

1) *E is a subalgebra;*

2) $\mathbf{1} \in E$;

3) *E separates points in K: that is, if* $t, t' \in K$ $(t \neq t')$, *then there exists a function* $x \in E$ *such that* $x(t) \neq x(t')$.

Then E is dense in $\mathbf{C}(K)$.

The classical result that the set of polynomials is dense in $\mathbf{C}[a, b]$ is easily derived from Theorem 1.

Theorem 1 is no longer valid if we consider a complex space $\mathbf{C}(K)$.

For let us take $K = \{z \in \mathbb{C}: |z| \leqslant 1\}$ to be the closed unit disc in the complex plane and consider the set E of complex polynomials on K: that is, $E = \mathscr{L}(1, z, z^2, \ldots)$. The set E satisfies the conditions of Theorem 1, but it is not dense in the complex $\mathbf{C}(K)$. For example, the function $x(z) = \operatorname{Re} z$ is not in $\overline{E}$, otherwise, by Weierstrass's Theorem, it would be analytic inside the disc. However, even in the complex case, the situation can easily be remedied.

THEOREM 2. *Let E be a subset of the complex B-space* $\mathbf{C}(K)$. *Suppose the conditions* 1)–3) *of Theorem* 1 *are satisfied, and in addition:*

4) *if* $x \in E$, *then* $\bar{x} \in E$, where $\bar{x}(t) = \overline{x(t)}$, *the bar denoting the complex conjugate.*

Then E is dense in $\mathbf{C}(K)$.

For proofs of Theorems 1 and 2 see, for example, Dunford and Schwartz-I; Schaefer-I.

LEMMA 1 (on partitions of unity). *If* $\{U_i\}_{i=1}^n$ *is a finite open cover of the compactum K, then there exists a system of continuous real-valued functions* $\{\phi_i(t)\}_{i=1}^n$ *on K satisfying the following conditions*:

1) $\phi_i(t) \geqslant 0 \quad (i = 1, 2, \ldots, n)$;
2) $\phi_i(t) = 0,\ t \notin U_i \quad (i = 1, 2, \ldots, n)$;
3) $\sum_{i=1}^{n} \phi_i(t) = 1$, *for every* $t \in K$.

Proof. We give this only for the case of a metric compactum K, having metric $r(t, s)$. Write $F_i = K \setminus U_i$ $(i = 1, 2, \ldots, n)$. For every $t \in K$, set

$$\phi_i(t) = \frac{r(t, F_i)}{\sum_{k=1}^{n} r(t, F_k)}.$$

Since $r(t, F_i) = 0$ is equivalent to $t \in F_i$, the system $\{\phi_i(t)\}_{i=1}^{n}$ satisfies 1)–3).

THEOREM 3. *If K is a metric compactum, then the B-space* $\mathbf{C}(K)$ *is separable.*

REMARK. It is easy to see that the converse is also true: if the B-space $\mathbf{C}(K)$ is separable, then the compactum K is metrizable.

Proof. With loss of generality we may consider a real $\mathbf{C}(K)$. By Corollary 1 to Theorem I.5.2, the compactum K is separable. Let M be a countable everywhere dense subset in K. By the compactness of K, there exists, for each $m \in \mathbb{N}$, a finite cover of K by open sets $U_i^m = \{s \in K : r(t_i, s) < 1/m\}$ $(i = 1, 2, \ldots, n(m))$, where $t_i \in M$. Let $\{\phi_i^m(t)\}_{i=1}^{n(m)}$ be the partition of unity constructed from $\{U_i^m\}_{i=1}^{n(m)}$ according to Lemma 1. We prove that the countable set E of functions of the form

$$\sum_{i=1}^{n(m)} r_i \phi_i^m (t_i) \qquad (m = 1, 2, \ldots),$$

where the r_i are arbitrary rational numbers, is dense in $\mathbf{C}(K)$. Choose $x \in \mathbf{C}(K)$ and $\varepsilon > 0$. Since the function $x(t)$ is uniformly continuous on K, there corresponds to ε a number $\delta > 0$ such that if $r(t, s) < \delta$, then $|x(t) - x(s)| < \varepsilon$. Choose $m \in \mathbb{N}$ such that $1/m < \delta$. Then we can choose rational numbers r_i satisfying $|x(t) - r_i| < 2\varepsilon$ for all $t \in U_i^m$ $(i = 1, 2, \ldots, n(m))$. Write

$$\tilde{x}(t) = \sum_{i=1}^{n(m)} r_i \phi_i^m (t).$$

Then $\tilde{x} \in E$, and if $t \in U_i^m$, we have

$$|x(t) - \tilde{x}(t)| = \left| \sum_{i=1}^{n(m)} x(t) \phi_i^m (t) - \sum_{i=1}^{n(m)} r_i \phi_i^m (t) \right| \leqslant \sum_{i=1}^{n(m)} |x(t) - r_i| \phi_i^m (t) < 2\varepsilon.$$

Since $\{U_i^m\}_{i=1}^{n(m)}$ is a cover of K, we conclude from this that $\|x - \tilde{x}\| < 2\varepsilon$, which proves that E is complete in $\mathbf{C}(K)$.

4.2. As we shall see below, the space of functions of bounded variation is closely connected with the space $\mathbf{C}[a, b]$.

Let $x(t)$ be a finite-valued function, defined in the interval $[a, b]$. Consider an arbitrary partition τ of $[a, b]$ $(a = t_0 < t_1 < \ldots < t_n = b)$ and form the expression

$$v_\tau = \sum_{k=1}^{n} |x(t_k) - x(t_{k-1})|. \tag{1}$$

If the set of sums v_τ corresponding to all possible partitions τ of $[a,b]$ is bounded, then $x(t)$ is said to be a *function of bounded variation* in $[a,b]$ and the quantity

$$\bigvee_a^b (x) = \sup_\tau v_\tau$$

is called the *total variation* of $x(t)$.

Consider the set **V** of all functions of bounded variation. **V** is a vector space. In fact, each function of bounded variation on $[a, b]$ is expressible as the difference of two non-decreasing functions (Fikhtengol'ts, vol. III):

$$x(t) = x_1(t) - x_2(t) \qquad (t \in [a, b]).$$

Conversely, it is easy to see that every function expressible as a difference of non-decreasing functions is a function of bounded variation. The linearity of **V** follows immediately from this remark.

If, for $x \in \mathbf{V}$, we write

$$\|x\| = |x(a)| + \bigvee_a^b (x),$$

then **V** becomes a normed space.

In fact, the first two axioms for a normed space are obviously satisfied, so we need only check the triangle inequality. Let $x, y \in \mathbf{V}$ and let $z = x + y$. Then, for any partition of $[a, b]$ $(a = t_0 < t_1 < \ldots < t_n = b)$, we have

$$|z(a)| = |x(a) + y(a)| \leqslant |x(a)| + |y(a)|$$

and

$$|z(t_k) - z(t_{k-1})| \leqslant |x(t_k) - x(t_{k-1})| + |y(t_k) - y(t_{k-1})|$$
$$(k = 1, 2, \ldots, n).$$

Thus

$$|z(a)| + \sum_{k=1}^n |z(t_k) - z(t_{k-1})| \leqslant |x(a)| + \sum_{k=1}^n |x(t_k) - x(t_{k-1})| + |y(a)| + \sum_{k=1}^n |y(t_k) - y(t_{k-1})|$$
$$\leqslant \|x\| + \|y\|,$$

and so the triangle inequality follows.

Let us prove that **V** is a *complete* space.

Let $\{x_n\}$ be a Cauchy sequence. First of all, since

$$|x_m(t) - x_n(t)| \leqslant |[x_m(t) - x_n(t)] - [x_m(a) - x_n(a)]| + |x_m(a) - x_n(a)| \leqslant |x_m(a) - x_n(a)|$$
$$+ \bigvee_a^b (x_m - x_n) = \|x_m - x_n\|,$$

the numerical sequence $\{x_n(t)\}$ converges for any $t \in [a, b]$. Write $x_0(t) = \lim_{n\to\infty} x_n(t)$. Let $\varepsilon > 0$ and let N_ε be so large that, when $m, n \geqslant N_\varepsilon$, we have

$$\|x_m - x_n\| < \varepsilon.$$

Then, for every partition $a = t_0 < t_1 < \ldots < t_r = b$, we have, *a fortiori*,

$$|x_m(a) - x_n(a)| + \sum_{k=1}^{r} |[x_m(t_k) - x_n(t_k)] - [x_m(t_{k-1}) - x_n(t_{k-1})]| < \varepsilon.$$

If we fix a partition and pass to the limit as $m \to \infty$, we find that

$$|x_0(a) - x_n(a)| + \sum_{k=1}^{r} |[x_0(t_k) - x_n(t_k)] - [x_0(t_{k-1}) - x_n(t_{k-1})]| \leqslant \varepsilon.$$

Because this is true for any partition, we have

$$|x_0(a) - x_n(a)| + \bigvee_a^b (x_0 - x_n) \leqslant \varepsilon \qquad (n > N_\varepsilon).$$

Hence, by the usual argument, we conclude: $x_0 \in \mathbf{V}$ and

$$\|x_0 - x_n\| \leqslant \varepsilon \qquad (n > N_\varepsilon),$$

that is, $x_n \to x_0$.

4.3. In addition to spaces of continuous functions and measurable functions, spaces of smooth functions and analytic functions play an important role in functional analysis. Examples of spaces of smooth functions are the spaces $\mathbf{C}^{(l)}(D)$ and the Sobolev spaces to be considered later. By way of an example, we give the definition of the Hardy spaces of analytic functions (for more details see Hoffman).

Denote by $\mathbf{H}^p (1 \leqslant p \leqslant \infty)$ the B-space of analytic functions $x(z)$ in the open unit disc D, such that the following norm is finite:

$$\|x\| = \begin{cases} \sup\limits_{0 \leqslant r < 1} \left(\int\limits_0^{2\pi} |x(re^{i\theta})|^p d\theta \right)^{1/p}, & 1 \leqslant p < \infty, \\ \sup\limits_{z \in D} |x(z)|, & p = \infty. \end{cases}$$

For $1 < p < \infty$, the spaces $\mathbf{H}^p$ can be embedded in a natural way as complemented subspaces of $\mathbf{L}^p(0, 2\pi)$. For $p = 1, \infty$, it is impossible to embed $\mathbf{H}^p$ as a complemented subspace of any $\mathbf{L}^p(T, \Sigma, \mu)$.

4.4. Let D be a bounded domain in $\mathbf{R}^n$. We shall say that a function $x(t)$, continuous in D, can be extended continuously to the closure $\bar{D}$, if there exists a continuous function $\tilde{x}(t)$ on $\bar{D}$ whose restriction to D coincides with $x(t)$ (such a function $\tilde{x}(t)$ is obviously unique).

We denote by $\mathbf{C}(D)$ the B-space of all functions $x(t)$ continuous in D and continuously extendable to $\bar{D}$, with the norm

$$\|x\| = \sup_{t \in D} |x(t)|.$$

It is easy to see that the mapping $x \to \tilde{x}$ is a linear isometry of the B-space $\mathbf{C}(D)$ onto $\mathbf{C}(\bar{D})$, and so $\mathbf{C}(D)$ is complete.

We shall say that $x(t)$ has l continuous derivatives on $\bar{D}$ if it has all the possible continuous partial derivatives of order $\leqslant l$ in D and these extend continuously to $\bar{D}$.

We denote by $\mathbf{C}^{(l)}(D)$ the B-space of all functions $x(t)$ on D having l continuous derivatives on $\bar{D}$, with the norm

$$\|x\| = \sum_{k=0}^{l} \sum_{(k)} \sup_{t \in D} \left| \frac{\partial^k x(t)}{\partial t_1^{k_1} \ldots \partial t_n^{k_n}} \right|,$$

where $\sum\limits_{(k)}$ is a summation over all the derivatives of order k.

The spaces $\mathbf{C}(D)$ and $\mathbf{C}^{(l)}(D)$ will play an important role in connection with embedding theorems (see Chapter XI and later).

§ 5. Hilbert space

5.1. In the class of finite-dimensional spaces, Euclidean space $\mathbf{R}^n$ is distinguished by having an inner product defined on it, which is connected with the norm by a simple relation: the square of the norm of an element is the inner product of that element with itself.

In the light of this example it is appropriate to investigate spaces having an "inner product" defined on them, from which the norm is derivable in the above manner.

A complex* vector space $\mathbf{H}$ is called an *inner product space* if for each pair of elements x, $y \in \mathbf{H}$ there is defined an inner product (x, y), which is a complex number satisfying the following conditions (axioms):

1) $(y, x) = \overline{(x, y)}$;
2) $(\lambda x_1 + \mu x_2, y) = \lambda(x_1, y) + \mu(x_2, y)$;
3) $(x, x) \geqslant 0$; $(x, x) = 0$ if and only if $x = \mathbf{0}$.

From the definition of an inner product space, it follows that:

a) $(x, \lambda y_1 + \mu y_2) = \bar{\lambda}(x, y_1) + \bar{\mu}(x, y_2)$ (axioms 1) and 2)).

b) $(x, \mathbf{0}) = (\mathbf{0}, y) = 0$. For we have $(x, 0.y) = 0.(x, y) = 0$.

c) $|(x, y)|^2 \leqslant (x, x)(y, y)$ (the *Cauchy–Buniakowski inequality*). To prove this, consider the expression

$$(x + \lambda y, x + \lambda y) = (x, x) + \bar{\lambda}(x, y) + \lambda(y, x) + |\lambda|^2 (y, y).$$

By axiom 3) this expression is non-negative, whatever value λ takes. Assuming $(y, y) > 0$ (otherwise $y = \mathbf{0}$ and the required inequality is obvious), we write $\lambda = -(x, y)/(y, y)$. By what has been said,

$$(x, x) - \frac{|(x, y)|^2}{(y, y)} - \frac{|(x, y)|^2}{(y, y)} + \frac{|(x, y)|^2}{(y, y)} \geqslant 0,$$

that is, $(x, x)(y, y) - |(x, y)|^2 \geqslant 0$, which is what we set out to prove.

If, in an inner product space $\mathbf{H}$, we set

$$\|x\| = \sqrt{(x, x)} \qquad (x \in \mathbf{H}), \tag{1}$$

then $\mathbf{H}$ is turned into a normed space. Indeed, with the exception of the triangle inequality, the axioms for a normed space follow immediately from the definition of $\mathbf{H}$. Let us prove

---

* Less often, inner product spaces are constructed from real vector spaces. In such cases, one assumes that the inner product is a real number, and condition 1) then becomes $(y, x) = (x, y)$.

When we speak of inner product spaces in the sequel we shall always take them to be complex spaces, unless we stipulate otherwise, although all the results, except the spectral theory of operators, are valid for the real case as well.

the triangle inequality. Let $x, y \in \mathbf{H}$. Using the Cauchy–Buniakowski inequality, we have

$$\|x+y\|^2 = (x+y, x+y) = |(x, x)+(x, y)+(y, x)+(y, y)| \leqslant \|x\|^2+2\|x\|\,\|y\|+\|y\|^2$$
$$= [\|x\|+\|y\|]^2.$$

A normed space $\mathbf{H}$ is said to be *unitary* if it is possible to introduce an inner product in $\mathbf{H}$ that is related to the norm by (1).

Since $\mathbf{H}$ is a normed space, it possesses all the properties of one. In particular, all the elementary facts mentioned in 1.1 apply to $\mathbf{H}$. We next record a few simple observations that apply specifically to inner product spaces.

1) *Continuity of the inner product.* If $x_n \to x$ and $y_n \to y$, then $(x_n, y_n) \to (x, y)$.

For, if we use the Cauchy–Buniakowski inequality, we have

$$|(x, y)-(x_n, y_n)| \leqslant |(x, y-y_n)|+|(x-x_n, y_n)| \leqslant \|x\|\,\|y-y_n\|+\|x-x_n\|\,\|y_n\|.$$

Since the convergent sequence $\{y_n\}$ is bounded, the right-hand side of this inequality tends to zero, and thus $(x_n, y_n) \to (x, y)$.

2) For any elements $x, y \in \mathbf{H}$, we have the equation*

$$\|x+y\|^2+\|x-y\|^2 = 2[\|x\|^2+\|y\|^2]. \tag{2}$$

For, by the definition of the norm, we have

$$\|x+y\|^2+\|x-y\|^2 = (x+y, x+y)+(x-y, x-y) = (x, x)+(x, y)+(y, x)+(y, y)$$
$$+(x, x)-(x, y)-(y, x)+(y, y) = 2[\|x\|^2+\|y\|^2].$$

The fact that $\mathbf{H}$ is a unitary space means that its norm has been introduced in a particular way, by means of an inner product. Now the completion $\overline{\mathbf{H}}$ of $\mathbf{H}$ is also a normed space (1.4). We show next that the inner product in $\mathbf{H}$ may be extended to $\overline{\mathbf{H}}$ in such a way that $\overline{\mathbf{H}}$ becomes a unitary space.

3) The completion $\overline{\mathbf{H}}$ of a unitary space $\mathbf{H}$ is itself unitary.

For let $\xi, \eta \in \overline{\mathbf{H}}$. There exist sequences $\{x_n\}, \{y_n\} \subset \mathbf{H}$ such that $x_n \to \xi$, $y_n \to \eta$. Repeating the argument in the proof of 1) almost word for word, we see that $\lim_{n\to\infty}(x_n, y_n)$ exists and does not depend on the choice of sequences $\{x_n\}$ and $\{y_n\}$ but is determined solely by ξ and η. Since 1) shows that $\lim (x_n, y_n) = (\xi, \eta)$ when $\xi, \eta \in \mathbf{H}$, it is also natural, for arbitrary ξ, $\eta \in \overline{\mathbf{H}}$, to set

$$(\xi, \eta) = \lim_{n\to\infty}(x_n, y_n).$$

As is easily verified, this definition makes $\overline{\mathbf{H}}$ into an inner product space. Bearing in mind the connection between the norms of $\mathbf{H}$ and $\overline{\mathbf{H}}$ mentioned in 1.4, we have

$$\|\xi\| = \lim_{n\to\infty}\|x_n\| = \lim \sqrt{(x_n, x_n)} = \sqrt{(\xi, \xi)}$$
$$(\xi \in \overline{\mathbf{H}},\ x_n \to \xi,\ x_n \in \mathbf{H}, \qquad n = 1, 2, \ldots),$$

* This equation expresses a generalization of the elementary geometrical fact that the sum of the squares of the diagonals of a parallelogram is equal to the sum of the squares of the two sides.

If (2) holds for all x, y in a normed space $\mathbf{X}$, then $\mathbf{X}$ is a unitary space: that is, it is possible to introduce an inner product in $\mathbf{X}$ such that the norm is given by (1).

which establishes that $\bar{\mathbf{H}}$ is unitary.

Let us record another completely obvious statement.

4) A subspace of a unitary space is also a unitary space.

5.2. Complete unitary spaces are of fundamental interest. Such spaces are called *Hilbert spaces* (after the German mathematician D. Hilbert).*

Hilbert space is an immediate generalization of Euclidean space, so its "geometry" approaches Euclidean geometry more closely than is the case for any other B-space. Hilbert space possesses many properties of Euclidean space which B-spaces in general do not possess (see, for example, equation (2)). This situation has made it possible to develop functional analysis based on Hilbert space much more extensively and completely than functional analysis based on general normed spaces and, as a result, Hilbert space theory has separated off into a large independent branch of functional analysis with its own results and methods, not confined within the same limits as the general functional analysis which forms the principal subject mater of this book. For this reason, Hilbert space will only be used, as a rule, to illustrate facts relating to the general theory, both here and in the sequel.

5.3. A concrete example of a Hilbert space is the space ℓ^2. An inner product is defined on this by

$$(x, y) = \sum_{k=1}^{\infty} \xi_k \overline{\eta_k} \quad (x = (\xi_1, \xi_2, \ldots, \xi_k, \ldots), \ y = (\eta_1, \eta_2, \ldots, \eta_k, \ldots)).$$

The norm derived from this inner product, in accordance with (1), namely,

$$\|x\| = \left[\sum_{k=1}^{\infty} |\xi_k|^2\right]^{1/2},$$

coincides with the norm defined in 2.3, which is why we retain the earlier notation ℓ^2 for this space. As we have proved, this is a complete space.

More generally, let (T, Σ, μ) be a measure space. $\mathbf{L}^2(T, \Sigma, \mu)$ becomes a Hilbert space if we set

$$(x, y) = \int_T x(t)\overline{y(t)}\, d\mu.$$

The norm in $\mathbf{L}^2$ coincides with that introduced by means of (1). Let us now consider a special case. Let $\phi \in \mathbf{L}^1(a, b)$ be a function such that $\phi(t) > 0$ a.e. (the interval here can also be infinite). Write

$$\nu(A) = \int_A \phi(t)\, dt$$

for all Lebesgue measurable sets A. We denote by $\mathbf{L}^2_\phi(a, b)$ the Hilbert space of all functions that are square summable relative to the measure ν. Here the inner product is, of course, defined by

$$(x, y) = \int_a^b \phi(t)[x(t)\overline{y(t)}]\, dt.$$

* It is sometimes further stipulated that the space be infinite-dimensional. Abstract Hilbert space was introduced by von Neumann [1].

If $\phi(t) = 1$, then $\mathbf{L}^2_\phi(a, b) = \mathbf{L}^2(a, b)$. As we have already shown, both ℓ^2 and $\mathbf{L}^2_\phi(a, b)$ are separable.

An example of an inseparable Hilbert space is the set $\mathbf{H}_c$ of functions defined on the whole real line and having at most a countable set of values different from zero. Here it is assumed that*

$$\sum_t |x(t)|^2 < \infty.$$

We define an inner product in $\mathbf{H}_c$ by

$$(x, y) = \sum_t x(t)\overline{y(t)}.$$

We leave it to the reader to prove that this space is complete, using the completeness of ℓ^2.

If we denote by x_τ the element of $\mathbf{H}_c$ given by:

$$x_\tau(t) = \begin{cases} 1, & t = \tau, \\ 0, & t \neq \tau, \end{cases}$$

then it is easy to verify that $\|x_\tau - x_{\tau'}\| = \sqrt{2}$ (if $\tau \neq \tau'$), which implies that $\mathbf{H}_c$ is inseparable.

5.4. In the study of Hilbert space the concept of orthogonality of elements turns out to be very important.

Elements x and y in a Hilbert space $\mathbf{H}$ are said to be *orthogonal* if $(x, y) = 0$. We then write $x \perp y$.

If a fixed element $x \in \mathbf{H}$ is orthogonal to each element of some set $E \subset \mathbf{H}$, then x is said to be *orthogonal* to E, and we write $x \perp E$.

Finally, if the elements of two sets E_1 and E_2 are pairwise orthogonal, then the sets are said to be *orthogonal* ($E_1 \perp E_2$).

Let us record a few simple facts relating to this concept.

a) If $x \perp y_1$ and $x \perp y_2$, then $x \perp \lambda y_1 + \mu y_2$.

b) If $x \perp y_n$ $(n = 1, 2, \ldots)$ and $y_n \to y$, then $x \perp y$.

This follows immediately from the continuity of the inner product (see 1) in 5.1).

c) If $x \perp E$, then $x \perp \overline{\mathscr{L}}(E)$.

To prove this, use a) and b).

d) If $x \perp E$, where E is a fundamental set in $\mathbf{H}$ (that is $\overline{\mathscr{L}}(E) = \mathbf{H}$), then $x = \mathbf{0}$.

For we then have $x \perp \mathbf{H}$, and therefore x is orthogonal to itself: that is, $(x, x) = 0$, which is equivalent to $x = \mathbf{0}$.

e) The collection of all elements orthogonal to a given set E is a subspace of $\mathbf{H}$; that is, it is a closed linear manifold.† This subspace is called the *orthogonal complement* of E.

The following theorem is of fundamental importance in the theory of Hilbert space.

THEOREM 1.‡ *Let $\mathbf{H}_1$ be a subspace of a Hilbert space $\mathbf{H}$ and $\mathbf{H}_2$ its orthogonal complement. Every element $x \in \mathbf{H}$ can be uniquely expressed in the form*

$$x = x' + x'' \qquad (x' \in \mathbf{H}_1,\ x'' \in \mathbf{H}_2). \tag{3}$$

* The function $x(t)$ is different from zero only for at most a countable set of values of t. Thus the sum occurring here may be regarded as an ordinary series.

† In the case of Hilbert space, the term "subspace" will in future always mean "closed linear manifold".

‡ This theorem was proved in its present form by Rellich [1].

Here the element x' realizes the distance from x to $\mathbf{H}_1$: that is,

$$\|x - x'\| = \rho(x, \mathbf{H}_1). \tag{4}$$

Proof. We write

$$d = \rho(x, \mathbf{H}_1)$$

and find elements $x_n \in \mathbf{H}_1$ such that

$$\|x - x_n\|^2 < d^2 + 1/n^2 \qquad (n = 1, 2, \dots). \tag{5}$$

By (2), we have

$$\|x_n - x_m\|^2 + \|(x - x_n) + (x - x_m)\|^2 = 2[\|x - x_n\|^2 + \|x - x_m\|^2]. \tag{6}$$

Now since $\frac{1}{2}(x_m + x_n) \in \mathbf{H}_1$, we have

$$\|(x - x_n) + (x - x_m)\|^2 = 4\left\|x - \frac{x_n + x_m}{2}\right\|^2 \geqslant 4d^2.$$

Therefore, using (6) together with (5) and these bounds, we obtain

$$\|x_n - x_m\|^2 \leqslant 2\left[d^2 + \frac{1}{n^2} + d^2 + \frac{1}{m^2}\right] - 4d^2 = \frac{2}{n^2} + \frac{2}{m^2}.$$

Thus the sequence $\{x_n\}$ is a Cauchy sequence. Since $\mathbf{H}$ is complete, $x' = \lim_{n \to \infty} x_n$ exists. Furthermore, $x' \in \mathbf{H}_1$ (as $\mathbf{H}_1$ is closed). Taking limits in (5), we find that $\|x - x'\| \leqslant d$, and since, for every element of $\mathbf{H}_1$, and in particular for x', we must have $\|x - x'\| \geqslant d$, it follows that

$$\|x - x'\| = d. \tag{7}$$

Now we prove that the element $x'' = x - x'$ is orthogonal to $\mathbf{H}_1$ and so belongs to $\mathbf{H}_2$. Choose a non-zero element $y \in \mathbf{H}_1$. For every λ, we have $x' + \lambda y \in \mathbf{H}_1$, so that

$$\|x'' - \lambda y\|^2 = \|x - (x' + \lambda y)\|^2 \geqslant d^2,$$

which may be rewritten, using (7), in the form

$$-\bar{\lambda}(x'', y) - \lambda(y, x'') + |\lambda|^2 (y, y) \geqslant 0.$$

In particular, when $\lambda = (x'', y)/(y, y)$, we conclude from this that

$$-\frac{|(x'', y)|^2}{(y, y)} - \frac{|(x'', y)|^2}{(y, y)} + \frac{|(x'', y)|^2}{(y, y)} \geqslant 0,$$

that is,

$$|(x'', y)|^2 \leqslant 0,$$

which can happen only when $(x'', y) = 0$, $x'' \perp y$.

Thus we have established that x is expressible in the form (3) and that (4) is true.

Let us prove that the representation in (3) is unique. In fact, if

$$x = x_1' + x_1'' \qquad (x_1' \in \mathbf{H}_1, \quad x_1'' \in \mathbf{H}_2),$$

then, comparing this with (3), we obtain

$$x' - x_1' = x_1'' - x''.$$

The element on the left-hand side of this equation belongs to $\mathbf{H}_1$, while that on the right-hand side belongs to $\mathbf{H}_2$, so that $x' - x_1' \perp x_1'' - x''$, yielding $x' - x_1' = x_1'' - x'' = \mathbf{0}$. This completes the proof of the theorem.

The elements x' and x'' uniquely determined by the element x are called the *projections* of x in the subspaces $\mathbf{H}_1$ and $\mathbf{H}_2$, respectively.

A system of elements $\{x_\alpha\}$ $(\alpha \in \mathrm{A})$ in a Hilbert space is said to be *complete* if the set $\{x_\alpha\}$ is fundamental in $\mathbf{H}$ (see II.3.2)—that is, if $\mathbf{H} = \overline{\mathscr{L}}(\{x_\alpha\})$.

COROLLARY. *A necessary and sufficient condition for a system of elements $\{x_\alpha\}$ in a space $\mathbf{H}$ to be complete is that there exist no non-zero element orthogonal to every element of the system.*

In fact, the necessity of the condition follows from d) in 5.4. The condition is sufficient because, if $\{x_\alpha\}$ were an incomplete system, then $\mathbf{H}_1 = \overline{\mathscr{L}}(\{x_\alpha\})$ would be a proper subspace of $\mathbf{H}$, and thus, by choosing $x \in \mathbf{H} \setminus \mathbf{H}_1$ and decomposing it into a sum of projections $x = x' + x''$ $(x' \in \mathbf{H}_1, x'' \in \mathbf{H}_2)$, we should have $x'' \neq \mathbf{0}$ and $x'' \perp x_\alpha$ $(\alpha \in \mathrm{A})$, contrary to our hypothesis.

5.5. A system of elements $\{x_\alpha\}$ $(\alpha \in \mathrm{A})$ in a Hilbert space $\mathbf{H}$ is called an *orthogonal system* if every two distinct elements of the system are orthogonal. If, in addition, the norm of each element x_α is equal to unity, then the system is called *orthonormal*.

An important example of an orthogonal system in $\mathbf{L}^2(a,b)$ is the system of trigonometric functions 1, $\cos\frac{2\pi(t-a)}{b-a}$, $\sin\frac{2\pi(t-a)}{b-a}, \ldots, \cos\frac{2\pi n(t-a)}{b-a}$, $\sin\frac{2\pi n(t-a)}{b-a}$. This is a *complete* system, since its linear hull—the set of all trigonometric polynomials—is dense in $\mathbf{L}^2(a,b)$. To see this, notice first that the set of all bounded (measurable) functions is dense in $\mathbf{L}^2(a,b)$. Next, using Luzin's Theorem, we see that the set of continuous periodic functions (that is, continuous functions taking equal values at the endpoints of the interval $[a,b]$) is dense in the former set. Finally, using Weierstrass's Theorem, we conclude that the set of trigonometric polynomials is dense in the set of continuous functions.

Another example of an orthogonal system in $\mathbf{L}^2(a,b)$ is the so-called *Rademacher system*, consisting of the functions

$$x_n(t) = (-1)^k$$

$$\left(\frac{k}{2^n}(b-a) < t-a < \frac{k+1}{2^n}(b-a); \quad k = 0, 1, \ldots, 2^n - 1; \; n = 0, 1, \ldots\right).$$

This system is not complete. Indeed the function $x(t) = (t-a)(t-b) + \frac{(b-a)^2}{6}$ is orthogonal to all functions in the system.

An orthogonal system in ℓ^2 is the system $\{x_n\}$, $x_n = (0, 0, \ldots, 0, 1, 0, \ldots)$. This system is clearly complete.
An orthogonal system in $\mathbf{H}_c$ is the collection $\{x_\tau\}$, where x_τ are the elements defined in 5.3 above.

If zero is not one of the elements of an orthogonal system, then the system is linearly independent. For if we had a relation of the form

$$\sum_{k=1}^{n} \lambda_k x_{\alpha_k} = \mathbf{0},$$

then, forming the inner product of each side of this equation with x_{α_j}, we should have $\lambda_j \|x_{\alpha_j}\|^2 = 0$.

It is easy to obtain an orthonormal system from an orthogonal one by dividing each element by its norm (assuming that the system does not contain the zero element). Going from an arbitrary system of elements to an orthonormal one is more complicated. Restricting ourselves to the case of a countable system, we next prove the following orthogonalization theorem (due to Schmidt).

THEOREM 2 (The orthogonalization theorem). *Let $\{y_n\}$ be a linearly independent system of elements in a Hilbert space* **H**. *There exists an orthonormal system $\{x_n\}$ such that*

$$x_n = \sum_{k=1}^{n} \lambda_k^{(n)} y_k \qquad (\lambda_n^{(n)} \neq 0; \quad n = 1, 2, \ldots). \tag{8}$$

We give two proofs of the theorem.

First proof. Write

$$x_1 = \lambda_1^{(1)} y_1 \qquad (\lambda_1^{(1)} = 1/\|y_1\|)$$

and assume that we have already constructed pairwise orthogonal and normalized elements $x_1, x_2, \ldots, x_{n-1}$, related to $y_1, y_2, \ldots, y_{n-1}$ by (8). Denote by $\mathbf{H}_n'$ the linear hull of the elements $y_1, y_2, \ldots, y_{n-1}$ and by $\mathbf{H}_n''$ its orthogonal complement. Since $y_n \notin \mathbf{H}_n'$, the projection y_n'' in the decomposition

$$y_n = y_n' + y_n'' \qquad (y_n' \in \mathbf{H}_n', \; y_n'' \in \mathbf{H}_n'')$$

is non-zero. Writing

$$x_n = \lambda_n^{(n)} y_n'' \qquad (\lambda_n^{(n)} = 1/\|y_n''\|),$$

we have $\|x_n\| = 1$; and, since $x_n \perp \mathbf{H}_n'$ and $x_k \in \mathbf{H}_k'$ $(k < n)$, we also have $x_n \perp x_k$ $(k < n)$.

Moreover,

$$x_n = \lambda_n^{(n)} y_n'' = \lambda_n^{(n)} y_n - \lambda_n^{(n)} y_n'.$$

As $y_n' \in \mathbf{H}_n'$, we have

$$y_n' = \sum_{k=1}^{n-1} \alpha_k^{(n)} y_k.$$

Setting $\lambda_k^{(n)} = -\alpha_k^{(n)} \lambda_n^{(n)}$ $(k < n)$, we obtain

$$x_n = \sum_{k=1}^{n} \lambda_k^{(n)} y_k.$$

The proof of the theorem now comes by induction.

REMARK 1. Since $\lambda_n^{(n)} \neq 0$, the elements y_n can be expressed in terms of $x_1, x_2, \ldots, x_n$:

$$y_n = \sum_{k=1}^{n} \mu_k^{(n)} x_k \quad \left(\mu_n^{(n)} = \frac{1}{\lambda_n^{(n)}}; \; n = 1, 2, \ldots\right). \tag{9}$$

REMARK 2. The elements x_n are not uniquely determined by the y_n. However, if we require that $\lambda_n^{(n)} > 0$ $(n = 1, 2, \ldots)$, then the system $\{x_n\}$ will be unique.

In fact, suppose there were two systems $\{x_n\}$ and $\{\tilde{x}_n\}$ satisfying the above requirements. Using (8) and (9) and the analogous equations for the system $\{\tilde{x}_n\}$, we can express the elements of the system $\{\tilde{x}_n\}$ in terms of the $\{x_n\}$—that is, we can obtain equations of the form

$$\tilde{x}_n = \sum_{k=1}^{n} \beta_k^{(n)} x_k. \tag{10}$$

Moreover, it is not hard to see that the coefficient $\beta_n^{(n)}$ is positive (it is equal to the product $\tilde{\lambda}_n^{(n)} \mu_n^{(n)}$, where $\tilde{\lambda}_n^{(n)}$ denotes the appropriate coefficient in the equation corresponding to (8) for the system $\{\tilde{x}_n\}$). Thus it is clear that $\tilde{x}_1 = x_1$. If we have already proved that $\tilde{x}_1 = x_1, \tilde{x}_2 = x_2, \ldots, \tilde{x}_{n-1} = x_{n-1}$, then, forming the inner product of each side of (10) with x_j, we have $0 = \beta_j^{(n)}$ $(j < n)$; that is, $\tilde{x}_n = \beta_n^{(n)} x_n$. But now, since $\|\tilde{x}_n\| = \|x_n\| = 1$, we have $\beta_n^{(n)} = 1$ and consequently $\tilde{x}_n = x_n$.

Second proof. We introduce the following notation:

$$\Delta_n = \begin{vmatrix} (y_1, y_1) & (y_1, y_2) & \ldots & (y_1, y_n) \\ (y_2, y_1) & (y_2, y_2) & \ldots & (y_2, y_n) \\ \ldots & \ldots & \ldots & \ldots \\ (y_n, y_1) & (y_n, y_2) & \ldots & (y_n, y_n) \end{vmatrix}, \qquad \Delta_0 = 1 \text{ (the \textit{Gram determinant})},$$

$$x_n' = \begin{vmatrix} (y_1, y_1) & (y_1, y_2) & \ldots & (y_1, y_{n-1}) & y_1 \\ (y_2, y_1) & (y_2, y_2) & \ldots & (y_2, y_{n-1}) & y_2 \\ \ldots & \ldots & \ldots & \ldots & \ldots \\ (y_n, y_1) & (y_n, y_2) & \ldots & (y_n, y_{n-1}) & y_n \end{vmatrix},$$

where the latter determinant is understood to be the sum of the products of the elements of the last column with the corresponding cofactors. Clearly,

$$(x_n', y_n) = \Delta_n, \tag{11}$$

$$(x_n', y_k) = 0 \quad (k = 1, 2, \ldots, n-1). \tag{12}$$

The latter is true because the determinant obtained on the left-hand side has two identical columns (the kth and the nth).

If we form the inner product of the last column of the determinant for x_n' with x_n', then, bearing in mind (11) and (12), we obtain a determinant in the last column of which all elements except the last are equal to zero, while the last is equal to Δ_n. Expanding this determinant by the elements of its last column, we obtain the equation

$$(x_n', x_n') = \Delta_n \Delta_{n-1} \quad (n = 1, 2, \ldots).$$

Since $x_n' \neq \mathbf{0}$ (otherwise the elements $y_1, y_2, \ldots, y_n$ would be linearly dependent), we see that $\Delta_n \neq 0$ $(n = 1, 2, \ldots)$.

Notice, finally, that the element x_k' $(k < n)$, being expressed as a linear combination of $y_1, y_2, \ldots, y_k$, will, by (12), be orthogonal to x_n'.

Bearing all this in mind, we obtain the required orthonormal sequence if we set

$$x_n = \frac{x_n'}{\|x_n'\|} = \frac{x_n'}{\sqrt{\Delta_n \Delta_{n-1}}} \quad (n = 1, 2, \ldots). \tag{13}$$

5.6. Let us use the results just obtained to construct a system of weighted orthogonal polynomials. As the $\{y_n\}$ in $\mathbf{L}^2_\phi(a, b)$ we now take

$$y_n(t) = t^n \quad (n = 0, 1, \ldots). \tag{14}$$

By orthogonalizing this sequence, we obtain a sequence $\{x_n\}$ such that

$$x_n(t) = \sum_{k=0}^{n} \lambda_k^{(n)} t^k \quad (\lambda_n^{(n)} \neq 0;\ n = 0, 1, \ldots),$$

and

$$(x_m, x_n) = \int_a^b \phi(t) x_m(t) \overline{x_n(t)}\, dt = \begin{cases} 0, & m \neq n, \\ 1, & m = n. \end{cases}$$

The polynomials $x_n(t)$ are said to be *orthogonal with weight* $\phi(t)$. They play a large role in the constructive theory of functions. Since the system (14) is complete, so is the system of orthogonal polynomials.

An explicit expression for the orthogonal polynomials can be obtained with the aid of (13). In this case,

$$\Delta_n = \begin{vmatrix} c_{00} & c_{01} & \cdots & c_{0n} \\ c_{10} & c_{11} & \cdots & c_{1n} \\ \cdots & \cdots & \cdots & \cdots \\ \cdots & \cdots & \cdots & \cdots \\ c_{n0} & c_{n1} & \cdots & c_{nn} \end{vmatrix}, \quad c_{jk} = \int_a^b \phi(t) t^{j+k}\, dt \quad (j, k = 0, 1, \ldots, n),$$

so that

$$x_n(t) = \frac{x'(t)}{\sqrt{\Delta_n \Delta_{n-1}}} = \frac{\begin{vmatrix} c_{00} & c_{01} & \cdots & c_{0\,n-1} & 1 \\ c_{10} & c_{11} & \cdots & c_{1\,n-1} & t \\ \cdots & \cdots & \cdots & \cdots & \cdots \\ \cdots & \cdots & \cdots & \cdots & \cdots \\ c_{n0} & c_{n1} & \cdots & c_{n\,n-1} & t^n \end{vmatrix}}{\sqrt{\Delta_n \Delta_{n-1}}} \quad (n = 0, 1, \ldots).$$

These equations show, among other things, that the polynomials x_n have real coefficients.

If $\phi(t) \equiv 1$; $a = -1$, $b = 1$, then the orthogonal polynomials are called *Legendre polynomials* and they are denoted by $P_n(t)$. It can be proved that

$$P_n(t) = \sqrt{\frac{2n+1}{2}} \frac{1}{(2n)!} \frac{d^n}{dt^n} [(t^2 - 1)^n] \quad (n = 0, 1, \ldots). \tag{15}$$

In the case where $\phi(t) = (1-t)^{-\alpha}(1+t)^{\beta}$; $a = -1$, $b = 1$, we obtain the *Jacobi polynomials*

$$J_n^{(\alpha, \beta)}(t) = k_n (1-t)^{-\alpha}(1+t)^{-\beta} \frac{d^n}{dt^n} [(1-t)^{\alpha+n}(1+t)^{\beta+n}], \tag{16}$$

where

$$k_n = (-1)^n \sqrt{2^{-(\alpha+\beta+2n+1)} \frac{\Gamma(\alpha+\beta+n+1)}{\Gamma(\alpha+n+1)\Gamma(\beta+n+1)} \frac{\alpha+\beta+2n+1}{n!}}.$$

In particular, when $\alpha = \beta = -\frac{1}{2}$, we have the *Chebyshev polynomials*, which are often encountered in various problems in the theory of functions.

When $\phi(t) = e^{-t}$; $a = 0$, $b = \infty$, we obtain the *Laguerre polynomials*

$$\mathrm{L}_n(t) = \frac{(-1)^n}{n!} e^t \frac{d^n}{dt^n} [e^{-t} t^n]. \tag{17}$$

Finally, if $\phi(t) = e^{-t^2}$; $a = -\infty$, $b = \infty$, then the polynomials, which in this case are denoted by $H_n(t)$, are called the Hermite polynomials,

$$H_n(t) = \frac{(-1)^n}{\sqrt{2^n n!\, \sqrt{\pi}}} e^{t^2} \frac{d^n}{dt^n} [e^{-t^2}]. \tag{18}$$

The reader will find a more complete information on orthogonal polynomials, and, in particular, proofs of the formulae (15)–(18), in Natanson-I (see also Szegö).

5.7. From now on we deal chiefly with separable Hilbert space. If we investigate particular orthogonal systems in specific separable spaces, we notice that they are all denumerable. This is not accidental. In fact, we have

THEOREM 3. *An orthogonal system* $\{x_\alpha\}$ $(\alpha \in \mathrm{A})$ *in a separable Hilbert space* $\mathbf{H}$ *is at most denumerable.*

Proof. Let D be a denumerable dense subset of $\mathbf{H}$. For each x_α, there exists $y_\alpha \in D$ such that

$$\|x_\alpha - y_\alpha\| < 1/2.$$

Further, the elements y_α and $y_{\alpha'}$ corresponding to distinct x_α and $x_{\alpha'}$ are distinct. For we have

$$\|y_\alpha - y_{\alpha'}\|^2 = \|(x_\alpha - x_{\alpha'}) + (x_{\alpha'} - y_{\alpha'}) + (y_\alpha - x_\alpha)\| \geq \|x_\alpha - x_{\alpha'}\| - [\|x_{\alpha'} - y_{\alpha'}\| + \|y_\alpha - x_\alpha\|]$$
$$> \sqrt{2} - 1 > 0,$$

because

$$\|x_\alpha - x_{\alpha'}\|^2 = (x_\alpha - x_{\alpha'}, x_\alpha - x_{\alpha'}) = (x_\alpha, x_\alpha) + (x_{\alpha'}, x_{\alpha'}) = 2.$$

Thus the set $\{x_\alpha\}$ is equipotent to a subset of the denumerable set D, and is therefore itself at most denumerable.

The next theorem is complementary to the one just proved.

THEOREM 4. *If a separable Hilbert space* $\mathbf{H}$ *is infinite-dimensional, then it has a countable complete orthonormal system.*

Proof. Starting with a countable dense subset $D = \{z_n\}$ of $\mathbf{H}$, we construct a linearly independent system $\{y_k\}$ that is complete in $\mathbf{H}$. First, assuming that $z_1 \neq \mathbf{0}$, we set $y_1 = z_1$. Next we take y_2 to be the element z_{n_2} having the least suffix $n_2 \geq 2$ for which y_1 and $y_2 = z_{n_2}$ are linearly independent. Proceeding in this way, we obtain the required system $\{y_k\}$. For $\{y_k\}$ is a complete system, as each z_n is obviously a linear combination of elements $y_j = z_{n_j}$ $(n_j < n)$, so that $D \subset \mathscr{L}(\{y_k\})$ and therefore $\overline{\mathscr{L}}(\{y_k\}) = \mathbf{H}$.

Applying the orthogonalization process to $\{y_n\}$, we obtain a complete orthonormal system, and this is evidently countable.

5.8. Suppose we are given an orthonormal system $\{x_k\}$ in a Hilbert space $\mathbf{H}$, and let $x \in \mathbf{H}$. The numbers

$$a_k = (x, x_k) \quad (k = 1, 2, \ldots)$$

are called the *Fourier coefficients* of x relative to the given orthonormal system, and the series

$$\sum_{k=1}^{\infty} a_k x_k$$

is called the *Fourier series* for x.

Let us form the subspace $\mathbf{H}_n = \mathscr{L}(\{x_1, x_2, \ldots, x_n\})$, whose elements are all the linear combinations of the first n elements of the given orthonormal system.

Then we have

THEOREM 5. *The partial sum* $s_n = \sum_{k=1}^{n} a_k x_k$ *of the Fourier series for the element* x *is the projection of* x *in the subspace* $\mathbf{H}_n$.

Proof. Since

$$x = s_n + (x - s_n)$$

and $s_n \in \mathbf{H}_n$, it is sufficient to prove that $x - s_n \perp \mathbf{H}_n$. But we clearly have $x - s_n \perp x_k$ ($k = 1, 2, \ldots, n$), and thus, by c) in 5.4, $x - s_n \perp \mathbf{H}_n$. This proves the theorem.

Using the conclusion of Theorem 1, we obtain from this

COROLLARY 1. *For any element* $z = \sum_{k=1}^{n} \alpha_k x_k \in \mathbf{H}_n$, *we have*

$$\|x - s_n\| = \rho(x, \mathbf{H}_n) \leqslant \|x - z\|.$$

Furthermore, since

$$\|x\|^2 = \|s_n\|^2 + \|x - s_n\|^2 \geqslant \|s_n\|^2, \tag{19}$$

$$\|s_n\|^2 = \sum_{k=1}^{n} |a_k|^2, \tag{20}$$

the following is also true.

COROLLARY 2. *We have the following inequality (Bessel's inequality):*

$$\sum_{k=1}^{n} |a_k|^2 \leqslant \|x\|^2.$$

If here we pass to the limit as $n \to \infty$, we obtain

$$\sum_{k=1}^{\infty} |a_k|^2 \leqslant \|x\|^2. \tag{21}$$

If, for some $x \in \mathbf{H}$, we have equality in (21), then x is said to satisfy *Parseval's identity.*

THEOREM 6. *The Fourier series of every element* $x \in \mathbf{H}$ *always converges, and the sum of the Fourier series is the projection of* x *in the subspace* $\mathbf{H}_0 = \overline{\mathscr{L}}(\{x_k\})$.

A necessary and sufficient condition for the sum of the Fourier series to be equal to the given element x *is that* x *should satisfy Parseval's identity.*

Proof. Inequality (21) implies that the series $\sum_{k=1}^{\infty} |a_k|^2$ converges. If, as before, we denote the partial sum of the Fourier series by s_n, then we have

$$\|s_{n+p} - s_n\|^2 = \sum_{k=n+1}^{n+p} |a_k|^2 \xrightarrow[n \to \infty]{} 0,$$

and so we conclude that the Fourier series is also convergent.

Let $s = \sum_{k=1}^{\infty} a_k x_k$. Since $s \in \mathbf{H}_0$ and

$$x = s + (x - s),$$

we can restrict ourselves, as in the proof of Theorem 5, to showing that $x - s \perp \mathbf{H}_0$. We again do this with the aid of c) in 5.4.

Finally, if we make use of (20) and rewrite (19) in the form

$$\|x - s_n\|^2 = \|x\|^2 - \sum_{k=1}^{n} |a_k|^2,$$

then we see easily that the second part of the theorem is true.

If the system $\{x_k\}$ is complete, then $\mathbf{H}_0 = \mathbf{H}$ and the projection of any element $x \in \mathbf{H}$ in $\mathbf{H}_0$ is x itself. This leads to the following

COROLLARY 1. *If the system $\{x_k\}$ is complete, then the Fourier series of every $x \in \mathbf{H}$ converges to x.*

An orthonormal system is said to be *closed* if each $x \in \mathbf{H}$ satisfies Parseval's identity. Taking the preceding results into account, we can now formulate

COROLLARY 2. *A system is closed if and only if it is complete.*

For, by Theorem 6, Parseval's identity is satisfied by all $x \in \mathbf{H}_0$ and only by these elements; therefore a system is closed precisely when $\mathbf{H}_0 = \mathbf{H}$, and this, in turn, is the case precisely when the system is complete.

REMARK. If Parseval's identity is satisfied for a fundamental set E in $\mathbf{H}$, then the given orthonormalized system is closed.

For, since $\mathbf{H}_0 \supset E$, we have $\mathbf{H}_0 = \overline{\mathscr{L}}(E) = \mathbf{H}$.

Thus if $\{x_n\}$ is a complete orthonormal system in $\mathbf{H}$, we can represent every element $x \in \mathbf{H}$ by its Fourier series:

$$x = \sum_{k=1}^{\infty} a_k x_k.$$

For this reason, a complete orthonormal system is called an *orthonormal basis* of the Hilbert space $\mathbf{H}$ (cf. XII.7.2).

THEOREM 7 (Riesz–Fischer). *Let $\{c_k\}$ be a numerical sequence for which the series $\sum_{k=1}^{\infty} |c_k|^2$ converges, and let $\{x_k\}$ be an orthonormal system in $\mathbf{H}$. Then there exists a unique element $x \in \mathbf{H}$ whose Fourier coefficients a_k are equal to the c_k, and which satisfies Parseval's identity.*

Proof. We see just as in the proof of the preceding theorem that the series $\sum_{k=1}^{\infty} c_k x_k$ converges. Let us denote its sum by x. Then

$$a_n = (x, x_n) = \lim_{m \to \infty} \left(\sum_{k=1}^{m} c_k x_k, x_n \right) = c_n \quad (n = 1, 2, \ldots),$$

since we have $\left(\sum_{k=1}^{m} c_k x_k, x_n \right) = c_n$ for $m \geqslant n$. The fact that x satisfies Parseval's identity now follows from the preceding theorem.

The uniqueness of the element x also follows from Theorem 6, because an element that satisfies Parseval's identity must be the sum of its Fourier series.

Let us go on to consider a generalization of Parseval's identity.

Suppose elements $x, y \in \mathbf{H}$ satisfy Parseval's identity. If $\{a_k\}$ and $\{b_k\}$ are corresponding sequences of Fourier coefficients of these elements, then

$$(x, y) = \sum_{k=1}^{\infty} a_k \bar{b}_k. \tag{22}$$

In fact, by Theorem 6,

$$x = \sum_{k=1}^{\infty} a_k x_k, \quad y = \sum_{k=1}^{\infty} b_k x_k;$$

therefore

$$(x, y) = \lim_{n\to\infty}\left(\sum_{k=1}^{n} a_k x_k, \sum_{k=1}^{n} b_k x_k\right) = \lim_{n\to\infty}\sum_{k=1}^{n} a_k \overline{b_k} = \sum_{k=1}^{\infty} a_k \overline{b_k}.$$

If the orthonormal system in question is complete, then any $x, y \in \mathbf{H}$ satisfy the generalized Parseval's identity.

If an infinite-dimensional separable Hilbert space $\mathbf{H}$ has an incomplete orthonormal system $\{x_k\}$ $(k = 1, 2, \ldots)$, then the subspace $\mathbf{H}_0$ spanned by $\{x_k\}$ is distinct from $\mathbf{H}$. Denote the orthogonal complement of $\mathbf{H}_0$ by $\tilde{\mathbf{H}}$. The subspace $\tilde{\mathbf{H}}$ is a separable Hilbert space and therefore we can find a countable complete orthonormal system $\{x_k\}$ $(k = 0, -1, -2, \ldots)$. Forming the union with the existing system, we obtain a system $\{x_k\}$ $(k = \ldots, -1, 0, 1, \ldots)$ which will be complete in the whole space $\mathbf{H}$. For an element $y \in \mathbf{H}$ orthogonal to all the x_k $(k = \ldots, -1, 0, 1, \ldots)$ must be orthogonal to $\mathbf{H}_0$; that is, we must have $y \in \tilde{\mathbf{H}}$. Then, since $\{x_k\}$ $(k = 0, -1, \ldots)$ is a complete system in $\tilde{\mathbf{H}}$, we have $y = \mathbf{0}$.

We note two facts from the theory of functions of a real variable, whose proofs are obtained using the general results presented above.

Since the trigonometric system is complete in $\mathbf{L}^2(a, b)$, Corollary 1 to Theorem 6 shows that every square-summable function has a decomposition into a Fourier (trigonometric) series, which converges in mean (of order 2).

It also follows from the Riesz–Fischer Theorem and inequality (21) that a necessary and sufficient condition for a given summable function to be an element of $\mathbf{L}^2$ is that the series of squares of the absolute values of its Fourier coefficients relative to the trigonometric system should be convergent.

It is interesting that the proofs of the analogous results for other classes of functions present great difficulties.

5.9. In conclusion, we prove that any two infinite-dimensional separable Hilbert spaces are linearly isometric (see 1.3); more precisely, one can establish a one-to-one correspondence between their elements which respects the linear operations and the inner product (and hence also the norm). Let $\mathbf{H}$ and $\mathbf{H}'$ be such spaces. By Theorem 4 there exist countable complete orthonormal systems $\{x_k\}$ in $\mathbf{H}$ and $\{x'_k\}$ in $\mathbf{H}'$. Let x be an arbitrary element in $\mathbf{H}$. Denote the sequence of Fourier coefficients of this element by $\{a_k\}$. Then, by Corollary 1 to Theorem 6,

$$x = \sum_{k=1}^{\infty} a_k x_k,$$

and, by inequality (21), $\sum_{k=1}^{\infty} |a_k|^2 < \infty$. Applying the Riesz–Fischer Theorem (in the space $\mathbf{H}'$) to the sequence $\{a_k\}$, we find an element $x' \in \mathbf{H}'$ having the same Fourier coefficients as x. Thus

$$x' = \sum_{k=1}^{\infty} a_k x'_k.$$

We associate this element x' with the given element x. This correspondence satisfies all the conditions of 1.3 (in place of equality of norms one needs to speak of equality of the inner products of the corresponding elements). Because of its simplicity, we shall not stop to verify that the algebraic operations are respected, but proceed to prove that if $x, y \in \mathbf{H}$ and

x' and y' are the corresponding elements in $\mathbf{H}'$, then $(x, y) = (x', y')$. In fact, since both the systems $\{x_k\}$ and $\{x'_k\}$ are complete, the elements x, y, and likewise x', y', satisfy the generalized Parseval's identity (22); that is,

$$(x, y) = \sum_{k=1}^{\infty} a_k \overline{b_k}, \quad (x', y') = \sum_{k=1}^{\infty} a_k \overline{b_k},$$

where $\{a_k\}$ denotes the sequence of Fourier coefficients of x (and x') and $\{b_k\}$ those of y (and y').

In particular, every infinite-dimensional separable Hilbert space $\mathbf{H}$ is linearly isometric (in the above sense) to the space ℓ^2. This result was proved by von Neumann [1].

Let us note without proof the analogous result for an arbitrary Hilbert space. It can be proved that the cardinality of a complete orthonormal system is an invariant of the space (its dimension) and that spaces of the same dimension are linearly isometric. For a proof of this fact, see, for example, Akhiezer and Glazman.

V

LINEAR OPERATORS AND FUNCTIONALS

§ 1. Spaces of operators and dual spaces

1.1. Since we shall sometimes need to consider non-linear operators as well as linear ones, in this section we introduce notation that will apply to both cases.

Let $\mathbf{X}$ and $\mathbf{Y}$ be two spaces (metric or normed) and let Ω be a subset of $\mathbf{X}$.

If with each element $x \in \Omega$ there is associated an element $y = U(x) \in \mathbf{Y}$, then we say that U is an *operator* mapping Ω into $\mathbf{Y}$ (or an operator from Ω into $\mathbf{Y}$). The set Ω is called the *domain of definition* of U, and is denoted by Ω_U. The set Δ_U of all elements of the form* $y = U(x)$ is called *the range of values* of U. If $\Omega_U = \mathbf{X}$ and $\Delta_U \subset \mathbf{X}$—that is, if U maps $\mathbf{X}$ into itself—then we say that U is an operator on $\mathbf{X}$.

If U is an operator from $\Omega \subset \mathbf{X}$ into $\mathbf{Y}$, then we shall write $U(E)$ $(E \subset \Omega)$ for the set of all $y \in \mathbf{Y}$ expressible in the form $y = U(x)$, where $x \in E$. In this situation, $U(E)$ is called the *image* of E.

Let us recall the definitions of the most important classes of operators.

a) Let $\mathbf{X}$ and $\mathbf{Y}$ be metric spaces and U an operator from $\Omega \subset \mathbf{X}$ into $\mathbf{Y}$. Then U is said to be *continuous at the point* $x_0 \in \Omega$ if, whenever $x_n \to x$ $(x_n \in \Omega)$, we have $U(x_n) \to U(x)$. If U is continuous at every point of a set $E \subset \Omega$, then we say simply that U is *continuous on* E.

b) If $\mathbf{X}$ and $\mathbf{Y}$ are normed spaces and Ω is a linear manifold contained in $\mathbf{X}$ then an operator U from Ω into $\mathbf{Y}$ is said to be *homogeneous* if

$$U(\lambda x) = \lambda U(x) \quad (x \in \Omega,\ \lambda \in \mathbb{K}).$$

U is said to be *additive* if

$$U(x_1 + x_2) = U(x_1) + U(x_2) \quad (x_1, x_2 \in \Omega).$$

c) U is said to be a *linear* operator if it is both additive and homogeneous on Ω_U.

Notice that if $\mathbf{X}$ and $\mathbf{Y}$ are normed spaces and U is an additive operator from $\Omega \subset \mathbf{X}$ into $\mathbf{Y}$ which is continuous at a particular point $x_0 \in \Omega$, then U is continuous on Ω, that is, continuous at every point of Ω.

For suppose that $x_n \to x$ $(x_n, x \in \Omega)$. Since

$$x_n = [x_0 + (x_n - x)] + (x - x_0)$$

and

$$x_0 + (x_n - x) \to x_0,$$

* Often we write Ux instead of $U(x)$ and, in the case of non-linear operators, we use the word "mapping" instead of "operator".

we have

$$U(x_n) = U(x_0 + (x_n - x)) + U(x - x_0) \to U(x_0) + U(x - x_0) = U(x).$$

1.2. We now present a continuity criterion for linear operators.

A linear operator U mapping a normed space $\mathbf{X}$ into a normed space $\mathbf{Y}$ is said to be *bounded* if there exists a constant C such that, for all $x \in \mathbf{X}$, we have

$$\|U(x)\| \leqslant C\|x\|. \tag{1}$$

THEOREM 1. *A necessary and sufficient condition for a linear operator U to be continuous is that it be bounded.*

Proof. Necessity. Let U be a continuous linear operator. We show that

$$C_0 = \sup_{\substack{\|x\|=1 \\ x\in\mathbf{X}}} \|U(x)\| < \infty.$$

If it were the case that $C_0 = \infty$, then we could find a sequence $\{x_n\}$ ($x_n \in \mathbf{X}$, $\|x_n\| = 1$) such that $\lambda_n = \|U(x_n)\| \to \infty$. Consider the sequence $\{x'_n\}$, $x'_n = x_n/\lambda_n$. Clearly $x'_n \to \mathbf{0}$; therefore, by the continuity of U we would have $U(x'_n) \to \mathbf{0}$, whereas in fact $\|U(x'_n)\| = 1$.

Now let $x \neq \mathbf{0}$ be any element of $\mathbf{X}$. Writing $x' = x/\|x\|$, we see that $\|x'\| = 1$. Therefore $\|U(x')\| \leqslant C_0$. But, by the homogeneity of U, we have $U(x') = (1/\|x\|)U(x)$, so that

$$\|U(x)\| \leqslant C_0\|x\|$$

and (1) is satisfied with $C = C_0$.

Sufficiency. It follows immediately from (1) that U is continuous at $\mathbf{0}$. By 1.1, U is continuous at every point of $\mathbf{X}$.

The theorem is thus proved.

Let us show that C_0 is the smallest constant satisfying (1). In fact, if $\|x\| = 1$, then $\|U(x)\| \leqslant C$, and therefore $C_0 \leqslant C$. But, on the other hand, C_0 satisfies (1).

REMARK. It is not hard to see that

$$C_0 = \sup_{\|x\| \leqslant 1} \|U(x)\|.$$

For we obviously have $C_0 \leqslant \sup_{\|x\| \leqslant 1} \|U(x)\|$. Moreover, for $x \in \mathbf{X}$ ($\|x\| \leqslant 1$, $x \neq \mathbf{0}$) we have $\|U(x)\| = \|x\| \left\|U\left(\frac{x}{\|x\|}\right)\right\| \leqslant C_0$, which gives the reverse inequality.

The number C_0 determined by the linear operator U is called the *norm** of U and is denoted by $\|U\|$. By the above remarks,

$$\|U\| = C_0 = \sup_{\|x\|=1} \|U(x)\| = \sup_{\|x\| \leqslant 1} \|U(x)\|.$$

Taking $C = C_0 = \|U\|$ in (1), we obtain

$$\|U(x)\| \leqslant \|U\|\,\|x\|.$$

Notice also that if an inequality of the form (1) holds for some C, then $\|U\| \leqslant C$.

* The norm of an operator in a Hilbert space was first defined by Hilbert (see Hilbert). For the more general definition, see Banach [1].

The concepts of linearity and continuity for operators and functionals had been introduced before this by several authors.

Finally, we draw attention to a simple geometric interpretation of $\|U\|$—it is the least upper bound of the coefficients of dilatation for vectors under the transformation induced by U.

1.3. Consider two normed spaces $\mathbf{X}$ and $\mathbf{Y}$. The set $L(\mathbf{X}, \mathbf{Y})$ of all linear operators mapping $\mathbf{X}$ into $\mathbf{Y}$ is a vector space (II.2.1). We denote by $B(\mathbf{X}, \mathbf{Y})$ the set of all continuous linear operators from $\mathbf{X}$ into $\mathbf{Y}$. We now verify that $B(\mathbf{X}, \mathbf{Y})$ is a linear manifold in $L(\mathbf{X}, \mathbf{Y})$ and that the operator norm $\|U\|$ is a norm on $B(\mathbf{X}, \mathbf{Y})$—that is, $B(\mathbf{X}, \mathbf{Y})$ is a normed space.

If $U_1, U_2 \in B(\mathbf{X}, \mathbf{Y})$, $U = U_1 + U_2$, then

$$\|U(x)\| \leqslant \|U_1(x)\| + \|U_2(x)\| \leqslant (\|U_1\| + \|U_2\|)\|x\|.$$

Thus $U \in B(\mathbf{X}, \mathbf{Y})$ and $\|U\| \leqslant \|U_1\| + \|U_2\|$.

If $U \in B(\mathbf{X}, \mathbf{Y})$, $\lambda \in \mathbb{K}$, $\tilde{U} = \lambda U$, then $\|\tilde{U}\| = \|\lambda U\| = |\lambda|\ \|U\|$.

If $\|U\| = 0$, then it clearly follows that $\|U(x)\| = 0$ for every $x \in \mathbf{X}$. Hence $U = \mathbf{0}$. We have thus proved that $B(\mathbf{X}, \mathbf{Y})$ is a normed space.

Let us prove that if the image space $\mathbf{Y}$ is complete, then $B(\mathbf{X}, \mathbf{Y})$ is also complete.

In fact, let $\{U_n\}$ be a Cauchy sequence of elements of $B(\mathbf{X}, \mathbf{Y})$. Taking any $\varepsilon > 0$, we have

$$\|U_m - U_n\| < \varepsilon \quad (m, n \geqslant N_\varepsilon),$$

that is, for every $x \in \mathbf{X}$,

$$\|U_m(x) - U_n(x)\| < \varepsilon\|x\|, \tag{2}$$

which implies that the sequence $\{U_n(x)\}$ of elements of $\mathbf{Y}$ is a Cauchy sequence, so that, by the completeness of $\mathbf{Y}$, we deduce the existence of

$$U(x) = \lim_{n \to \infty} U_n(x) \quad (x \in \mathbf{X}).$$

Clearly $U \in L(\mathbf{X}, \mathbf{Y})$. Letting $m \to \infty$ in (2), we obtain

$$\|U(x) - U_n(x)\| = \lim_{m \to \infty} \|U_m(x) - U_n(x)\| \leqslant \varepsilon\|x\| \; (n \geqslant N_\varepsilon), \tag{3}$$

that is, the operator V defined by:

$$V(x) = U(x) - U_n(x) (x \in \mathbf{X})$$

is an element of $B(\mathbf{X}, \mathbf{Y})$. Therefore, so is $U = V + U_n$. Moreover, (3) implies that

$$\|U - U_n\| \leqslant \varepsilon \quad (n \geqslant N_\varepsilon),$$

which shows that $U_n \to U$ in $B(\mathbf{X}, \mathbf{Y})$, as required.

The above remarks give us the following theorem.

THEOREM 2. *If* $\mathbf{X}$ *is a normed space and* $\mathbf{Y}$ *is a B-space, then* $B(\mathbf{X}, \mathbf{Y})$ *is a B-space.*

Since the space of scalars $\mathbb{K}$ is complete, it follows that the space of all continuous linear functionals $B(\mathbf{X}, \mathbb{K})$, which we now denote by $\mathbf{X}^*$, is a B-space. This B-space $\mathbf{X}^*$ is said to be *dual* (or *conjugate*) to $\mathbf{X}$. If $f \in \mathbf{X}^*$, then

$$\|f\| = \sup_{\|x\| \leqslant 1} |f(x)|.$$

We recall (II.2.2) that if we are dealing with a complex space $\mathbf{X}$ then multiplication by a complex number is defined in $\mathbf{X}^*$ by

$$(\lambda f)(x) = \bar{\lambda} f(x) \quad (f \in \mathbf{X}^*, \; x \in \mathbf{X}). \tag{4}$$

The reason for the appropriateness of the definition (4) will be given later (see 3.2).

§ 2. Some functionals and operators on specific spaces

2.1. We consider the following functional on $\mathbf{C}[a, b]$:

$$f(x) = \sum_{k=1}^{n} c_k x(t_k), \tag{1}$$

where $t_1, t_2, \ldots, t_n$ is some system of points in the interval $[a, b]$. Examples of such functionals include: the value of the function at a fixed point, the finite differences of a function, Riemann sums, and weighted sums with respect to some system of nodes.

Let us show that the functional defined by (1) is linear, and that

$$\|f\| = \sum_{k=1}^{n} |c_k|. \tag{2}$$

The linearity of f is obvious:

$$f(\lambda x_1 + \mu x_2) = \sum_{k=1}^{n} c_k[\lambda x_1(t_k) + \mu x_2(t_k)] = \sum_{k=1}^{n} c_k \lambda x_1(t_k) + \sum_{k=1}^{n} c_k \mu x_2(t_k) = \lambda f(x_1) + \mu f(x_2).$$

Moreover, it is clear from the inequality

$$|f(x)| = \left|\sum_{k=1}^{n} c_k x(t_k)\right| \leqslant \max_{a \leqslant t \leqslant b} |x(t)| \sum_{k=1}^{n} |c_k| = \sum_{k=1}^{n} |c_k| \|x\|$$

that f is continuous, and that

$$\|f\| \leqslant \sum_{k=1}^{n} |c_k|.$$

We now consider the piecewise linear function $\tilde{x}$ on $[a, b]$ taking the values

$$\tilde{x}(t_k) = \operatorname{sign} c_k \quad (k = 1, 2, \ldots, n),$$

at $t_1, t_2, \ldots, t_n$, and linear in the intervals $[t_k, t_{k+1}]$ $(k = 1, 2, \ldots, n-1)$, and constant in $[a, t_1]$ and $[t_n, b]$.

Clearly,

$$|\tilde{x}(t)| \leqslant 1,$$

that is,

$$\|\tilde{x}\| \leqslant 1.$$

Thus

$$\|f\| = \sup_{\|x\| \leqslant 1} |f(x)| \geqslant f(\tilde{x}) = \sum_{k=1}^{n} c_k \tilde{x}(t_k) = \sum_{k=1}^{n} c_k \operatorname{sign} c_k = \sum_{k=1}^{n} |c_k|,$$

which, together with the opposite inequality already established above, yields (2).

2.2. Consider the functional on $\mathbf{C}[a, b]$ defined by

$$f(x) = \int_a^b \phi(t) x(t) dt, \tag{3}$$

where ϕ is a given summable function.

The following are examples of such functionals: the integral of the function over the whole interval, or over part of it, the moments of the function, its Fourier coefficients, and so on.

Let us show that the functional (3) is linear, and that

$$\|f\| = \int_a^b |\phi(t)|dt. \tag{4}$$

Obviously, f is defined for all $x \in \mathbf{C}[a, b]$, and even for $x \in \mathbf{L}^\infty(a, b)$; and it is obviously also linear. The inequality

$$|f(x)| \leqslant \int_a^b |\phi(t)x(t)|dt \leqslant \max_{a \leqslant t \leqslant b} |x(t)| \int_a^b |\phi(t)|dt = \|x\| \int_a^b |\phi(t)|dt$$

implies that f is continuous, and yields a bound for its norm:

$$\|f\| \leqslant \int_a^b |\phi(t)|\,dt.$$

To establish the reverse inequality, we first consider the case where ϕ is continuous. Choose any $\varepsilon > 0$ and divide the interval $[a, b]$ up by choosing points $a = t_0 < t_1 < \ldots < t_n = b$ such that the oscillation of ϕ is less than ε in each $[t_k, t_{k+1}]$. We now assign each of these subintervals to one of two groups. To the first group we assign the intervals $\Delta'_1, \Delta'_2, \ldots, \Delta'_r$ on each of which the values of ϕ have the same sign (though the sign may vary from interval to interval). The remaining intervals $\Delta''_1, \Delta''_2, \ldots, \Delta''_s$ are all assigned to the second group. Notice that, since ϕ changes sign in Δ''_k $(k = 1, 2, \ldots, s)$, it takes the value zero there; and thus, remembering that its oscillation is less than ε, we deduce that

$$|\phi(t)| < \varepsilon \quad (t \in \Delta''_k;\ k = 1, 2, \ldots, s).$$

Next we define a function $\tilde{x}$ in $\mathbf{C}[a, b]$. On the intervals of the first group we set

$$\tilde{x}(t) = \operatorname{sign} \phi(t) \quad (t \in \Delta'_j;\ j = 1, 2, \ldots, r).$$

At the rest of the points of $[a, b]$ we take the function to be linear. Moreover, if a (or b) is an endpoint of an interval in the second group, we take $\tilde{x}(a) = 0$ (respectively, $\tilde{x}(b) = 0$).

Let us find a lower bound for

$$f(\tilde{x}) = \int_a^b \phi(t)\tilde{x}(t)dt.$$

Taking into account the fact that $|(\tilde{x}(t)| \leqslant 1$ $(a \leqslant t \leqslant b)$, we have

$$\int_a^b \phi(t)\tilde{x}(t)dt = \sum_{j=1}^r \int_{\Delta'_j} \phi(t)\tilde{x}(t)dt + \sum_{k=1}^s \int_{\Delta''_k} \phi(t)\tilde{x}(t)dt \geqslant \sum_{j=1}^r \int_{\Delta'_j} |\phi(t)|dt$$
$$- \sum_{k=1}^s \int_{\Delta''_k} |\phi(t)|dt = \int_a^b |\phi(t)|dt - 2\sum_{k=1}^s \int_{\Delta''_k} |\phi(t)|dt > \int_a^b |\phi(t)|dt - 2\varepsilon(b-a).$$

Moreover, since $\|\tilde{x}\| \leqslant 1$,

$$\|f\| \geqslant f(\tilde{x}) > \int_a^b |\phi(t)|\,dt - 2\varepsilon(b-a).$$

Letting $\varepsilon \to 0$ here, we obtain the required inequality.

Now take ϕ to be any summable function. Since the set of all continuous functions is dense in the space $\mathbf{L}^1$, we can find a continuous function $\tilde{\phi}$ such that

$$\|\phi-\tilde{\phi}\|_{\mathbf{L}^1}=\int_a^b|\phi(t)-\tilde{\phi}(t)|dt<\varepsilon.$$

Starting from the function $\tilde{\phi}$ we construct a function $\tilde{x}\in\mathbf{C}[a,b]$ with $\|\tilde{x}\|\leqslant 1$ in the manner described above: that is, such that

$$\int_a^b\tilde{\phi}(t)\tilde{x}(t)dt\geqslant\int_a^b|\tilde{\phi}(t)|dt-2\varepsilon(b-a)=\|\tilde{\phi}\|_{\mathbf{L}^1}-2\varepsilon(b-a).$$

However,

$$\|\tilde{\phi}\|_{\mathbf{L}^1}\geqslant\|\phi\|_{\mathbf{L}^1}-\|\phi-\tilde{\phi}\|_{\mathbf{L}^1}>\|\phi\|_{\mathbf{L}^1}-\varepsilon,$$

and so

$$f(\tilde{x})=\int_a^b\phi(t)\tilde{x}(t)dt=\int_a^b\tilde{\phi}(t)\tilde{x}(t)dt+\int_a^b[\phi(t)-\tilde{\phi}(t)]\tilde{x}(t)dt>\|\phi\|_{\mathbf{L}^1}-\varepsilon-2\varepsilon(b-a)-\int_a^b|\phi(t)-\tilde{\phi}(t)|dt>\|\phi\|_{\mathbf{L}^1}-2\varepsilon(b-a+1).$$

As before, we conclude from this that

$$\|f\|\geqslant f(\tilde{x})>\|\phi\|_{\mathbf{L}^1}-2\varepsilon(b-a+1),$$

so that, as $\varepsilon\to 0$,

$$\|f\|\geqslant\|\phi\|_{\mathbf{L}^1},$$

as we required to prove.

2.3. One can also consider functionals of the form (3) on the space $\mathbf{L}^p(a,b)$. More generally, let (T,Σ,μ) be a σ-finite measure space. We prove* that

$$f(x)=\int_T x(t)y(t)d\mu, \tag{5}$$

where $y\in\mathbf{L}^q(T,\Sigma,\mu)$ $(1\leqslant p\leqslant\infty,\ 1/p+1/q=1)$, is a continuous linear functional on $\mathbf{L}^p(T,\Sigma,\mu)$, and that

$$\|f\|=\|y\|_{\mathbf{L}^q}=\begin{cases}\left[\int_T|y(t)|^q d\mu\right]^{1/q}, & 1<p\leqslant\infty,\\ \operatorname{vrai\,sup}_{t\in T}|y(t)|, & p=1.\end{cases} \tag{6}$$

1) Consider first the case $1<p<\infty$. The integral (5) is defined for all $x\in\mathbf{L}^p$. This follows from Hölder's inequality for integrals (see IV.2.3). Using the same inequality we find that

$$|f(x)|=\left|\int_T x(t)y(t)d\mu\right|\leqslant\left[\int_T|y(t)|^q d\mu\right]^{1/q}\left[\int_T|x(t)|^p d\mu\right]^{1/p}=\|y\|_{\mathbf{L}^q}\|x\|,$$

and so

$$\|f\|\leqslant\|y\|_{\mathbf{L}^q}. \tag{7}$$

* We shall prove later that the functionals expressible in the form (5) exhaust the set of linear functionals on $\mathbf{L}^p(1\leqslant p<\infty)$: that is, (5) is a general form for any linear functional on $\mathbf{L}^p$ (see Chapter VI, § 2).

To prove the reverse inequality, consider the function

$$\tilde{x}(t) = |y(t)|^{q-1} \operatorname{sign} y(t).$$

Since

$$|\tilde{x}(t)|^p = |y(t)|^{p(q-1)} = |y(t)|^q,$$

we have $\tilde{x} \in \mathbf{L}^p$; moreover,

$$\|x\| = \left[\int_T |\tilde{x}(t)|^p \, d\mu\right]^{\frac{1}{p}} = \left[\int_T |y(t)|^q \, d\mu\right]^{\frac{1}{q}\cdot\frac{q}{p}} = [\|y\|_{\mathbf{L}^q}]^{\frac{q}{p}},$$

and consequently

$$\|f\| \geqslant f\left(\frac{\tilde{x}}{\|\tilde{x}\|}\right) = \frac{1}{\|\tilde{x}\|}\int_a^b y(t)\tilde{x}(t)\, d\mu = \frac{1}{\|\tilde{x}\|}\int_a^b |y(t)|^q \, d\mu = [\|y\|_{\mathbf{L}^q}]^{q\left(1-\frac{1}{p}\right)} = \|y\|_{\mathbf{L}^q}.$$

Together with (7), this yields the equation we require.

2) Now suppose $p = 1$. Write

$$\|y\|_{\mathbf{L}^\infty} = \operatorname*{vrai\,sup}_{t \in T} |y(t)| = Q.$$

Obviously, $\|f\| \leqslant Q$. On the other hand, choose $\varepsilon > 0$ and let A be the set

$$A = \{t \in T\colon\ y(t) > Q - \varepsilon\}.$$

Note that $\mu(A) > 0$. Define

$$\tilde{x}(t) = \begin{cases} \operatorname{sign} y(t), & t \in A, \\ 0, & t \notin A. \end{cases}$$

Clearly, $\tilde{x} \in \mathbf{L}^1$ and $\|\tilde{x}\| = \int_a^b |\tilde{x}(t)| \, d\mu = \mu(A)$. Furthermore,

$$\|f\| \geqslant f\left(\frac{\tilde{x}}{\|\tilde{x}\|}\right) = \frac{1}{\mu(A)}\int_T y(t)\tilde{x}(t)\, d\mu = \frac{1}{\mu(A)}\int_T |\phi(t)|\, d\mu \geqslant \frac{1}{\mu(A)}[Q-\varepsilon]\mu(A) = Q - \varepsilon.$$

As ε was arbitrary, it follows from this that $\|f\| \geqslant Q$, which, with the above, gives $\|f\| = Q$.

3) Finally, suppose $p = \infty$. The inequality $\|f\| \leqslant \|y\|_{\mathbf{L}^1}$ is obvious. On the other hand, the function $\tilde{x}(t) = \operatorname{sign} y(t)$ belongs to $\mathbf{L}^\infty$ and $\|\tilde{x}\|_{\mathbf{L}^\infty} \leqslant 1$. From the inequality

$$\|f\| \geqslant |f(\tilde{x})| = \int_T \operatorname{sign} y(t) \cdot y(t)\, d\mu = \int_T |y(t)|\, d\mu = \|y\|_{\mathbf{L}^1}$$

we deduce that $\|f\| \geqslant \|y\|_{\mathbf{L}^1}$ and thus finally we have $\|f\| = \|y\|_{\mathbf{L}^1}$.

REMARK. If we are considering the complex space $\mathbf{L}^p$, then it is more convenient to express the functional (5) in the form

$$f(x) = \int_T x(t)\overline{y(t)}\, d\mu.$$

Then (6) is clearly unchanged (cf. 1.3).

2.4. Consider the integral operator $y = U(x)$ defined by

$$y(s) = \int_a^b K(s, t)x(t)\,dt, \tag{8}$$

where $K(s, t)$ is a continuous function. As examples of operators of the form (8), we mention the Dirichlet and Fejér integrals. Let us show that U is a continuous linear operator from $\mathbf{C}[a, b]$ into $\mathbf{C}[a, b]$, and that

$$\|U\| = \max_{a \leqslant s \leqslant b} \int_a^b |K(s, t)|\,dt = M. \tag{9}$$

It is obvious that the operator (8) is linear, and the inequality

$$\|U(x)\| = \max_{a \leqslant s \leqslant b} \left| \int_a^b K(s, t)x(t)\,dt \right| \leqslant \|x\| \max_{a \leqslant s \leqslant b} \int_a^b |K(s, t)|\,dt = M\|x\|$$

shows that it is continuous, and that $\|U\| \leqslant M$. We now establish the reverse inequality. Since the integral

$$\int_a^b |K(s, t)|\,dt$$

is a continuous function of s, there exists an $s_0 \in [a, b]$ such that

$$M = \max_{a \leqslant s \leqslant b} \int_a^b |K(s, t)|\,dt = \int_a^b |K(s_0, t)|\,dt.$$

Consider the functional of type (3) on $\mathbf{C}[a, b]$ with $\phi(t) = K(s_0, t)$; that is, .

$$f(x) = \int_a^b K(s_0, t)x(t)\,dt \quad (x \in \mathbf{C}[a, b]).$$

By the above (see 2.2), we can find an element $\tilde{x} \in \mathbf{C}[a, b]$ with $\|\tilde{x}\| \leqslant 1$ such that

$$f(\tilde{x}) \geqslant \|f\| - \varepsilon = \int_a^b |K(s_0, t)|\,dt - \varepsilon = M - \varepsilon.$$

Setting $\tilde{y} = U(\tilde{x})$, we clearly have

$$\|U\| \geqslant \|U(\tilde{x})\| = \|\tilde{y}\| \geqslant \tilde{y}(s_0) = \int_a^b K(s_0, t)\tilde{x}(t)\,dt = f(\tilde{x}) \geqslant M - \varepsilon$$

and, as ε was arbitrary, this leads to (9).

2.5. We now show that the integral operator (8) can also be considered as a linear operator from $\mathbf{L}^1(a, b)$ into $\mathbf{L}^1(a, b)$; however, in this case

$$\|U\| = \max_{a \leqslant t \leqslant b} \int_a^b |K(s, t)|\,ds = M'. \tag{10}$$

The additivity of U is clear in this case too. Using Fubini's Theorem on interchanging the order of integration, we have the inequality

$$\|U(x)\| = \int_a^b \left| \int_a^b K(s, t)x(t)\,dt \right| ds \leqslant \int_a^b \left[\int_a^b |K(s, t)|\,ds \right] |x(t)|\,dt \leqslant M'\|x\|,$$

so that

$$\|U\| \leqslant M'.$$

Next, if as above we choose $t_0 \in [a, b]$ such that

$$M' = \max_{a \leqslant t \leqslant b} \int_a^b |K(s, t)|\, ds = \int_a^b |K(s, t_0)| ds,$$

and take any $\varepsilon > 0$, then by the uniform continuity of $K(s, t)$, we can find $\delta > 0$ such that

$$|K(s', t') - K(s, t)| < \varepsilon,$$

provided $|s - s'| < \delta, |t - t'| < \delta$. Now surround t_0 by an interval $d = [t_1, t_2]$ $(a \leqslant t_1 \leqslant t_0 \leqslant t_2 \leqslant b)$ having length less than $\delta (t_2 - t_1 < \delta)$. We set

$$\tilde{x}(t) = \begin{cases} \dfrac{1}{t_2 - t_1}, & t \in d, \\ 0, & t \notin d \end{cases} \qquad (\|\tilde{x}\| = 1).$$

Writing $\tilde{y} = U(\tilde{x})$, we find that

$$\|U\| \geqslant \|U(\tilde{x})\| = \|\tilde{y}\| = \int_a^b |\tilde{y}(s)|\, ds = \int_a^b \left| \int_a^b K(s, t)\tilde{x}(t)\, dt \right| ds = \frac{1}{t_2 - t_1} \int_a^b \left| \int_{t_1}^{t_2} K(s, t)\, dt \right| ds \geqslant$$

$$\geqslant \frac{1}{t_2 - t_1} \int_a^b \left| \int_{t_1}^{t_2} K(s, t_0)\, dt \right| ds - \frac{1}{t_2 - t_1} \int_a^b \left[\int_{t_1}^{t_2} |K(s, t) - K(s, t_0)|\, dt \right] ds$$

$$= M' - \varepsilon(b - a),$$

which gives (10), as required.

The reader will convince himself without difficulty that if this operator (8) is regarded as an operator from $\mathbf{L}^1(a, b)$ into $\mathbf{C}[a, b]$, then

$$\|U\| = \max_{\substack{a \leqslant s \leqslant b \\ a \leqslant t \leqslant b}} |K(s, t)|.$$

2.6. Now we regard the operator (8) as an operator from $\mathbf{L}^2(a, b)$ into $\mathbf{L}^2(a, b)$. Here we relax the requirement on the kernel $K(s, t)$ of the operator. Namely, instead of assuming it continuous, we assume it square-measurable, and that

$$\int_a^b \int_a^b |K(s, t)|^2\, ds dt = N^2 < \infty. \tag{11}$$

Let us prove that, in this situation, the expression (8) defines a continuous operator from $\mathbf{L}^2$ into $\mathbf{L}^2$, and that

$$\|U\| \leqslant N. \tag{12}$$

The continuity of the operator (8) and the bound (12) for the norm are obtained as a corollary to Buniakowski's inequality:

$$\|U(x)\|^2 = \int_a^b \left| \int_a^b K(s,t)x(t)dt \right|^2 ds \leqslant \int_a^b \left[\int_a^b |K(s,t)|^2 dt \int_a^b |x(t)|^2 dt \right] ds = N^2 \|x\|^2.$$

Let us prove that, in the case of a symmetric kernel ($K(s,t) = K(t,s)$), we have the precise equation

$$\|U\| = \frac{1}{|\lambda_1|}, \tag{13}$$

where λ_1 is the eigenvalue of $K(s,t)$ having least absolute value.

If, however, the kernel is asymmetric, then the norm of U is given by a more complicated expression, namely

$$\|U\| = \frac{1}{\sqrt{\Lambda_1}}, \tag{14}$$

where Λ_1 is the least eigenvalue of the kernel

$$K^*(s,t) = \int_a^b K(\tau, s)K(\tau, t)\,d\tau.$$

To establish equations (13) and (14), we shall need to use certain facts from the theory of integral equations.* As is well known, an integral equation with symmetric kernel has a complete system of (normalized) eigenfunctions and characteristic values: $\omega_1(t), \omega_2(t), \ldots, \lambda_1, \lambda_2, \ldots$, namely, the set of linearly independent solutions of the homogeneous equation and those λ for which the homogeneous equation has non-zero solutions:

$$\omega_k(s) = \lambda_k \int_a^b K(s,t)\omega_k(t)dt \qquad (k = 1, 2, \ldots).$$

By the Hilbert–Schmidt Theorem,* every function expressible by means of a kernel—in particular, $y(s)$—is a value of an operator expressible in terms of eigenfunctions by a series that converges in mean (in $\mathbf{L}^2$):

$$y(s) = \sum_k \frac{h_k}{\lambda_k}\omega_k(s) \quad \left(h_k = \int_a^b x(t)\omega_k(t)dt; \quad k = 1, 2, \ldots \right).$$

Hence (cf. Theorem IV.5.6)

$$\|y\|^2 = \int_a^b |y(s)|^2 ds = \sum_k \frac{h_k^2}{\lambda_k^2} \leqslant \frac{1}{\lambda_1^2} \sum_k h_k^2 \leqslant \frac{1}{\lambda_1^2}\|x\|^2.$$

---

* See, for example, Petrovskii.

This shows that $\|U\| \leqslant \frac{1}{|\lambda_1|}$. Taking $\tilde{x} = \omega_1$, we have $\tilde{y} = U(\tilde{x}) = \frac{1}{\lambda_1}\tilde{x}$, and so

$$\|U\| \geqslant \|U(\tilde{x})\| = \frac{1}{|\lambda_1|}\|\tilde{x}\| = \frac{1}{|\lambda_1|}.$$

This establishes (13).

Now turn to the case of an arbitrary kernel. Using Fubini's Theorem, we have

$$\begin{aligned}\|y\|^2 &= \int_a^b |y(s)|^2\,ds = \int_a^b \left[\int_a^b K(s,t)x(t)\,dt \int_a^b K(s,\tau)x(\tau)\,d\tau\right] ds \\ &= \int_a^b x(\tau)\,d\tau \int_a^b \left[\int_a^b K(s,t)K(s,\tau)\,ds\right] x(t)\,dt \\ &= \int_a^b\int_a^b K^*(t,\tau)x(t)x(\tau)\,dt\,d\tau.\end{aligned}$$

Denoting the characteristic values and eigenfunctions of $K^*(t, \tau)$ by Λ_k, $\Omega_k(t)$, we find, by applying the Hilbert–Schmidt Theorem again, that

$$\|y\|^2 = \int_a^b x(\tau)d\tau \int_a^b K^*(t,\tau)x(t)\,dt = \int_a^b x(\tau)d\tau \sum_k \frac{H_k}{\Lambda_k}\Omega_k(\tau) = \sum_k \frac{H_k^2}{\Lambda_k} \leqslant \frac{1}{\Lambda_1}\|x\|$$

$$\left(H_k = \int_a^b x(t)\,\Omega_k(t)\,dt, \quad k = 1, 2, \ldots\right).$$

It is thus clear that $\|U\| \leqslant \frac{1}{\sqrt{\Lambda_1}}$; exactly as in the preceding case, we see, by taking $\tilde{x} = \Omega_1$, that we have equality here: that is, that (14) is true.

2.7. Analogous arguments enable one also to define the norm of an operator U in the case where there is a term outside the integral. Let

$$y = U(x), \quad y(s) = x(s) - \lambda\int_a^b K(s,t)\,x(t)\,dt. \tag{15}$$

We restrict ourselves to the case of a symmetric kernel. We again apply the Hilbert–Schmidt Theorem, first completing the system of eigenfunctions by adjoining functions ω_0, ω_{-1}, ω_{-2} .... The function x will have Fourier coefficients relative to this system, which we denote by $\ldots, h_{-1}, h_0, h_1, h_2, \ldots$. The function y has Fourier coefficients

$$\eta_k = \int_a^b y(s)\,\omega_k(s)\,ds = \int_a^b \left[x(s) - \lambda\int_a^b K(s,t)\,x(t)\,dt\right]\omega_k(s)\,ds = h_k\left(1 - \frac{\lambda}{\lambda_k}\right)$$

$$(k = 1, 2, \ldots),\ \eta_k = h_k\ \ (k = 0, -1, -2, \ldots).$$

Notice that we can use the first of these formulae in all cases if we take $\lambda_k = \infty$ for $k = 0, -1, -2, \ldots$. As a result, we have

$$\|y\|^2 = \sum_{k=-\infty}^{\infty} |\eta_k|^2 = \sum_{k=-\infty}^{\infty} |h_k|^2 \left|1 - \frac{\lambda}{\lambda_k}\right|^2 \leqslant L^2 \sum |h_k|^2 = L^2 \|x\|^2,$$

where we write

$$L = \sup_k \left|1 - \frac{\lambda}{\lambda_k}\right|.$$

Evidently,

$$\|U\| = L.$$

2.8. Now we consider operators on finite-dimensional spaces. Let **X** and **Y** be finite-dimensional B-spaces having dimensions m and n respectively. Let U be a linear operator mapping **X** into **Y**. Denote by $e_k (k = 1, 2, \ldots, m)$ the element of **X** whose coordinates are all zero except the k-th, which is unity. Let g_k be the analogous elements in **Y** ($k = 1, 2, \ldots, n$). If $x = (\xi_1, \xi_2, \ldots, \xi_m) \in \mathbf{X}$ and $y = (\eta_1, \eta_2, \ldots, \eta_n) \in \mathbf{Y}$ then

$$x = \sum_{k=1}^{m} \xi_k e_k, \quad y = \sum_{k=1}^{n} \eta_k g_k.$$

Hence, if $y = U(x)$, then we have

$$y = \sum_{k=1}^{m} \xi_k U(e_k).$$

Denoting the coordinates of the element $U(e_k)$ $(k = 1, 2, \ldots, m)$ by $a_{1k}, a_{2k}, \ldots, a_{nk}$, we deduce from this that

$$\|y\| = \sum_{j=1}^{n} \eta_j g_j = \sum_{k=1}^{m} \xi_k \sum_{j=1}^{n} a_{jk} g_j = \sum_{j=1}^{n} \left(\sum_{k=1}^{m} a_{jk} \xi_k \right) g_j.$$

Consequently, since the elements $g_1, g_2, \ldots, g_n$ are linearly independent, we have

$$\eta_j = \sum_{k=1}^{m} a_{jk} \xi_k \quad (j = 1, 2, \ldots, n),$$

that is, the coordinates of $y = U(x)$ are obtained from those of x by transformation according to the matrix

$$A = \begin{pmatrix} a_{11} & a_{12} & & a_{1m} \\ a_{21} & a_{22} & \cdots & a_{2m} \\ \cdots & \cdots & \cdots & \cdots \\ a_{n1} & a_{n2} & \cdots & a_{nm} \end{pmatrix}.$$

Conversely, every matrix of this form determines a linear operator from **X** into **Y**, as we shall leave the reader to convince himself.* The norm of U depends on what metrics we use in the spaces **X** and **Y**.

a) Let us regard U as an operator from ℓ_m^∞ into ℓ_n^∞ (see IV.3.5). Then

$$\|y\| = \|U(x)\| = \max_j |\eta_j| \leqslant \max_j \sum_{k=1}^m |a_{jk}| \, |\xi_k| \leqslant \|x\| \max_j \sum_{k=1}^m |a_{jk}|,$$

that is,

$$\|U\| \leqslant \max_j \sum_{k=1}^m |a_{jk}| = L.$$

We choose j_0 such that

$$\sum_{k=1}^m |a_{j_0 k}| = L,$$

and construct $\tilde{x} = (\tilde{\xi}_1, \tilde{\xi}_2, \ldots, \tilde{\xi}_m)$ by setting

$$\tilde{\xi}_k = \operatorname{sign} a_{j_0 k} \quad (k = 1, 2, \ldots m).$$

Then

$$\|U\| \geqslant \|U(\tilde{x})\| = \max_j \left| \sum_{k=1}^m a_{jk} \tilde{\xi}_k \right| \geqslant \sum_{k=1}^m a_{j_0 k} \tilde{\xi}_k = \sum_{k=1}^m |a_{j_0 k}| = L.$$

Thus

$$\|U\| = L = \max_j \sum_{k=1}^m |a_{jk}|.$$

b) If U is regarded as an operator from ℓ_m^1 into ℓ_n^1 (see IV.3.5), then

$$\|U\| = \max_k \sum_{j=1}^n |a_{jk}| = L'. \tag{16}$$

In fact, we clearly have $\|U\| \leqslant L'$. Moreover, there exists a k_0 such that

$$\sum_{j=1}^n |a_{jk_0}| = L'.$$

It is now sufficient to take $\tilde{x} = (0, 0, \ldots, 1, \ldots, 0)$ (the 1 in the k_0-th place). Then

$$\|U\| \geqslant \|U(\tilde{x})\| = \sum_{j=1}^n \left| \sum_{k=1}^m a_{jk} \tilde{\xi}_k \right| = \sum_{j=1}^n |a_{jk_0}| = L',$$

so that $\|U\| = L'$.

c) Consider the same operator, regarding the spaces as Euclidean, that is, with $\mathbf{X} = \ell_m^2$,

* Consequently all linear operators between finite-dimensional B-spaces are continuous.

$\mathbf{Y} = \ell_n^2$. Then, if $m = n$ and the matrix A is symmetric,

$$\|U\| = |\lambda_1|.$$

Here λ_1 is the eigenvalue of A having least absolute value.

In the general case,

$$\|U\| = \sqrt{\Lambda_1},$$

where Λ_1 is the eigenvalue of the matrix A^*A (A^* being the Hermitian conjugate* of A) having least absolute value.

Let us look at the first case once again. The matrix A has real eigenvalues $\lambda_1, \lambda_2, \ldots, \lambda_n$ and linearly independent eigenvectors $x_1, x_2, \ldots, x_n$, which we may assume normalized and pairwise orthogonal. If $x = c_1x_1 + c_2x_2 + \ldots + c_nx_n$, then

$$y = U(x) = U(c_1x_1 + c_2x_2 + \ldots + c_nx_n) = c_1\lambda_1x_1 + c_2\lambda_2x_2 + \ldots + c_n\lambda_nx_n,$$

$$\|y\|^2 = |c_1\lambda_1|^2 + |c_2\lambda_2|^2 + \ldots + |c_n\lambda_n|^2 \leqslant \lambda_1^2(|c_1|^2 + |c_2|^2 + \ldots + |c_n|^2) = \lambda_1^2\|x\|^2,$$

and we have equality for $x = x_1$.

REMARK. Note that we also have another expression for $\|U\|$, namely

$$\|U\| = \sup_{\|x\|=1} |(Ax, x)| = \sup_{\|x\|=1} \left| \sum_{j=1}^{n} \left(\sum_{k=1}^{n} a_{jk}\xi_k \right) \bar{\xi}_j \right|.$$

For

$$|(Ax, x)| \leqslant \|Ax\| \, \|x\| = \|U(x)\| \cdot \|x\| \leqslant \|U\| = |\lambda_1| \quad (\|x\| = 1).$$

But for $x = x_1$ we have equality. Therefore $\sup\limits_{\|x\|=1} |(Ax, x)| = \|U\|$.

Now let A be an arbitrary matrix. In this case we have

$$\|U\|^2 = \sup_{\|x\|=1} \|U(x)\|^2 = \sup_{\|x\|=1} (Ax, Ax) = \sup_{\|x\|=1} \sum_{j=1}^{n} \left(\sum_{k=1}^{m} a_{jk}\xi_k \sum_{s=1}^{m} \bar{a}_{js}\bar{\xi}_s \right) =$$

$$= \sup_{\|x\|=1} \sum_{s=1}^{m} \left[\sum_{k=1}^{m} \left(\sum_{j=1}^{n} a_{jk}a_{sj}^* \right) \xi_k \right] \bar{\xi}_s = \sup_{\|x\|=1} (A^*Ax, x) = \Lambda_1,$$

where, as we mentioned in the Remark, Λ_1 is the largest eigenvalue of A^*A.

For $n = 1$, the operator U becomes a functional. In this case, the matrix A degenerates into a single row. By a change of notation we may suppose that every linear functional f on the m-dimensional space $\mathbf{X}$ is determined by a system $(\phi_1, \phi_2, \ldots, \phi_m)$ of m numbers:

$$f(x) = \sum_{k=1}^{m} \phi_k\xi_k \quad (x = (\xi_1, \xi_2, \ldots, \xi_m) \in \mathbf{X}). \tag{17}$$

Since it is also true, conversely, that every system of numbers $(\phi_1, \phi_2, \ldots, \phi_\mu)$ determines a functional on $\mathbf{X}$ according to (17), we can identify the functional f with the system

* The elements a_{jk}^* of A^* are defined by $a_{jk}^* = \bar{a}_{kj}$ ($j = 1, 2, \ldots, m$; $k = 1, 2, \ldots, n$). In particular, if A is a real matrix, then $a_{jk}^* = a_{kj}$.

determining it, and write

$$f = (\phi_1, \phi_2, \ldots, \phi_m). \tag{18}$$

The norm of the functional (18) in various specific spaces is defined as follows:

$$\mathbf{X} = \ell_m^\infty, \quad \|f\| = \sum_{k=1}^{m} |\phi_k|, \tag{19}$$

$$\mathbf{X} = \ell_m^1, \quad \|f\| = \max_k |\phi_k|, \tag{20}$$

$$\mathbf{X} = \ell_m^2, \quad \|f\| = \sqrt{\sum_{k=1}^{m} |\phi_k|^2}. \tag{21}$$

REMARK. If $\mathbf{X}$ is a complex space, then it is more convenient to write (17) in the form

$$f(x) = \sum_{k=1}^{m} \overline{\phi}_k \xi_k,$$

identifying f as before with the system $(\phi_1, \phi_2, \ldots, \phi_m)$ (cf. the Remark in 2.3).

§ 3. Linear functionals and operators on Hilbert space

3.1. An operator* on a separable Hilbert space admits a matrix representation similar to that for operators on finite-dimensional spaces (see 2.8).

In fact, let U be an operator on a separable (infinite-dimensional) Hilbert space $\mathbf{H}$. Choose any complete orthonormal system $x_1, x_2, \ldots, x_n, \ldots$. Each element $x \in \mathbf{H}$ can be expressed in the form

$$x = \sum_{k=1}^{\infty} c_k x_k,$$

where $c_k = (x, x_k)$ is the Fourier coefficient of x. Therefore, as U is continuous,

$$y = Ux = \sum_{k=1}^{\infty} c_k U x_k.$$

Hence, comparing the Fourier coefficients on the left- and right-hand sides, we find that

$$d_j = \sum_{k=1}^{\infty} a_{jk} c_k, \tag{1}$$

where the d_j are the Fourier coefficients of y and the a_{jk} are those of Ux_k.

Thus the sequence of Fourier coefficients of $y = Ux$ is obtained from the sequence of

* When we are dealing with Hilbert space, the term "operator" will always mean "continuous linear operator". Also we write Ux instead of $U(x)$.

Fourier coefficients of x by means of the transformation with matrix

$$A = \begin{pmatrix} a_{11} & a_{12} & \dots & a_{1k} & \dots \\ a_{21} & a_{22} & \dots & a_{2k} & \dots \\ \dots & \dots & \dots & \dots & \dots \\ a_{j1} & a_{j2} & \dots & a_{jk} & \dots \\ \dots & \dots & \dots & \dots & \dots \end{pmatrix}. \tag{2}$$

Therefore U has a matrix representation, as in the finite-dimensional case.

However, it should be kept in mind that, unlike the finite-dimensional case, by no means every matrix of the form (2) determines a linear operator. A necessary and sufficient condition for this is that there exist a constant C such that, for $m, n = 1, 2, \dots,$

$$\sum_{j=1}^{n} \left| \sum_{k=1}^{m} a_{jk} c_k \right|^2 \leqslant C^2 \sum_{k=1}^{m} |c_k|^2, \tag{3}$$

whatever values we give to $c_1, c_2, \dots, c_m$.

The necessity of this condition is almost obvious. Taking x to be an element whose Fourier coefficients, from the $(m+1)$-st onwards, are zero, we have

$$\sum_{j=1}^{n} \left| \sum_{k=1}^{m} a_{jk} c_k \right|^2 \leqslant \sum_{j=1}^{\infty} \left| \sum_{k=1}^{m} a_{jk} c_k \right|^2 = \|y\|^2 = \|Ux\|^2 \leqslant \|U\|^2 \|x\|^2 = \|U\|^2 \sum_{k=1}^{m} |c_k|^2,$$

from which it is clear that we may take C to be $\|U\|$.

Conversely, if the inequality (3) holds, then, letting m and then n tend to infinity, we obtain

$$\|y\|^2 = \sum_{j=1}^{\infty} \left| \sum_{k=1}^{\infty} a_{jk} c_k \right|^2 \leqslant C^2 \sum_{k=1}^{\infty} |c_k|^2 = C^2 \|x\|^2.$$

The above criterion is difficult to verify, and for this reason we mention a simpler sufficient condition, which strongly restricts the class of operators. Namely, if

$$D^2 = \sum_{j=1}^{\infty} \sum_{k=1}^{\infty} |a_{jk}|^2 < \infty,$$

then the matrix (2) determines a linear operator on $\mathbf{H}$ by (1).

For

$$\|y\|^2 = \sum_{j=1}^{\infty} \left| \sum_{k=1}^{\infty} a_{jk} c_k \right|^2 \leqslant \sum_{j=1}^{\infty} \sum_{k=1}^{\infty} |a_{jk}|^2 \sum_{k=1}^{\infty} |c_k|^2 = \sum_{j=1}^{\infty} \sum_{k=1}^{\infty} |a_{jk}|^2 \|x\|^2,$$

which implies that U is bounded, with $\|U\| \leqslant D$.

We also note that the matrix representation depends on the choice of the orthonormal system $x_1, x_2, \dots, x_n, \dots$, and that by changing the latter we also change the matrix (2) representing U. The problem of finding the conditions under which two matrices of the form (2) determine the same operator presents great difficulties, and we shall not go into it, referring the reader to a specialist monograph.*

* See, e.g., Akhiezer and Glazman.

3.2. Let us now clarify our overall picture of linear functionals on Hilbert space.

By considering the inner product (x, y) in $\mathbf{H}$ for a fixed $y \in \mathbf{H}$ we convince ourselves without difficulty that this is a linear functional. For let

$$f(x) = (x, y). \tag{4}$$

The additivity and homogeneity of f amount simply to axiom 2 for an inner product space, while the boundedness is a consequence of the Cauchy–Buniakowski inequality:

$$|f(x)| = |(x, y)| \leqslant \|x\| \|y\|,$$

from which we see that

$$\|f\| \leqslant \|y\|. \tag{5}$$

We now prove that functionals of the form (4) exhaust the linear functionals on $\mathbf{H}$, and that we have equality in (5).

THEOREM 1. *For every continuous linear functional f on a Hilbert space $\mathbf{H}$, there exists an element $y \in \mathbf{H}$, uniquely determined by f, such that* (4) *is true for each $x \in \mathbf{H}$. In addition, we have the equation*

$$\|f\| = \|y\|. \tag{6}$$

Proof. First let us establish the existence of y. Denote by $\mathbf{H}_0$ the set of all $x \in \mathbf{H}$ such that $f(x) = 0$. As f is linear and continuous, this set is a closed subspace. If in addition we have $\mathbf{H}_0 = \mathbf{H}$, then we can take $y = \mathbf{0}$.

Consider the case where $\mathbf{H}_0 \neq \mathbf{H}$. Let $y_0 \notin \mathbf{H}_0$. Write y_0 in the form

$$y_0 = y' + y'' \qquad (y' \in \mathbf{H}_0, y'' \perp \mathbf{H}_0)$$

(Theorem IV.5.1). Clearly, $y'' \neq \mathbf{0}, f(y'') \neq 0$, so we may assume that $f(y'') = 1$. Choose any element $x \in \mathbf{H}$ and write $f(x) = \alpha$. The element $x' = x - \alpha y''$ belongs to $\mathbf{H}_0$, because

$$f(x') = f(x) - \alpha f(y'') = \alpha - \alpha = 0.$$

Therefore

$$(x, y'') = (x' + \alpha y'', y'') = \alpha(y'', y''),$$

so that

$$f(x) = \alpha = \left(x, \frac{y''}{(y'', y'')}\right)$$

and we may take $y = y''/(y'', y'')$.

We have thus established the existence of y.

The uniqueness of this element is very simply established. For if

$$(x, y) = (x, y_1),$$

for all $x \in \mathbf{H}$, then $(x, y - y_1) = 0$ and hence $y - y_1 \perp \mathbf{H}$, which is only possible if $y = y_1$.

Moreover,

$$\|f\| \geqslant f\left(\frac{y}{\|y\|}\right) = \frac{(y, y)}{\|y\|} = \|y\|,$$

which, together with (5), yields equation (6).*

The proof of the theorem is therefore complete.

In particular, a linear functional on $\mathbf{L}^2(T, \Sigma, \mu)$ has the general form

$$f(x) = (x, y) = \int_T x(t)\overline{y(t)}\,d\mu \qquad (x, y \in \mathbf{L}^2).$$

On ℓ^2, we have

$$f(x) = (x, y) = \sum_{k=1}^{\infty} \xi_k \bar{\eta}_k \quad (x = (\xi_1, \xi_2, \ldots), \quad y = (\eta_1, \eta_2, \ldots) \in \ell^2).$$

The general form for a functional on $\mathbf{L}^2(a, b)$ was described by Fréchet [2], that for a functional on an abstract Hilbert space, in the separable case, by von Neumann [1] and, in the general case, by Rellich [1].

In 1.3 we introduced scalar multiplication for functionals on a complex normed space by the formula

$$(\lambda f)\ (x) = \bar{\lambda} f(x). \tag{7}$$

To see the appropriateness of this definition, consider the functionals on a Hilbert space $\mathbf{H}$. The theorem on the general form of a functional on a Hilbert space provides a justification for identifying a functional $f \in \mathbf{H}^*$ with the element $y \in \mathbf{H}$ that determines it:

$$f(x) = (x, y) \qquad (x \in \mathbf{H}). \tag{8}$$

Moreover, if f_1 is determined by y_1, and f_2 by y_2, then clearly $f_1 + f_2$ is determined by $y_1 + y_2$. As λf is defined by (7), we may also speak of the functional λf and the element λy. In fact,

$$(\lambda f)\ (x) = \bar{\lambda} f(x) = \bar{\lambda}(x, y) = (x, \lambda y). \tag{9}$$

Since, moreover, $\|f\| = \|y\|$, the above remarks provide grounds for identifying the space $\mathbf{H}^*$ of functionals f with the space $\mathbf{H}$ of elements y—these are linearly isometric.

3.3. Let U be a (bounded) linear operator on a Hilbert space $\mathbf{H}$. Consider the functional

$$f(x) = (Ux, y),$$

where y is a fixed element of $\mathbf{H}$. Clearly f is a linear functional, and, since

$$|f(x)| = |(Ux, y)| \leqslant \|Ux\|\,\|y\| \leqslant \|U\|\,\|x\|\,\|y\|,$$

f is bounded and we have

$$\|f\| \leqslant \|U\|\,\|y\|. \tag{10}$$

Therefore, by Theorem 1, there exists an element $y^* \in \mathbf{H}$ such that

$$f(x) = (x, y^*) \qquad (x \in \mathbf{H}),$$

that is,

$$(Ux, y) = (x, y^*) \qquad (x \in \mathbf{H}). \tag{11}$$

The element y^* is uniquely determined by f, and therefore ultimately by y. Thus, by associating y^* with y, we obtain an operator on $\mathbf{H}$. This operator is denoted by U^* and is

---

* If $y = \mathbf{0}$, then (6) follows immediately from (4).

said to be adjoint to U. If we replace y^* in (11) by U^*y, we obtain the equation defining the adjoint operator,

$$(Ux, y) = (x, U^*y) \qquad (x, y \in \mathbf{H}).$$

The operator U^* is additive since, by combining the equations

$$(Ux, y_1) = (x, U^*y_1), \qquad (Ux, y_2) = (x, U^*y_2),$$

which hold for all $x \in \mathbf{H}$, we obtain

$$(Ux, y_1 + y_2) = (x, U^*y_1 + U^*y_2),$$

so that

$$U^*(y_1 + y_2) = U^*y_1 + U^*y_2.$$

Analogously, we have

$$(Ux, iy) = -i(Ux, y) = -i(x, U^*y) = (x, iU^*y) \ (x, y \in \mathbf{H}),$$

so that

$$U^*(iy) = iU^*y.$$

Therefore $U^*(\lambda y) = \lambda U^*y$ for every λ, and U^* is thus a linear operator.

Furthermore, since $\|f\| = \|y^*\|$, equation (10), rewritten in the form

$$\|y^*\| = \|U^*y\| \leqslant \|U\| \, \|y\|,$$

shows that U^* is bounded, and that $\|U^*\| \leqslant \|U\|$.

It is not hard to see that the operator $U^{**} = (U^*)^*$ coincides with U. In fact, for all $x, y \in \mathbf{H}$, we have the equations

$$(U^*x, y) = (x, U^{**}y)$$

and

$$(U^*x, y) = (x, Uy),$$

so that $(x, U^{**}y) = (x, Uy)$, and thus $U^{**}y = Uy$ for every $y \in \mathbf{H}$.

This fact enables us to show that the norms of U and U^* are equal. For, if we use the inequality we have already proved, we have $\|U\| = \|U^{**}\| \leqslant \|U^*\|$, which, with the preceding inequality, yields $\|U^*\| = \|U\|$.

An operator U that coincides with its adjoint is said to be *self-adjoint*. A self-adjoint operator is characterized by the equation

$$(Ux, y) = (x, Uy) \qquad (x, y \in \mathbf{H}).$$

Self-adjoint operators play a fundamental role in the theory of operators on Hilbert space, since the deepest facts in the theory relate precisely to the self-adjoint operators. Furthermore, it turns out to be possible to consider unbounded operators also.

The concept of an adjoint operator will later (see Chapter IX) be generalized to the case of a linear operator on an arbitrary B-space. However, the adjoint operator will, generally speaking, then be on another space, so the concept of a self-adjoint operator cannot be carried over to the general case.

Let $\mathbf{H} = \mathbf{L}^2(a, b)$ and let U be the integral operator defined in 2.6; that is, $y = Ux$ is to mean

$$y(s) = \int_a^b K(s, t)x(t)dt \qquad \left(\int_a^b \int_a^b |K(s, t)|^2 \, ds\, dt < +\infty\right). \tag{12}$$

It is not hard to check that the adjoint operator U^* is also an integral operator. For if $y^* = U^*y$, then

$$y^*(s) = \int_a^b K^*(s, t)y(t)dt \qquad (K^*(s, t) = \overline{K(t, s)}). \tag{13}$$

In fact, for every $x \in \mathbf{L}^2$ we must have

$$(Ux, y) = (x, y^*),$$

that is,

$$\int_a^b \left[\int_a^b K(s, t)x(t)dt\right]\overline{y(s)}\, ds = \int_a^b x(t)\overline{y^*(t)}\, dt.$$

Therefore

$$\int_a^b x(t)\left[\overline{y^*(t) - \int_a^b \overline{K(s, t)}y(s)ds}\right] = 0,$$

so that, as x was arbitrary, we find that

$$y^*(t) = \int_a^b \overline{K(s, t)}\, y(s)ds,$$

which, apart from the notation, is equation (13).

The operator (12) will be self-adjoint if the kernel $K(s, t)$ is complex-symmetric—that is, if

$$K(t, s) = \overline{K(s, t)}.$$

Now suppose the operator U is defined by the matrix (2). Denoting the Fourier coefficients of y by $d_1, d_2, \ldots$, we have

$$(Ux, y) = \sum_{j=1}^{\infty}\left(\sum_{k=1}^{\infty} a_{jk}c_k\right)\overline{d_j} = \sum_{j=1}^{\infty} c_k \overline{\sum_{j=1}^{\infty} \bar{a}_{jk}d_j} = (x, y^*),$$

where y^* is the element whose Fourier coefficients are

$$d_j^* = \sum_{k=1}^{\infty} a_{jk}^* d_k \qquad (j = 1, 2, \ldots; a_{jk}^* = \bar{a}_{jk}^*).$$

Since $y^* = U^*y$, this shows that U^* is defined by the matrix

$$A^* = \begin{pmatrix} a_{11}^* & a_{12}^* & \dots & a_{1k}^* & \dots \\ a_{21}^* & a_{22}^* & \dots & a_{2k}^* & \dots \\ \dots & \dots & \dots & \dots & \dots \\ a_{j1}^* & a_{j2}^* & \dots & a_{jk}^* & \dots \\ \dots & \dots & \dots & \dots & \dots \end{pmatrix},$$

the Hermitian conjugate of the matrix (2). In particular, $U^* = U$ means that $a_{jk} = \bar{a}_{kj}$.

3.4. An important class of operators on Hilbert space is composed of the so-called projections.

Let $\mathbf{H}_0$ be a subspace of a Hilbert space $\mathbf{H}$. By Theorem IV.5.1, there corresponds to each $x \in \mathbf{H}$, in a unique manner, its projection in the subspace $\mathbf{H}_0$. In this way we define an operator $P = P_{\mathbf{H}_0}$, which is called the *projection operator* (onto $\mathbf{H}_0$).

Let us note some obvious properties of projections.

a) For any $x \in \mathbf{H}$, the elements Px and $x - Px$ are orthogonal.

b) $x \in \mathbf{H}_0$ is equivalent to $Px = x$.

c) $x \perp \mathbf{H}_0$ is equivalent to $Px = \mathbf{0}$.

Further, since $\|x\|^2 = \|Px\|^2 + \|x - Px\|^2$, we have $\|Px\| \leqslant \|x\|$, so that $\|P\| \leqslant 1$. Provided $\mathbf{H}_0$ does not reduce to the zero element alone, we can choose $\bar{x} \in \mathbf{H}_0$ with $\|\bar{x}\| = 1$, and deduce that $\|P\| \geqslant \|P\bar{x}\| = \|\bar{x}\| = 1$. Thus we deduce

d) $\|P\| = 1$.

The class of projections is characterized in the following theorem.

THEOREM 2. *A necessary and sufficient condition for a linear operator P on $\mathbf{H}$ to be a projection is that:*

1) *P be self-adjoint;*

2) *$P(Px) = Px$, for every $x \in \mathbf{H}$.*

Proof. Necessity. Let P be a projection onto a subspace $\mathbf{H}_0$. Choose any $x, y \in \mathbf{H}$ and write them in the form

$$x = x' + x'' \qquad (x' = Px \in \mathbf{H}_0, \quad x'' \perp \mathbf{H}_0),$$

$$y = y' + y'' \qquad (y' = Py \in \mathbf{H}_0, \quad y'' \perp \mathbf{H}_0).$$

Then we have

$$(Px, y) = (x', y' + y'') = (x', y') = (x' + x'', y') = (x, Py),$$

so that P is self-adjoint.

Further, using b), we have

$$P(Px) = Px' = x' = Px,$$

which shows the necessity of condition 2).

Sufficiency. Let P be a linear operator satisfying both the conditions in the theorem. Denote the set of all $x \in \mathbf{H}$ such that $Px = x$ by $\mathbf{H}_0$. It is easily verified that $\mathbf{H}_0$ is a subspace. We show that P is the projection of $\mathbf{H}$ on $\mathbf{H}_0$. Take any $x \in \mathbf{H}_0$ and express it in the form $x = Px + (x - Px)$. We want to show that $Px \in \mathbf{H}_0$ and $x - Px \perp \mathbf{H}_0$.

The first of these is clear, since $P(Px) = Px$. Also, if $u \in \mathbf{H}_0$, then $Pu = u$ and hence

$$(x - Px, u) = (x, u) - (Px, u) = (x, u) - (x, Pu) = 0;$$

then $x - Px \perp \mathbf{H}_0$, and the proof of the theorem is complete.

To obtain a useful matrix representation for a projection operator P, we must choose an orthonormal system $x_1, x_2, \ldots$ such that, for each $k = 1, 2, \ldots$, we have either $x_k \in \mathbf{H}_0$ or $x_k \perp \mathbf{H}_0$.*

With such a choice, we have

$$a_{jk} = (Px_k, x_j) = \begin{cases} 0, & j \neq k, \\ 0, & j = k, \quad x_k \perp \mathbf{H}_0, \\ 1, & j = k, \quad x_k \in \mathbf{H}_0. \end{cases}$$

Thus all the entries of the matrix (2) representing a projection P are zero except those on the principal diagonal, which may be equal to 1.

Adjoint operators and projections were introduced for the case of a separable Hilbert space by von Neumann [1], and for the general case by Rellich [1].

3.5. Let us establish the relationship between the projection operators introduced in 3.4 and projections in the sense of the theory of normed spaces (IV.1.9). For the time being, we shall call the former orthoprojections and the latter projections. Let $\mathbf{H}$ be a Hilbert space. An orthoprojection is clearly a projection; however, the converse is false. Theorem IV.5.1 shows that every subspace of a Hilbert space has an orthoprojection onto it, and that the subspace is complemented (in the sense of IV.1.9). It turns out that this property characterizes a Hilbert space, within the class of B-spaces.

THEOREM 3. *If every closed subspace of a B-space $\mathbf{X}$ has a complement, then $\mathbf{X}$ is isomorphic to a Hilbert space.*

Theorem 3 was proved by J. Lindenstrauss and L. Tzafriri—a proof (and also specific examples of non-complemented subspaces in various B-spaces) may be found in Kadets and Mityagin [1]. The most-well-known specific examples of non-complemented subspaces are as follows: $\mathbf{c}_0$ is not complemented in ℓ^∞, and $\mathbf{C}[0, 1]$ is not complemented in $\mathbf{L}^\infty(0, 1)$.

§4. Rings of operators

4.1. It is possible to multiply linear operators together, as well as add them. Let $\mathbf{X}$, $\mathbf{Y}$ and $\mathbf{Z}$ be three normed spaces, and let U and V be linear operators, from $\mathbf{X}$ into $\mathbf{Y}$ and from $\mathbf{Y}$ into $\mathbf{Z}$, respectively ($U \in B(\mathbf{X}, \mathbf{Y})$, $V \in B(\mathbf{Y}, \mathbf{Z})$). The *product* of V and U is the operator $W = VU$ from $\mathbf{X}$ into $\mathbf{Z}$ defined as follows:†

$$W(x) = V(U(x)) \quad (x \in \mathbf{X}).$$

Since

$$W(x_1 + x_2) = V(U(x_1 + x_2)) = V(U(x_1) + U(x_2)) = V(U(x_1)) + V(U(x_2)) = \\ = W(x_1) + W(x_2)$$

and

$$\|W(x)\| = \|V(U(x))\| \leqslant \|V\| \|U(x)\| \leqslant \|U\| \|V\| \|x\|,$$

* This is accomplished as follows: choose an orthonormal system in $\mathbf{H}_0$ and one in its orthogonal complement, each complete in its subspace. Then unite these two systems into one (cf. IV.5.8).

† This definition makes sense for non-linear operators as well.

W is a linear operator, and we have

$$\|W\| = \|VU\| \leqslant \|V\|\|U\|. \tag{1}$$

If $\mathbf{X}$, $\mathbf{Y}$ and $\mathbf{Z}$ are finite-dimensional spaces (of dimension m, n, p, respectively) and A, B, C are the matrices determining U, V, W, respectively, then $C = BA$. For $z = W(x)$ implies

$$\zeta_j = \sum_{k=1}^{m} c_{jk}\xi_k \ (j = 1, 2, \ldots, p;\ x = (\xi_1, \xi_2, \ldots, \xi_m);\ z = (\zeta_1, \zeta_2, \ldots, \zeta_p)).$$

On the other hand, $z = V(y)$, where $y = U(x)$; that is,

$$\zeta_j = \sum_{m=1}^{n} b_{jm}\eta_m,\ \eta_m = \sum_{k=1}^{m} a_{mk}\xi_k$$

$$(j = 1, 2, \ldots, p;\ m = 1, 2, \ldots, n;\ y = (\eta_1, \eta_2, \ldots, \eta_n)).$$

Comparing this equation with the preceding one, and bearing in mind that x was arbitrary, we deduce that

$$c_{jk} = \sum_{m=1}^{n} b_{jm}a_{mk} \ (j = 1, 2, \ldots, p;\ k = 1, 2, \ldots, m),$$

which is what we wished to prove.

It is known from algebra that the product of two matrices can be a matrix all of whose entries are zero even if the matrix factors are non-zero. This shows that we cannot replace the inequality sign by an equality sign in (1), even in the case of finite-dimensional spaces.

However, several properties of products of numbers are also true for operators. We mention, for example, omitting the trivial proofs, the distributivity and associativity of multiplication. Furthermore, there are operators that play the role which unity (the "identity" element) plays in numerical multiplication. However, unlike numerical multiplication, there are two "identities"—a left identity and a right identity. Namely, the operator $I \in B(\mathbf{X}, \mathbf{X})$, which is the identity mapping on $\mathbf{X}$:

$$I(x) = x \ (x \in \mathbf{X}), \tag{2}$$

is a right identity:

$$UI = U,$$

while the analogous operator I_1 on $\mathbf{Y}$ is a left identity:

$$I_1U = U.$$

We should particularly stress that multiplication of operators is not commutative. Furthermore, the operator UV is, in general, not even defined, so the equation $UV = VU$ has no meaning. And even where UV does make sense (this will be so, clearly, if $\mathbf{Z} = \mathbf{X}$), the operators VU and UV are defined on different spaces—the former on $\mathbf{X}$, the latter on $\mathbf{Y}$.

For this reason the question of the permutability of U and V can arise only when $\mathbf{X} = \mathbf{Y} = \mathbf{Z}$: that is, when both operators are on the same space $\mathbf{X}$—or, in other words, are elements of $B(\mathbf{X}, \mathbf{X})$.

However, even in this case the equation $UV = VU$ does not hold for arbitrary operators $U, V \in B(\mathbf{X}, \mathbf{X})$, unless $\mathbf{X}$ is a one-dimensional space.

4.2. Among the spaces $B(\mathbf{X}, \mathbf{Y})$, the space $B(\mathbf{X}, \mathbf{X})$ of continuous linear operators mapping $\mathbf{X}$ into itself occupies a special position, for only in this one does the product of two arbitrary elements make sense.

A system on which there are defined two operations, addition and multiplication, subject to the usual rules for the operations on numbers (with the exception of commutativity and invertibility of multiplication) is called a *ring*. If a multiplication is defined for elements of a normed space, then the normed space is called a *normed ring*.* In this case, to connect the product operation with the metric of the space, one also demands that the product be continuous.

Thus the space $B(\mathbf{X}, \mathbf{X})$ is a normed ring (the continuity of the product follows from inequality (1)).†

In a normed ring, and in particular in $B(\mathbf{X}, \mathbf{X})$, one can define powers of any element. Let $U \in B(\mathbf{X}, \mathbf{X})$. By definition, we set

$$U^{\circ} = I, \; U^n = U^{n-1}U \quad (n = 1, 2, \ldots),$$

where I is the identity operator on $\mathbf{X}$ (see (2)).

Since this definition in no way differs from the corresponding one for numbers, it follows that for any positive integers m and n we have

$$U^m U^n = U^{m+n},$$

from which we deduce that all the powers of a particular operator U commute with one another.

Furthermore, applying inequality (1) successively, we find that

$$\|U^n\| \leqslant \|U\|^n \quad (n = 0, 1, \ldots). \tag{3}$$

Here equality does not hold, in general.

In the sequel we assume that the space $\mathbf{X}$ is complete. Then, by Theorem 1.2, so is the space $B(\mathbf{X}, \mathbf{X})$. With this assumption, we consider the "geometric progression":

$$I + U + U^2 + \ldots + U^n + \ldots. \tag{4}$$

Let us determine the conditions under which this series converges. It follows immediately from (3) that it always converges when

$$\|U\| < 1, \tag{5}$$

since the series (4) is majorized by the numerical series

$$1 + \|U\| + \|U\|^2 + \ldots + \|U\|^n + \ldots$$

and hence converges, in view of the completeness of $B(\mathbf{X}, \mathbf{X})$ (see IV.1.5). However, in contrast to the numerical case, condition (5) is not a necessary condition for the convergence of (3).

---

* In the literature, the terms *algebra* and *normed algebra*, respectively, are often used—and also *Banach algebra* when the space is complete.

† We shall apply the theory of normed rings only in connection with the spaces $B(\mathbf{X}, \mathbf{X})$. We refer the reader wishing to obtain further knowledge about normed rings to the book by M. A. Naimark (see Naimark).

THEOREM 1. *For any operator* $U \in B(\mathbf{X}, \mathbf{X})$,

$$\lim_{n \to \infty} \sqrt[n]{\|U^n\|} = c_U$$

exists.

Furthermore, if $c_U < 1$, *then the series* (4) *is convergent, and if* $c_U > 1$ *then it is divergent.*

Proof. We write

$$a = \inf_n \sqrt[n]{\|U^n\|}$$

and show that $c_U = \lim_{n \to \infty} \sqrt[n]{\|U\|^n} = a$. Let $\varepsilon > 0$. Choose m such that

$$\sqrt[m]{\|U^m\|} < a + \varepsilon.$$

Also let

$$M = \max[1, \|U\|, \ldots, \|U^{m-1}\|].$$

Now consider any n and write it in the form $n = k_n m + l_n (0 \leqslant l_n \leqslant m-1)$. Then by (3) we have

$$\sqrt[n]{\|U^n\|} \leqslant \sqrt[n]{\|U^{l_n}\| \, \|U^m\|^{k_n}} \leqslant M^{1/n} \|U^m\|^{k_n/n} < M^{1/n}(a+\varepsilon)^{(n-l_n)/n}.$$

Since

$$\lim_{n \to \infty} M^{1/n}(a+\varepsilon)^{(n-l_n)/n} = a + \varepsilon,$$

there exists N_ε such that

$$M^{1/n}(a+\varepsilon)^{(n-l_n)/n} < a + 2\varepsilon \quad (n \geqslant N_\varepsilon).$$

Therefore, for $n \geqslant N_\varepsilon$,

$$a \leqslant \sqrt[n]{\|U^n\|} < a + 2\varepsilon,$$

and hence it follows, as required, that $\lim_{n \to \infty} \sqrt[n]{\|U^n\|} = a$ does exist.

We now deduce that the series (4) converges (if $c_U < 1$)—or diverges (if $c_U > 1$)—by applying Cauchy's test for convergence to the series $\sum_{n=0}^{\infty} \|U^n\|$.

COROLLARY. *A necessary and sufficient condition for the series* (4) *to converge is that for some k we have*

$$\|U^k\| < 1. \tag{6}$$

For if (4) is convergent, then $\|U^n\| \to 0$ and so (6) is true for large enough k. Conversely, if (6) is true, then $c_U = \inf_n \sqrt[n]{\|U\|^n} \leqslant \sqrt[k]{\|U\|^k} < 1$, and so (4) converges.

4.3. We now consider the problem of generalizing the multiplication operation for operators. Suppose U is a linear operator mapping a normed space $\mathbf{X}$ into a normed space $\mathbf{Y}$. We say that U *has an inverse* (or is *invertible*) if there exists an operator V mapping $\mathbf{Y}$

into $\mathbf{X}$ such that

$$VU = I_{\mathbf{X}}, \ UV = I_{\mathbf{Y}}, \tag{7}$$

where $I_{\mathbf{X}}$ and $I_{\mathbf{Y}}$ are the operators that act identically on $\mathbf{X}$ and $\mathbf{Y}$ respectively.* The operator V is said to be inverse to U and is denoted by $V = U^{-1}$.

It follows immediately from the definition that the inverse operator U^{-1}, like the given operator, is linear. For if $y_1, y_2 \in \mathbf{Y}$ then, by the second of the equations (7),

$$y_k = U(x_k) \ (x_k = V(y_k); \ k = 1, 2).$$

Hence, by the first of the equations (7), we have

$$V(y_1 + y_2) = V(U(x_1 + x_2)) = x_1 + x_2 = V(y_1) + V(y_2).$$

The homogeneity of U^{-1} is proved similarly.

It also follows from the definition that the operator U is inverse to U^{-1}; that is, $(U^{-1})^{-1} = U$.

The definition of the inverse operator just given has a formal nature. To clarify its essential meaning, we prove that, if an inverse operator U^{-1} exists, then U is a one-to-one mapping of $\mathbf{X}$ onto $\mathbf{Y}$.

In fact, choose $x_1 \neq x_2$ $(x_1, x_2 \in \mathbf{X})$. If we had $U(x_1) = U(x_2)$ then by the first of the equations (7) we should have

$$x_1 = VU(x_1) = VU(x_2) = x_2.$$

Moreover, every $y \in \mathbf{Y}$ is the image of some $x \in \mathbf{X}$ (and, by what we have proved, there can only be one such x). For take $x = V(y)$. Then, by the second of the equations (7),

$$U(x) = UV(y) = y.$$

Conversely, suppose the operator U yields a one-to-one mapping from $\mathbf{X}$ onto $\mathbf{Y}$. With each element $y \in \mathbf{Y}$ we associate its inverse image: that is, the element $x \in \mathbf{X}$ such that $U(x) = y$. This defines an operator V mapping $\mathbf{Y}$ onto $\mathbf{X}$. It is easy to check that V is linear, and that $V = U^{-1}$.

In fact, the equations $x = V(y)$ and $y = U(x)$ are equivalent, so

$$VU(x) = V(y) = x, \ UV(y) = U(x) = y.$$

From what has been said it follows, among other things, that an inverse operator, if it exists, must be unique.

The above remarks enable us to study the concept of an inverse operator from another point of view. Suppose we are given an equation

$$U(x) = y, \tag{8}$$

where y is an arbitrary, but fixed, element of $\mathbf{Y}$ and x is an unknown element of $\mathbf{X}$.

Clearly, if the inverse operator U^{-1} exists, then (8) has a unique solution, for each $y \in \mathbf{Y}$, and this solution will be $x = U^{-1}(y)$.

The unique solubility of equation (8) is not enough to guarantee the existence of a continuous inverse operator U^{-1}.

* In future we shall use the notations $I_{\mathbf{X}}$, $I_{\mathbf{Y}}$, and so on, without specifically mentioning them each time. We shall sometimes omit the subscripts indicating the spaces on which they are defined, when this can be done without ambiguity.

THEOREM 2. *Suppose equation (8) has a solution, for each $y \in \mathbf{Y}$, and suppose that there exists a positive number m such that, for every $x \in \mathbf{X}$,*

$$\|U(x)\| \geqslant m\|x\|. \tag{9}$$

Then U has a continuous inverse U^{-1}, and

$$\|U^{-1}\| \leqslant \frac{1}{m}. \tag{10}$$

Proof. Clearly the mapping induced by such an operator U is one-to-one, since if $x_1 \neq x_2$ then

$$\|U(x_1) - U(x_2)\| \geqslant m\|x_1 - x_2\| > 0.$$

Furthermore, as (8) is soluble for each $y \in \mathbf{Y}$, we must have $U(\mathbf{X}) = \mathbf{Y}$.

Hence, as we showed above, the linear operator U^{-1} exists. Since $x = V(y)$ implies that $y = U(x)$, we see from (9) that

$$\|y\| \geqslant m\|V(y)\| \quad (y \in \mathbf{Y})$$

or

$$\|U^{-1}(y)\| = \|V(y)\| \leqslant \frac{1}{m}\|y\| \quad (y \in \mathbf{Y}). \tag{11}$$

In other words, U^{-1} is bounded, and it is therefore a continuous operator. The bound (10) follows immediately from (11).

REMARK. The conditions of the theorem are in fact also necessary for the existence of a continuous inverse operator U^{-1}.

The first condition is obviously necessary. The necessity of the second is easily verified by taking $m = 1/\|U^{-1}\|$.

As an illustration of the theorem just proved, let us consider the operator U on $\mathbf{L}^2(a, b)$ defined in 2.7:

$$y = U(x), \ y(s) = x(s) - \lambda \int_a^b K(s, t)x(t)\, dt$$

(we assume that the kernel $K(s, t)$ is symmetric).

Assuming that λ is not an eigenvalue of the kernel, we conclude, from well-known facts in the theory of integral equations, that, in this situation, (8) has a solution, for every $y \in \mathbf{L}^2$ (see Petrovskii). Furthermore, it was shown in 2.7 (we keep to the previous notation) that

$$\|y\|^2 = \|U(x)\|^2 = \sum_{-\infty}^{\infty} h_k^2 \left(1 - \frac{\lambda}{\lambda_k}\right)^2.$$

Hence

$$\|U(x)\|^2 \geqslant m^2 \sum_{-\infty}^{\infty} h_k^2 = m^2\|x\|^2,$$

where

$$m = \inf \left| 1 - \frac{\lambda}{\lambda_k} \right| \quad (k = \ldots, -1, 0, 1, \ldots),$$

and where the λ_k denote the eigenvalues of $K(s, t)$ (for $k = 0, -1, -2, \ldots$, we must assume that $\lambda_k = \infty$). Since $\lambda_k \neq \lambda$ and $\lambda_k \to \infty$ as $k \to \infty$, we have $m > 0$. Therefore there exists a continuous operator U^{-1} and we have

$$\|U^{-1}\| \leqslant \frac{1}{m} = \sup \left| 1 + \frac{\lambda}{\lambda_k - \lambda} \right| \quad (k = \ldots, -1, 0, 1, \ldots).$$

4.4. If we retain only one of the equations (7), we obtain the concept of a left inverse or a right inverse, respectively, of an operator. To be precise, an operator V_l from $U(\mathbf{X})$ into $\mathbf{X}$ is said to be a *left inverse* to an operator U if

$$V_l U = I_{\mathbf{X}}.$$

Likewise, an operator V_r from $\mathbf{Y}$ into $\mathbf{X}$ is said to be a *right inverse* to U if

$$U V_r = I_{\mathbf{Y}}.$$

We shall denote a left inverse operator by the symbol U_l^{-1} and a right inverse operator by the symbol U_r^{-1}; we shall sometimes omit the suffices when it is clear from the context what kind of inverse operator is meant. The argument given above shows that a left inverse operator is necessarily linear.

The existence of a left or right inverse operator enables one to draw conclusions about the solubility of equation (8).

Namely, if a left inverse operator U_l^{-1} exists, then a solution to (8), if one exists, is unique. In other words, if there is a left inverse operator, then U is a one-to-one mapping from $\mathbf{X}$ onto $U(\mathbf{X})$. For when we verified this in the case of the inverse operator we used only the first of the equations (7), which is true for left inverse operators also.

One should bear in mind that, in general, $U(\mathbf{X})$ will not coincide with $\mathbf{Y}$, so that U_l will not be defined on the whole of $\mathbf{Y}$.

Similarly, one can verify that the existence of a right inverse operator implies that equation (8) is soluble (though, generally speaking, not uniquely) for each $y \in \mathbf{Y}$ (the solution being $x = U_r^{-1}(y)$)—that is, we have $U(\mathbf{X}) = \mathbf{Y}$ in this case.

It follows from what has been said that, if there exist both a left universe U_l^{-1} and a right inverse U_r^{-1}, then these are equal; furthermore, the inverse $U^{-1} = U_l^{-1} = U_r^{-1}$ then exists. For this reason, we sometimes call inverse operators two-sided inverses.

Finally we note that a necessary and sufficient condition for the existence of a continuous left inverse operator is that U satisfy equation (9).

4.5. The case where $\mathbf{Y} = \mathbf{X}$ is interesting since if an operator $U \in B(\mathbf{X}, \mathbf{X})$ has a continuous inverse, then this will also be an element of the space $B(\mathbf{X}, \mathbf{X})$.

The arguments that we applied in 4.2 to the series (4):

$$I + U + U^2 + \ldots + U^n + \ldots,$$

yield the following theorem (see Banach [2]).

THEOREM 3 (Banach). *Let* $\mathbf{X}$ *be a B-space and let* $U \in B(\mathbf{X}, \mathbf{X})$. *If*

$$\|U\| \leqslant q < 1, \tag{12}$$

then the operator $I-U$ has a continuous inverse, and

$$\|(I-U)^{-1}\| \leqslant \frac{1}{1-q}. \tag{13}$$

Proof. It was shown in 4.2 that, under the conditions of the theorem—that is, under the conditions (5)—the series (4) is convergent. Denote the sum of this series by V. Then we have

$$\begin{aligned} V(I-U) &= (I+U+\ldots+U^n+\ldots)(I-U) \\ &= (I+U+\ldots+U^n+\ldots)-(U+U^2+\ldots+U^{n+1}) = I \end{aligned} \tag{14}$$

and similarly

$$(I-U)V = I. \tag{15}$$

Hence $V=(I-U)^{-1}$.

Furthermore, by (3),

$$\|V\| \leqslant \|I\|+\|U\|+\ldots+\|U^n\|+\ldots \leqslant 1+q+\ldots+q^n+\ldots = \frac{1}{1-q},$$

giving the required bound (13).

REMARK. Since (14) and (15) are always true if the series (4) is convergent, it follows from Theorem 1 and its corollary that the continuous linear operator $(I-U)^{-1}$ will also exist whenever

$$\lim_{n\to\infty} \sqrt[n]{\|U^n\|} < 1, \tag{16}$$

or if, for some $k=1,2,\ldots$, we have

$$\sqrt[k]{\|U^k\|} < 1. \tag{17}$$

4.6. Banach's theorem shows that an operator $I-U$ that differs by a small amount from the identity operator I—which has a continuous inverse ($I^{-1}=I$)—will itself have a continuous inverse. This fact lends itself to generalization.

THEOREM 4. *Let $U_0 \in B(\mathbf{X},\mathbf{Y})$, where $\mathbf{X}$ and $\mathbf{Y}$ are B-spaces, and suppose U_0 has an inverse $U_0^{-1} \in B(\mathbf{Y},\mathbf{X})$. If an operator $U \in B(\mathbf{X},\mathbf{Y})$ satisfies the condition*

$$\|U\| < \frac{1}{\|U_0^{-1}\|}, \tag{18}$$

then the operator $V=U_0+U$ has a continuous inverse V^{-1}, and

$$\|V^{-1}\| \leqslant \frac{\|U_0^{-1}\|}{1-\|U_0^{-1}U\|} \leqslant \frac{\|U_0^{-1}\|}{1-\|U_0^{-1}\|\|U\|}.$$

Proof. Consider the operator

$$W = U_0^{-1}V = I_{\mathbf{X}} + U_0^{-1}U.$$

Since

$$\|U_0^{-1}U\| \leqslant \|U_0^{-1}\|\|U\| < 1,$$

Banach's theorem shows that W has a continuous inverse W^{-1}. Furthermore, we have

$$\|W^{-1}\| \leqslant \frac{1}{1-\|U_0^{-1}U\|} \leqslant \frac{1}{1-\|U_0^{-1}\|\|U\|}. \tag{19}$$

Also, we have

$$U_0^{-1}VW^{-1} = I_{\mathbf{X}},$$

so that

$$VW^{-1} = U_0$$

and hence

$$VW^{-1}U_0^{-1} = I_{\mathbf{Y}}.$$

On the other hand,

$$W^{-1}U_0^{-1}V = I_{\mathbf{X}}.$$

From the last two equations we conclude that the operator $W^{-1}U_0^{-1}$ is a continuous inverse to V.

From the equation $V^{-1} = W^{-1}U_0^{-1}$ we see by (19) that we have the bound

$$\|V^{-1}\| \leqslant \|W^{-1}\|\|U_0^{-1}\| \leqslant \frac{\|U_0^{-1}\|}{1-\|U_0^{-1}U\|} \leqslant \frac{\|U_0^{-1}\|}{1-\|U_0^{-1}\|\|U\|}.$$

§ 5. The method of successive approximations

5.1. Consider the equation

$$x - U(x) = y, \tag{1}$$

where U is a continuous linear operator on a B-space $\mathbf{X}$, y is a given element of $\mathbf{X}$ and x is an unknown element of $\mathbf{X}$.

One of the prevalent methods of determining solutions to (1) is the so-called *method of successive approximations*, in which one starts with an arbitrary given element $x_0 \in \mathbf{X}$—the *initial approximation*—and constructs a sequence of *approximate solutions*

$$x_{n+1} = y + U(x_n) \qquad (n = 0, 1, \ldots). \tag{2}$$

If one obtains a convergent sequence whose limit is a solution of the given equation, then one says that the successive approximation process for equation (1), starting from x_0, is convergent (to a solution of (1)). Since $U \in B(\mathbf{X}, \mathbf{X})$, the mere fact that $\{x_n\}$ converges implies that $x^* = \lim_{n\to\infty} x_n$ is a solution of (1). To convince oneself of this, it is sufficient to let $n \to \infty$ in (2).

The question of the convergence of the successive approximation process for equation (1) turns out to be connected with the convergence of the series

$$I + U + \ldots + U^n + \ldots, \tag{3}$$

whose sum (when the series does converge) is $(I-U)^{-1}$ (see the Remark following Banach's theorem—Theorem 4.3).

THEOREM 1. *If the series* (3) *is convergent, then for any initial approximation* x_0 *the successive approximation process converges to a unique solution* x^* *of equation* (1). *We then have the following bound*:

$$\|x^*-x_n\| \leqslant \|(I-U)^{-1}\| \|U^n\| \|x_1-x_0\| \qquad (n=1,2,\dots). \tag{4}$$

In particular, if the conditions of Banach's theorem are satisfied, we can replace this bound by

$$\|x^*-x_n\| \leqslant \frac{q^n}{1-q}\|x_1-x_0\| \qquad (n=1,2,\dots).$$

Proof. Applying (2) successively, we have

$$x_n = y+U(y)+\dots+U^{n-1}(y)+U^n(x_0) \qquad (n=1,2,\dots). \tag{5}$$

Hence it is clear that, if (3) is convergent, then, since $U^n(x_0)\to \mathbf{0}$ in this case, we can infer the existence of

$$x^* = \lim_{n\to\infty} x_n = \sum_{k=1}^{\infty} U^k(y) = (I-U)^{-1}(y).$$

Since x^* is obviously a solution to (1), the first part of the theorem is proved.

To obtain the bound (4), substitute x^* for x_0 in (5). Then, as is clear from (2), we have $x_n = x^*$ $(n = 1, 2, \dots)$. Thus we arrive at the equation

$$x^* = y+U(y)+\dots+U^{n-1}(y)+U^n(x^*) \qquad (n=1,2,\dots).$$

Subtracting equation (5) from this and taking norms, we have

$$\|x^*-x_n\| \leqslant \|U^n\| \|x^*-x_0\| \qquad (n=1,2,\dots). \tag{6}$$

Write $\tilde{x} = x^*-x_0$. Taking into account the fact that x^* is a solution of (1) and consequently $x^*-U(x^*) = y$, we have

$$(I-U)(\tilde{x}) = \tilde{x}-U(\tilde{x}) = x^*-U(x^*)-x_0+U(x_0) = y+U(x_0)-x_0 = x_1-x_0;$$

and from this we find that

$$\tilde{x} = (I-U)^{-1}(x_1-x_0).$$

Using this in (6), we obtained the required bound.

5.2. Theorem 1 has various applications. Let us now dwell on some of these.

First let $\mathbf{X}$ be an m-dimensional space. In this case the linear operator U is determined by a square matrix $A = (a_{jk})$, and equation (1) may be written in an explicit form as a system of equations:

$$\xi_j - \sum_{k=1}^{m} a_{jk}\xi_k = \eta_j \ (j=1,2,\dots,m;\ x=(\xi_1,\xi_2,\dots,\xi_m); \quad y=(\eta_1,\eta_2,\dots,\eta_m)). \tag{7}$$

The successive approximations $x_n = (\xi_1^{(n)}, \xi_2^{(n)}, \dots, \xi_m^{(n)})$ $(n = 0, 1, \dots)$ to a solution are found from the formula

$$\xi_j^{(n+1)} = \sum_{k=1}^{m} a_{jk}\xi_k^{(n)} + \eta_j \qquad (j = 1, 2, \ldots, m, \quad n = 0, 1, \ldots).$$

The condition (12) of the preceding subsection, guaranteeing that the successive approximation process converges to a solution, depends on the definition of the norm in $\mathbf{X}$. Thus, if $\mathbf{X} = \ell_m^\infty$, then since in that situation we then have (see 2.8)

$$\|U\| = \max_j \sum_{k=1}^{m} |a_{jk}|,$$

we see that, provided

$$\sum_{k=1}^{m} |a_{jk}| < 1 \qquad (j = 1, 2, \ldots, m) \tag{8}$$

is satisfied, $\lim_{n\to\infty} \xi_j^{(n)} = \xi_j^*$ exists $(j = 1, 2, \ldots, m)$, and $\xi_1 = \xi_1^*, \xi_2 = \xi_2^*, \ldots, \xi_m = \xi_m^*$ is a (unique) solution to the system (7).

Taking $\mathbf{X} = \ell_m^1$, we obtain another condition

$$\sum_{j=1}^{m} |a_{jk}| < 1 \qquad (k = 1, 2, \ldots, m). \tag{9}$$

Finally, if we take $\mathbf{X} = \ell_m^2$, then, since we have the bound

$$\|U\| \leqslant \left[\sum_{j=1}^{m}\sum_{k=1}^{m} |a_{jk}|^2\right]^{1/2},$$

we obtain the condition

$$\sum_{j=1}^{m}\sum_{k=1}^{m} |a_{jk}|^2 < 1. \tag{10}$$

It is interesting to note that if the matrix A is symmetric and $\mathbf{X} = \ell_m^2$, then the condition $\|U\| < 1$ is not only sufficient, but also necessary, for the successive approximation process to converge.

For in this case, as we showed in 2.8,

$$\|U\| = |\lambda_1|,$$

where λ_1 is the eigenvalue of A having greatest absolute value. Taking y in the system (7) to be the corresponding eigenvector, and setting $x_0 = \mathbf{0}$, we obtain

$$x_1 = y;\ x_2 = U(x_1) + y(\lambda_1 + 1)y;\ \ldots;$$

$$x_{n+1} = U(x_n) + y = (\lambda_1^n + \ldots + \lambda_1 + 1)y \quad (n = 1, 2, \ldots),$$

and thus it follows that the sequence $\{\xi_j^{(n)}\}$ does not tend to a limit as $n \to \infty$ if $|\lambda_1| \geqslant 1\ (\eta_j \neq 0)$.

Thus, when the matrix A is symmetric, the successive approximation process converges to a solution if and only if the eigenvalues of A are less than unity in absolute value.

In Chapter XIII we shall prove a general result which implies that this is true even without the assumption that A is symmetric. We leave the immediate verification of this to the reader.

Let us now indicate further bounds for the eigenvalue λ_1 of A of greatest absolute value. Denoting the eigenvector in $\mathbf{X}$ corresponding to λ_1 by x_1, we have $U(x_1) = \lambda_1 x_1$, and therefore

$$|\lambda_1| \|x_1\| = \|U(x_1)\| \leqslant \|U\| \|x_1\|,$$

that is, $|\lambda_1| \leqslant \|U\|$. Taking $\mathbf{X}$ to be ℓ_m^∞, we deduce from this that

$$|\lambda_1| \leqslant \max_j \sum_{k=1}^m |a_{jk}|.$$

Taking $\mathbf{X} = \ell_m^1$, we obtain the bound

$$|\lambda_1| \leqslant \max_k \sum_{j=1}^m |a_{jk}|.$$

Finally, if $\mathbf{X} = \ell_m^2$, then*

$$|\lambda_1| \leqslant \|U\| \leqslant \left[\sum_{j=1}^m \sum_{k=1}^m |a_{jk}|^2 \right]^{1/2}.$$

5.3. Let us turn to a study of infinite systems. Consider the system of equations

$$\xi_j - \sum_{k=1}^\infty a_{jk}\xi_k = \eta_j \quad (j = 1, 2, \ldots). \tag{11}$$

By a *solution* of the system we mean a numerical sequence $\{\xi_j^*\}$ such that the series on the left-hand sides of (12) are convergent for $\xi_j = \xi_j^*$ and all the equations (11) then vanish identically.

Systems of this type occur in the study of boundary problems for equations of mathematical physics and integral equations.†

We assume first that the infinite matrix

$$A = \begin{pmatrix} a_{11} & a_{12} & \cdots & a_{1k} & \cdots \\ a_{21} & a_{22} & \cdots & a_{2k} & \cdots \\ \cdots & \cdots & \cdots & \cdots & \cdots \\ a_{j1} & a_{j2} & \cdots & a_{jk} & \cdots \\ \cdots & \cdots & \cdots & \cdots & \cdots \end{pmatrix}$$

of the system (11) satisfies the condition

$$\sum_{j=1}^\infty \sum_{k=1}^\infty |a_{jk}|^2 < 1. \tag{12}$$

---

* Concerning iterative methods of solution for finite systems, see Faddeev and Faddeeva, Ch. III.

† For infinite systems, see Kantorovich [4]. The book by Kantorovich and Krylov has a bibliography for infinite systems.

In this case, as we showed in III.3.1, the matrix A determines a linear operator U on the space ℓ^2:

$$z = U(x), \qquad \zeta_j = \sum_{k=1}^{\infty} a_{jk}\xi_k$$

$$(j = 1, 2, \ldots; \quad x = (\xi_1, \xi_2, \ldots),\ z = (\zeta_1, \zeta_2, \ldots)).$$

Condition (12) yields

$$\|U\| \leqslant \left[\sum_{j=1}^{\infty}\sum_{k=1}^{\infty} |a_{jk}|^2\right]^{1/2} < 1,$$

so Theorem 1 is applicable to the system (11), expressed in the form of a single equation of the form (1):

$$x - U(x) = y \quad (x = (\xi_1, \xi_2, \ldots), \quad y = (\eta_1, \eta_2, \ldots)),$$

and this shows that, for each $y \in \ell^2$, there exists a unique solution (in ℓ^2) $x^* = \{x_k^*\}$, which is obtainable by the method of successive approximations:

$$\xi_j^{(n+1)} = \sum_{k=1}^{\infty} a_{jk}\xi_k^{(n)} + \eta_j \quad (j = 1, 2, \ldots; \ n = 0, 1, \ldots),$$

$$\lim_{n\to\infty} \sum_{j=1}^{\infty} |\xi_j^* - \xi_j^{(n)}|^2 = 0$$

($\{\xi_k^{(0)}\}$ being any sequence in ℓ^2).

We now replace condition (12) by a weaker condition:

$$\sum_{j=1}^{\infty}\sum_{k=1}^{\infty} |a_{jk}|^2 < \infty. \tag{13}$$

Although the matrix A here also determines a linear operator U on ℓ^2, the inequality $\|U\| < 1$ does not hold in general, so we cannot apply Theorem 1.

Let us show that the study of the system (11) subject to the condition (13) reduces to a study of finite systems (we assume that the sequence $\{\eta_j\}$ of right-hand sides belongs to ℓ^2).

Choose n_0 such that

$$\sum_{j=n_0+1}^{\infty}\ \sum_{k=n_0+1}^{\infty} |a_{jk}|^2 < 1.$$

Fix the first n_0 unknowns $\xi_1, \xi_2, \ldots, \xi_{n_0}$ and consider the system

$$\xi_j - \sum_{k=n_0+1}^{\infty} a_{jk}\xi_k = \eta_j + \sum_{k=1}^{n_0} a_{jk}\xi_k \quad (j = n_0+1,\ n_0+2, \ldots). \tag{14}$$

Since

$$\sum_{j=n_0+1}^{\infty} |a_{jk}\xi_k|^2 = |\xi_k|^2 \sum_{j=n_0+1}^{\infty} |a_{jk}|^2 \leqslant |\xi_k|^2 \sum_{j=1}^{\infty}\sum_{s=1}^{\infty} |a_{js}|^2 < \infty \quad (k = 1, 2, \ldots, n_0),$$

the sequence of right-hand sides of the system (14) belongs to ℓ^2, so, by what was said above, the system has a unique solution $\tilde{\xi}_{n_0+1}, \tilde{\xi}_{n_0+2}, \ldots$ in ℓ^2, depending on the fixed values $\xi_1, \xi_2, \ldots, \xi_{n_0}$. To determine the nature of this dependence, consider the systems

$$\xi_j - \sum_{k=n_0+1}^{\infty} a_{jk}\xi_k = a_{js} \quad (j = n_0+1, n_0+2, \ldots; \; s = 1, 2, \ldots, n_0)$$

and

$$\xi_j - \sum_{k=n_0+1}^{\infty} a_{jk}\xi_k = \eta_j \quad (j = n_0+1, \; n_0+2, \ldots),$$

whose solutions (which are unique, as we have remarked) we denote by $\{c_{ks}\}$ and $\{\tilde{\eta}_k\}$ $(k = n_0+1, n_0+2, \ldots; s = 1, 2, \ldots, n_0)$, respectively. We then have the equations

$$c_{js} - \sum_{k=n_0+1}^{\infty} a_{jk}c_{ks} = a_{js} \quad (j = n_0+1, \; n_0+2, \ldots; \; s = 1, 2, \ldots, n_0), \tag{15}$$

$$\tilde{\eta}_j - \sum_{k=n_0+1}^{\infty} a_{jk}\tilde{\eta}_k = \eta_j \quad (j = n_0+1, \; n_0+2, \ldots). \tag{16}$$

Multiplying the first equations by ξ_s, summing and adding them to the second set, we obtain

$$\left(\tilde{\eta}_j + \sum_{s=1}^{n_0} c_{js}\xi_s\right) - \sum_{k=n_0+1}^{\infty} a_{jk}\left(\tilde{\eta}_k + \sum_{s=1}^{n_0} c_{ks}\,\xi_s\right) = \eta_j + \sum_{s=1}^{n_0} a_{js}\xi_s$$

$$(j = n_0+1, \; n_0+2 \ldots),$$

which shows that the sequence $\left\{\tilde{\eta}_k + \sum_{s=1}^{n_0} c_{ks}\xi_s\right\}$ is a solution of the system (14). As the solution is unique, we have

$$\tilde{\xi}_k = \tilde{\eta}_k + \sum_{s=1}^{n_0} c_{ks}\xi_s \quad (k = n_0+1, \; n_0+2, \ldots).$$

This result may be formulated as follows: the given system (11) is equivalent to the system

$$\begin{aligned} \xi_j - \sum_{k=1}^{\infty} a_{jk}\xi_k &= \eta_j \quad (j = 1, 2, \ldots, n_0), \\ \xi_j - \sum_{k=1}^{n_0} c_{jk}\xi_k &= \tilde{\eta}_j \quad (j = n_0+1, \; n_0+2, \ldots). \end{aligned} \tag{17}$$

Now substitute for the ξ_k with $k > n_0$ in the first group of equations the values obtained for these from the second group:

$$\xi_j - \sum_{k=1}^{n_0} a_{jk}\xi_k - \sum_{k=n_0+1}^{\infty} a_{jk}\left(\tilde{\eta}_k + \sum_{s=1}^{n_0} a_{ks}\xi_s\right) = \eta_j \quad (j = 1, 2, \ldots, n_0), \tag{18}$$

or, if we write (cf. (15) and (16))

$$c_{jk} = a_{jk} + \sum_{s=n_0+1}^{\infty} a_{js}c_{sk}, \quad \tilde{\eta}_j = \eta_j + \sum_{k=n_0+1}^{\infty} a_{jk}\tilde{\eta}_k \quad (j, k = 1, 2, \ldots, n_0),^*$$

then we can express (18) in the form

$$\xi_j - \sum_{k=1}^{n_0} c_{jk}\xi_k = \tilde{\eta}_j \quad (j = 1, 2, \ldots, n_0). \tag{19}$$

Now adjoin to these equations the equations of the second group in the system (17):

$$\xi_j - \sum_{k=1}^{n_0} c_{jk}\xi_k = \tilde{\eta}_j \quad (j = n_0 + 1, n_0 + 2, \ldots), \tag{20}$$

and we obtain an infinite system of equations, equivalent to the given system (11).

The solution of the system thus constructed can be carried out in two steps: first, find the values of the first n_0 unknowns from (19); second, find the values of the other unknowns by substituting these first values in (20).

Since we can always carry out the second step once we have carried out the first, we have thus reduced the solution of the infinite system (11) to that of the finite system (19).

Using this, one can show that a system (11) subject to the condition (13) has all the properties of a finite system. Since we shall obtain this result in Chapter XIII from other general considerations, we mention here only one such property: if the homogeneous system corresponding to (11) (that is, the system obtained from (11) when $\eta_1 = \eta_2 = \ldots = 0$) has a unique solution in ℓ^2 (obviously, it must be zero), then the given system (11) has a unique solution, whatever sequence $\{\eta_j\}$ appears on the right-hand sides.

For if the system (11) is homogeneous, then so is the system (19). But for a finite system the above assertion is true, so an inhomogeneous system (19) has a unique solution, and the required result thus follows.

We leave it to the reader to verify that a homogeneous system (11) has a finite number of linearly independent solutions, and also to deduce a condition for the inhomogeneous system to be soluble when this number is greater than zero.

5.4. We now assume that the matrix A satisfies the condition

$$\sum_{k=1}^{\infty} |a_{jk}| \leqslant 1 - \rho \quad (j = 1, 2, \ldots), \tag{21}$$

where $\rho > 0$.

A system (11) satisfying this condition is said to be *completely regular*.

---

* The series that appear on the right-hand side are convergent since $\sum_{s=n_0+1}^{\infty} |a_{js}|^2$, $\sum_{s=n_0+1}^{\infty} |c_{sk}|$ and $\sum_{k=n_0+1}^{\infty} |\tilde{\eta}_k|^2$ are all convergent series.

On the space ℓ^∞ we introduce an operator U:

$$z = U(x),$$

$$\zeta_j = \sum_{k=1}^{\infty} a_{jk}\xi_k \quad (j = 1, 2, \ldots; \quad x = (\xi_1, \xi_2, \ldots); \ z = (\zeta_1, \zeta_2, \ldots)).$$

Since by (21) we have

$$|\zeta_j| \leqslant \sum_{k=1}^{\infty} |a_{jk}||\xi_k| \leqslant \|x\| \sum_{k=1}^{\infty} |a_{jk}| \leqslant (1-\rho)\|x\| \quad (j = 1, 2, \ldots), \tag{22}$$

$z = U(x)$ is meaningful for any $x \in \ell^\infty$ and is an element of ℓ^∞. Moreover,

$$\|z\| = \|U(x)\| \leqslant (1-\rho)\|x\|.$$

Hence, as U is obviously linear, we conclude that U is a continuous linear operator on ℓ^∞, and that

$$\|U\| \leqslant 1 - \rho. \tag{23}$$

As in 3.3, we can now describe the system (11) as a single equation of type (1):

$$x - U(x) = y \quad (x = (\xi_1, \xi_2, \ldots), \quad y = (\eta_1, \eta_2, \ldots)). \tag{24}$$

The inequality (23) ensures that we can apply Theorem 1 to the equation (24), and, accordingly, this equation has, for any $x \in \ell^\infty$, a unique solution (in ℓ^∞), which can be found by the method of successive approximations.

Thus a completely regular system has a unique bounded solution, whatever bounded sequence $\{\eta_j\}$ appears on the right-hand sides.

REMARK. The above remarks guarantee a unique bounded solution for a completely regular system. Here we cannot exclude the possibility of the system having other, unbounded, solutions. For instance, the system

$$\xi_j - \frac{1}{j+1}\xi_{j+1} = 0 \quad (j = 1, 2, \ldots)$$

has infinitely many solutions

$$\xi_1 = a, \quad \xi_2 = 2a, \ldots, \quad \xi_k = k!a, \ldots,$$

where a is an arbitrary constant. However, if $a \neq 0$, these solutions are unbounded.

Similar remarks may be made about the system considered in 5.3.

If we consider a system for which there exist n_0 and M such that

$$\sum_{k=n_0+1}^{\infty} |a_{jk}| \leqslant 1 - \rho, \quad \sum_{k=1}^{\infty} |a_{jk}| \leqslant M \quad (j = 1, 2, \ldots, \rho > 0),$$

then everything that we established in 5.3 for systems satisfying (13) can be asserted here; the arguments carry over without any essential changes.

5.5. Let us consider the integral equation

$$x(s) - \lambda \int_a^b K(s, t)x(t)\,dt = y(s),$$

the kernel $K(s, t)$ being assumed continuous.

Introducing the integral operator U on $\mathbf{C}[a, b]$ or on $\mathbf{L}^2(a, b)$ (see 2.4 or 2.6), we can write the integral equation in the form

$$x - \lambda U(x) = y. \tag{25}$$

If

$$|\lambda| < \frac{1}{\|U\|}, \tag{26}$$

then, by Theorem 4.3, the operator $I - \lambda U$ has the continuous inverse

$$(I - \lambda U)^{-1} = I + \lambda U + \lambda^2 U^2 + \ldots + \lambda^n U^n + \ldots.$$

Hence the unique solution x^* of (25) has the form

$$x^* = (I - \lambda U)^{-1}(y) = y + \lambda U(y) + \lambda^2 U^2(y) + \ldots + \lambda^n U^n(y) + \ldots. \tag{27}$$

This series is called the *Neumann series.*

Let us show that the operators U^n, like U, are integral operators. In fact, $v = U^2(x)$ means that $v = U(z)$, where $z = U(x)$; that is,

$$v(s) = \int_a^b K(s, t)z(t)\,dt, \quad z(t) = \int_a^b K(t, u)x(u)\,du.$$

Hence

$$v(s) = \int_a^b K(s, t)\left[\int_a^b K(t, u)x(u)\,du\right]dt =$$

$$= \int_a^b \left[\int_a^b K(s, t)K(t, u)\,dt\right]x(u)\,du = \int_a^b K_2(s, u)\,x(u)\,du$$

$$\left(K_2(s, u) = \int_a^b K(s, t)K(t, u)\,dt\right).$$

By induction, it can be shown that $v = U^n(x)$ means

$$v(s) = \int_a^b K_n(s, u)x(u)\,du \quad (n = 2, 3, \ldots),$$

where $K_n(s, u)$ is determined by the recurrence relation

$$K_n(s, u) = \int_a^b K_{n-1}(s, t)K(t, u)\,dt \quad (n = 2, 3, \ldots),$$

which, when expanded, yields

$$K_n(s, u) = \int_a^b \dots \int_a^b K(s, t_1)K(t_1, t_2) \dots K(t_{n-1}, u)dt_1, dt_2 \dots dt_{n-1}.$$

The functions $K_n(s, u)$ are known as the *iterated kernels*.

The Neumann series can now be expressed in the expanded form:

$$x^*(s) = y(s) + \lambda \int_a^b K(s, t)y(t)\,dt + \lambda^2 \int_a^b K_2(s, t)y(t)\,dt + \dots$$

$$\dots + \lambda^n \int_a^b K_n(s, t)y(t)\,dt + \dots.$$

The nature of the convergence of this series depends on the space on which we regard the operator U as acting.

We obtain conditions for the Neumann series to converge from (26) if we write out the expression for $\|U\|$. Thus, in the space $\mathbf{C}[a, b]$, we have (see 2.4)

$$\|U\| = \max_s \int_a^b |K(s, t)|\,dt \leqslant M(b-a) \quad \left(M = \max_{s,\,t} |K(s, t)|\right)$$

and (26) takes the form

$$|\lambda| < \frac{1}{\max_s \int_a^b |K(s, t)|\,dt},$$

or, more simply,

$$|\lambda| < \frac{1}{M(b-a)}.$$

For $\mathbf{L}^2(a, b)$ we have the bound (see 2.6)

$$\|U\| \leqslant \left[\int_a^b \int_a^b |K(s, t)|^2\,ds\,dt\right]^{1/2}.$$

Hence a sufficient condition for the Neumann series to converge (in $\mathbf{L}^2$) is

$$|\lambda| < \frac{1}{\left[\int_a^b \int_a^b |K(s, t)|^2\,ds\,dt\right]^{1/2}}.$$

If the kernel $K(s, t)$ is symmetric, then we showed in 2.6 that one has an exact equation

$$\|U\| = \frac{1}{|\lambda_1|},$$

where λ_1 is the eigenvalue of $K(s, t)$ having greatest absolute value. Thus, in the case of a symmetric kernel, the Neumann series will converge (again in $\mathbf{L}^2$) when

$$|\lambda| < |\lambda_1|.$$

One can verify, exactly as we did in 5.2 for finite systems, that the Neumann series no longer converges in the present situation for any $y \in \mathbf{L}^2$ if $|\lambda| \geqslant |\lambda_1|$. Hence, in this case, condition (26) is not only sufficient, but also necessary, for the Neumann series to converge.

Similarly, using the inequality $|\lambda_1| \geqslant \dfrac{1}{\|U\|}$, we obtain (lower) bounds for the eigenvalue of the kernel having least absolute value:

$$|\lambda_1| \geqslant \frac{1}{\max \int_a^b |K(s,t)|\,dt} \geqslant \frac{1}{M(b-a)}, \quad |\lambda_1| \geqslant \frac{1}{\left[\int_a^b \int_a^b |K(s,t)|^2\,ds\,dt\right]^{1/2}}.$$

Finally, we note the following fact. By our remark on Banach's theorem (4.5), the Neumann series converges if and only if, for some $n = 1, 2, \ldots$,

$$|\lambda| < \frac{1}{\sqrt[n]{\|U^n\|}},$$

that is, (for $\mathbf{C}[a, b]$),

$$|\lambda| < \frac{1}{\left[\max_s \int_a^b |K_n(s,t)|\,dt\right]^{1/n}},$$

or (for $\mathbf{L}^2(a, b)$)

$$|\lambda| < \frac{1}{\left[\int_a^b \int_a^b |K_n(s,t)|^2\,ds\,dt\right]^{1/2n}}$$

§ 6. The ring of operators on a Hilbert space

6.1. Let us consider in more detail the ring $B(\mathbf{H}, \mathbf{H})$ of linear operators on a Hilbert space $\mathbf{H}$. We first explain how the operation of passing to the adjoint operator is connected with the algebraic operations of the ring. We have the following propositions:

a) $$[U_1 + U_2]^* = U_1^* + U_2^*. \tag{1}$$

For the following statements hold for any $x, y \in \mathbf{H}$:

$$([U_1 + U_2]x, y) = (x, [U_1 + U_2]^* y),$$

$$(U_1 x, y) = (x, U_1^* y),$$

$$(U_2 x, y) = (x, U_2^* y).$$

Adding the second and third of these, we find that

$$([U_1 + U_2]x, y) = (x, U_1^* y + U_2^* y) = (x, [U_1^* + U_2^*] y),$$

so that, as x was arbitrary,

$$[U_1 + U_2]^* y = [U_1^* + U_2^*] y,$$

giving (1).

b)
$$[\lambda U]^* = \lambda U^*. \tag{2}$$

We have

$$([\lambda U]x, y) = (x, [\lambda U]^* y) \quad (x, y \in \mathbf{H}). \tag{3}$$

But $[\lambda U]x = \lambda U x$, so

$$([\lambda U]x, y) = \lambda(Ux, y) = \lambda(x, U^* y) = (x, [\lambda U^*]y).$$

Comparing this with (3), we obtain (2).

c)
$$[U_1 U_2]^* = U_2^* U_1^*. \tag{4}$$

For we have

$$([U_1 U_2]x, y) = (x, [U_1 U_2]^* y) \quad (x, y \in \mathbf{H}).$$

On the other hand,

$$([U_1 U_2]x, y) = (U_1(U_2 x), y) = (U_2 x, U_1^* y) = (x, U_2^*(U_1^* y))$$
$$= (x, [U_2^* U_1^*]y).$$

Proceeding as above, we obtain (4).

d) If an operator U has a linear inverse U^{-1}, then the adjoint operator U^* has a linear inverse,* and

$$(U^*)^{-1} = (U^{-1})^*. \tag{5}$$

For, since $U^{-1}U = UU^{-1} = I$, it follows from c) that

$$U^*(U^{-1})^* = (U^{-1})^* U^* = I^* = I.$$

The assertion now follows from this.

In the situation where U_1 and U_2 are self-adjoint, a)–c) reduce to the following:

e) If λ, μ are real numbers, then the operator $\lambda U_1 + \mu U_2$ is self-adjoint.

f) The product $U_1 U_2$ is self-adjoint if and only if $U_1 U_2 = U_2 U_1$—that is, if and only if U_1 and U_2 commute.

For, as U_1 and U_2 are self-adjoint, c) shows that

$$[U_1 U_2]^* = U_2 U_1,$$

which gives the required result.

Now suppose we have a sequence of operators $\{U_n\}$ and an operator U. We say that the

* Recall that, for a Hilbert space, "linear operator" means "continuous linear operator" (see p. 141)

sequence $\{U_n\}$ is *weakly convergent to* U if, for any $x, y \in \mathbf{H}$,

$$\lim_{n\to\infty} (U_n x, y) = (Ux, y).$$

Clearly, the limit of a weakly convergent sequence of self-adjoint operators is a self-adjoint operator.

In exactly the same way, if $U_n x \to Ux$ for each $x \in \mathbf{H}$ (in this case, we say that the sequence of operators $\{U_n\}$ *converges* to U on $\mathbf{H}$) and the U_n are self-adjoint, then so is U.

Finally, if $U_n \to U$ in the space $B(\mathbf{H}, \mathbf{H})$, then U will again be self-adjoint whenever the U_n are.

Both the last two results become obvious if we note that, in each case, $\{U_n\}$ is weakly convergent to U.

6.2. An operator U on a Hilbert space $\mathbf{H}$ is said to be *positive* ($U \geqslant \mathbf{0}$) if

$$(Ux, x) \geqslant 0$$

for each $x \in \mathbf{H}$.

It is easy to see that a positive operator is self-adjoint. In fact, suppose (Ux, x) is real for each $x \in \mathbf{H}$. Since

$$\begin{aligned}(Ux, y) = \tfrac{1}{4}\{&[(U(x+y), x+y) - (U(x-y), x-y)] + \\ &+ i[(U(x+yi), x+yi) - (U(x-yi), x-yi)]\}\end{aligned}$$

and since the expressions in square brackets are real, if we interchange x and y, we have

$$\begin{aligned}(Uy, x) &= \tfrac{1}{4}\{[(U(y+x), y+x) - (U(y-x), y-x)] + \\ &\quad + i[(U(y+xi), y+xi) - (U(y-xi), y-xi)]\} \\ &= \tfrac{1}{4}\{[(U(x+y), x+y) - (U(x-y), x-y)] - \\ &\quad - i[(U(x+yi), x+yi) - (U(x-yi), x-yi)]\} = \overline{(Ux, y)}.\end{aligned}$$

Hence we find that

$$(Ux, y) = \overline{(Uy, x)} = (x, Uy),$$

as we required to prove.

REMARK. We have actually proved that the operator U is self-adjoint if (Ux, x) is real for each $x \in \mathbf{H}$. It is easy to see that this is also a necessary condition for U to be self-adjoint.

We say that an operator U_1 is *greater* than an operator U_2 ($U_1 \geqslant U_2$) if the difference $U_1 - U_2$ is a positive operator.

Notice that the operator $U^* U$ (or $U U^*$) will be positive, whatever operator we take for U. For

$$(U^*Ux, x) = (Ux, Ux) \geqslant 0.$$

In particular, if $U^* = U$ (that is, if U is self-adjoint), then $U^2 \geqslant \mathbf{0}$.

It is also clear that a sum of positive operators is positive.

We further note that a linear combination of positive operators in which the coefficients are real and non-negative is itself a positive operator.

Moreover, every power U^n of a positive operator U is positive. For if $n = 2m$ is even,*

* The statement is obviously also true if $n = 0$, in which case $U^n = I$.

then

$$(U^n x, x) = (U^m x, U^m x) = \|U^n x\|^2 \geqslant 0 \quad (x \in \mathbf{H}).$$

If $n = 2m+1$ is odd, then

$$(U^n x, x) = (U(U^m x), U^m x) = (Uy, y) \geqslant 0 \quad (x \in \mathbf{H}, y = U^m x).$$

It now follows that a linear combination of powers of a positive operator with non-negative coefficients, that is, an operator of the form

$$\phi(U) = a_0 U^n + a_1 U^{n-1} + \ldots + a_n I \quad (a_0, a_1, \ldots a_n \geqslant 0),$$

is a positive operator. We call $\phi(U)$ an *operator polynomial* in U.

Let U be a positive operator. We have the inequality

$$|(Ux, y)|^2 \leqslant (Ux, x)(Uy, y) \quad (x, y \in \mathbf{H}), \tag{6}$$

which is a generalization of Buniakowski's inequality (the latter is obtained from (6) when $U = I$). The proof of (6) is almost a word-for-word repetition of the proof of Buniakowski's inequality presented in IV.5.1, so we leave it to the reader.

6.3. Using inequality (6), we now prove the remarkable "monotone sequence theorem".

THEOREM 1. *Let $\{U_n\}$ be an increasing sequence of self-adjoint operators. If $\sup_n \|U_n\| = A < \infty$, then there exists a linear operator U such that, for each $x \in \mathbf{H}$,*

$$Ux = \lim_{n \to \infty} U_n x,$$

and we have

$$\|U\| \leqslant A. \tag{7}$$

Proof. Choose $m \geqslant n$. The operator $U_m - U_n$ is positive, so for each $x \in \mathbf{H}$,

$$(U_m x, x) - (U_n x, x) = ([U_m - U_n]x, x) \geqslant 0,$$

that is, the numerical sequence $\{(U_n x, x)\}$ is increasing. Since

$$|(U_n x, x)| \leqslant \|U_n x\| \, \|x\| \leqslant A\|x\|^2, \tag{8}$$

the limit $\lim_{n \to \infty} (U_n x, x)$ exists and is finite.

Also, by applying (6) to the operator $U_m - U_n$ $(m \geqslant n)$, we can use (8) to write

$$\begin{aligned} |(U_m x - U_n x, y)|^2 &\leqslant [(U_m x, x) - (U_n x, x)][(U_m y, y) - (U_n y, y)] \\ &\leqslant 2A\|y\|^2[(U_m x, x) - (U_n x, x)]. \end{aligned}$$

Now set

$$y = U_m x - U_n x.$$

Then we have

$$\|U_m x - U_n x\|^2 = 2A[(U_m x, x) - (U_n x, x)] \qquad (x \in \mathbf{H}).$$

By what we have proved, the expression on the right tends to zero as $m, n \to \infty$. We therefore deduce the existence of

$$Ux = \lim_{n \to \infty} U_n x.$$

The operator U thus defined is additive and homogeneous. And, since

$$\|U_n x\| \leqslant \|U_n\| \|x\| \leqslant A \|x\|,$$

we have, in the limit,

$$\|Ux\| \leqslant A \|x\|,$$

so that U is a bounded operator and $\|U\| \leqslant A$.

This proves the theorem.

REMARK. The theorem is also clearly true for decreasing sequences of operators.

6.4. Let U be a positive operator. A positive operator V is called a *square root* of U if $V^2 = U$. We then write $V = \sqrt{U}$ or $V = U^{1/2}$.

THEOREM 2. *Let U be a positive operator. Then U has a unique square root $V = \sqrt{U}$. Further, V commutes with every operator that commutes with U.*

Proof. We may assume, without losing any generality, that $\|U\| \leqslant 1$. Write $U_0 = I - U$. Then U_0 is positive, since

$$(U_0 x, x) = (x, x) - (Ux, x) \geqslant \|x\|^2 - \|U\| \|x\|^2 \geqslant 0 \quad (x \in \mathbf{H}).$$

By (6) we have

$$|(U_0 x, y)|^2 \leqslant (U_0 x, x)(U_0 y, y) \leqslant \|x\|^2 \|y\|^2.$$

With $y = U_0 x$ we have the inequality

$$\|U_0 x\| \leqslant \|x\|.$$

Hence

$$\|U_0\| \leqslant 1. \tag{9}$$

Now define a sequence of operators $\{V_n\}$ by setting

$$V_1 = 0, \; V_{n+1} = \tfrac{1}{2}(U_0 + V_n^2) \quad (n = 1, 2, \ldots). \tag{10}$$

Using the inequality (9), one can verify without difficulty, by induction, that

$$\|V_n\| \leqslant 1 \quad (n = 1, 2, \ldots). \tag{11}$$

We now show that the operators V_n and $V_{n+1} - V_n$ are operator polynomials in U_0 with non-negative coefficients. For $n = 1$ this is obvious. Also, since operations on operator polynomials are carried out exactly as on ordinary polynomials, it follows from (10) that, if V_n is an operator polynomial in U_0 with non-negative coefficients, then V_{n+1} will also be such a polynomial. If, using (10) again, we write

$$\begin{aligned} V_{n+1} - V_n &= \tfrac{1}{2}(U_0 + V_n^2) - \tfrac{1}{2}(U_0 + V_{n-1}^2) = \tfrac{1}{2}(V_n^2 - V_{n-1}^2) \\ &= \tfrac{1}{2}(V_n + V_{n-1})(V_n - V_{n-1}), \end{aligned}$$

then, assuming it has already been proved that $V_n - V_{n-1}$ is an operator polynomial with non-negative coefficients, and bearing in mind that $V_n + V_{n-1}$ is such a polynomial, by what we have proved, we see that $V_{n+1} - V_n$, being a product of operator polynomials in U_0 with non-negative coefficients, is itself an operator polynomial in U_0 with non-negative coefficients.

Using the fact that operator polynomials with non-negative coefficients are positive (see 6.2), we can now deduce that $V_n \geqslant \mathbf{0}$ and $V_{n+1} - V_n \geqslant \mathbf{0}$. In other words, the sequence $\{V_n\}$ is increasing, and the V_n, being positive, will also be self-adjoint. Since the sequence $\{V_n\}$ is bounded, by (11), Theorem 1 shows that there exists an operator V_0 such that

$$V_0 x = \lim_{n\to\infty} V_n x.$$

Furthermore, since the V_n are positive, so is V_0, and by (11) we have

$$\|V_0\| \leqslant 1. \tag{12}$$

Let $\bar{U}$ be an operator that commutes with U, and therefore also with U_0. The operators V_n also clearly commute with $\bar{U}$, and so

$$\bar{U}V_0 x = \bar{U}\Big(\lim_{n\to\infty} V_n x\Big) = \lim_{n\to\infty} \bar{U}V_n x = \lim_{n\to\infty} V_n\bar{U}x = V_0\bar{U}x \quad (x\in\mathbf{H}),$$

that is, V_0 commutes with $\bar{U}$. In particular, V_0 commutes with each of the operators V_n. Hence $V_0^2 - V_n^2 = (V_0 + V_n)\,(V_0 - V_n)$, and therefore

$$\|V_0^2 x - V_n^2 x\| \leqslant \|V_0 + V_n\|\,\|V_0 x - V_n x\| \leqslant 2\|V_0 x - V_n x\| \to 0 \quad (x\in\mathbf{H}),$$

so that

$$V_0^2 x = \lim_{n\to\infty} V_n^2 x \quad (x\in\mathbf{H}).$$

Passing to the limit in the equation

$$V_{n+1}x = \tfrac{1}{2}(U_0 x + V_0^2 x) \quad (n = 1, 2, \ldots,\ x\in\mathbf{H}),$$

as $n\to\infty$, we see that

$$V_0 x = \frac{1}{2}(U_0 x + V_0^2 x) \quad (x\in\mathbf{H}).$$

Hence, if we set $V = I - V_0$, we have

$$V^2 = I - 2V_0 + V_0^2 = [I - 2V_0] + [2V_0 - U_0] = I - U_0 = U.$$

Bearing (12) in mind, we can prove that V is positive in exactly the same way we used to prove that U_0 was positive. Hence $V = \sqrt{U}$.

Notice also that V, like V_0, commutes with every operator that commutes with U.

We now prove the uniqueness of the square root. Let V' be a square root of U. Since

$$UV' = V'^3 = V'U,$$

V' commutes with U and so V also commutes with V'. Choose any $x\in\mathbf{H}$ and write $y = V'x - Vx$. Then

$$(V'y, y) + (Vy, y) = ([V' + V]\,[V' - V]x, y) = ([V'^2 - V^2]x, y)$$
$$= ([U - U]x, y) = 0.$$

Hence, as $(Vy, y) \geqslant 0$ and $(V'y, y) \geqslant 0$,

$$(Vy, y) = (V'y, y) = 0.$$

Let W denote any square root of V. Since

$$\|Wy\|^2 = (Wy, Wy) = (W^2 y, y) = (Vy, y) = 0,$$

we have $Wy = \mathbf{0}$ and, *a fortiori*, $Vy = W(Wy) = \mathbf{0}$. Similarly, $V'y = \mathbf{0}$. But we then have

$$\|V'x - Vx\|^2 = ([V' - V]^2 x, x) = ([V' - V]y, x) = 0$$

and $V'x = Vx$ for any $x \in H$. Consequently $V' = V$.

The proof of the theorem is now complete.*

COROLLARY. *If U_1 and U_2 are positive operators that commute with one another, then their product is also positive.*

For the operator $\sqrt{U_2} = V$ commutes with U_1, so $(U_1 U_2 x, x) = (U_1 V^2 x, x) = (VU_1 Vx, x) = (U_1(Vx), Vx) \geqslant 0$ $(x \in \mathbf{H})$.

6.5. When U is a self-adjoint operator, Theorem V.4.1, on the domain of convergence of the series

$$I + U + \ldots + U^n + \ldots, \tag{13}$$

can be simplified.

THEOREM 3. *If U is a self-adjoint operator, then a necessary and sufficient condition for the series* (13) *to converge is that $\|U\| < 1$.*

Proof. We first show that

$$\|U^2\| = \|U\|^2. \tag{14}$$

In fact, we have

$$\|U\|^2 = \sup_{x \neq \mathbf{0}} \frac{\|Ux\|^2}{\|x\|^2} = \sup_{x \neq \mathbf{0}} \frac{(U^2 x, x)}{\|x^2\|} \leqslant \sup_{x \neq \mathbf{0}} \frac{\|U^2 x\|}{\|x\|} = \|U^2\|,$$

and the reverse inequality $\|U^2\| \leqslant \|U\|^2$ is true for every operator U.

It follows from (14) that

$$\|U^{2^m}\| = \|U\|^{2^m} \qquad (m = 1, 2, \ldots) \tag{15}$$

and that

$$c_U = \lim_{n \to \infty} \|U^n\|^{1/n} = \lim_{m \to \infty} \|U^{2^m}\|^{1/2^m} = \|U\|. \tag{16}$$

If we take into account Theorem 4.1 and (16), it is sufficient to verify that (13) diverges if $c_U = \|U\| = 1$. But in this case, by (15), we have

$$\|U^{2^m}\| = 1 \qquad (m = 1, 2, \ldots),$$

and the general term of (13) does not tend to zero.

6.6. The family of projections (3.4) plays a large role in the normed ring $B(\mathbf{H}, \mathbf{H})$ of operators on a Hilbert space $\mathbf{H}$. As we shall show in Chapter IX, § 5, every self-adjoint operator can be reduced to a projection by means of a special construction. This accounts for the importance of the theorems on projections presented below.

Two projections P_1 and P_2 are said to be *orthogonal* if $P_1 P_2 = \mathbf{0}$.† Denote the subspaces onto which P_1 and P_2 project by $\mathbf{H}_1$ and $\mathbf{H}_2$, respectively. A necessary and

---

* The proof we have given is due to Visser (existence) and Szoekefalvi–Nagy (uniqueness). See Riesz and Sz.-Nagy.

† Since $\mathbf{0}$ is a self-adjoint operator, it follows from f) in 6.1 that P_1 and P_2 are permutable, so $P_2 P_1 = P_1 P_2 = \mathbf{0}$, and the definition does not depend on the order in which the projections are multiplied.

sufficient condition for P_1 and P_2 to be orthogonal is that $\mathbf{H}_1$ and $\mathbf{H}_2$ be orthogonal subspaces. For suppose $P_1P_2 = \mathbf{0}$, and let $x' \in \mathbf{H}_1$, $x'' \in \mathbf{H}_2$. Using b) in 3.4 and Theorem 3.2, we have

$$(x', x'') = (P_1x', P_2x'') = (x', P_1P_2x'') = 0.$$

Conversely, if $\mathbf{H}_1 \perp \mathbf{H}_2$ and $x \in \mathbf{H}$, then $P_2x \in \mathbf{H}_2$ and hence $P_2x \perp \mathbf{H}_1$, whence, by c) of 3.4, we have $P_1P_2x = \mathbf{0}$, so that $P_1P_2 = \mathbf{0}$.

THEOREM 4. *Let $P_1, P_2, \ldots, P_n$ be projections. A necessary and sufficient condition for $P = P_1 + P_2 + \ldots + P_n$ to be a projection is that $P_1, P_2, \ldots P_n$ be pairwise orthogonal.*

Proof. Necessity. Assume that P is a projection. By Theorem 3.2,

$$\|Px\|^2 = (Px, Px) = (P^2x, x) = (Px, x) \quad (x \in \mathbf{H}) \tag{17}$$

and similarly

$$\|P_kx\|^2 = (P_kx, x) \quad (x \in \mathbf{H};\ k = 1, 2, \ldots, n).$$

Hence

$$\|P_1x\|^2 + \|P_2x\|^2 \leqslant \sum_{k=1}^{n} \|P_kx\|^2 = \sum_{k=1}^{n} (P_kx, x) = (Px, x) = \|Px\|^2 \leqslant \|x\|^2$$

$$(x \in \mathbf{H}). \tag{18}$$

Consider any element $y \in \mathbf{H}$ and put $x = P_1y$ in (18). Since $P_1x = P_1^2y = P_1y$, this gives

$$\|P_1y\|^2 + \|P_2P_1y\|^2 \leqslant \|P_1y\|^2.$$

It follows from this that $P_2P_1y = \mathbf{0}$ and $P_2P_1 = \mathbf{0}$, so P_1 and P_2 are orthogonal. The proof that the other projections are pairwise orthogonal is exactly the same.

Sufficiency. We verify that $P = P_1 + P_2 + \ldots + P_n$ satisfies the conditions of Theorem 3.2 and so is a projection. Since P is a sum of self-adjoint operators, it is itself self-adjoint (6.1, d). Let us check that $P^2 = P$. Since the P_k $(k = 1, 2, \ldots, n)$ are pairwise orthogonal projections—that is, since

$$P_jP_k = \begin{cases} \mathbf{0} & (j \neq k), \\ P_k & (j = k) \end{cases} \quad (j, k = 1, 2, \ldots, n),$$

we have

$$P^2 = \left(\sum_{k=1}^{n} P_k\right)^2 = \sum_{j=1}^{n}\sum_{k=1}^{n} P_jP_k = \sum_{k=1}^{n} P_k = P.$$

This proves the theorem.

Assuming that the conditions of the theorem are satisfied, we now describe the structure of the subspace $\tilde{\mathbf{H}}$ onto which $P = P_1 + P_2 + \ldots + P_n$ projects. Denote the subspace corresponding to P_k $(k = 1, 2, \ldots, n)$ by $\mathbf{H}_k$. Let $x \in \tilde{\mathbf{H}}$. Then

$$x = Px = P_1x + P_2x + \ldots + P_nx = x_1 + x_2 + \ldots + x_n,$$

where $x_k = P_kx \in \mathbf{H}_k$ $(k = 1, 2, \ldots, n)$.

Conversely, suppose x is an element admitting a representation of the form

$$x = x_1 + x_2 + \dots + x_n \quad (x_k \in \mathbf{H}_k;\ k = 1, 2, \dots, n). \tag{19}$$

Since $PP_k = \sum_{j=1}^{n} P_j P_k = P_k$ and $x_k = P_k x_k = PP_k x_k$, we have

$$Px = \sum_{k=1}^{n} Px_k = \sum_{k=1}^{n} PP_k x_k = \sum_{k=1}^{n} x_k = x,$$

that is, $x \in \tilde{\mathbf{H}}$. Since $P_k x_j = \mathbf{0}$ $(j \neq k)$,

$$x_k = \sum_{j=1}^{n} P_k x_j = P_k\left(\sum_{j=1}^{n} x_j\right) = P_k x \quad (k = 1, 2, \dots, n),$$

from which it follows, in particular, that the representation (19) for x is unique.

To summarize, the subspace $\tilde{\mathbf{H}}$ consists of all those elements x that admit a representation of the form (19). $\tilde{\mathbf{H}}$ is said to be the *orthogonal sum* of the subspaces $\mathbf{H}_1, \mathbf{H}_2, \dots, \mathbf{H}_n$, and is denoted by

$$\tilde{\mathbf{H}} = \mathbf{H}_1 \oplus \mathbf{H}_2 \oplus \dots \oplus \mathbf{H}_n = \oplus \sum_{k=1}^{n} \mathbf{H}_k.$$

6.7. We now investigate differences and products of projections. As a preliminary, we first prove the following lemma.

LEMMA 1. *Let P_1 and P_2 be projections, projecting onto subspaces $\mathbf{H}_1$ and $\mathbf{H}_2$, respectively. The following four conditions are equivalent:*

a) $P_1 \geqslant P_2$; b) $\mathbf{H}_1 \supset \mathbf{H}_2$; c) $P_1 P_2 = P_2$; d) $P_2 P_1 = P_2$.

Proof. a) implies b). Let $x \in \mathbf{H}$. Since $P_1 x \perp x - P_1 x$, we have

$$\|x\|^2 = \|P_1 x\|^2 + \|x - P_1 x\|^2. \tag{20}$$

If $x \in \mathbf{H}_2$, then $P_2 x = x$ and $(P_2 x, x) = (x, x) = \|x\|^2$. *A fortiori*,

$$\|x\|^2 \geqslant \|P_1 x\|^2 = (P_1 x, P_1 x) = (P_1 x, x) \geqslant (P_2 x, x) = \|x\|^2,$$

so from (20) we conclude that $x = P_1 x$, which means that $x \in \mathbf{H}_1$.

b) implies c). Let $x \in \mathbf{H}$. Recalling that $P_2 x \in \mathbf{H}_2$ and so $P_2 x \in \mathbf{H}_1$, we have $P_1 P_2 x = P_2 x$.

c) implies d). Since the product of the self-adjoint operators P_1 and P_2 is a projection, and thus a self-adjoint operator, P_1 and P_2 are permutable: $P_2 P_1 = P_1 P_2 = P_2$.

d) implies a). For $x \in \mathbf{H}$,

$$(P_2 x, x) = \|P_2 x\|^2 = \|P_2 P_1 x\|^2 \leqslant \|P_1 x\|^2 = (P_1 x, x).$$

THEOREM 5. *Let P_1 and P_2 be projections. A necessary and sufficient condition for $P = P_1 - P_2$ to be a projection is that $P_1 \geqslant P_2$.*

Proof. Necessity. Assume that P is a projection. Since $P_1 = P + P_2$, the preceding theorem shows that $PP_2 = \mathbf{0}$, so that $(P_1 - P_2)P_2 = P_1 P_2 - P_2 = \mathbf{0}$ and condition c) of the lemma is satisfied.

Sufficiency. Obviously P is a self-adjoint operator. Also, by conditions c) and d) of the lemma,

$$P^2 = P_1^2 - P_1P_2 - P_2P_1 + P_2^2 = P_1 - P_2 - P_2 + P_2 = P;$$

it remains only to apply Theorem 3.2.

Denote the subspace onto which P projects by $\check{\mathbf{H}}$. We leave it to the reader to prove that $\check{\mathbf{H}}$ consists of all elements $x \in \mathbf{H}_1$ orthogonal to $\mathbf{H}_2$. The subspace $\check{\mathbf{H}}$ is called the *orthogonal complement* of H_2 in $\mathbf{H}_1$ and is denoted by $\check{\mathbf{H}} = \mathbf{H}_1 \ominus \mathbf{H}_2$.

REMARK. If $P_1 = I$, then the hypothesis of the theorem is obviously satisfied for any P_2. Hence $P = I - P_2$ is a projection, whatever projection P_2 we take. Here P projects $\mathbf{H}$ onto the orthogonal complement $\mathbf{H} \ominus \mathbf{H}_2$ of the subspace $\mathbf{H}_2$.

We now prove a theorem about products of projections.

THEOREM 6. *Let P_1 and P_2 be projections. A necessary and sufficient condition for $P = P_1P_2$ to be a projection is that P_1 and P_2 be permutable.*

Proof. For the necessity of the condition we need only use the fact that P is self-adjoint. To establish its sufficiency, we again use Theorem 3.2, noting that P is self-adjoint and

$$P^2 = P_1P_2P_1P_2 = P_1^2P_2^2 = P.$$

We leave it to the reader to verify that, in this case, the subspace onto which P projects is the intersection of $\mathbf{H}_1$ and $\mathbf{H}_2$.

6.8. By applying Theorem 1 to increasing (or decreasing) sequences $\{P_n\}$ of projections, we obtain the following result.

THEOREM 7. *Let $\{P_n\}$ be a monotone sequence of projections. For each $x \in \mathbf{H}$, the limit*

$$Px = \lim_{n \to \infty} P_n x \tag{21}$$

exists and the operator P is a projection onto the subspace $\check{\mathbf{H}}$ defined by

$$\check{\mathbf{H}} = \overline{\bigcup_{n=1}^{\infty} \mathbf{H}_n},$$

if $\{P_n\}$ is an increasing sequence, and by

$$\check{\mathbf{H}} = \bigcap_{n=1}^{\infty} \mathbf{H}_n,$$

if it is decreasing. Here the $\mathbf{H}_n$ are the subspaces corresponding to the projections P_n ($n = 1, 2, \ldots$).

Proof. Assume that $\{P_n\}$ is an increasing sequence. The existence of the limit (25) is a consequence of Theorem 1, since $\|P_n\| \leqslant 1$ ($n = 1, 2, \ldots$). We prove that P is a projection onto $\check{\mathbf{H}}$. For this it is sufficient to verify that, if $x \in \check{\mathbf{H}}$, then $Px = x$, while if $x \perp \check{\mathbf{H}}$, then $Px = \mathbf{0}$. Write $\Omega = \bigcup_{n=1}^{\infty} \mathbf{H}_n$ and choose $x \in \Omega$. There exists m such that $x \in \mathbf{H}_n$ ($n \geqslant m$). This means that $P_n x = x$ ($n \geqslant m$) and so $Px = \lim_{n \to \infty} P_n x = x$. Remembering that $\check{\mathbf{H}} = \overline{\Omega}$, we see that $Px = x$ also for $x \in \check{\mathbf{H}}$. Now suppose $x \perp \check{\mathbf{H}}$. Then, *a fortiori*, $x \perp \mathbf{H}_n$ ($n = 1, 2, \ldots$). Therefore $P_n x = \mathbf{0}$ and, taking limits, we have $Px = \mathbf{0}$.

Similar arguments give the result when $\{P_n\}$ is a decreasing sequence.

REMARK. The theorem remains true if, instead of a sequence $\{P_n\}$, we consider a family of projections $\{P_\lambda\}$ depending on a continuously variable real parameter.

COROLLARY. *Let $\{P_k\}$ be a family of pairwise orthogonal projections. Then for each $x \in \mathbf{H}$ the series*

$$Px = \sum_{k=1}^{\infty} P_k x \tag{22}$$

is convergent and P is a projection.

For we need only apply the above theorem to be increasing sequence $\{P^{(n)}\}$ of projections, where $P^{(n)} = \sum_{k=1}^{n} P_k$, $n = 1, 2, \ldots$.

We leave it to the reader to verify that P projects $\mathbf{H}$ onto the subspace $\tilde{\mathbf{H}}$ consisting of all $x \in \mathbf{H}$ having a representation of the form

$$x = \sum_{k=1}^{\infty} x_k \quad \left(x_k \in \mathbf{H}_k;\ k = 1, 2, \ldots;\ \sum_{k=1}^{\infty} \|x_k\|^2 < \infty \right).$$

The subspace $\tilde{\mathbf{H}}$ is called the *orthogonal sum* (cf. 6.6) of the subspaces $\mathbf{H}_1, \mathbf{H}_2, \ldots$ and is denoted by

$$\tilde{\mathbf{H}} = \mathbf{H}_1 \oplus \mathbf{H}_2 \oplus \cdots = \oplus \sum_{k=1}^{\infty} \mathbf{H}_k.$$

§ 7. The weak topology and reflexive spaces

7.1. In III.2.2 we established the Hahn–Banach Theorem for LCSs and presented a number of its most important corollaries, relating to the existence of sufficiently many continuous linear functionals. Now we present some results on this theme in a formulation that is more suited to normed spaces.

THEOREM 1 (Hahn–Banach). *Let $\mathbf{X}$ be a normed space and f_0 a continuous linear functional defined on a subspace $\mathbf{X}_0 \subset \mathbf{X}$. Then there exists a continuous linear functional f, defined on the whole of $\mathbf{X}$ and extending f_0; furthermore, $\|f\| = \|f_0\|$.*

Proof. Since f_0 is continuous on $\mathbf{X}_0$, we have $|f_0(x)| \leqslant \|f_0\|\,\|x\|$ $(x \in \mathbf{X}_0)$. By Theorem II.4.2, there exists an extension f of f_0 on $\mathbf{X}$ such that $|f(x)| \leqslant \|f_0\|\,\|x\|$ $(x \in \mathbf{X})$, and so $\|f\| \leqslant \|f_0\|$. On the other hand, we have $\|f_0\| \leqslant \|f\|$ because $\mathbf{X}_0 \subset \mathbf{X}$, and the theorem is therefore proved.

By corollary 2 to Theorem III.2.4, we have

THEOREM 2. *For each non-zero element x_0 of a normed space $\mathbf{X}$, there exists a functional $f \in \mathbf{X}^*$ such that*

$$\|f\| = 1, \quad f(x_0) = \|x_0\|,$$

that is, such that the relation $|f(x)| \leqslant \|f\|\,\|x\|$ holds with equality for the element x_0.

COROLLARY. *For each $x \in \mathbf{X}$ we have*

$$\|x\| = \sup\{|f(x)| : f \in \mathbf{X}^*,\ \|f\| \leqslant 1\}.$$

As a generalization of Theorem 2, we have

THEOREM 3. *Let Ω be a linear manifold in a normed space $\mathbf{X}$ and let $x_0 \in \mathbf{X}$ be an element having distance $d > 0$ from Ω ($\rho(x_0, \Omega) = d$).*

Then there exists a functional $f \in \mathbf{X}^*$ such that

$$f(x) = 0 \quad (x \in \Omega), \quad \|f\| = 1, \quad f(x_0) = d.$$

Proof. Let $\mathbf{X}_0$ denote the simple linear extension of Ω obtained by adjoining x_0 to Ω: that is, let $\mathbf{X}_0 = \mathscr{L}(\Omega, x_0)$. Since $x \notin \Omega$, the expression for an element $x \in \mathbf{X}$ in the form

$$x = \lambda x_0 + x' \quad (x' \in \Omega)$$

is unique. For $x \in \mathbf{X}_0$, set

$$f_0(x) = \lambda d.$$

Clearly f_0 is a linear functional. Also $f_0(x_0) = d$ and $f_0(x) = 0$ for $x \in \Omega$. Furthermore,

$$\|x\| = \|\lambda x_0 + x'\| = |\lambda| \left\| x_0 + \frac{x'}{\lambda} \right\| \geqslant |\lambda| \rho(x_0, \Omega) = |\lambda| d = |f_0(x)|.$$

Therefore the functional f_0 is bounded, with $\|f_0\| \leqslant 1$. To obtain the reverse inequality, take $x' \in \Omega$ such that $\|x_0 - x'\| < d + \varepsilon$. Then

$$d = f_0(x_0) = f_0(x_0 - x') \leqslant \|f_0\| \, \|x_0 - x'\| < \|f_0\| (d + \varepsilon),$$

so that

$$\|f_0\| < \frac{d}{d + \varepsilon},$$

and, as ε was arbitrary, we have $\|f_0\| = 1$.

Applying Theorem 1 on the extension of functionals to f_0, we thus obtain a functional f having the required properties.

REMARK 1. Theorem 2, on the existence of sufficiently many functionals, is a special case of Theorem 3. For if $\Omega = \{\mathbf{0}\}$, then $\rho(x_0, \Omega) = \|x_0\|$.

REMARK 2. In Theorems 2 and 3 it is sometimes more useful to give the alternative formulation obtained by replacing f by the functional defined by:

$$\tilde{f}(x) = \frac{1}{d} f(x).$$

The functional $\tilde{f}$ has the properties:

$$\tilde{f}(x) = 0 \quad (x \in \Omega), \quad \|\tilde{f}\| = \frac{1}{d}, \quad \tilde{f}(x_0) = 1.$$

Let us give an application of Theorem 3. We say that a system $\{x_\alpha\}$ $(\alpha \in \mathrm{A})$ in a normed space $\mathbf{X}$ admits biorthogonalization if there exists a system of functionals $\{f_\alpha\}$ $(\alpha \in \mathrm{A})$, $f_\alpha \in \mathbf{X}^*$, such that

$$f_{\alpha'}(x_\alpha) = \begin{cases} 0, & \alpha' \neq \alpha, \\ 1, & \alpha' = \alpha. \end{cases}$$

THEOREM 4. *A necessary and sufficient condition for a system $\{x_\alpha\}$ $(\alpha \in \mathrm{A})$ to admit biorthogonalization is that it be minimal—that is, no element x_k belongs to the closed linear hull of the set of remaining elements.*

Proof. Necessity. Let $\{x_\alpha\}$ be a system admitting biorthogonalization. If, for some α_0, we were to have

$$x_{\alpha_0} = \lim_{n\to\infty} \sum_{k=1}^{n} \lambda_k^{(n)} x_{\alpha_k} \quad (\alpha_k \neq \alpha_0),$$

then

$$f_{\alpha_0}(x_{\alpha_0}) = \lim_{n\to\infty} \sum_{k=1}^{n} \lambda_k^{(n)} f_{\alpha_0}(x_{\alpha_k}) = 0,$$

whereas, in fact, $f_{\alpha_0}(x_{\alpha_0}) = 1$.

Sufficiency. Now suppose $\{x_\alpha\}$ is a minimal system. Denote the closed linear hull of the set of elements $\{x_\alpha\}$ $(\alpha \neq \alpha')$ by $\mathbf{X}_{\alpha'}$. Since $x_{\alpha'} \notin \mathbf{X}_{\alpha'}$ and $\mathbf{X}_{\alpha'}$ is a closed linear manifold, it follows from Theorem 3 (more precisely, from the Remark after it) that there exists a functional $f_{\alpha'}$ mapping $\mathbf{X}_{\alpha'}$, to zero (that is, $f_{\alpha'}(x_\alpha) = 0$ for $\alpha \neq \alpha'$) and such that $f_{\alpha'}(x_{\alpha'}) = 1$. The family of functionals $\{f_{\alpha'}\}$ now satisfies the requirements of the definition, so the system $\{x_\alpha\}$ admits biorthogonalization.

REMARK. Every finite system $x_1, x_2, \ldots, x_n$ of linearly independent elements is minimal. For $x_j \notin \mathscr{L}(\{x_k\})$ $(k \neq j)$, by the linear independence, and, as a finite-dimensional linear manifold is closed, x_j also does not belong to the closure of $\mathscr{L}(\{x_k\})$.

If follows from this Remark that, for each finite system of linearly independent elements, there is a system of functionals that is biorthogonal to it.

By applying Theorem 4 to a Hilbert space, taking into account the general form for linear functionals on Hilbert spaces, we obtain the following result.

Let $\{x_n\}$ be any sequence of elements in a Hilbert space $\mathbf{H}$. A necessary and sufficient condition for the existence of a sequence $\{y_n\}$ biorthogonal to the given one—that is, satisfying

$$(x_m, y_n) = \begin{cases} 0, & m \neq n, \\ 1, & m = n, \end{cases}$$

is that $\{x_n\}$ be minimal.

The extension theorem for linear functionals was proved in the real case by Hahn. It had been obtained in essence by Banach [2]. Corollaries to the extension theorem are mentioned in Banach.

7.2. In III.3.1 we defined the weak topology corresponding to any dual pairing of vector spaces. Next we shall be interested in the following situation. Let $\mathbf{X}$ be a normed space and $\mathbf{X}^*$ its dual. We consider the topology $\sigma(\mathbf{X}, \mathbf{X}^*)$ on $\mathbf{X}$, which we shall call the *weak topology* of $\mathbf{X}$, and the topology $\sigma(\mathbf{X}^*, \mathbf{X})$ on $\mathbf{X}^*$, which we shall call the *weak* topology* of $\mathbf{X}^*$. Hence on $\mathbf{X}^*$ there are two weak topologies: the weak* topology $\sigma(\mathbf{X}^*, \mathbf{X})$ and the weak topology $\sigma(\mathbf{X}^*, \mathbf{X}^{**})$; moreover, we have $\sigma(\mathbf{X}^*, \mathbf{X}) \leqslant \sigma(\mathbf{X}^*, \mathbf{X}^{**})$ (this inequality is strict in the non-reflexive case; see below). Convergence in the weak topology (resp., weak* topology) will be called weak convergence (resp. weak* convergence), while convergence in norm will be called strong convergence. The following terms are to be understood analogously: weak compactness, weak boundedness, weak closure, etc.

LEMMA 1. *Let $\langle \mathbf{X}, \mathbf{Y} \rangle$ be a dual pairing of vector spaces such that $\mathbf{Y}$ contains a countable total subset $\{f_n\}_{n=1}^{\infty}$ of $\mathbf{X}$. Then the topology $\sigma(\mathbf{X}, \mathbf{Y})$ is metrizable on weakly compact sets.*

Proof. Let K be a $\sigma(\mathbf{X}, \mathbf{Y})$-compact set in $\mathbf{X}$. Define a metric on K by:

$$\rho(x, y) = \sum_{n=1}^{\infty} (1/2^n)\,|f_n(x-y)| \qquad (x, y \in K).$$

Since K is weakly compact, it is weakly bounded, so $\rho(x, y) < \infty$. If $\rho(x, y) = 0$, then $f_n(x - y) = 0$ for all n. Therefore, as $\{f_n\}$ is total, we deduce that $x = y$. The remaining properties of a metric obviously hold for ρ.

Consider the identity mapping i: $(K, \sigma(\mathbf{X}, \mathbf{Y})) \to (K, \rho)$, where (K, ρ) is the metric space having the metric ρ. Obviously i is continuous, so, by Corollary 3 to Theorem I.2.5, it is a homeomorphism, which proves that the space $(K, \sigma(\mathbf{X}, \mathbf{Y}))$ is metrizable.

The class of separable B-spaces plays an important role in various problems. We now record some results on this class.

COROLLARY. *If* $\mathbf{X}$ *is a separable normed space, then the topology* $\sigma(\mathbf{X}, \mathbf{X}^*)$ *is metrizable on weakly compact sets.*

Proof. We verify that the condition of Lemma 1 is satisfied. Let $\{x_n\}$ be a countable everywhere dense subset of $\mathbf{X}$. Theorem 2 shows that for each $n \in \mathbb{N}$ there exists an $f_n \in \mathbf{X}^*$, $\|f_n\| = 1$, such that $f_n(x_n) = \|x_n\|$. Assume that $f_n(x) = 0$ for all $n \in \mathbb{N}$. Let us show that $x = \mathbf{0}$. Choose a sequence $\{x_{n_m}\}$ with $x_{n_m} \to x$. Then

$$\|x_{n_m}\| = |f_{n_m}(x_{n_m})| = |f_{n_m}(x_{n_m}) - f_{n_m}(x)| \leqslant \|f_{n_m}\| \, \|x_{n_m}\| \, \|x_{n_m} - x\| \to 0,$$

so that $x_{n_m} \to \mathbf{0}$, and therefore $x = \mathbf{0}$.

THEOREM 5. *If a normed space* $\mathbf{X}$ *has a separable dual* $\mathbf{X}^*$, *then* $\mathbf{X}$ *is separable.*

Proof. Let $\{f_n\}_{n=1}^{\infty}$ be an everywhere dense subset of $\mathbf{X}^*$. By the definition of the norm of a functional, there exists an $x_n \in \mathbf{X}$ for each $n \in \mathbb{N}$, such that $\|x_n\| \leqslant 1$ and $|f_n(x_n)| \geqslant \frac{1}{2}\|f_n\|$. The set E of linear combinations of the elements x_n with rational coefficients is countable. Let us show that E is dense in $\mathbf{X}$. Obviously $\bar{E}$ is a closed linear manifold. If $\bar{E} \neq \mathbf{X}$, then by Theorem 2 there exists $f \in \mathbf{X}^*$, $\|f\| = 1$, such that $f(x) = 0$, $x \in \bar{E}$. Assume that $f_{n_k} \to f$ in the norm. Then

$$(1/2)\|f_{n_k}\| \leqslant |f_{n_k}(x_{n_k})| = |(f_{n_k} - f)(x_{n_k})| \leqslant \|f_{n_k} - f\| \to 0,$$

so that $f_{n_k} \to \mathbf{0}$, $f = \mathbf{0}$, contradicting the equation $\|f\| \leqslant 1$.

Let $\mathbf{X}$ be a normed space. The polar of the closed unit ball* $B_{\mathbf{X}}$ in $\mathbf{X}$ is the closed unit ball $B_{\mathbf{X}^*}$ in the dual space. By the Alaoglu–Bourbaki Theorem (Theorem III.3.7), $B_{\mathbf{X}^*}$ is $\sigma(\mathbf{X}^*, \mathbf{X})$-compact. We note that $B_{\mathbf{X}^*}$ is not always weak* sequentially compact (for example, if $\mathbf{X} = \ell^{\infty}$; see X.3.4); however, if $\mathbf{X}$ is separable, then $B_{\mathbf{X}^*}$ is sequentially compact.

THEOREM 6. *If* $\mathbf{X}$ *is a separable normed space, then the unit ball* $B_{\mathbf{X}^*}$ *with the weak* topology is metrizable, and consequently also sequentially compact.*

Proof. Consider $\mathbf{X}$ as a set of linear functionals on $\mathbf{X}^*$ and form the dual pair $\langle \mathbf{X}^*, \mathbf{X} \rangle$. A countable everywhere dense subset of $\mathbf{X}$ is clearly a total subset on $\mathbf{X}^*$. Now apply Lemma 1.

REMARK. If $B_{\mathbf{X}^*}$ is metrizable in the weak* topology, then $\mathbf{X}$ is separable.

7.3. Let $\mathbf{X}$ be a normed space. Since $\mathbf{X}^*$ is in turn a normed space, it makes sense to speak of the space $\mathbf{X}^{**} = (\mathbf{X}^*)^*$ dual to $\mathbf{X}^*$—this is the collection of all continuous linear functionals on $\mathbf{X}^*$. In exactly the same way, one can consider the space $\mathbf{X}^{***}$, and so on.

Now consider the canonical embedding $\pi_{\mathbf{X}^*}: \mathbf{X} \to \mathbf{X}^{**}$ (see III.3.2), which we shall in future denote simply by π. We recall that π assigns to each $x \in \mathbf{X}$ the functional F_x on $\mathbf{X}^*$ defined by

$$F_x(f) = \overline{f(x)} \qquad (f \in \mathbf{X}^*).$$

---

* Here, and in what follows, $B_{\mathbf{X}} = \{x \in \mathbf{X}: \|x\| \leqslant 1\}$.

Clearly,

$$|F_x(f)| = |f(x)| \leqslant \|x\| \, \|f\|,$$

so that $F_x \in \mathbf{X}^{**}$. Hence π maps $\mathbf{X}$ into $\mathbf{X}^{**}$. By the Corollary to Theorem 2,

$$\|\pi(x)\| = \|F_x\| = \sup\{|F_x(f)| : f \in \mathbf{X}^*, \ \|f\| \leqslant 1\} = \sup\{|f(x)| : f \in \mathbf{X}^*, \ \|f\| \leqslant 1\} = \|x\|,$$

and so we see that π is a linear isometry from $\mathbf{X}$ on the subspace $\pi(\mathbf{X})$ of $\mathbf{X}^{**}$.

A B-space $\mathbf{X}$ is said to be *reflexive* if $\pi(\mathbf{X}) = \mathbf{X}^{**}$, that is, if $\mathbf{X}$ is isometric to $\mathbf{X}^{**}$ under the canonical embedding π. Since the correspondence which determines the isometry in this situation is of a special form, the existence of a linear isometry between $\mathbf{X}$ and $\mathbf{X}^{**}$ is not enough to imply that $\mathbf{X}$ is reflexive (a relevant example is given by James [1]).

It is easy to see that every finite-dimensional B-space is reflexive. We leave it to the reader to prove, using Theorem 3.1 on the general form of a linear functional on a Hilbert space, that Hilbert spaces are reflexive. As we shall see below, the $\mathbf{L}^p$ spaces $(1 < p < \infty)$ are reflexive, but $\mathbf{L}^1$, $\mathbf{L}^\infty$, $\mathbf{C}[0, 1]$, $\mathbf{c}_0$ are not.

THEOREM 7. *A necessary and sufficient condition for a B-space* $\mathbf{X}$ *to be reflexive is that the closed unit ball* $B_{\mathbf{X}}$ *be weakly compact.*

Proof. Necessity. As $\mathbf{X}$ is reflexive, $B_{\mathbf{X}}$ coincides with $B_{\mathbf{X}^{**}}$ in $\mathbf{X}^{**}$, which is $\sigma(\mathbf{X}^{**}, \mathbf{X}^*)$-compact by the Alaoglu–Bourbaki Theorem.

Sufficiency. If we show that the set $\pi(B_{\mathbf{X}})$ coincides with $B_{\mathbf{X}^{**}}$, then we shall have $\pi(\mathbf{X}) = \mathbf{X}^{**}$. By the Bipolar Theorem, $B_{\mathbf{X}}^{\circ\circ}$ is the $\sigma(\mathbf{X}^{**}, \mathbf{X}^*)$-closure of $\pi(B_{\mathbf{X}})$. Since $B_{\mathbf{X}}$ is weakly compact, $\pi(B_{\mathbf{X}})$ is $\sigma(\mathbf{X}^{**}, \mathbf{X}^*)$-closed, so that $B_{\mathbf{X}}^{\circ\circ} = \pi(B_{\mathbf{X}})$. On the other hand, by the definition of the bipolar, $B_{\mathbf{X}}^{\circ\circ} = B_{\mathbf{X}^{**}}$, so we obtain the equation $\pi(B_{\mathbf{X}}) = B_{\mathbf{X}^{**}}$, as required.

COROLLARY 1. *Every closed subspace of a reflexive B-space* $\mathbf{X}$ *is reflexive.*

Proof. Let $\mathbf{Y}$ be a closed subspace of $\mathbf{X}$ and let $B_{\mathbf{Y}}$ be the closed unit ball in $\mathbf{Y}$. Since $B_{\mathbf{X}}$ is weakly compact and $B_{\mathbf{Y}} = B_{\mathbf{X}} \cap \mathbf{Y}$, Corollary 3 to Theorem III.3.2 shows that $B_{\mathbf{Y}}$ is weakly compact in $\mathbf{Y}$. Applying Theorem 7 again, we find that $\mathbf{Y}$ is reflexive.

COROLLARY 2. *A B-space* $\mathbf{X}$ *is reflexive if and only if its dual* $\mathbf{X}^*$ *is reflexive.*

Proof. If $\mathbf{X}$ is reflexive, then the topologies $\sigma(\mathbf{X}^*, \mathbf{X})$ and $\sigma(\mathbf{X}^*, \mathbf{X}^{**})$ coincide on $\mathbf{X}^*$. Since $B_{\mathbf{X}^*}$ is weak* compact by the Alaoglu–Bourbaki Theorem, this implies that $B_{\mathbf{X}^*}$ is weakly compact and, by Theorem 7, $\mathbf{X}^*$ is reflexive.

If $\mathbf{X}^*$ is reflexive, then, as we have already shown, $\mathbf{X}^{**}$ is reflexive. Since $\pi(\mathbf{X})$ is a closed subspace of $\mathbf{X}^{**}$, it follows from Corollary 1 that $\pi(\mathbf{X})$, and hence also $\mathbf{X}$, is reflexive.

COROLLARY 3. *The weak topology and the weak* topology coincide on the unit ball* $B_{\mathbf{X}^*}$ *if and only if* $\mathbf{X}$ *is reflexive.*

Proof. If the weak topology and the weak* topology coincide on $B_{\mathbf{X}^*}$, then $B_{\mathbf{X}^*}$ is weakly compact, and so $\mathbf{X}^*$ is reflexive. Hence, by Corollary 2, $\mathbf{X}$ is reflexive. The converse is obvious.

REMARK. If B-spaces $\mathbf{X}$ and $\mathbf{Y}$ are isomorphic and $\mathbf{X}$ is reflexive, then $\mathbf{Y}$ is reflexive.

We shall become acquainted with further examples of reflexive B-spaces below.

7.4. Here we formulate, without proofs, a few important results connected with the weak topology.

Let $\mathbf{X}$ be a B-space. How can we describe the set $\pi(\mathbf{X})$? It consists, of course, of all the functionals $F \in \mathbf{X}^{**}$ that are $\sigma(\mathbf{X}^*, \mathbf{X})$-continuous. However, verification of weak* continuity on unrestricted nets of functionals is always difficult. In this connection, the following theorem merits attention.

THEOREM 8 (A. Grothendieck). *If* $\mathbf{X}$ *is a B-space, then a functional F belongs to* $\pi(\mathbf{X})$ *if and only if*

$$f_\alpha \underset{A}{\to} \mathbf{0}\ (\sigma(\mathbf{X}^*, \mathbf{X})) \quad \text{and} \quad \|f_\alpha\| \leqslant 1 \quad (\alpha \in A) \tag{$*$}$$

imply that $F(f_\alpha) \to 0$.

If $\mathbf{X}$ is a separable B-space, then we can substitute sequences for nets in $(*)$.

The inconvenience of working with unrestricted nets of functionals is also the reason for the importance of the following theorem.

THEOREM 9 (Krein–Shmul'yan). *If* $\mathbf{X}$ *is a B-space, then a convex subset E of* $\mathbf{X}$ *is weak* closed if the set* $\{x \in E : \|x\| \leqslant n\}$ *is weak* closed for each* $n \in \mathbb{N}$.

In Theorems 8 and 9, the completeness of $\mathbf{X}$ is essential. For proofs, see Schaefer-I (Chapter IV, §6).

§8. Extensions of linear operators

8.1. The Hahn–Banach Theorem asserts that every continuous linear functional defined on a subspace of a normed space $\mathbf{X}$ has a continuous extension to the whole of $\mathbf{X}$. The analogous statement for operators is no longer true. In this subsection we consider what positive results in this direction can be obtained.

Let us agree on some notation. Let U_0 and U be two operators mapping sets Ω_0 and Ω in a space $\mathbf{X}$, respectively, into a space $\mathbf{Y}$. If $\Omega_0 \subset \Omega$ and we have $U(x) = U_0(x)$ for $x \in \Omega_0$, then U is said to be an *extension* (or *prolongation*) of U_0. In this case we write $U_0 \subset U$.

THEOREM 1.* *Let* $\mathbf{X}$ *and* $\mathbf{Y}$ *be normed spaces and let* U_0 *be an operator from* $\Omega \subset \mathbf{X}$ *into* $\mathbf{Y}$. *A necessary and sufficient condition for* U_0 *to have a continuous extension U to the set* $\mathscr{L}(\Omega)$ *is that there exist a constant C such that*

$$\left\| \sum_{k=1}^{n} \lambda_k U_0(x_k) \right\| \leqslant C \left\| \sum_{k=1}^{n} \lambda_k x_k \right\| \tag{1}$$

for any $x_1, x_2, \ldots, x_n$ *in* Ω *and any numbers* $\lambda_1, \lambda_2, \ldots, \lambda_n$.

Proof. Necessity. If the extension U exists, then, since the element $x = \sum_{k=1}^{n} \lambda_k x_k$ belongs to $\mathscr{L}(\Omega)$, we have

$$U(x) = \sum_{k=1}^{n} \lambda_k U(x_k) = \sum_{k=1}^{n} \lambda_k U_0(x_k).$$

But, as U is linear and bounded, we have

$$\left\| \sum_{k=1}^{n} \lambda_k U_0(x_k) \right\| = \|U(x)\| \leqslant \|U\| \, \|x\| = \|U\| \left\| \sum_{k=1}^{n} \lambda_k x_k \right\|,$$

that is, (1) is satisfied for $C = \|U\|$.

Sufficiency. It follows immediately from (1) that, if a linear combination of the elements $x_1, x_2, \ldots, x_n$ is equal to the zero element

$$\sum_{k=1}^{n} \alpha_k x_k = \mathbf{0}, \text{ then also } \sum_{k=1}^{n} \alpha_k U_0(x_k) = \mathbf{0}.$$

---

* See F. Riesz [2].

Now take any $x \in \mathscr{L}(\Omega)$:

$$x = \sum_{k=1}^{n} \lambda_k x_k \qquad (x_k \in \Omega;\ k = 1, 2, \ldots, n). \tag{2}$$

Write

$$U(x) = \sum_{k=1}^{n} \lambda_k U_0(x_k). \tag{3}$$

We prove that, in spite of the possible multiplicity of representations (2), the value of the operator (3) is uniquely determined by x. For if $x = \sum_{k=1}^{n} \lambda_k x_k = \sum_{k=1}^{n} \lambda'_k x_k$,* then $\sum_{k=1}^{n} (\lambda_k - \lambda'_k) x_k = \mathbf{0}$. Hence, as we remarked above, $\sum_{k=1}^{n} (\lambda_k - \lambda'_k) U_0(x_k) = \mathbf{0}$. Therefore

$$\sum_{k=1}^{n} \lambda_k U_0(x_k) = \sum_{k=1}^{n} \lambda'_k U_0(x_k).$$

There is no difficulty in verifying that U is linear. For if $x = \sum_{k=1}^{n} \lambda_k x_k$ and $y = \sum_{k=1}^{n} \mu_k x_k$, then $x + y = \sum_{k=1}^{n} (\lambda_k + \mu_k) x_k$ and

$$U(x+y) = \sum_{k=1}^{n} (\lambda_k + \mu_k) U_0(x_k)$$

$$= \sum_{k=1}^{n} \lambda_k U_0(x_k) + \sum_{k=1}^{n} \mu_k U_0(x_k) = U(x) + U(y).$$

The boundedness of U follows from (1), since

$$\|U(x)\| \leqslant C\|x\| \qquad (x \in \mathscr{L}(\Omega)).$$

It is also obvious that U is an extension of U_0.

The proof of the theorem is therefore complete.

8.2. The following theorem demonstrates that it is possible to extend the domain of a continuous linear operator from a given subset to its closure.

THEOREM 2 (Extension by continuity). *Let* **X** *and* **Y** *be normed spaces, and suppose* **Y** *is complete. Every continuous linear operator* U_0 *from* $\Omega \subset \mathbf{X}$ *into* **Y** *has a unique continuous linear extension* U *to the closure* $\bar{\Omega}$ *of* Ω, *and* $\|U\| = \|U_0\|$.

Proof. Let $x \in \bar{\Omega}$. Then there exists a sequence $\{x_n\}$ of elements of Ω converging to x. Consider the sequence $\{U_0(x_n)\}$ in **Y**. Since $\|U_0(x_n) - U_0(x_m)\| \leqslant \|U_0\|\,\|x_n - x_m\|$, this is a Cauchy sequence; hence, as **Y** is complete, $\lim_{n\to\infty} U_0(x_n)$ exists. We prove that this limit is independent of the choice of the sequence $\{x_n\}$. In fact, if $x'_n \in \Omega$ and $x'_n \to x$, then

$$\|U_0(x'_n) - U_0(x_n)\| \leqslant \|U_0\|\,\|x'_n - x_n\| \to 0.$$

* We do not lose any generality by assuming the same elements x_k occur in both sums, because we can add to each sum a term composed of the missing x_k with zero coefficients.

Hence $\lim_{n\to\infty} U_0(x'_n) = \lim_{n\to\infty} U_0(x_n)$.

Now set

$$U(x) = \lim_{n\to\infty} U_0(x_n).$$

Obviously U is linear, and, taking limits in the inequality $\| U_0(x_n)\| \leqslant \| U_0 \| \; \| x_n \|$, we find that $\| U(x) \| \leqslant \| U_0 \| \; \| x \|$; that is, U is bounded, and we have $\| U \| \leqslant \| U_0 \|$. On the other hand,

$$\|U\| = \sup_{\substack{\|x\| \leqslant 1 \\ x\in\Omega}} \|U(x)\| \geqslant \sup_{\substack{\|x\| \leqslant 1 \\ x\in\Omega}} \|U_0(x)\| = \|U_0\|,$$

so that $\| U \| = \| U_0 \|$.

Finally, if V is a continuous linear operator from $\bar{\Omega}$ into $\mathbf{Y}$ with $V \supset U_0$, then, by taking limits in the equation $U_0(x_n) = V(x_n)$ $(x_n \in \Omega,\ x_n \to x)$ and using the continuity of V, we find that $U(x) = V(x)$: that is, U and V coincide.

This proves the theorem.

COROLLARY. *If a continuous linear operator U from a normed space $\mathbf{X}$ into a B-space $\mathbf{Y}$ maps a dense subset of $\mathbf{X}$ to zero, then $U(x) = \mathbf{0}$ for all $x \in \mathbf{X}$.*

8.3. If a linear operator U_0 is defined on a dense subset Ω of space $\mathbf{X}$, then by Theorem 2 one can extend it to the whole of $\mathbf{X}$, preserving its linearity and its norm. However, if we omit the stipulation that Ω be dense in $\mathbf{X}$, then an extension of this type will, in general, not be possible.

A solution to the problem of the existence of an extension of an operator U_0 having the same norm as U_0 depends on properties of the image space $\mathbf{Y}$.

For convenience in what follows, we make the following definition: a system of sets $\mathfrak{A} = \{A\}$ will be called *overlapping* if each pair of sets in the system have non-empty intersection.

We shall say that a normed space $\mathbf{Y}$ is a *space of type* $\mathfrak{M}$ if it is a normed space in which every overlapping family of closed balls has non-empty intersection.

The simplest examples of a space of type $\mathfrak{M}$ is the real line. In fact, a closed sphere in this space is just a closed interval. We therefore consider an overlapping set $\{[a_\xi, b_\xi]\}$ $(\xi \in \Xi)$ of closed intervals and verify that this has non-empty intersection. For this we fix some $\xi_0 \in \Xi$. Since the intervals $[a_\xi, b_\xi]$, $[a_{\xi_0}, b_{\xi_0}]$ have points in common whatever $\xi \in \Xi$ we choose, we have

$$a_{\xi_0} \leqslant b_\xi \qquad (\xi \in \Xi).$$

Therefore

$$a_{\xi_0} \leqslant \inf_{\xi\in\Xi} b_\xi.$$

Since ξ_0 was arbitrary, this yields

$$\bar{a} = \sup_{\xi\in\Xi} a_\xi \leqslant \inf_{\xi\in\Xi} b_\xi = \bar{b}.$$

Thus every point of $[\bar{a}, \bar{b}]$ belongs to each interval $[a_\xi, b_\xi]$, which proves our assertion.

By similar arguments, which we leave to the reader, one can prove that the real spaces $\ell^\infty(T)$ and $\mathbf{L}^\infty(T, \Sigma, \mu)$, where μ is σ-finite, are also spaces of type $\mathfrak{M}$.

Note that the complex plane, regarded as a (complex) normed space, is not a space of type $\mathfrak{M}$, since it is easy to specify a system of three disks in the plane such that every two intersect but the intersection of all three is empty.

In the rest of this subsection all spaces are assumed to be real.

The class of spaces of type $\mathfrak{M}$ is fairly narrow. The reader can convince himself that, apart from $\ell^\infty(T)$ and $\mathbf{L}^\infty(T, \Sigma, \mu)$, none of the specific normed spaces considered above is a space of type $\mathfrak{M}$.

Finally, we note that every space $\mathbf{Y}$ of type $\mathfrak{M}$ is complete.

For let $\{y_n\}$ be a Cauchy sequence of elements of $\mathbf{Y}$. Write

$$r_n = \sup_{m \geqslant n} \|y_m - y_n\|$$

and consider the family of closed balls $\{B_{r_n}(y_n)\}$ $(n = 1, 2, \ldots,)$. This is an overlapping family, since for $m > n$

$$\|y_m - y_n\| \leqslant r_n$$

and hence $y_m \in B_{r_n}(y_n) \cap B_{r_m}(y_m)$. If $\mathbf{Y}$ is a space of type $\mathfrak{M}$, then there exists an element $y \in \mathbf{Y}$ belonging to each of the balls $B_{r_n}(y_n)$, that is, satisfying

$$\|y - y_n\| \leqslant r_n \qquad (n = 1, 2, \ldots),$$

and, since $r_n \to 0$, we have $y_n \to y$. This shows that $\mathbf{Y}$ is complete.

The fundamental Theorems 3 and 4, which now follow, were proved by Nachbin [1]. In this paper, the author gave a characterization of spaces of type $\mathfrak{M}$ from the point of view of the theory of ordered spaces. A refinement of this characterization was given by Kelley [1]. On extensions of linear operators, see also Akilov [1], [2], Kakutani [1] and Kantorovich, Vulikh and Pinsker.

THEOREM 3. *Let $\mathbf{X}$ be a normed space and $\mathbf{X}_0$ a linear manifold* in $\mathbf{X}$. Let U_0 be a continuous linear operator mapping $\mathbf{X}_0$ into a space $\mathbf{Y}$ of type $\mathfrak{M}$. Then there exists a continuous linear extension U of U_0, mapping $\mathbf{X}$ into $\mathbf{Y}$, and such that $\|U\| = \|U_0\|$.*

Proof. Let $\mathbf{X}_0$ be a linear manifold in $\mathbf{X}$. By a simple extension of $\mathbf{X}_0$ we shall mean a linear manifold $\mathbf{X}_1$ consisting of all the elements of the form

$$x' = \lambda x_1 + x \qquad (x \in \mathbf{X}_0), \tag{4}$$

where x_1 is a given element in $\mathbf{X}$.

We shall prove that, when the hypotheses of the theorem are satisfied, every continuous linear operator U_0 mapping $\mathbf{X}_0 \subset \mathbf{X}$ into $\mathbf{Y}$ has a linear extension U_1, with the same norm, to every simple extension† $\mathbf{X}_1$ of $\mathbf{X}_0$.

Obviously, if such an operator U_1 exists, it is completely determined by specifying $y_0 = U_1(x_1)$. Since the extended operator must have the same norm, this element must satisfy

$$\|y_0 - U_0(x)\| = \|U_1(x_1) - U_1(x)\| \leqslant \|U_1\|\,\|x_1 - x\| = \|U_0\|\,\|x_1 - x\|.$$

Thus a necessary condition for the existence of U_1 is that all the closed balls $B^{(x)} = B_r(y)$ centred at $y = U_0(x)$ and with radius $r = r_x = \|U_0\|\;\|x_1 - x\|$ have a common point.

We shall prove that this condition is also sufficient. Writing $\mathfrak{K}$ for the collection of all balls of this type, assume that there is a point y_0 belonging to all balls in $\mathfrak{K}$. Set

$$U_1(x_1) = y_0.$$

Then for $x' \in \mathbf{X}_1$, we have, by (4),

$$U_1(x') = \lambda U_1(x_1) + U_1(x) = \lambda y_0 + U_0(x).$$

As the representation (4) is unique, this shows that U_1 is linear; and, as (for $\lambda \neq 0$), we have the condition

$$\|U_1(x')\| = |\lambda|\left\|y_0 + U_0\left(\frac{x}{\lambda}\right)\right\| \leqslant r_{-x/\lambda}|\lambda| = |\lambda|\,\|U_0\|\left\|x_1 + \frac{x}{\lambda}\right\| = \|U_0\|\,\|x'\|,$$

we see that U_1 is a continuous linear operator with $\|U_1\| \leqslant \|U_0\|$. The opposite inequality is obvious, so we conclude that $\|U_1\| = \|U_0\|$.

Thus we need only prove that the balls $B^{(x)}$ in the system $\mathfrak{K}$ have non-empty intersection, and, in view of the hypotheses of the theorem, it is therefore sufficient to verify that $\mathfrak{K}$ is an overlapping system.

For this, we take any two balls $B(x_1')$ and $B(x_2')$ in $\mathfrak{K}$. Then

$$\begin{aligned} r_1 + r_2 = r_{x_1'} + r_{x_2'} &= \|U_0\|\,[\,\|x_1 - x_1'\| + \|x_1 - x_2'\|\,] \\ &\geqslant \|U_0\|\,\|(x_1 - x_1') - (x_1 - x_2')\| \geqslant \|U_0(x_1' - x_2')\| = \|y_1' - y_2'\|, \end{aligned}$$

* If we bear in mind Theorem 2 and the fact that $\mathbf{Y}$ is complete, we can assume that $\mathbf{X}_0$ is a closed manifold.

† Here we are concerned with proper extensions—that is, extensions with $x_1 \notin \mathbf{X}_0$. It is easy to see that in such a case the representation (4) is unique.

that is, the sum of the radii of the balls is not less than the distance between their centres; consequently, the balls $B(x_1')$ and $B(x_2')$ intersect,* which shows, since these balls were arbitrarily chosen, that $\mathfrak{K}$ is overlapping.

We have thus proved that we can carry out simple extensions. To complete the proof, we use Zorn's Lemma (I.1.2). Let A denote the collection of all linear operators extending U_0 (and having the same norm). We introduce an order in A as follows: if $V', V'' \in$ A, we put $V' \leqslant V''$ if $V' \subset V''$. Let us check that the conditions of Zorn's Lemma are satisfied. Let A_0 be a totally ordered subset of A. Write

$$\check{\mathbf{X}} = \bigcup_{V \in \mathrm{A}_0} \Omega_V.$$

Since, for any two of the sets comprising this union, one is necessarily contained in the other, $\check{\mathbf{X}}$ is a linear manifold. Consider any element $x \in \check{\mathbf{X}}$. There exists $V' \in \mathrm{A}_0$ such that $x \in \Omega_{V'}$. Define

$$\check{V}(x) = V'(x).$$

This definition does not depend on the choice of V', since if $x \in \Omega_{V''}$, then either $V' \leqslant V''$ or $V'' \leqslant V'$, so $V'(x) = V''(x)$. It is also easy to see that $\check{V}$ is linear and that $\| \check{V} \| = \| U_0 \|$. Further, $\check{V}$ is obviously an extension of every $V \in \mathrm{A}_0$ and hence also of U_0. Therefore $\check{V} \in$ A and $V \leqslant \check{V}$ $(V \in \mathrm{A}_0)$. Applying Zorn's Lemma, we can find an operator U that is a maximal element of A. This U is the operator we are looking for, since if we had $\Omega_U \neq \mathbf{X}$, then we could find a simple extension of Ω_U, larger than Ω_U, and, by what we have proved, we could then extend the operator U, preserving its norm. Denoting the resulting operator by U_1, we should then have $U_1 \in$ A and $U \leqslant U_1$, which is impossible since $U \neq U_1$.

This completes the proof.

8.4. The condition in Theorem 3 for a linear operator to have an extension is also necessary. Before we prove this result, we make the following definition.

A B-space $\mathbf{Y}$ is called a P_1*-space* if, whatever B-space $\mathbf{X}$ we take, every continuous linear operator U_0 mapping an arbitrary subspace $\mathbf{X}_0$ of $\mathbf{X}$ into $\mathbf{Y}$ has a linear extension U to the whole space $\mathbf{X}$, with the same norm.

THEOREM 4. *The following conditions on a B-space $\mathbf{Y}$ are equivalent*:

1) $\mathbf{Y}$ *is a* P_1*-space*,

2) *if a B-space $\mathbf{X}$ contains $\mathbf{Y}$ as a subspace, then there exists a continuous linear projection from $\mathbf{X}$ onto $\mathbf{Y}$ with norm* 1;

3) $\mathbf{Y}$ *is a space of type* $\mathfrak{M}$.

Proof. 1) $\Rightarrow$ 2). Suppose $\mathbf{Y}$ is a subspace of a B-space $\mathbf{X}$. Then the identity operator $U_0 : \mathbf{Y} \to \mathbf{Y}$ has an extension $U : \mathbf{X} \to \mathbf{Y}$, $\| U \| = \| U_0 \| = 1$. Obviously U is the required projection.

2) $\Rightarrow$ 3). Let T be the ball $B_{\mathbf{Y}^*}$ in the dual space. As we showed in 7.3, there exists a linear isometry from $\mathbf{Y}$ onto a subspace of the B-space $\ell^\infty(T)$. By the hypothesis, there is a projection of norm 1 from $\ell^\infty(T)$ onto $\mathbf{Y}$. We have already remarked that $\ell^\infty(T)$ is a space of type $\mathfrak{M}$. The property of being a space of type $\mathfrak{M}$ is clearly preserved under a projection with norm 1, so $\mathbf{Y}$ is a space of type $\mathfrak{M}$.

3) $\Rightarrow$ 1) was proved in Theorem 3.

It can be shown (see Kelley [1], Nachbin [1]) that a B-space $\mathbf{Y}$ is a P_1-space if and only if it is linearly isometric to $\mathbf{C}(Q)$, where Q is an extremally disconnected compactum (these compacta are defined on p. 279).

The extension problem for linear operators may also be approached from a somewhat different point of view. Let $\mathbf{X}$ be a given normed space, $\mathbf{X}_0$ any subspace, and $\mathbf{Y}$ any B-space. Let U_0 be a continuous linear operator mapping $\mathbf{X}_0$ into $\mathbf{Y}$. A necessary and sufficient condition for the existence of a linear operator U, with $U \supset U_0$, $\| U \| = \| U_0 \|$, mapping $\mathbf{X}$ into $\mathbf{Y}$ is that $\mathbf{X}$ be a unitary space. This result was proved by Kakutani [1].

* If $B_{r_1}(y_1)$ and $B_{r_2}(y_2)$ are two closed balls in a space $\mathbf{Y}$ and $r_1 + r_2 \geqslant \| y_1 - y_2 \|$, then their intersection is non-empty; for example, the element $\dfrac{r_2}{r_1 + r_2} y_1 + \dfrac{r_1}{r_1 + r_2} y_2$ belongs to $B_{r_1}(y_1) \cap B_{r_2}(y_2)$.

VI

THE ANALYTIC REPRESENTATION OF FUNCTIONALS

IN APPLICATIONS of the general theory it is very valuable to know a general form for the linear functionals on specific spaces. By a general form for the linear functionals in a given class (the class of all continuous functionals on a given space is the one most often considered) we mean an analytic expression, containing parameters of various kinds (numbers, functions, etc.), that yields a functional in the class for fixed values of the parameters, and moreover is such that every functional in the class is expressible in this way.

In this chapter we determine a general form for the linear functionals on a number of the specific spaces considered above.

§ 1. Integral representations for functionals on spaces of measurable functions

1.1. In § 3 of Chapter IV we preceded the study of specific spaces with an account of the theory of general spaces of measurable functions. In this chapter we proceed in precisely the same way for the study of general forms for functionals on these spaces.

Let $\mathbf{X}$ be an IS on (T, Σ, μ), where the measure μ is σ-finite. A linear functional f on $\mathbf{X}$ is said to be *order continuous* if the conditions $x_n, x \in \mathbf{X}$, $x_n(t) \to 0$ a.e. and $|x_n(t)| \leqslant x(t)$ a.e. imply that* $f(x_n) \to 0$. The set of all (o)-continuous functionals on $\mathbf{X}$, which we shall denote by $\mathbf{X}_n^{\sim}$, is a vector space.

Denote by $\mathbf{X}'$ the set of all $x' \in \mathbf{S}(T, \Sigma, \mu)$ such that supp $x' \subset$ supp $\mathbf{X}$ (mod μ), $\int |x x'| \, d\mu < \infty$ for every $x \in \mathbf{X}$.

It is easy to see that $\mathbf{X}'$ is an IS (it may happen that $\mathbf{X}' = \{\mathbf{0}\}$). The IS $\mathbf{X}'$ is said to be *dual* to $\mathbf{X}$. For each $x' \in \mathbf{X}'$ one can construct a linear functional $f_{x'}$ on $\mathbf{X}$ according to the formula

$$f_{x'}(x) = \int_T x(t)\overline{x'(t)} \, d\mu \quad (x \in \mathbf{X}). \tag{1}$$

It is clear from Lebesgue's Theorem that $f_{x'} \in \mathbf{X}_n^{\sim}$. We shall show that the integral functionals of the form (1) exhaust the space $\mathbf{X}_n^{\sim}$.

THEOREM 1. *Equation* (1) *gives a general form for* (o)*-continuous functionals on an* IS $\mathbf{X}$. *The mapping* $x' \in \mathbf{X}' \to f_{x'} \in \mathbf{X}_n^{\sim}$ *is a linear isomorphism; moreover,* $x' \geqslant \mathbf{0}$ *if and only if* $f_{x'}(x) \geqslant 0$ *for each* $x \in \mathbf{X}_+$.

---

* It is easy to see that this remains unchanged if convergence everywhere is replaced by convergence in measure.

Proof. First assume that $\mathbf{X}$ is real and $\mathbf{L}^\infty(T, \Sigma, \mu) \subset \mathbf{X}$. We shall prove that every functional $f \in \mathbf{X}_n^\sim$ has a representation of the form (1). For each $A \in \Sigma$, set

$$\phi(A) = f(\chi_A).$$

If $\mu(A_n) \to 0$, then $\phi(A_n) = f(\chi_{A_n}) \to 0$, so ϕ is a countably-additive set function on Σ, absolutely continuous with respect to μ. By the Radon–Nikodym Theorem (see Theorem I.6.10), there exists a function $x' \in \mathbf{L}^1(T, \Sigma, \mu)$ such that

$$f(\chi_A) = \phi(A) = \int_A x' \, d\mu. \tag{2}$$

We show that, for each $x \in \mathbf{X}$, we have

$$f(x) = \int_T x x' \, d\mu. \tag{3}$$

Let $B_+ = \{t \in T: x'(t) \geqslant 0\}$, $B_- = \{t \in T: x'(t) < 0\}$. Write

$$f_+(x) = f(x\chi_{B_+}), \quad f_-(x) = f(x\chi_{B_-}) \ (x \in \mathbf{X}).$$

Since

$$f_+(\chi_A) = \int_A x'_+ \, d\mu, \quad f_-(\chi_A) = \int_A x'_- \, d\mu$$

and

$$f_+, f_- \in \mathbf{X}_n^\sim, \quad f = f_+ - f_-,$$

we may assume in proving (3) that $x' \geqslant 0$ and that (2) is satisfied.

Let $x \in \mathbf{X}_+$. Then there exists a sequence $\{x_n\}$ of simple functions such that $\mathbf{0} \leqslant x_n \uparrow x$.

Now from the fact that $f \in \mathbf{X}_n^\sim$ and from Beppo Levi's Theorem we obtain (3). For an arbitrary function $x \in \mathbf{X}$, we obtain (3) from the equation $x = x_+ - x_-$. It follows from (3) that $x' \in \mathbf{X}'$.

Let us now dispense with the assumption that $\mathbf{L}^\infty \subset \mathbf{X}$. First we note that, by Corollary 2 to Lemma IV.3.1, a function x' satisfying (3) is unique (to within equivalence).

By Corollary 1 to Lemma IV.3.1, there is a non-decreasing sequence $\{A_n\} \subset \Sigma$ such that $\chi_{A_n} \in \mathbf{X}$ $(n \in \mathbb{N})$ and $\bigcup_{n=1}^{\infty} A_n = \operatorname{supp} \mathbf{X}$. Consider the ISs $\mathbf{X}_n = \{x \in \mathbf{X}: \operatorname{supp} x \subset A_n\}$ and the functionals $f_n(x) = f(x)$ $(x \in \mathbf{X}_n)$ on these. By what we have proved, there exist $x'_n \in \mathbf{X}'_n$ $(n \in \mathbb{N})$ such that we have

$$f_n(x) = \int_T x x'_n \, d\mu \quad (x \in \mathbf{X}_n). \tag{4}$$

By considering the functionals f_n and f_{n+1} on $\mathbf{X}_{n+1}$ and using the uniqueness of the representing function in (3), we see that $x'_n(t) = x'_{n+1}(t)$ for almost all $t \in A_n$. Hence we can set

$$x'(t) = \begin{cases} x'_n(t), & t \in A_n, \\ 0, & t \notin \operatorname{supp} \mathbf{X}. \end{cases}$$

Let us prove (3) in this situation. As above, we may assume that $x' \geqslant \mathbf{0}$ and that (4) is satisfied. If $x_n = x\chi_{A_n}$ $(x \in \mathbf{X}_+)$, then $f(x_n) \to f(x)$, while, on the other hand, by Beppo Levi's Theorem we have

$$\int_T x x'_n \, d\mu = \int_T x_n x' \, d\mu \to \int_T x x' \, d\mu,$$

and so we deduce (3) for $x \in \mathbf{X}_+$, and therefore also for all $x \in \mathbf{X}$.

If $\mathbf{X}$ is a complex space, $f \in \mathbf{X}_n^{\sim}$, then write $f(x) = \operatorname{Re} f(x) + i \operatorname{Im} f(x)$. Both $\operatorname{Re} f$ and $\operatorname{Im} f$ are linear functionals on the real IS $\mathbf{X}_{\mathbb{R}}$, and $\operatorname{Re} f, \operatorname{Im} f \in (\mathbf{X}_{\mathbb{R}})_n^{\sim}$. By what has been proved, there exist $x_1', x_2' \in (\mathbf{X}_{\mathbb{R}})'$ such that

$$\operatorname{Re} f(x) = \int x x_1' \, d\mu, \ \operatorname{Im} f(x) = \int x x_2' \, d\mu \quad (x \in \mathbf{X}_{\mathbb{R}}).$$

Write $x' = x_1' - i x_2'$. Then for every $x \in \mathbf{X}$ we have

$$f(x) = f(\operatorname{Re} x) + i f(\operatorname{Im} x) = \int (\operatorname{Re} x) x_1' \, d\mu + i \int (\operatorname{Re} x) x_2' \, d\mu + i \int (\operatorname{Im} x) x_1' \, d\mu$$
$$- \int (\operatorname{Im} x) x_2' \, d\mu = \int x \bar{x}' \, d\mu,$$

and so we see also that $x' \in \mathbf{X}'$.

The statement that $f_{x'}(x) \geqslant 0$ for all $x \in \mathbf{X}_+$ is equivalent to $x' \geqslant \mathbf{0}$, as is clear from Corollary 2 to Lemma IV.3.1. Thus the proof of the theorem is now complete.

Theorem 1 was in essence already contained in the monograph Kantorovich, Vulikh and Pinsker, but it appeared explicitly in print only in the mid-1960s, simultaneously in papers by several authors.

1.2. It is most tempting to obtain an integral representation for all the continuous linear functionals on a BIS $\mathbf{X}$. Here we consider when this is possible.

Let $\mathbf{X}$ be an NIS on (T, Σ, μ). Write

$$\mathbf{X}^{\times} = \{x' \in \mathbf{X}' \colon f_{x'} \in \mathbf{X}^*\}.$$

For every $x' \in \mathbf{X}^{\times}$ we set

$$\|x'\| = \|f_{x'}\|_{\mathbf{X}^*} = \sup\left\{ \left|\int_T x \bar{x}' \, d\mu\right| : x \in \mathbf{X}, \ \|x\| \leqslant 1 \right\}.$$

THEOREM 2. *$\mathbf{X}^{\times}$ is a* BIS *satisfying conditions* (B) *and* (C).

Proof. To verify the monotonicity of the norm, it is enough to establish that

$$\|x'\| = \sup\left\{ \int_T x |x'| \, d\mu \colon x \in \mathbf{X}_+, \ \|x\| \leqslant 1 \right\}. \tag{5}$$

Equation (5) follows from the fact that, if

$$\|x'\| < \left|\int x x' \, d\mu\right| + \varepsilon, \ \|x\| \leqslant 1,$$

then $x_1(t) = \operatorname{sign} x(t) \, . \, x(t)$ satisfies

$$x_1 \in \mathbf{X}, \ \|x_1\| \leqslant 1 \quad \text{and} \quad \left|\int_T x x' \, d\mu\right| \leqslant \int_T |x x'| d\mu = \int_T x_1 |x'| d\mu.$$

To verify (B) and (C) it is sufficient, by Lemma IV.3.5, to establish that the statements $x_n' \in \mathbf{X}^{\times}, x' \in \mathbf{S}, \|x_n'\| \leqslant 1 \ (n \in \mathbb{N}), x_n' \to x(\mu)$ imply that $x' \in \mathbf{X}^{\times}$ and $\|x'\| \leqslant 1$. For, by Fatou's Lemma, we have, for any $x \in \mathbf{X}_+$,

$$|f_{x'}(x)| \leqslant \int_T x |x'| \, d\mu \leqslant \sup_n \int_T x |x_n'| \, d\mu \leqslant \|x\| \sup_n \|x_n'\| \leqslant \|x\|.$$

Hence $f_{x'} \in \mathbf{X}^*$, $x' \in \mathbf{X}^{\times}$ and $\|x'\| \leqslant 1$, which completes the proof.

THEOREM 3. *If* $\mathbf{X}$ *is a* BIS, *then* $\mathbf{X}_n^{\sim} \subset \mathbf{X}^*$.

Proof. Let $f \in \mathbf{X}_n^{\sim}$. If we assume that $f \notin \mathbf{X}^*$, then there exists a sequence $\{x_n\}$ with $x_n \to \mathbf{0}$

in norm, such that $|f(x_n)| \geqslant \varepsilon > 0$ $(n \in \mathbb{N})$. By Lemma IV.3.2 and the fact that $f \in \mathbf{X}^{\sim}_n$, some sequence $\{f(x_{n_k})\}$ converges to 0, which is a contradiction.

Hence if $\mathbf{X}$ is a BIS, then $\mathbf{X}^{\times} = \mathbf{X}'$ (in this case, we use the notation $\mathbf{X}'$).

THEOREM 4. *If* $\mathbf{X}$ *is an* NIS, *then* $\mathbf{X}^{\sim}_n \supset \mathbf{X}^*$ *if and only if* $\mathbf{X}$ *satisfies condition* (A).

Proof. Assume that $\mathbf{X}^{\sim}_n \supset \mathbf{X}^*$. For each $x \in \mathbf{X}$ we have

$$\|x\| = \sup\{|f_{x'}(|x|)|: x' \in \mathbf{X}^{\times},\ \|x'\| \leqslant 1\}. \tag{6}$$

Equation (6) is deduced in a similar manner to (5).

Assume that $\mathbf{X}$ does not satisfy condition (A). Then there exists a sequence $\{x_n\}$, with $\mathbf{0} \leqslant x_n \downarrow \mathbf{0}$, in $\mathbf{X}$ such that $\|x_n\| \geqslant \varepsilon > 0$ $(n \in \mathbb{N})$. By (6), there exists a sequence $\{x'_n\} \subset \mathbf{X}'_+$, $\|x'_n\| \leqslant 1$, such that $\int x_n x'_n \, d\mu \geqslant \varepsilon$ $(n \in \mathbb{N})$. Since a ball in the space $\mathbf{X}^*$ is weak* compact, the sequence $\{f_{x'_n}\}$ has a weak* limit point $f = f_{x'} \in \mathbf{X}^{\times}$, $\|x'\| \leqslant 1$. If $x \geqslant \mathbf{0}$, then $f_{x'}(x) = \lim f_{x'_{n_k}}(x) \geqslant \mathbf{0}$ and so $x' \geqslant \mathbf{0}$.*

Since $x_n \downarrow \mathbf{0}$, there exists $k \in \mathbb{N}$ such that

$$\int_T x_k x' \, d\mu < \varepsilon/4.$$

As f is a limit point, there exists $m \geqslant k$ such that

$$\left|\int_T x_k x'_m \, d\mu - \int_T x_k x' \, d\mu\right| < \varepsilon/4.$$

Using the fact that $x_m \leqslant x_k$, we deduce that

$$\int_T x_m x'_m \, d\mu \leqslant \int_T x_k x'_m \, d\mu < \varepsilon/2,$$

which gives a contradiction.

Conversely, suppose $\mathbf{X}$ satisfies (A). Choose $f \in \mathbf{X}^*$ and $\{x_n\}$ with $x_n \to \mathbf{0}$ a.e., $|x_n| \leqslant x \in \mathbf{X}$. By condition (A), $x_n \to \mathbf{0}$ in norm, and so $f(x_n) \to 0$. Hence $\mathbf{X}^* \subset \mathbf{X}^{\sim}_n$. By Theorems 3 and 4, we have

COROLLARY 1. *If* $\mathbf{X}$ *is a* BIS, *then* $\mathbf{X}^{\sim}_n = \mathbf{X}^*$ *if and only if* $\mathbf{X}$ *satisfies condition* (A).

From Corollary 1 and Theorem 1 we deduce

COROLLARY 2. *If* $\mathbf{X}$ *is a* BIS, *then* (1) *is a general form for a continuous linear functional on* $\mathbf{X}$ *if and only if* $\mathbf{X}$ *satisfies condition* (A).

Corollary 2 shows that every continuous linear functional on a BIS $\mathbf{X}$ has an integral representation if and only if $\mathbf{X}$ satisfies condition (A). We defer the application of this result to the next section, and now consider the case where condition (A) is not satisfied.

1.3. THEOREM 5. *If* $\mathbf{X}$ *is an* NIS, *then* supp $\mathbf{X}'$ = supp $\mathbf{X}^{\times}$ = supp $\mathbf{X}$, *and hence the set* $\mathbf{X}^* \cap \mathbf{X}^{\sim}_n$ *separates points on* $\mathbf{X}$.

Proof. Assume that supp $\mathbf{X}^{\times} \neq$ supp $\mathbf{X}$. Then there exists $A \in \Sigma(\mu)$, $\mu(A) > 0$, $\chi_A \in \mathbf{X}$, such that $x'\chi_A = 0$ for every $x' \in \mathbf{X}^{\times}$, or—what is the same—$f(\chi_A) = 0$ for every $f \in \mathbf{X}^* \cap \mathbf{X}^{\sim}_n$. By the Corollary to Theorem IV.3.1, the unit ball $B_{\mathbf{X}}$ is bounded in the TVS $\mathbf{S}(T, \Sigma, \mu)$, and hence the set E, equal to the closure of $B_{\mathbf{X}} \cap \mathbf{L}^2(T, \Sigma, \mu)$ in $\mathbf{L}^2$, is also bounded in $\mathbf{S}(T, \Sigma, \mu)$. The boundedness of E means that there exists a number $\lambda > 0$ such that $\lambda\chi_A \notin E$. By Theorem III.2.6, there exists a continuous linear functional f_0 on $\mathbf{L}^2$ such

* Generally speaking, a subsequence must be chosen for every x.

that

$$\sup\{|f_0(x)|: x\in E\} \leqslant 1 < f_0(\lambda\chi_A). \tag{7}$$

By Theorem V.3.1, there exists a function $z_0 \in \mathbf{L}^2$ such that

$$f_0(x) = \int_T xz_0\,d\mu \quad (x\in \mathbf{L}^2).$$

Since supp $\mathbf{X}$ = supp $(\mathbf{X}\cap\mathbf{L}^2)$, there exists, for each $x\in B_{\mathbf{X}}$, a sequence $\{x_n\}\subset B_{\mathbf{X}}\cap\mathbf{L}^2$, $x_n\to x$ a.e. By Fatou's Theorem and (7), we see that

$$\int |xz_0|\,d\mu \leqslant \sup_n \int |x_n z_0|\,d\mu \leqslant 1.$$

Therefore the functional

$$f(x) = \int xz_0\,d\mu \quad (x\in\mathbf{X})$$

is bounded on the whole of $\mathbf{X}$, and $f\in\mathbf{X}^*\cap\mathbf{X}_n^{\sim}$. By (7), we have $f(\lambda\chi_A) > 1$, and so we obtain a contradiction to $f(\chi_A) = 0$. Thus supp $\mathbf{X}^{\times}$ = supp $\mathbf{X}'$ = supp $\mathbf{X}$.

Now we show that $\mathbf{X}^*\cap\mathbf{X}_n^{\sim}$ separates points on $\mathbf{X}$; that is, for every $x\in\mathbf{X}$, $x\neq\mathbf{0}$, there exists $f\in\mathbf{X}^*\cap\mathbf{X}_n^{\sim}$, $f(x)\neq 0$. We may assume that $x > \mathbf{0}$. Since supp $\mathbf{X}^{\times}$ = supp $\mathbf{X}$, there exists $A\in\Sigma(\mu)$, with $\mu(A) > 0$, $A\subset$ supp x, such that $\chi_A\in\mathbf{X}^{\times}$. Then we have $\int_T x\chi_A\,d\mu > 0$. This completes the proof.

We have shown that $\mathbf{X}^*\cap\mathbf{X}_n^{\sim}$ is a total subset on $\mathbf{X}$. We next consider the question of when the norm on $\mathbf{X}$ is retrievable from this set.

THEOREM 6 (Nakano–Amemiya–Mori). *If* $\mathbf{X}$ *is an* NIS, *then the following statements are equivalent*:

1) $\|x\| = \sup\{|f(x)|: f\in\mathbf{X}^*\cap\mathbf{X}_n^{\sim}, \|f\|\leqslant 1\}$, *for each* $x\in\mathbf{X}$;

2) $\mathbf{X}$ *satisfies condition* (C).

Proof. 1) $\Rightarrow$ 2). If $x_n, x\in\mathbf{X}$, $x_n\to x(\mu)$, $\|x_n\|\leqslant 1$, then, by Fatou's Theorem, $\|x\|\leqslant 1$. Hence $\mathbf{X}$ satisfies (C), by Lemma IV.3.4.

2) $\Rightarrow$ 1). First assume that $\mathbf{X}\subset\mathbf{L}^1(T,\Sigma,\mu)$. For each $n\in\mathbb{N}$, write

$$\|x\|_n = \inf\Big\{\max\Big\{\|y\|, n\int_T z(t)d\mu\Big\}: y, z\in\mathbf{X}_+, |x| = y+z\Big\} \quad (x\in\mathbf{X}).$$

We verify that $\|\cdot\|_n$ is a monotone norm on $\mathbf{X}$. The only non-trivial part is that $\|x\|_n = 0$ implies $x = \mathbf{0}$. Choose $y_k, z_k\in\mathbf{X}_+$ such that

$$|x| = y_k + z_k, \quad \|y_k\|\to 0, \quad n\int |z_k|\,d\mu \xrightarrow[k\to\infty]{} 0.$$

Then, by Theorem IV.3.1, $y_k\to\mathbf{0}(\mu)$ and $z_k\to\mathbf{0}(\mu)$. Therefore $x = \mathbf{0}$.

Note that

$$\|x\| \geqslant \|x\|_n, \quad n\int |x|\,d\mu \geqslant \|x\|_n.$$

Moreover, it is clear that

$$\|x\|_1 \leqslant \|x\|_2 \leqslant \dots \leqslant \|x\|_n \leqslant \|x\|_{n+1} \leqslant \dots.$$

We shall prove that

$$\|x\|_n \to \|x\|. \tag{8}$$

Choose $R > \|x\|_n$ ($n \in \mathbb{N}$). Then there exist $y_n, z_n \in \mathbf{X}_+$ such that

$$|x| = y_n + z_n, \quad \|y_n\| < R, \qquad \int_T |z_n|\, d\mu < R/n.$$

Hence $z_n \to \mathbf{0}$ (μ) and so $y_n = |x| - z_n \to |x|$ (μ). Since $\mathbf{0} \leqslant y_n \leqslant |x|$, condition (C) implies that

$$\|x\| \leqslant \sup \|y_n\| \leqslant R.$$

Consequently (8) is established. Let us now prove (1). We may assume that $x \geqslant \mathbf{0}$. It follows from (8) that, for each $\varepsilon > 0$, there exists $n \in \mathbb{N}$ such that $\|x\|_n > \|x\| - \varepsilon$.

Consider the NIS $(\mathbf{X}, \|\cdot\|_n)$. Since $\|x\|_n \leqslant n \int |x|\, d\mu$, it follows that $\|\cdot\|_n$ is (o)-continuous, so that, by Theorem 4, $(\mathbf{X}, \|\cdot\|_n)^* \subset \mathbf{X}^{\sim}_n$. Thus there exists $f \in (\mathbf{X}, \|\cdot\|_n)^*$ such that

$$\|f\|_{(\mathbf{X}, \|\cdot\|_n)^*} \leqslant 1, \qquad \|x\|_n < |f(x)| + \varepsilon.$$

Hence

$$\|f\|_{\mathbf{X}^*} \leqslant \|f\|_{(\mathbf{X}, \|\cdot\|_n)^*} \leqslant 1, \qquad \|x\| < \|x\|_n + \varepsilon < |f(x)| + 2\varepsilon,$$

which proves 1) in the case where $\mathbf{X} \subset \mathbf{L}^1$. The general case is reduced to this one by means of Corollary 1 to Lemma IV.3.1.

Let $\mathbf{X}$ be an IS. Write $\mathbf{X}'' = (\mathbf{X}')'$. Clearly $\mathbf{X} \subset \mathbf{X}''$. We now consider the question of when we have $\mathbf{X} = \mathbf{X}''$. If $\mathbf{X}$ is an NIS, then we set $\mathbf{X}^{\times\times} = (\mathbf{X}^{\times})^{\times}$. Clearly $\mathbf{X} \subset \mathbf{X}^{\times\times}$. We denote the norm on $\mathbf{X}^{\times\times}$ by $\|\cdot\|^{\times\times}$. We have the inequality $\|x\| \geqslant \|x\|^{\times\times}$ ($x \in \mathbf{X}$).

THEOREM 7. *The following statements are equivalent for an* NIS $\mathbf{X}$:

1) $\mathbf{X}$ *satisfies conditions* (B) *and* (C);

2) $\mathbf{X} = \mathbf{X}^{\times\times}$ *and* $\|\cdot\| = \|\cdot\|^{\times\times}$.

Proof. 1) $\Rightarrow$ 2). Since $\mathbf{X}$ satisfies condition (C), Theorem 6 shows that $\|x\| = \|x\|^{\times\times}$ for $x \in \mathbf{X}$. Let us show that $\mathbf{X} = \mathbf{X}^{\times\times}$. Since $\operatorname{supp} \mathbf{X} = \operatorname{supp} \mathbf{X}^{\times} = \operatorname{supp} \mathbf{X}^{\times\times}$, there exists, for each $x \in (\mathbf{X}^{\times\times})_+$, a sequence $\{x_n\} \subset \mathbf{X}_+$, $\mathbf{0} \leqslant x_n \uparrow x$ (Lemma IV.3.1). Using (B), we see that $x \in \mathbf{X}$.

2) $\Rightarrow$ 1). This holds since $\mathbf{X}^{\times\times}$ satisfies conditions (B) and (C), by Theorem 2.

Applications of the results just obtained to specific spaces will be presented in the next section. For references to the literature, see Chapter X.

§ 2. The spaces $\mathbf{L}^p(T, \Sigma, \mu)$

2.1. We now apply the theory developed in the last section to some specific spaces—first of all the spaces $\mathbf{L}^p(T, \Sigma, \mu)$, where we assume the measure μ is σ-finite (though we shall not need this when $1 < p < \infty$).

THEOREM 1. *Let* $1 \leqslant p < \infty$ *and suppose* $1/p + 1/q = 1$. *The following formula yields a general form for the continuous linear functionals on* $\mathbf{L}^p(T, \Sigma, \mu)$:

$$f(x) = \int_T x(t)\overline{y(t)}\, d\mu, \quad x \in \mathbf{L}^p, \tag{1}$$

where y *is any element of* $\mathbf{L}^q(T, \Sigma, \mu)$. *Furthermore, we have*

$$\|f\| = \|y\|_{\mathbf{L}^q}. \tag{2}$$

Proof. Since the spaces $\mathbf{L}^p$ $(1 \leqslant p < \infty)$ satisfy condition (A) (Theorem IV.3.5), Corollary 2 to Theorem 1.4 shows that (1) gives a general form for the continuous linear functionals. Equation (2) follows from V.2.3.

Now consider the space $\mathbf{L}^\infty(T, \Sigma, \mu)$. Since in non-trivial cases this does not satisfy condition (A) (see IV.3.3), equation (1) (with $y \in \mathbf{L}^1(T, \Sigma, \mu)$) no longer gives a general form for continuous linear functionals.

Write $\mathbf{ba}(\Sigma, \mu)$ for the set of "bounded additive" functions on Σ—that is, the set of (real or complex) additive functions ϕ on Σ such that

1) $\mu(A) = 0$ implies $\phi(A) = 0$;
2) the total variation $|\phi|(T)$ is finite.

If we introduce linear operations on $\mathbf{ba}(\Sigma, \mu)$ by the rules

$$(\lambda\phi)(A) = \lambda\phi(A), \quad (\phi_1 + \phi_2)(A) = \phi_1(A) + \phi_2(A), \quad A \in \Sigma,$$

then $\mathbf{ba}(\Sigma, \mu)$ becomes a vector space. The equation $\|\phi\| = |\phi|(T)$ defines a norm on $\mathbf{ba}(\Sigma, \mu)$, and with this definition $\mathbf{ba}(\Sigma, \mu)$ becomes a B-space.

We now define $\int_T x(t)d\phi$ for $x \in \mathbf{L}^\infty(T, \Sigma, \mu)$. Consider the set Ω of simple measurable functions of the form

$$x(t) = \sum_{i=1}^{k} \lambda_i \chi_{A_i}(t), \tag{3}$$

where $\lambda_i \in \mathbb{R}$ or $\mathbb{C}$, $A_i \in \Sigma$ $(i = 1, 2, \ldots, k)$, and where the A_i are pairwise disjoint. We set

$$\int_T x(t)d\phi = \sum_{i=1}^{k} \lambda_i \phi(A_i).$$

We leave it to the reader to verify that $\int_T x\,d\phi$ does not depend on the representation for x given by (3). By Theorem I.6.3, Ω is dense in $\mathbf{L}^\infty(T, \Sigma, \mu)$. Clearly $\int_T x\,d\phi$ is a linear functional on Ω, and

$$\left|\int_T x\,d\phi\right| \leqslant \sum_{i=1}^{k} |\lambda_i||\phi(A_i)| \leqslant \max_{i=1}^{k} |\lambda_i||\phi|(T) \leqslant \|x\|_{L^\infty}\|\phi\|.$$

By Theorem V.8.2, $\int_T x\,d\phi$ can be linearly extended by continuity to the whole of $\mathbf{L}^\infty(T, \Sigma, \mu)$; we denote the extension by the same symbol.

THEOREM 2. *The formula*

$$f(x) = \int_T x(t)\,d\phi \tag{4}$$

*gives a general form for a continuous linear functional f on $\mathbf{L}^\infty(T, \Sigma, \mu)$, where ϕ is an arbitrary element of** $\mathbf{ba}(\Sigma, \mu)$. *Furthermore, $\|f\| = \|\phi\|$ and $\phi(A) \geqslant 0$ for every $A \in \Sigma$ if and only if $f(x) \geqslant 0$ for every $x \in \mathbf{L}^\infty(T, \Sigma, \mu)$, $x \geqslant \mathbf{0}$.*

---

* If $\mathbf{L}^\infty(T, \Sigma, \mu)$ is complex, then one must use the complex space $\mathbf{ba}(\Sigma, \mu)$ while if $\mathbf{L}^\infty(T, \Sigma, \mu)$ is real, one can take $\mathbf{ba}(\Sigma, \mu)$ to be real.

Proof. In constructing $\int x\, d\phi$ we showed that (4) determines an $f \in (\mathbf{L}^{\infty})^*$ and that $\|f\| \leqslant \|\phi\|$. Let us prove the opposite inequality. For every $\varepsilon > 0$, there exist $A_i \in \Sigma$ such that

$$\|\psi\| \leqslant \sum_{i=1}^{k} |\phi(A_i)| + \varepsilon, \quad A_i \cap A_j = \varnothing \ (i \neq j), \quad \bigcup_{i=1}^{k} A_i = T.$$

Write $\tilde{x} = \sum_{i=1}^{k} \operatorname{sign} \phi(A_i) \chi_{A_i}$. Then $\|\tilde{x}\| \leqslant 1$ and $\int_T \tilde{x}\, d\phi = \sum_{i=1}^{k} |\phi(A_i)|$, so that $\|\phi\| \leqslant \int_T \tilde{x}\, d\phi + \varepsilon \leqslant \|f\| + \varepsilon$. Therefore $\|f\| = \|\phi\|$.

It remains to prove that every $f \in (\mathbf{L}^{\infty})^*$ has a representation (4). Write $\phi(A) = f(\chi_A)$ $(A \in \Sigma)$.

Then $|\phi|(T) \leqslant \|f\|$ (see the above argument) and $\phi \in \mathbf{ba}(\Sigma, \mu)$. By the construction of $\int_T x\, d\phi$ it is clear that $f(x) = \int_T x\, d\phi$ $(x \in \mathbf{L}^{\infty})$. The assertion that ϕ is positive is also clear from the construction, so the proof is complete.

Notice that the mapping $y \in \mathbf{L}^q \to f \in (\mathbf{L}^p)^*$ (respectively, $\phi \in \mathbf{ba} \to f \in (\mathbf{L}^{\infty})^*$) determined by (1) (respectively, (4)) is a linear isometry from $\mathbf{L}^q$ onto $(\mathbf{L}^p)^*$ (respectively, $\mathbf{ba}$ onto $\mathbf{L}^{\infty}$). Therefore one often says that $\mathbf{L}^q$ is dual to $\mathbf{L}^p$ $(1 \leqslant p < \infty)$ and writes $(\mathbf{L}^p)^* = \mathbf{L}^q$. We shall use similar abbreviated notation also for the representations to be obtained below.

One cannot conclude from the resulting equation $(\mathbf{L}^p)^{**} = (\mathbf{L}^q)^* = \mathbf{L}^p (1 < p < \infty)$ that the space $\mathbf{L}^p$ is reflexive, as the equation $(\mathbf{L}^p)^{**} = \mathbf{L}^p$ must be understood as a linear isometry, which need not be of the special type required in the definition of reflexivity (see V.7.3). Nevertheless one can deduce, by investigating the specific isometry in Theorem 1, that the space $\mathbf{L}^p (1 < p < \infty)$ is reflexive. For let us take $F \in (\mathbf{L}^p)^{**}$ and show that there exists $x \in \mathbf{L}^p$ such that

$$F(f) = f(x), \ f \in (\mathbf{L}^p)^*. \tag{5}$$

Denote by α the linear isometry that associates to each $y \in \mathbf{L}^q$ the functional $f \in (\mathbf{L}^p)^*$ given by (1). If we set $F_1(y) = F(\alpha y)$, $y \in \mathbf{L}^q$, then $F_1 \in (\mathbf{L}^q)^*$. Hence by Theorem 1 there exists $x \in \mathbf{L}^p$ such that

$$F_1(y) = \int y(t)\overline{x(t)}\, d\mu, \ y \in \mathbf{L}^q.$$

Let us verify that (5) holds for this $x \in \mathbf{L}^p$. If $f \in (\mathbf{L}^p)^*$ and $y = \alpha^{-1}(f) \in \mathbf{L}^q$, then

$$F(f) = F_1(y) = \int y(t)\overline{x(t)}\, d\mu = \overline{\int x(t)\overline{y(t)}\, d\mu} = f(x).$$

For $p = 1$ the canonical embedding $\pi: \mathbf{L}^1 \to (\mathbf{L}^1)^{**}$ maps $\mathbf{L}^1$ into the set of functionals on $\mathbf{L}^{\infty}$ having an integral representation (1), which, as we have already remarked, is a proper subset of $(\mathbf{L}^1)^{**} = (\mathbf{L}^{\infty})^*$ when $\mathbf{L}^1$ is infinite-dimensional. Thus $\mathbf{L}^1$ and $\mathbf{L}^{\infty} = (\mathbf{L}^1)^*$ are non-reflexive in the infinite-dimensional case.

A general form for linear functionals on $\mathbf{L}^p(a, b)$ $(1 < p < \infty)$ was described by F. Riesz, and for linear functionals on $\mathbf{L}^1(a, b)$ by Steinhaus (see F. Riesz [2] and Steinhaus [1], respectively); the generalization to abstract measure spaces was obtained by Nikodym.

2.2. Let us devote a little space to the special case of sequence spaces. From Theorem 1 we have

THEOREM 3. *Let* $1 \leqslant p < \infty$ *and suppose* $1/p + 1/q = 1$. *A general form for a linear* *functional f on the space ℓ^p is given by*

$$f(x)=\sum_{k=1}^{\infty}\xi_k\overline{\eta_k},\qquad x=\{\xi_k\}_{k=1}^{\infty}\in\ell^p,$$

where $y=\{\eta_k\}_{k=1}^{\infty}$ is an arbitrary element of ℓ^q. Moreover, we have $\|f\|=\|y\|_{\ell^q}$.

Now we consider the space $\mathbf{c}_0$ (see IV.3.4). We shall prove that $(\mathbf{c}_0)^*=\ell^1$.

THEOREM 4. *A general form for a linear functional f on the space $\mathbf{c}_0$ is given by*

$$f(x)=\sum_{k=1}^{\infty}\xi_k\overline{\eta_k},\qquad x=\{\xi_k\}_{k=1}^{\infty}\in\mathbf{c}_0,\tag{6}$$

where $y=\{\eta_k\}_{k=1}^{\infty}$ is an arbitrary element of ℓ^1. Moreover, we have $\|f\|=\|y\|_{\ell^1}$.

Proof. Since $\mathbf{c}_0$ satisfies condition (A), a general form for functionals is given by (6). Let $f\in(\mathbf{c}_0)^*$. Write $\tilde{x}=\{\tilde{\xi}_k\}$, where

$$\tilde{\xi}_k=\begin{cases}\operatorname{sign}\eta_k, & k\leqslant n,\\ 0, & k>n.\end{cases}$$

Since $\tilde{x}\in\mathbf{c}_0$ and $\|\tilde{x}\|\leqslant 1$, we have

$$f(\tilde{x})=\sum_{k=1}^{n}|\eta_k|\leqslant\|f\|.$$

Letting $n\to\infty$, we have

$$\sum_{k=1}^{\infty}|\eta_k|\leqslant\|f\|,\tag{7}$$

so that $y\in\ell^1$ and $\|y\|_{\ell^1}\leqslant\|f\|$.

If $y\in\ell^1$, then

$$\left|\sum_{k=1}^{\infty}\eta_k\xi_k\right|\leqslant\max|\xi_k|\sum_{k=1}^{\infty}|\eta_k|=\|y\|_{\ell^1}\cdot\|x\|.$$

From this we deduce in the usual way that f is a linear functional and that $\|f\|\leqslant\|y\|_{\ell^1}$; and, by (7), this leads to the equation $\|f\|=\|y\|_{\ell^1}$.

Although the space $\mathbf{c}$ is not a BIS, certain additional considerations allow us to determine a general form for a linear functional on it.

Consider the element $e_0=(1,1,\ldots,1,\ldots)$ in $\mathbf{c}$. If $x=\{\xi_k\}$ is an arbitrary element in $\mathbf{c}$ and $\xi_0=\lim_{k\to\infty}\xi_k$, then clearly $x-\xi_0e_0\in\mathbf{c}_0$.

Let f be a continuous linear functional on $\mathbf{c}$; by considering it as acting only on $\mathbf{c}_0$, we also obtain a continuous linear functional f_0. Suppose (for $x\in\mathbf{c}_0$) that

$$f_0(x)=\sum_{k=1}^{\infty}\eta_k\xi_k\quad(y=\{\eta_k\}\in\ell^1).$$

Writing $\eta_0=f(e_0)$, we find that, for any $x\in\mathbf{c}$,

$$f(x)=f(\xi_0e_0)+f_0(x-\xi_0e_0)=\eta_0\xi_0+\sum_{k=1}^{\infty}\eta_k(\xi_k-\xi_0).$$

Setting $\alpha = \eta_0 - \sum_{k=1}^{\infty} \eta_k$, we finally obtain

$$f(x) = \alpha\xi_0 + \sum_{k=1}^{\infty} \eta_k\xi_k = \alpha \lim_{k\to\infty} \xi_k + \sum_{k=1}^{\infty} \eta_k\xi_k. \tag{8}$$

We leave it to the reader to prove that (8) gives a general form for a continuous linear functional on **c** and that

$$\|f\| = |\alpha| + \sum_{k=1}^{\infty} |\eta_k|.$$

We leave it to the reader to prove that the *B*-spaces $\mathbf{c}_0$ and **c** are non-reflexive.

2.3. Let us dwell briefly on the analytic representation of functionals on the other spaces of measurable functions introduced in § 3 of Chapter IV.

First we consider Orlicz spaces. Here we have the equations

$$(\mathbf{E}_M)' = (\mathbf{L}_M)' = \mathbf{L}_{M^*}.$$

THEOREM 5. *A general form for an (o)-continuous linear functional f on the space $\mathbf{L}_M(T, \Sigma, \mu)$ is given by*

$$f(x) = \int_T x(t)\overline{y(t)}\, d\mu, \qquad x \in \mathbf{L}_M, \tag{9}$$

where y is an arbitrary element of $\mathbf{L}_{M^*}(T, \Sigma, \mu)$. Moreover,

$$\|f\|_{(\mathbf{L}_M, \|\cdot\|_1)^*} = \|y\|_2, \quad \|f\|_{(\mathbf{L}_M, \|\cdot\|_2)^*} = \|y\|_1. \tag{10}$$

A general form for an $f \in (\mathbf{E}_M)^*$ is given by (9); also, the analogues of equations (10) (with $\mathbf{E}_M$ in place of $\mathbf{L}_M$) are true. Therefore (9) gives a general form for a continuous linear functional on $\mathbf{L}_M$ if and only if M satisfies the Δ_2-condition. The Orlicz space $\mathbf{L}_M$ is reflexive if and only if M and M^* satisfy the Δ_2-condition. For proofs of these results, see Krasnosel'skii and Rutickii. T. Ando has succeeded in obtaining a representation for all the continuous linear functionals on an arbitrary Orlicz space.

A connection between Lorentz and Marcinkiewicz spaces is given by the following formulae, where equality is to be interpreted as referring to the elements constituting the spaces and to the norms:

$$\mathbf{M}(\psi)' = \Lambda(\psi), \quad \Lambda(\psi)' = \mathbf{M}(\psi).$$

Since $\Lambda(\psi)$ satisfies condition (A), we have $\Lambda(\psi)^* = \mathbf{M}(\psi)$. The space $\mathbf{M}(\psi)^*$ is substantially larger than $\Lambda(\psi)$.

§ 3. A general form for linear functionals on the space C(K)

3.1. Let K be a compactum. Denote by **rca** (K) the set of all (real or complex) regular countably additive functions ϕ, defined on the σ-algebra $\mathscr{B}$ of all Borel sets in K and having finite total variation $|\phi|(K) < \infty$.

If we introduce linear operations in **rca**(K) in a manner analogous to that used for **ba**(Σ, μ), and write $\|\phi\| = |\phi|(K)$, then **rca**(K) is turned into a *B*-space.

THEOREM 1. *A general form for a continuous linear functional f on the space* **C**(K) *is given by*

$$f(x) = \int_K x(t)\, d\phi, \qquad x \in \mathbf{C}(K),$$

*where ϕ is an arbitrary element of** **rca**(K). *Moreover, $\|f\| = \|\phi\|$ and $\phi(A) \geq 0$ for every $A \in \mathscr{B}$ if and only if $f(x) \geq 0$ for every $x \in \mathbf{C}(K)$.*

* If **C**(K) is complex, then we must take **rca**(K) to be a complex space, while if **C**(K) is real, we must take **rca**(K) to be real.

The correspondence determined by Theorem 1 is a linear isometry between $\mathbf{C}(K)^*$ and $\mathbf{rca}(K)$. The proof of Theorem 1 is rather long and requires further use of results from measure theory and topology, so we omit it (see Dunford and Schwartz-I; Zaanen-II). Theorem 1 was proved for $K = [0, 1]$ by F. Riesz (hence it is often called Riesz's Theorem). The general formulation is due to A. A. Markov and S. Kakutani.

We now reformulate Theorem 1 for the case of the real space $\mathbf{C}[a, b]$. First we add some remarks about functions of bounded variation (see IV.4.1).

Consider an increasing function $\phi(t)$ on the interval $[a, b]$. Let $a < t_1 < t_2 < \ldots < t_n < b$ be any system of points. Choose t'_k and t''_k such that $a < t'_1 < t_1 < t''_1 < t'_2 < t_2 < t''_2 < \ldots < t''_n < b$. Clearly, we have

$$\sum_{k=1}^{n} [\phi(t''_k) - \phi(t'_k)] \leqslant \phi(b) - \phi(a)$$

and if in this expression we let $t'_k \to t_k - 0$ and $t''_k \to t_k + 0$, then in the limit we obtain

$$\sum_{k=1}^{n} [\phi(t_k + 0) - \phi(t_k - 0)] \leqslant \phi(b) - \phi(a).$$

From this it follows that a monotone function can have only a finite number of discontinuities at which the jump exceeds a given ε. Consequently the set of all discontinuities of a monotone function is at most countable.

Since every function of bounded variation is expressible as a difference of two increasing functions, the above property also holds for functions of bounded variation.

It follows from this that the set of points of continuity of a function of bounded variation is dense in $[a, b]$.

Let $g(t)$ be a function of bounded variation. Consider the function $\tilde{g}(t)$ defined by

$$\tilde{g}(t) = \tfrac{1}{2}[g(t+0) + g(t-0)] \qquad (a < t < b),$$

$$\tilde{g}(b) = g(b), \qquad \tilde{g}(a) = g(a).$$

Thus $\tilde{g}(t)$ coincides with $g(t)$ at all points of continuity of the latter, and at $t = a, b$. The function $\tilde{g}(t)$ is said to be *regulated*.

If a function $g(t)$ coincides with its regulated function, then we shall say it is *regular*. We denote the set of all regular functions vanishing at $t = a$ by $\mathbf{V}_0$. Clearly $\mathbf{V}_0$ is a linear subspace of the space $\mathbf{V}$ of all functions of bounded variation. It is easy to verify that it is a closed subspace. Hence $\mathbf{V}_0$, being a closed subspace of the B-space $\mathbf{V}$, is itself a B-space.

For every function $x \in \mathbf{C}[a, b]$, one can define the Stieltjes integral (see Vulikh-III, Chap. XI, § 5) $\int_a^b x(t)\,dg(t)$ with respect to a function $g(t)$ of bounded variation.

THEOREM 2. *A general form for a continuous linear function on the space* $\mathbf{C}[a, b]$ *is given by the Stieltjes integral*

$$f(x) = \int_a^b x(t)\,dg(t),$$

where $g(t)$ is an arbitrary function of bounded variation. If, moreover, $g(t)$ is a regular function, then

$$\|f\| = \bigvee_a^b (g).$$

The functional f determines the function $g \in \mathbf{V}_0$ uniquely.

Since a Stieltjes integral of a continuous function can be reduced to an integral with respect to a countably additive set function (see Vulikh-III, Chap. XI, § 5), Theorem 2 is easily obtained from Theorem 1. We leave it to the reader to check the details (for a proof independent of Theorem 1, see Kolmogorov and Fomin). The space $\mathbf{C}(K)$ is not reflexive in the infinite-dimensional case (that is, when K is infinite). We leave it to the reader to verify that the functional $F_t(g) = g(t)$, $g \in \mathbf{V}_0$, where $t \in [0, 1]$ is fixed, does not belong to the image of $\mathbf{C}[0, 1]$ under the canonical embedding.

3.2. Let us consider an application of the theorem on the general form of linear functionals on $\mathbf{C}[a, b]$ to the so-called problem of moments.

Suppose we are given a sequence $\{x_n\}$ of linearly independent elements in a normed space $\mathbf{X}$. Choosing an arbitrary linear functional $f \in \mathbf{X}^*$, we form the numerical sequence

$$f(x_n) = \mu_n \qquad (n = 0, 1, \dots). \tag{1}$$

The *problem of moments* in the wide sense is the problem of determining f from the sequence $\{\mu_n\}$. With the problem formulated in this generality, one cannot make much progress towards a solution. However, even in the general case, we can still state conditions for the problem of moments to be soluble and for it to have a unique solution.

If $\{x_n\}$ is a fundamental set, then, by what was said in III.3.2, the functional f (if we assume that it exists) is uniquely determined by the sequence $\{\mu_n\}$. It is not difficult to see that being a fundamental set is also a necessary condition for the problem of moments to have a unique solution.

Furthermore, if we denote by ϕ the (possibly non-additive) functional defined on the set $\{x_n\}$ by:

$$\phi(x_n) = \mu_n \qquad (n = 0, 1, \dots), \tag{2}$$

then, since the existence of the functional f means that one can find a linear extension of f on the linear hull $\mathscr{L}(\{x_n\})$, Theorem V.8.1 shows that a necessary and sufficient condition for the problem of moments to be soluble is that there exist a constant $M > 0$ such that, for any $\lambda_0, \lambda_1, \dots, \lambda_n$, we have

$$\left| \sum_{k=1}^{n} \lambda_k \mu_k \right| \leqslant M \left\| \sum_{k=0}^{n} \lambda_k x_k \right\| \tag{3}$$

or, in other words,

$$\sup \frac{\left| \sum_{k=0}^{n} \lambda_k \mu_k \right|}{\left\| \sum_{k=0}^{n} \lambda_k x_k \right\|} < \infty, \tag{4}$$

where the supremum is taken over all values of $\lambda_0, \lambda_1, \dots, \lambda_n$ $(n = 0, 1, \dots)$.

From among the whole variety of specific problems of moments, we shall consider only the *problem of power moments*, when $\mathbf{X} = \mathbf{C}[a, b]$ and

$$x_n(t) = t^n \qquad (n = 0, 1, \dots). \tag{5}$$

Bearing in mind the general form for a linear functional on $\mathbf{C}[a, b]$, we can formulate the problem in this case as follows: determine conditions under which there exists a function of bounded variation $g(t)$ such that

$$\int_a^b t^n \, dg(t) = \mu_n \qquad (n = 0, 1, \ldots). \tag{6}$$

Note that, since in the present case the system $\{x_n\}$ is complete in $\mathbf{C}[a, b]$, a solution of the problem of moments, if it exists at all, is unique.

Condition (4) for the problem of moments to be soluble can be rewritten in the following form:

$$\sup \frac{\left| \sum_{k=0}^{n} \lambda_k \mu_k \right|}{\max\limits_{a \leqslant t \leqslant b} \left| \sum_{k=0}^{n} \lambda_k t^k \right|} < \infty. \tag{7}$$

Finally, we present a more concrete result relating to the problem of power moments on the interval $[0, 1]$.

THEOREM 3 (Hausdorff). *A necessary and sufficient condition for the existence of a function of bounded variation $g(t)$ such that*

$$\int_0^1 t^n \, dg(t) = \mu_n \qquad (n = 0, 1, \ldots) \tag{8}$$

is that

$$\sum_{k=0}^{n} C_n^k |\Delta^{n-k} \mu_k| \leqslant M \qquad (n = 0, 1, \ldots), \tag{9}$$

where the C_n^k are binomial coefficients and $\Delta^m \mu_k$ are the m-th differences for the sequence $\{\mu_n\}$, defined inductively by

$$\Delta^{m+1} \mu_k = \Delta^m \mu_k - \Delta^m \mu_{k+1}, \qquad \Delta^0 \mu_k = \mu_k$$
$$(m = 0, 1, \ldots; \qquad k = 0, 1, \ldots). \tag{10}$$

Proof. Necessity. Suppose the problem of moments (8) is soluble. Let f denote the linear functional on $\mathbf{C}[0, 1]$ induced by $g(t)$. Also write

$$x_k^{(m)}(t) = t^k (1 - t)^m \qquad (m, k = 0, 1, \ldots). \tag{11}$$

Since

$$x_k^{(m+1)}(t) = t^k (1-t)^{m+1} = t^k (1-t)^m - t^{k+1} (1-t)^m = x_k^{(m)}(t) - x_{k+1}^{(m)}(t),$$

we have

$$f(x_k^{(m+1)}) = f(x_k^{(m)}) - f(x_{k+1}^{(m)}) \qquad (m, k = 0, 1, \ldots).$$

Furthermore,

$$f(x_k^{(0)}) = \mu_k.$$

Using (10) it is easy to verify (by induction) that

$$f(x_k^{(m)}) = \Delta^m \mu_k \qquad (m, k = 0, 1, \ldots).$$

Now let $\theta_k^{(n)} = \text{sign}\, \Delta^{n-k}\mu_k\, (k = 0, 1, \ldots, n)$. Consider the function

$$\tilde{x}(t) = \sum_{k=1}^{n} \theta_k^{(n)} C_n^k x_k^{(n-k)}(t).$$

In view of the fact that $x_k^m(t) \geqslant 0$ on $[0, 1]$, we have

$$|\tilde{x}(t)| \leqslant \sum_{k=0}^{n} C_n^k x_k^{(n-k)}(t) = \sum_{k=0}^{n} C_n^k t^k (1-t)^{n-k} = [t + (1-t)]^n = 1.$$

Therefore $\|\tilde{x}\| < 1$. Hence

$$\sum_{k=0}^{n} C_n^k |\Delta^{n-k}\mu_k| = \sum_{k=0}^{n} C_n^k \theta_k^{(n)} \Delta^{n-k}\mu_k$$
$$= \sum_{k=0}^{n} C_n^k \theta_k^{(n)} f(x_k^{(n-k)}) = f(\tilde{x}) \leqslant \|f\|$$
$$(n = 0, 1, \ldots),$$

which establishes the necessity of the condition, when we set $M = \|f\|$.

Sufficiency. Let ϕ denote the functional defined on $\{x_n\}$ by (2), where we take $x_n(t) = t^n$. Extend ϕ to the linear hull of $\{x_n\}$—that is, to the set of all polynomials. Explicitly, if $x(t) = \lambda_0 + \lambda_1 t + \ldots + \lambda_n t^n$, then we set

$$f_0(x) = \lambda_0\mu_0 + \lambda_1\mu_1 + \ldots + \lambda_n\mu_n.$$

This defines f_0 uniquely, since the functions $x_n(t)$ are linearly independent.

The functional f_0 thus defined is clearly additive and homogeneous. We prove that the condition (10) ensures that f_0 is continuous.

Notice that, independently of this condition, f_0 is continuous on the set $\mathbf{H}_m$ of polynomials of degree not exceeding m, since $\mathbf{H}_m$ is a finite-dimensional space (the coefficients of a polynomial being the coordinates), and so convergence in $\mathbf{H}_m$ is coordinatewise.

Keeping the previous notation, we have as before

$$f_0(x_k^{(s)}) = \Delta^s \mu_k \qquad (s, k = 0, 1, \ldots).$$

Now consider an arbitrary polynomial $x(t)$. Suppose it has degree m. We introduce the sequence of associated Bernstein polynomials

$$x_n(t) = B_n(x; t) = \sum_{k=0}^{n} C_n^k x\left(\frac{k}{n}\right) t^k (1-t)^{n-k}. \tag{12}$$

It is well known that the degree of $x_n(t)$, for every $n = 1, 2, \ldots$, is not more than m,* and, since the $x_n(t)$ converge uniformly to $x(t)$ as $n \to \infty$ (see Natanson-I), the above remark shows that

$$f_0(x_n) \to f_0(x),$$

while

$$|f_0(x_n)| \leqslant \sum_{k=0}^{n} C_n^k \left| x\left(\frac{k}{n}\right) \right| |f_0(x_n^{(n-k)})| \leqslant \|x\| \sum_{k=0}^{n} C_n^k |\Delta^{n-k}\mu_k| \leqslant M\|x\|.$$

Letting $n \to \infty$ in the left-hand side, we find that

$$|f_0(x)| \leqslant M\|x\|.$$

As x was an arbitrary polynomial, this proves the continuity of f_0. We now need only to extend f_0 to all of $\mathbf{C}[0, 1]$ by continuity (see IV.8.2) and apply the theorem on the general form of linear functionals on $\mathbf{C}[0, 1]$ to the resulting functional f, giving

$$f(x) = \int_0^1 x(t)\, dg(t) \qquad (x \in \mathbf{C}[0, 1]),$$

where $g(t)$ is a function of bounded variation.

This completes the proof of the theorem.

* Here is a proof of this fact. Differentiating expression (12) and replacing the index of summation k by $k+1$ in the first sum, we have

$$\frac{dx_n}{dt} = \sum_{k=0}^{n} C_n^k [kt^{k-1}(1-t)^{n-k} - (n-k)t^k(1-t)^{n-k-1}] x\left(\frac{k}{n}\right)$$
$$= n \sum_{k=0}^{n-1} C_{n-1}^k t^k (1-t)^{n-1-k} \left[x\left(\frac{k+1}{n}\right) - x\left(\frac{k}{n}\right) \right].$$

By induction, we obtain

$$\frac{d^s x_n}{dt^s} = n(n-1)\ldots(n-s+1) \sum_{k=0}^{n-s} C_{n-s}^k t^k (1-t)^{n-s-k} \Delta^s x\left(\frac{k}{n}\right).$$

Since $x(t)$ is a polynomial of degree m, we see that $\Delta^{m+1} x\left(\frac{k}{n}\right) = 0$. Therefore $\frac{d^{m+1} x_n}{dt^{m+1}} = 0$, from which it follows that $x_n(t)$ is a polynomial whose degree is at most m.

VII

SEQUENCES OF LINEAR OPERATORS

A SIGNIFICANT number of concrete mathematical processes can be included in an abstract scheme that can be described with the aid of sequences of operators. Such problems include, for example, the study of the convergence of Fourier series and interpolation polynomials, the study of the formulae of mechanical quadrature, the theory of singular integrals, and so on.

Moreover, in abstract form the study of a problem generally reduces either to establishing the convergence of a sequence of linear operators, or to proving the boundedness of the norms of the operators in this sequence, or to other similar problems.

§ 1. Basic theorems

1.1. The following theorem plays an essential role among the theorems on sequences of linear operators.

THEOREM 1. *If a sequence of continuous linear operators $\{U_n\}$, mapping a B-space* **X** *into a normed space* **Y**. *is bounded at each point, that is, if*

$$\sup_n \|U_n(x)\| < \infty \qquad (x \in \mathbf{X}), \tag{1}$$

then the norms of the operators are bounded in aggregate:

$$\|U_n\| \leqslant M \qquad (n = 1, 2, \ldots).$$

Proof. Notice first of all that if we know a bound for the values of a linear operator in some ball:

$$\|U(x)\| \leqslant B \qquad (x \in B_\delta(x_0)),$$

then we can bound its norm; in fact,

$$\|U\| \leqslant 2B/\delta.$$

For if we take any x' with $\|x'\| < 1$, then we have

$$x = x_0 + \delta x' \in B_\delta(x_0).$$

Therefore

$$\|U(x_0) + \delta U(x')\| \leqslant B,$$

and consequently

$$\|U(x')\| = \frac{1}{\delta}\|U(\delta x')\| \leqslant \frac{1}{\delta}[\|U(x_0) + \delta U(x')\| + \|U(x_0)\|] \leqslant \frac{2B}{\delta},$$

which implies that $\|U\| \leqslant 2B/\delta$, as claimed.

We now turn to the proof of the theorem. Assume that the sequence $\{\|U_n\|\}$ is unbounded. We introduce the functional

$$p(x) = \sup_n \|U_n(x)\|.$$

This functional is unbounded in every ball, for if we had $p(x) \leqslant B$ in $B_\delta(x_0)$, then, as we have remarked, we should have $\|U_n\| \leqslant 2B/\delta$ for every $n = 1, 2, \ldots$.

It follows from this that the set $E_k = \{x \in \mathbf{X} : p(x) > k\}$ is dense in $\mathbf{X}$. Also E_k is open, for if $x_0 \in E_k$, that is, if $p_0(x_0) > k$, then for some n_0 we have $\|U_{n_0}(x_0)\| > k$, and so by the continuity of $\|U_{n_0}(x)\|$ we also have $\|U_{n_0}(x)\| > k$ for x sufficiently close to x_0.

The set E_k, being a dense open subset of $\mathbf{X}$, is a residual (see I.4.7). But the intersection of a countable system of residuals is a residual, and so is non-empty. If $x_0 \in \bigcap_{k=1}^{\infty} E_k$, then

$$\sup_n \|U_n(x_0)\| = \infty,$$

contrary to our assumption.

REMARK. The condition in the theorem can be weakened, as it is sufficient to require (1) to hold for a set of second category, rather than for the whole of $\mathbf{X}$ (we are not assuming that $\mathbf{X}$ is complete).

The theorem can also be stated as follows (the *principle of location of singularities*): if

$$\sup_n \|U_n\| = \infty,$$

then there exists an element $x_0 \in \mathbf{X}$ such that

$$\sup_n \|U_n(x_0)\| = \infty. \tag{2}$$

Moreover, the set of elements satisfying (2) is a residual.

The latter formulation enables us to generalize the stated result somewhat. Let $\{U_n^{(k)}\}$ $(k, n = 1, 2, \ldots)$ be continuous linear operators from $\mathbf{X}$ into $\mathbf{Y}$, such that

$$\sup_n \|U_n^{(k)}\| = \infty \qquad (k = 1, 2, \ldots).$$

Then there exists an element x_0 such that

$$\sup_n \|U_n^{(k)}(x_0)\| = \infty \qquad (k = 1, 2, \ldots). \tag{3}$$

For the set A_k of those $x \in \mathbf{X}$ for which

$$\sup_n \|U_n^{(k)}(x)\| = \infty,$$

is a residual; hence the intersection $A_0 = \bigcap_{k=1}^{\infty} A_k$ is non-empty. We may clearly take x_0 to be any element of A_0.

This result is sometimes called the *principle of condensation of singularities*.

1.2. By applying Theorem 1 in the case where the sequence of operators $\{U_n\}$ converges on $\mathbf{X}$—that is, where

$$U(x) = \lim_{n\to\infty} U_n(x) \tag{4}$$

exists for each $x \in \mathbf{X}$—we obtain an important result on convergent sequences of linear operators. Namely, under the above hypotheses on $\mathbf{X}$ and $\mathbf{Y}$, we have

THEOREM 2. *If a sequence of continuous linear operators $\{U_n\}$ converges on $\mathbf{X}$ to an operator U, then U is a continuous linear operator, and*

$$\|U\| \leqslant \lim_{n\to\infty} \|U_n\|. \tag{5}$$

Proof. The operator U is obviously linear. Further, since $\lim_{n\to\infty} \|U_n(x)\| = \|U(x)\| < \infty$, we also have $\sup_n \|U_n(x)\| < \infty$, and by Theorem 1 the sequence of norms $\{\|U_n\|\}$ is bounded. Hence

$$\|U(x)\| = \lim_{n\to\infty} \|U_n(x)\| \leqslant \varliminf_{n\to\infty} \|U_n\| \|x\|,$$

which establishes the continuity of U and equation (5), as required.

REMARK. We emphasize that the U_n do not necessarily converge in norm to U in $B(\mathbf{X}, \mathbf{Y})$, as one can easily see in the following example: $\mathbf{X} = \ell^1$, $\mathbf{Y} = \mathbf{R}^1$, $U_n = f_n$, where

$$f_n(x) = \xi_n \qquad (x = \{\xi_k\} \in \ell^1; \qquad n = 1, 2, \ldots).$$

Although $f_n(x) \to 0$, we have $\|f_n\| = 1$ $(n = 1, 2, \ldots)$.

Boundedness of the sequence of norms is also, in a certain sense, a sufficient condition for a sequence of linear operators to converge. More precisely, we have the following result (see Banach and Steinhaus [1]).

THEOREM 3 (Banach–Steinhaus). *The following conditions are together necessary and sufficient for a sequence of continuous linear operators $\{U_n\}$, mapping a B-space $\mathbf{X}$ into a B-space $\mathbf{Y}$, to converge on $\mathbf{X}$ to a linear operator:*

1) *the norms of the U_n are bounded in aggregate:*

$$\|U_n\| \leqslant M \qquad (n = 1, 2, \ldots);$$

2) *$\{U_n(x')\}$ is a Cauchy sequence, for each element x' in some dense subset D of $\mathbf{X}$.*

Proof. Necessity. The first condition is necessary by Theorem 2. The second is obviously necessary.

Sufficiency. Take any $x \in \mathbf{X}$ and choose $x' \in D$ such that

$$\|x - x'\| < \varepsilon.$$

For large enough m, n,

$$\|U_m(x') - U_n(x')\| < \varepsilon.$$

Therefore

$$\|U_m(x) - U_n(x)\| \leqslant \|U_m(x') - U_n(x')\| + \|U_m(x) - U_m(x')\| +$$
$$+ \|U_n(x) - U_n(x')\| < \varepsilon + (\|U_m\| + \|U_n\|)\|x - x'\| < (2M + 1)\varepsilon.$$

Hence, as $\mathbf{Y}$ is complete, $U(x) = \lim_{n \to \infty} U_n(x)$ exists, and so, by Theorem 2, U is a continuous linear operator.

REMARK 1. The theorem (or at least the sufficiency part) remains true if we replace the condition that $\{U_n(x')\}$ be a Cauchy sequence by the condition that it converge to $U(x')$, where U is a given continuous linear operator. Here we do not need $\mathbf{Y}$ to be complete.

REMARK 2. Condition 2) may be replaced by requiring convergence on a fundamental subset D of $\mathbf{X}$. For convergence on D implies convergence on $\mathscr{L}(D)$, and the latter is dense in $\mathbf{X}$.

REMARK 3. Theorem 1 shows that we can take (1) in place of the first condition of Theorem 3.

COROLLARY. *The set of points of convergence of a sequence of continuous linear operators is either a set of the first category or the whole space.*

For if the set of points of convergence is a set of second category, then the sequence of norms is bounded, by the Remark following Theorem 1. On the other hand, the set of points of convergence, being a set of second category, is dense in some ball; and since it is clearly a linear manifold, it must also be dense in the whole space $\mathbf{X}$. Now apply Theorem 3.

REMARK 4. Theorems 1 and 2 and Theorem 3 (the sufficiency part) remain true if we replace the sequences of operators by nets of operators.

§ 2. Some applications to the theory of functions

The theorems of the preceding section have various applications. Let us look at some of these.

2.1. First we consider the question of the convergence of *mechanical quadrature formulae.*

For the approximate evaluation of integrals one usually makes use of mechanical quadrature formulae having the form

$$\int_a^b x(t)\,dt \cong \sum_{k=0}^{n} A_k x(t_k) \qquad (a \leqslant t_0 < t_1 \ldots < t_n \leqslant b).$$

The rectangular, trapezium and Simpson formulae are examples. More complicated examples of exactly the same type are the Newton–Cotes and Gauss formulae. A general theory of *cubic formulae* has been developed by S. L. Sobolev (see Sobolev-II).

Since we cannot ensure a desired level of accuracy from a single formula, it is natural to consider sequences of formulae

$$\int_a^b x(t)\,dt \cong \sum_{k=0}^{n} A_k^{(n)} x(t_k^{(n)}) \tag{1}$$

$$(a \leqslant t_0^{(n)} < t_1^{(n)} < \ldots < t_n^{(n)} \leqslant b; \qquad n = 0, 1, \ldots)$$

and to pose the question: under what conditions will the error in calculating integrals by these formulae tend to zero as $n \to \infty$? If this does happen for a given function x, we shall say that the mechanical quadrature formulae (1) *converge* for x.

One answer to the question posed above is given by

THEOREM 1 (Szegö). *The following conditions are necessary and sufficient for the mechanical quadrature formulae* (1) *to converge for every continuous function*:

1) $\sum_{k=0}^{n} |A_k^{(n)}| \leqslant M \qquad (n = 0, 1, \ldots);$

2) *the formulae converge for every polynomial.*

Proof. Consider the following functionals on the space $\mathbf{C}[a,b]$:

$$f_n(x) = \sum_{k=0}^{n} A_k^{(n)} x(t_k^{(n)}) \qquad (n = 0, 1, \ldots),$$

$$f(x) = \int_a^b x(t)\,dt.$$

As we showed in V.2.1,

$$\| f_n \| = \sum_{k=0}^{n} |A_k^{(n)}| \qquad (n = 0, 1, \ldots).$$

Thus condition 1) means that the norms of the f_n are bounded in aggregate, and condition 2) that $f_n(x) \to f(x)$ for x belonging to the dense subset of all polynomials in $\mathbf{C}[a, b]$. Hence the stated result is a special case of the Banach–Steinhaus Theorem (if we take the Remark following it into account).

REMARK 1. If the coefficients $A_k^{(n)}$ are positive for all k and n, then the first condition follows from the second.

For, taking $x(t) \equiv 1$, we deduce the convergence of the formulae from the second condition: that is, we have

$$b - a = \int_a^b dt = \lim_{n \to \infty} \sum_{k=0}^{n} A_k^{(n)},$$

and from this it also follows that the sums $\sum_{k=0}^{\infty} |A_k^{(n)}| = \sum_{k=0}^{n} A_k^{(n)}$ are bounded.

REMARK 2. In the second condition, the set of all polynomials can be replaced by another dense subset of $\mathbf{C}[a, b]$, for example the set of all piecewise linear functions, or even by a set which is complete in $\mathbf{C}[a, b]$, for example, the set of powers of the independent variable (see Remark 2 following the Banach–Steinhaus Theorem).

REMARK 3. What we have said above about the formulae (1) carries over without any change to the more general case of formulae

$$\int_a^b p(t)x(t)\,dt \cong \sum_{k=0}^{n} A_k^{(n)} x(t_k^{(n)}) \tag{2}$$

$$(a \leqslant t_0^{(n)} < t_1^{(n)} < \ldots < t_n^{(n)} \leqslant b; \qquad n = 0, 1, \ldots),$$

where $p(t)$ is a fixed summable function, called the weight function.

The following is one of the basic methods of obtaining mechanical quadrature formulae.

For $n = 0, 1, \ldots,$ we specify values $t_0^{(n)}, t_1^{(n)}, \ldots, t_n^{(n)}$ in $[a, b]$ in any manner and construct the Lagrange interpolation polynomials $P_n(x; t)$ associated with $x(t)$, coinciding with $x(t)$ at $t_0^{(n)}, t_1^{(n)}, \ldots, t_n^{(n)}$. It is well known that

$$P_n(x; t) = \sum_{k=0}^{n} l_k^{(n)}(t) x(t_k^{(n)}),$$

where

$$l_k^{(n)}(t) = \frac{\omega_n(t)}{(t - t_k^{(n)})\omega_n'(t_k^{(n)})}$$

$$(\omega_n(t) = (t - t_0^{(n)})(t - t_1^{(n)}) \ldots (t - t_n^{(n)}); \qquad k = 0, 1, \ldots, n; \; n = 0, 1, \ldots).$$

If we replace $x(t)$ in the integral $\int_a^b p(t) x(t) \, dt$ by its interpolation polynomial, we obtain the mechanical quadrature formulae

$$\int_a^b p(t) x(t) \, dt \cong \sum_{k=0}^{n} A_k^{(n)} x(t_k^{(n)}) \quad \left(A_k^{(n)} = \int_a^b p(t) l_k^{(n)}(t) \, dt \right). \tag{3}$$

Formulae obtained by this method are called *interpolation formulae* (see Natanson-I).

If $x(t)$ is a polynomial of degree at most n, then it coincides with its interpolation polynomial, so in this case (3) is an exact formula. Thus, if $x(t)$ is an arbitrary polynomial, then the error of the formula is zero for sufficiently large n—that is, interpolation formulae for mechanical quadrature always converge on the set of all polynomials. Hence the first condition of Theorem 1 is, by itself, a necessary and sufficient condition for such formulae to converge for all continuous functions. In particular, the convergence is guaranteed when all the coefficients $A_k^{(n)}$ are non-negative, by Remark 1.

This latter situation arises, for example, when the weight function is positive and the points $t_0^{(n)}, t_1^{(n)}, \ldots, t_n^{(n)}$ are chosen such that the polynomials $\omega_n(t)$ form an orthogonal system with respect to the weight $p(t)$. The quadrature formulae thus obtained are called formulae of *Gaussian type*. They are distinguished from other interpolation formulae for mechanical quadrature by being exact for polynomials of degree $2n + 1$ (see Natanson-I).

2.2. Now we consider the space $\tilde{\mathbf{C}}$, whose elements are the continuous periodic functions defined on the whole real line and having the same period (which, for definiteness, we take to be 2π). Every such function may clearly be regarded as a function defined on some interval $[a, a + 2\pi]$ of length 2π and satisfying $x(a + 2\pi) = x(a)$. This enables us to identify $\tilde{\mathbf{C}}$ with a closed subspace of $\mathbf{C}[a, a + 2\pi]$.

It follows from this that the operator $y = U(x)$ given by

$$y(s) = \int_a^{a+2\pi} K(s, t) x(t) \, dt \qquad (s \in [a, a + 2\pi]), \tag{4}$$

where the kernel $K(s, t)$ is continuous, is a linear operator from $\tilde{\mathbf{C}}$ into $\mathbf{C}$. We leave it to the reader to check that the norm of U is given by the expression

$$\|U\| = \max_s \int_a^{a+2\pi} |K(s, t)| \, dt \tag{5}$$

(cf. V.2.4).

In general, the operator (4) maps periodic functions into non-periodic ones. An obvious necessary and sufficient condition for U to be an operator from $\tilde{\mathbf{C}}$ into $\tilde{\mathbf{C}}$ is that $K(s, t)$

have period 2π in its first argument; that is, $K(s+2\pi, t) = K(s, t)$. Finally, if $K(s, t)$ is defined in the whole plane and has period 2π in its second argument also, then the integration in (4) and (5) can be carried out over any interval of length 2π.

Let us form the Fourier series of the continuous 2π-periodic function $x(t)$:

$$x(t) \sim \frac{a_0}{2} + \sum_{k=1}^{\infty} (a_k \cos kt + b_k \sin kt),$$

$$a_k = \frac{1}{\pi} \int_0^{2\pi} x(t) \cos kt\, dt, \qquad b_k = \frac{1}{\pi} \int_0^{2\pi} x(t) \sin kt\, dt$$

$$(k = 0, 1, \ldots).$$

If we associate with each function $x(t)$ in $\tilde{C}$ the partial sum $S_n(x)$ of its Fourier series, we obtain an operator S_n mapping $\tilde{C}$ into $\tilde{C}$. It is well known that this sum is expressible as a Dirichlet integral:

$$y = S_n(x), \qquad y(s) = \frac{1}{2\pi} \int_0^{2\pi} x(t) \frac{\sin (2n+1)\frac{t-s}{2}}{\sin \frac{t-s}{2}}\, dt,$$

that is, S_n is of the form (4), and furthermore, by the continuity of the kernel, S_n is a continuous linear operator. Let us show that $\|S_n\| \to \infty$ as $n \to \infty$.

In fact, using (5) and the periodicity of the kernel, we have

$$\|S_n\| = \frac{1}{2\pi} \int_0^{2\pi} \left| \frac{\sin (2n+1)\frac{t-s}{2}}{\sin \frac{t-s}{2}} \right| dt = \frac{1}{2\pi} \int_0^{2\pi} \left| \frac{\sin \frac{2n+1}{2} t}{\sin \frac{t}{2}} \right| dt$$

$$= \frac{1}{\pi} \int_0^{\pi} \left| \frac{\sin mt}{\sin t} \right| dt,$$

where $m = 2n+1$. Using the following well-known inequalities from analysis,

$$|\sin t| \leqslant |t|, \qquad \sin s \geqslant \frac{2}{\pi} s \qquad \left(0 \leqslant s \leqslant \frac{\pi}{2}\right),$$

we have also

$$\|S_n\| = \frac{1}{\pi} \int_0^{\pi} \left| \frac{\sin mt}{\sin t} \right| dt > \frac{1}{\pi} \sum_{k=1}^{m-1} \int_{\frac{k\pi}{m}}^{\frac{(2k+1)\pi}{2m}} \frac{|\sin mt|}{\sin t}\, dt =$$

$$= \frac{1}{\pi} \sum_{k=1}^{m-1} \int_0^{\frac{\pi}{2m}} \frac{|\sin mt|}{\sin\left(t + \frac{k\pi}{m}\right)}\, dt \geqslant$$

$$\geqslant \frac{1}{\pi}\sum_{k=1}^{m-1}\int_0^{\frac{\pi}{2m}}\frac{\frac{2}{\pi}mt}{t+\frac{k\pi}{m}}dt \geqslant \frac{1}{\pi}\sum_{k=1}^{m-1}\int_{\frac{\pi}{4m}}^{\frac{\pi}{2m}}\frac{\frac{2}{\pi}mt}{t+\frac{k\pi}{m}}dt \geqslant$$

$$\geqslant \frac{1}{\pi}\sum_{k=1}^{m-1}\frac{\frac{2}{\pi}m\frac{\pi}{4m}}{\frac{\pi}{4m}+\frac{k\pi}{m}}\frac{\pi}{4m} \geqslant \frac{1}{8\pi}\sum_{k=2}^{m-1}\frac{1}{k} \geqslant \frac{1}{8\pi}\sum_{k=2}^{m-1}\int_k^{k+1}\frac{dt}{t} =$$

$$= \frac{1}{8\pi}\ln\frac{m}{2} \geqslant \frac{1}{8\pi}\ln n.$$

Hence

$$\|S_n\| \geqslant \frac{1}{8\pi}\ln n,$$

from which we obtain the required result.

From Theorem 1.3 we conclude that there exists a continuous periodic function whose Fourier series does not converge uniformly to any function.

The arguments we have given also enable us to establish the existence of a continuous periodic function whose Fourier series diverges at an arbitrary preassigned point.

For this we consider the sequence of functionals f_n on the space $\tilde{\mathbf{C}}$, defined by

$$f_n(x) = S_n(x)(t_0) = \frac{1}{2\pi}\int_0^{2\pi}\frac{\sin(2n+1)\frac{t-t_0}{2}}{\sin\frac{t-t_0}{2}}x(t)\,dt.$$

Exactly as in V.2.2, the norm of f_n is determined by the equation

$$\|f\| = \frac{1}{2\pi}\int_0^{2\pi}\left|\frac{\sin(2n+1)\frac{t-t_0}{2}}{\sin\frac{t-t_0}{2}}\right|dt = \frac{1}{2\pi}\int_0^{2\pi}\left|\frac{\sin(2n+1)\frac{t}{2}}{\sin\frac{t}{2}}\right|dt = \|S_n\|$$

and hence $\|f_n\| \to \infty$ as $n + \infty$. Therefore by Theorem 1.1 there exists an $x_0 \in \tilde{\mathbf{C}}$ such that $\sup_n |f_n(x_0)| = \infty$, which is what we required to prove.

Now take an arbitrary countable set $e = \{t_k\}$ on the real line and form the functionals $f_n^{(k)}$, given by

$$f_n^{(k)}(x) = S_n(x)(t_k) \qquad (k, n = 1, 2, \ldots).$$

Applying the principle of condensation of singularities to these, we find an element $x_0 \in \tilde{\mathbf{C}}$ such that

$$\sup_n |f_n^{(k)}(x_0)| = \infty \qquad (k = 1, 2, \ldots),$$

that is, we have a function $x_0(t)$ whose Fourier series diverges at each point of the set e.

An example of a continuous function whose Fourier series is nowhere convergent was first given by du Bois Reymond (see Zygmund).

If S_n is regarded as a linear operator from $\mathbf{L}^1$ into $\mathbf{L}^1$, then, in view of the symmetry of the kernel, $\|S_n\|$ keeps its former value; and thus we conclude from Theorem 1.3 that there exists a summable function whose Fourier series does not converge in mean on $\mathbf{L}^1$.

2.3. We can obtain a wide generalization of the preceding results if we introduce the concept of a polynomial operator.

We again consider the space $\tilde{\mathbf{C}}$ of continuous periodic functions (with period 2π) and denote by $\tilde{\mathbf{H}}_n$ the subspace consisting of all trigonometric polynomials of degree at most n.

A continuous linear operator U on $\tilde{\mathbf{C}}$ is called a (trigonometric) *polynomial operator of degree n* if

1) $U(x) \in \tilde{\mathbf{H}}_n$ for every $x \in \tilde{\mathbf{C}}$;
2) $U(x) = x$ for every $x \in \tilde{\mathbf{H}}_n$.

In other words, a polynomial operator assigns to each 2π-periodic function a trigonometric polynomial of degree at most n and leaves these polynomials themselves fixed.

The simplest example of a polynomial operator is the operator S_n studied in 2.2. Another example is provided by the operator that associates with a function one of its (trigonometric) interpolation polynomials, constructed with respect to a fixed system of weights.

Let us introduce the following notation. If $y = U(x)$, then the value of the function y for a given s will be denoted by $U(x; s)$. For example, $S_n(x; s) = S_n(x)(s)$. Further, if $x(t)$ is a function in $\tilde{\mathbf{C}}$, then we shall denote by $x^h(t)$ the function obtaining from $x(t)$ by translating the argument:

$$x^h(t) = x(t+h).$$

Clearly $x^h \in \tilde{\mathbf{C}}$, for every h. Notice also that, as $x \in \tilde{\mathbf{C}}$ is uniformly continuous,

$$\|x^h - x\| = \max_t |x^h(t) - x(t)| \to 0 \qquad \text{when} \qquad h \to 0. \tag{6}$$

It is possible to establish some very important general facts concerning polynomial operators and sequences of polynomial operators. These are all based on the following lemma, which connects an arbitrary polynomial operator with the simplest one, namely S_n.

LEMMA 1. *If U is a polynomial operator of degree n, then we have the identity*

$$\frac{1}{2\pi}\int_0^{2\pi} U(x^\tau;\ s-\tau)\,d\tau = S_n(x; s) \qquad (x \in \tilde{\mathbf{C}}) \tag{7}$$

(*the Zygmund–Marcinkiewicz–Berman identity*).

Proof. First suppose that $x \in \tilde{\mathbf{H}}_n$, so that $x^\tau \in \tilde{\mathbf{H}}_n$ also. Then

$$U(x^\tau; s-\tau) = x^\tau(s-t) = x(s).$$

But, since S_n is a polynomial operator of degree n, we have also

$$S_n(x; s) = x(s),$$

which proves the identity in this case.

Now suppose that $x(t) = \cos mt$ or $x(t) = \sin mt$, where $m > n$. Restricting ourselves for definiteness to the first case, we have

$$x^{\tau}(t) = \cos m(t+\tau) = \cos mt \cos m\tau - \sin mt \sin m\tau$$
$$= x_1(t)\cos m\tau + x_2(t)\sin m\tau.$$

Therefore, if we set

$$y_1 = U(x_1), \qquad y_2 = U(x_2)$$

(y_1, y_2 being elements of $\tilde{\mathbf{H}}_n$), then

$$U(x^{\tau}; s-\tau) = y_1(s-\tau)\cos m\tau + y_2(s-\tau)\sin m\tau.$$

But $y_1(s-\tau)$ and $y_2(s-\tau)$ are trigonometric polynomials in τ of degree not exceeding n, so they are orthogonal to the functions $\cos m\tau$ and $\sin m\tau$. Hence

$$\frac{1}{2\pi}\int_0^{2\pi} U(x^{\tau}; s-\tau)\,d\tau = 0.$$

The right-hand side of (7) is clearly also zero.

Thus identity (7) is proved in this case, and therefore—since both sides are additive in x—also for arbitrary trigonometric polynomials.

Now we consider the left-hand side of (7) and prove that, for fixed s, it is a continuous linear functional on $\tilde{\mathbf{C}}$.

Denoting the left-hand side of (7) by $f_s(x)$, we verify that the functional f_s makes sense for any element $x \in \tilde{\mathbf{C}}$. We do this by showing that the integrand is continuous in τ. We have

$$\begin{aligned}|U(x^{\tau+h}; s-\tau-h) - U(x^{\tau}; s-\tau)| &\leqslant |U(x^{\tau+h}; s-\tau-h) - U(x^{\tau}; s-\tau-h)| + \\ + |U(x^{\tau}; s-\tau-h) - U(x^{\tau}; s-\tau)| &\leqslant \|U(x^{\tau+h} - x^{\tau})\| + |U(x^{\tau}; s-\tau-h) - \\ - U(x^{\tau}; s-\tau)| &\leqslant \|U\|\,\|x^{\tau+h} - x^{\tau}\| + |U(x^{\tau}; s-\tau-h) - \\ &\quad - U(x^{\tau}; s-\tau)|.\end{aligned}$$

For sufficiently small h, the first term here is as small as we please, by (6), and the same is true of the second term, since $U(x^{\tau}, t)$ is continuous.

The functional f_s is obviously additive, and, since

$$|f_s(x)| \leqslant \frac{1}{2\pi}\int_0^{2\pi} |U(x^{\tau}; s-\tau)|\,d\tau \leqslant \frac{1}{2\pi}\int_0^{2\pi} \|U\|\,\|x^{\tau}\|\,d\tau = \|U\|\,\|x\|,$$

it is also a bounded functional.

Finally, if we write $g_s(x) = S_n(x; s)$, then (7) amounts to the assertion that f_s and g_s coincide. But it was shown earlier that they coincide on the dense subset of trigonometric polynomials in $\tilde{\mathbf{C}}$. Since they are continuous functions, it follows from this that they coincide on the whole of $\tilde{\mathbf{C}}$, which is what we required to prove.

THEOREM 2. *Among all trigonometric polynomial operators of degree n, the operator S_n has the least norm: that is, we always have*

$$\|U\| \geqslant \|S_n\| > A \ln n.$$

For, by (7),

$$\|S_n(x)\| = \max_s |S_n(x; s)| \leqslant \frac{1}{2\pi} \int_0^{2\pi} \max_s |U(x^\tau; s-\tau)| \, d\tau \leqslant \|U\| \, \|x\|,$$

so that $\|S_n\| \leqslant \|U\|$.

THEOREM 3 (Lozinskii–Kharshiladze). *If $\{U_n\}$ is a sequence of trigonometric polynomial operators, where U_n has degree n, then the norms of these operators tend to infinity. In particular, no such sequence can be convergent on the whole space $\tilde{\mathbf{C}}$.*

The first statement follows immediately from the last theorem. The second is obtained using Theorem 1.1.

As a special case of the theorem just stated, we note the following important fact. For any given points of interpolation, there exists a continuous 2π-periodic function whose associated sequence of interpolation polynomials is not uniformly convergent (Faber's Theorem).

We note without proof that the statements of Theorems 2 and 3 apply also to the space $\mathbf{L}^1$.

One can, in an analogous fashion, consider (algebraic) polynomial operators on $\mathbf{C}[0, 1]$, by which we mean continuous linear operators that map $\mathbf{C}[0, 1]$ into the subspace $\mathbf{H}_n$ of all algebraic polynomials of degree at most n and leave the elements of $\mathbf{H}_n$ fixed. However, it is more convenient to study the algebraic case by reducing it to the trigonometric case (see Natanson-I).

Let us consider the (algebraic) polynomial operator U_n assigning to a continuous function the n-th partial sum of its Fourier series with respect to a given system of orthogonal polynomials. By applying the algebraic analogue of Theorem 3 to sequences of operators, we obtain the result (Nikolaev [1]): *for any system of orthogonal polynomials, there exists a continuous function whose Fourier series with respect to this system is not uniformly convergent.*

Theorems 2 and 3 were established by S. M. Lozinskii and F. I. Kharshiladze (see Lozinskii [1], [2]). S. M. Lozinskii has made a far-reaching development of these ideas.

2.4. Let us consider the problem of representing functions by singular integrals.

Suppose we are given a sequence of functions $\{K_n(s, t)\}$ on the square $[a, b; a, b]$. A function $x(s)$ is said to be representable by a singular integral if the sequence

$$x_n(s) = \int_a^b K_n(s, t) x(t) \, dt \qquad (n = 1, 2, \ldots) \tag{8}$$

converges to $x(s)$ in one sense or another.

Singular integrals occur regularly in various problems in analysis. By way of example we mention the Dirichlet integral, the Fejér integral, the de la Vallé–Poussin integral, the Hilbert integral, etc.

Let us prove a theorem on the convergence in mean of singular integrals in $\mathbf{L}^1(a, b)$, restricting ourselves to continuous kernels.

THEOREM 4. *A necessary and sufficient condition for the sequence* (8) *to converge to x in the space $\mathbf{L}^1$, for every summable function x, is that*

$$1) \qquad \int_a^b \left| \int_a^b K_n(s, t) x(t) \, dt - x(s) \right| ds \xrightarrow[n \to \infty]{} 0 \tag{9}$$

for every x in a complete subset D of $\mathbf{L}^1$*; and*

2) $$\int_a^b |K_n(s,t)|\,ds \leqslant M \qquad (t \in [a,b]; \quad n = 1, 2, \ldots). \tag{10}$$

For if U_n denotes the operator on $\mathbf{L}^1$ associated with the kernel $K_n(s,t)$:

$$y = U_n(x), \qquad y(s) = \int_a^b K_n(s,t)x(t)\,dt,$$

then the first condition amounts to requiring that $U_n(x) \to x$ for $x \in D$, and the second that $\|U_n\| \leqslant M$, by V.2.5. Thus the stated result is a special case of the Banach–Steinhaus Theorem (or, more precisely, of Remark 2 following it) (see 1.2).

REMARK 1. For the sufficiency part, condition 1) can be replaced by two simpler conditions, namely

$$\int_\alpha^\beta K_n(s,t)\,dt \xrightarrow[n\to\infty]{} 1 \qquad (s \in (\alpha,\beta) \subset [a.b]), \tag{11}$$

$$\int_a^b |K_n(s,t)|\,dt \leqslant M \qquad (s \in [a,b]; \quad n = 1, 2, \ldots). \tag{12}$$

It is easy to verify that in this case one can take D to be the collection of characteristic functions of intervals contained in $[a,b]$.

In fact, if $\tilde{x}$ is the characteristic function of $[\alpha,\beta]$, then, for $s \in (\alpha,\beta)$,

$$\tilde{x}_n(s) = \int_\alpha^\beta K_n(s,t)\,dt \xrightarrow[n\to\infty]{} 1 = \tilde{x}(s);$$

while if $s \notin [\alpha,\beta]$ and we have, say, $a \leqslant s < \alpha$, then

$$\tilde{x}_n(s) = \int_\alpha^\beta K_n(s,t)\,dt = \int_\alpha^\beta K_n(s,t)\,dt - \int_\alpha^\alpha K_n(s,t)\,dt \xrightarrow[n\to\infty]{} 1 - 1 = 0 = \tilde{x}(s).$$

Thus $\tilde{x}_n(s) \to \tilde{x}(s)$ for all $s \neq \alpha, \beta$; that is, almost everywhere. Condition (12) ensures that we can take the limit under the integral sign in $\int_a^b \left| \int_a^b K_n(s,t)\tilde{x}(t)\,dt - \tilde{x}(s) \right| ds$, from which we see that this limit is zero.

REMARK 2. If the kernels $K_n(s,t)$ are symmetric, the condition (12) coincides with condition 2) of the theorem; hence conditions (11) and (12) are in this case both necessary and sufficient.

If the operators U_n associated with the kernels $K_n(s,t)$ are regarded as operators on $\mathbf{C}[a,b]$, then one obtains, in an analogous way, conditions for the sequence (8) to converge uniformly to a continuous function $x(s)$.

The reader can obtain more detailed information on singular integrals in Dunford and Schwartz-II and in Natanson [1].

2.5. In conclusion, we consider the problem of generalized summation of series (see Zygmund).

Suppose we have a numerical sequence

$$a_1 + a_2 + \ldots + a_n + \ldots, \tag{13}$$

and let $\{s_k\}$ be the sequence of its partial sums. We introduce the infinite matrix

$$\begin{pmatrix} \alpha_{11} & \alpha_{12} & \dots & \alpha_{1k} & \dots \\ \alpha_{21} & \alpha_{22} & \dots & \alpha_{2k} & \dots \\ \dots & \dots & \dots & \dots & \dots \\ \alpha_{n1} & \alpha_{n2} & \dots & \alpha_{nk} & \dots \\ \dots & \dots & \dots & \dots & \dots \end{pmatrix} \tag{14}$$

and form the expression

$$\sigma_n = \sum_{k=1}^{\infty} \alpha_{nk} s_k \qquad (n = 1, 2, \dots), \tag{15}$$

assuming that all the series on the right-hand side converge.

The series (13) is said to be generalized-summable by means of the matrix (14) if the sequence $\{\sigma_n\}$ has a finite limit. The value

$$\sigma = \lim_{n \to \infty} \sigma_n$$

of this limit is called the *generalized sum* of the series (13).

For example, the *Cesàro summation method*, where

$$\sigma_n = \frac{1}{n} \sum_{k=1}^{n} s_k,$$

is characterized by the matrix

$$\begin{pmatrix} 1 & 0 & \dots & 0 & \dots \\ \frac{1}{2} & \frac{1}{2} & \dots & 0 & \dots \\ \dots & \dots & \dots & \dots & \dots \\ \frac{1}{n} & \frac{1}{n} & \dots & \frac{1}{n} & \dots \\ \dots & \dots & \dots & \dots & \dots \end{pmatrix}.$$

Instead of talking of generalized sums of series, one can consider the problem of defining the generalized limit of a sequence, associating with a given numerical sequence $x = \{\xi_k\}$ the sequence

$$\sigma_n(x) = \sum_{k=1}^{\infty} \alpha_{nk} \xi_k \qquad (n = 1, 2, \dots)$$

and studying its behaviour as $n \to \infty$. Clearly, one way of formulating the problem is easily reduced to the other. Since the sequence point of view turns out to be more convenient, we shall restrict ourselves to this, though we retain the term "summation method".

It is natural to consider only those summation methods that are applicable to every sequence which converges in the usual sense, and that assign the usual limit as the generalized limit of such a sequence. Summation methods having this property are called *permanent* or *regular*.

Conditions for the permanence of the summation method defined by (14) are formulated in the following theorem.

THEOREM 5 (Toeplitz). *A necessary and sufficient condition for the summation method defined by the matrix* (14) *to be permanent is that*

1) $\lim\limits_{n\to\infty} \alpha_{nk} = 0 \qquad (k = 1, 2, \ldots);$

2) $\lim\limits_{n\to\infty} \sum\limits_{k=1}^{\infty} \alpha_{nk} = 1;$

3) $\sum\limits_{k=1}^{\infty} |\alpha_{nk}| \leqslant M \qquad (n = 1, 2, \ldots).$

Proof. Consider the following functionals on the space **c** of convergent sequences:

$$\sigma_n(x) = \sum_{k=1}^{\infty} \alpha_{nk}\xi_k \qquad (x = \{\xi_k\}; \quad n = 1, 2, \ldots)$$

and

$$\sigma(x) = \lim_{k\to\infty} \xi_k.$$

It was shown in VI.2.2 that

$$\|\sigma_n\| = \sum_{k=1}^{\infty} |\alpha_{nk}| \qquad (n = 1, 2, \ldots),$$

so condition 3) means that the functionals σ_n have bounded norms.

We also introduce the sequences

$$x_0 = (1, 1, \ldots, 1, \ldots) \qquad \text{and} \qquad x_k = (0, \ldots, 0, 1, 0, \ldots)$$

(with unity in the k-th place).

Since

$$\sigma_n(x_0) = \sum_{k=1}^{\infty} \alpha_{nk}, \qquad \sigma_n(x_k) = \alpha_{nk} \quad (n, k = 1, 2, \ldots),$$

conditions 1) and 2) can be expressed in the form

$$\sigma_n(x_k) \xrightarrow[n\to\infty]{} 0 = \sigma(x_k), \qquad \sigma_n(x_0) \xrightarrow[n\to\infty]{} 1 = \sigma(x_0).$$

Since the collection of elements $x_0, x_1, \ldots, x_k, \ldots$ is complete in **c**, the conditions of the theorem agree with those of the Banach–Steinhaus Theorem (taking into account Remarks 1 and 2).

Hence 1)–3) are necessary and sufficient conditions for

$$\sigma_n(x) \xrightarrow[n\to\infty]{} \sigma(x),$$

and this is just the required permanence condition for the summation method.

The matrix associated with the Cesàro summation method obviously satisfies the conditions of the theorem. From this follows the well-known fact that arithmetic mean (Cesàro) summation is a permanent summation method.

VIII

THE WEAK TOPOLOGY IN A BANACH SPACE

In Chapters III and V we have already had to deal with the simplest properties of the weak topology. In this chapter we shall consider some of the deeper properties of this topology. In §4 we study a problem in mathematical economics whose solution uses properties of the weak topology and a theorem on the general form of continuous linear operators on a space of continuous functions.

§1. Weakly bounded sets

1.1. We begin by formulating some important consequences of the Banach–Steinhaus Theorem for the weak and weak* topologies. First of all we consider some criteria for weak and weak* convergence of sequences.

Suppose we have a sequence of continuous linear functionals $\{f_n\}$, defined on a B-space $\mathbf{X}$. If the limit

$$\lim_{n\to\infty} f_n(x) = f(x) \tag{1}$$

exists for each $x \in \mathbf{X}$, then by Theorem VII.1.2 the functional f will also be continuous and linear. Also $f_n \to f(\sigma(\mathbf{X}^*, \mathbf{X}))$.

The Banach–Steinhaus Theorem yields the following criterion for weak* convergence.

Theorem 1. *Let* $\mathbf{X}$ *be a B-space. The following conditions are necessary and sufficient for the statement*

$$f_n \to f(\sigma(\mathbf{X}^*, \mathbf{X})) \tag{2}$$

to hold in $\mathbf{X}$:

1) *the sequence of norms* $\{\|f_n\|\}$ *is bounded;*
2) *equation* (1) *holds for all* x *belonging to a dense subset* D *of* $\mathbf{X}$.

Remark. In condition 2) we need only require that D be a fundamental subset of $\mathbf{X}$ (cf. Remark 2 following the Banach–Steinhaus Theorem).

The norm is not continuous with respect to weak convergence—that is, (2) does not imply that $\|f_n\| \to \|f\|$. All we can say is that

$$\|f\| \leqslant \varliminf_{n\to\infty} \|f_n\|$$

(see Theorem VII.1.2).

Next we consider weak convergence in $\mathbf{X}$. Note first of all that the canonical embedding π of a normed space $\mathbf{X}$ into $\mathbf{X}^{**}$ is a linear isomorphism from the LCS $(\mathbf{X}, \sigma(\mathbf{X}, \mathbf{X}^*))$ onto the LCS $\pi(\mathbf{X})$, viewed as a subspace of $(\mathbf{X}^{**}, \sigma(\mathbf{X}^{**}, \mathbf{X}^*))$—in fact, in both spaces convergence of a net of elements is the same as convergence for all $f \in \mathbf{X}^*$.

In a number of cases this simple observation reduces the study of the weak topology on $\mathbf{X}$ to that of the weak* topology on $\mathbf{X}^{**}$. For example, bearing in mind that π is an isometry, we see from the Remark following Theorem 1 that, if $x_n \to x$ $(\sigma(\mathbf{X}, \mathbf{X}^*))$, then $\|x\| \leqslant \underline{\lim}_{n\to\infty} \|x_n\|$. By means of similar considerations, using Theorem 1, we obtain

THEOREM 2. *The following conditions are necessary and sufficient for $x_n \to x$ $(\sigma(\mathbf{X}, \mathbf{X}^*))$ in a normed space $\mathbf{X}$:*

1) $\sup_n \|x_n\| < \infty$;

2) $f(x_n) \to f(x)$ *for a dense (fundamental) set of functionals f in $\mathbf{X}^*$.*

1.2. Next we consider weak boundedness of subsets. A set E in a normed space $\mathbf{X}$ is said to be weakly bounded if for each $f \in \mathbf{X}^*$ we have $\sup\{|f(x)| : x \in E\} < \infty$. A set E in $\mathbf{X}^*$ is said to be weak* bounded if for each $x \in \mathbf{X}$ we have $\sup\{|f(x)| : f \in E\} < \infty$. Obviously, a set $E \subset \mathbf{X}$ is weakly bounded if and only if it is bounded in the LCS $(\mathbf{X}, \sigma(\mathbf{X}, \mathbf{X}^*))$, while a set $E \subset \mathbf{X}^*$ is weak* bounded if and only if it is bounded in the LCS $(\mathbf{X}^*, \sigma(\mathbf{X}^*, \mathbf{X}))$.

THEOREM 3. *If $\mathbf{X}$ is a B-space, then a set $E \subset \mathbf{X}^*$ is weak* bounded if and only if it is bounded relative to the norm of $\mathbf{X}^*$.*

Proof. We need only establish that a weak* bounded E is bounded relative to the norm. Assume the contrary. Then we can choose a sequence $\{f_n\}$ in E such that $\|f_n\| \geqslant n^2$ $(n \in \mathbb{N})$. Since

$$\left|\frac{1}{n} f_n(x)\right| \leqslant \frac{1}{n} \sup\{|f(x)| : f \in E\} \to 0 \qquad (x \in \mathbf{X}),$$

we have $\left(\frac{1}{n}\right) f_n \to 0$ $(\sigma(\mathbf{X}^*, \mathbf{X}))$, which is in contradiction to condition 1) of Theorem 1, as $\|(1/n) f_n\| \geqslant n$ $(n \in \mathbb{N})$.

In exactly the same way as we deduced Theorem 3 from Theorem 1, we deduce the next theorem from Theorem 2.

THEOREM 4. *If $\mathbf{X}$ is a normed space, then a set $E \subset \mathbf{X}$ is weakly bounded if and only if it is bounded relative to the norm of $\mathbf{X}$.*

We promised a proof of this theorem as far back as Chapter III (see Theorem III.3.3), but it is only now that the Banach–Steinhaus Theorem enables us to give it. Since weakly compact sets are weakly bounded, it follows from Theorem 4 that every weakly compact set is bounded relative to the norm.

We also recall that the weak closure and the closure relative to the norm coincide for convex subsets of a normed space $\mathbf{X}$ (see Theorem III.3.2.). We shall show below that, in spite of this property, the weak topology and the norm topology are distinct for an infinite-dimensional B-space.

The reasoning at the beginning of 1.1 shows that, if $\mathbf{X}$ is a B-space, then $\mathbf{X}^*$ is weak* sequentially complete: that is, if the numerical sequence $\{f_n(x)\}$ $(f_n \in \mathbf{X}^*)$ has a limit for each $x \in \mathbf{X}$, then there exists an $f \in \mathbf{X}^*$ with $f_n \to f(\sigma(\mathbf{X}^*, \mathbf{X}))$.

The space $\mathbf{X}$ itself does not always have the analogous property. A B-space $\mathbf{X}$ is said to be *weakly sequentially complete* if the LCS $(\mathbf{X}, \sigma(\mathbf{X}, \mathbf{X}^*))$ is sequentially complete, that is, if the following condition is satisfied: if a numerical sequence $\{f(x_n)\}$ $(x_n \in \mathbf{X})$ has a limit for each $f \in \mathbf{X}^*$ then there exists an $x \in \mathbf{X}$ such that $x_n \to x$ $(\sigma(\mathbf{X}, \mathbf{X}^*))$. It follows from the fact that $\mathbf{X}^*$ is weak* sequentially complete that a reflexive B-space $\mathbf{X}$ is weakly sequentially complete. In Chapter X we shall see that the space $\mathbf{c}_0$ is not weakly sequentially complete, while the non-reflexive space $\mathbf{L}^1[0, 1]$ is weakly sequentially complete (see Theorem X.4.9). An essential difference between the weak topology and the strong topology is apparent from

THEOREM 5. *A necessary and sufficient condition for the weak topology and the norm topology to coincide in a normed space* $\mathbf{X}$ *is that* $\mathbf{X}$ *be finite-dimensional.*

Proof. The sufficiency follows from Theorem IV.1.2. Let us prove the necessity. It follows from III.3.3 that in this case the ball $B_{\mathbf{X}^*}$ in the dual space is contained in the absolutely convex hull of a finite number of functionals, so that $\mathbf{X}^*$, and hence also $\mathbf{X}$, is finite-dimensional.

§2. Eberlein–Shmul'yan theory

If E is a subset of a topological space $\mathbf{X}$, then there are three properties close to the property of compactness that E may possess (see I.2.7). Let us recall these.

1) E is relatively compact; that is, its closure is compact;

2) E is relatively sequentially compact; that is, every sequence in E contains a subsequence that converges to a point of $\mathbf{X}$;

3) E is relatively countably compact; that is, every sequence in E has a cluster point in $\mathbf{X}$.

The implications 1)$\Rightarrow$ 3) and 2)$\Rightarrow$3) are valid in general. In I.5.1 we proved that, in the case of a metric space, properties 1)–3) are equivalent. Here we shall establish that one has a similar picture if $\mathbf{X}$ is a B-space endowed with the weak topology. We note at once that a weakly compact set is not necessarily metrizable. For consider any inseparable reflexive B-space $\mathbf{X}$ (for example, a Hilbert space with an uncountable basis). The ball $B_{\mathbf{X}}$ in $\mathbf{X}$ is weakly compact, but if it were metrizable in the weak topology then, by Corollary 1 to Theorem I.5.2, $B_{\mathbf{X}}$ would be weakly separable and thus, as $B_{\mathbf{X}}$ is convex, also strongly separable. From this it would follow that $\mathbf{X}$ is separable, contradicting the choice of $\mathbf{X}$.

The equivalence of 1)–3) for the weak topology was proved as a result of efforts made over several years by a number of mathematicians, the basic step being made by V. L. Shmul'yan and Eberlein (in 1940), so that this important division of B-space theory is naturally called Eberlein–Shmul'yan theory.* In our exposition we follow R. Wheatley and H. Cohen.

LEMMA 1. *Let* $\mathbf{X}$ *be a* B*-space, and* $\mathbf{Y}$ *a finite-dimensional subspace of* $\mathbf{X}^{**}$. *Then there exists a finite set* $\{f_m\}_{m=1}^n$, $\|f_m\| = 1$ $(m = 1, 2, \ldots, n)$, *of elements of* $\mathbf{X}^*$ *such that, for every* $F \in \mathbf{Y}$, *we have*

$$\max\{|F(f_m)| : 1 \leqslant m \leqslant n\} \geqslant \tfrac{1}{2}\|F\|. \tag{1}$$

Proof. Since the unit ball in $\mathbf{Y}$ is compact, it has a finite $\frac{1}{4}$-net $\{F_m\}_{m=1}^n$ ($\|F_m\| = 1$, $m = 1, 2, \ldots, n$). Choose elements $f_m \in \mathbf{X}^*$, $\|f_m\| = 1$ in $\mathbf{X}^*$, such that $|F_m(f_m)| > \frac{3}{4}$ $(m = 1, 2, \ldots, n)$. Taking $F \in \mathbf{Y}$, let us prove (1). We may assume that $F \neq \mathbf{0}$. Then, corresponding to the functional $F/\|F\|$, we can find an F_m such that $\|F/\|F\| - F_m\| < \frac{1}{4}$. We have

$$|F(f_m)| \geqslant \|F\|\,|F_m(f_m)| - \|F\|\,|F_m(f_m) - (F/\|F\|)(f_m)| \geqslant \|F\|\,|F_m(f_m)| - \|F\|\,\|F/\|F\| - F_m\|$$
$$> \tfrac{3}{4}\|F\| - \tfrac{1}{4}\|F\| = \tfrac{1}{2}\|F\|,$$

and so we obtain (1).

---

* For historical remarks, see Dunford and Schwartz-I.

LEMMA 2. *Let E be a relatively weakly countably compact subset of a B-space* $\mathbf{X}$. *Let* E_1 *denote the* $\sigma(\mathbf{X}^{**}, \mathbf{X})$*-closure of* $\pi(E)$, *where* $\pi: \mathbf{X} \to \mathbf{X}^{**}$ *is the canonical embedding. Then* E_1 *is* $\sigma(\mathbf{X}^{**}, \mathbf{X})$*-compact and, for each point* $F \in E_1$, *there exists a sequence* $\{x_n\} \subset E$ *such that* $\pi(x) = F$ *for every weak cluster point* $x \in \mathbf{X}$ *of this sequence.*

REMARK. It is clear that the sequence $\{x_n\}$ in the statement of Lemma 2 has a unique cluster point.

Proof. Since E is relatively countably compact, the set $f(E)$ is a relatively compact set of scalars, for each $f \in \mathbf{X}^*$. Hence E is weakly bounded, and therefore also strongly bounded (Theorem 1.4). Hence E_1 *is* $\sigma(\mathbf{X}^{**}, \mathbf{X}^*)$-compact by the Alaoglu–Bourbaki Theorem.

Let $F \in E_1$. We construct the required sequence $\{x_n\}$ by induction. Choose $f_1 \in \mathbf{X}^*$, $\|f_1\| = 1$. Since $F \in E_1$, there exists $x_1 \in E$ such that

$$|(F - \pi(x_1))(f_1)| < 1.$$

The space $\mathbf{Y}_2$ spanned by the functionals F and $F - \pi(x_1)$ is finite-dimensional. By Lemma 1, there exist vectors $f_2, \ldots, f_{k(2)}$ of norm 1 in $\mathbf{X}^*$ such that

$$\max\{|G(f_m)| : 2 \leqslant m \leqslant k(2)\} \geqslant \tfrac{1}{2}\|G\|$$

for every $G \in \mathbf{Y}_2$. Using the fact that $F \in E_1$, we can find an $x_2 \in E$ such that

$$\max\{|(F - \pi(x_2))(f_m)| : 1 \leqslant m \leqslant k(2)\} < 1/2.$$

Using Lemma 1 for $\mathbf{Y}_3 = \mathscr{L}(F - \pi(x_2), F - \pi(x_1), F)$, we find points $f_{k(2)+1}, \ldots, f_{k(3)}$ of norm 1 in $\mathbf{X}^*$ such that

$$\max\{|G(f_m)| : k(2) < m \leqslant k(3)\} \geqslant \tfrac{1}{2}\|G\|$$

for every $G \in \mathbf{Y}_3$.

Since $F \in E_1$, we can find $x_3 \in E$ such that

$$\max\{|(F - \pi(x_3))(f_m)| : 1 \leqslant m \leqslant k(3)\} < 1/3.$$

Continuing this process, we obtain a sequence $\{x_n\} \subset E$ satisfying the condition

$$\max\{|G(f_m)| : k(n-1) < m \leqslant k(n)\} \geqslant \tfrac{1}{2}\|G\| \tag{2}$$

for each $G \in \mathbf{Y}_n = \mathscr{L}(F - \pi(x_n), F - \pi(x_{n-1}), \ldots, F)$, where $f_m \in \mathbf{X}^*$, $\|f_m\| = 1$, and

$$\max\{|(F - \pi(x_n))(f_m)| : 1 \leqslant m \leqslant k(n)\} < 1/n. \tag{3}$$

Let $x \in \mathbf{X}$ be a cluster point of the sequence $\{x_n\}$ in the weak topology (such points exist because E is relatively weakly countably compact). Since the closure (in the norm) of the linear hull $\mathscr{L}(\{x_n\})$ is weakly closed, we have $x \in \overline{\mathscr{L}}(\{x_n\})$. Hence $F - \pi(x) \in F - \overline{\mathscr{L}}(\{\pi(x_n)\}) \subset \overline{\mathscr{L}}(\{F, F - \pi(x_1), \ldots, F - \pi(x_n), \ldots\})$, where the closure is formed relative to the norm of $\mathbf{X}^{**}$.

By (2), each $G \in \mathscr{L}(\{F, F - \pi(x_1), \ldots, F - \pi(x_n), \ldots\})$ satisfies

$$\sup_m |G(f_m)| \geqslant \tfrac{1}{2}\|G\|.$$

Therefore the same is true of each G in the closure of this subspace, and, in particular, of $F - \pi(x)$. By (3), we have, for fixed m,

$$|(F - \pi(x_n))(f_m)| < 1/p \qquad \text{for} \qquad n \geqslant k(p) \geqslant m.$$

Thus

$$|(F-\pi(x))(f_m)| \leqslant |(F-\pi(x_n))(f_m)| + |f_m(x_n - x)| \leqslant 1/p + |f_m(x_n - x)|$$

for $n \geqslant k(p) \geqslant m$. Since x is a weak cluster point of $\{x_n\}$, there exists, for each $N > m$, a suffix n such that

$$|f_m(x_n - x)| < 1/N \quad \text{and} \quad n \geqslant k(N) \geqslant m.$$

For such an x_n we have (since $k(N) \geqslant m$, we can take $p = N$ above)

$$|(F-\pi(x_n))(f_m)| + |f_m(x_n - x)| < 2/N.$$

Therefore, as N was arbitrary, we see that $(F-\pi(x))(f_m) = 0$ for all $m \in \mathbb{N}$. Since

$$\tfrac{1}{2}\|F-\pi(x)\| \leqslant \sup_m |(F-\pi(x))(f_m)| = 0,$$

we have $F = \pi(x)$, as we wished to show.

THEOREM 1. *Let E be a subset of a B-space $\mathbf{X}$. The following statements are equivalent:*

1) *E is relatively weakly compact;*
2) *E is relatively weakly sequentially compact;*
3) *E is relatively weakly countably compact.*

Proof. 1) $\Rightarrow$ 2). If $\{x_n\}$ is a sequence of elements of E, then denote the closed linear hull of the set $\{x_n\}$ by $\mathbf{Y}$. By Corollary 3 to Theorem III.3.2 we see that $E \cap \mathbf{Y}$ is relatively compact in the weak topology of the separable space $\mathbf{Y}$. By the Corollary to Lemma V.7.1, the weak closure of $E \cap \mathbf{Y}$ is weakly metrizable. Hence, by Theorem I.5.2, there exists a subsequence x_{n_k} with $x_{n_k} \to x$ $(\sigma(\mathbf{Y}, \mathbf{Y}^*))$, and hence also $x_{n_k} \to x$ $(\sigma(\mathbf{X}, \mathbf{X}^*))$.

2) $\Rightarrow$ 3) is obvious.

3) $\Rightarrow$ 1). Denote the $\sigma(\mathbf{X}^{**}, \mathbf{X}^*)$-closure of $\pi(E)$ by E_1 and the $\sigma(\mathbf{X}, \mathbf{X}^*)$-closure of E by E_2. By Lemma 2, E_1 is contained in $\pi(\mathbf{X})$ and is $\sigma(\mathbf{X}^{**}, \mathbf{X}^*)$-compact. Since π is a homeomorphism in the weak and weak* topologies (see 1.1), we have $\pi(E_2) = E_1$, and hence E_2 is weakly compact.

REMARK. If E is relatively weakly countably compact, then for each x in the weak closure of E there exists a sequence $\{x_n\} \subset E$ such that $x_n \to x$ $(\sigma(\mathbf{X}, \mathbf{X}^*))$.

For by Lemma 2 there exists a sequence $\{x_n\} \subset E$ having x as its unique cluster point. By Theorem 1, E_2 is weakly compact, and a sequence with a unique cluster point in a compact space converges to that point (Corollary to Lemma I.2.2). Hence $x_n \to x$ $(\sigma(\mathbf{X}, \mathbf{X}^*))$.

COROLLARY. *Let E be a subset of a B-space $\mathbf{X}$. The following statements are equivalent:*

1) *E is weakly compact;*
2) *E is weakly sequentially compact;*
3) *E is weakly countably compact.*

Proof. 1) $\Rightarrow$ 2), by Theorem 1, while 2) $\Rightarrow$ 3) is obvious.

3) $\Rightarrow$ 1). Since a convergent sequence has a unique cluster point (namely its limit), the Remark following Theorem 1 shows that E is weakly closed. It now follows from Theorem 1 that E is weakly compact.

§ 3. Weak convergence in specific spaces

In this section we explain the significance of weak convergence for sequences of elements in the Banach spaces $\mathbf{L}^p(T, \Sigma, \mu)$ and $\mathbf{C}(K)$.

3.1. We first prove a lemma.

LEMMA 1. *Suppose that $1 < p \leqslant 2$. There exists a positive constant c such that, for all real u,*

$$|1+u|^p \geqslant 1 + pu + c\theta(u), \tag{1}$$

where

$$\theta(u) = \begin{cases} |u|^2, & |u| < 1, \\ |u|^p, & |u| \geqslant 1. \end{cases} \tag{2}$$

Proof. We introduce the functions

$$\chi(u) = |1+u|^p - 1 - pu$$

and

$$\psi(u) = \frac{\chi(u)}{\theta(u)}.$$

Since

$$\lim_{u \to 0} \psi(u) = \lim_{u \to 0} \frac{(1+u)^p - 1 - pu}{u^2} = \frac{p(p-1)}{1 \cdot 2},$$

$$\lim_{u \to \infty} \psi(u) = \lim_{u \to \infty} \frac{(1+u)^p - 1 - pu}{|u|^p} = 1,$$

there exist $\delta > 0$, $\Delta > 0$ and $c > 0$ such that

$$\psi(u) \geqslant c \qquad \text{for} \qquad |u| \leqslant \delta \qquad \text{or} \qquad |u| \geqslant \Delta. \tag{3}$$

Furthermore, we have

$$\chi'(u) = p|1+u|^{p-1}\operatorname{sign}(1+u) - p, \ \chi''(u) = p(p-1)|1+u|^{p-2}$$
$$(u \neq -1).$$

Since $\chi''(u) > 0$ and $\chi'(0) = 0$, the function $\chi(u)$ has a unique minimum at $u = 0$. But $\chi(0) = 0$, so for $\delta \leqslant |u| \leqslant \Delta$ we have $\chi(u) > 0$ and therefore $\psi(u) > 0$ for such u. Replacing the c found above by a smaller number if necessary, we may assume that

$$\psi(u) \geqslant c \qquad (\delta \leqslant |u| \leqslant \Delta).$$

Taking this together with (3), we obtain the required result.

3.2. In certain spaces weak convergence of a sequence $\{x_n\}$ to x and convergence of the sequence of norms: $\|x_n\| \to |x|$ together guarantee that $\{x_n\}$ converges in norm to x.

THEOREM 1. *The following conditions are necessary and sufficient for a sequence $\{x_n\}$ to converge in norm to x in the real space $\mathbf{L}^p(T, \Sigma, \mu)$ $(1 < p < \infty)$:*

1) $x_n \to x$ *weakly;*

2) $\|x_n\| \to \|x\|$.

Proof. The necessity of the two conditions is obvious. Let us prove the sufficiency. Recall that, since $(\mathbf{L}^p)^* = \mathbf{L}^q$ $(1/p + 1/q = 1)$, the weak convergence $x_n \to x$ means that, for every $y \in \mathbf{L}^q$,

$$\int x_n(t)y(t)\,d\mu \to \int x(t)y(t)\,d\mu.$$

Write $A_0 = \{t \in T\colon\ x(t) = 0\}$. Assuming that $1 < p \leqslant 2$, we make use of inequality (1) in Lemma 1. Replacing u in this inequality by $\frac{x_n(t) - x(t)}{x(t)}$, for $t \notin A_0$, we obtain

$$\left|\frac{x_n(t)}{x(t)}\right|^p \geqslant 1 + p\frac{x_n(t)-x(t)}{x(t)} + c\theta\left(\frac{x_n(t)-x(t)}{x(t)}\right).$$

Multiply this inequality through by $|x(t)|^p$ and integrate:

$$\int_T |x_n(t)|^p d\mu \geqslant \int_{A_0} |x_n(t)|^p d\mu + \int_T |x(t)|^p d\mu + p\int_T |x(t)|^{p-1} \operatorname{sign} x(t)(x_n(t)-x(t))d\mu +$$

$$+c\int_{T\setminus A_0} |x(t)|^p \theta\left(\frac{x_n(t)-x(t)}{x(t)}\right) d\mu. \tag{4}$$

Writing $\tilde{y}$ for the element of $\mathbf{L}^q$ defined by

$$\tilde{y}(t) = |x(t)|^{p-1} \operatorname{sign} x(t)$$

and writing $\tilde{f}$ for the corresponding linear functional on $\mathbf{L}^p$, we can rewrite inequality (4) in the form

$$c\int_{T\setminus A_0} |x(t)|^p\theta\left(\frac{x_n(t)-x(t)}{x(t)}\right)d\mu + \int_{A_0} |x_n(t)|^p d\mu \leqslant [\|x_n\|^p - \|x\|^p + p(\tilde{f}(x) - \tilde{f}(x_n))],$$

from which we see, remembering the conditions of the theorem, that the expression on the right-hand side of the last inequality, and consequently also both integrals on the left-hand side, tend to zero.

If $A_n' = \{t \in T\setminus A_0 : |x_n(t)-x(t)| \geqslant |x(t)|\}$, $A_n'' = \{t \in T\setminus A_0 : |x_n(t)-x(t)| < |x(t)|\}$, then, bearing in mind the definition of the function θ (see (2)), we can write

$$\int_{T\setminus A_0} |x(t)|^p\theta\left(\frac{x_n(t)-x(t)}{x(t)}\right)d\mu = \int_{A_n'} |x_n(t)-x(t)|^p d\mu + \int_{A_n''} |x(t)|^{p-2}|x_n(t)-x(t)|^2 d\mu. \tag{5}$$

As we have already remarked, the left-hand side of (5) tends to zero as $n \to \infty$. Therefore both integrals on the right-hand side of (5) tend to zero.

On the other hand, if we keep in mind that the inequality $|x_n(t)-x(t)| < |x(t)|$ holds on A_n'', we see, using Hölder's inequality, that

$$\int_{A_n''} |x_n(t)-x(t)|^p d\mu \leqslant \int_{A_n''} |x(t)|^{p-1}|x_n(t)-x(t)|\, d\mu = \int_{A_n''} [|x(t)|^{p/2-1}|x_n(t)-$$

$$-x(t)|]\,|x(t)|^{p/2} d\mu \leqslant \left[\int_{A_n''} |x(t)|^{p-2}|x_n(t)-x(t)|^2 d\mu\right]^{1/2}\left[\int_{A_n''} |x(t)|^p d\mu\right]^{1/2},$$

and so, since the first factor tends to zero and the second is bounded,

$$\int_{A_n''} |x_n(t)-x(t)|^p d\mu \underset{n\to\infty}{\to} 0.$$

Therefore

$$\int_{T\setminus A_0} |x_n(t)-x(t)|^p d\mu = \int_{A_n'} + \int_{A_n''} |x_n(t)-x(t)|^p d\mu \underset{n\to\infty}{\longrightarrow} 0.$$

Finally we have

$$\int_T |x_n(t)-x(t)|^p d\mu = \int_{T\setminus A_0} |x_n(t)-x(t)|^p d\mu + \int_{A_0} |x_n(t)|^p d\mu \underset{n\to\infty}{\longrightarrow} 0,$$

and so we have proved that $\{x_n\}$ converges to x.

If $p > 2$, then instead of the inequality (1) we must use the inequality

$$|1+u|^p \geqslant 1+pu+c\,|u|^p, \tag{1'}$$

which can be proved by imitating the proof of (1). If, for $t \notin A_0$, we substitute $u = \frac{x_n(t)-x(t)}{x(t)}$ in (1′), multiply by $|x(t)|^p$ and integrate, we obtain

$$\int_T |x_n(t)|^p\,d\mu \geqslant \int_{A_0} |x_n(t)|^p\,d\mu + \int_T |x(t)|^p\,d\mu + p\int_T |x(t)|^{p-1}\,\text{sign}\,x(t)(x_n(t)-x(t))\,d\mu +$$
$$+c\int_{T\setminus A_0} |x_n(t)-x(t)|^p\,d\mu.$$

With $\tilde{x}$ and $\tilde{f}$ as above, and writing $c' = 1/\min(1, c)$, we find that

$$\int_T |x_n(t)-x(t)|^p\,d\mu \leqslant c'[\,\|x_n\|^p - \|x\|^p + p(\tilde{f}(x)-\tilde{f}(x_n))],$$

from which it follows that $x_n \to x$ in norm in $\mathbf{L}^p$.

3.3. We now explain the meaning of weak convergence in $\mathbf{L}^p(T, \Sigma, \mu)$ $(1 \leqslant p < \infty)$.

THEOREM 2. *The following conditions are necessary and sufficient for a sequence $\{x_n\}$ to converge weakly to x in $\mathbf{L}^p(T, \Sigma, \mu)$ $(1 \leqslant p < \infty)$:*

1) $\sup_n \|x_n\| < \infty$;

2) $\int_A x_n(t)d\mu \to \int_A x(t)\,d\mu$ *for every* $A \in \Sigma(\mu)$.

Proof. Since $\{\chi_A : A \in \Sigma(\mu)\}$ is a fundamental set in $\mathbf{L}^p$ (see IV.3.4), Theorem 2 is a special case of Theorem 1.2.

It is clear from the proof that condition 2) does not need to be verified for all $A \in \Sigma(\mu)$, but only for a collection of sets whose characteristic functions form a fundamental set in $\mathbf{L}^p$. In particular, in the case of $\mathbf{L}^p(a, b)$, it is enough to consider the collection of sets $A = [a, s]$, where $a \leqslant s \leqslant b$; and in the case of ℓ^p we can take the collection of one-point sets A, so that in this case condition 2) amounts to coordinatewise convergence.

3.4. We now consider weak convergence in $\mathbf{C}(K)$.

THEOREM 3. *The following conditions are necessary and sufficient for a sequence $\{x_n\}$ to converge weakly to x_0 in $\mathbf{C}(K)$:*

1) $|x_n(t)| \leqslant M$ *for all* $t \in K$ $(n \in \mathbb{N})$;

2) $x_n(t) \to x_0(t)$ *for each* $t \in K$.

Proof. The necessity of the two conditions is almost obvious. For, by Theorem 1.2, $\sup\|x_n\| = M < \infty$, and so $\|x_n(t)\| \leqslant M$.

The necessity of the second condition is verified as follows: for a fixed $t \in K$, consider the functional f_t on $\mathbf{C}(K)$ defined by

$$f_t(x) = x(t).$$

Since x_n converges weakly to x_0, we must have

$$f_t(x_n) \to f_t(x_0),$$

which is just condition 2).

Sufficiency. We use Theorem VI.3.1 on the structure of $\mathbf{C}(K)^*$. By this theorem we need to prove that, for every function $\phi \in \mathbf{rca}(K)$, we have

$$\int_K x_n(t)\,d\phi \to \int_K x_0(t)\,d\phi. \tag{6}$$

Let us now verify that (6) does hold.

We have the bound

$$\left| \int_K x_n(t)\,d\phi - \int_K x_0(t)\,d\phi \right| \leqslant \int_K |x_n(t) - x_0(t)|\,d|\phi|, \tag{7}$$

where $|\phi|$ is the total variation of ϕ, which is a measure on K. Since this measure is finite, the function identically equal to M is measurable with respect to it. By Lebesgue's Theorem, the right-hand side of (7) tends to zero, so we obtain (6).

Theorems III.3.2 and 3 yield the interesting

COROLLARY. *Suppose the sequence $\{x_n(t)\}$ of continuous functions is bounded and converges at each point of a compactum K to a continuous function $x_0(t)$. Then there exist convex combinations $y_n(t) = \sum_{k=1}^{n} \lambda_k^{(n)} x_k(t)$ such that the sequence $\{y_n(t)\}$ converges uniformly to $x_0(t)$.*

3.5. As we showed above (see Theorem 1.5), if the weak topology on a B-space coincides with the norm topology, then it is finite-dimensional. Next we present an example of an infinite-dimensional space in which weak convergence of a sequence implies strong convergence. This shows that an investigation of sequences is not sufficient for studying the weak topology.

THEOREM 4 (Schur). *For sequences, weak convergence in ℓ^1 coincides with convergence in norm.*

Proof. Suppose $x_n \to x_0$ weakly. Replacing x_n by $x_n - x_0$, we may assume that $x_n \to \mathbf{0}$ weakly. Then we need to prove that also $\|x_n\| \to 0$. Assume the contrary. Suppose there exists a subsequence $\{x_{n_m}\}$ such that

$$\lim_{k\to\infty} \|x_{n_m}\| = l > 0. \tag{8}$$

Weak convergence is not destroyed by passing to subsequences; further, by replacing x_{n_m} by $x_{n_m}/\|x_{n_m}\|$ if necessary, we obtain a sequence which, as before, converges weakly to zero, and, moreover, consists of elements having norm 1.

Thus we may assume that the given sequence $\{x_n\}$ satisfies the following conditions:

$$x_n \to \mathbf{0} \text{ weakly in } \ell^1 \tag{9}$$

and

$$\|x_n\| = 1 \qquad (n = 1, 2, \ldots). \tag{10}$$

Suppose $x_n = (\xi_1^{(n)}, \xi_2^{(n)}, \ldots, \xi_k^{(n)}, \ldots)$. Introduce functionals f_k such that

$$f_k(x) = \xi_k \qquad (x = \{\xi_k\}; \quad k = 1, 2, \ldots).$$

In view of (9), we must have $f_k(x_n) \to 0$ as $n \to \infty$; that is,

$$\xi_k^{(n)} \xrightarrow[n\to\infty]{} 0 \qquad (k = 1, 2, \ldots). \tag{11}$$

Now set $n_1 = 1$. We have

$$\sum_{k=1}^{\infty} |\xi_k^{(n_1)}| = \|x_{n_1}\| = 1.$$

Therefore there exists a suffix $p_1 > 0$ such that

$$\sum_{k=1}^{p_1} |\xi_k^{(n_1)}| > 3/4.$$

Assume that we have already chosen integers $1 = n_1 < n_2 < \ldots < n_j$ and $0 = p_0 < p_1 < \ldots < p_j$ such that

$$\sum_{k=1}^{p_{s-1}} |\xi_k^{(n_s)}| < 1/4 \qquad (s = 1, 2, \ldots, j) \tag{12}$$

and

$$\sum_{k=p_{s-1}+1}^{p_s} |\xi_k^{(n_s)}| > 3/4 \quad (s = 1, 2, \ldots, j). \tag{13}$$

Then, by (9), there exists an $n_{j+1} > n_j$ such that

$$\sum_{k=1}^{p_j} |\xi_k^{(n_{j+1})}| < 1/4.$$

Using this inequality and (10), we obtain

$$\sum_{k=p_j+1}^{\infty} |\xi_k^{(n_{j+1})}| = \sum_{k=1}^{\infty} |\xi_k^{(n_{j+1})}| - \sum_{k=1}^{p_j} |\xi_k^{(n_{j+1})}| > 3/4$$

and thus we can find a $p_{j+1} > p_j$ such that

$$\sum_{k=p_j+1}^{p_{j+1}} |\xi_k^{(n_{j+1})}| > 3/4.$$

The argument just given shows that there exist two sequences $1 = n_1 < n_2 < \ldots$ and $0 = p_0 < p_1 < \ldots$ of integers such that (12) and (13) are true for each $s = 1, 2, \ldots$.

Now write

$$\eta_k = \operatorname{sign} \xi_k^{(n_s)} \qquad (p_{s-1} < k \leqslant p_s;\ k, s = 1, 2, \ldots).$$

The sequence $\{\eta_k\}$ belongs to ℓ^∞, so we can consider the linear functional f_0 on ℓ^1 given by

$$f_0(x) = \sum_{k=1}^{\infty} \eta_k \xi_k \qquad (x = \{\xi_k\}).$$

Let us find a lower bound for $f_0(x_{n_s})$. Remembering that $|\eta_k| \leqslant 1$, we have

$$|f_0(x_{n_s})| = \left| \sum_{k=1}^{\infty} \eta_k \xi_k^{(n_s)} \right| \geqslant \left| \sum_{k=p_{s-1}+1}^{p_s} \eta_k \xi_k^{(n_s)} \right| - \sum_{k=1}^{p_{s-1}} |\eta_k \xi_k^{(n_s)}| -$$

$$- \sum_{k=p_s+1}^{\infty} |\eta_k \xi_k^{(n_s)}| \geqslant \sum_{k=p_{s-1}+1}^{p_s} |\xi_k^{(n_s)}| - \sum_{k=1}^{p_{s-1}} |\xi_k^{(n_s)}| -$$

$$- \sum_{k=p_s+1}^{\infty} |\xi_k^{(n_s)}| = 2 \sum_{k=p_{s-1}+1}^{p_s} |\xi_k^{(n_s)}| - \|x_{n_s}\|.$$

Therefore, by (10) and (13), $f_0(x_{n_s}) > \frac{1}{2}$, contradicting (9).

3.6. Now we consider weak convergence in Hilbert space **H**.

Since every linear functional f on H has the form

$$f(x) = (x, y) \qquad (x \in \mathbf{H})$$

(see V.3.2), weak convergence of x_n to x_0 means that, for each $y \in \mathbf{H}$,

$$(x_n, y) \xrightarrow[n \to \infty]{} (x_0, y).$$

Moreover, in Hilbert space it is very easy to prove that, as we have already shown above in the case of $\mathbf{L}^p$, weak convergence and convergence of norms together imply convergence in norm. For if $x_n \to x_0$ weakly and $\|x_n\| \to \|x_0\|$, then

$$\|x_n - x_0\|^2 = (x_n - x_0, x_n - x_0) = (x_n, x_n) + (x_0, x_0) - (x_n, x_0) - \overline{(x_n, x_0)}.$$

But $(x_n, x_0) \to (x_0, x_0)$. Therefore $\|x_n - x_0\| \to 0$.

Concerning the contents of this section as a whole, see Banach. Weak convergence in $\mathbf{L}^2$ was already considered by Hilbert. Theorem 3 on weak convergence in $\mathbf{C}[a, b]$ is due to F. Riesz [2].

§4. The problem of translocation of mass and the normed space it generates

4.1. In a paper [5] by Kantorovich an important class of finite-dimensional extremal problems were studied in connection with the analysis of certain questions of production organization and planning. The methods of classical mathematical analysis proved to be of little effectiveness in application to these problems. These investigations in the end led to the birth of a new mathematical discipline, which was given the name linear programming.

The problem of interest to us—the problem of translocation of mass—is directly connected with the following very simple problem in linear programming, which is widely used at the present time in the practical planning of rail, road, air and sea transportation.

The transportation problem. The components of a given vector

$$\phi = (\phi_1, \ldots, \phi_m) \tag{1}$$

(or, more precisely, their absolute values) represent volumes of production (when $\phi_k \leqslant 0$) or consumption (when $\phi_k > 0$) of some uniform product at m points, labelled by an index k in $K = \{1, 2, \ldots, m\}$. It is further assumed that the total volume of consumption coincides with the total volume of production, that is, that

$$\sum_{k \in K} \phi_k = 0. \tag{2}$$

A transportation plan is determined by choosing a matrix

$$\psi = [\psi_{ij}]_{i,j \in K}, \qquad \psi_{ij} \geqslant 0, \qquad i \in K, \qquad j \in K, \tag{3}$$

whose elements indicate the planned volumes of transportation from each point i to each point j. In carrying out this transportation it is clear that, at each point $k \in K$, $\sum_{i \in K} \psi_{ik}$ units of the product under consideration will be imported and $\sum_{j \in K} \psi_{kj}$ units exported. This means that the matrix (3) determines an admissible transportation plan if the following balancing conditions are satisfied:

$$\sum_{i \in K} \psi_{ik} - \sum_{j \in K} \psi_{kj} = \phi_k, \quad k \in K. \tag{4}$$

The total expenditure involved in putting each transportation plan into effect is determined, according to the model we are considering, by the formula

$$\tau(\psi) = \sum_{i\in K}\sum_{j\in K} r_{ij}\psi_{ij}, \tag{5}$$

where the r_{ij} are non-negative numbers representing the outlay for transferring one unit of the product from point i to point j.

Hence the collection of admissible transportation plans is determined by the set Ψ_ϕ of non-negative solutions (3) of the system of linear equations (4). And the most economical plan $\psi \in \Psi_\phi$ we are seeking—the so-called optimal transportation plan—is characterized by the fact that the total expenditure (5) attains a minimum.

With the aid of some general results from the theory of linear programming (or by direct arguments) it is easily verified that the extremal problem posed here is always soluble. Moreover, an admissible transportation plan (3) is optimal if and only if there exists a vector $u = (u_1, u_2, \ldots, u_m)$ such that

$$u_j - u_i \leqslant r_{ij}, \quad i \in K, \quad j \in K, \tag{6}$$

where $\psi_{ij}(u_j - u_i - r_{ij}) = 0, i \in K, j \in K$. The latter means that if a non-zero quantity ψ_{ij} is to be transported from point i to point j according to the given admissible plan, then the corresponding condition in (6) must be satisfied with equality. It is important to note that this test for optimality enables one to construct an effective algorithm,* by means of which one can, on a modern computer, solve transportation problems with several thousand production and consumption points.

To conclude this introductory subsection, we make some remarks concerning the transportation problem we have just been considering, and prove a proposition that we shall need below for studying the problem of translocation of mass.

The components of the given vector (1) may clearly be expressed in the form $\phi_k = \phi_k^+ - \phi_k^-$, $k \in K$, where

$$\phi_k^+ = \max\{0, \phi_k\}, \qquad \phi_k^- = -\min\{0, \phi_k\}, \qquad \phi_k^+ + \phi_k^- = |\phi_k|.$$

Then, by (2), the quantities $\phi^+(K) = \sum_{k\in K} \phi_k^+$, $\phi^-(K) = \sum_{k\in K} \phi_k^-$ coincide and are equal to half the sum of the absolute values of the components of the vector (1).

Now we consider the function

$$v(\psi) = \sum_{i\in K}\sum_{j\in K} \psi_{ij} \tag{7}$$

and show that, for each matrix $\psi \in \Psi_\phi$, we have the inequality

$$v(\psi) \geqslant \phi^+(K), \tag{8}$$

which becomes an equality if and only if

$$\sum_{i\in K} \psi_{ik} = \phi_k^+, \quad \sum_{j\in K} \psi_{kj} = \phi_k^-, \qquad k \in K. \tag{9}$$

For equation (4) and the non-negativity of the summands imply that

$$\sum_{i\in K} \psi_{ik} \geqslant \phi_k^+, \qquad \sum_{j\in K} \psi_{kj} \geqslant \phi_k^-, \qquad k \in K. \tag{10}$$

Summing the first or second of these inequalities over all $k \in K$, we obtain the required inequality (8). Further, it is clear that we have equality if and only if the equations (9) hold.

Notice also that the lower bounds for the function (7) on Ψ_ϕ given by (8) are attained: that is, the set of matrices (3) satisfying the conditions (9) is non-empty. For example, these conditions are satisfied by the matrix

$$\psi^* = [\psi_{ij}^*]_{i,j\in K}, \qquad \psi_{ij}^* = \frac{\phi_i^- \cdot \phi_j^+}{\phi^+(K)}, \qquad i \in K, \quad j \in K. \tag{11}$$

LEMMA 1. *If the given non-negative numbers r_{ij} in the transportation problem under consideration satisfy the triangle inequality*

$$r_{ij} \leqslant r_{ik} + r_{kj}, \quad i \in K, \quad j \in K, \quad k \in K, \tag{12}$$

then there exists an optimal transportation plan (3) *for which condition* (8) *is satisfied with equality.*

---

* The first algorithm of this kind was put forward in a paper by Kantorovich and Gavurin [1] (this paper was written in 1940 and is cited in Kantorovich [6]).

Proof. Consider the non-empty closed bounded sets

$$\Psi_\phi^\nu = \{\psi \in \Psi_\phi | v(\psi)| \leqslant \nu\}, \qquad \nu \in [\phi^+(K), +\infty).$$

As the functions (5) and (7) are continuous, there exists for each $\nu \geqslant \phi^+(K)$ a matrix $\psi_\nu \in \Psi_\phi^\nu$ such that

$$\tau(\psi_\nu) = \min\{\tau(\psi): \psi \in \Psi_\phi^\nu\},$$

$$v(\psi_\nu) = \min\{v(\psi): \psi \in \Psi_\phi^\nu, \ \tau(\psi) = \tau(\psi_\nu)\}.$$

We need to show that when (12) is satisfied for all these matrices $\psi_\nu \in \Psi_\phi^\nu$, we have the equation

$$v(\psi_\nu) = \phi^+(K). \tag{13}$$

Assume that we have strict inequality in (8) for some matrix $\psi = \psi_\nu$. Then for some $k_0 \in K$ we must have strict inequalities in (10). But then there exist $i_0 \in K$ and $j_0 \in K$ such that $\varepsilon = \min\{\psi_{i_0k_0}, \psi_{k_0j_0}\} > 0$.

Consider the matrix $\psi' \in \Psi_\phi$ obtained from the original matrix $\psi = \psi_\nu$ by the following substitution for three elements

$$\psi'_{i_0j_0} = \psi_{i_0j_0} + \varepsilon, \qquad \psi'_{i_0k_0} = \psi_{i_0k_0} - \varepsilon, \qquad \psi'_{k_0j_0} = \psi_{k_0j_0} - \varepsilon.$$

For this matrix we have

$$\tau(\psi') = \tau(\psi) + \varepsilon(r_{i_0j_0} - r_{i_0k_0} - r_{k_0j_0}) \leqslant \tau(\psi),$$

$$v(\psi') = v(\psi) - \varepsilon,$$

and this contradicts the choice of $\psi_\nu \in \Psi_\phi^\nu$. This contradiction shows that the required equation (13) holds for all $\nu \in [\phi^+(K), +\infty)$, and the lemma is proved.

REMARK. It follows from this lemma that, if the original numbers r_{ij} satisfy (12), then condition (4) can be replaced by the more stringent requirement (9).

4.2. The infinite-dimensional analogue of the transportation problem that we present below was first considered in a paper by Kantorovich [6] (see also Kantorovich [8], Kantorovich and Rubinshtein [1], [2]).

The problem of translocation of mass. The finite set of points is here replaced by an arbitrary metric compactum K with metric $r(t, s)$, representing the outlay involved in transferring one unit of mass from any point $t \in K$ to any point $s \in K$. The analogue of the vector (1) satisfying (2) is a countably-additive function ϕ defined on the system $\mathscr{B}$ of all Borel sets of K, whose positive variation $\phi_+(K)$ coincides with its negative variation $\phi_-(K)$—that is, which satisfies

$$\phi(K) = \phi_+(K) - \phi_-(K) = 0. \tag{14}$$

We recall (see I.6.3) that, for $e \in \mathscr{B}$,

$$\phi_+(e) = \sup\{\phi(e'): \ e' \in \mathscr{B}, e' \subset e\},$$

$$\phi_-(e) = \sup\{-\phi(e'): \ e' \in \mathscr{B}, e' \subset e\}.$$

For each $e \in \mathscr{B}$, the values of $\phi_+(e)$ and $\phi_-(e)$ are interpreted as the required amount and the existing amount of mass at e, respectively. Thus (14) has the same meaning as (2) for the case of a finite number of points.

A plan for translocation of mass on K is specified by choosing a finite measure ψ defined on the σ-algebra $\tilde{\mathscr{B}}$ of Borel sets on the compactum $\tilde{K} = K \times K$. Here the measure of a set $e \times e'$, where $e, e' \in \mathscr{B}$, which we now denote by $\phi(e, e')$, indicates the quantity of mass to be translocated from e to e'. Such a plan is admissible if it satisfies the balancing conditions

$$\psi(K, e) - \psi(e, K) = \phi(e), \qquad e \in \mathscr{B}, \tag{15}$$

analogous to the conditions (4) of the transportation problem.

As above, the collection of admissible translocations ψ will be denoted by Ψ_ϕ. Moreover, the total outlay involved in putting such a translocation into effect is now naturally calculated by means of the double integral

$$\tau(\psi) = \int_{\tilde{K}} r(t, s)\, d\psi(t, s). \tag{16}$$

Therefore the optimal translocation that we seek is characterized by a function $\psi \in \Psi_\phi$ for which the corresponding value of (16) attains a minimum.

It will be shown below that the extremal problem we have posed is always soluble. Moreover the test for an optimal translocation is here, in essence, little different from that presented for the transportation problem in the last subsection. It turns out that an admissible translocation $\Psi = \Psi_\phi$ is optimal if and only if there exists a function $u: K \to \mathbb{R}$ such that

$$u(s) - u(t) \leqslant r(t, s), \qquad t \in K, \quad s \in K, \tag{17}$$

where $u(s)-u(t) = r(t,s)$ if (t,s) belongs to the support* supp ψ of ψ, that is, if we have the strict inequality $\psi(e_t, e_s) > 0$ for all neighbourhoods e_t, e_s of t and s.

Let us now devote some space to a problem with almost 200 years' history (see Kantorovich [8]).

In 1781 the prominent French mathematician G. Monge posed the following problem in connection with a study of the most efficient transfer of earth from an embankment to a cutting: divide two equal volumes into infinitesimal particles and put these into correspondence with one another such that the sum of the products of the lengths of the paths joining them by the volumes of the particles being transferred is a minimum.

The geometric theory of congruences was developed by Monge in connection with his study of this problem. In connection with this same problem he stated his hypothesis that the required paths of translocation of mass were the normals to a certain one-parameter family of surfaces.

Other well-known mathematicians and specialists in mechanics afterwards took up Monge's problem. However, a rigorous proof of the truth of the hypothesis was given only after a hundred years, in a 200-page memoir by P. Appel in 1884. Although the author afterwards succeeded in simplifying his proof somewhat, it nevertheless remained very lengthy and complicated, and was based on deep theorems that had already been developed at that time in the calculus of variations.

One can obtain a proof of Monge's hypothesis which is also applicable to an essentially wider class of problems of translocation of mass as a simple consequence of the test for optimality of transfer presented above. As a preliminary, let us prove an auxiliary statement.

LEMMA 2. *Let $u: K \to \mathbb{R}$ be a function satisfying condition (17) and let t_0, s_0 and z_0 be points of K such that*

$$u(s_0)-u(t_0) = r(t_0, s_0) = r(t_0, z_0)+r(z_0, s_0). \tag{18}$$

Then the set

$$U_{z_0} = \{z \in K \mid u(z) = u(z_0)\} \tag{19}$$

is situated between two balls passing through z_0 and having centres at t_0 and s_0. In other words, for each point $z \in U_{z_0}$, we have the inequalities

$$r(t_0, z) \geqslant r(t_0, z_0), \qquad r(z, s_0) \geqslant r(z_0, s_0).$$

Proof. By (17), we have

$$u(z_0)-u(t_0) \leqslant r(t_0, z_0), \quad u(s_0)-u(z_0) \leqslant r(z_0, s_0).$$

From this and (18) we obtain

$$u(z_0)-u(t_0) = r(t_0, z_0), \quad u(s_0)-u(z_0) = r(z_0, s_0).$$

But then, by condition (17), we have, for any $z \in U_{z_0}$,

$$r(t_0, z) \geqslant u(z)-u(t_0) = u(z_0)-u(t_0) = r(t_0, z_0),$$
$$r(z, s_0) \geqslant u(s_0)-u(z) = u(s_0)-u(z_0) = r(z_0, s_0),$$

as we required to prove.

If we now bear in mind the test for optimality mentioned above, we can assert that the Monge–Appel Theorem is true for any problem of translocation of mass in a convex compactum in an arbitrary Euclidean or Hilbert space. Moreover, if the admissible translocation defined by a measure $\psi \in \Psi_\phi$ is optimal, then we can take the required family of surfaces to be the one-parameter family $u(z) = \text{const}$, corresponding to the function $u: K \to \mathbb{R}$ in the test for optimality. For if there is translocation from the point t_0 to the point s_0—that is, if $(t_0, s_0) \in \operatorname{supp} \psi$—then by Lemma 2 the open interval (t_0, s_0) coincides, for any $z_0 \in (t_0, s_0)$, with the normal to the corresponding surface (19).

Another peculiarity of the problem of translocation of mass considered by G. Monge is that the original countably additive function $\phi: \mathscr{B} \to \mathbb{R}$ satisfying (14) is taken to be

$$\phi(e) = \mu(e \cap N)-\mu(e \cap M), \qquad e \in \mathscr{B},$$

where μ is a fixed measure (Lebesgue measure) and M and N are given measurable sets such that $\mu(M) = \mu(N)$. However, this peculiarity, as we have seen, is not used in the proof we have given of the Monge–Appel Theorem.

4.3. For the proof of our assertions about the problem of translocation of mass, we first study, following Kantorovich and Rubinshtein [1], [2], an associated normed linear space which has some independent interest. In particular, we shall obtain at the same time an important property of weak* convergence of linear functionals on a space of continuous functions on a metric compactum.

Consider an arbitrary metric compactum K with metric $r(t,s)$ and the system $\mathscr{B}$ of Borel subsets of K. In the B-space **rca**(K) of all countably additive functions† $\phi: \mathscr{B} \to \mathbb{R}$ we single out a narrower set $\Phi_0(\mathscr{B})$, consisting of the

* For the definition of the support of a Borel measure, see Halmos.

† We note that any finite countably-additive set function defined on the σ-algebra $\mathscr{B}$ of Borel sets of a compactum K is regular (see Halmos).

functions that satisfy (14). With each finite Borel measure ψ on $\tilde{K}(=K \times K)$ we associate a function

$$\phi(e) = \psi(K, e) - \psi(e, K), \qquad e \in \mathscr{B}, \tag{20}$$

which obviously satisfies (14) and therefore belongs to $\Phi_0(\mathscr{B})$. On the other hand, if for any $\phi \in \Phi_0(\mathscr{B})$ we define the function

$$\psi^*(e, e') = \frac{\phi_-(e) \cdot \phi_+(e')}{\phi_+(K)}, \qquad e \in \mathscr{B}, \quad e' \in \mathscr{B}, \tag{21}$$

we have:

$$\psi^*(K, e) - \psi^*(e, K) = \phi_+(e) - \phi_-(e) = \phi(e), \qquad e \in \mathscr{B},$$

so that the set Ψ_ϕ of measures $\psi \in \Psi(\tilde{\mathscr{B}})$ satisfying (20) is non-empty.

It is not hard to check that these sets Ψ_ϕ satisfy the following relations:

$$\Psi_{\phi_1} + \Psi_{\phi_2} = \{\psi_1 + \psi_2 \,|\, \psi_1 \in \Psi_{\phi_1}, \psi_2 \in \Psi_{\phi_2}\} \subset \Psi_{\phi_1 + \phi_2}, \tag{22}$$

$$0 \in \Psi_0, \qquad \lambda \Psi_\phi = \{\lambda \psi \in \Psi_\phi\} = \Psi_{\lambda\phi}, \qquad \lambda > 0. \tag{23}$$

Also, to obtain the set $\Psi_{-\phi}$ one need only replace each ψ by the "transposed" measure ψ^T, where $\psi^T(E)$ $E^T = \{(t, s) : (s, t) \in E\}$, for any Borel set $E \subset \tilde{K}$. In particular,

$$\psi^{\mathrm{T}}(e, e') = \psi(e', e), \qquad e \in \mathscr{B}, \quad e' \in \mathscr{B}.$$

Hence

$$\Psi_{-\phi} = \{\psi^{\mathrm{T}} \,|\, \psi \in \Psi_\phi\} = (\Psi_\phi)^{\mathrm{T}}. \tag{24}$$

Now we can easily show that the non-negative function defined on $\Phi_0(\mathscr{B})$ by

$$\|\phi\|_\tau = \inf_{\psi \in \Psi_\phi} \tau(\psi) = \inf_{\psi \in \Psi_\phi} \int_{\tilde{K}} r(t, s)\, d\psi(t, s) \tag{25}$$

satisfies the axioms for a seminorm:

$$\|\lambda\phi\|_\tau = |\lambda| \cdot \|\phi\|_\tau, \tag{26}$$

$$\|\phi_1 + \phi_2\|_\tau \leqslant \|\phi_1\|_\tau + \|\phi_2\|_\tau. \tag{27}$$

In fact, (27) is a consequence of (22) and the additivity of the function (16). Also, if $\lambda \geqslant 0$, then (26) follows from (23) and the homogeneity of (16), while if $\lambda < 0$ then it follows from (24) provided we observe that

$$\tau(\psi^{\mathrm{T}}) = \int_{\tilde{K}} r(t, s)\, d\psi(s, t) = \int_{\tilde{K}} r(s, t)\, d\psi(s, t) = \tau(\psi).$$

It will be shown below (see the Remark 1 following Theorem 1) that (25) also satisfies the condition

$$\|\phi\|_\tau = 0 \Rightarrow \phi = 0, \tag{28}$$

that is, it is a norm on $\Phi_0(\mathscr{B})$. However, it will be convenient to prove this fact after establishing certain properties of the associated space $\Phi_0^\tau(\mathscr{B})$, which for the present we consider as a topological space, having the topology induced by the semi-metric* $\rho_\tau(\phi_1, \phi_2) = \|\phi_1 - \phi_2\|\tau$.† Although it will be proved below that we obtain in this way the topology of a normed space, we cannot strictly assume that this is so at present. Hence we must reach an agreement concerning terminology for the present case.

We shall say that a sequence $\{\phi_n\}$ converges strongly to ϕ (and write $\phi_n \xrightarrow{\tau} \phi$) if $\|\phi_n - \phi\|_\tau \to 0$. We write $(\Phi_0^\tau(\mathscr{B}))^*$ for the set of all linear functionals on $\Phi_0(\mathscr{B})$ that are continuous with respect to strong convergence. We shall say that a sequence $\{\phi_n\}$ converges weakly to ϕ (and write $\phi_n \dashrightarrow \phi$) if $L(\phi_n) \to L(\phi)$ for each $L \in (\Phi_0^\tau(\mathscr{B}))^*$.

4.4. In this subsection some auxiliary facts will be proved.

We recall that the support supp ψ of a measure ψ consists of the $(t, s) \in K \times K$ such that $\psi(e_t, e_s) > 0$ for all neighbourhoods e_t and e_s of the relevant points. Also the support supp ϕ of a function $\phi \in \mathbf{rca}(K)$ consists of those $t \in K$ such that $\max\{\phi_+(e_t), \phi_-(e_t)\} > 0$ for every neighbourhood e_t of t. This implies that

$$\operatorname{supp} \phi = (\operatorname{supp} \phi_+) \cup (\operatorname{supp} \phi_-)$$

and this is the smallest closed set F in K for which $\phi(e) = \phi(e \cap F)$, $e \in \mathscr{B}$.

---

* A semi-metric satisfies all the axioms for a metric except the one stating that $\rho(x, y) = 0$ implies $x = y$. The topology induced by a semi-metric is defined exactly like that induced by a metric.

† (*Editor's note*) This metric is often called the Kantorovich–Rubinshtein metric.

LEMMA 3. *For every function* $\phi \in \Phi_0^\xi(\mathscr{B})$, *we have the following bounds*:

$$\|\phi\|_\tau \leqslant \phi_+(K)\cdot\max\{r(t,s): t\in\operatorname{supp}\phi_-, s\in\operatorname{supp}\phi_+\},$$
$$\|\phi\|_\tau \leqslant \phi_+(K)\operatorname{diam}(\operatorname{supp}\phi),$$

where diam (supp ϕ) *denotes the diameter of* supp ϕ *in* K.

Proof. It is sufficient to note that the measures $\psi^*\in\Psi_\phi$ defined by (21) satisfy:

$$\psi^*(K,K) = \phi_-(K) = \phi_+(K),$$
$$\operatorname{supp}\psi^* = (\operatorname{supp}\phi_-)\times(\operatorname{supp}\phi_+).$$

Therefore

$$\|\phi\|_\tau = \inf_{\psi\in\Psi_\phi}\tau(\psi) \leqslant \tau(\psi^*) = \int_K\int_K r(t,s)\,d\psi^*(t,s) \leqslant \phi_+(K)\cdot\max\{r(t,s): t\in\operatorname{supp}\phi_-,$$
$$s\in\operatorname{supp}\phi_+\} \leqslant \phi_+(K)\operatorname{diam}(\operatorname{supp}\phi),$$

as we sought to prove.

Using these bounds, we now establish some important properties of the sets

$$S_0^\nu = \{\phi\in\Phi_0(\mathscr{B}): \phi_+(K)+\phi_-(K)\leqslant\nu\}, \tag{29}$$

whose union, over all $\nu\geqslant 0$, clearly coincides with $\Phi_0(\mathscr{B})$. To this end, we single out the subset $\tilde{\Phi}(\mathscr{B})$ of **rca** (K) consisting of all functions with finite support—in other words, the linear hull of the set of simple functions

$$\phi_t(e) = \begin{cases}1 & \text{when } t\in e,\\ 0 & \text{when } t\notin e,\end{cases}$$

which have one-point sets for their supports.

Now consider the sets

$$\tilde{\Phi}_0(\mathscr{B}) = \Phi_0(\mathscr{B})\cap\tilde{\Phi}(\mathscr{B}), \qquad \tilde{S}_0^\nu = S_0^\nu\cap\tilde{\Phi}(\mathscr{B}).$$

The simplest functions in $\tilde{\Phi}_0(\mathscr{B})$ are the functions

$$\phi_{ts} = \phi_s - \phi_t, \tag{30}$$

for which we have, if $t\neq s$, supp $\phi_{ts} = \{t,s\}$, $(\phi_{ts})_+ = \phi_s$, $(\phi_{ts})_- = \phi_t$ so that, by Lemma 3,

$$\|\phi_{ts}\|_\tau \leqslant (\phi_{ts})_+(K)\operatorname{diam}(\operatorname{supp}\phi_{ts}) = r(t,s). \tag{31}$$

Furthermore the linear hull of these functions (30) is obviously the whole of $\tilde{\Phi}_0(\mathscr{B})$.

LEMMA 4. *For any given* $\nu>0$ *and* $\varepsilon>0$, *the set* $\tilde{S}_0^\nu$ *has a finite* ε*-net relative to the seminorm** (25) *in* S_0^ν.

Proof. We express the given compactum K in the form $K = \bigcup_{i=1}^m e_i$, where the e_i are non-empty Borel sets such that $e_i\cap e_j = \varnothing$ when $i\neq j$ and $\max(\operatorname{diam} e_i) < \varepsilon/(2\nu)$. Fix a point t_i in each e_i. Also choose a further point $t_0\in K$ and a natural number

$$q > \frac{2m}{\varepsilon}\operatorname{diam} K.$$

Now we consider the set P of integer vectors

$$p = (p_1,\ldots,p_m) \tag{32}$$

satisfying the condition $\sum_{i=1}^m |p_i| + |\sum_{i=1}^m p_i| \leqslant q\nu$, and show that the corresponding functions

$$\tilde{\phi} = \sum_{i=1}^m \frac{pi}{q}\phi_{t_0t_1}, \qquad p\in P, \tag{33}$$

which clearly belong to $\tilde{S}_0^\nu$, constitute the required ε-net.

First we associate with each $\phi\in S_0^\nu$ the function $\tilde{\phi}\in\tilde{S}_0^\nu$ given by

$$\tilde{\phi} = \sum_{i=1}^m \phi(e_i)\phi_{t_i} = \sum_{i=1}^m \phi(e_i)\phi_{t_0t_i} \tag{34}$$

* An ε-net relative to a semi-metric is defined exactly like an ε-net relative to a metric.

and show that

$$\|\phi-\tilde{\phi}\|_\tau \leqslant \varepsilon/2. \tag{35}$$

To do this we consider the functions in $\Phi_0(\mathscr{B})$ defined by

$$\phi_i(e) = \phi(e \cap e_i) - \phi(e_i)\phi_{t_i}(e), \quad e \in \mathscr{B}, \quad i = 1, \ldots, m.$$

The supports of these functions are contained in the closures of the corresponding e_i, so that

$$\operatorname{diam}(\operatorname{supp} \phi_i) \leqslant \operatorname{diam} e_i < \frac{\varepsilon}{2v}.$$

Furthermore, $(\phi_i)_+(K) = (\phi_i)_+(e_i) \leqslant \phi_+(e_i) + \phi_-(e_i)$, and hence, by the lemma, we have

$$\|\phi_i\|_\tau \leqslant \frac{\varepsilon}{2v}[\phi_+(e_i)+\phi_-(e_i)].$$

But then

$$\|\phi-\tilde{\phi}\|_\tau = \left\| \sum_{i=1}^m \phi_i \right\|_\tau \leqslant \sum_{i=1}^m \|\phi_i\|_\tau \leqslant \frac{\varepsilon}{2v} \sum_{i=1}^m [\phi_+(e_i)+\phi_-(e_i)] = \frac{\varepsilon}{2v}[\phi_+(K)+\phi_-(K)] \leqslant \frac{\varepsilon}{2},$$

so the bounds given by (35) are established.

It now remains to associate with the function (34) a vector $p \in P$ such that the corresponding function (33) satisfies

$$\|\tilde{\phi}-\tilde{\tilde{\phi}}\|_\tau < \frac{\varepsilon}{2}. \tag{36}$$

Write $p_i = s_i c_i$, $i = 1, 2, \ldots, m$, where $s_i = \operatorname{sign} \phi(e_i)$ and the c_i are the integral parts of the non-negative numbers $q|\phi(e_i)|$. Then

$$\sum_{i=1}^m |p_i| + \left| \sum_{i=1}^m p_i \right| = \sum_{i=1}^m c_i + \left| \sum_{i=1}^m [q\phi(e_i) - p_i] \right| \leqslant$$

$$\leqslant \sum_{i=1}^m c_i + \sum_{i=1}^m |q\phi(e_i) - p_i| = \sum_{i=1}^m c_i + \sum_{i=1}^m [q|\phi(e_i)| - c_i] =$$

$$= q \sum_{i=1}^m |\phi(e_i)| \leqslant q \sum_{i=1}^m [\phi_+(e_i)+\phi_-(e_i)] \leqslant qv,$$

that is, the associated vector (32) belongs to P. Let us show that the function (33) corresponding to this vector satisfies the required condition (36). In fact, we have

$$\|\tilde{\phi}-\tilde{\tilde{\phi}}\|_\tau = \left\| \sum_{i=1}^m \left[\phi(e_i) - \frac{p_i}{q}\right] \phi_{t_0 t_i} \right\|_\tau \leqslant \sum_{i=1}^m \left| \phi(e_i) - \frac{p_i}{q} \right| \|\phi_{t_0 t_i}\|_\tau \leqslant$$

$$\leqslant \frac{1}{q} \operatorname{diam} K \sum_{i=1}^m [q|\phi(e_i)| - c_i] \leqslant \frac{m}{q} \operatorname{diam} K < \frac{\varepsilon}{2}.$$

This completes the proof of the lemma.

COROLLARY 1. *The linear manifold* $\tilde{\Phi}_0(\mathscr{B})$ *is everywhere dense in* $\Phi_0^\tau(\mathscr{B})$.

COROLLARY 2. *Two continuous linear functionals on* $\Phi_0^\tau(\mathscr{B})$ *coincide if they take equal values on all the functions* (30).

LEMMA 5. *Strong and weak convergence in* $\Phi_0^\tau(\mathscr{B})$ *are the same for functions belonging to* S_0^v. *More precisely, if* $\{\phi_n\}_{n=1}^\infty \subset S_0^v$, *then* $\phi_n \xrightarrow{\tau} \phi_0$ *is equivalent to* $\phi_n \dashrightarrow \phi_0$.

Proof. Strong convergence obviously implies weak convergence, without any additional assumptions. Assume now that $\phi_n \dashrightarrow \phi_0$, where $\phi_n \in S_0^v$. If the functions ϕ_n here did not converge strongly to ϕ_0, then for some $\varepsilon > 0$ there would exist a subsequence ϕ_{n_k} such that

$$\|\phi_{n_k} - \phi_0\|_\tau \geqslant 2\varepsilon, \qquad k = 1, 2, \ldots.$$

If we take Lemma 4 into account, we may assume without any loss of generality that this subsequence is a Cauchy sequence. Hence there exists a k_0 such that

$$\|\phi_{n_k} - \phi_{n_{k_0}}\|_\tau \leqslant \varepsilon$$

for all $k \geqslant k_0$.* By the Hahn–Banach Theorem there exists a functional $L \in (\Phi_0^\xi(\mathscr{B}))^*$ such that

$$\|L\| = 1, \qquad L(\phi_{nk_0} - \phi_0) = \|\phi_{nk_0} - \phi_0\|_\tau.$$

Then, for all $k \geqslant k_0$, we have:

$$L(\phi_{n_k} - \phi_0) = L(\phi_{nk_0} - \phi_0) + L(\phi_{nk} - \phi_{nk_0}) \geqslant 2\varepsilon - \varepsilon = \varepsilon,$$

contradicting the assumption that ϕ_n converges weakly to ϕ_0. This contradiction completes the proof of the lemma.

LEMMA 6. *For all $\phi \in \Phi_0(\mathscr{B})$ and $\psi \in \Psi_\phi$ we have*

$$\psi(K, K) \geqslant \phi_+(K) = \phi_-(K),$$

with equality if and only if

$$\psi(K, e) = \phi_+(e), \qquad \psi(e, K) = \phi_-(e), \qquad e \in \mathscr{B}.$$

Proof. The analogous statement for the transportation problem was proved in 4.1. We obtain the present result by repeating the arguments given before, replacing the individual points by the sets $e \in \mathscr{B}$ and the finite sums by integrals.

4.5. Let us now establish a general form for linear functionals on $\Phi_0^\xi(\mathscr{B})$. With this aim in mind, we consider the space $\mathbf{Lip}^1(K)$ consisting of functions $u: K \to \mathbb{R}$ satisfying a Lipschitz condition—that is, such that

$$\|u\|_{\mathbf{Lip}} = \sup_{t \neq s} \frac{u(s) - u(t)}{r(t, s)} < \infty.$$

This is obviously a semi-normed space; moreover,

$$\|u\|_{\mathbf{Lip}} = 0 \Leftrightarrow u(t) = \text{const} \qquad x \in K. \tag{37}$$

To obtain a normed space we must factor out the subspace of constant functions or else restrict our attention to a particular linear subspace, for example

$$\mathbf{Lip}^1(K, t_0) = \{u \in \mathbf{Lip}^1(K) \mid u(t_0) = 0\}.$$

The following theorem shows that the normed space $\mathbf{Lip}^1(K, t_0)$ is linearly isometric to the space $(\Phi_0^\xi(\mathscr{B}))^*$ of continuous linear functionals on $\Phi_0^\xi(\mathscr{B})$.

THEOREM 1. *For each function $u \in \mathbf{Lip}^1(K)$, the corresponding additive and homogeneous functional*

$$L_u(\phi) = \int_K u(t)\, d\phi(t), \qquad \phi \in \Phi_0^\xi(\mathscr{B}), \tag{38}$$

is continuous, and

$$\|L_u\| = \|u\|_{\mathbf{Lip}}.\dagger \tag{39}$$

Conversely, for any given continuous linear functional L in $\Phi_0^\xi(\mathscr{B})$, there exists a function $u \in \mathbf{Lip}^1(K)$ such that $L = L_u$. Moreover this function u is determined to within a constant factor.

Proof. For all $u \in \mathbf{Lip}^1(K)$, $\phi \in \Phi_0^\xi(\mathscr{B})$, $\psi \in \Psi_\phi$, we have

$$L_u(\phi) = \int_K u(t)\, d\phi = \int_K u(t)\, d\psi(K, t) - \int_K u(t)\, d\psi(t, K) =$$

$$= \int_{\tilde{K}} u(s)\, d\psi(t, s) - \int_{\tilde{K}} u(t)\, d\psi(t, s) = \int_{\tilde{K}} (u(s) - u(t))\, d\psi(t, s) \leqslant$$

$$\leqslant \|u\|_{\mathbf{Lip}} \int_{\tilde{K}} r(t, s)\, d\psi(t, s) = \|u\|_{\mathbf{Lip}}\, \tau(\psi).$$

Therefore

$$L_u(\phi) \leqslant \|u\|_{\mathbf{Lip}} \inf_{\psi \in \Psi_\phi} \tau(\psi) = \|u\|_{\mathbf{Lip}} \|\phi\|_\tau. \tag{40}$$

Hence the functional (38) is linear and we have

$$\|L_u\| \leqslant \|u\|_{\mathbf{Lip}}. \tag{41}$$

---

* This can be proved exactly as for metric spaces.

† $\|L_u\| = \sup\{|L_u(\phi)| : \|\phi\|_\tau \leqslant 1\}$.

To obtain the reverse inequality we need only consider the simplest functions (30), for which we have

$$L_u(\phi_{ts}) = \int_K u(z)\,d\phi_{ts}(z) = u(s) - u(t).$$

Remembering (31), we deduce that

$$\|u\|_{\mathbf{Lip}} = \sup_{t \neq s} \frac{u(s)-u(t)}{r(t,s)} \leqslant \sup_{t \neq s} \frac{L_u(\phi_{ts})}{\|\phi_{ts}\|_\tau} \leqslant \|L_u\|. \tag{42}$$

The required equation (39) now follows from (41) and (42).

To prove the second part of the theorem we fix a point $t_0 \in K$ and associate with each functional $L \in (\Phi_0^\tau(\mathscr{B}))^*$ the function

$$u(t) = L(\phi_{t_0 t}), \qquad t \in K.$$

Now for all t and s in K we have

$$L(\phi_{ts}) = L(\phi_{t_0 s} - \phi_{t_0 t}) = u(s) - u(t),$$

$$u(s) - u(t) = L(\phi_{ts}) \leqslant \|L\|\,\|\phi_{ts}\|_\tau \leqslant \|L\|\,r(t,s).$$

Hence $u \in \mathbf{Lip}^1(K)$ and

$$L(\phi_{ts}) = L_u(\phi_{ts}) = u(s) - u(t), \qquad t \in K,\ s \in K.$$

But now L and L_u must coincide by Corollary 2 to Lemma 3.

The final statement regarding the freedom of choice for u follows from (37) and (39).

REMARK 1. Using inequality (40), we can now show that the semi-norm (25) is a norm.

To do this we consider an arbitrary non-zero function $\phi \in \Phi_0(\mathscr{B})$ and the corresponding decomposition of the compactum K into disjoint Borel sets K_ϕ^+ and K_ϕ^- such that*

$$\phi_+(e) = \phi(e \cap K_\phi^+), \qquad \phi_-(e) = -\phi(e \cap K_\phi^-), \qquad e \in \mathscr{B}.$$

Since ϕ is non-zero, $\phi_+(K) = \phi_+(K_\phi^+) > 0$. As ϕ_+ is regular, there exists a closed set $F \subset K_\phi^+$ such that $\phi_+(F) > 0$. Choose ε with $0 < \varepsilon < \frac{1}{2}\phi_+(F)$; then, using the fact that $\phi_-(F) = 0$ and the regularity of ϕ_+ and ϕ_-, we can find a $\delta = \delta(\varepsilon) > 0$ such that the δ-neighbourhood

$$F_\delta = \left\{ t \in K;\ r(t,F) = \min_{s \in F} r(t,s) < \delta \right\}$$

of the set F satisfies

$$\phi_-(F_\delta) < \varepsilon, \qquad \phi_+(F_\delta \setminus F) < \varepsilon.$$

It is easy to see that the function

$$u(t) = \begin{cases} \delta - r(t,F) & \text{for } t \in F_\delta, \\ 0 & \text{for } t \notin F_\delta \end{cases}$$

satisfies the Lipschitz condition, and that $\|u\|_{\mathbf{Lip}} \leqslant 1$. Thus, by (40), we have

$$\|\phi\|_\tau \geqslant \|u\|_{\mathbf{Lip}}\|\phi\|_\tau \geqslant L_u(\phi) = \int_K u(t)\,d\phi = \int_{F_\delta} u(t)\,d\phi = \int_{F_\delta} u(t)\,d\phi_+ - \int_{F_\delta} u(t)\,d\phi_-.$$

Further, since $r(t,F) = 0$ for $t \in F$ and $r(t,F) < \delta$ for $t \in F_\delta \setminus F$, we find that

$$\int_{F_\delta} u(t)\,d\phi = \delta\phi_+(F_\delta) - \int_{F_\delta} r(t,F)\,d\phi_+ \geqslant \delta\phi_+(F_\delta) - \delta\phi_+(F_\delta \setminus F) > \delta[\phi_+(F_\delta) - \varepsilon]$$

$$\int_{F_\delta} u(t)\,d\phi_- \leqslant \delta\phi_-(F_\delta) < \delta\varepsilon.$$

Therefore

$$\|\phi\|_\tau \geqslant \delta[\phi_+(F_\delta) - 2\varepsilon] > 0,$$

as we set out to prove.

REMARK 2. If M is any subset of K, then every function $\tilde{u} \in \mathbf{Lip}^1(M)$ is the restriction to M of some function $u \in \mathbf{Lip}^1(K)$ having the same norm.

For $\tilde{u}$ can be extended to the closure $\overline{M}$ of M by continuity. Now the linear functional on

$$E = \{\phi \in \Phi_0^\tau(\mathscr{B}):\ \operatorname{supp} \phi \subseteq \overline{M}\}$$

determined by the resulting function can be extended to all of $\Phi_0^\tau(\mathscr{B})$ keeping the norm unchanged. As a result, we obtain a function in $\mathbf{Lip}^1(K)$ which differs from the one we require by a constant multiple.

REMARK 3. In addition to the metric compactum K with metric $r(t,s)$ we now consider the metric compactum

* See Theorem I.6.2.

$K' = K$ with the metric $r'(t, s) = [r(t, s)]^\alpha$, where $\alpha \in (0, 1]$. The corresponding normed space $\mathbf{Lip}^1(K', t_0)$ obviously coincides with the space $\mathbf{Lip}^\alpha(K, t_0)$, where

$$\| u \|_{\mathbf{Lip}^\alpha} = \sup_{t \neq s} \frac{u(s) - u(t)}{[r(t, s)]^\alpha}.$$

Hence the latter is linearly isometric to the space $(\Phi_0^\xi(\mathscr{B}_{K'}))^*$ of linear functionals on $\Phi_0^\xi(\mathscr{B}_{K'})$.

We now establish the criterion for optimal translocation stated in 4.2.

THEOREM 2. *We have equality*

$$\tau(\psi) = \| \phi \|_\tau \tag{43}$$

for a measure $\psi \in \Psi_\phi$ *if and only if there exists a function* $u: K \to \mathbb{R}$ *such that*

$$u(s) - u(t) \leqslant r(t, s), \qquad t \in K, \ s \in K, \tag{44}$$

and

$$u(s) - u(t) = r(t, s) \text{ when } (t, s) \in \operatorname{supp} \psi. \tag{45}$$

Proof. For all measures $\psi \in \Psi_\phi$ and all functions $u \in \mathbf{Lip}^1(K)$ with $\| u \|_{\mathbf{Lip}} \leqslant 1$ we have

$$\tau(\psi) = \int_{\hat{K}} r(t, s)\, d\psi(t, s) \geqslant \int_{\hat{K}} (u(s) - u(t))\, d\psi(t, s) =$$

$$= \int_K u(s)\, d\psi(K, s) - \int_K u(t)\, d\psi(t, K) = \int_K u(t)\, d\phi = L_u(\phi), \tag{46}$$

$$L_u(\phi) \leqslant \| \phi \|_\tau \leqslant \tau(\psi). \tag{47}$$

Now assume that, given a measure $\psi \in \Psi_\phi$, we can find a function $u: K \to \mathbb{R}$ satisfying (44) and (45). Then we have equality in (46) for this function. From this and (47) it follows that (43) is true.

Conversely, assume that (43) holds for the measure $\psi \in \Psi_\phi$. By Theorem 1 and the Hahn–Banach Theorem, there exists a function $u \in \mathbf{Lip}^1(K)$ with $\| u \|_{\mathbf{Lip}} = 1$ such that $L_u(\phi) = \| \phi \|_\tau$. In other words, this function satisfies (44), and the left-hand inequality in (47) holds as an equality for this function. Hence both parts of (47) hold with equality; consequently we also have equality in (46). But this means that u satisfies (45). Thus the theorem is proved.

This theorem implies, in particular, that the τ-norm of an elementary function $\phi_{t,s} \in \Phi_0^\xi(\mathscr{B})$ coincides with the distance $r(t, s)$ between the corresponding points. In fact, for any t_0 and s_0 in K, the function

$$\psi_{t_0 s_0}(e, e') = \phi_{t_0}(e)\phi_{s_0}(e'), \qquad e \in \mathscr{B}, \quad e' \in \mathscr{B},$$

in $\Psi_{\phi_{t_0 s_0}}$ satisfies

$$\operatorname{supp} \psi_{t_0 s_0} = \{(t_0, s_0)\}, \qquad \tau(\psi_{t_0 s_0}) = \int_K \int_K r(t, s)\, d\psi_{t_0, s_0}(t, s) = r(t_0, s_0),$$

and the conditions of the theorem just proved hold for the function

$$u(t) = r(t_0, t), \qquad t \in K.$$

However, more important than this are the corollaries stated below, which are proved by means of Remark 2 and Lemma 1 of 4.1.

COROLLARY 1. *Let* $\bar{K} \subset K$ *be a closed set and let* $\tilde{\phi} \in \Phi_0^\xi(\mathscr{B}_{\bar{K}})$. *Suppose there exists a measure* $\tilde{\psi} \in \Psi_{\tilde{\phi}}$ *such that*

$$\tau(\tilde{\psi}) = \int_{\bar{K}} \int_{\bar{K}} r(t, s)\, d\tilde{\psi}(t, s) = \| \tilde{\phi} \|_\tau.$$

Then the function $\phi(e) = \tilde{\phi}(e \cap \bar{K})$, $e \in \mathscr{B}_K$, *in* $\Phi_0^\xi(\mathscr{B}_K)$, *and the measure*

$$\psi(e, e') = \tilde{\psi}(e \cap \bar{K}, e' \cap \bar{K}), \quad e \in \mathscr{B}_K, \ e' \in \mathscr{B}_K,$$

in Ψ_ϕ *satisfy the analogous inequality*

$$\tau(\psi) = \int_K \int_K r(t, s)\, d\psi(t, s) = \| \phi \|_\tau,$$

and we have $\| \phi \|_\tau = \| \tilde{\phi} \|_\tau$.

COROLLARY 2. *If the function $\phi \in \Phi_0^\tau(\mathscr{B})$ has finite support then there is always a measure ψ in Ψ_ϕ such that $\tau(\psi) = \|\phi\|_\tau$, and*

$$\psi(K, K) = \phi_+(K) = \phi_-(K),$$

that is, $\psi(K, e) = \phi_+(e)$, $\psi(e, K) = \phi_-(e)$, $e \in \mathscr{B}$.

The last statement is in fact also true for functions $\phi \in \Phi_0^\tau(\mathscr{B})$ having infinite support. However, to prove this we first need to establish the connection between the τ-norm and weak* convergence of linear functionals on $\mathbf{C}(K)$.

4.6. We already know (see Theorem VI.3.1) that the space $\mathbf{C}(K)^*$ of continuous linear functionals on $\mathbf{C}(K)$ is linearly isometric to the B-space $\mathbf{rca}(K)$. When a sequence ϕ_n in $\mathbf{rca}(K)$ is weak* convergent to ϕ_0, we shall write $\phi_n \overset{(*)}{\dashrightarrow} \phi_0$. Notice that weak* convergence of ϕ_n to ϕ_0 clearly implies that $\phi_n(K) \to \phi_0(K)$. Also we denote the norm in $\mathbf{rca}(K)$ by $\|\phi\|$: that is, $\|\phi\| = \phi_+(K) + \phi_-(K)$. It follows from Theorem 1.1 that if $\phi_n \overset{(*)}{\dashrightarrow} \phi_0$, then

$$\|\phi_0\| \leqslant \underline{\lim}\, \|\phi_n\| \leqslant \overline{\lim}\, \|\phi_n\| < \infty. \tag{48}$$

Moreover, every weak* closed set $M \subset \mathbf{rca}(K)$ is weak* sequentially complete (see 1.2); that is, if $\phi_n \in M$ are functions such that the integrals $\int_K x\, d\phi_n$ converge for each $x \in \mathbf{C}(K)$, then these functions are weak* convergent to some $\phi_0 \in M$. In particular, this applies to the set $\mathbf{rca}_+(K)$ of all measures $\phi \in \mathbf{rca}(K)$, to the subspace $\Phi_0(\mathscr{B}) \subset \mathbf{rca}(K)$, and also to the closed balls

$$S^\nu = \{\phi \in \mathbf{rca}\,(K): \|\phi\| \leqslant \nu\}$$

of arbitrary radii $\nu \in [0, +\infty)$ and to the closed balls

$$S_0^\nu = \{\phi \in \Phi_0(\mathscr{B}): \|\phi\| \leqslant \nu\}$$

in $\Phi_0(\mathscr{B})$, which we encountered in 4.4. Moreover, since $\mathbf{C}(K)$ is separable (see Theorem IV.4.3), these balls S^ν and S_0^ν are weak* sequentially compact (see Theorem I.7.6).

The space $\mathbf{rca}(K)$ is widely used in probability theory, geometry and other mathematical disciplines. Moreover, weak* convergence in this space often plays a significantly larger role than weak convergence. This is explained firstly by the weak* compactness of the balls S^ν, already referred to above, and also by the fact that for many problems the norm in $\mathbf{rca}(K)$ turns out to be too coarse. This norm is not directly linked to the original metric $r(t, s)$ and the induced topology in K. Because of this the measure of closeness defined by this norm for functions in $\mathbf{rca}(K)$ does not always accord with one's intuitive ideas. For example, if $t \neq t_0$, the ν-norm of the difference $\phi_t - \phi_{t_0}$ is equal to 2, regardless of the distance $r(t, t_0)$ between the points. At the same time, it is not always convenient in applications to work with the weak* topology in $\mathbf{rca}(K)$, as this is not metrizable. In such cases the τ-norm may turn out to be useful.

For functions $\phi \in \Phi_0(\mathscr{B})$, besides strong convergence $\overset{\nu}{\to}$ and weak* convergence $\overset{(*)}{\dashrightarrow}$ in $\mathbf{rca}(K)$ we have also strong convergence $\overset{\tau}{\to}$ and weak convergence $\dashrightarrow$ in $\Phi_0^\tau(\mathscr{B})$. The latter is weaker than weak* convergence since not every continuous function satisfies a Lipschitz condition. Therefore, for functions in $\Phi_0(\mathscr{B})$, taking (48) into account, we have:

$$\phi_n \overset{(*)}{\dashrightarrow} \phi_0 \Rightarrow \sup \|\phi_n\| < +\infty, \quad \phi_n \dashrightarrow \phi_0. \tag{49}$$

In addition, by Lemma 5 of 4.4, we have:

$$\{\phi_n\} \subset S_0^\nu,\ \phi_n \dashrightarrow \phi_0 \Leftrightarrow \{\phi_n\} \subset S_0^\nu,\ \phi_n \overset{\tau}{\to} \phi_0. \tag{50}$$

LEMMA 7. *If a sequence of functions $\phi_n \in S_0^\nu$ is a Cauchy sequence relative to the τ-norm, then it is weak* convergent to some function $\phi_0 \in S_0^\nu$.*

Proof. Since S_0^ν is weak* sequentially compact, to prove the lemma we need only show that the given sequence $\{\phi_n\} \subset S_0^\nu$ cannot have more than one cluster point. It follows from (49) and (50) that every weak* cluster point of the sequence $\{\phi_n\}$ is also a τ-cluster point. But since the sequence is a Cauchy sequence in the τ-norm it cannot have more than one τ-cluster point, so the lemma is proved.

THEOREM 3. *For any $\nu \in [0, +\infty)$, the set S_0^ν is τ-compact. Moreover, the three types of convergence coincide for the functions in S_0^ν occurring in* (49) *and* (50)*: that is,*

$$\phi_n \overset{(*)}{\dashrightarrow} \phi_0 \Leftrightarrow \phi_n \dashrightarrow \phi_0 \Leftrightarrow \phi_n \overset{\tau}{\to} \phi_0.$$

Proof. The assertions in the theorem follow immediately from Lemma 7 and (49) and (50).

COROLLARY. *A sequence of functions $\phi_n \in \mathbf{rca}(K)$ is weak* convergent to a function ϕ_0 if and only if* $\sup \|\phi_n\|$

$< \infty$, $\phi_n(K) \to \phi_0(K)$ *and, for some (and hence also for any) point* $t_0 \in K$, *the functions* $\phi'_n = \phi_n - \phi_n(K)\phi_{t_0}$ *in* $\Phi_0(\mathscr{B})$ *are* τ-*convergent to* $\phi'_0 = \phi_0 - \phi_0(K)\phi_{t_0}$.

REMARK. In view of Lemma 3, which was proved in 4.4, every function $\phi \in \Phi_0(\mathscr{B})$ satisfies:

$$\|\phi\|_\tau \leq \tfrac{1}{2} \operatorname{diam} K \|\phi\|.$$

At the same time the norms $\|\phi\|$ and $\|\phi\|_\tau$ are clearly equivalent only for finite compacta K. It thus follows that, if a compactum K contains an infinite number of points, then the normed space $\Phi_0^\tau(\mathscr{B}_k)$ is not complete. However, since S_0^ν is τ-complete, it is possible in many applications to manage without the former space being complete.

In the proof to be given below of the solubility of the problem of translocation of mass for every given function $\phi \in \Phi_0(\mathscr{B})$, essential use will be made of the property of weak* convergence that we have referred to. It is easy to see that weak* convergence $\psi_n \xrightarrow{(*)} \psi_0$ in $\mathbf{rca}(\tilde{K})$, where $\tilde{K} = K \times K$, implies that $\psi_n(K, .) \xrightarrow{(*)} \psi_0(K, .)$ and $\psi_n(. , K) \xrightarrow{(*)} \psi_0(. , K)$ in $\mathbf{rca}(K)$. In what follows, $\|\psi\|$ will denote the norm of the measure ψ in $\mathbf{rca}(\tilde{K})$.

THEOREM 4. *For any function* $\phi \in \Phi_0(\mathscr{B})$, *there is a measure* ψ *in* Ψ_ϕ *such that* $\tau(\psi) = \|\phi\|_\tau$, *and*

$$\psi(K, K) = \phi_+(K) = \phi_-(K) = \tfrac{1}{2}\|\phi\|, \tag{51}$$

that is

$$\psi(K, e) = \phi_+(e), \qquad \psi(e, K) = \phi_-(e), \qquad e \in \mathscr{B}. \tag{52}$$

Proof. Consider an arbitrary function $\phi \in \Phi_0(\mathscr{B})$. By Lemma 4, we can construct a sequence of functions $\phi_n \in \Phi(\mathscr{B})$, having finite supports, such that

$$\|\phi_n\| \leq \|\phi\|, \qquad \phi_n \xrightarrow{\tau} \phi. \tag{53}$$

By Theorem 3 this sequence will also be weak* convergent to ϕ.

Since the ϕ_n have finite supports, Corollary 2 to Theorem 2 shows that, for each ϕ_n, there exists a measure $\psi_n \in \Psi_{\phi_n}$ such that $\tau(\psi_n) = \|\phi_n\|_\tau$ and

$$\|\psi_n\| = \psi_n(K, K) = (\phi_n)_+(K) = (\phi_n)_-(K) = \tfrac{1}{2}\|\phi_n\|, \tag{54}$$

that is,

$$\psi_n(K, e) = (\phi_n)_+(e), \qquad \psi_n(e, K) = (\phi_n)_-(e), \qquad e \in \mathscr{B}.$$

As $\tilde{K}$ is a metric compactum, $\mathbf{C}(\tilde{K})$ is separable (see Theorem IV.4.3). Therefore bounded sets in $\mathbf{rca}(\tilde{K}) = \mathbf{C}(\tilde{K})^*$ are relatively compact and metrizable in the weak* topology (see Theorem V.7.6). Thus, by (54), we can choose a subsequence $\{\psi_{n_k}\}$ of $\{\psi_n\}$ that is weak* convergent to a measure ψ satisfying

$$\|\psi\| \leq \varliminf_{k \to \infty} \|\psi_{n_k}\| \leq \tfrac{1}{2}\|\phi\|. \tag{55}$$

We show that ψ satisfies the requirements of the theorems.

Firstly, as we have mentioned, the weak* convergence of the ψ_{n_k} to ψ implies that $\psi_{n_k}(K, .)$ and $\psi_{n_k}(. , K)$ are weak* convergent in $\mathbf{rca}(K)$ to $\psi(K, .)$ and $\psi(. , K)$, respectively. Also $\phi_{n_k} \xrightarrow{(*)} \phi$. But then, if we pass to the weak* limit in the equations

$$\psi_{n_k}(K, \cdot) - \psi_{n_k}(\cdot, K) = \phi_{n_k},$$

which follow from the definition of $\Psi_{\phi_{n_k}}$, we obtain

$$\psi(K, \cdot) - \psi(\cdot, K) = \phi,$$

so that the measure ψ belongs to Ψ_ϕ. Equations (51) and (52) follow from (55) and Lemma 6. Finally, we obtain the equation

$$\tau(\psi) = \|\phi\|_\tau$$

from the equations $\tau(\psi_{n_k}) = \|\phi_{n_k}\|_\tau$ by passing to the limit and using (53). The theorem is therefore completely proved.

REMARK. It follows from this theorem, in particular, that if condition (15) in the original problem of translocation of mass is replaced by the more stringent requirement (53), thus restricting the class of admissible translocations, the problem remains in essence unchanged.

In conclusion we note that the results we have presented concerning the problem of translocation of mass and the associated normed space had already been obtained by the 1950s.

Later they were used repeatedly, especially in information theory and ergodic theory, and were also supplemented and extended in different directions by several authors (see, for example, S. G. Krein and Yu. I. Petunin [1], Rubinshtein [1–3], Vershik [1], [2], Rvachev [1]). Interesting results concerning the existence of optimal transports with densities in the problem of translocation of Lebesgue measure have been obtained by Sudakov [1] and Ekhlakov [1]. In papers by V. L. Levin [1], [3], [4] (see also the earlier paper cited in these) the author considers the case where K is an arbitrary compactum and $r(t, s)$ is a continuous non-negative function on $K \times K$. Here the functional (25) is, in general, no longer a norm on $\Phi_0(\mathscr{B})$; nevertheless, several results remain true even in this more general situation. Let us note some of these. In addition to the original problem of translocation of mass we consider the dual problem, which consists in maximizing the functional

$$\int_K u(t)\,d\phi \tag{56}$$

under the restrictions (17) on $u \in \mathbf{C}(K)$. Let us write A for the optimal value for the problem of translocation of mass—that is, the greatest lower bound for the functional (16) on the set of measures ψ satisfying (15). Let us write B for the optimal value for the dual problem—that is, the least upper bound for the functional (56) subject to the restrictions (17). It is easy to see that the criterion for optimality of translocation contained in Theorem 2 (in the case where r is a metric) can be reformulated as the duality relation $A = B$. In the general case, an optimal translocation may not exist; however, this duality relation always makes sense and if $r(t, t) = 0$ $(t \in K)$ then it is satisfied. Furthermore, if $r(t, t) = 0$ $(t \in K)$ and r satisfies the triangle inequality, then an optimal translocation of mass ψ exists and Theorem 4 holds—that is, one can manage without transit transport. The study of this last case is also the subject of a paper by Alsynbaev, Imomnazarov and Rubinshtein [1], where an auxiliary space with an asymmetric norm is constructed and all the required results are established, with its help, exactly as we did for the classical case we were concerned with above.

IX

COMPACT AND ADJOINT OPERATORS

§ 1. Compact sets in normed spaces

1.1. In I.5.2 we mentioned a compactness criterion for a set in the space $\mathbf{C}(K)$. We now state some compactness conditions for a set in $\mathbf{L}^p(D)$ ($1 < p < \infty$, D is a measurable bounded set in Euclidean space $\mathbf{R}^n$).

Consider a family $\{\omega_h\}$ of functions of a real argument, depending on a positive real parameter and having the following properties:

1) $0 \leqslant \omega_h(\xi) \leqslant M$ $(\xi \geqslant 0, h > 0)$;
2) $\omega_h(\xi) = 0$ for $\xi \geqslant h$;
3) ω_h is continuous for $\xi < h$;
4) if t denotes a vector in $\mathbf{R}^n$ and $|t|$ its length, then

$$\int_{\mathbf{R}^n} \omega_h(|t|)\,dt = v_h,$$

for all $h > 0$, where v_h is the volume of an n-dimensional ball of radius h; on account of condition 2) the integration can be carried out over the n-dimensional ball G_h with centre at the origin.

We extend functions in $\mathbf{L}^p(D)$ to all of $\mathbf{R}^n$, taking them to be zero outside the set D. We also denote by $G_h(s)$ the n-dimensional ball of radius h with centre at $s \in \mathbf{R}^n$. For $x \in \mathbf{L}^p(D)$, we write

$$x_h(s) = \frac{1}{v_h} \int_{\mathbf{R}^n} \omega_h(s-t)x(t)\,dt = \frac{1}{v_h} \int_{G_h(s)} \omega_h(s-t)x(t)\,dt \quad (s \in \mathbf{R}^n), \tag{1}$$

where, for simplicity, we put $\omega_h(s-t) = \omega_h(|s-t|)$.

The function x_h is called the *Steklov average* function of x or the *mean function* (with respect to x); the family of functions $\{\omega_h\}$ is called the *kernel* of the averaging process. If we take ω_h to be the function

$$\omega_h(\xi) = \begin{cases} 1, & \xi < h, \\ 0, & \xi \geqslant h, \end{cases}$$

then $x_h(s)$ is the mean value of x on the ball $G_h(s)$, in the usual sense.

LEMMA 1. *Let $x \in \mathbf{L}^p(D)$. Then for every $h > 0$ the mean function x_h is continuous. Furthermore,*

$$|x_h(s)| \leqslant \frac{M}{(v_h)^{1/p}} \left\{ \int_{G_h(s)} |x(t)|^p\,dt \right\}^{1/p} \leqslant \frac{M}{(v_h)^{1/p}} \|x\| \qquad (s \in \mathbf{R}^n). \tag{2}$$

Proof. We have

$$x_h(s) - x_h(s') = \frac{1}{v_h} \int\limits_{G_h(s)} \omega_h(s-t)x(t)\,dt - \frac{1}{v_h} \int\limits_{G_h(s')} \omega_h(s'-t)x(t)\,dt \quad (s, s' \in \mathbf{R}^n).$$

Let $0 < \lambda < 1$. Write

$$A'_\lambda = G_{\lambda h}(s) \cap G_{\lambda h}(s'), \qquad A''_\lambda = [G_h(s) \cup G_h(s')] \setminus A'_\lambda.$$

Using Hölder's inequality, we can write

$$|x_h(s) - x_h(s')| \leqslant \frac{1}{v_h}\left[\int\limits_{A'_\lambda} |\omega_h(s-t) - \omega_h(s'-t)|\,|x(t)|\,dt + 2M \int\limits_{A''_\lambda} |x(t)|\,dt\right] \leqslant$$

$$\leqslant \frac{1}{v_h}\left[\left(\int\limits_{A'_\lambda} |\omega_h(s-t) - \omega_h(s'-t)|^q\,dt\right)^{1/q}\left(\int\limits_{A'_\lambda} |x(t)|^p\,dt\right)^{1/p} + \right.$$

$$\left. + 2M\left(\int\limits_{A''_\lambda} |x(t)|^p\,dt\right)^{1/p} (\text{mes } A''_\lambda)^{1/q}\right] \leqslant \frac{\|x\|}{v_h}\left\{\left[\int\limits_{A'_\lambda} |\omega_h(s-t)\right.\right.$$

$$\left.\left. - \omega_h(s'-t)|^q\,dt\right]^{1/q} + 2M[\text{mes } A''_\lambda]^{1/q}\right\}.$$

If $|s'-s| \to 0$ and $\lambda \to 1$ then obviously mes $A''_\lambda \to 0$. Therefore, given $\varepsilon > 0$, we can find $\delta > 0$ and $\lambda_0 < 1$ such that the second term inside the curly brackets is less than ε when $|s' - s| < \delta$ and $\lambda = \lambda_0$.

The function $\omega_h(s-t)$ is uniformly continuous, as a function of t, on $G_{\lambda_0 h}(s)$, so we can find a $\delta_0 \leqslant \delta$ such that

$$|\omega_h(s-t) - \omega_h(s'-t)| \leqslant \varepsilon \quad (|s'-s| < \delta_0, \; t \in A'_{\lambda_0}).$$

These considerations now enable us, by augmenting the bound, to write

$$|x_h(s) - x_h(s')| \leqslant \frac{\varepsilon\|x\|}{v_h}(v_h + 1) \qquad (|s-s'| \leqslant \delta_0). \tag{3}$$

The continuity of x_h is thus proved.

The bound in (2) is obtained from (1) by applying Holder's inequality.

REMARK. The number δ_0 appearing in the proof of the lemma does not depend on x.

LEMMA 2. *If the function x is uniformly continuous in $\mathbf{R}^n$, then the mean functions x_h converge uniformly to x as $h \to 0$.*

Proof. Since

$$x(s) = \frac{1}{v_h} \int\limits_{G_h(s)} \omega_h(s-t)x(s)\,dt,$$

we have

$$|x_h(s)-x(s)| \leqslant \frac{1}{v_h} \int\limits_{G_h(s)} \omega_h(s-t)|x(t)-x(s)|\,dt.$$

Since $|x(t)-x(s)| < \varepsilon$ for sufficiently small h, we have

$$|x_h(s)-x(s)| \leqslant \frac{\varepsilon}{v_h} \int\limits_{G_h(s)} \omega_h(s-t)\,dt = \varepsilon$$

simultaneously for all $s \in \mathbf{R}^n$.

If we associate with each element $x \in \mathbf{L}^p(D)$ the mean function x_h, we obtain a linear operator, defined on $\mathbf{L}^p(D)$. Let us denote this by U_h: thus $U_h(x) = x_h$. It is not hard to see that U_h is a continuous operator from $\mathbf{L}^p(D)$ into $\mathbf{C}(Q)$, where Q is any closed bounded set containing D. For, by Lemma 1, x_h is continuous in $\mathbf{R}^n$ and so *a fortiori* in Q, and by (2) we find that

$$\|x_h\|_{\mathbf{C}(Q)} = \max_{s \in Q} |x_h(s)| \leqslant \frac{M}{(v_h)^{1/p}} \|x\|_{\mathbf{L}^p},$$

and so

$$\|U_h\| \leqslant \frac{M}{v_h^{1/p}}.$$

If U_h is regarded as a linear operator on $\mathbf{L}^p(D)$ then the bound for its norm can be substantially improved.

LEMMA 3. *If U_h is regarded as an operator on $\mathbf{L}^p(D)$, then*

$$\|U_h\| \leqslant M \qquad (h > 0). \tag{4}$$

Proof. By (2) we have

$$\|x_h\|_{\mathbf{L}^p(D)} = \left[\int\limits_D |x_h(s)|^p\,ds\right]^{1/p} \leqslant \left\{\frac{M^p}{v_h} \int\limits_D \left[\int\limits_{G_h(s)} |x(t)|^p\,dt\right] ds\right\}^{1/p}.$$

Changing the order of integration, we find that

$$\|x_h\|_{\mathbf{L}^p(D)} \leqslant \left\{\frac{M^p}{v_h} \int\limits_D \left[\int\limits_{G_h(t)} |x(t)|^p\,ds\right] dt\right\}^{1/p} = M\left[\int\limits_D |x(t)|^p\,dt\right]^{1/p},$$

which gives (4).

LEMMA 4. *In $\mathbf{L}^p(D)$, we have $x_h \to x$ as $h \to 0$.*

Proof. Let D_0 be a closed cube in $\mathbf{R}^n$ that contains D and is large enough that there exists a closed cube D_1, also containing D and itself contained in the interior of D_0. We shall prove that the set $\mathring{\mathbf{C}}(D_0)$ of all functions continuous on D_0 and vanishing on the boundary of D_0 is dense in $\mathbf{L}^p(D)$. Let $x \in \mathbf{L}^p(D_0)$. Since the set of almost everywhere bounded functions is dense in $\mathbf{L}^p(D_0)$, we can assume that x is bounded on D_0. Assume that $|x(t)|$

$\leqslant K$ for $t \in D_0$. Choose any $\varepsilon > 0$. By Luzin's Theorem, there exists a function ϕ which is continuous on D_0 and coincides with x everywhere except on a set $A \subset D_0$ such that mes $A \leqslant \varepsilon$; also $|\phi(t)| \leqslant K$ for $t \in D_0$. Choosing D_1 such that mes $(D_0 \setminus D_1) < \varepsilon$ and using the theorem on extending continuous functions (Natanson-II), we construct a function x_0, continuous on D_0, coinciding with ϕ on D_1 and equal to zero on the boundary of D_0. By the theorem just mentioned, we can choose this function such that $|x_0(t)| \leqslant K$ for $t \in D_0$. Let us find a bound for the norm (in $\mathbf{L}^p(D_0)$) of $x - x_0$. We have

$$\|x - x_0\| \leqslant \|x - \phi\| + \|\phi - x_0\|.$$

However,

$$\|x - \phi\| = \left[\int_{D_0} |x(t) - \phi(t)|^p\, dt\right]^{1/p} = \left[\int_A |x(t) - \phi(t)|^p\, dt\right]^{1/p} \leqslant 2K\varepsilon^{1/p},$$

$$\|\phi - x_0\| = \left[\int_{D_0} |\phi(t) - x_0(t)|^p\, dt\right]^{1/p} = \left[\int_{D_0 \setminus D_1} |\phi(t) - x_0(t)|^p\, dt\right]^{1/p} \leqslant 2K\varepsilon^{1/p},$$

and so $\|x - x_0\| \leqslant 4K\varepsilon^{1/p}$. It remains to note that $x_0 \in \mathring{\mathbf{C}}(D_0)$.

If x is a function in $\mathring{\mathbf{C}}(D_0)$, then, by setting it equal to zero outside D_0, we can conclude from Lemma 2 that $x_h \to x$ uniformly on $\mathbf{R}^n$ and hence also on D_0. *A fortiori*, $x_h \to x$ in $\mathbf{L}^p(D_0)$. By Lemma 3, the norms of the operators U_h are bounded in aggregate, and since, by what has been proved, $\mathring{\mathbf{C}}(D_0)$ is dense in $\mathbf{L}^p(D_0)$, the Banach–Steinhaus Theorem (VII.1.2) shows that $x_h \to x$ on all of $\mathbf{L}^p(D_0)$. But a function in $\mathbf{L}^p(D)$ which is extended so that it is zero outside D is an element of $\mathbf{L}^p(D_0)$, and thus the convergence also holds on $\mathbf{L}^p(D)$. This proves the lemma.

1.2. We now obtain a compactness criterion for a set in $\mathbf{L}^p(D)$.

THEOREM 1 (Kolmogorov). *The following conditions are together necessary and sufficient for a set $E \subset \mathbf{L}^p(D)$ to be relatively compact*:

1) *E is bounded*;

2) *the mean functions converge (in $\mathbf{L}^p(D)$) uniformly to their generating functions on E; that is,*

$$x_h \to x \text{ uniformly on } E.$$

Proof. The necessity of the first condition is obvious. Let us prove the necessity of the second. Since E is relatively compact, it has a finite ε-net. Let this be $x^{(1)}, x^{(2)}, \dots, x^{(m)}$. For sufficiently small h ($h \leqslant h_\varepsilon$), we have $\|x_h^{(k)} - x^{(k)}\| \leqslant \varepsilon$ ($k = 1, 2, \dots, m$), by Lemma 4.

For each $x \in E$, there exists an $x^{(k)}$ such that

$$\|x - x^{(k)}\| \leqslant \varepsilon.$$

But then we have

$$\|x_h - x\| = \|U_h(x) - x\| \leqslant \|U_h(x - x^{(k)})\| + \|x_h^{(k)} - x^{(k)}\| + \|x - x^{(k)}\| < 2\varepsilon + M\varepsilon.$$

Sufficiency. Write E_h for the set of elements of the form $U_h(x)$, where $x \in E$. By Lemma 1, $E_h \subset \mathbf{C}(\bar{D})$. We show that E_h is relatively compact in $\mathbf{C}(\bar{D})$. The first condition for a set in $\mathbf{C}(\bar{D})$ to be compact—namely, that it be bounded—is satisfied, in view of the inequality (2). The equicontinuity of the functions in E_h follows from (3).

Thus E_h is relatively compact in $\mathbf{C}(\bar{D})$, for every h, and so *a fortiori* in $\mathbf{L}^p(D)$. By condition 2), there exists an h_0 sufficiently small that $\|x - x_{h_0}\| = \|x - U_{h_0}x\| < \varepsilon$ for arbitrary $x \in E$.

Hence E_h is a relatively compact ε-net for E. By Hausdorff's Theorem (see I.5.1), E is relatively compact.

The theorem is therefore proved.

REMARK. The compactness criterion given in this theorem is also valid in $\mathbf{L}^1(D)$ (see Tulaikov [1]).

1.3. We now give another compactness criterion for a set in $\mathbf{L}^p(D)$. For simplicity we restrict ourselves to the case where D is the interval $[a, b]$.

THEOREM 2 (M. Riesz). *The following conditions are together necessary and sufficient for a set $E \subset \mathbf{L}^p(a, b)$ $(1 < p < \infty)$ to be relatively compact:*

1) *E is bounded;*

2) *the translates of a function in E converge to the given function, uniformly with respect to $x \in E$, i.e. $\int_a^b |x(t+\tau) - x(t)|^p \, dt \to 0$ as $\tau \to 0$.*

Proof. Necessity. We introduce a translation operator V_τ on $\mathbf{L}^p(a, b)$ by:

$$y = V_\tau(x), \quad y(t) = x(t+\tau).$$

Clearly V_τ is a linear operator on $\mathbf{L}^p(a, b)$ and

$$\|V_\tau\| \leqslant 1. \tag{5}$$

Let us show that

$$V_\tau(x) \xrightarrow[\tau \to 0]{} x \quad (\text{in } \mathbf{L}^p(a, b)). \tag{6}$$

In fact, if x is a continuous function, then

$$\int_a^b |x(t+\tau) - x(t)|^p \leqslant \int_a^b \varepsilon^p \, dt = \varepsilon^p(b-a) \quad (\tau \leqslant \tau_\varepsilon)$$

and (6) is therefore satisfied. Since the set of continuous functions is dense in $\mathbf{L}^p(a, b)$, the Banach–Steinhaus Theorem is applicable, in view of (5), and hence (6) is true for every $x \in \mathbf{L}^p(a, b)$.

The rest of the argument proceeds exactly as in the proof of Theorem 1.

Sufficiency. We prove that, if the conditions of the present theorem are satisfied, then so are those of Theorem 1, provided we take the kernel of the averaging process to be

$$\omega_h(\xi) = \begin{cases} 1, & \xi < h, \\ 0, & \xi \geqslant h \end{cases} \quad (h > 0).$$

Retaining the previous notation, we can write

$$\|U_h(x) - x\|^p = \frac{1}{(2h)^p} \int_a^b \left| \int_{-h}^h [x(t+\tau) - x(t)] \, d\tau \right|^p dt.$$

Applying Hölder's inequality to the inner integral and then changing the order of integration, we obtain

$$\|U_h(x) - x^p\| \leqslant \frac{1}{(2h)^p} \int_a^b \left\{ \left[\int_{-h}^h 1^q \, d\tau \right]^{p/q} \left[\int_{-h}^h |x(t+\tau) - x(t)|^p \, d\tau \right] \right\} dt = \frac{1}{2h} \int_{-h}^h d\tau \int_a^b |x(t+\tau) -$$

$$- x(t)|^p \, dt = \frac{1}{2h} \int_{-h}^h \|V_\tau(x) - x\|^p \, d\tau.$$

If the second condition of the theorem is satisfied, then for any $\varepsilon > 0$ there exists an $h_0 > 0$ such that, when $|\tau| \leqslant h_0$, we have $\|V_\tau(x)-x\|^p < \varepsilon$ for every $x \in E$; but then, by the inequality we established above, we have

$$\|U_h(x)-x\|^p \leqslant \frac{1}{2h}\int_{-h}^{h} \varepsilon d\tau = \varepsilon.$$

Hence the second condition of Theorem 1 is satisfied. Since the first condition is the same in both theorems, we can apply Theorem 1 to E to show that it is relatively compact.

1.4. Finally we state a general compactness criterion for a set in a normed space (see Gel'fand [1]).

THEOREM 3 (Gel'fand). *For a set E in a B-space $\mathbf{X}$ to be relatively compact it is necessary—and, if $\mathbf{X}$ is separable, also sufficient—that, for every sequence of functionals $\{f_n\}$ that is weak* convergent to zero, we have*

$$f_n(x) \underset{n\to\infty}{\longrightarrow} 0 \tag{7}$$

uniformly on E: that is, to each $\varepsilon > 0$ there corresponds an n_ε such that

$$|f_n(x)| < \varepsilon$$

for all $x \in E$ and all $n > n_\varepsilon$.

Proof. Sufficiency. Let us first prove that E is bounded. In fact, if it were not, then we could find a sequence of elements $x_n \in E$ $(n = 1, 2, \ldots)$ such that $\|x_n\| = \gamma_n^2 \to \infty$ as $n \to \infty$. By the theorem on the existence of sufficiently many functionals, there would then exist functionals f_n' with $\|f_n'\| = 1$ and $f_n'(x_n) = \|x_n\|$. Setting $f_n = \dfrac{1}{\gamma_n} f_n'$, we would obtain a sequence $\{f_n\}$ that converged to zero, but for which

$$f_n(x_n) = \gamma_n \underset{n\to\infty}{\longrightarrow} \infty.$$

Hence the convergence $f_n(x) \to 0$ would not be uniform.

Let B denote the closed unit ball in the space $\mathbf{X}^*$, endowed with the weak* topology. We already know (III.3.1) that we can view every element $x \in \mathbf{X}$ as a continuous function on $(\mathbf{X}^*, \sigma(\mathbf{X}^*, \mathbf{X}))$, and hence also on B. Denote the resulting embedding of $\mathbf{X}$ into $\mathbf{C}(B)$ by ϕ. We shall show that $\phi(E)$ is relatively compact in the B-space $\mathbf{C}(B)$. Since it is easy to see that ϕ is a linear isometry, this will also prove that E is relatively compact. By Theorem V.7.6 B is a metric compactum (with metric r). We now verify that $\phi(E)$ satisfies the conditions of the Arzela–Ascoli Theorem (see Theorem I.5.4). Since we have proved that E is bounded, we need only show that it is equicontinuous. Assume the contrary. This implies that there exists $\varepsilon > 0$ such that, whatever $\delta > 0$ we take, there exist $x_\delta \in E$ and $f_n, f_n' \in B$ for which we have $r(f_n, f_n') < \delta$ and

$$|f_n(x_\delta) - f'_n(x_\delta)| \geqslant \varepsilon.$$

If we choose successively $\delta = 1, 1/2, \ldots, 1/n, \ldots$, then we obtain a sequence x_n of elements of E and two sequences of functionals $\{f_n\}, \{f_n'\} \subset B$ such that $r(f_n, f_n') < 1/n$ and

$$|f_n(x_n) - f_n'(x_n)| \geqslant \varepsilon \qquad (n \in \mathbb{N}). \tag{8}$$

Since B is a metric compactum, we may assume (by passing to a subsequence if necessary) that $f_n \to f_0$ $(\sigma(\mathbf{X}^*, \mathbf{X}))$, $f_0 \in B$. Since $r(f_n, f_n') < 1/n$, we have also $f_n' \to f_0$ $(\sigma(\mathbf{X}^*, \mathbf{X}))$. Now $f_n - f_n' \to \mathbf{0}$ in the weak* topology, so by (7) $f_n(x) - f_n'(x) \to 0$ as $n \to \infty$ uniformly on E, which contradicts (8).

Sufficiency. We show that if $\{U_n\}$ is a sequence of continuous linear operators mapping the given space $\mathbf{X}$ into a normed space $\mathbf{Y}$, such that, for $x \in \mathbf{X}$,

$$U_n(x) \xrightarrow[n \to \infty]{} \mathbf{0}, \tag{9}$$

then (9) holds uniformly on E when E is relatively compact.

As $\mathbf{X}$ is a B-space, (9) implies that the norms of the operators U_n are bounded (see VII.1.1): thus $\|U_n\| \leqslant M$, say. Since E is compact, it has a finite $\varepsilon/2M$-net: $x_1, x_2, \ldots, x_m$. For sufficiently large n, we have

$$\|U_n(x_k)\| < \frac{\varepsilon}{2} \qquad (k = 1, 2, \ldots, m). \tag{10}$$

Also, for arbitrary $x \in E$, there exists an x_k such that

$$\|x - x_k\| < \frac{\varepsilon}{2M}. \tag{11}$$

Hence, for these n, we have

$$\|U_n(x)\| \leqslant \|U_n(x_k)\| + \|U_n(x - x_k)\| < \frac{\varepsilon}{2} + \|U_n\| \|x - x_k\| \leqslant \varepsilon,$$

that is, $U_n(x) \to \mathbf{0}$ uniformly on E.

REMARK. The sufficiency part of Theorem 3 is in general false if $\mathbf{X}$ is not separable (see Bukhvalov [1]; this also contains generalizations of Theorems 3 and V.7.8).

§ 2. Compact operators

2.1. As we stated in Chapter V, every linear operator from one finite-dimensional space into another is determined by a rectangular matrix. The study of such operators is therefore a relatively easy task, since the properties of finite matrices are well known from algebra. If, however, we consider operators on an arbitrary normed space, then it is by no means always possible to establish for these all the properties analogous to those of the "finite-dimensional" ones. In this respect the so-called compact operators approach closest of all to the latter condition. The fundamental properties of compact operators, which bring them close to "finite-dimensional" operators, will be established only in Chapter XIII. Here we shall just give the definition, mention the simplest consequences that can be deduced from it, and confine our more detailed remarks to compact operators on Hilbert space.

An arbitrary operator U mapping one normed space $\mathbf{X}$ into another normed space $\mathbf{Y}$ is said to be *compact* if it maps every bounded set in $\mathbf{X}$ into a relatively compact set in $\mathbf{Y}$. If, in addition, U is continuous, then it is said to be *completely continuous*.

In what follows we shall apply this definition almost exclusively to the case where U is a linear operator. It is not hard to see that every compact linear operator is completely continuous. For the unit ball in $\mathbf{X}$ is mapped by the operator into a relatively compact, and in particular bounded, set in $\mathbf{Y}$, so that $\sup_{\|x\| \leqslant 1} \|U(x)\| < \infty$; that is, U is a bounded operator.

Finally, we note that for this case one can restrict oneself in the definition to the requirement that the unit ball (or a ball of any other radius) be transformed into a relatively compact set. In future, unless we stipulate otherwise, all our compact operators will be linear.

The definition of a compact operator and the simple properties of this class of operators to be considered in this section were given by Hilbert (see Hilbert) in the case of $\mathbf{L}^2$. For the general case, see Banach.

The finite-dimensional operators mentioned above are, of course, compact. Another trivial example of a compact operator is a continuous linear functional on a space $\mathbf{X}$, considered as an operator from $\mathbf{X}$ into the space of scalars.

A more interesting example of a compact operator is the operator U_h introduced in § 1 (see 1.1). For, as can easily be checked, the compactness of $U_h(E)$ is a consequence of the boundedness of E alone.

Finally, we consider the following integral operator with continuous kernel:

$$y = U(x), \qquad y(s) = \int_a^b K(s, t)x(t)\,dt.$$

We regard U as an operator from $\mathbf{C}[a, b]$ into $\mathbf{C}[a, b]$. Let us show that U is compact. In fact, if E is a bounded set in $\mathbf{C}[a, b]$ (say $\|x\| \leqslant M$ for $x \in E$), then $U(E)$ is clearly also bounded ($\|y\| \leqslant M\|U\|$ for $y \in U(E)$). Further, if $x \in \mathbf{X}$, then (for $y = U(x)$) we have

$$|y(s') - y(s)| \leqslant \int_a^b |K(s', t) - K(s, t)| \; |x(t)|\,dt \leqslant M \int_a^b |K(s', t) - K(s, t)|\,dt.$$

If s and s' are sufficiently close, then the right-hand side is as small as we please, independently of $x \in E$. This shows that the functions in $U(E)$ are equicontinuous. Hence $U(E)$ is relatively compact in $\mathbf{C}[a, b]$.

Using Theorem IX.1.2, it can be shown that U will also be compact if it is regarded as an operator from $\mathbf{L}^p(a, b)$ into $\mathbf{L}^r(a, b)$ $(1 < p, r < \infty)$.

All these facts, and their proofs also, carry over to the spaces $\mathbf{C}(K)$ and $\mathbf{L}^p(D)$.

2.2. We now prove three simple theorems on compact operators. Let $\mathbf{X}$ and $\mathbf{Y}$ be normed spaces.

THEOREM 1. *If U is a compact operator from $\mathbf{X}$ into $\mathbf{Y}$, then the image space $\mathbf{Y}_0 = \Delta_U$ of U is separable.*

Proof. Write B_n for the ball in $\mathbf{X}$ with centre at $\mathbf{0}$ and radius n, and write Q_n for the set $U(B_n)$ in $\mathbf{Y}$, that is, the set of elements of the form $y = U(x)$ $(x \in B_n)$. Since $\mathbf{X} = \bigcup_{n=1}^{\infty} B_n$, we have $\mathbf{Y}_0 = \bigcup_{n=1}^{\infty} Q_n$. But $\overline{Q}_n$ is compact and therefore separable (I.5.1). Therefore $\mathbf{Y}_0$ is also separable.

THEOREM 2. a) *A linear combination $U = \alpha U_1 + \beta U_2$ of compact operators U_1 and U_2, mapping $\mathbf{X}$ into $\mathbf{Y}$, is a compact operator.*

b) *If $U \in B(\mathbf{X}, \mathbf{Y})$ and $V \in B(\mathbf{Y}, \mathbf{Z})$, and if one of these operators is compact, then the product VU is also compact.*

Proof. a) Let E be a bounded subset of $\mathbf{X}$ and suppose $\{y_n\} \subset U(E)$. We have

$$y_n = \alpha U_1(x_n) + \beta U_2(x_n) \qquad (x_n \in E, n \in \mathbb{N}).$$

As U_1 is compact, we can choose a convergent subsequence $\{U_1(x_{n_k})\}$ of the sequence $\{U_1(x_n)\}$. For the same reason we can choose a convergent subsequence of the sequence $\{U_2(x_{n_k})\}$. Suppose this subsequence is $\{U_2(x_{n_{k_j}})\}$. Obviously $\{U(x_{n_{k_j}})\}$ is convergent, so $U(E)$ is relatively compact.

b) This follows immediately from the fact that a continuous linear operator maps a bounded set into a bounded set, and a compact set into a compact set.

THEOREM 3. *Let $\{U_n\}$ be a sequence of continuous linear operators from* $\mathbf{X}$ *into a B-space* $\mathbf{Y}$, *converging to an operator U (in $\mathbf{B}(\mathbf{X}, \mathbf{Y})$). If the U_n ($n = 1, 2, \ldots$) are compact, then so is U.*

Proof. Let B be the unit ball in $\mathbf{X}$. We show that $U(B)$ is relatively compact. Choose U_{n_0} such that

$$\|U_{n_0} - U\| < \varepsilon.$$

The set $U_{n_0}(B)$ is compact, and for $y \in U(B)$ we have

$$\|y - y_{n_0}\| = \|U(x) - U_{n_0}(x)\| \leqslant \|U - U_{n_0}\|\|x\| < \varepsilon$$

$$(y_{n_0} = U_{n_0}(x),\ y = U(x),\ x \in B),$$

so $U_{n_0}(B)$ forms an ε-net for $U(B)$. By the Remark following Hausdorff's Theorem, $U(B)$ is relatively compact, and this implies that U is compact.

REMARK. In the conditions of the theorem it is not possible to replace convergence in the space of operators—that is, convergence in norm—by convergence on $\mathbf{X}$. For, as we have already remarked, the operators U_h (see IX.1.1) are compact and

$$U_h(x) \to x \quad \text{on} \quad \mathbf{L}^p (x \in \mathbf{L}^p).$$

However, the identity operator into $\mathbf{L}^p$ is not compact. In fact, we can reformulate Theorem IV.1.3 as follows: the identity operator on a normed space $\mathbf{X}$ is compact if and only if $\mathbf{X}$ is finite-dimensional.

§ 3. Adjoint operators

3.1. Suppose we have a continuous linear operator U mapping a normed space $\mathbf{X}$ into a normed space $\mathbf{Y}$. Let g be a continuous linear functional on $\mathbf{Y}$, in other words, an element of the dual space $\mathbf{Y}^*$. For any $x \in \mathbf{X}$, write

$$f(x) = g(U(x)). \tag{1}$$

Thus the functional f is the product of the functional g, regarded as a linear operator from $\mathbf{X}$ into the space of real (or complex) scalars, and the operator U: that is, $f = gU$. Therefore f is a continuous linear functional, and

$$\|f\| \leqslant \|U\|\|g\|. \tag{2}$$

Thus (1) associates with each functional $g \in \mathbf{Y}^*$ a functional $f \in \mathbf{X}^*$. The operator that effects this correspondence is called the *adjoint* of the given operator U and is denoted by U^*. That is, $f = U^*g$ is the same as (1); in other words, $U^*(g) = gU$.

Hence to every $U \in B(\mathbf{X}, \mathbf{Y})$ there corresponds a dual operator U^* mapping $\mathbf{Y}^*$ into $\mathbf{X}^*$. We shall show that $U^* \in B(\mathbf{Y}^*, \mathbf{X}^*)$.

THEOREM 1. *The adjoint operator U^* is a continuous linear operator mapping $\mathbf{Y}^*$ into $\mathbf{X}^*$, and*

$$\|U^*\| = \|U\|. \tag{3}$$

Proof. Let us check that U^* is linear. If $g = \lambda g_1 + \mu g_2$ and $f = U^*(g)$, then

$$f(x) = g(U(x)) = \bar{\lambda} g_1(U(x)) + \bar{\mu} g_2(U(x)) = \bar{\lambda} U^*(g_1)(x) + \bar{\mu} U^*(g_2)(x).$$

Hence

$$U^*(g) = \lambda U(g_1) + \mu U(g_2).$$

The boundedness of U^* follows from (2), which shows that

$$\|U^*\| \leqslant \|U\|.$$

Next choose any $x \in \mathbf{X}$ and let $y = U(x)$. We construct a functional $g \in \mathbf{Y}^*$ such that

$$g(y) = \|y\|, \qquad \|g\| = 1.$$

Then

$$\|U(x)\| = \|y\| = g(y) = (gU)(x) = U^*(g)(x) \leqslant \|U^*(g)\| \|x\| \leqslant \|U^*\| \|x\|,$$

from which it is clear that

$$\|U\| \leqslant \|U^*\|.$$

Notice also that, if U_1 and U_2 are two continuous linear operators from $\mathbf{X}$ into $\mathbf{Y}$ and

$$U = \lambda U_1 + \mu U_2,$$

then

$$U^* = \bar{\lambda} U_1^* + \bar{\mu} U_2^*.$$

For we have (with $g \in \mathbf{Y}^*$, $x \in \mathbf{X}$)

$$\begin{aligned} U^*(g)(x) &= \lambda[g(U_1(x))] + \mu[g(U_2(x))] = (\lambda g)(U_1(x)) + (\mu g)(U_2(x)) \\ &= U_1^*(\bar{\lambda} g)(x) + U_2^*(\bar{\mu} g)(x) = [\bar{\lambda} U_1^*(g) + \bar{\mu} U_2^*(g)](x). \end{aligned}$$

Let V and U be continuous linear operators from $\mathbf{X}$ into $\mathbf{Y}$ and from $\mathbf{Y}$ into $\mathbf{Z}$, respectively. If $W = UV$, then $W^* = V^* U^*$. For, if we set $f = W^*(g)$, $f_1 = U^*(g)$ $(g \in \mathbf{Z}^*)$, we can write

$$f(x) = g(W(x)) = g(UV(x)) = f_1(V(x)) = V^* U^*(g)(x) \qquad (x \in \mathbf{X}).$$

3.2. Consider real finite-dimensional spaces $\mathbf{X}$ and $\mathbf{Y}$, of dimensions m and n respectively, and a linear operator U from $\mathbf{X}$ into $\mathbf{Y}$. The operator U is determined by a matrix

$$A = \begin{pmatrix} a_{11} & a_{12} & \dots & a_{1m} \\ a_{21} & a_{22} & \dots & a_{2m} \\ \dots & \dots & \dots & \dots \\ a_{n1} & a_{n2} & \dots & a_{nm} \end{pmatrix}$$

by means of the formula

$$\eta_j = \sum_{k=1}^{m} a_{jk} \xi_k \qquad (j = 1, 2, \dots, n), \tag{4}$$

where

$$x = (\xi_1, \xi_2, \dots, \xi_m) \in \mathbf{X}, \qquad y = (\eta_1, \eta_2, \dots, \eta_n) \in \mathbf{Y}.$$

Let $g = (\psi_1, \psi_2, \ldots, \psi_n)$ be the functional on $\mathbf{Y}$ given by

$$g(y) = \sum_{j=1}^{n} \psi_i \eta_i.$$

The functional $f = U^*g$ will have the form

$$f(x) = g(U(x)) = \sum_{j=1}^{n} \psi_j \sum_{k=1}^{m} a_{jk}\xi_k = \sum_{k=1}^{m} \left(\sum_{j=1}^{n} a_{jk}\psi_j \right) \xi_k,$$

that is, $f = (\phi_1, \phi_2, \ldots, \phi_m)$, where

$$\phi_k = \sum_{j=1}^{n} a_{jk}\psi_j = \sum_{j=1}^{n} a^*_{kj}\psi_j \qquad (a^*_{kj} = a_{jk}; j = 1, 2, \ldots, n; k = 1, 2, \ldots, m).$$

Hence U^* is determined by the matrix

$$A^* = \begin{pmatrix} a^*_{11} & a^*_{12} & \ldots & a^*_{1n} \\ a^*_{21} & a^*_{22} & \ldots & a^*_{2n} \\ \cdots & \cdots & \cdots & \cdots \\ a^*_{m1} & a^*_{m2} & \ldots & a^*_{mn} \end{pmatrix} = \begin{pmatrix} a_{11} & a_{21} & \ldots & a_{n1} \\ a_{12} & a_{22} & \ldots & a_{n2} \\ \cdots & \cdots & \cdots & \cdots \\ a_{1m} & a_{2m} & \ldots & a_{nn} \end{pmatrix},$$

which is obtained from A by interchanging the rows and columns.

Now let U be the integral operator with continuous kernel defined by

$$y = U(x), \qquad y(s) = \int_a^b K(s, t)x(t)\,dt.$$

We shall regard U as an operator on $\mathbf{L}^p(a, b)$ $(1 < p < \infty)$.

If we form the functional $g \in (\mathbf{L}^p)^*$ given by

$$g(y) = \int_a^b \psi(s)y(s)\,ds \qquad \left(y \in \mathbf{L}^p,\ \psi \in \mathbf{L}^q,\ \frac{1}{p} + \frac{1}{q} = 1\right),$$

then we have

$$g(U(x)) = \int_a^b \psi(s)\left[\int_a^b K(s, t)x(t)\,dt\right]ds = \int_a^b \left[\int_a^b K(s, t)\psi(s)\,ds\right]x(t)\,dt = \int_a^b \phi(t)x(t)\,dt.$$

Thus the adjoint operator U^* is also an integral operator, with kernel $K^*(s, t) = K(t, s)$:

$$f = U^*g, \qquad \phi(t) = \int_a^b K^*(t, s)\psi(s)\,ds = \int_a^b K(s, t)\psi(s)\,ds.$$

REMARK. If we consider complex finite-dimensional spaces, then the adjoint operator is determined by the matrix

$$A^* = \begin{pmatrix} a^*_{11} & a^*_{12} & \ldots & a^*_{1n} \\ a^*_{21} & a^*_{22} & \ldots & a^*_{2n} \\ \cdots & \cdots & \cdots & \cdots \\ a^*_{m1} & a^*_{m2} & \ldots & a^*_{mn} \end{pmatrix} \qquad (a^*_{jk} = \overline{a_{kj}}; j = 1, 2, \ldots, m; k = 1, 2, \ldots, n).$$

To obtain this result, it is enough to recall that the functionals $f = (\phi_1, \phi_2, \ldots,$

$\phi_m) \in \mathbf{X}^*$ and $g = (\psi_1, \psi_2, \ldots, \psi_n) \in \mathbf{Y}^*$ are defined by the formulae

$$f(x) = \sum_{k=1}^{m} \overline{\phi_k} \xi_k \qquad (x = (\xi_1, \xi_2, \ldots, \xi_m) \in \mathbf{X}),$$

$$g(y) = \sum_{k=1}^{n} \overline{\psi_k} \eta_k \qquad (y = (\eta_1, \eta_2, \ldots, \eta_n) \in \mathbf{Y}).$$

Similarly, in the case of the complex space $\mathbf{L}^p$, the operator dual to an integral operator will also be an integral operator with kernel

$$K^*(s, t) = \overline{K(t, s)}.$$

3.3. Regarding the operator U^* as given, we can form its adjoint, which we shall call the *second adjoint* of the operator U and denote by U^{**}. By Theorem 1, this will be a continuous linear operator from $\mathbf{X}^{**}$ into $\mathbf{Y}^{**}$. Identifying $\mathbf{X}$ with a subspace of $\mathbf{X}^{**}$ and $\mathbf{Y}$ with a subspace of $\mathbf{Y}^{**}$, we now show that U^{**} coincides with U on $\mathbf{X}$. Let π_1 be the canonical embedding of $\mathbf{X}$ into $\mathbf{X}^{**}$ and π_2 that of $\mathbf{Y}$ into $\mathbf{Y}^{**}$.

THEOREM 2. *For every $x \in \mathbf{X}$, we have*

$$U^{**}(\pi_1(x)) = \pi_2(U(x)). \tag{5}$$

Proof. If $g \in \mathbf{Y}^*$, then

$$[U^{**}(\pi_1(x))](g) = [\pi_1(x)](U^*(g)) = \overline{[U^*(g)](x)} = \overline{g(U(x))} = [\pi_2(U(x))](g),$$

from which (5) follows, as claimed.

COROLLARY. *If $\mathbf{X}$ is reflexive, then $U^{**} = U$.* (Equality must be understood as meaning that corresponding elements of $\mathbf{X}$ and $\mathbf{X}^{**}$, as well as those of $\mathbf{Y}$ and $\mathbf{Y}^{**}$, are to be identified.)

3.4. The property of compactness is preserved when we go from an operator to its adjoint.

THEOREM 3. *Let U be a continuous linear operator mapping a normed space $\mathbf{X}$ into a B-space $\mathbf{Y}$. If one of the operators U and U^* is compact, then so is the other.*

Proof. Assume first that U is compact, and choose a bounded sequence $\{g_n\}$ of functionals on $\mathbf{Y}$ (say $\|g_n\| \leqslant M$; $n = 1, 2, \ldots$). We show that we can select a convergent subsequence in the sequence of functionals $\{U^* g_n\}$. This will prove that U^* is compact.

Let $\mathbf{Y}_0$ denote the closure of $U(\mathbf{X})$. By Theorem IX.2.1, $\mathbf{Y}_0$ is a separable space. Let us consider the functionals g_n on $\mathbf{Y}_0$ only. The sequence of functionals $\{g_n\}$ is bounded, so by Theorem V.7.6 we can choose a weak* convergent subsequence. Let us assume that this has already been done—that is, assume that

$$\lim_{n \to \infty} g_n(y) = g(y) \qquad (y \in \mathbf{Y}_0).$$

Here g will be a linear functional on $\mathbf{Y}_0$. Since we can extend it to the whole of $\mathbf{Y}$ and leave the norm unchanged, we can assume that $g \in \mathbf{Y}^*$. Notice that

$$\|g\| \leqslant \underline{\lim}_{n \to \infty} \|g_n\| \leqslant M.$$

If we set $f = U^*(g)$, $f_n = U^*(g_n)$, then, for every $x \in \mathbf{X}$, we have

$$f_n(x) = g_n(U(x)) \to g(U(x)) = f(x),$$

because $U(x) \in \mathbf{Y}_0$. Hence we see that $f_n \to f\,(\sigma(\mathbf{X}^*, \mathbf{X}))$. Let us show that we also have convergence in norm. Assume the contrary. By passing to a subsequence if necessary, we may assume that

$$\|f_n - f\| \geqslant m > 0 \qquad (n = 1, 2, \ldots).$$

For each $n = 1, 2, \ldots$, we find an element x_n such that

$$\|x_n\| = 1, \; |f_n(x_n) - f(x_n)| > \tfrac{1}{2}\|f_n - f\| \geqslant \tfrac{1}{2}m \qquad (n = 1, 2, \ldots),$$

that is, if we set $y_n = U(x_n)$,

$$|g_n(y_n) - g(y_n)| \geqslant \tfrac{1}{2}m \qquad (n = 1, 2, \ldots). \tag{6}$$

Now U is compact and $\|x_n\| = 1$ $(n = 1, 2, \ldots)$, so we can choose a convergent subsequence $\{y_{n_k}\}$ of the sequence $\{y_n\}$. Assume that $y_{n_k} \to y_0$. Remembering that $y_0 \in \mathbf{Y}_0$, we have

$$|g_{n_k}(y_{n_k}) - g(y_{n_k})| \leqslant |g_{n_k}(y_{n_k}) - g_{n_k}(y_0)| + |g_{n_k}(y_0) - g(y_0)| + |g(y_0) - g(y_{n_k})| \leqslant 2M\,\|y_{n_k} - y_0\| + |g_{n_k}(y_0) - g(y_0)| \underset{k \to \infty}{\longrightarrow} 0,$$

which contradicts (6).

Now we show that if U^* is compact, then so is U.

Consider the operator U^{**}. By what we have proved, this is a compact operator, and since, by Theorem 2, we can regard it as an extension of U, it will also be compact. For if E is a bounded subset of $\mathbf{X}$, then $U(E)$ is relatively compact in $\mathbf{Y}^{**}$. As $\mathbf{Y}$ $(= \pi_2(\mathbf{Y}))$ is closed in $\mathbf{Y}^{**}$, $U(E)$ will also be relatively compact in $\mathbf{Y}$.

3.5. The concept of an adjoint operator enables us to state a convenient compactness criterion.

THEOREM 4. *Let U be a continuous linear operator mapping a normed space $\mathbf{X}$ into a separable B-space $\mathbf{Y}$. A necessary and sufficient condition for U to be compact is that the adjoint operator U^* map every sequence of functionals $\{g_n\}$ that is weak* convergent to zero in $\mathbf{Y}$ to a sequence $\{f_n\}$ of functionals in $\mathbf{X}$ that converges to zero in norm.*

Proof. Necessity. Assume that $\{f_n\}$ does not converge in norm to zero. We may suppose that

$$\|f_n\| \geqslant m > 0 \qquad (n = 1, 2, \ldots). \tag{7}$$

By the preceding theorem, U^* is a compact operator, and, since the sequence $\{g_n\}$ is bounded, we can select in $\{f_n\}$ a convergent subsequence $\{f_{n_k}\}$. Suppose that

$$f_{n_k} \to f. \tag{8}$$

For every $x \in \mathbf{X}$ we have

$$f(x) = \lim_{k \to \infty} f_{n_k}(x) = \lim_{k \to \infty} g_{n_k}(U(x)) = 0,$$

since $g_{n_k} \to \mathbf{0}$ $(\sigma(\mathbf{Y}^*, \mathbf{Y}))$. Consequently $f = \mathbf{0}$ and thus (8) contradicts (7).

Sufficiency. By Theorem 3 it is enough to check that U^* is a compact operator. Let B be the unit ball in $\mathbf{Y}^*$ and suppose that $\{g_n\} \subset B$. As $\mathbf{Y}$ is separable, Theorem V.7.6 shows that $g_{n_k} \to g$ $(\sigma(\mathbf{Y}^*, \mathbf{Y}))$. Hence, by the hypothesis, $U^*(g_{n_k}) \to U^*(g)$ in norm, which proves that $U^*(B)$ is relatively compact and so also that U^* is compact.

3.6. For an operator U on a Hilbert space $\mathbf{H}$ we already defined the concept of an adjoint operator, in V.3.3. Let us show that the new definition leads, in essence, to the same idea.

Let g be a linear functional on $\mathbf{H}$. By V.3.2, there exists an element $z \in \mathbf{H}$ such that

$$g(y) = (y, z) \qquad (y \in \mathbf{H}). \tag{9}$$

Now write $f = U^*g$, and suppose the functional f is defined by the element z^*:

$$f(x) = (x, z^*) \qquad (x \in \mathbf{H}). \tag{10}$$

Using (9) and (10), we rewrite the equation $f(x) = g(U(x))$ in the form

$$(x, z^*) = (Ux, z) \qquad (x \in \mathbf{H}). \tag{11}$$

The operator U^* is defined on $\mathbf{H}^*$. Since $\mathbf{H}^*$ is linearly isometric to $\mathbf{H}$, we can consider U^* as an operator on $\mathbf{H}$ as well (also clearly linear), which we also denote by the symbol U^*. Since g corresponds to the element z and f to the element z^*, we have $z^* = U^*z$, and (11) can be rewritten as $(Ux, z) = (x, U^*z)$ $(x \in \mathbf{H})$, which agrees with the previous definition of the adjoint operator.

A definition of the adjoint operator was given for specific operators in $\mathbf{L}^2$ by Hilbert (see Hilbert). For other particular cases see Riesz [3]. For the general definition, see Schauder [2], Hildebrandt [1], Banach.

§ 4. Compact self-adjoint operators on Hilbert space

4.1. In this section we give a detailed analysis of the structure of a compact self-adjoint operator on a Hilbert space $\mathbf{H}$. It turns out that the structure of such an operator is reminiscent of the structure of a symmetric matrix. As in the case of matrices, eigenvalues and related concepts play an important role in determining the properties of a compact self-adjoint operator.

The results in this section were established for integral operators in $\mathbf{L}^2$ by Hilbert and Schmidt, and in the general case by von Neumann [1] (for separable $\mathbf{H}$) and Rellich [1] (arbitrary $\mathbf{H}$).

An *eigenvalue* of an operator U is a number λ such that there exists an element $x_0 \neq \mathbf{0}$ with the property that

$$Ux_0 = \lambda x_0.$$

An element x for which the equation $Ux = \lambda x$ holds is called an *eigenvector* belonging to (or corresponding to) the given eigenvalue λ. The eigenvectors belonging to a given eigenvalue λ form the *eigenspace* $\mathbf{H}_\lambda$ corresponding to λ. It is not hard to verify that $\mathbf{H}_\lambda$ is in fact a subspace of $\mathbf{H}$.*

4.2. Let us note some very simple facts concerning eigenvalues and eigenvectors of a self-adjoint operator U on a Hilbert space $\mathbf{H}$.

I. *(Ux, x) is real, for every $x \in \mathbf{H}$* (cf. V.6.2).

This follows from

$$(Ux, x) = (x, Ux) = \overline{(Ux, x)}.$$

* All the above definitions can be carried over also to the case where one considers an arbitrary normed space $\mathbf{X}$ and an operator U on it. In this case we shall denote the eigenspace by $\mathbf{X}_\lambda$.

II. *We have the equation*

$$\|U\| = \sup_{\|x\|=1} |(Ux, x)|.$$

Write $Q = \sup_{\|x\|=1} |(Ux, x)|$. Since, when $\|x\| = 1$, we have

$$|(Ux, x)| \leqslant \|Ux\| \|x\| \leqslant \|Ux\| \leqslant \|U\|,$$

it follows that $Q \leqslant \|U\|$.

Also, by the easily verified identity (cf. V.4.2)

$$(Ux, y) = \tfrac{1}{4}\{[(U(x+y), x+y) - (U(x-y), x-y)] + \\ + i[(U(x+iy), x+iy) - (U(x-iy), x-iy)]\} \tag{1}$$

we see from I that

$$\operatorname{Re}(Ux, y) = \tfrac{1}{4}[(U(x+y), x+y) - (U(x-y), x-y)] \leqslant \tfrac{1}{4}Q[\|x+y\|^2 + \|x-y\|^2] \\ = \tfrac{1}{2}Q[\|x\|^2 + \|y\|^2]$$

(for the last equality, see IV.5.1). If we take an x with $\|x\| = 1$ and $y = \dfrac{Ux}{\|Ux\|}$, we obtain

$$\|Ux\| = \operatorname{Re}(Ux, y) \leqslant Q,$$

from which we obtain

$$\|U\| \leqslant Q.$$

III. *The eigenvalues of U are real.*

For if λ is an eigenvalue and x is a non-zero eigenvector corresponding to it, then

$$\lambda = \frac{(Ux, x)}{(x, x)}.$$

Since U is self-adjoint, the numerator in this expression is real, and hence so is λ.

IV. *Eigenspaces $\mathbf{H}_{\lambda_1}$ and $\mathbf{H}_{\lambda_2}$ corresponding to distinct eigenvalues λ_1 and λ_2 of U are orthogonal.*

For let x and y be elements of $\mathbf{H}_{\lambda_1}$ and $\mathbf{H}_{\lambda_2}$ respectively. Since $Ux = \lambda_1 x$, $Uy = \lambda_2 y$, we have, assuming, say, that $\lambda_1 \neq 0$,

$$(x, y) = \frac{1}{\lambda_1}(Ux, y) = \frac{1}{\lambda_1}(x, Uy) = \frac{\lambda_2}{\lambda_1}(x, y),$$

which is possible only if $(x, y) = 0$.

4.3. Now assume that the operator U is not merely self-adjoint but also compact. We have

THEOREM 1. *The operator U has at least one eigenvalue.*

Proof. Since the assertion of the theorem is obvious when $U = \mathbf{0}$, we assume that $U \neq \mathbf{0}$. Let us consider the *bounds* of the operator, namely

$$m = \inf_{\|x\|=1} (Ux, x), \qquad M = \sup_{\|x\|=1} (Ux, x).$$

By II, we have $\|U\| = \max[|m|, M]$. We prove that

$$\lambda_1 = \begin{cases} m, & \text{if } \|U\| = |m|, \\ M, & \text{if } \|U\| = M, \end{cases}$$

is an eigenvalue of U.

In fact, let us consider, say, the case where $\|U\| = M$. By the definition of the number M, there is a sequence of normalized elements $\{x_n\}$ such that

$$(Ux_n, x_n) \underset{n\to\infty}{\longrightarrow} M = \lambda_1. \tag{2}$$

Since $\{x_n\}$ is a bounded sequence, the compactness of U means that we can choose a convergent subsequence of $\{Ux_n\}$. Assume that this has already been done—that is, assume that $\{Ux_n\}$ is convergent. Assume, say, that $Ux_n \to y_0$. If we recall that $\lambda_1^2 = \|U\|^2$ and use the relations

$$\|Ux_n - \lambda_1 x_n\|^2 = \|Ux_n\|^2 - 2\lambda_1(Ux_n, x_n) + \lambda_1^2 \leqslant \|U\|^2 + \lambda_1^2 - 2\lambda_1(Ux_n, x_n),$$

we see, by (2), that

$$Ux_n - \lambda_1 x_n \underset{n\to\infty}{\longrightarrow} \mathbf{0}.$$

But then

$$x_n = \frac{1}{\lambda_1}[Ux_n - (Ux_n - \lambda_1 x_n)]$$

also has a limit. Namely, we have $x_n \to x_0 = 1/\lambda_1\, y_0$. Since $Ux_n \to Ux_0 = y_0$, we have $y_0 = Ux_0 = \lambda_1 x_0$. As $x_0 \neq \mathbf{0}$ ($\|x_0\| = 1$), λ_1 is an eigenvalue.

The theorem is therefore proved.

COROLLARY. *If U has no non-zero eigenvalues, then $U = \mathbf{0}$.*

REMARK. The eigenvalue of U that was found in the theorem is the one with largest absolute value.

For if λ is an eigenvalue and x an eigenvector corresponding to it—which we may take to be normalized—then

$$|\lambda| = |\lambda|(x, x) = |(Ux, x)| \leqslant \|U\| = |\lambda_1|.$$

4.4. Let P_λ denote the projection (see V.3.4) onto the eigenspace $\mathbf{H}_\lambda$. The following theorem is a substantial refinement of Theorem 1.

THEOREM 2. *The set of eigenvalues of a compact self-adjoint operator U is at most countable. Moreover,*

$$U = \sum_k \lambda_k P_{\lambda_k}, \tag{3}$$

where $\lambda_1, \lambda_2, \ldots$ are the distinct eigenvalues of U (convergence of the series being understood to mean convergence in norm in the space of operators).

Proof. Let λ be an eigenvalue of U. The following relations hold:

$$\lambda P_\lambda = UP_\lambda = P_\lambda U. \tag{4}$$

For, since $P_\lambda x \in \mathbf{H}_\lambda$ for every $x \in \mathbf{H}$, we have $UP_\lambda x = \lambda P_\lambda x$. The permutability of P_λ and U follows from the fact that the product $UP_\lambda = \lambda P_\lambda$ is a self-adjoint operator (see V.6.1).

Consider the operator

$$U_2 = U_1 - \lambda_1 P_{\lambda_1}, \qquad (U_1 = U). \tag{5}$$

In view of (4), we can write

$$U_2 = \tilde{P}_1 U_1 = U_1 \tilde{P}_1, \tag{6}$$

where we have put $\tilde{P}_1 = I - P_{\lambda_1}$. It follows from this that U_2 is a compact self-adjoint operator (the compactness comes from Theorem IX.2.2). Also (6) shows that

$$\|U_2\| \leqslant \|\tilde{P}_1\| \|U_1\| \leqslant \|U_1\|.$$

Applying the preceding theorem to U_2, we can find an eigenvalue λ_2 for it. Further, we have

$$|\lambda_1| \geqslant |\lambda_2|,$$

as $|\lambda_1| = \|U_1\|$, $|\lambda_2| = \|U_2\|$.

We show that λ_1 is not an eigenvalue of U_2. In fact, there would otherwise be an element $x \neq \mathbf{0}$ such that

$$U_2 x = \lambda_1 x$$

or, in view of (5),

$$U_1 x - \lambda_1 P_{\lambda_1} x = \lambda_1 x. \tag{7}$$

Applying P_{λ_1} to both sides of this equation, we obtain, in view of (4),

$$\lambda_1 P_{\lambda_1} x = U P_{\lambda_1} x - \lambda_1 P_{\lambda_1} x = \mathbf{0}.$$

Therefore, by (7),

$$U_1 x = \lambda_1 x,$$

that is, $x \in \mathbf{H}_{\lambda_1}$, and so $x = P_{\lambda_1} x = \mathbf{0}$, giving a contradiction.

Let us show further that every non-zero eigenvalue of U_2 is at the same time also an eigenvalue of U_1, and the corresponding eigenspaces coincide.

In fact, let $\lambda \neq 0$ be an eigenvalue of U_2 and let $x \neq \mathbf{0}$ be an element such that $U_2 x = \lambda x$. By (6), we have

$$U_1 \tilde{P}_1 x = \lambda x, \tag{8}$$

so that

$$\tilde{P}_1 U_1 \tilde{P}_1 x = \lambda \tilde{P}_1 x.$$

On the other hand,

$$\tilde{P}_1 U_1 \tilde{P}_1 x = U_1 \tilde{P}_1^2 x = U_1 \tilde{P}_1 x = \lambda x,$$

and so $\tilde{P}_1 x = x$. Thus, by (8), we have

$$U_1 x = \lambda x,$$

that is, λ is an eigenvalue of U_1. If we now regard x as an eigenvector of U_1 belonging to the eigenvalue λ, then, since $\lambda \neq \lambda_1$, we see that $\mathbf{H}_\lambda$ and $\mathbf{H}_{\lambda_1}$ are orthogonal (see III in 4.3), so $P_{\lambda_1} x = \mathbf{0}$ and, according to (5), we have

$$U_2 x = U_1 x - \lambda_1 P_{\lambda_1} x = U_1 x = \lambda x.$$

Thus x is an eigenvector of U_2.

If U_2 is not identically zero, then we can construct from it the operator $U_3 = U_2 - \lambda_2 P_{\lambda_2}$, and so on. Repeating this process, we construct compact self-adjoint operators $U_1 = U, U_2, \ldots, U_n$ and eigenvalues of these, $\lambda_1, \lambda_2, \ldots, \lambda_n$, such that

$$U_{k+1} = U_k - \lambda_k P_{\lambda_k} = U - \sum_{j=1}^{k} \lambda_j P_{\lambda_j} \qquad (k = 1, 2, \ldots, n-1),$$

$$|\lambda_1| \geqslant |\lambda_2| \geqslant \ldots \geqslant |\lambda_n|, \qquad \|U_k\| = |\lambda_k| \qquad (k = 1, 2, \ldots, n).$$

Further, by what has been proved, the λ_k will be pairwise distinct eigenvalues of $U_1 = U$.
Assume that $U_n = \mathbf{0}$. Then

$$U = \sum_{j=1}^{n-1} \lambda_j P_{\lambda_j}. \tag{9}$$

However, if $U_n \neq \mathbf{0}$ for every $n = 1, 2, \ldots$, then the process leads to a sequence of operators $U_1, U_2, \ldots$ and eigenvalues $\lambda_1, \lambda_2, \ldots$. We show that, in this case, $\lambda_n \to 0$. For otherwise, for all $n = 1, 2, \ldots$, we should have

$$|\lambda_n| \geqslant \lambda_0 > 0.$$

Choose a normalized element x_n in $\mathbf{H}_{\lambda_n}$. The different elements x_n are pairwise orthogonal, so

$$\|Ux_m - Ux_n\|^2 = \|\lambda_m x_m - \lambda_n x_n\|^2 = |\lambda_m|^2 + |\lambda_n|^2 \geqslant 2\lambda_0^2$$
$$(m \neq n),$$

from which it follows that neither the sequence $\{Ux_n\}$ nor any of its subsequences converges, which contradicts the assumption that U is compact. Since $\|U_n\| = |\lambda_n|$, it follows from what we have proved that $U_n \to \mathbf{0}$ as $n \to \infty$, and therefore

$$U = \sum_{k=1}^{\infty} \lambda_k P_{\lambda_k}.$$

Hence we have established that U has a representation in the form (3).

Let us prove that the operator has no other non-zero eigenvalues, apart from $\lambda_1, \lambda_2, \ldots, \lambda_n, \ldots$. In fact, if λ were such an eigenvalue, then for some $x \neq \mathbf{0}$ we should have $Ux = \lambda x$; that is,

$$\lambda x = \sum_k \lambda_k P_{\lambda_k} x.$$

The elements $P_{\lambda_k} x$ belong to $\mathbf{H}_{\lambda_k}$ so they are pairwise orthogonal. Hence

$$\lambda P_{\lambda_m} x = \lambda_m P_{\lambda_m} x \qquad (m \in \mathbb{N}).$$

Since $\lambda \neq \lambda_m$, we have $P_{\lambda_m} x = \mathbf{0}$. Hence $x = \mathbf{0}$.

This proves the theorem.

REMARK 1. It is not hard to verify that the eigenspace $\mathbf{H}_{\lambda_k}$ corresponding to a non-zero eigenvalue is finite-dimensional.

For every bounded set E in $\mathbf{H}_{\lambda_k}$ is the image of a bounded set $\tilde{E}$ ($\tilde{E}$ consists of elements z of the form $z = (1/\lambda_k)x$, where $x \in E$). Therefore, as U is compact, E is relatively compact, and this is possible only if $\mathbf{H}_{\lambda_k}$ is finite-dimensional (see Theorem IV.1.3).

REMARK 2. It follows from (6) that the eigenvalue λ_k may be defined by

$$\lambda_k = \pm \sup_{\substack{\|x\|=1 \\ x \perp \mathbf{H}_{\lambda_j}}} |(Ux, x)| \qquad (j = 1, 2, \ldots, k-1).$$

4.5. Let $\mathbf{H}_0$ denote the eigenspace corresponding to the eigenvalue zero, and let $\tilde{\mathbf{H}}$ be its orthogonal complement. Since

$$(Ux, y) = (x, Uy) = 0,$$

for $x \in \mathbf{H}$, $y \in \mathbf{H}_0$, we have $Ux \perp \mathbf{H}_0$; that is, $U(\mathbf{H}) \subset \tilde{\mathbf{H}}$. Conversely, if $y \perp U(\mathbf{H})$, then, for every $x \in \mathbf{H}$,

$$0 = (Ux, y) = (x, Uy),$$

which shows that $Uy = \mathbf{0}$, that is, $y \in \mathbf{H}_0$. Hence $\tilde{\mathbf{H}} = \overline{U(\mathbf{H})}$.

Let us choose a complete orthogonal system in each of the subspaces $\mathbf{H}_{\lambda_k}$. In view of the above remark, this will be finite, for each $k = 1, 2, \ldots$. Putting these systems together, we obtain an orthonormal system $x_1, x_2, \ldots$ consisting of eigenvalues of U. We denote the eigenvalue corresponding to x_k by λ_k, as before. In this procedure we shall, in general, have repetitions among the eigenvalues.* It follows from Theorem 2 that the resulting orthonormal system will be complete in $\tilde{\mathbf{H}}$, and so every element $z \in \mathbf{H}$ can be expressed in the form

$$z = x_0 + \tilde{x}, \tag{10}$$

where

$$x_0 \in \mathbf{H}_0, \qquad \tilde{x} = \sum_k c_k x_k \in \tilde{\mathbf{H}}$$

$$(c_k = (\tilde{x}, x_k) = (z, x_k), \quad k \in \mathbb{N}).$$

In particular, if $z = Ux$, then, as we have mentioned above, $z \in \tilde{\mathbf{H}}$, and so

$$z = \sum_k c_k x_k \qquad (c_k = (z, x_k) = (Ux, x_k) = (x, Ux_k) = \lambda_k (x, x_k)).$$

Let us consider the equation

$$x - \mu U x = y, \tag{11}$$

in which y is a fixed element of $\mathbf{H}$ and μ is a numerical parameter.

Expressing y and the unknown element x in the form (10), we obtain the equation

$$x_0 + \sum_k c_k x_k - \mu \sum_k c_k \lambda_k x_k = y_0 + \sum_k d_k x_k$$

$$(c_k = (x, x_k), \qquad d_k = (y, y_k)),$$

so that

$$x_0 = y_0, \qquad c_k(1 - \mu\lambda_k) = d_k \qquad (k = 1, 2, \ldots)$$

---

* The convenience of this notation consists in the fact that not only does an eigenvector determine an eigenvalue, but also, conversely, an eigenvalue uniquely determines (to within a multiple) an eigenvector.

and therefore

$$x_0 = y_0, \qquad c_k = \frac{d_k}{1 - \mu\lambda_k} \qquad (k = 1, 2, \ldots), \tag{12}$$

provided

$$1 - \mu\lambda_k \neq 0$$

for $k = 1, 2, \ldots$.

Moreover,

$$\sum |c_k|^2 \leqslant \frac{1}{\min_k |1 - \mu\lambda_k|^2} \sum_k |d_k|^2 < \infty,$$

and a solution exists and is unique. Denoting the solution we seek by x^*, we have, in this case,

$$x^* = y_0 + \sum_k \frac{d_k}{1 - \mu\lambda_k} x_k.$$

However, if $\mu\lambda_k = 1$, then a solution exists only if the corresponding coefficients $d_k = 0$, that is, only if the right-hand side of (11)—namely, the element y—is orthogonal to all the eigenvectors belonging to the eigenvalue $1/\mu$. Here the solution is not unique, since the coefficients c_k may be chosen arbitrarily: if x^* is a solution, then so is $x^* + \bar{x}$, where $\bar{x}$ is any element of $\mathbf{H}_{1/\mu}$.

The numbers

$$\mu_k = \frac{1}{\lambda_k} \qquad (k = 1, 2, \ldots)$$

will be called the *characteristic values* of equation (11), and the eigenvectors belonging to an eigenvalue λ_k will be called the *characteristic vectors* corresponding to the characteristic value μ_k.

If we take U to be an integral operator

$$y = Ux, \qquad y(s) = \int_a^b K(s, t)x(t)dt$$

in $\mathbf{L}^2$ with continuous symmetric kernel, then the general results stated above become well-known results in the theory of integral equations, whose formulation we shall not go into.

Also, by considering a symmetric matrix as an operator on a finite-dimensional Hilbert space, one can derive known theorems in the theory of matrices from what has been said above.

4.6. Let $\mathbf{H}$ be a separable space. For the matrix representation (see V.3.1) of a compact self-adjoint operator U, we choose a complete orthonormal system as follows: we adjoin an orthonormal system in $\mathbf{H}_0$ to the system constructed above for $\hat{\mathbf{H}}$. If we denote the resulting system by $\{x_k\}$, and the corresponding eigenvalues by $\{\lambda_k\}$ (so that λ_k may be equal to zero), we see that the matrix representing U is

$$A = \begin{pmatrix} \lambda_1 & 0 & 0 & \ldots \\ 0 & \lambda_2 & 0 & \ldots \\ 0 & 0 & \lambda_3 & \ldots \\ \ldots & \ldots & \ldots & \ldots \end{pmatrix}. \tag{13}$$

The diagonal elements of this matrix are the eigenvalues, and all other elements are zero.

It can be shown that, if U admits a matrix representation by means of a matrix of the form (13), where the λ_k are real and $\lambda_k \to 0$, then U is a compact self-adjoint operator.

§ 5. Integral representations of self-adjoint operators

In this section we establish a representation for an arbitrary self-adjoint operator in the form of an abstract integral of Stieltjes type, which reduces to formula (3) of the preceding section when the operator in question is compact.

Such a representation was found for operators on specific spaces by Hilbert (see Hilbert), and in the general case by von Neumann [1]. The idea behind the arguments presented below is due to F. Riesz. See also Akhiezer and Glazman for the material in this section.

The basic tool for the present investigation is the concept of a function on operators. We encountered special cases of this concept earlier: for example, we defined powers of operators on normed spaces with integral exponents, and square roots of positive operators on Hilbert space. The correspondence between functions of a real argument and operators will be systematically extended below to all continuous functions.

5.1. Let U be a self-adjoint operator on a Hilbert space $\mathbf{H}$. As in the preceding section, we denote its bounds by

$$m = \inf_{\|x\|=1} (Ux, x), \qquad M = \sup_{\|x\|=1} (Ux, x).$$

Now let $\phi(t) = c_0 + c_1 t + c_2 t^2 + \ldots + c_n t^n$. By definition, we set

$$\phi(U) = c_0 I + c_1 U + \ldots + c_n U^n.$$

The operator $\phi(U)$ is called an *operator polynomial*.

We note some properties of operator polynomials.

I. $[\phi(U)]^* = \bar{\phi}(U)$. In particular, if $\phi(t)$ is a real polynomial, then $\phi(U)$ is a self-adjoint operator.

II. If $\phi(t) = \alpha\phi_1(t) + \beta\phi_2(t)$, then $\phi(U) = \alpha\phi_1(U) + \beta\phi_2(U)$.

III. If $\phi(t) = \phi_1(t)\phi_2(t)$, then $\phi(U) = \phi_1(U)\phi_2(U)$.

IV. An operator polynomial is permutable with any operator that permutes with U: that is, $UV = VU$ implies $\phi(U)V = V\phi(U)$.

V. If $\phi(t) \geqslant 0$ for $t \in [m, M]$, then $\phi(U)$ is a positive operator.

As the first four properties are obvious, we shall prove only the last one.

Since the polynomial $\phi(t)$ is positive in the interval $[m, M]$, it cannot have roots of odd multiplicities inside this interval; moreover, since the sign of a polynomial changes at a root of odd multiplicity and the sign of $\phi(t)$ coincides with that of its highest coefficient c_n for sufficiently large t, the number of roots greater than or equal to M and of odd multiplicity is even if $c_n > 0$ and odd if $c_n < 0$. Bearing all this in mind, one can convince oneself that $\phi(t)$ is expressible in the form

$$\phi(t) = \phi_1(t)\phi_2(t) \ldots \phi_s(t),$$

where the factors $\phi_k(t)$ have one of the following forms:

$$\phi_k(t) = \begin{cases} a & (a > 0), \\ (t+\alpha_k)^2 + \beta_k & (\alpha_k \text{ real}, \beta_k > 0), \\ t - t_k & (t_k \leqslant m), \\ t_k - t & (t_k \geqslant M). \end{cases}$$

It is not hard to check that in every case $\phi_k(U) \geqslant \mathbf{0}$.

For if, say, $\phi_k(t) = t - t_k$ $(t_k \leqslant m)$, then $\phi_k(U) = U - t_k I$ and

$$(\phi_k(U)x, x) = (Ux, x) - t_k(x, x) \geqslant (m - t_k)(x, x) \geqslant 0.$$

By III, we have

$$\phi(U) = \phi_1(U)\phi_2(U) \ldots \phi_s(U).$$

But the operators $\phi_k(U)$ are permutable with one another (property IV), and so we can apply the corollary to Theorem V.6.2, which shows that the product $\phi(U)$ is also a positive operator.

Let us note another two properties of operator polynomials that follow from what we have said above.

VI. If $\phi_1(t) \leqslant \phi_2(t)$ in the interval $[m, M]$, then $\phi_1(U) \leqslant \phi_2(U)$.

VII. $\|\phi(U)\| \leqslant \max_{t \in [m, M]} |\phi(t)|$.

For

$$\|\phi(U)\|^2 = \sup_{\|x\| = 1} (\phi(U)x, \phi(U)x) = \sup(\bar{\phi}(U)\phi(U)x, x) = \sup_{\|x\| = 1} (\psi(U)x, x),$$

where we have set $\psi(t) = |\phi(t)|^2$. If $l = \max_{t \in [m, M]} |\phi(t)|$, then

$$0 \leqslant \psi(t) \leqslant l^2 \qquad (t \in [m, M])$$

and so

$$0 \leqslant \psi(U) \leqslant l^2 I.$$

Therefore

$$\sup_{\|x\| = 1} (\psi(U)x, x) \leqslant l^2.$$

5.2. Suppose the function $\phi(t)$ is continuous in the interval $[m, M]$. There is a sequence of polynomials $\{\phi_n(t)\}$ that converges uniformly to $\phi(t)$ in $[m, M]$. It is not hard to see that the sequence of operators $\{\phi_n(U)\}$ will converge to an operator. For

$$\|\phi_{n+p}(U) - \phi_n(U)\| \leqslant \max_{t \in [m, M]} |\phi_{n+p}(t) - \phi_n(t)| \underset{n \to \infty}{\longrightarrow} 0.$$

If we have another sequence of polynomials also converging uniformly to $\phi(t)$ (in $[m, M]$), then the sequence of corresponding operator polynomials will converge to the same limiting operator as the original sequence. One can convince oneself of this by uniting the two sequences into one. These arguments provide grounds for the notation

$$\phi(U) = \lim_{n \to \infty} \phi(U).$$

The operator $\phi(U)$ will be called an *operator function.*

Some properties of operator functions are formulated in the following theorem.

THEOREM 1. a) $[\phi(U)]^* = \bar{\phi}(U)$. *In particular, if* $\phi(t)$ *is real, for* $t \in [m, M]$, *then* $\phi(U)$ *is a self-adjoint operator.*

b) *If* $\phi(t) = \alpha\phi_1(t) + \beta\phi_2(t)$, *then* $\phi(U) = \alpha\phi_1(U) + \beta\phi_2(U)$.

c) *If* $\phi(t) = \phi_1(t)\phi_2(t)$, *then* $\phi(U) = \phi_1(U)\phi_2(U)$.

d) *The operator* $\phi(U)$ *is permutable with every operator that permutes with* U.

e) *If* $\phi(t) \geqslant \psi(t)$ $(t \in [m, M])$, *then* $\phi(U) \geqslant \psi(U)$.

f) $\|\phi(U)\| \leqslant \max_{t \in [m, M]} |\phi(t)|$.*

g) *If a sequence* $\{\phi_n(t)\}$ *of continuous functions converges uniformly to a function* $\phi(t)$ *in the interval* $[m, M]$, *then* $\|\phi_n(U) - \phi(U)\| \to 0$.

h) *If P is a projection that is permutable with U, then*

$$P\phi(U) = P\phi(PU).$$

Proof. The truth of a)–g) is established without difficulty, using properties of operator polynomials.

As regards h), if $\phi(t)$ is a polynomial, then the relation we require to prove is easily verified, when we note that $P^2 = P$ and so $P^k = P$ $(k = 1, 2, \ldots)$. In the case where $\phi(t)$ is an arbitrary continuous function, we need to use a limit argument.

5.3. To determine the operator $\phi(U)$ uniquely it is not necessary to know the values of $\phi(t)$ on the whole interval $[m, M]$. It turns out to be sufficient to specify $\phi(t)$ on a certain closed set $S_U \subset [m, M]$, called the spectrum of U.

We shall say that a number λ is a spectral point of a self-adjoint operator U if there exists a sequence $\{x_n\}$ such that

$$\|x_n\| = 1 \qquad (n = 1, 2, \ldots), \qquad Ux_n - \lambda x_n \underset{n \to \infty}{\longrightarrow} 0. \tag{1}$$

In other words, λ is a spectral point if

$$\inf_{\|x\| = 1} \|Ux - \lambda x\| = 0. \tag{2}$$

The collection of all spectral points is called the *spectrum*† of U and is denoted by S_U.

Clearly every eigenvalue belongs to the spectrum; however, the spectrum may contain points that are not eigenvalues. For instance, when U is compact and **H** is infinite-dimensional, $\lambda = 0$ will be a spectral point, although it is easy to see that it is not always an eigenvalue.

Let λ be a spectral point. By (1), we have

$$\lambda = \lim_{n \to \infty} (Ux_n, x_n).$$

It thus follows that the spectrum of U is located on the real line and contained in the interval $[m, M]$. Let us show that *the bounds of an operator U are spectral points.*

We assume that‡ $0 \leqslant m \leqslant M$, and consider the point $\lambda = M$.

If $\|x\| = 1$, then, bearing in mind that $\|U\| = \lambda$, we have

$$\|Ux - \lambda x\|^2 = \|Ux\|^2 - 2\lambda(Ux, x) + \lambda^2 \leqslant 2\lambda[\lambda - (Ux, x)],$$

---

* Below (in 5.3) we shall establish a more precise expression for $\|\phi(U)\|$.

† This definition makes sense formally even when U is an arbitrary linear operator on an arbitrary normed space. However, we shall give another definition of the spectrum in the general case, which is equivalent to this one only when U is a self-adjoint operator on a Hilbert space.

‡ It is always possible to arrange that this condition is satisfied by adding an operator of the form μI to U; then the spectrum and the bounds are both shifted a distance μ to the right along the real axis.

so that

$$\inf_{\|x\|=1} \|Ux-\lambda x\|^2 \leqslant 2\lambda\left[\lambda - \sup_{\|x\|=1} (Ux, x)\right] = 0,$$

and, by (2), this means that $\lambda \in S_U$.

We can now show, further, that S_U is a closed set.

For if $\lambda_0 \notin S_U$ then, by (2),

$$\inf_{\|x\|=1} \|Ux-\lambda_0 x\| = d > 0.$$

Thus if $|\lambda-\lambda_0| < d/2$, we have

$$\inf_{\|x\|=1} \|Ux-\lambda x\| \geqslant \inf_{\|x\|=1} \|Ux-\lambda_0 x\| - \sup_{\|x\|=1} \|\lambda_0 x-\lambda x\| > d-\frac{d}{2} = \frac{d}{2} > 0$$

and hence the complement of S_U is open.

Before stating the fundamental result, we prove two auxiliary propositions.

LEMMA 1. *Let $\phi(t)$ be a real polynomial. Then $S_\phi(U) = \phi(S_U)$: that is, the spectrum of $\phi(U)$ consists of all μ expressible in the form $\mu = \phi(\lambda)$ $(\lambda \in S_U)$.*

Proof. Let μ be a real number and let $t_1, t_2, \ldots, t_n$ be all the roots of the equation

$$\phi(t) = \mu.$$

The operator $\phi(U)-\mu I$ is obviously expressible as a product

$$\phi(U)-\mu I = c(U-t_1 I)(U-t_2 I) \ldots (U-t_s I). \tag{3}$$

Choose $\lambda \in S_U$. There exists a sequence $\{x_n\}$ of normalized elements such that $Ux_n - \lambda x_n \to 0$ as $n \to \infty$. Set $\mu = \phi(\lambda)$, $t_s = \lambda$ in (3). Then

$$\phi(U)x_n - \mu x_n = c(U-t_1 I)(U-t_2 I) \ldots (Ux_n - \lambda x_n) \underset{n\to\infty}{\longrightarrow} 0,$$

that is, $\mu \in S_{\phi(U)}$.

Conversely, if none of the t_k belong to the spectrum of U, then

$$\inf_{\|x\|=1} \|Ux-t_s x\| = \rho_s > 0,$$

$$\inf_{\|x\|=1} \|(U-t_{s-1}I)\ (Ux-t_s x)\| \geqslant \inf_{\|y\|\geqslant \rho_s} \|Uy-t_{s-1}y\| = \rho_{s-1} > 0$$

and, proceeding in this way, we find that

$$\inf_{\|x\|=1} \|\phi(U)x-\mu x\| = \rho_1 > 0.$$

Hence $\mu = \phi(t_k)$ $(k = 1, 2, \ldots, s)$ does not belong to $S_{\phi(U)}$.

LEMMA 2. *If $\phi(t)$ is a polynomial, then*

$$\|\phi(U)\| = \max_{t\in S_U} |\phi(t)|.$$

For

$$\|\phi(U)\|^2 = \sup_{\|x\|=1} (\phi(U)x, \phi(U)x) = \sup_{\|x\|=1} (\bar{\phi}(U)\phi(U)x, x) = \sup_{\|x\|=1} (\psi(U)x, x), \tag{4}$$

where we have set $\psi(t) = |\phi(t)|^2$.

Thus $\|\phi(U)\|^2$ is the supremum of $\psi(U)$. But the supremum of the positive operator $\psi(U)$ coincides with the supremum of its spectrum:

$$M_{\psi(U)} = \sup S_{\psi(U)}. \tag{5}$$

In accordance with Lemma 1, we have

$$\sup S_{\psi(U)} = \sup \psi(S_U) = \sup_{t \in S_U} \psi(t) = \left[\sup_{t \in S_U} |\phi(t)|\right]^2.$$

Since S_U is closed, we may replace the word "supremum" here by "maximum". Comparing the last equation with (4) and (5), we obtain the required result.

The fundamental theorem can now be proved without any difficulty.

THEOREM 2. *If the function $\phi(t)$ is continuous in $[m, M]$, then*

$$\|\phi(U)\| = \max_{t \in S_U} |\phi(t)|. \tag{6}$$

For let $\{\phi_n(t)\}$ be a sequence of polynomials converging uniformly to $\phi(t)$. By Lemma 2,

$$\|\phi_n(U)\| = \max_{t \in S_U} |\phi_n(t)| = \|\phi_n\|_{\mathbf{C}}.$$

Letting $n \to \infty$ here, we obtain (6).

COROLLARY. *If the functions ϕ_1 and ϕ_2 are continuous in $[m, M]$ and coincide on the spectrum S_U, then $\phi_1(U) = \phi_2(U)$.*

Suppose $\phi(t)$ is defined and continuous on the spectrum of an operator U. If we extend it continuously to the whole interval $[m, M]$, we obtain a function $\tilde{\phi}(t)$, continuous in $[m, M]$. By definition, we set

$$\phi(U) = \tilde{\phi}(U).$$

By what was shown above, $\phi(U)$ does not depend on how $\phi(t)$ was extended and is determined by the values of ϕ on S_U alone. Clearly the properties of operator functions stated in Theorem 1 carry over to the functions just defined (in the statement of Theorem 1 the interval $[m, M]$ must be replaced by S_U throughout).

The extension of the concept of an operator function enables us to prove an important theorem characterizing the spectrum of a self-adjoint operator.

We shall say that a complex number λ is a *regular value* of an operator U if it does not belong to the spectrum of U.

THEOREM 3. *A necessary and sufficient condition that λ be a regular value of an operator U is that there exist an inverse linear operator**

$$R_\lambda = [U - \lambda I]^{-1}$$

on **H**.

Proof. Necessity. Suppose λ is a regular value. Define the function ρ_λ on S_U by

$$\rho_\lambda(t) = \frac{1}{t - \lambda}$$

and write $R_\lambda = \rho_\lambda(U)$. As

$$(t - \lambda)\rho_\lambda(t) = 1 \qquad (t \in S_U),$$

* It is in fact this theorem that provides the basis for the definition of the spectrum in the general case of an arbitrary linear operator on an arbitrary normed space (see Chapter XIII).

Theorem 1 shows that

$$(U-\lambda I)R_\lambda = R_\lambda(U-\lambda I) = I,$$

and so

$$R_\lambda = [U-\lambda I]^{-1}.$$

Sufficiency. If there exists a linear inverse (or even left inverse) operator

$$R_\lambda = [U-\lambda I]^{-1},$$

then, for $\|x\| = 1$,

$$\|R_\lambda(U-\lambda I)x\| = \|x\| = 1.$$

Therefore

$$\inf_{\|x\|=1} \|Ux-\lambda x\| \geqslant \frac{1}{\|R_\lambda\|} > 0.$$

The theorem we have just proved enables us to extend the result of Lemma 1 to an arbitrary continuous real-valued function.

THEOREM 4. *Let $\phi(t)$ be a real-valued function continuous on S_U. Then*

$$S_{\phi(U)} = \phi(S_U).$$

Proof. Suppose $\mu \notin \phi(S_U)$. The function

$$\psi(t) = \frac{1}{\phi(t)-\mu} \qquad (t \in S_U)$$

is continuous on S_U, so the operator $\psi(U)$ makes sense. Clearly

$$\psi(U) = [\phi(U)-\mu I]^{-1}.$$

Hence, by Theorem 3, $\mu \notin S_{\phi(U)}$.

Now assume that $\mu = \phi(\lambda)$, where $\lambda \in S_U$. Let $\{\phi_n(t)\}$ be a sequence of polynomials converging uniformly on S_U to $\phi(t)$. We have

$$\|\phi(U)x-\mu x\| \leqslant \|\phi_n(U)x-\phi_n(\lambda)x\| + \|\phi(U)-\phi_n(U)\|\,\|x\| + |\mu-\phi_n(\lambda)|\,\|x\|.$$

Now Lemma 1 shows that

$$\inf_{\|x\|=1} \|\phi_n(U)x-\phi_n(\lambda)x\| = 0,$$

so

$$\inf_{\|x\|=1} \|\phi(U)x-\mu x\| \leqslant \|\phi(U)-\phi_n(U)\| + |\mu-\phi_n(\lambda)|.$$

If we let $n \to \infty$ here, we find that

$$\inf_{\|x\|=1} \|\phi(U)x-\mu x\| = 0,$$

that is, $\mu \in S_{\phi(U)}$.

5.4. With every self-adjoint operator U one can associate a family of projections which enable one to construct representations both for U and for operator functions as Stieltjes type integrals.

Let U be a self-adjoint operator. We consider the function

$$\phi_\lambda^+(t) = \begin{cases} 0 & (t \leqslant \lambda), \\ t-\lambda & (t > \lambda) \end{cases}$$

and introduce the notation

$$U_\lambda^+ = \phi_\lambda^+(U).$$

Also let $\mathbf{H}_\lambda^+$ denote the set of all elements x such that $U_\lambda^+ x = \mathbf{0}$; in other words, $\mathbf{H}_\lambda^+$ is the eigenspace of U_λ^+ corresponding to the eigenvalue zero. Finally, denote the projection onto $\mathbf{H}_\lambda^+$ by I_λ.

As the theorem below shows, the properties of the projections I_λ are closely connected with the spectrum of U, and for this reason the family of projections I_λ is called the *spectral function of* U.

THEOREM 5. a) *If* $\lambda < m$, *then* $I_\lambda = \mathbf{0}$; *if* $\lambda > M$, *then* $I_\lambda = I$.

b) *If* $\lambda \leqslant \mu$, *then* $I_\lambda \leqslant I_\mu$.

c) *A spectral function is continuous to the right as a function of* λ, *in the sense that*

$$I_\lambda = I_{\lambda+0} = \lim_{\mu \to \lambda+0} I_\mu \qquad \text{on } \mathbf{H}.$$

d) *The projection* I_λ $(-\infty < \lambda < \infty)$ *is permutable with every operator that permutes with* U.

e) *Every real regular value* λ *of* U *is a point of constancy of the spectral function; that is, there exists* $\delta > 0$ *such that* $I_{\lambda-\delta} = I_{\lambda+\delta}$. *Conversely, the points of constancy of the spectral function are regular values of* U.

f) *A real number* λ *is a regular value of* U *if and only if* $I_\lambda = I_{\lambda+0} \neq I_{\lambda-0}$, *where* $I_{\lambda-0} = \lim_{\mu \to \lambda-0} I_\mu$ *on* $\mathbf{H}$. *In this case the operator* $P_\lambda = I_\lambda - I_{\lambda-0}$ *is the projection onto the eigenspace corresponding to the eigenvalue* λ.

g) *If* $\phi(t) \geqslant 0$ *in* $[\lambda, \mu]$, *then* $[I_\mu - I_\lambda]\phi(U) \geqslant \mathbf{0}$.*

Proof. a) If $\lambda < m$, then $U_\lambda^+ = U - \lambda I$. Therefore, for $x \neq \mathbf{0}$,

$$(U_\lambda^+ x, x) = (Ux, x) - \lambda(x, x) \geqslant (m-\lambda)(x, x) > 0,$$

so that $\mathbf{H}_\lambda^+$ consists just of the zero element.

If $\lambda > M$ we have $U_\lambda^+ = \mathbf{0}$ and $\mathbf{H}_\lambda^+ = \mathbf{H}$.

b) If $\lambda \leqslant \mu$, then $\phi_\lambda^+(t) = \phi_\mu^+(t)$ and so by Theorem 1 $U_\lambda^+ \geqslant U_\mu^+$. Further, $U_\mu^+ \geqslant \mathbf{0}$. Hence

$$0 \leqslant (U_\mu^+ x, x) \leqslant (U_\lambda^+ x, x) \qquad (x \in \mathbf{H}).$$

If $x \in \mathbf{H}_\lambda^+$, then $U_\lambda^+ x = \mathbf{0}$ and so

$$(U_\mu^+ x, x) = 0.$$

From this it follows that† $U_\mu^+ x = \mathbf{0}$, that is, $x \in \mathbf{H}_\mu^+$. Hence $\mathbf{H}_\lambda^+ \subset \mathbf{H}_\mu^+$, which conforms to the required relation $I_\lambda \leqslant I_\mu$ between the projections (V.6.7, Lemma).

c) Since I_λ decreases as λ does, Theorem V.6.7 shows that $I_{\lambda+0} = \lim_{\mu \to \lambda+0} I_\mu$ exists on $\mathbf{H}$ and is a projection. Denote the corresponding subspace by $\mathbf{H}_{\lambda+0}^+$. Since $I_\lambda \leqslant I_{\lambda+0}$, we have $\mathbf{H}_\lambda^+ \subset \mathbf{H}_{\lambda+0}^+$. Let $x \in \mathbf{H}_{\lambda+0}^+$. Since, clearly, $x \in \mathbf{H}_\mu^+$ $(\mu > \lambda)$, we have $I_\mu x = x$. Therefore

* It is obviously enough to require only that $\phi(t) \geqslant 0$ at points of the spectrum that belong to $[\lambda, \mu]$.

† If $(Vx, x) = 0$ for a positive operator V and some $x \in \mathbf{H}$ then $Vx = \mathbf{0}$. For $0 = (Vx, x) = ([\sqrt{V}]^2 x, x) = (\sqrt{V}x, \sqrt{V}x)$, i.e. $\sqrt{V}x = \mathbf{0}$ and so also $Vx = \sqrt{V}(\sqrt{V}x) = 0$.

$U_\mu^+ x = \mathbf{0}$, but

$$\|U_\mu^+ - U_\lambda^+\| = \max_{t \in S_U} |\phi_\mu^+(t) - \phi_\lambda^+(t)| \leqslant \mu - \lambda,$$

so that

$$U_\lambda^+ x = \lim_{\mu \to \lambda + 0} U_\mu^+ x = \mathbf{0},$$

and hence $x \in \mathbf{H}_\lambda^+$. Thus $\mathbf{H}_{\lambda+0}^+ \subset \mathbf{H}_\lambda^+$, which, together with the above, yields $\mathbf{H}_{\lambda+0}^+ = \mathbf{H}_\lambda^+$, or in other words $I_{\lambda+0} = I_\lambda$.

d) Let V be an operator permutable with U. By Theorem 1, V is also permutable with U_λ^+. Using this fact, we deduce that, for $x \in \mathbf{H}_\lambda^+$,

$$U_\lambda^+ Vx = VU_\lambda^+ x = \mathbf{0},$$

that is, $Vx \in \mathbf{H}_\lambda^+$. If x is an arbitrary element of $\mathbf{H}$ then by what has been proved $VI_\lambda x \in \mathbf{H}_\lambda^+$, so

$$VI_\lambda = I_\lambda VI_\lambda. \tag{7}$$

The operator V^* is also permutable with U, so we can replace V in (7) by V^*:

$$V^* I_\lambda = I_\lambda V^* I_\lambda.$$

Taking adjoints of both sides of the last equation, we have

$$I_\lambda V = I_\lambda VI_\lambda.$$

Comparing this with (7), we obtain

$$I_\lambda V = VI_\lambda.$$

e) We introduce the operator

$$U_\lambda^- = \phi_\lambda^-(U),$$

where

$$\phi_\lambda^-(t) = \begin{cases} t - \lambda & (t \leqslant \lambda), \\ 0 & (t > \lambda), \end{cases}$$

and prove that

$$U_\lambda^- = I_\lambda(U - \lambda I). \tag{8}$$

In fact, as

$$U - \lambda I = U_\lambda^+ + U_\lambda^-, \tag{9}$$

we have, for $x \in \mathbf{H}_\lambda^+$,

$$Ux - \lambda x = U_\lambda x.$$

If now x is an arbitrary element of $\mathbf{H}$, then we can write

$$I_\lambda(Ux - \lambda x) = (U - \lambda I) I_\lambda x = U_\lambda^- I_\lambda x = I_\lambda U_\lambda^- x. \tag{10}$$

But $U_\lambda^+ U_\lambda^- = \mathbf{0}$, i.e. $U_\lambda^+ (U_\lambda^- x) = \mathbf{0}$, so $U_\lambda^- x \in \mathbf{H}_\lambda^+$, and therefore $I_\lambda U_\lambda^- x = U_\lambda^- x$, which yields (8).

Now assume that λ_0 is a point of constancy of the spectral function; that is, for some interval (λ, μ) containing λ_0 we have $I_\lambda = I_\mu$. Consider the operator

$$P = \frac{U_\lambda^- - U_\mu^-}{\mu - \lambda}.$$

Using (8), we have

$$P = \frac{I_\lambda(U - \lambda I) - I_\mu(U - \mu I)}{\mu - \lambda} = I_\lambda,$$

so P is a projection. Further,

$$P = \sigma(U),$$

where $\sigma(t)$ is a function equal to 1 for $t \leqslant \lambda$, equal to zero for $t \geqslant \mu$, and linear in $[\lambda, \mu]$.

Assume that the interval (λ, μ) contains points of the spectrum of U; let $\tilde{\lambda}$, say, be one of these. Then

$$Q = P - P^2 = \mathbf{0}.$$

On the other hand, by Theorem 2,

$$\|Q\| = \max_{t \in S_U} |\sigma(t) - \sigma^2(t)| \geqslant \sigma(\tilde{\lambda}) - \sigma^2(\tilde{\lambda}) > 0.$$

Thus λ_0, being a point of (λ, μ), is a regular value.

Now suppose that λ is a point of growth of the spectral function. This means that, for every $\delta > 0$,

$$I_{\lambda+\delta} - I_{\lambda-\delta} \neq \mathbf{0}.$$

Consider an $x \in \mathbf{H}_{\lambda+\delta}^+ \ominus \mathbf{H}_{\lambda-\delta}^+$. Since $x \in \mathbf{H}_{\lambda+\delta}^+$, we have $U_{\lambda+\delta}^+ x = \mathbf{0}$. But as

$$\|U_{\lambda+\delta}^+ - U_\lambda^+\| \leqslant \delta,$$

we have

$$\|U_\lambda^+ x\| \leqslant \delta \|x\|.$$

Further, $x \perp H_{\lambda-\delta}^+$; that is, $I_{\lambda-\delta} x = \mathbf{0}$. Therefore

$$I_{\lambda-\delta} U_{\lambda-\delta}^- x = U_{\lambda-\delta}^- I_{\lambda-\delta} x = \mathbf{0},$$

from which we see that $U_{\lambda-\delta}^- x \perp \mathbf{H}_{\lambda-\delta}^+$. But, as we have already noted above, we always have $U_{\lambda-\delta}^- x \in \mathbf{H}_{\lambda-\delta}^+$. Hence, in the present case, $U_{\lambda-\delta}^- x = \mathbf{0}$. As before, we have

$$\|U_{\lambda-\delta}^- - U_\lambda^-\| \leqslant \delta,$$

so that

$$\|U_\lambda^- x\| \leqslant \delta \|x\|.$$

In view of (9), we deduce that

$$\|Ux - \lambda x\| \leqslant \|U_\lambda^+ x\| + \|U_\lambda^- x\| \leqslant 2\delta \|x\|. \tag{11}$$

Choosing a sequence $\{\delta_n\}$ with $\delta_n \to 0$ and a corresponding sequence of normalized elements $x_n \in \mathbf{H}_{\lambda+\delta_n}^+ \ominus \mathbf{H}_{\lambda-\delta_n}^-$, we find that

$$\|Ux_n - \lambda x_n\| \xrightarrow[n \to \infty]{} 0,$$

and this shows that λ belongs to the spectrum of U.

f) We note first of all that, as the spectral function is monotonic, the limit

$$I_{\lambda-0} = \lim_{\mu\to\lambda-0} I_\mu$$

exists on $\mathbf{H}$, and $I_{\lambda-0}$ is a projection. It is also clear that $I_{\lambda-0} \leqslant I_\lambda$.

Now assume that λ is a point such that

$$I_{\lambda-0} \neq I_{\lambda+0} = I_\lambda.$$

Writing $\mathbf{H}^+_{\lambda-0}$ for the subspace corresponding to the projection $I_{\lambda-0}$, we choose an $x \in \mathbf{H}^+_\lambda \ominus \mathbf{H}^+_{\lambda-0}$. Clearly $x \in \mathbf{H}^+_{\lambda+\delta} \ominus \mathbf{H}^+_{\lambda-\delta}$, for any $\delta > 0$. Now (11) is true for any $\delta > 0$ and its left-hand side is independent of δ, so

$$\|Ux - \lambda x\| = 0,$$

that is, λ is an eigenvalue of U, and x is a corresponding eigenvector. If, as usual, we write $\mathbf{H}_\lambda$ for the eigenspace and P_λ for the projection on $\mathbf{H}_\lambda$, then it follows from what we have proved that

$$\mathbf{H}_\lambda \supset \mathbf{H}^+_\lambda - \mathbf{H}^+_{\lambda-0} \qquad (P_\lambda \geqslant I_\lambda - I_{\lambda-0}).$$

Next we consider the projection

$$I'_\mu = P_\lambda - I_\mu P_\lambda \qquad (-\infty < \mu < \infty)$$

and prove that $I'_\mu x = \mathbf{0}$ if and only if $P_\lambda U^+_\mu x = \mathbf{0}$. In fact, if $I'_\mu x = \mathbf{0}$, then $P_\lambda x = I_\mu P_\lambda x$, so that $P_\lambda U^+_\mu x = U^+_\mu P_\lambda x = \mathbf{0}$. Conversely, if $P_\lambda U^+_\mu x = \mathbf{0}$, then $P_\lambda x \in \mathbf{H}^+_\mu$ and $I_\mu P_\lambda x = P_\lambda x$, that is, $I'_\mu x = \mathbf{0}$.

Using part h) of Theorem 1, we have

$$P_\lambda U^+_\mu = P_\lambda \phi^+_\lambda(P_\lambda U).$$

However,

$$P_\lambda U = \lambda P_\lambda,$$

so we deduce immediately from the definition of an operator function that

$$P_\lambda U^+_\mu = \begin{cases} (\lambda-\mu)P_\lambda, & \mu < \lambda, \\ \mathbf{0}, & \mu \geqslant \lambda. \end{cases}$$

It thus follows that for $\mu < \lambda$ the equation $I'_\mu x = \mathbf{0}$ is equivalent to $P_\lambda x = \mathbf{0}$: that is, for $\mu < \lambda$, we have $I'_\mu = P_\lambda$. However, if $\mu \geqslant \lambda$, then $I'_\mu = \mathbf{0}$. Hence

$$I_\mu P_\lambda = \begin{cases} \mathbf{0}, & \mu < \lambda, \\ P_\lambda, & \mu \geqslant \lambda. \end{cases}$$

But

$$I_\mu = I_\mu P_\lambda + I_\mu (I - P_\lambda).$$

Therefore

$$I_{\lambda-0} = \lim_{\mu\to\lambda-0} I_\mu = I_{\lambda-0}(I - P_\lambda).$$

Subtracting this from the equation

$$I_\lambda = I_\lambda P_\lambda + I_\lambda(I - P_\lambda) = P_\lambda + I_\lambda(I - P_\lambda),$$

we obtain

$$I_\lambda - I_{\lambda-0} = P_\lambda + (I_\lambda - I_{\lambda-0})\ (I - P_\lambda) \geqslant P_\lambda.$$

Now $P_\lambda \neq \mathbf{0}$, so certainly $I_\lambda - I_{\lambda-0} \neq \mathbf{0}$; that is, the spectral function has a jump at the point λ. Moreover, if we take into account the relation obtained earlier, we find that

$$P_\lambda = I_\lambda - I_{\lambda-0}.$$

g) We assume first that $\phi(t) = 0$ in $[\lambda, \mu]$, and prove that $[I_\mu - I_\lambda]\phi(U) = \mathbf{0}$. To this end we consider the function $\tilde{\phi}(t)$, which coincides with $\phi(t)$ at the points $\lambda_1, \lambda_2, \ldots, \lambda_r$ ($\lambda_k \leqslant \lambda$, $\lambda_r = \lambda$) and $\mu_1, \mu_2, \ldots, \mu_s$ ($\mu_k \geqslant \mu$, $\mu_1 = \mu$) and is linear in the intervals between these points. Clearly,

$$\tilde{\phi}(t) = \sum_{k=1}^{r} \alpha_k \phi_{\lambda_k}^{-}(t) + \sum_{k=1}^{s} \beta_k \phi_{\mu_k}^{+}(t)$$

and therefore

$$\tilde{\phi}(U) = \sum_{k=1}^{r} \alpha_k U_{\lambda_k}^{-} + \sum_{k=1}^{s} \beta_k U_{\mu_k}^{+}.$$

Let us verify that

$$[I_\mu - I_\lambda]\tilde{\phi}(U) = 0. \tag{12}$$

We write

$$[I_\mu - I_\lambda]\tilde{\phi}(U) = \sum_{k=1}^{r} \alpha_k [I_\mu - I_\lambda] U_{\lambda_k}^{-} + \sum_{k=1}^{s} \beta_k [I_\mu - I_\lambda] U_{\mu_k}^{+}$$

and consider the individual terms separately. Using (8), we have

$$[I_\mu - I_\lambda]U_{\lambda_k}^{-} = [I_\mu - I_\lambda] I_{\lambda_k}(U - \lambda_k I) = [I_\mu I_{\lambda_k} - I_\lambda I_{\lambda_k}](U - \lambda_k I).$$

However, $I_{\lambda_k} \leqslant I_\lambda \leqslant I_\mu$, so $I_\mu I_{\lambda_k} = I_\lambda I_{\lambda_k} = I_{\lambda_k}$, and from this it follows that $[I_\mu - I_\lambda]U_{\lambda_k}^{-} = \mathbf{0}$. Also $[I_\mu - I_\lambda]U_{\mu_k}^{+} = I_\mu [I - I_\lambda] U_{\mu_k}^{+} = [I - I_\lambda] U_{\mu_k}^{+} I_\mu = \mathbf{0}$. Thus (12) is proved.

The function $\phi(t)$ may be expressed as a limit of a sequence of piecewise-linear functions of the type just considered that converges uniformly on $[m, M]$. Since each function in the sequence satisfies an equation like (12), we obtain the required equation $[I_\mu - I_\mu]\phi(U) = \mathbf{0}$ by passing to the limit, which we can do by part g) of Theorem 1.

Now assume that $\phi(t)$ satisfies the conditions of the theorem; that is, $\phi(t) \geqslant 0$ for $t \in [\lambda, \mu]$. In addition to $\phi(t)$, we consider the function

$$\psi(t) = \begin{cases} \phi(\lambda), & t < \lambda, \\ \phi(t), & \lambda \leqslant t \leqslant \mu, \\ \phi(\mu), & t < \mu. \end{cases}$$

Since $\psi(t) \geqslant 0$ for all t, we have $\psi(U) \geqslant \mathbf{0}$ and so, *a fortiori*, $[I_\mu - I_\lambda]\psi(U) \geqslant \mathbf{0}$. But $\phi(t)$ and $\psi(t)$ coincide in $[\lambda, \mu]$, so by what was proved above

$$[I_\mu - I_\lambda][\phi(U) - \psi(U)] = \mathbf{0},$$

from which we see that

$$[I_\mu - I_\lambda]\phi(U) = [I_\mu - I_\lambda]\psi(U) \geqslant \mathbf{0}.$$

This completes the proof of the theorem.

5.5. We now turn to integral representations of operator functions. Let $\phi(t)$ be a continuous, bounded, real-valued function defined on* $(-\infty, +\infty)$. We consider a partition of the real line, $-\infty = \lambda_0 < \lambda_1 < \ldots < \lambda_n < \lambda_{n+1} = \infty$ ($\lambda_1 < m$, $\lambda_n > M$), and form the sums

$$s = \sum_{k=1}^{n-1} l_k[I_{\lambda_{k+1}} - I_{\lambda_k}]; \qquad S = \sum_{k=1}^{n-1} L_k[I_{\lambda_{k+1}} - I_{\lambda_k}], \tag{13}$$

where

$$l_k = \inf_{t\in[\lambda_k, \lambda_{k+1}]} \phi(t), \qquad L_k = \sup_{t\in[\lambda_k, \lambda_{k+1}]} \phi(t).$$

Since $\phi(t)$ satisfies the condition $l_k \leqslant \phi(t) \leqslant L_k$ in $[\lambda_k, \lambda_{k+1}]$, part g) of Theorem 5 yields

$$l_k[I_{\lambda_{k+1}} - I_{\lambda_k}] \leqslant [I_{\lambda_{k+1}} - I_{\lambda_k}]\phi(U) \leqslant L_k[I_{\lambda_{k+1}} - I_{\lambda_k}].$$

Adding all these inequalities, we obtain

$$s \leqslant \phi(U) \leqslant S. \tag{14}$$

On the other hand,

$$S - s = \sum_{k=1}^{n-1} (L_k - l_k)\ [I_{\lambda_{k+1}} - I_{\lambda_k}] \leqslant \delta I,$$

where δ denotes $\max\limits_{k=1,2,\ldots,n-1} (\lambda_{k+1} - \lambda_k)$. Hence $\|S - s\| \to 0$ as $\delta \to 0$. In view of the fact that $\mathbf{0} \leqslant S - \phi(U) \leqslant S - s$, we have $S \to \phi(U)$ as $\delta \to 0$, and similarly $s \to \phi(U)$ as $\delta \to 0$.

Now form the "integral" sum

$$\sigma = \sum_{k=1}^{n-1} \phi(\xi_k)\ [I_{\lambda_{k+1}} - I_{\lambda_k}] \qquad (\lambda_k \leqslant \xi_k \leqslant \lambda_{k+1}). \tag{15}$$

We verify without any difficulty that

$$s \leqslant \sigma \leqslant S$$

and so

$$\lim_{\delta\to 0} \sigma = \lim_{\delta\to 0} S = \phi(U).$$

But we can regard $\lim\limits_{\delta\to 0} \sigma$ as the integral

$$\lim_{\delta\to 0} \sigma = \int_{-\infty}^{+\infty} \phi(t)\, dI_t. \tag{16}$$

Hence we obtain the formula

$$\phi(U) = \int_{-\infty}^{+\infty} \phi(t)\, dI_t. \tag{17}$$

In particular, if $\phi(t) = t$, then (17) gives an *integral representation* for U:

$$U = \int_{-\infty}^{+\infty} t\, dI_t. \tag{18}$$

* We can assume that $\phi(t)$ was originally defined on the spectrum of U, and then extended to the whole real line, retaining all the above properties.

The properties of operator functions expressed in Theorem 1 and 5.3 can be formulated in terms of the operator integral (16). We also mention a formula connecting the concept of an operator function with the usual Stieltjes integral,

$$(\phi(U)x, y) = \int_{-\infty}^{+\infty} \phi(t)\, d(I_t x, y)$$

and the resulting formula

$$\|\phi(U)\|^2 = \int_{-\infty}^{+\infty} |\phi(t)|^2\, d(I_t x, x).$$

The simple deduction of this is left to the reader.

We can also consider integrals of the form (16) over an interval $[\lambda, \mu]$. It is not difficult to verify that

$$\int_{\lambda}^{\mu} \phi(t)\, dI_t = [I_\mu - I_\lambda]\phi(U).$$

In particular, if $\lambda < m$, $\mu \geqslant M$, then

$$\int_{\lambda}^{\mu} \phi(t)\, dI_t = \phi(U).$$

5.6. To construct integrals of the form (16) we do not need to assume that the family of projections I_λ is the spectral function of a self-adjoint operator. It is sufficient for the family of projections to satisfy the conditions laid down in parts a) and b) of Theorem 5.

Let $\{E_\lambda\}$ be a family of projections depending on a real parameter and satisfying the conditions:

1) $E_\lambda \leqslant E_\mu$ if $\lambda \leqslant \mu$;

2) there exist numbers m and M such that $E_\lambda = \mathbf{0}$ for $\lambda < m$ and $E_\lambda = I$ for $\lambda > M$.

In this case we call the family $\{E_\lambda\}$ a *resolution of the identity*. With these assumptions, the complex function $(E_t x, y)$ of a real argument will be a function of bounded variation, whatever elements $x, y \in \mathbf{H}$ we take. This follows immediately from the formula

$$(E_t x, y) = \frac{1}{4}\{[(E_t(x+y), x+y) - (E_t(x-y), x-y)] + i[(E_t(x+yi), x+yi) - (E_t(x-yi), x-yi)]\}.$$

Hence, if $\phi(t)$ is a continuous bounded function, we can assign a meaning to the integral

$$J(\phi; x, y) = \int_{-\infty}^{+\infty} \phi(t)\, d(E_t x, y) \quad (x, y \in \mathbf{H}).$$

It is not hard to see that the functional $J(\phi; x, y)$ is additive and homogeneous* in both x and y. Moreover, since

$$|J(\phi; x, y)| \leqslant \sup|\phi(t)| \mathop{\mathbf{V}}_{-\infty}^{+\infty} [(E_t x, y)] \leqslant \sup|\phi(t)| \cdot \frac{1}{4}[\|x+y\|^2 + \|x-y\|^2 + \|x+yi\|^2 + \|x-yi\|^2] = \sup|\phi(t)|[\|x\|^2 + \|y\|^2], \tag{19}$$

$J(\phi; x, y)$ is continuous in x, y.

---

* The homogeneity in y is of the "second kind": $J(\phi; x, \alpha y) = \bar{\alpha} J(\phi; x, y)$.

For a fixed x, we consider the following functional in y:

$$f_x(y) = \overline{J(\phi; x, y)}.$$

Bearing in mind the general form for a linear functional on a Hilbert space, we have

$$f_x(y) = (y, \tilde{x}),$$

where the element $\tilde{x} \in \mathbf{H}$ is determined by the functional f_x alone, and so ultimately by x alone. Setting $\tilde{x} = J(\phi)x$, we can write

$$J(\phi; x, y) = (J(\phi)x, y).$$

It is obvious that $J(\phi)$ is additive and homogeneous. Furthermore, using (19) we see that

$$\| J(\phi)x\| \left(J(\phi)x, \frac{J(\phi)x}{\| J(\phi)x\|} \right) \leqslant \sup_{\|y\|=1} (J(\phi)x, y) \leqslant \sup |\phi(t)|[\|x\|^2 + 1],$$

from which it follows that $J(\phi)$ is bounded.

In exactly the same way we can define the integral over any interval $[\lambda, \mu]$.

Now let us prove that $J(\phi)$ is the limit of integral sums of the form (15):

$$\sigma = \sum_{k=1}^{n-1} \phi(\xi_k) \, [E_{\lambda_{k+1}} - E_{\lambda_k}] \qquad (\lambda_1 < m, \lambda_n > M).$$

We have

$$(J(\phi)x, y) - (\sigma x, y) = \sum_{k=1}^{n-1} \int_{\lambda_k}^{\lambda_{k+1}} \phi(t) \, d(E_t x, y) - \sum_{k=1}^{n-1} \int_{\lambda_k}^{\lambda_{k+1}} \phi(\xi_k) \, d(E_t x, y) =$$

$$= \sum_{k=1}^{n-1} \int_{\lambda_k}^{\lambda_{k+1}} [\phi(t) - \phi(\xi_k)] \, d(E_t x, y).$$

Hence

$$\|J(\phi) - \sigma\| = \sup_{\substack{\|x\|=1 \\ \|y\|=1}} |(J(\phi)x, y) - (\sigma x, y)| \leqslant \sup_{\substack{\|x\|=1 \\ \|y\|=1}} \sum_{k=1}^{n-1} \int_{\lambda_k}^{\lambda_{k+1}} \left| [\phi(t) - \phi(\xi_k)] d(E_t x, y) \right|.$$

Writing ω for the maximum oscillation of $\phi(t)$ in the intervals $[\lambda_1, \lambda_2]$, $[\lambda_2, \lambda_3]$, $\dots$, $[\lambda_{n-1}, \lambda_n]$, we can extend this inequality:

$$\|J(\phi) - \sigma\| \leqslant \omega \sup_{\substack{\|x\|=1 \\ \|y\|=1}} \sum_{k=1}^{n-1} \mathop{\mathbf{V}}_{\lambda_k}^{\lambda_{k+1}} [(E_t x, y)] \leqslant \omega \sup_{\substack{\|x\|=1 \\ \|y\|=1}} \mathop{\mathbf{V}}_{-\infty}^{+\infty} [(E_t x, y)] \leqslant 2\omega.$$

If $\delta = \max\limits_{k=1, 2, \dots, n-1} (\lambda_{k+1} - \lambda_k) \to 0$, then $\omega \to 0$, and hence $\lim\limits_{\delta \to 0} \sigma = J(\phi)$. This provides the basis for the notation $J(\phi) = \int\limits_{-\infty}^{+\infty} \phi(t) \, dE_t$.

Using what we have proved, let us record some properties of this operator:

I. $[J(\phi)]^* = J(\overline{\phi})$.

II. If $\phi(t) = \alpha\phi_1(t) + \beta\phi_2(t)$, then $J(\phi) = \alpha J(\phi_1) + \beta J(\phi_2)$.

III. If $\phi(t) = \phi_1(t)\phi_2(t)$, then

$$J(\phi) = J(\phi_1) J(\phi_2) = J(\phi_2) J(\phi_1). \tag{20}$$

Let us dwell on the proof of the last property. For some partition we form the integral sums

$$\sigma = \sum_{k=1}^{n-1} \phi(\xi_k)[E_{\lambda_{k+1}} - E_{\lambda_k}],$$

$$\sigma' = \sum_{k=1}^{n-1} \phi_1(\xi_k)[E_{\lambda_{k+1}} - E_{\lambda_k}], \qquad \sigma'' = \sum_{k=1}^{n-1} \phi_2(\xi_k)[E_{\lambda_{k+1}} - E_{\lambda_k}].$$

Since for $j < k$ we have $\lambda_j < \lambda_{j+1} \leqslant \lambda_k < \lambda_{k+1}$, it follows that

$$[E_{\lambda_{k+1}} - E_{\lambda_k}][E_{\lambda_{j+1}} - E_{\lambda_j}] = E_{\lambda_{k+1}}E_{\lambda_{j+1}} - E_{\lambda_{k+1}}E_{\lambda_j} - E_{\lambda_k}E_{\lambda_{j+1}} + E_{\lambda_k}E_{\lambda_j}$$

$$= E_{\lambda_{j+1}} - E_{\lambda_j} - E_{\lambda_{j+1}} + E_{\lambda_j} = \mathbf{0}.$$

Hence in the product

$$\sigma'\sigma'' = \sum_{j,k=1}^{n-1} \phi_1(\xi_j)\phi_2(\xi_k)\ [E_{\lambda_{j+1}} - E_{\lambda_j}][E_{\lambda_{k+1}} - E_{\lambda_k}]$$

only the terms with $j = k$ can be different from zero. Hence

$$\sigma'\sigma'' = \sigma.$$

Taking limits, we obtain (20).

IV. We have the upper bound

$$\|J(\phi)\| \leqslant \max_{t\in[m,M]} |\phi(t)|. \tag{21}$$

For

$$\|J(\phi)x\|^2 = (J(\phi)x, J(\phi)x) = (J(|\phi|^2)x, x) = \int_{-\infty}^{+\infty} |\phi(t)|^2\, d(E_t x, x).$$

Since

$$(E_t x, x) = \begin{cases} 0, & t < m, \\ \|x\|^2, & t > M, \end{cases}$$

we have

$$\|J(\phi)x\|^2 = \int_{m-\varepsilon}^{M+\varepsilon} |\phi(t)|^2\, d(E_t x, x) \leqslant \max_{t\in[m-\varepsilon, M+\varepsilon]} |\phi(t)|^2 \|x\|^2.$$

Since this inequality is true for any $\varepsilon > 0$, in the limit we have (remembering that $\phi(t)$ is continuous)

$$\|J(\phi)x\|^2 \leqslant \max_{t\in[m,M]} |\phi(t)|^2 \|x\|^2,$$

which is equivalent to (21).

V. If a sequence of continuous functions $\{\phi_n\}$ converges uniformly in $[m, M]$ to a function ϕ, then $J(\phi_n) \to J(\phi)$.

The properties of the operator integrals $J(\phi)$ coincide with the properties of operator functions. The reasons for this will become clear from the following theorem.

THEOREM 6. *Let $\{E_\lambda\}$ be a resolution of the identity that is continuous to the right as a function of the parameter λ. Then $\{E_\lambda\}$ is the spectral function of the operator* *and*

$$U = \int_{-\infty}^{-\infty} t\,dE_t,$$

$$J(\phi) = \phi(U). \tag{22}$$

Proof. The truth of (22) may be verified without difficulty using properties II, III and V. In particular,

$$U_\lambda^+ = \phi_\lambda^+(U) = J(\phi_\lambda^+),$$

so that

$$(U_\lambda^+ x, x) = \int_{-\infty}^{+\infty} \phi_\lambda^+(t)\,d(E_t x, x) = \int_{\lambda}^{\infty} (t-\lambda)\,d(E_t x, x). \tag{23}$$

If $I_\lambda x = x$, that is, if $U_\lambda^+ x = \mathbf{0}$, then

$$\int_{\lambda}^{\infty} (t-\lambda)\,d(E_t x, x) = 0,$$

which can only be so in the case where

$$(E_t x, x) = \text{const for } t > \lambda.$$

Hence

$$(E_t x, x) = (E_\infty x, x) = (x, x) \qquad (t > \lambda)$$

and therefore

$$E_t x = x \quad (t > \lambda).$$

If we let $t \to \lambda + 0$ here, we find that

$$E_\lambda x = E_{\lambda+0} x = x,$$

that is, $I_\lambda \leqslant E_\lambda$.

Conversely, if $E_\lambda x = x$, then certainly $E_t x = x$ for $t > \lambda$. Thus from (23) we obtain

$$(U_\lambda^+ x, x) = 0.$$

Therefore*

$$U_\lambda^+ x = \mathbf{0},$$

and this shows that $I_\lambda x = x$. Hence $I_\lambda \geqslant E_\lambda$, which with the preceding observation yields the equation $I_\lambda = E_\lambda$.

REMARK. If the resolution of the identity does not satisfy the requirement of continuity to the right, then it can be shown that

$$I_\lambda = E_{\lambda+0}.$$

For the inequality $I_\lambda \leqslant E_{\lambda+0}$ was in fact obtained in the proof of the theorem. Furthermore, if we take limits in the inequality $I_t \geqslant E_t$ as $t \to \lambda + 0$, we obtain the reverse inequality

$$I_\lambda = I_{\lambda+0} \geqslant E_{\lambda+0}.$$

---

* See the note on p. 264.

Let U be a compact self-adjoint operator and let $\{\lambda_k\}$ be the set of its eigenvalues. As before, we write $\mathbf{H}_{\lambda_k}$ for the eigenspaces and P_{λ_k} for the projections on these subspaces; and we now introduce operators

$$E_\lambda = \sum_{\lambda_k \leqslant \lambda} P_{\lambda_k} \qquad (-\infty < \lambda < \infty). \tag{24}$$

It is not hard to check that the family $\{E_\lambda\}$ is a resolution of the identity which is continuous to the right. Hence $\{E_\lambda\}$ is the spectral function of the operator

$$\tilde{U} = \int_{-\infty}^{+\infty} t\, dE_t.$$

But we can convince ourselves by an immediate check that

$$\int_{-\infty}^{+\infty} t\, dE_t = \sum_k \lambda_k P_{\lambda_k};$$

so that, bearing in mind equation (3) of the preceding section, we have

$$\tilde{U} = U.$$

Hence the projections (24) form a spectral function for U and the integral representation for U given by (18) coincides with formula (3) of the preceding section.

5.7. The integral representation (17) gives a way of further extending the concept of an operator function. Consider a function $\phi(t)$ of a type for which the Lebesgue–Stieltjes integral

$$J(\phi; x, y) = \int_{-\infty}^{+\infty} \Phi(t) d(I_t x, y),$$

where $\{I_\lambda\}$ is the spectral function of U, exists (and is finite) for all (x, y). As above, it can be shown that the functional $J(\phi; x, y)$ is expressible in the form

$$J(\phi; x, y) = (J(\phi)x, y) \qquad (x, y \in \mathbf{H}),$$

where $J(\phi)$ is a linear operator. Bearing in mind (22), it is natural to set

$$\phi(U) = J(\phi).$$

One can also reach the generalization of the operator function concept somewhat differently. Starting from a spectral function, one can define an operator measure exactly as one defines an additive set function for a real monotonic function. If $\phi(t)$ is the characteristic function of a set e that is measurable relative to the operator measure, then $\phi(U)$ is the value of the operator measure on e. The meaning of $\phi(U)$ is also clear when $\phi(t)$ is a simple measurable function. A function that is measurable relative to an operator measure is the limit of a uniformly convergent sequence of simple measurable functions, which makes it possible to define $\phi(U)$ for an arbitrary measurable function $\phi(t)$.

X

ORDERED NORMED SPACES

FOR CERTAIN specific investigations of importance in the analysis of vector spaces of functions and sequences the concepts of positivity and inequality play an essential role, along with the concept of a positive operator. However, these concepts, and the facts relating to them, are not reflected in any way in Banach's theory of normed spaces, which we have been studying up to now. Consequently it has been impossible to include certain essential problems of classical and applied analysis within the scope of the methods of functional analysis. It was this that led to the construction, in 1935–1937, of the theory of partially ordered linear spaces, in papers by L. V. Kantorovich (see [2]–[4a]). These spaces—which are also known as vector lattices—are endowed, in a specific way, with both a vector space structure and a compatible order structure. In the papers just mentioned, the author laid the axiomatic foundations for vector lattices, constructed a theory of linear operators on them, and gave various applications of the theory to problems in the theory of functions and functional equations.

Later this theory had a fruitful development in papers by the Leningrad school on partially ordered spaces (B. Z. Vulikh, A. G. Pinsker, A. I. Yudin, G. P. Akilov, and their followers, A. I. Veksler, D. A. Vladimirov, G. Ya. Rotkovich). An important division in this area of functional analysis—namely, the theory of normed lattices—owes its development partly to a significant contribution by L. Ya. Lozanovskii.

Closely allied to the theory of vector lattices are the investigations of M. G. Krein on normed spaces with a cone of positive elements defined in them, which were initiated in the late thirties and extended by the Voronezh school (M. A. Krasnosel'skii and his followers). These investigations are principally concerned with the spectral theory of operators and the solution of operator equations (non-linear analysis). In this book we shall not study this theory, and refer the reader instead to the survey article by Krein and Rutman [1] and to Krasnosel'skii-II. For connections between the theory of vector lattices and Krein's theory, see Vulikh-IV. Also connected with the theory of vector lattices is the theory of topological near-fields (see Antonovskii, Boltyanskii and Sarymsakov).

The development of the theory of vector lattices was influenced by the report on lattices of functionals by F. Riesz [4] at the Bologna mathematical congress in 1929. During the 1940s the Japanese mathematician H. Nakano, extending earlier work, constructed a systematic theory of partially ordered spaces, which contained many new avenues of investigation (see Nakano). In the development of the theory of vector lattices, important roles were also played by other Japanese mathematicians (T. Ogasavara, I. Amemiya, T. Ando, K. Yosida, S. Kakutani, T. Shimogaki) and by Americans (G. Birkhoff, S. Bochner, M. Stone, etc.), Dutchmen (H. Freudenthal, V. Luxemburg, A. Zaanen) and Frenchmen (J. Dieudonné).

In this chapter we consider normed lattices—that is, vector lattices that are also normed spaces, the norm and the ordering being related in a natural manner.

We remark that one important class of vector lattices is the class of ideal spaces already discussed earlier in the book. In fact, even if applications to ISs are all one has in mind, one can obtain new and interesting results through the axiomatic approach (see § 5).

In order to present facts of any significance from the theory of normed lattices, we shall require a number of purely algebraic results from the theory of vector lattices. Although these results will be stated precisely, we shall present them without proofs, referring the reader to Vulikh-I. The results in this chapter from the theory of normed lattices are essentially complementary to Vulikh-I. The fundamental books on the theory of vector lattices are: Kantorovich, Vulikh and Pinsker; Vulikh-I; Nakano; Luxemburg and Zaanen; Fremlin; Schaefer-II. A good deal of space is devoted to the theory of normed lattices in a series of papers by Luxemburg [1] and Luxemburg and Zaanen [1], which are essentially survey articles. The majority of the results in §§ 1–4 are well known and covered in the sources listed above, to which we also refer the reader for references concerning priorities.

We remark that, owing to the development of the theory of vector lattices, we have often needed to revise the terminology and notation employed in Vulikh-I. All the changes we have made are noted explicitly.

§ 1. Vector lattices

1.1. A real vector space $\mathbf{X}$ is called a *vector lattice* (VL for short)* if it is at the same time a lattice – that is, a partially ordered set in which there exist a supremum $x \vee y$ and an infimum $x \wedge y$ for every two elements x and y, subject to the following compatibility conditions relating the algebraic operations and the ordering:

1) $x \leqslant y$ implies $x+z \leqslant y+z$, for every $z \in \mathbf{X}$;

2) if $x \geqslant \mathbf{0}$ and $\lambda \geqslant 0$ is a scalar, then $\lambda x \geqslant \mathbf{0}$.

The set $\mathbf{X}_+ = \{x \in \mathbf{X} : x \geqslant 0\}$ is called the *cone* of positive elements of $\mathbf{X}$. In a VL the supremum and infimum clearly exist for any finite set of elements.

For every $x \in \mathbf{X}$ the element $x_+ = x \vee \mathbf{0}$ is called the *positive part* of x, the element $x_- = (-x) \vee \mathbf{0} = (-x)_+$ is called the *negative part* of x, and the element $|x| = x_+ + x_-$ is called the *modulus* of x. For every $x \in \mathbf{X}$, we have $x = x_+ - x_-$.

An *order interval* (or simply *interval*) in a VL $\mathbf{X}$ is any set of the form $[x_1, x_2] = \{x \in \mathbf{X} : x_1 \leqslant x \leqslant x_2\}$, where $x_1 \leqslant x_2$ $(x_1, x_2 \in \mathbf{X})$.

Two elements x, y in a VL $\mathbf{X}$ are said to be *disjoint* (notation: $x \mathrm{~d~} y$) if $|x| \wedge |y| = \mathbf{0}$. Two sets $E_1, E_2 \subset \mathbf{X}$ are said to be *disjoint* ($E_1 \mathrm{~d~} E_2$) if every element E_1 is disjoint from every element $y \in E_2$. Finally, an element $x \in \mathbf{X}$ is said to be *disjoint* from a set $E \subset \mathbf{X}$ ($x \mathrm{~d~} E$) if $x \mathrm{~d~} y$ for all $y \in E$. If E is an arbitrary set in a VL $\mathbf{X}$ then its *disjunctive complement* is the set E^{d} consisting of all $x \in \mathbf{X}$ that are disjoint from E. We also write $E^{\mathrm{dd}} = (E^{\mathrm{d}})^{\mathrm{d}}$. It is easy to see that $(E^{\mathrm{d}})^{\mathrm{dd}} = E^{\mathrm{d}}$ for every E.

We shall say that a set E_1 is *wider* than a set E_2 ($E_1, E_2 \subset \mathbf{X}$) if $E_1^{\mathrm{dd}} \supset E_2^{\mathrm{dd}}$; E_1 is said to be *of the same width* as E_2 if $E_1^{\mathrm{dd}} = E_2^{\mathrm{dd}}$. An element $x \in \mathbf{X}_+$ is called a (*weak*) *unit* if $\{x\}^{\mathrm{dd}} = \mathbf{X}$.

A subset E of a VL $\mathbf{X}$ is said to be *solid* if the statements $x \in \mathbf{X}$, $y \in E$, $|x| \leqslant |y|$ imply that $x \in E$. If E is an arbitrary subset of $\mathbf{X}$, then the smallest solid set containing E is called the

* In Vulikh-I vector lattices are called linear lattices; in Bourbaki-IV and in Luxemburg and Zaanen they are called Riesz spaces. The term vector lattice was introduced by G. Birkhoff.

solid hull of E and is denoted by sol (E). It is easy to see that sol (E) is the set of all $x \in \mathbf{X}$ such that there exists $y \in E$ with $|x| \leqslant |y|$.

1.2. A linear manifold $\mathbf{Y}$ in a VL $\mathbf{X}$ is called a *vector sublattice** of $\mathbf{X}$ if for all $y_1, y_2 \in \mathbf{Y}$ we have $y_1 \vee y_2, y_1 \wedge y_2 \in \mathbf{Y}$. A solid linear manifold $\mathbf{Y}$ in a VL $\mathbf{X}$ is called an *ideal*. An ideal $\mathbf{Y}$ of a VL $\mathbf{X}$ is called a *foundation* if $\mathbf{Y}$ is of the same width as $\mathbf{X}$, that is, if $\mathbf{Y}^d = \{\mathbf{0}\}$.

If $u \in \mathbf{X}_+$, then the smallest ideal of $\mathbf{X}$ containing u is called a *principal ideal* or *u-ideal* of $\mathbf{X}$, and is denoted by $\mathbf{X}(u)$. It is easy to see that $\mathbf{X}(u)$ consists of all $x \in \mathbf{X}$ such that $|x| \leqslant \lambda u$ for some λ, $0 \leqslant \lambda < +\infty$.

If a VL $\mathbf{X}$ coincides with the principal ideal $\mathbf{X}(u)$ for some choice of $u \in \mathbf{X}_+$, then $\mathbf{X}$ is said to be a *vector lattice of bounded elements*, and the element u is called a *strong unit* (in $\mathbf{X}$).

A *band* in a VL $\mathbf{X}$ is any set $\mathbf{Y} \subset \mathbf{X}$ that is the disjunctive complement of some set $E \subset \mathbf{X}$. In other words, a set $\mathbf{Y} \subset \mathbf{X}$ is said to be a *band* if $\mathbf{Y} = \mathbf{Y}^{dd}$.

We shall say that a VL $\mathbf{X}$ is *Archimedean* if, from the fact that $nx \leqslant y \in \mathbf{X}$ for all $n \in \mathbb{N}$, it follows that $x = \mathbf{0}$. All the types of VL of interest for functional analysis are Archimedean.

If $\mathbf{X}$ is an Archimedean VL then a set $\mathbf{Y} \subset \mathbf{X}$ is a band if and only if $\mathbf{Y}$ is an ideal in $\mathbf{X}$ and the following condition is satisfied (the so-called *regularity condition*): if $E \subset \mathbf{Y}$ and sup E (inf E) exists in $\mathbf{X}$, then sup $E \in \mathbf{Y}$ (inf $E \in \mathbf{Y}$).

1.3. A VL $\mathbf{X}$ in which every countable set that is bounded above has a supremum is called a *K_σ-space* (or σ-complete VL). A VL $\mathbf{X}$ in which every set that is bounded above has a supremum is called a *K-space* (or a complete VL). It is easy to see (cf. the proof of Corollary 1 to Theorem I.6.17) that, in a K_σ-space, any countable set that is bounded below has an infimum, while, in a K-space, any set whatsoever that is bounded below has an infimum. A K_σ-space is an Archimedean VL. An ideal in a K_σ-space or a K-space is a K_σ-space, or a K-space, respectively.

The most important property of bands in a K-space $\mathbf{X}$ is that one can project $\mathbf{X}$ onto them in a canonical way. Let $\mathbf{Y}$ be a band in a K-space $\mathbf{X}$, and let $x \in \mathbf{X}_+$. Set

$$[\mathbf{Y}]x = \sup\{y \in \mathbf{Y}_+ : y \leqslant x\}.$$

By the definition of a K-space, this supremum exists in $\mathbf{X}$, and, by the regularity condition, $[\mathbf{Y}]x \in \mathbf{Y}$. For an arbitrary $x \in \mathbf{X}$, we write

$$[\mathbf{Y}]x = [\mathbf{Y}]x_+ - [\mathbf{Y}]x_-.$$

It is clear that $[\mathbf{Y}]$ is a linear operator mapping $\mathbf{X}$ onto $\mathbf{Y}$ and leaving elements of $\mathbf{Y}$ fixed. It is easy to see that

$$|[\mathbf{Y}]x| \leqslant |x| \quad \text{and} \quad |[\mathbf{Y}]x| = [\mathbf{Y}](|x|) \quad (x \in \mathbf{X}).$$

The operator $[\mathbf{Y}]$ is called a *projection onto the band* $\mathbf{Y}$. Every $x \in \mathbf{X}$ is uniquely expressible in the form $x = y + z$, where $y \in \mathbf{Y}$, $z \in \mathbf{Y}^d$, and $y = [\mathbf{Y}]x$, $z = [\mathbf{Y}^d]x$ (see Vulikh-I, Theorem IV.3.2).

For an arbitrary set $E \subset \mathbf{X}$, we denote by $[E]$ the projection $[E^{dd}]$ on the band E^{dd} generated by E. If $u \in \mathbf{X}$, then the smallest band $\{u\}^{dd}$ in $\mathbf{X}$ containing u is called a *principal band* and denoted by $\mathbf{X}_u$. In accordance with the notation introduced above, the projection onto $\mathbf{X}_u$ is denoted by $[u]$.

A VL $\mathbf{X}$ is said to be *of countable type* if every bounded family of non-zero pairwise disjoint elements is at most countable. The class of VLs of countable type is convenient

---

* In Vulikh-I, a linear sublattice.

because all bounds are realized on countable sets. More precisely, an Archimedean VL $\mathbf{X}$ is of countable type if and only if the following condition is satisfied:

(*) for every set $E \subset \mathbf{X}$ such that $\sup E$ exists, there is an at most countable subset $E_0 \subset E$ such that $\sup E_0 = \sup E$; or the analogous condition obtained by replacing sup by inf in (*).

Let us introduce the following notation. Let $\{x_\alpha\}$ $(\alpha \in \mathrm{A})$ be a net in a VL $\mathbf{X}$. The notation $x_\alpha \uparrow$ means that $x_\alpha \geqslant x_{\alpha'}$ whenever $\alpha, \alpha' \in \mathrm{A}$ and $\alpha \geqslant \alpha'$. The notation $x_\alpha \uparrow x$ means that $x_\alpha \uparrow$ and $\sup x_\alpha = x$. We define $x_\alpha \downarrow$ and $x_\alpha \downarrow x$ analogously.

The study of Archimedean VLs often can be reduced to the study of K-spaces by means of K-completions. For every Archimedean VL $\mathbf{X}$, there exists a unique K-space $k\mathbf{X}$, called the *K-completion* of $\mathbf{X}$, having the following properties:

1) $\mathbf{X}$ is a vector sublattice of $k\mathbf{X}$;
2) if $E \subset \mathbf{X}$ and $x = \sup E$ exists in $\mathbf{X}$, then the same x is the supremum of E in $k\mathbf{X}$;
3) for every $\hat{x} \in k\mathbf{X}$, there exist nets $\{y_\alpha\}$ and $\{y'_\beta\}$ in $\mathbf{X}$ such that $y_\alpha \uparrow \hat{x}$, $y'_\beta \downarrow \hat{x}$.

The K-completion is obtained by the usual method of cuts, analogous to the construction of the real numbers by Dedekind (see Vulikh-I, Theorem IV.11.1).

1.4. We consider two types of convergence in a VL $\mathbf{X}$.

A net $\{x_\alpha\}$ $(\alpha \in \mathrm{A})$ is said to be *(o)-convergent* to the limit $x \in \mathbf{X}$ $(x_\alpha \xrightarrow{(o)} x$ or $x = (o)\text{-}\lim x_\alpha)$ if there exists a net $\{y_\beta\}$ $(\beta \in \mathrm{B})$ such that

1) $y_\beta \downarrow \mathbf{0}$;
2) for every $\beta \in \mathrm{B}$, there exists an $\alpha_0 \in \mathrm{A}$ such that $|x_\alpha - x| \leqslant y_\beta$ whenever $\alpha \geqslant \alpha_0$.

In the case of a sequence $\{x_n\}$ we shall also use another definition of order convergence.

A sequence is said to be *$(o\sigma)$-convergent* to the limit $x \in \mathbf{X}$ $(x_n \xrightarrow{(o\sigma)} x$ or $x = (o\sigma)\text{-}\lim x_n)$ if there exists a sequence $\{y_n\}$ such that

1) $y_n \downarrow \mathbf{0}$;
2) $|x_n - x| \leqslant y_n$, $n \in \mathbb{N}$.

If $\mathbf{X}$ is a K_σ-space or an Archimedean VL of countable type, then the two definitions of order convergence for sequences coincide. In general this is not so.

Let $\mathbf{X}$ be a VL, and suppose $\{x_n\}_{n=1}^{\infty} \subset \mathbf{X}$. We shall say that the series $\sum\limits_{n=1}^{\infty} x_n$ is (o)-convergent to x, and write $x = (o)\text{-}\sum\limits_{n=1}^{\infty} x_n$, if the sequence of its partial sums is (o)-convergent to x. The definition of $(o\sigma)$-convergence is analogous.

If $\mathbf{X}$ is an Archimedean VL of countable type and $\{x_\alpha\}$ $(\alpha \in \mathrm{A})$ is a net with $x_\alpha \xrightarrow{(o)} x$, then there exists an increasing sequence of suffices $\alpha_n \in \mathrm{A}$ $(n \in \mathbb{N})$ such that $x_{\alpha_n} \xrightarrow{(o\sigma)} x$.

Note that the lattice operations and the linear operations in a VL are continuous with respect to order convergence.

Let $\mathbf{X}$ be an Archimedean VL. A sequence $\{x_n\}$ is said to be *(r)-convergent* to the limit $x \in \mathbf{X}$ $(x_n \xrightarrow{(r)} x)$ if there exists an element $z \in \mathbf{X}_+$, a so-called *convergence regulator*, such that $|x - x_n| \leqslant \varepsilon_n z$ $(n \in \mathbb{N})$, where $\varepsilon_n \in [0, +\infty)$ and $\varepsilon_n \to 0$. A sequence $\{x_n\}$ is said to be (r)-fundamental if there exist a $z \in \mathbf{X}_+$ and a numerical sequence $\varepsilon_n \downarrow 0$ such that $|x_n - x_m| \leqslant \varepsilon_n z$ whenever $m \geqslant n$. The property of (r)-convergence implies that of $(o\sigma)$-convergence.

An Archimedean VL $\mathbf{X}$ is said to be (r)-*complete* if every r-fundamental sequence of its elements is (r)-convergent. Every K_σ-space is (r)-complete (Vulikh-I, Lemma V.3.1).

1.5. An important method of studying VLs is by means of theorems on representing them as spaces of continuous functions.

Let $\mathbf{X}$ and $\mathbf{Y}$ be VLs. A linear mapping U from $\mathbf{X}$ into $\mathbf{Y}$ is said to be a *lattice homomorphism* if, for $x_1, x_2 \in \mathbf{X}$, we have

$$U(x_1 \vee x_2) = U(x_1) \vee U(x_2).$$

Obviously, in this case the formulae $U(x_1 \wedge x_2) = U(x_1) \wedge U(x_2)$ and $|U(x)| = U(|x|)$ will also be satisfied. In particular, if $x \geqslant \mathbf{0}$, then $U(x) \geqslant \mathbf{0}$.

We shall say that U *preserves bounds* if the existence of the supremum of a set E in $\mathbf{X}$ implies that of the supremum of $U(E)$ in $\mathbf{Y}$, and if we have

$$\sup U(E) = U(\sup E).$$

Then the analogous equation for infima is also true.

A one-to-one lattice homomorphism U, mapping $\mathbf{X}$ onto $\mathbf{Y}$, is called a *lattice isomorphism* or an *order isomorphism*. If U is a lattice isomorphism then the mappings U and U^{-1} preserve bounds. We shall say that two VLs $\mathbf{X}$ and $\mathbf{Y}$ are *order isomorphic* if there exists a lattice isomorphism from $\mathbf{X}$ onto $\mathbf{Y}$. It is clear that a mapping $U: \mathbf{X} \to \mathbf{Y}$ will be a lattice isomorphism if and only if U is vector space isomorphism from $\mathbf{X}$ onto $\mathbf{Y}$ and $U(\mathbf{X}_+) = \mathbf{Y}_+$.

A compactum Q is said to be *extremally disconnected* if the closure of every open set in Q is open-and-closed. It can be shown that the VL of continuous functions $\mathbf{C}(K)$ is a K-space if and only if the compactum K is extremally disconnected (see Vulikh-I).

Now let Q be an extremally disconnected compactum. We denote by $\mathbf{C}_\infty(Q)$ the set of all continuous extended-real-valued functions on Q—that is, functions that may take the values $+\infty$ and $-\infty$ on nowhere dense sets. The space $\mathbf{C}_\infty(Q)$ is turned into a vector space in a natural way. In fact, it can be shown that, if $x_1, x_2 \in \mathbf{C}_\infty(Q)$ and we set $y(t) = x_1(t) + x_2(t)$, for all $t \in Q$ such that $x_1(t)$ and $x_2(t)$ are finite, then the function y admits of a unique extension to an element of $\mathbf{C}_\infty(Q)$, which we take to be the sum $x_1 + x_2$. No difficulties arise in defining multiplication by scalars. We introduce an ordering in $\mathbf{C}_\infty(Q)$, putting $x_1 \geqslant x_2$ if $x_1(t) \geqslant x_2(t)$ for all $t \in Q$. Then $\mathbf{C}_\infty(Q)$ becomes a K-space (Vulikh-I, Theorem V.2.2). The following theorem plays a most important role (see Vulikh-I, Theorem V.4.2).

THEOREM 1. *Every K-space $\mathbf{X}$ is order isomorphic to some foundation of the K-space $\mathbf{C}_\infty(Q)$ constructed on an extremally disconnected compactum* Q. If E is a fixed set of pairwise disjoint positive elements of $\mathbf{X}$, then this isomorphism can be chosen such that the elements of E correspond to the characteristic functions of the open-and-closed sets in Q.*

Since an Archimedean VL can be embedded in its K-completion, and the K-completion can, by Theorem 1, be represented as a K-space of functions in which the linear operations and the lattice operations, applied to a finite number of elements, are calculated pointwise, we arrive at A. I. Yudin's useful *principle of invariance of relations*. Suppose we have two expressions $u(x_1, \ldots, x_n)$ and $v(x_1, \ldots, x_n)$ composed of a finite number of linear operations and lattice operations, where the variables $x_1, \ldots, x_n$ can take values in an arbitrary VL. Then the inequality $u \geqslant v$ holds in an Archimedean VL $\mathbf{X}$ for all values of the

* This compactum is unique to within homeomorphism.

arguments if and only if it holds when we replace $x_1, \ldots, x_n$ by arbitrary real numbers. (The infinite analogue of this principle is false!) We note that it can be shown by a different method that the principle of invariance of relations is true for an arbitrary VL. We leave it to the reader to convince himself of the convenience of the principle by proving the following inequalities, for all $x, y, z \in \mathbf{X}$,

$$|(x \vee z) - (y \vee z)| \leqslant |x - y|, \qquad |(x \wedge z) - (y \wedge z)| \leqslant |x - y|,$$

first using the principle, and then from the definition of a VL.

Let us now characterize the K-spaces $\mathbf{X}$ for which we obtain the whole K-space $\mathbf{C}_\infty(Q)$ under the representation of Theorem 1. A K-space $\mathbf{W}$ is said to be *extended* if every set of pairwise disjoint elements in $\mathbf{W}$ has a supremum. The K-space $\mathbf{C}_\infty(Q)$ is extended, and conversely every extended K-space is isomorphic, under the isomorphism of Theorem 1, to the whole of $\mathbf{C}_\infty(Q)$. If a K-space $\mathbf{X}$ is order isomorphic to a foundation of some extended K-space $\mathbf{W}$, then $\mathbf{W}$ is called a *maximal extension* of $\mathbf{X}$ and is denoted by $\mathfrak{M}(\mathbf{X})$. This space is unique to within an order isomorphism. The existence of a maximal extension is guaranteed by Theorem 1.

A modern approach to the realization of VLs may be found in a paper by Veksler [1].

1.6. Let us illustrate the concepts introduced above for the case of ideal spaces, with which we are already somewhat familiar.

First of all we extend the concept of an ideal space, retaining the earlier definition but waiving the condition that measure be σ-finite, and replacing this by the weaker direct sum property (I.6.10). It follows from Theorem I.6.17 that a real space $\mathbf{S}(T, \Sigma, \mu)$ is a K-space and that the elements denoted by $x \wedge y$, $x \vee y$, $|x|$, x_+, x_- in 1.1 coincide with those of IV.3.1. An IS is an ideal in $\mathbf{S}(T, \Sigma, \mu)$ and an FS is a foundation in $\mathbf{S}(T, \Sigma, \mu)$—hence their names. A set E_1 is wider than E_2 if $\operatorname{supp} E_1 \supset \operatorname{supp} E_2 \pmod{\mu}$. A unit in an IS $\mathbf{X}$ is a function $x \in \mathbf{X}$ such that $x(t) > 0$ a.e. An example of a vector sublattice of $\mathbf{S}(T, \Sigma, \mu)$ that is not an ideal is the set of all simple measurable functions. Clearly $\mathbf{L}^\infty(T, \Sigma, \mu)$ is a u-ideal in $\mathbf{S}(T, \Sigma, \mu)$, where $u(t) \equiv 1$. The set of bands in an FS $\mathbf{X}$ is in one-to-one correspondence with the elements of Σ (sets that are equal mod μ are identified). Here $A \in \Sigma$ corresponds to the band $\mathbf{X}_A = \{x \in \mathbf{X} : \operatorname{supp} x \subset A \pmod{\mu}\}$. The projection $[\mathbf{X}_A]$ onto the band $\mathbf{X}_A$ is the operator of multiplication by the characteristic function of A: that is, $[\mathbf{X}_A]x = x\chi_A$.

It is easy to verify that the K-space $\mathbf{S}(T, \Sigma, \mu)$ is a K-space of countable type if and only if the measure μ is σ-finite. Because of this, we have been able to restrict ourselves to sequences up to now in our study of ISs.

If $\mathbf{X}$ is an IS, then $x_n \xrightarrow{(o)} x$ in $\mathbf{X}$ if and only if $x_n(t) \to x(t)$ a.e. and $|x_n| \leqslant y \in \mathbf{X}$ $(n \in \mathbb{N})$. This type of convergence is one we have already encountered (see VI.1.1). In the case of $\mathbf{S}(0, 1)$, the extremally disconnected compactum Q of Theorem 1 is much more exotic in its topological properties than $[0, 1]$, but nevertheless the representation of Theorem 1 even in this case often provides a key to new results.* Finally, we note that $\mathbf{S}(T, \Sigma, \mu)$ is an extended K-space, the maximal extension of any FS on (T, Σ, μ).

1.7. Often facts that we have proved for ISs carry over to abstract VLs. However, it should be said that the theory of ISs was developed later than the general theory of VLs. Later on, we shall require the following generalization of Lemma IV.3.1.

---

* Note that in this case Q is a Gel'fand compactum of the Banach algebra $\mathbf{L}^\infty(0, 1)$ (see Naimark).

LEMMA 1. *If* $\mathbf{X}$ *is an Archimedean* VL *and* $\mathbf{Y}$ *is a foundation in* $\mathbf{X}$, *then, for each* $x \in \mathbf{X}_+$, *there is a net* $\{y_\alpha\}$ *with* $y_\alpha \uparrow x$, $\mathbf{0} \leqslant y_\alpha \in \mathbf{Y}$.

Proof. First we note that it is sufficient to prove that, for every $x \in \mathbf{X}_+$,

$$x = \sup\{y \in \mathbf{Y}: 0 \leqslant y \leqslant x\}. \tag{1}$$

Since $\mathbf{Y}$ contains, together with every two of its elements, also their supremum, the set $\mathrm{A} = \{y \in \mathbf{Y}: \mathbf{0} \leqslant y \leqslant x\}$, with the ordering induced from $\mathbf{X}$, is a directed set. Therefore, if every element $y \in \mathrm{A}$ is taken as its own index, then A is a net, and $y \uparrow x$ $(y \in \mathrm{A})$.

Assume that (1) is false for some $x \in \mathbf{X}_+$. Then there exists $y_0 \in \mathbf{X}$ such that $y \leqslant y_0$ for all $y \in \mathrm{A}$, but not $x \leqslant y_0$. Taking $y_1 = x \wedge y_0$, we see that $y \leqslant y_1$ for all $y \in \mathrm{A}$ and $y_1 < x$. As $\mathbf{Y}$ is a foundation in $\mathbf{X}$, there exists a $z_0 \in \mathbf{Y}$, corresponding to the element $x - y_1 > \mathbf{0}$, such that $(x - y_1) \wedge z_0 > \mathbf{0}$. The element $z_1 = (x - y_1) \wedge z_0 \in \mathbf{Y}$ satisfies $\mathbf{0} < z_1 \leqslant x - y_1$. Therefore, for every $y \in \mathrm{A}$ we have $y + z_1 \leqslant x - y_1 + y \leqslant x$. As $y + z_1 \in \mathbf{Y}$, we conclude, by induction, that $y + nz_1 \leqslant x$ $(n \in \mathbb{N})$, contrary to the Archimedean axiom.

§ 2. Linear operators and functionals

2.1. Let $\mathbf{X}$ be a VL, and $\mathbf{Y}$ a K-space. In this chapter we shall only be considering linear operators and functionals, so we shall omit the word "linear" in all subsequent definitions.

An operator $U: \mathbf{X} \to \mathbf{Y}$ is said to be *positive* if $U(x) \geqslant \mathbf{0}$ for every $x \geqslant \mathbf{0}$. An operator U is said to be *regular* if $U = U_1 - U_2$, where the U_i are positive linear operators $(i = 1, 2)$. We shall denote the set of all regular operators from $\mathbf{X}$ into $\mathbf{Y}$ by $L^\sim(\mathbf{X}, \mathbf{Y})$.

THEOREM 1. *A necessary and sufficient condition for an operator* $U: \mathbf{X} \to \mathbf{Y}$ *to be regular is that the image* $U(E)$ *of every bounded subset* $E \subset \mathbf{X}$ *be a bounded subset of* $\mathbf{Y}$.

Operators that satisfy the criterion of Theorem 1 are said to be *order-bounded.** Theorem 1 states that, if $\mathbf{Y}$ is a K-space, then the class of regular operators coincides with the class of (o)-bounded operators. In general this is not so.

We introduce an ordering in $L^\sim(\mathbf{X}, \mathbf{Y})$ as follows: $U \geqslant \mathbf{0}$ if U is a positive operator; $U_1 \geqslant U_2$ if $U_1 - U_2 \geqslant \mathbf{0}$. It is an extremely important fact that the set $L^\sim(\mathbf{X}, \mathbf{Y})$, with the ordering defined in this way, is a K-space (Vulikh-I, Theorem VIII.2.1). In this situation, if $\{U_\xi\}$ $(\xi \in \Xi)$ is a set of operators that is bounded above, then $U = \sup U_\xi$ is calculated, for each $x \in \mathbf{X}_+$, from the formula

$$U(x) = \sup\left\{U_{\xi_1}(x_1) + \ldots + U_{\xi_n}(x_n): x = \sum_{i=1}^{n} x_i,\ x_i \geqslant \mathbf{0}, \xi_i \in \Xi\ (1 \leqslant i \leqslant n)\right\}. \tag{1}$$

The infimum is calculated analogously. It is simple to show from (1) that if $U \in L^\sim(\mathbf{X}, \mathbf{Y})$, then for $x \in \mathbf{X}_+$ we have

$$U_+(x) = \sup\{U(x'): 0 \leqslant x' \leqslant x\},\ U_-(x) = -\inf\{U(x'): 0 \leqslant x' \leqslant x\},$$
$$|U|(x) = \sup\{|U(x')|: |x'| \leqslant x\}.$$

This implies that, for every $x \in \mathbf{X}$,

$$|U(x)| \leqslant |U|(|x|). \tag{2}$$

* As before, we shall shorten the word "order" to (o), and write, for example, (o)-bounded operator.

It follows from (1) that, if U_α, $U \in L^\sim(\mathbf{X}, \mathbf{Y})$ and $U_\alpha \xrightarrow{(o)} U$, then for each $x \in \mathbf{X}$ we have $U_\alpha(x) \xrightarrow{(o)} U(x)$ (the converse is false).

We now introduce some further classes of operators.

An operator $U \in L^\sim(\mathbf{X}, \mathbf{Y})$ is said to be *order continuous** or *normal* if, for every net $\{x_\alpha\}$, the statement $x_\alpha \xrightarrow{(o)} x$ implies that $U(x_\alpha) \xrightarrow{(o)} U(x)$. The set $L_n^\sim(\mathbf{X}, \mathbf{Y})$ of all (o)-continuous operators is a band in the K-space $L^\sim(\mathbf{X}, \mathbf{Y})$.

An operator $U \in L^\sim(\mathbf{X}, \mathbf{Y})$ is said to be *order σ-continuous*† if $x_n \xrightarrow{(o\sigma)} x$, for a sequence $\{x_n\}$, implies that $U(x_n) \xrightarrow{(o)} U(x)$. The set $L^\sim_{n\sigma}(\mathbf{X}, \mathbf{Y})$ of all $(o\sigma)$-continuous operators is also a band in $L^\sim(\mathbf{X}, \mathbf{Y})$. Obviously, we have $L_n^\sim(\mathbf{X}, \mathbf{Y}) \subset L^\sim_{n\sigma}(\mathbf{X}, \mathbf{Y})$. For proofs of these facts, see Vulikh-I, Chapter VIII, §§ 3, 4.

2.2. We now dwell in some detail on the case of functionals—that is, the case where $\mathbf{Y} = \mathbf{R}^1$. In this case we write

$$\mathbf{X}^\sim = \mathbf{L}^\sim(\mathbf{X}, \mathbf{R}^1), \quad \mathbf{X}_n^\sim = L_n^\sim(\mathbf{X}, \mathbf{R}^1), \quad \mathbf{X}^\sim_{n\sigma} = L^\sim_{n\sigma}(\mathbf{X}, \mathbf{R}^1).$$

We note that, if f is a linear functional on an Archimedean VL $\mathbf{X}$, then $f \in \mathbf{X}^\sim_{n\sigma}$ if $x_n \xrightarrow{(o\sigma)} x$ implies $f(x_n) \to f(x)$, which simplifies verification of (o)-continuity. The K-space $\mathbf{X}_n^\sim$ is said to be *(o)-dual* to $\mathbf{X}$. If $\mathbf{X}$ is a K-space of countable type, then, by 1.4, $\mathbf{X}_n^\sim = \mathbf{X}^\sim_{n\sigma}$.

A functional $f \in \mathbf{X}^\sim$ is said to be *singular*‡ if there exists a foundation $\mathbf{G}$ in $\mathbf{X}$ on which f vanishes. It is easy to see that the set $\mathbf{X}_s^\sim$ of all singular functionals is an ideal in $\mathbf{X}^\sim$.

LEMMA 1. *If* $\mathbf{X}$ *is an Archimedean* VL, *then* $\mathbf{X}_n^\sim = (\mathbf{X}_s^\sim)^d$.

Proof. Assume the contrary. Then there exists a functional $f \in \mathbf{X}_s^\sim$ such that for some $g \in \mathbf{X}_n^\sim$ we have $|f| \wedge |g| > \mathbf{0}$. Since $\mathbf{X}_s^\sim$ and $\mathbf{X}_n^\sim$ are ideals, $f_1 = |f| \wedge |g| \in \mathbf{X}_s^\sim \cap \mathbf{X}_n^\sim$. By the definition of a singular functional, there exists a foundation $\mathbf{Y}$ in $\mathbf{X}$ on which f_1 vanishes. Lemma 1.1 shows that, for each $x \in \mathbf{X}_+$, there exists a net $\{x_\alpha\}$ with $x_\alpha \uparrow x$, $\mathbf{0} \leqslant x_\alpha \in \mathbf{Y}$. Since $f_1 \in \mathbf{X}_n^\sim$, we have $f_1(x) = \lim f_1(x_\alpha) = 0$. Hence $f_1 = \mathbf{0}$, and we have a contradiction to the assumption that $f_1 > \mathbf{0}$.

We shall prove below that one often has the equation $\mathbf{X}_s^\sim = (\mathbf{X}_n^\sim)^d$.

2.3. Now we record some facts connected with the canonical embedding of a VL $\mathbf{X}$ in its second (o)-dual space.

Let $\mathbf{X}$ be a VL and $\mathbf{Y}$ an ideal in $\mathbf{X}^\sim$ that is total on $\mathbf{X}$. Denote by π the canonical embedding of $\mathbf{X}$ into $\mathbf{Y}^\sim$. First note that $\pi(\mathbf{X}) \subset \mathbf{Y}_n^\sim$. For if $x \in \mathbf{X}$ and $f_\alpha \xrightarrow{(o)} \mathbf{0}$ in the K-space $\mathbf{Y}$, then $f_\alpha \xrightarrow{(o)} \mathbf{0}$ in $\mathbf{X}^\sim$, so that

$$\pi(x)(f_\alpha) = f_\alpha(x) \to 0,$$

which proves that $\pi(x) \in \mathbf{Y}_n^\sim$.

---

* In Vulikh-I, completely linear.

† In Vulikh-I, (o)-linear.

‡ In Vulikh-I, anormal.

THEOREM 2. 1) *The mapping* π *is a lattice homomorphism from the* VL **X** *onto a vector sublattice of the K-space* $\mathbf{Y}_n^{\sim}$.

2) *The mapping* π *preserves bounds if and only if* $\mathbf{Y} \subset \mathbf{X}_n^{\sim}$.

3) *The vector sublattice* $\pi(\mathbf{X})$ *is an ideal in* $\mathbf{Y}^{\sim}$ *if and only if* $\mathbf{Y} \subset \mathbf{X}_n^{\sim}$ *and* **X** *is a K-space.*

Proof. 1) is contained in the proof of Theorem IX.5.1 in Vulikh-I (see also Schaefer-I, V.1.6). That π preserves bounds and $\pi(\mathbf{X})$ is an ideal if $\mathbf{Y} \subset \mathbf{X}_n^{\sim}$ is proved in Lemma IX.5.1, Vulikh-I. The converse part of 2) is obvious, while 3) is proved in a similar fashion to Theorem IX.7.2, Vulikh-I.

Notice also that, by the proof of Lemma IX.5.1, Vulikh-I, $\pi(\mathbf{X})$ is always of the same width as $\mathbf{Y}_n^{\sim}$.

COROLLARY 1. *If an element* $x \in \mathbf{X}$ *is such that* $f(x) \geqslant 0$ for every $f \in \mathbf{Y}_+$, *then* $x \geqslant \mathbf{0}$.

COROLLARY 2. *If* $x_\alpha \downarrow$ *in* **X** *and* $x_\alpha \to x$ $(\sigma(\mathbf{X}, \mathbf{Y}))$, *then* $x_\alpha \downarrow x$.

Proof. If $\alpha' \geqslant \alpha$, then $f(x_\alpha - x_{\alpha'}) \geqslant 0$ for all $f \in \mathbf{Y}_+$, so, taking limits with respect to α', we see that $f(x_\alpha - x) \geqslant 0$. Hence $x_\alpha \geqslant x$ for all α, by Corollary 1. On the other hand, if $y \leqslant x_\alpha$ for all α, then $f(y) \leqslant f(x_\alpha) \to f(x)$ for $f \in \mathbf{Y}_+$, and so, by Corollary 1 again, we have $x \geqslant y$. Hence $x_\alpha \downarrow x$.

Let **X** be a K-space for which $\mathbf{X}_n^{\sim}$ is total, and let $\kappa: \mathbf{X} \to (\mathbf{X}_n^{\sim})_n^{\sim}$ be the canonical embedding. Then **X** is said to be *(o)-reflexive* if $\kappa(\mathbf{X}) = (\mathbf{X}_n^{\sim})_n^{\sim}$.

THEOREM 3. *Let* **X** *be a K-space for which* $\mathbf{X}_n^{\sim}$ *is total. The following statements are equivalent*:

1) **X** *is (o)-reflexive*;

2) $\kappa(\mathbf{X})$ *is a band in* $(\mathbf{X}_n^{\sim})_n^{\sim}$;

3) *if* $\{x_\alpha\}$ *is a net in* **X** *with* $0 \leqslant x_\alpha \uparrow$, *such that* $\sup f(x_\alpha) < +\infty$ *for all* $f \in \mathbf{X}_n^{\sim}, f \geqslant \mathbf{0}$, *then there exists* $x \in \mathbf{X}$ *such that* $x_\alpha \uparrow x$.

The equivalence 1) $\Leftrightarrow$ 3) is proved in Theorem IX.6.1, Vulikh-I, while 1) $\Rightarrow$ 2) is obvious. 2) $\Rightarrow$ 1), since $\kappa(\mathbf{X})$ is a foundation in $(\mathbf{X}_n^{\sim})_n^{\sim}$, as we remarked after Theorem 2.

COROLLARY. *If* **X** *is a* VL *and* **Y** *is a band in* $\mathbf{X}^{\sim}$, *then* **Y** *is (o)-reflexive.*

Proof. The set $\{\pi(x): x \in \mathbf{X}\}$ is obviously a total set of (o)-continuous functionals on **Y**. Suppose $\mathbf{0} \leqslant f_\alpha \uparrow$ in **Y** and suppose $\sup F(f_\alpha) < +\infty$ for all $F \in \mathbf{Y}_n^{\sim}$, $F \geqslant \mathbf{0}$. In particular,

$$\sup_\alpha \pi(x)(f_\alpha) = \sup_\alpha f_\alpha(x) < +\infty$$

for all $x \in \mathbf{X}_+$. By the definition of the supremum in $\mathbf{X}^{\sim}$, there exists a functional $f \in \mathbf{X}^{\sim}$ such that $f_\alpha \uparrow f$. As **Y** is a band, $f \in \mathbf{Y}$, and hence 3) of Theorem 3 is satisfied.

In particular, $\mathbf{X}^{\sim}$ and $\mathbf{X}_n^{\sim}$ are (o)-reflexive.

2.4. Let us again consider the case of an ideal space **X**, assuming, as in 1.6, that the measure has the direct sum property. Since in the case of a σ-finite measure the space $S(T, \Sigma, \mu)$ and all its ideals are K-spaces of countable type, the class of order continuous functionals coincides with that of order σ-continuous functionals, as mentioned in 2.2; hence in Chapter VI we were actually dealing with order continuous functionals.

Theorem VI.1.1, on the representation of $\mathbf{X}_n^{\sim}$ by means of its dual $\mathbf{X}'$, carries over without any change to the case where the measure μ has the direct sum property. Moreover, the mapping $x' \in \mathbf{X}' \to f_{x'} \in \mathbf{X}_n^{\sim}$ is an order isomorphism between the K-spaces $\mathbf{X}'$ and $\mathbf{X}_n^{\sim}$. Hence, if

$$f(x) = \int_T x(t)x'(t)\,d\mu \qquad (x \in \mathbf{X}),$$

then

$$|f|(x) = \int_T x(t)|x'(t)|\,d\mu,$$

$$f_+(x) = \int_T x(t)x'_+(t)\,d\mu, \qquad f_-(x) = \int_T x(t)x'_-(t)\,d\mu.$$

If $\mathbf{X}$ is an IS, then, using Theorem VI.1.1, it is easy to see that $\mathbf{X}$ is (o)-reflexive if and only if $\mathbf{X} = \mathbf{X}''$.

2.5. The class of ideal spaces (together with the VL $\mathbf{C}(K)$) is the most important class of VLs, so it is important to obtain an abstract characterization of it in the class of all VLs. It turns out that a K-space is order isomorphic to some IS if and only if it contains a foundation $\mathbf{Y}$ for which $\mathbf{Y}_n^{\sim}$ is total. We shall conclude this section with a proof of this. First of all we need the following lemma.

LEMMA 2. *If $\mathbf{X}$ is a K-space, then for each $f \in \mathbf{X}_n^{\sim}$, $f \geqslant \mathbf{0}$, there exists a band $\mathbf{C}_f$, called the band of essential positivity of f, having the following properties:*

1) $f(x) > 0$ for all $x > \mathbf{0}$ in $\mathbf{C}_f$;

2) $f(x) = 0$ for all $x \in (\mathbf{C}_f)^{\mathrm{d}}$.

Proof. Write

$$\mathbf{N}_f = \{x \in \mathbf{X}: f(|x|) = 0\}.$$

Since $f \in \mathbf{X}_n^{\sim}$, we see that $\mathbf{N}_f$ is a band. Obviously the band we are looking for is $\mathbf{C}_f = (\mathbf{N}_f)^{\mathrm{d}}$.

LEMMA 3. *If $\mathbf{X}$ is a K-space and $f, g \geqslant 0$ are (o)-continuous functionals on $\mathbf{X}$, then $f\,\mathrm{d}\,g$ if and only if $\mathbf{C}_f\,\mathrm{d}\,\mathbf{C}_g$.*

Proof. Since $\mathbf{C}_f$ and $\mathbf{C}_g$ are bands, the statement $\mathbf{C}_f\,\mathrm{d}\,\mathbf{C}_g$ is equivalent to $\mathbf{C}_f \cap \mathbf{C}_g = \{\mathbf{0}\}$. If $f\,\mathrm{d}\,g$ and $\mathbf{0} \leqslant x \in \mathbf{C}_f \cap \mathbf{C}_g$, then, as $f \wedge g = \mathbf{0}$, we have

$$\inf\{f(x_1)+g(x_2): x = x_1+x_2; \quad x_1, x_2 \geqslant \mathbf{0}\} = 0.$$

Hence we can construct two sequences $x_1^{(n)}, x_2^{(n)} \geqslant \mathbf{0}$ $(n \in \mathbb{N})$ such that $x = x_1^{(n)} + x_2^{(n)}$ and

$$f(x_1^{(n)})+g(x_2^{(n)}) < 1/2^n.$$

Write

$$y_i^{(n)} = \sup_{m \geqslant n} x_i^{(m)} \qquad (i = 1, 2).$$

Then $y_i^{(n)} \downarrow y_i \geqslant \mathbf{0}$ $(i = 1, 2)$ and

$$f(y_1^{(n)}) \leqslant \sum_{m=n}^{\infty} f(x_1^{(m)}) \leqslant 1/2^{n-1}, \qquad g(y_2^{(n)}) \leqslant 1/2^{n-1}.$$

Hence $f(y_1) = g(y_2) = 0$. Now $\mathbf{0} \leqslant y_i \leqslant x \in \mathbf{C}_f \cap \mathbf{C}_g$, so $y_i = \mathbf{0}$ $(i = 1,\ 2)$ by the definition of a band of essential positivity. Thus $x = x_1^{(n)} + x_2^{(n)} \xrightarrow{(o)} \mathbf{0}$, and so $x = \mathbf{0}$.

Conversely, suppose that $\mathbf{C}_f\,\mathrm{d}\,\mathbf{C}_g$. If $\mathbf{Y} = (\mathbf{C}_f \cup \mathbf{C}_g)^{\mathrm{d}}$, then by 1.3 we have, for every $x \in \mathbf{X}_+$,

$$x = [\mathbf{C}_f]x + [\mathbf{C}_g]x + [\mathbf{Y}]x.$$

Thus $g([\mathbf{C}_f]x) = f([\mathbf{C}_g]x) = f([\mathbf{Y}]x) = g([\mathbf{Y}]x) = 0$. Consequently

$$0 \leqslant (f \wedge g)\ (x) \leqslant f([\mathbf{C}_g]x + [\mathbf{Y}]x) + g([\mathbf{C}_f]x) = 0,$$

and so $f\ \mathrm{d}\,g$.

A functional $f \in \mathbf{X}_n^{\sim}$, $f \geqslant \mathbf{0}$, is said to be *essentially positive* if $\mathbf{C}_f = \mathbf{X}$.

LEMMA 4. *If* $\mathbf{X}$ *is a K-space for which* $\mathbf{X}_n^{\sim}$ *is total, then there exists a system* $\{f_\alpha\}$ $(\alpha \in \mathrm{A})$ *of pairwise disjoint functionals in* $\mathbf{X}_n^{\sim}$ *such that* $f_\alpha > \mathbf{0}$ $(\alpha \in \mathrm{A})$, $\left(\bigcup_{\alpha \in \mathrm{A}} \mathbf{C}_{f_\alpha}\right)^{\mathrm{dd}} = \mathbf{X}$ *and each of the bands* $\mathbf{C}_{f_\alpha}$ *is principal.*

Proof. Let $\mathfrak{M}$ denote the set of all systems $\mathfrak{m} = \{f_\beta\}$ $(\beta \in \mathrm{B})$ of pairwise disjoint functionals in $\mathbf{X}_n^{\sim}, f_\beta > \mathbf{0}$ $(\beta \in \mathrm{B})$, such that each band $\mathbf{C}_{f_\beta}$ is principal. We introduce an ordering in $\mathfrak{M}$, setting $\mathfrak{m} \geqslant \mathfrak{n}$, where $\mathfrak{m} = \{f_\beta\}$ $(\beta \in \mathrm{B}), \mathfrak{n} = \{g_\gamma\}$ $(\gamma \in \Gamma)$, if for each $\gamma \in \Gamma$ there exists a $\beta \in \mathrm{B}$ such that $f_\beta = g_\gamma$. By Zorn's Lemma, $\mathfrak{M}$ contains a maximal system $\mathfrak{m}_0 = \{f_\alpha\}$ $(\alpha \in \mathrm{A})$. Let us show that $\{f_\alpha\}$ $(\alpha \in \mathrm{A})$ is the required system. Assume that $\mathbf{Y} = \left(\bigcup_{\alpha \in \mathrm{A}} \mathbf{C}_{f_\alpha}\right)^{\mathrm{dd}} \neq \mathbf{X}$. Write $\mathbf{Z} = \mathbf{Y}^{\mathrm{d}}$. As $\mathbf{Z} \neq \{\mathbf{0}\}$, there exists $x_0 \in \mathbf{Z}$ such that $x_0 > \mathbf{0}$. Because $\mathbf{X}_n^{\sim}$ is total on $\mathbf{X}$, there exists a functional $g \in \mathbf{X}_n^{\sim}$ such that $g(x_0) \neq \mathbf{0}$. Since $|g(x_0)| \leqslant |g|\,(x_0)$, we may assume that $g > \mathbf{0}$. For each $x \in \mathbf{X}$, write

$$f(x) = g([x_0]x).$$

Clearly $f \in \mathbf{X}_n^{\sim}$; and $f(x_0) = g(x_0) > 0$, so $f > \mathbf{0}$. Furthermore, $\mathbf{C}_f \subset \mathbf{X}_{x_0} \subset \mathbf{Z}$, so $\mathbf{C}_f \,\mathrm{d}\, \mathbf{C}_{f_\alpha}$, for all $\alpha \in \mathrm{A}$. By Lemma 3, $f \,\mathrm{d}\, f_\alpha$ for all $\alpha \in \mathrm{A}$. This gives a contradiction to the maximality of $\mathfrak{m}_0$, so the lemma is proved.

Now we present a lemma on spaces of continuous functions. Even in the VL $\mathbf{C}[0, 1]$ it is easy to check that the bands of infinite sets are not calculated pointwise.

LEMMA 5. *Let* K *be a compactum,* E *a set of non-negative functions in the VL* $\mathbf{C}(K)$, *and* $z(t) = \inf\{x(t) : x \in E\}$ $(t \in K)$. *A necessary and sufficient condition for* $\inf E = \mathbf{0}$ *is that the set* $P = \{t \in K : z(t) > 0\}$ *be of first category in* K.

Proof. Necessity. For each $\varepsilon > 0$, write $P(\varepsilon) = \{t \in K : z(t) \geqslant \varepsilon\}$. Then $P = \bigcup_{n=1}^{\infty} P(1/n)$. It is therefore sufficient to establish that, for each $\varepsilon > 0$, the set $P(\varepsilon)$ is nowhere dense. Since

$$P(\varepsilon) = \bigcap_{x \in E} \{t \in K : x(t) \geqslant \varepsilon\},$$

$P(\varepsilon)$ is closed. If $P(\varepsilon)$ is not nowhere dense, then there exists a point $t_0 \in K$ and an open neighbourhood V of t_0 with $V \subset P(\varepsilon)$. Since a compactum is completely regular (see Dunford and Schwartz-I, Chapter I, § 5), there exists a function $x_0 \in \mathbf{C}(K)$ such that

1) $0 \leqslant x_0(t) \leqslant \varepsilon$, $t \in K$;
2) $x_0(t_0) = \varepsilon$;
3) $x_0(t) = 0$, $t \in K \setminus V$.

Thus $\mathbf{0} < x_0 \leqslant x$ for every $x \in E$, which contradicts $\inf E = \mathbf{0}$.

Sufficiency. Suppose P is of first category. Assume that $\mathbf{0}$ is not the infimum of E. Then there exists a function $x_0 \in \mathbf{C}(K)$ such that $\mathbf{0} < x_0 \leqslant x$ for every $x \in E$; hence there exists a point $t_0 \in K$ such that $x_0(t_0) > 0$. Thus $V = \{t \in K : x_0(t) > 0\}$ is a non-empty open set. Also $z(t) \geqslant x_0(t) > 0$ for all $t \in V$, so V is contained in P. Hence V is a set of first category. But it is well known from topology that a non-empty open set in a compactum cannot be of first category (for a proof of this one need only slightly modify the proof of Theorem I.4.2).

We now consider measures on extremally disconnected compacta. From now on let Q be an extremally disconnected compactum, $\mathscr{E}_Q$ the collection of all open-and-closed subsets of Q, and $\mathcal{N}_Q$ the collection of all sets of first category in Q. Denote by $\mathfrak{B}_Q$ the collection of all symmetric differences $G \Delta N$, where $G \in \mathcal{E}_Q$, $N \in \mathcal{N}_Q$.

LEMMA 6. *$\mathfrak{B}_Q$ is a σ-algebra containing the Borel σ-algebra of Q.*

Proof. It is easy to see that a set A belongs to $\mathfrak{B}_Q$ if and only if there exists $G \in \mathcal{E}_Q$ such that the sets

$$N_1 = A \setminus G \quad \text{and} \quad N_2 = G \setminus A \tag{3}$$

are of first category. For if $A = G \Delta N$, $G \in \mathcal{E}_Q$, $N \in \mathcal{N}_Q$, then $N_1 = N \setminus G$ and $N_2 = N \cap G$ are sets of first category. Conversely, if A is such that (3) holds, then write $N = N_1 \cup N_2 \in \mathcal{N}_Q$. Clearly, $A = G \Delta N$. It is therefore sufficient to prove that the sets satisfying (3) form a σ-algebra. If $\{A_n\}$ is a sequence of such sets, and $G_n \in \mathcal{E}_Q$ are the sets corresponding to these, then we set $G = \bigcup_{n=1}^{\infty} G_n$. Since G is open, the set $\bar{G} \setminus G$ is nowhere dense. Further, $A = \bigcup_{n=1}^{\infty} A_n$ satisfies

$$A \setminus \bar{G} \subset A \setminus G \subset \bigcup_{n=1}^{\infty} (A_n \setminus G_n) \in \mathcal{N}_Q,$$

$$G \setminus A \subset \bigcup_{n=1}^{\infty} (G_n \setminus A_n) \in \mathcal{N}_Q.$$

As $\bar{G} \setminus A = [(\bar{G} \setminus G) \setminus A] \cup (G \setminus A) \in \mathcal{N}_Q$, we have $A \in \mathfrak{B}_Q$. If $A = G \Delta N \in \mathfrak{B}_Q$, then $Q \setminus A = (Q \setminus G) \Delta N$. Since $Q \setminus G \in \mathcal{E}_Q$, it follows that $Q \setminus A \in \mathfrak{B}_Q$. Thus we have proved that $\mathfrak{B}_Q$ is a σ-algebra.

Since, for every open set U, the sets $\bar{U} \in \mathcal{E}_Q$ and $\bar{U} \setminus U$ are nowhere dense, we see that $U \in \mathfrak{B}_Q$, and so the Borel σ-algebra is contained in $\mathfrak{B}_Q$.

A measure μ defined on the σ-algebra $\mathfrak{B}_Q$ is said to be *normal* if:

1) $N \in \mathcal{N}_Q$ implies $\mu(N) = 0$;

2) for every $G \in \mathcal{E}_Q$ with $\mu(G) = +\infty$, there exists $G_1 \in \mathcal{E}_Q$ such that $G_1 \subset G$ and $0 < \mu(G_1) < +\infty$.

A normal measure μ is said to be *strictly positive* if

3) $G \in \mathcal{E}_Q$, $G \neq \varnothing$, implies $\mu(G) > 0$.

An extremally disconnected compactum on which there is a strictly positive normal measure is called a *hyper-Stone* compactum. Let Q be a hyper-Stone compactum with the strictly positive normal measure μ. We note the simplest consequences of the definitions.

a) $\mu(G \Delta N) = \mu(G)$, if $G \in \mathcal{E}_Q$, $N \in \mathcal{N}_Q$.

For $\mu(G \Delta N) = \mu(G \setminus N) + \mu(N \setminus G) = \mu(G \setminus N) = \mu(G \setminus N) + \mu(N \cap G) = \mu(G)$.

b) The measure μ is complete.

If $A \subset G \Delta N$, $G \in \mathcal{E}_Q$, $N \in \mathcal{N}_Q$, and $\mu(G \Delta N) = 0$, then by a) and 3) we have $G = \varnothing$, so $A \subset N$ and hence $A \in \mathcal{N}_Q$.

c) The measure μ is locally finite.

If $\mu(G \Delta N) = +\infty$, $G \in \mathcal{E}_Q$, $N \in \mathcal{N}_Q$, then by 1) $\mu(G) = +\infty$, and thus, by 2), there exists a $G_1 \in \mathcal{E}_Q$ such that $G_1 \subset G$ and $0 < \mu(G_1) < +\infty$. Hence $G_1 \Delta N \subset G \Delta N$ and $\mu(G_1 \Delta N) = \mu(G_1) < +\infty$, which proves c).

d) The compactum Q contains a system $\{Q_\alpha\} \subset \mathcal{E}_Q$ of pairwise disjoint sets such that $0 < \mu(Q_\alpha) < +\infty$, for every α, and the set $Q \setminus \bigcup Q_\alpha$ is nowhere dense.

First we note that, for every open set $V \subset Q$, the set $\bar{V} \setminus V$ belongs to $\mathcal{N}_Q$. It is easy to deduce d) from this, using Zorn's Lemma and 2).

e) Let $\{Q_\alpha\}$ be the system in d). If $A \cap Q_\alpha \in \mathfrak{B}_Q$ for every α then $A \in \mathfrak{B}_Q$.

Suppose $A \cap Q_\alpha = G_\alpha \Delta N_\alpha$, where $G_\alpha \in \mathscr{E}_Q$, $N_\alpha = \bigcup_{n=1}^{\infty} N_\alpha^{(n)}$, and the sets $N_\alpha^{(n)}$ are nowhere dense. We may assume here that $G_\alpha, N_\alpha \subset Q_\alpha$. Write $N^{(n)} = \bigcup_\alpha N_\alpha^{(n)}$ $(n \in \mathbb{N})$. Let us show that $N^{(n)}$ is nowhere dense. Assume the contrary.

Assume that the non-empty open set V is contained in $\overline{N^{(n)}}$. Since $Q \setminus \bigcup_\alpha Q_\alpha$ is closed and nowhere dense, the set $U = V \cap \left(\bigcup_\alpha Q_\alpha \right)$ is non-empty and open. As the sets Q_α are pairwise disjoint and open-and-closed, we have $\overline{N^{(n)}} \cap Q_\alpha = \overline{N_\alpha^{(n)}}$, for every α. Therefore $U \cap Q_\alpha \subset \overline{N_\alpha^{(n)}}$, and so $U \cap Q_\alpha$ is both open and nowhere dense. Hence $U \cap Q_\alpha = \varnothing$ for all α, which means that $U = \varnothing$.

Thus $N^{(n)}$ is nowhere dense, $N = \bigcup_{n=1}^{\infty} N^{(n)} \in \mathscr{N}_Q$, $G = \bigcup G_\alpha$ is open, and so, by Lemma 6, $G \in \mathfrak{B}_Q$.

It is now clear that

$$A = (G \Delta N) \cup (A \cap (Q \setminus \bigcup Q_\alpha)) \in \mathfrak{B}_Q.$$

We see from b), c) and e) that $(Q, \mathfrak{B}_Q, \mu)$ satisfies all the axioms for a measure space (see I.6.2).

f) The space $(Q, \mathfrak{B}_Q, \mu)$ has the direct sum property.

We show that the direct sum condition is satisfied for the system $\{Q_\alpha\}$ in d). It is obviously sufficient to consider the case where $G \in \mathscr{E}_Q$, $0 < \mu(G) < \infty$.

We express G in the form

$$G = \bigcup_\alpha (G \cap Q_\alpha) \cup \left[G \cap \left(Q \setminus \bigcup_\alpha Q_\alpha \right) \right].$$

For the proof of f) it is clear enough to verify that $G \cap Q_\alpha \neq \varnothing$ for at most a countable set of suffices α. If we assume the contrary, then there exist an $\varepsilon > 0$ and a countable set of suffices α_n such that $\mu(G \cap Q_{\alpha_n}) \geqslant \varepsilon$. This contradicts the inequality

$$\sum_{n=1}^{\infty} \mu(G \cap Q_{\alpha_n}) \leqslant \mu(G) < +\infty.$$

THEOREM 4. *Let μ be a strictly positive normal measure on $\mathfrak{B}_Q$.*

1) *If $x \in \mathbf{C}_\infty(Q)$, then x is a measurable a.e. finite function.*

2) *If $x \neq y$ in $\mathbf{C}_\infty(Q)$, then x and y are not equivalent functions.*

3) *If y is a measurable a.e. finite function on Q, then there exists $x \in \mathbf{C}_\infty(Q)$ such that $x(t) = y(t)$ a.e.*

Proof. 1) This is obvious.

2) If $x \neq y$, then the set $P = \{t \in Q : x(t) = y(t)\}$ is open and non-empty, so that $\mu(P) = \mu(\overline{P}) > 0$.

3) Let us show that the identity mapping $x \in \mathbf{C}_\infty(Q) \to x \in \mathbf{S}(Q, \mathfrak{B}_Q, \mu)$ is an order isomorphism between the extended K-spaces $\mathbf{C}_\infty(Q)$ and $\mathbf{S}(Q, \mathfrak{B}_Q, \mu)$. First we show that this mapping yields an isomorphism from $\mathbf{C}(Q)$ to $\mathbf{L}^\infty(Q, \mathfrak{B}_Q, \mu)$. To do this we need only verify that for each $y \in \mathbf{L}^\infty(Q, \mathfrak{B}_Q, \mu)$ there exists $x \in \mathbf{C}(Q)$ such that $x(t) = y(t)$ a.e. If

$G \Delta N \in \mathfrak{B}_Q$, then

$$\chi_{G \Delta N}(t) = \chi_G(t) \text{ a.e.} \tag{4}$$

and $\chi_G \in \mathbf{C}(Q)$, since G is open-and-closed. We may assume that y is a bounded function. Then by Theorem I.6.3 there exists a sequence of simple measurable functions $\{y_n\}$ such that $y_n \to y$ uniformly on Q. In view of (4), each y_n determines an $x_n \in \mathbf{C}(Q)$ with $x_n(t) = y_n(t)$ a.e. Thus $x_n(t) \to y(t)$ uniformly on a set A that is the complement to some set of first category. Hence the restriction of y to A is continuous and $\overline{A} = Q$. On the other hand, $\{x_n\}$ is a Cauchy sequence in $\mathbf{C}(Q)$ since we clearly have

$$\|x_n - x_m\|_{\mathbf{C}(Q)} = \|y_n - y_m\|_{\mathbf{L}^\infty(Q, \mathfrak{B}_Q, \mu)}.$$

As $\mathbf{C}(Q)$ is complete, $x_n \to x$ in norm in $\mathbf{C}(Q)$ and $x(t) = y(t)$ a.e. Thus we have established that $\mathbf{C}(Q)$ and $\mathbf{L}^\infty(Q, \mathfrak{B}_Q, \mu)$ are isomorphic.

Let $y \in \mathbf{S}(Q, \mathfrak{B}_Q, \mu)$. We may assume that the function $y(t)$ is everywhere finite. For $n = 0, 1, 2, \ldots$, we set

$$y_n(t) = \begin{cases} y(t) & \text{if } n \leqslant |y(t)| < n+1, \\ 0 & \text{otherwise.} \end{cases}$$

Then $y_n \in \mathbf{L}^\infty(Q, \mathfrak{B}_Q, \mu)$. By what has been proved, there exists $x_n \in \mathbf{C}(Q)$ with $x_n(t) = y_n(t)$ a.e. As $|y_n| \wedge |y_m| = \mathbf{0}$ $(n \neq m)$ in $\mathbf{L}^\infty(Q, \mathfrak{B}_Q, \mu)$, we have $|x_n| \wedge |x_m| = \mathbf{0}$ $(n+m)$ in $\mathbf{C}(Q)$, and so also in $\mathbf{C}_\infty(Q)$. Now $\mathbf{C}_\infty(Q)$ is an extended K-space, so $x = \sup x_n$ exists in $\mathbf{C}_\infty(Q)$. Clearly the pointwise supremum satisfies $\sup x_n(t) = y(t)$ a.e. and so, by Lemma 5, $x(t) = \sup x_n(t)$ except on a set of first category—that is, almost everywhere. This proves, therefore, that $y(t) = x(t)$ a.e.

In the course of the proof we established

COROLLARY. *The mapping $x \in \mathbf{C}_\infty(Q) \to x \in \mathbf{S}(Q, \mathfrak{B}_Q, \mu)$ is an order isomorphism.*

Now we can prove the fundamental theorem.

THEOREM 5. *A K-space $\mathbf{X}$ is order isomorphic to an ideal space on some measure space (T, Σ, μ) with the direct sum property if and only if there exists a foundation $\mathbf{Y}$ in $\mathbf{X}$ with $\mathbf{Y}_n^\sim$ total.*

Proof. If $\mathbf{X}$ is an IS on (T, Σ, μ), then $\mathbf{X} \cap \mathbf{L}^1(T, \Sigma, \mu)$ is a foundation in $\mathbf{X}$, on which the set of (o)-continuous functionals is total, since it is total on $\mathbf{L}^1(T, \Sigma, \mu)$.

Let us now prove the converse. We may assume that $\mathbf{X}_n^\sim$ is total on $\mathbf{X}$, since if $\mathbf{Y}$ is isomorphic to a foundation in $\mathbf{S}(T, \Sigma, \mu)$, then $\mathbf{X}$ is isomorphic to a foundation in $\mathbf{S}(T, \Sigma, \mu)$, in view of the uniqueness of the maximal extension. Let $\{f_\alpha\}$ $(\alpha \in \mathrm{A})$ be the system of functionals in Lemma 4, and let x_α be the element generating the band $\mathbf{C}_{f_\alpha}$. By Lemma 3, the elements x_α are pairwise disjoint. Thus, by Theorem 1.1, the maximal extension $\mathfrak{M}(\mathbf{X})$ can be represented in the form $\mathbf{C}_\infty(Q)$, such that the elements x_α correspond to characteristic functions of pairwise disjoint open-and-closed sets Q_α. Since $\{x_\alpha : \alpha \in \mathrm{A}\}^{dd} = \mathbf{X}$, it follows that $\bigcup Q_\alpha = Q$. We now define a measure on $\mathfrak{B}_Q$. If $A = G \Delta N$, $G \in \mathscr{E}_Q$, $N \in \mathscr{N}_Q$, where A is contained in some Q_α, then we set

$$\mu(A) = f_\alpha(\chi_G).$$

(Here we identify $\mathbf{X}$ with its isomorphic image in $\mathbf{C}_\infty(Q)$.) For every $A \in \mathfrak{B}_Q$, we set

$$\mu(A) = \sup\left\{ \sum_{i=1}^n \mu(A \cap Q_{\alpha_i}) : \{\alpha_i\}_{i=1}^n \subset \mathrm{A} \right\}.$$

It is easy to see that, to prove μ is measure, it is enough to verify that its restriction to every σ-algebra $\mathfrak{B}_{Q_\alpha}$ is countably additive. First we note that, if

$$G \Delta N \subset G_1 \Delta N_1; G, G_1 \in \mathscr{E}_Q; N, N_1 \in \mathscr{N}_Q,$$

then $G \subset G_1$. For otherwise there would be a non-empty open set U contained in G such that $U \cap G_1 = \varnothing$. Then

$$U \setminus N \subset G \setminus N \subset G_1 \Delta N_1 = (G_1 \setminus N_1) \cup (N_1 \setminus G_1),$$

so that $U \setminus N \subset N_1 \setminus G_1$, as $U \cap G_1 = \varnothing$. Therefore $U \subset N \cup N_1$; but we have already remarked that an open set in a compactum cannot be of first category.

Thus we now prove that the restriction of μ to the σ-algebra $\mathfrak{B}_{Q_\alpha}$ is countably additive. Let $A_n = G_n \Delta N_n \in \mathfrak{B}_{Q_\alpha}$, $A_n \downarrow \varnothing$. It follows from what we have proved that $G_n \downarrow$. The set $\bigcap G_n$ is clearly of first category. Hence, by Lemma 5, $\chi_{G_n} \downarrow \mathbf{0}$ in the K-space $\mathbf{X}$, so that $\mu(A_n) = f_\alpha(\chi_{G_n}) \to 0$ as $n \to \infty$.

We now verify that μ is a normal measure. Properties 1) and 2) are obvious from the definition of μ. Property 3) follows from the fact that for each $G \in \mathscr{E}_Q$ there exists a Q_α such that $G \cap Q_\alpha \neq \varnothing$ and the fact that f_α is essentially positive on $\mathbf{C}_{f_\alpha}$.

§ 3. Normed lattices

3.1. A norm $\|\cdot\|$ on a VL $\mathbf{X}$ is said to be *monotone* if $|x| \leqslant |y|$ implies $\|x\| \leqslant \|y\|$. A VL equipped with a monotone norm is called a *normed lattice* (NL, for short). An NL that is complete in its norm is called a *Banach lattice* (BL, for short). If an NL (or BL) $\mathbf{X}$ is a K-space then $\mathbf{X}$ is said to be a *normed K-space* (or *Banach K-space*, respectively).

Obviously, an NIS is a normed K-space, and a BIS is a Banach K-space. Properties 1)–4) of IV.3.1 for convergence in norm are clearly valid for any NL $\mathbf{X}$. Property 5) amounts to continuity of the projection operator in a normed K-space $\mathbf{X}$: if $x_n \to x$ in norm, then $[E]x_n \to [E]x$ in norm (moreover the norm of $[E]$ is at most 1). Note also that every NL is Archimedean.

Let us consider the connection between convergence in norm and convergence with respect to a regulator. If $\mathbf{X}$ is an NL and $x_n \xrightarrow{(r)} x$, then $x_n \to x$ in norm, since $\|x_n - x\| \leqslant \varepsilon_n \|u\| \to 0$ (where u is the convergence regulator). As a rule, the converse statement is false. However, since the proof of Lemma IV.3.2 can be repeated word for word in the case of a BL $\mathbf{X}$, we see that, if $x_n \to x$ in norm, then there is a subsequence $\{x_{n_k}\}$ with $x_{n_k} \xrightarrow{(r)} x$. Now we see, as in Theorem IV.3.2, that any two monotone norms on a VL $\mathbf{X}$ that turn it into a BL are equivalent.

THEOREM 1. 1) *If* $\mathbf{X}$ *is a* NL *then* $\mathbf{X}^* \subset \mathbf{X}^\sim$.

2) *If* $\mathbf{X}$ *is a* BL, *then* $\mathbf{X}^* = \mathbf{X}^\sim$.

Proof. 1) As the norm is monotone, the order intervals in a VL $\mathbf{X}$ are bounded in norm, and, as a functional $f \in \mathbf{X}^*$ is bounded on any set that is bounded in norm, Theorem 2.1 shows that $f \in \mathbf{X}^\sim$.

2) Suppose $f \in \mathbf{X}^\sim$ but $f \notin \mathbf{X}^*$. Then there exists a sequence $\{x_n\}$ such that $x_n \to \mathbf{0}$ in norm, and $f(x_n) \geqslant \varepsilon > 0$ $(n \in \mathbb{N})$. On the other hand, there exists a subsequence $\{x_{n_k}\}$ with $x_{n_k} \xrightarrow{(r)} \mathbf{0}$, so that $|f(x_{n_k})| \leqslant |f|(|x_{n_k}|) \leqslant \varepsilon_{n_k} |f|(u) \to 0$. This contradiction proves that $f \in \mathbf{X}^*$.

If $\mathbf{X}$ is an NL, then it is easy to see that the Banach dual $\mathbf{X}^*$, with the usual norm and ordering induced from $\mathbf{X}^\sim$, is a Banach K-space (and an ideal in $\mathbf{X}^\sim$).

3.2. For the study of VLs there is a further class of representations, in addition to the one considered in 1.5, that is convenient.

Let $\mathbf{X}$ be an Archimedean VL, and let $u \in \mathbf{X}_+$. We turn the u-ideal $\mathbf{X}(u)$ into an NL, taking the norm to be the Minkowski functional of the order interval $[-u, u]$. This norm will be denoted by $\|\cdot\|_u$ and called the *u-norm*. It is clear that a VL $\mathbf{X}$ is (r)-complete if and only if each NL $\mathbf{X}(u)$ is complete $(u \in \mathbf{X}_+)$. The importance of (r)-complete lattices is underlined by the fact that a theorem of M. G. and S. G. Krein and S. Kakutani (see Vulikh-I, Theorem VII.5.1) shows that every u-ideal $\mathbf{X}(u)$ that is complete in the u-norm is order isomorphic and isometrically isomorphic to the BL $\mathbf{C}(K)$ on some compactum K. In view of the results of 3.1 on the connection between convergence in norm and (r)-convergence in BLs, we deduce that every BL is (r)-complete.

In the theory of operators it is often natural to consider vector spaces (and vector lattices) over the field of complex numbers $\mathbb{C}$. Since in general the supremum is undefined for two functions in, say, the complex space $\mathbf{L}^p(0, 1)$ with the natural ordering (two complex numbers have no maximum), but the modulus of a function is defined, it is the modulus that is used as the basis for generalizing the concept of a VL to the complex case.

Suppose the complex vector space $\mathbf{X}$ is the complexification* of a real vector space $\mathbf{X}_0$ which is an (r)-complete VL. We now define a modulus $|z|$ in $\mathbf{X}_0$ for the element $z = x + iy$ $(x, y \in \mathbf{X}_0)$. For this purpose we consider the principal ideal $\mathbf{X}_0(|x| + |y|)$, which, in view of the (r)-completeness of $\mathbf{X}_0$, is isomorphic to the real space $\mathbf{C}(K)$. We extend this isomorphism in a natural way to an isomorphism from $\mathbf{X}_0(|x|+|y|) + i\mathbf{X}_0(|x|+|y|)$ onto the complex space $\mathbf{C}_{\mathbb{C}}(K)$. If $z = x + iy$, $x, y \in \mathbf{C}(K)$, then the function $|z(t)|$ $(t \in K)$ can be calculated by $|z| = \sup\limits_{0 \leqslant \theta < 2\pi} |(\cos\theta)x + (\sin\theta)y|$, where the supremum is formed in the VL $\mathbf{C}(K)$. Hence, for every $z \in \mathbf{X}$, we have established the existence of

$$|z| = \sup_{0 \leqslant \theta < 2\pi} |(\cos\theta)x + (\sin\theta)y|, \tag{1}$$

which completes the construction of the modulus in $\mathbf{X}$.

A complex vector space $\mathbf{X}$ satisfying the above conditions is called a *complex vector lattice* if the modulus of an element is defined by (1). Almost all the concepts from the theory of VLs and BLs carry over to the complex case (see Schaefer-II).

3.3. We now look at some completeness criteria for NLs.

THEOREM 2 (Amemiya). *Let* $\mathbf{X}$ *be an NL. The following statements are equivalent*:

1) $\mathbf{X}$ *is complete in norm*;

2) *if* $\mathbf{0} \leqslant x_n \uparrow$ *is a Cauchy sequence in* $\mathbf{X}$, *then* $x_n \to x \in \mathbf{X}$ *in norm*;

3) *if* $\mathbf{0} \leqslant x_n \uparrow$ *is a Cauchy sequence in* $\mathbf{X}$, *then the supremum* $x = \sup x_n$ *exists in* $\mathbf{X}$.

Proof. Obviously 1) $\Rightarrow$ 2) $\Rightarrow$ 3).

2) $\Rightarrow$ 1). If $\{x_n\}$ is a Cauchy sequence in $\mathbf{X}$, then, by passing to a subsequence if necessary, we may assume that

$$\sum_{n=1}^{\infty} \|x_{n+1} - x_n\| < +\infty. \tag{2}$$

* That is, $\mathbf{X}$ is the set of all elements of the form $x + iy$ $(x, y \in \mathbf{X}_0)$ with linear operations defined in the natural way (see §2 of Chapter XIII).

Write

$$y_n = \sum_{k=1}^{n} (x_{k+1} - x_k)_+, \qquad z_n = \sum_{k=1}^{n} (x_{k+1} - x_k)_-.$$

Then $\mathbf{0} \leqslant y_n \uparrow$, $\mathbf{0} \leqslant z_n \uparrow$. If $n < m$, then

$$\|y_m - y_n\| = \left\| \sum_{k=n+1}^{m} (x_{k+1} - x_k)_+ \right\| \leqslant \sum_{k=n+1}^{m} \|(x_{k+1} - x_k)_+\| \leqslant \sum_{k=n+1}^{m} \|x_{k+1} - x_k\| \underset{\substack{n \to \infty \\ m \to \infty}}{\longrightarrow} \mathbf{0}$$

by (2). Hence $\{y_n\}$ is an increasing Cauchy sequence. Therefore $y_n \to y$ in norm. Similarly, $z_n \to z$ in norm. Thus $y_n - z_n \to y - z$; but

$$y_n - z_n = \sum_{k=1}^{n} (x_{k+1} - x_k) = x_{n+1} - x_1,$$

so that $x_n \to y - z + x_1$ in norm.

3) $\Rightarrow$ 2). Let $\{x_n\}$ be a Cauchy sequence with $\mathbf{0} \leqslant x_n \uparrow$. By the hypothesis, $x = \sup x_n$ exists. Without losing any generality, we may assume that

$$\|x_{n+1} - x_n\| \leqslant 1/n^3 \qquad (n \in \mathbb{N}). \tag{3}$$

Consider

$$y_n = x_1 + \sum_{k=1}^{n} k(x_{k+1} - x_k).$$

We have $\mathbf{0} \leqslant y_n \uparrow$, and, if $n < m$,

$$\|y_m - y_n\| = \left\| \sum_{k=n+1}^{m} k(x_{k+1} - x_k) \right\| \leqslant \sum_{k=n+1}^{m} 1/k^2 \underset{\substack{n \to \infty \\ m \to \infty}}{\longrightarrow} \mathbf{0}$$

by (3). Hence $y = \sup y_n$ exists, by hypothesis. Furthermore,

$$n(x - x_n) = \sup_{m > n} \sum_{k=n}^{m} n(x_{k+1} - x_k) \leqslant \sup_{m > n} \sum_{k=n}^{m} k(x_{k+1} - x_k) \leqslant y,$$

so that

$$\mathbf{0} \leqslant x - x_n \leqslant y/n.$$

Since the norm is monotone, $x_n \to x$. This completes the proof.

3.4. We conclude this section by considering functionals on VLs.

Let $\mathbf{X}(u)$ be a u-ideal with the u-norm in a K-space $\mathbf{X}$.

THEOREM 3 (Grothendieck [1]). *If a sequence $\{f_n\}$ is weak* convergent to $\mathbf{0}$ in $(\mathbf{X}(u))^*$, then $f_n \to \mathbf{0}$ weakly.*

The proofs of Theorem 3 and its generalizations (see Schaefer-II) are non-trivial and fairly lengthy, so we shall omit them.

Theorem 3 gives interesting results, in particular, when $\mathbf{X}(u) = \mathbf{L}^\infty(T, \Sigma, \mu)$. We prove, as a corollary, that, if $\mathbf{X} = \ell^\infty$, then the closed ball B in the space $\mathbf{X}^*$ is not sequentially $\sigma(\mathbf{X}^*, \mathbf{X})$-compact (see p. 179). For otherwise, by Theorem 3 it would be sequentially weakly compact in $\mathbf{X}^*$, and thus by Theorem VIII.2.1 B would be weakly compact. Hence $\mathbf{X}^*$, and so also ℓ^∞, would be reflexive, which is not the case (see VI.2.1).

LEMMA 1. *Let* $\mathbf{X}$ *be an* (r)*-complete* VL, *and let* $\{f_\alpha\}$ $(\alpha \in \mathrm{A})$ *be a net in* $\mathbf{X}^*$ *such that*

1) $\sup_{\alpha \in \mathrm{A}} |f_\alpha(x)| < \infty$, *for each* $x \in \mathbf{X}$;

2) $\lim f_\alpha(x) = f(x)$ *exists and is finite, for each* $x \in \mathbf{X}$.

Then $f \in \mathbf{X}^\sim$.

Proof. As $\mathbf{X}$ is (r)-complete, all u-ideals are complete. By the Banach–Steinhaus Theorem (see the Remark on p. 204), the restriction of f to each u-ideal $\mathbf{X}(u)$ is continuous; and this means that f is bounded on every interval. Hence $f \in \mathbf{X}^\sim$.

COROLLARY. *Let* $\mathbf{X}$ *be a* K*-space for which* $\mathbf{X}^\sim$ *is total. Then*

1) *every* $\sigma(\mathbf{X}^\sim, \mathbf{X})$*-bounded set* M *is relatively compact in this topology;*

2) *if* $\mathbf{Y}$ *is an ideal in* $\mathbf{X}^\sim$, *then every interval in* $\mathbf{Y}$ *is compact in the topology of* $\sigma(\mathbf{Y}, \mathbf{X})$.

Proof. 1) By Lemma III.3.5 it is sufficient to prove that the $\sigma(\mathbf{X}^+, \mathbf{X})$-closure of M is contained in $\mathbf{X}^\sim$. However, this follows from Lemma 1.

2) follows from 1) and Corollary 1 to Theorem 2.2.

THEOREM 4. *Suppose* $\mathbf{X}$ *is a* K*-space,* $\{f_m\}_{m=1}^\infty \subset \mathbf{X}_n^\sim$, *and let* f *be a linear functional on* $\mathbf{X}$. *If* $f_m(x) \to f(x)$ *for every* $x \in \mathbf{X}$, *then* $f \in \mathbf{X}_n^\sim$.

Proof. By Lemma 1, $f \in \mathbf{X}^\sim$. It is clearly sufficient to prove that, for each $u \in \mathbf{X}_+$, the restriction of f to the u-ideal $\mathbf{X}(u)$, which we denote by g, is (o)-continuous. Write $g_m = f_m \mid \mathbf{X}(u)$. Then $g_m \in \mathbf{X}(u)_n^\sim$, $g \in \mathbf{X}(u)^\sim = \mathbf{X}(u)^*$. Since g_m is weak* convergent to g in $\mathbf{X}(u)^*$, Theorem 3 shows that $g_m \to g$ weakly. By Corollary 2 to Theorem III.3.2, there are convex combinations g'_m of the elements g_m which converge in norm to g in $\mathbf{X}(u)^*$. Now $g'_m \in \mathbf{X}(u)_n^\sim$ and $\mathbf{X}(u)_n^\sim$ is a band in $\mathbf{X}(u)^*$; and since bands in a VL are closed relative to the norm, it follows that $g \in \mathbf{X}(u)_n^\sim$, which is what we set out to prove.

From Theorem 4 we immediately deduce

COROLLARY 1. *If* $\mathbf{X}_n^\sim$ *is total on the* K*-space* $\mathbf{X}$, *then the space* $(\mathbf{X}_n^\sim, \sigma(\mathbf{X}_n^\sim, \mathbf{X}))$ *is sequentially complete.*

COROLLARY 2. *If the* K*-space* $\mathbf{X}$ *is* (o)*-reflexive, then the space* $(\mathbf{X}, \sigma(\mathbf{X}, \mathbf{X}_n^\sim))$ *is sequentially complete.*

THEOREM 5. *If* $\mathbf{X}$ *is an* NL *for which* $\mathbf{X}_n^\sim$ *is total, then* $\mathbf{X}^* \cap \mathbf{X}_n^\sim$ *separates points on* $\mathbf{X}$.

We shall prove this result only for K-spaces. First note that the proof of Theorem VI.1.5 carries on to the case of a σ-finite measure; thus, for a K-space $\mathbf{X}$, Theorem 5 follows from the representation in Theorem 2.5.

We now turn to an important theorem on the structure of $\mathbf{X}^\sim$.

THEOREM 6. *If* $\mathbf{X}$ *is a* VL *on which* $\mathbf{X}_n^\sim$ *is total, then* $\mathbf{X}_s^\sim = (\mathbf{X}_n^\sim)^{\mathrm{d}}$: *that is,* $\mathbf{X}^\sim$ *decomposes into a sum of two disjoint bands* $\mathbf{X}_n^\sim$ *and* $\mathbf{X}_s^\sim$.

Proof. We prove Theorem 6 in the case where $\mathbf{X}$ is a K-space. By Lemma 2.1, $\mathbf{X}_s^\sim \subset (\mathbf{X}_n^\sim)^{\mathrm{d}}$. Conversely, let $\phi \in (\mathbf{X}_n^\sim)^{\mathrm{d}}$, $\phi \geqslant \mathbf{0}$; $\mathbf{N}_\phi = \{x \in \mathbf{X} : \phi(|x|) = 0\}$. Clearly $\mathbf{N}_\phi$ is an ideal in $\mathbf{X}$. We shall prove that $\mathbf{N}_\phi$ is of the same width as $\mathbf{X}$, from which it will follow that $\phi \in \mathbf{X}_s^\sim$. Assume that this is not the case. Then, writing $\mathbf{Y} = \mathbf{N}_\phi^{\mathrm{d}}$, we have $\mathbf{Y} \neq \{\mathbf{0}\}$. For each $y \in \mathbf{Y}$, we set $\|y\| = \phi(|y|)$. Then $\mathbf{Y}$ is an NL. Now $\mathbf{X}_n^\sim$ is total on $\mathbf{X}$, so $\mathbf{Y}_n^\sim$ is total on $\mathbf{Y}$. By Theorem 5, $\mathbf{Y}^* \cap \mathbf{Y}_n^\sim$ is total on $\mathbf{Y}$. Therefore there exists a functional f such that $f \in \mathbf{Y}^* \cap \mathbf{Y}_n^\sim$, $f > \mathbf{0}$ and $f(y) \leqslant \|y\| = \phi(y)$ for every $y \in \mathbf{Y}$.

For each $x \in \mathbf{X}$, we set

$$f_1(x) = f([\mathbf{Y}]x).$$

Then $\mathbf{0} < f_1 \in \mathbf{X}_n^\sim$, $0 \leqslant f_1 \leqslant \phi \in (\mathbf{X}_n^\sim)^{\mathrm{d}}$. Hence $f_1 = \mathbf{0}$, giving a contradiction.

In particular, we see from Theorem 6 that every continuous functional on an NIS decomposes in a natural way as a sum of a functional having an integral representation and a singular functional. This result is important in various applications (especially in optimal control and convex analysis: see Dubovitskii and Milyutin [2]). The following classical theorem of Yosida and Hewitt* on the representation of functionals on $\mathbf{L}^\infty(T, \Sigma, \mu)$ is a special case of Theorem 6.

THEOREM 7 (Yosida–Hewitt). *If the measure μ is finite, then every continuous linear functional f on $\mathbf{L}^\infty(T, \Sigma, \mu)$ is uniquely decomposable as a sum $f = f_1 + f_2$, where*

$$f_1(x) = \int_T x(t)y(t)\,d\mu$$

$$(y \in \mathbf{L}^1(T, \Sigma, \mu)),$$

and f_2 has the property that, for each $\varepsilon > 0$, there exists $A \in \Sigma$, $\mu(T \setminus A) < \varepsilon$, such that $f_2(x) = 0$ when $\operatorname{supp} x \subset A \pmod{\mu}$.

§ 4. *KB*-spaces

4.1. In this section we briefly consider BLs satisfying conditions (A), (B) and (C) analogous to those introduced in IV.3.2.†

Let $\mathbf{X}$ be a normed lattice.

The norm in $\mathbf{X}$ is said to be *order continuous*, or $\mathbf{X}$ is said to satisfy condition (A), if

$$\mathbf{0} \leqslant x_\alpha \downarrow \mathbf{0} \text{ implies } \|x_\alpha\| \to 0.$$

The norm in $\mathbf{X}$ is said to be *order semicontinuous*, or $\mathbf{X}$ is said to satisfy condition (C) if

$$\mathbf{0} \leqslant x_\alpha \uparrow x \in \mathbf{X} \text{ implies } \sup \|x_\alpha\| = \|x\|.$$

The norm in $\mathbf{X}$ is said to be *monotone complete*, or $\mathbf{X}$ is said to satisfy condition (B), if

$$\mathbf{0} \leqslant x_\alpha \uparrow,\ x_\alpha \in \mathbf{X},\ \sup \|x_\alpha\| < \infty \text{ imply that there exists } x \in \mathbf{X} \text{ such that } x_\alpha \uparrow x.$$

If the analogues of these conditions are satisfied for sequences only, then we shall speak of *order σ-continuity* or condition (A_σ), *order σ-semicontinuity* or condition (C_σ), and *monotone σ-completeness* or condition (B_σ), respectively. As we shall see below, the conditions for sequences are equivalent to those for nets in the case of an NIS on a σ-finite measure space—that is, the above terminology is compatible with that of IV.3.2.

4.2. Let us consider condition (A). Let $\mathbf{X}$ be a normed K-space. In Vulikh-I, p. 188, it is proved that a K-space satisfying condition (A) is of countable type, so that $(A) \Leftrightarrow (A_\sigma)$.‡ It is also clear that, if $\mathbf{X}$ satisfies condition (A), then $x_\alpha \xrightarrow{(o)} x$ implies that $x_\alpha \to x$ in norm.

We first consider condition (A) in connection with the class of (o)-continuous functionals. We presented some results in this direction for BISs in § 1 of Chapter VI; here we generalize those results and supplement them with some new theorems.

---

* The Yosida–Hewitt theorem has been generalized many times. Theorem 6 was proved in its final form independently by G. Ya. Lozanovskii and, in a joint paper, by V. Luxemburg and A. Zaanen [1]. See also Lozanovskii [3].

† Conditions (A) and (B) were introduced by L. V. Kantorovich, and condition (C) by H. Nakano.

‡ What is more, if $\mathbf{X}$ is a normed K_σ-space satisfying condition (A_σ), then $\mathbf{X}$ is a K-space satisfying condition (A). In general, (A_σ) does not imply (A).

LEMMA 1. *If* **X** *is an* NL *and* $\{x_\alpha\}$ *is a net in* **X** *such that* $x_\alpha \downarrow$ *and* $x_\alpha \to x$ *in the weak topology, then* $x_\alpha \to x$ *in norm.*

Proof. Assume the contrary. Then we may assume that, for all α,

$$\|x_\alpha - x\| \geqslant \varepsilon > 0. \tag{1}$$

Since $x_\alpha \downarrow x$, by Corollary 2 to Theorem 2.2, we have $x_\alpha - x \geqslant \mathbf{0}$ for each α. Now $x_\alpha \to x$ $(\sigma(\mathbf{X}, \mathbf{X}^*))$, so there exists an element $y = \sum_{i=1}^{n} \lambda_i x_{\alpha_i}$, where $0 \leqslant \lambda_i \leqslant 1$, $\sum_{i=1}^{n} \lambda_i = 1$, such that $\|y - x\| < \frac{1}{2}\varepsilon$. Choose $\alpha \geqslant \alpha_i$ $(1 \leqslant i \leqslant n)$. Then

$$\mathbf{0} \leqslant x_\alpha - x \leqslant \sum_{i=1}^{n} \lambda_i x_{\alpha_i} - x,$$

from which we have $\|x_\alpha - x\| < \frac{1}{2}\varepsilon$, contradicting (1).

THEOREM 1. *If* **X** *is a* NL, *then* $\mathbf{X}_n^\sim \supset \mathbf{X}^*$ *if and only if* **X** *satisfies condition* (*A*).

Proof. If $\mathbf{X}_n^\sim \supset \mathbf{X}^*$ and $\mathbf{0} \leqslant x_\alpha \downarrow \mathbf{0}$, then $x_\alpha \to \mathbf{0}$ in the weak topology, so $x_\alpha \to x$ in norm, by Lemma 1.

If **X** satisfies (A), $f \in \mathbf{X}^*$ and $x_\alpha \xrightarrow{(o)} \mathbf{0}$, then $x_\alpha \to \mathbf{0}$ in norm, so that $f(x_\alpha) \to 0$. Hence $f \in \mathbf{X}_n^\sim$.

THEOREM 2. *Let* **X** *be a Banach K-space. The following conditions are equivalent*:

1) **X** *satisfies condition* (*A*);

2) *intervals in* **X** *are weakly compact*;

3) *if* $\pi: \mathbf{X} \to \mathbf{X}^{**}$ *is the canonical embedding, then* $\pi(\mathbf{X})$ *is an ideal in* $\mathbf{X}^{**}$.

Proof. 1) ⇒ 3). Since **X** satisfies (A), Theorems 1 and 3.1 show that $\mathbf{X}^* = \mathbf{X}_n^\sim$. Hence $\pi(\mathbf{X})$ is an ideal in $\mathbf{X}^{**}$ by Theorem 2.2.

3) ⇒ 2). Take any interval I in **X**. Since $\pi(\mathbf{X})$ is an ideal in $\mathbf{X}^{**}$, so is $\pi(I)$ in $\mathbf{X}^{**}$. By the Corollary to Lemma 3.1, the interval $\pi(I)$ is weak* compact, so I is weakly compact.

2) ⇒ 1). In view of Theorem 1, it is sufficient to prove that $\mathbf{X}^* \subset \mathbf{X}_n^\sim$. Let $f \in \mathbf{X}^*$. Assume that $f \notin \mathbf{X}_n^\sim$. Then there exist a net $x_\alpha \downarrow \mathbf{0}$ and a number $\varepsilon > 0$ such that

$$|f(x_\alpha)| \geqslant \varepsilon. \tag{2}$$

Since we may assume that $x_1 \geqslant x_\alpha$ for all α, the weak compactness of intervals means that there exists a subnet $\{x_\beta\}$ with $x_\beta \to x$ $(\sigma(\mathbf{X}, \mathbf{X}^*))$. By Corollary 2 to Theorem 2.2, $x = \mathbf{0}$, and so $f(x_\beta) \to 0$, contradicting (2).

Now we consider an interesting property of Banach K-spaces satisfying condition (A), which is formulated in terms of B-spaces only, without any use of the ordering. As a preliminary, we record an auxiliary theorem.

Let **X** be an NL. We shall say that a sequence $\{x_n\}$ in **X** *decreases sideways to zero*, and write $x_n \xrightarrow{\quad}\!\!\downarrow \mathbf{0}$, if $x_n \downarrow \mathbf{0}$ and $x_n - x_{n+1} \,\mathrm{d}\, x_{n+1}$ $(n \in \mathbb{N})$.

THEOREM 3. *If* **X** *is a normed K-space in which* $x_n \xrightarrow{\quad}\!\!\downarrow \mathbf{0}$ *implies* $\|x_n\| \to 0$, *then* **X** *satisfies condition* (*A*).

Proof. Suppose $x_n \downarrow \mathbf{0}$. Fix $\varepsilon > 0$ and write $q_n = (x_n - \varepsilon x_1)_+$. Then $q_n \downarrow \mathbf{0}$ and $[q_n]x_1 \xrightarrow{\quad}\!\!\downarrow \mathbf{0}$. Consequently, $\|[q_n]x_1\| \to 0$. As

$$x_n = [q_n]x_n + (x_n - [q_n]x_n) \leqslant [q_n]x_1 + \varepsilon x,$$

we have

$$\|x_n\| \leqslant \|[q_n]x_1\| + \varepsilon \|x_1\|,$$

so that $\|x_n\| \to 0$: that is, **X** satisfies condition (A_σ). But (A_σ) implies (A), as we have already remarked.

LEMMA 2. *If* $\mathbf{X}$ *is a Banach K-space not satisfying condition (A), then there is a sequence* $\{z_n\} \subset \mathbf{X}_+$ *such that*

1) $z_n \mathrm{d}\, z_m$ $(n \neq m)$;
2) $\|z_n\| = 1 \quad (n \in \mathbb{N})$;
3) $\sup z_n = u$ *exists in* $\mathbf{X}$.

Proof. As $\mathbf{X}$ does not satisfy condition (A), the hypothesis of Theorem 3 is not satisfied in $\mathbf{X}$. Thus there exists sequence $\{x_n\}$ with $x_n \downarrow \mathbf{0}$, $\|x_n\| \geqslant \varepsilon > 0$. This sequence is not a Cauchy sequence. Otherwise $x_n \to x$ in norm, and since $x_n \downarrow 0$, this gives $x = \mathbf{0}$, contradicting $\|x_n\| \geqslant \varepsilon > 0$. Hence there exist $\delta > 0$ and $n_1 < n_2 < \ldots < n_k < \ldots$ such that $\|x_{n_k} - x_{n_{k+1}}\| \geqslant \delta$. Write

$$z_k = \frac{x_{n_k} - x_{n_{k+1}}}{\|x_{n_k} - x_{n_{n+1}}\|}.$$

Then 1) and 2) are obvious, while 3) follows from the inequality $z_k \leqslant x_1/\delta$.

LEMMA 3. *If* $\mathbf{X}$ *is a Banach K-space satisfying condition (A) and having a weak unit, then the closed unit ball* $B_{\mathbf{X}^*}$ *in* $\mathbf{X}^*$ *is weak* sequentially compact.*

Proof. Let x_0 be a weak unit in $\mathbf{X}$, and let $I = [-x_0, x_0]$. We have $x \wedge nx_0 \uparrow x$ for every $x \in \mathbf{X}_+$ (see Vulikh-I, Lemma IV.2.1), so, as $\mathbf{X}$ satisfies condition (A), we see that the linear hull of the interval is dense in norm (and so also weakly dense) in $\mathbf{X}$. Hence the functional

$$\rho(f) = \sup\{|f(x)| : x \in I\}, \qquad f \in \mathbf{X}^*,$$

will be a norm on $\mathbf{X}^*$. Denote the resulting normed space by $\mathbf{X}^*_\rho$. As $\mathbf{X}$ satisfies condition (A), the interval I is weakly compact, by Theorem 2. Hence, by the Mackey–Arens Theorem (Theorem III.3.8), the dual of the normed space $\mathbf{X}^*_\rho$ can be identified with the linear hull of I. Therefore the topology $\sigma(\mathbf{X}^*, \mathbf{X})$ is stronger than the weak topology of $\mathbf{X}^*_\rho$, and so the weak topologies $\sigma(\mathbf{X}^*, \mathbf{X})$ and $\sigma(\mathbf{X}^*_\rho, (\mathbf{X}^*_\rho)^*)$ coincide on the ball $B_{\mathbf{X}^*}$. Hence we see, by applying Theorem VIII.2.1 to the normed space $\mathbf{X}^*_\rho$, that $B_{\mathbf{X}^*}$ is sequentially compact in both weak topologies.*

Now we can state and prove the criterion mentioned above, which was obtained by G. Ya. Lozanovskii [1].

THEOREM 4. *A necessary and sufficient condition for a K-space* $\mathbf{X}$ *to satisfy condition (A) is that* $\mathbf{X}$ *contain no closed subspace* $\mathbf{Y}$ *isomorphic as a B-space to* ℓ^∞.

Proof. Necessity. Assume the contrary. Then there exists an isomorphism $U : \ell^\infty \to \mathbf{Y}$, where $\mathbf{Y}$ is a closed subspace of $\mathbf{X}$. Let $\mathbf{Y}_0 = U(\mathbf{c}_0)$ and let $\mathbf{Z}$ be the band generated by $\mathbf{Y}_0$ in $\mathbf{X}$. Since $\mathbf{Y}_0$ is separable, $\mathbf{Z}$ contains a weak unit. In fact, if $\{x_n\}$ is a countable dense subset of $\mathbf{Y}_0$, then it is easy to see that the element $x_0 = \sum_{n=1}^{\infty} \frac{1}{2^n \|x_n\|} |x_n|$ is a weak unit.

Let $\{e_k\}_{k=1}^{\infty}$ be the sequence of unit vectors in $\mathbf{c}_0$. Then every element of $\mathbf{c}_0$ is expressible in the form $\sum_{k=1}^{\infty} \xi_k e_k$, where $\xi_k \to 0$. Write

$$f_n\left(\sum_{k=1}^{\infty} \xi_k e_k\right) = \xi_n, \qquad n \in \mathbb{N}.$$

* Theorem VIII.2.1 was stated for B-spaces, but the implication 1) ⇒2) is valid for any normed space, as is clear from passing to the completion.

These f_n are functionals in $(\mathbf{c}_0)^*$ and $\|f_n\| = 1$ $(n \in \mathbb{N})$. Define functionals g_n on $\mathbf{Y}_0$ by

$$g_n(y) = f_n(U^{-1}(y)), \qquad y \in \mathbf{Y}_0.$$

Then $\|g_n\| \leqslant \|f_n\| \|U^{-1}\| = \|U^{-1}\|$ $(n \in \mathbb{N})$, so the set $\{g_n\}$ is bounded in norm in $\mathbf{Y}_0^*$. We extend each functional g_n to $\mathbf{Z}$, preserving the norm. Denote the resulting functionals by $\tilde{g}_n$. Then $\{\tilde{g}_n\}$ is a subset of $\mathbf{Z}^*$ bounded in norm. By Lemma 3 we can choose a subsequence $\{\tilde{g}_{n_m}\}$ such that $\tilde{g}_{n_m} \to \tilde{g} \in \mathbf{Z}^*$ in the topology $\sigma(\mathbf{Z}^*, \mathbf{Z})$. For all $x \in \mathbf{X}$, we set

$$\tilde{\tilde{g}}_{n_m}(x) = \tilde{g}_{n_m}([\mathbf{Z}]x), \qquad \tilde{\tilde{g}}(x) = \tilde{g}([\mathbf{Z}]x).$$

Then $\tilde{\tilde{g}}_{n_m} \to \tilde{\tilde{g}}(\sigma(\mathbf{X}^*, \mathbf{X}))$.

If we now write $\phi_{n_m} = U^*\tilde{\tilde{g}}_{n_m}$ and $\phi = U^*\tilde{\tilde{g}}$, then $\phi_{n_m} \to \phi$ in the weak* topology of $(\ell^\infty)^*$. By Theorem 3.3, $\phi_{n_m} \to \phi$ in the weak topology of $(\ell^\infty)^*$. In the natural embedding, the space ℓ^1 is a band in $(\ell^\infty)^*$ (since ℓ^1 is obviously (o)-reflexive). Let us project our functionals onto this band. Then it is easy to see that $[\ell^1]\phi_{n_m} \to [\ell^1]\phi$ in the weak topology of ℓ^1. For, since $[\ell^1]$ is an operator in $(\ell^\infty)$the operator $[\ell^1]^*$ is an operator in $(\ell^\infty)^{**}$, and if $\psi \in (\ell^1)^* = (\ell^\infty)^{**}$, then $[\ell^1]^*(\psi) \in (\ell^\infty)^{**}$, so that

$$([\ell^1]\phi_{n_m})\ (\psi) = \phi_{n_m}([\ell^1]^*\psi) \to \phi([\ell^1]^*\psi) = ([\ell^1]\phi)\ (\psi).$$

Thus, in view of Theorem VIII.3.4, $[\ell^1]\phi_{n_m} \to [\ell^1]\phi$ in norm in ℓ^1. Since $(\mathbf{c}_0)^* = \ell^1$, we can regard $[\ell^1]\phi_{n_m}$ and $[\ell^1]\phi$ as functionals on $\mathbf{c}_0$, and it is easy to see that these are simply the restrictions of the corresponding functionals ϕ_{n_m} and ϕ from ℓ^∞ to $\mathbf{c}_0$. We now show that $[\ell^1]\phi_{n_m} = f_{n_m}$. In fact, for each unit vector e_k, we have

$$([\ell^1]\phi_{n_m})\ (e_k) = \phi_{n_m}(e_k) = (U^*\tilde{\tilde{g}}_{n_m})\ (e_k) = \tilde{g}_{n_m}\ (Ue_k) = g_{n_m}(Ue_k) = f_{n_m}(e_k),$$

and so we obtain the required result. Thus we have shown that the sequence $\{f_{n_m}\}$ converges in norm in ℓ^1. But this is impossible, for $\|f_n - f_m\| = 1$ when $n \neq m$. This contradiction completes the proof of the necessity.

Sufficiency. Assuming $\mathbf{X}$ does not satisfy condition (A), choose the sequence $\{z_n\}$ according to Lemma 2. Define an operator $U: \ell^\infty \to \mathbf{X}$ as follows: if $\xi = \{\xi_k\}_{k=1}^\infty$, then

$$U(\xi) = (o)\text{-}\sum_{k=1}^{\infty} \xi_k z_k.$$

Then, for each $n \in \mathbb{N}$, we have

$$|\xi_n| z_n \leqslant |U(\xi)| \leqslant \|\xi\|_{\ell^\infty} u_1$$

so that

$$\|\xi\|_{\ell^\infty} \leqslant \|U(\xi)\|_{\mathbf{X}} \leqslant \|\xi\|_{\ell^\infty} \|u\|.$$

Hence $\mathbf{Y} = U(\ell^\infty)$ is a subspace isomorphic to ℓ^∞. This contradiction completes the proof.

COROLLARY. *If a Banach K-space* $\mathbf{X}$ *is separable, then it satisfies condition (A).*

Proof. If $\mathbf{X}$ did not satisfy condition (A), then by Theorem 4 it would have a subspace $\mathbf{Y}$ isomorphic to ℓ^∞. As ℓ^∞ is inseparable, $\mathbf{Y}$ would then be inseparable, contradicting the hypothesis on $\mathbf{X}$.

From this Corollary we obtain the part of Theorem IV.3.3 that we did not prove in Chapter IV.

4.3. We now consider conditions (B) and (C). If $\mathbf{X}$ is a K-space of countable type, then, obviously, (C) $\Leftrightarrow$ (C_σ). Hence these conditions coincide for an NIS on a σ-finite measure space. In general, this is not so.

Conditions (B) and (B_σ) are distinct even in the class of K-spaces of countable type; however, in the case of an NIS on a σ-finite measure space, (B_σ) $\Leftrightarrow$ (B), and thus our terminology agrees with that of IV.3.2. For suppose such an NIS $\mathbf{X}$ satisfies (B_σ), and let $\{x_\alpha\}$ be a net, $\mathbf{0} \leqslant x_\alpha \uparrow$, which is bounded in norm in $\mathbf{X}$. Then, by the Corollary to Theorem IV.3.1, this net is a bounded set in the TVS $\mathbf{S}(T, \Sigma, \mu)$. The proof of Lemma III.1.1 shows that $\sup_\alpha x_\alpha(t) < +\infty$ a.e., so by Corollary 2 to Theorem I.6.17, $x_0 = \sup x_\alpha$ exists in $\mathbf{S}(T, \Sigma, \mu)$ and there is a sequence $\{x_{\alpha_n}\}$ with $x_{\alpha_n} \uparrow x_0$. As $\{x_{\alpha_n}\}$ is also bounded in norm, (B_σ) implies that $x_0 \in \mathbf{X}$ and thus $\mathbf{X}$ satisfies (B).

By 3) in Theorem 3.2, we have

THEOREM 5. *If a normed K-space $\mathbf{X}$ satisfies (B_σ) then it is complete in its norm.*

The following theorem shows that spaces satisfying conditions (B) and (C) exist in profusion.

THEOREM 6. *If $\mathbf{X}$ is an* NL, *then $\mathbf{X}^*$ satisfies conditions* (B) *and* (C).

Proof. Let $\{f_\alpha\}$ be a net in $\mathbf{X}^*$, $0 \leqslant f_\alpha \uparrow$, and suppose $\{f_\alpha\}$ is bounded in norm. For $x \in \mathbf{X}_+$, we set $f(x) = \sup f_\alpha(x) < +\infty$; and for $x \in \mathbf{X}$, we set $f(x) = f(x_+) - f(x_-)$. Then f is a linear functional on $\mathbf{X}$. Let us prove that $f \in \mathbf{X}^*$.

Choose any $\varepsilon > 0$. For each $x \in \mathbf{X}_+$, $\|x\| \leqslant 1$, there exists α_n such that $f(x) \leqslant f_{\alpha_x}(x) + \varepsilon \leqslant \|f_{\alpha_x}\| + \varepsilon \leqslant \sup \|f_\alpha\| + \varepsilon$, and so $\|f\| \leqslant \sup \|f_\alpha\| < \infty$. Therefore $f \in \mathbf{X}^*$, $f_\alpha \uparrow f$ and $\|f\| = \sup \|f_\alpha\|$; that is, $\mathbf{X}^*$ satisfies (C) and (B). As $\mathbf{X}^* \cap \mathbf{X}_n^\sim$ is a band in $\mathbf{X}^*$, it also satisfies (B) and (C).

The most important property of spaces satisfying condition (C) is contained in the following theorem.

THEOREM 7 (Nakano–Amemiya–Mori). *If $\mathbf{X}$ is a normed K-space on which $\mathbf{X}_n^\sim$ is total, then the following statements are equivalent*:

1) $\mathbf{X}$ *satisfies condition* (C);

2) $\|x\| = \sup\{|f(x)| : f \in \mathbf{X}^* \cap \mathbf{X}_n^\sim, \|f\| \leqslant 1\}$, *for each* $x \in \mathbf{X}$;

3) *the canonical embedding* $\pi: \mathbf{X} \to (\mathbf{X}_n^\sim \cap \mathbf{X}^*)^*$ *preserves the norm.*

Proof. 1) $\Rightarrow$ 2). Theorem VI.1.6 carries over to NISs without the σ-finiteness condition on the measure. Now we can apply Theorem 2.5.

2) $\Rightarrow$ 3) is obvious.

3) $\Rightarrow$ 1), since $(\mathbf{X}_n^\sim \cap \mathbf{X}^*)^*$ satisfies (C) by Theorem 6.

If $\mathbf{X}$ is complete, then $\mathbf{X}_n^\sim \subset \mathbf{X}^*$, and we can consider the norm on $\mathbf{X}_n^\sim$ induced from $\mathbf{X}^*$. Thus the Banach K-spaces $(\mathbf{X}_n^\sim)^*$ and $(\mathbf{X}_n^\sim)_n^\sim$ are defined.

THEOREM 8. *If $\mathbf{X}$ is a Banach K-space on which $\mathbf{X}_n^\sim$ is total, then the following statements are equivalent*:

1) $\mathbf{X}$ *satisfies conditions* (B) *and* (C);

2) *the canonical embedding* $\kappa: \mathbf{X} \to (\mathbf{X}_n^\sim)_n^\sim$ *is an isometry from* $\mathbf{X}$ *onto* $(\mathbf{X}_n^\sim)_n^\sim$;

3) *if* $\pi: \mathbf{X} \to \mathbf{X}^{**}$ *is the natural embedding, then there exists a projection P of unit norm from* $\mathbf{X}^{**}$ *onto* $\pi(\mathbf{X})$;

4) *every system of closed balls in* $\mathbf{X}$ *with the finite intersection property has non-empty intersection.*

Proof. 1) $\Rightarrow$ 2). By Theorem 7, κ preserves the norm. Let us show that $\kappa(\mathbf{X}) = (\mathbf{X}_n^\sim)_n^\sim$. As we remarked in 2.3, $\kappa(\mathbf{X})$ is a foundation in $(\mathbf{X}_n^\sim)_n^\sim$. Therefore, for each $F > \mathbf{0}$ in $(\mathbf{X}_n^\sim)_n^\sim$, there exists a net $\{x_\alpha\} \subset \mathbf{X}_+$ such that $\kappa(x_\alpha) \uparrow F$. Then we have $\|x_\alpha\| = \|\kappa(x_\alpha)\| \leqslant \|F\|$ and by condition (B) $x = \sup x_\alpha$ exists. Since κ preserves bonds, $F = \kappa(x) \in \kappa(\mathbf{X})$.

2) $\Rightarrow$ 1). This follows from Theorems 6 and 2.2.

2) $\Rightarrow$ 3). For $F \in \mathbf{X}^{**}$ we let F_1 denote the restriction of F to $\mathbf{X}_n^\sim$. Write $F_2 = [(\mathbf{X}_n^\sim)_n^\sim](F_1)$. Then $F_2 \in (\mathbf{X}_n^\sim)_n^\sim = \kappa(\mathbf{X})$ and so there exists $x_F \in \mathbf{X}$ such that $F_2 = \kappa(x_F)$. We set $P(F) = \pi(x_F)$ and show that P is the projection we require. Let us show that, for $F = \pi(x)$, we have

$$P(F) = F.$$

Since $\mathbf{X}_n^\sim$ is total on $\pi(\mathbf{X})$, it is sufficient to prove that $x = x_F$. Now

$$F_1(f) = f(x) \qquad (f \in \mathbf{X}_n^\sim),$$

so we have $F_2 = \kappa(x)$ and so $x = x_F$.

It remains for us to check that $\|P\| = 1$. Let $F \in \mathbf{X}^{**}$, $\|F\| \leqslant 1$, where $P(F) = \pi(x_F)$. We have the obvious relations

$$\|P(F)\| = \|\pi(x_F)\| = \|x_F\| = \|\kappa(x_F)\| = \|F_2\| \leqslant \|F_1\| \leqslant \|F\|.$$

Hence $\|P\| \leqslant 1$. As we have already proved that P leaves elements of $\pi(\mathbf{X})$ fixed, it follows that $\|P\| = 1$.

3) $\Rightarrow$ 4). Let $\{B_\xi\}$ be a system of closed balls in $\mathbf{X}$ having the finite intersection property, and let $\bar{B}_\xi$ be the weak* closure of $\pi(B_\xi)$ in $\mathbf{X}^{**}$. Since the balls are weak* compact, $\bigcap \bar{B}_\xi$ is non-empty in $\mathbf{X}^{**}$. Let $F \in \bigcap \bar{B}_\xi$. Then $P(F) \in \bigcap B_\xi$.

4) $\Rightarrow$ 1). This is a recent and interesting result of A. A. Sedaev [1]; we refer the reader to the paper for the proof. We emphasize that property 4) is clearly an isometric invariant.

Some interesting properties of spaces satisfying conditions (B) and (C) have also been given in a paper by Lozanovskii [2].

4.4. A Banach K-space satisfying conditions (A_σ) and (B_σ) is called a *KB-space*. As we remarked in 4.2, a KB-space satisfies condition (A). In view of Theorem VII.6.3 (Vulikh-I), a KB-space also satisfies condition (B), so that, by Theorem 8, a KB-space is (o)-reflexive.

THEOREM 9. *If* $\mathbf{X}$ *is a* BL, *then the following statements are equivalent*:

1) $\mathbf{X}$ *is a KB-space*;

2) *if* $\pi: \mathbf{X} \to \mathbf{X}^{**}$ *is the canonical embedding, then* $\pi(\mathbf{X})$ *is a band in* $\mathbf{X}^{**}$;

3) $\mathbf{X}$ *is weakly sequentially complete*;

4) $\mathbf{X}$ *contains no closed subspace* $\mathbf{Y}$ *isomorphic as a B-space to* $\mathbf{c}_0$;

5) *if* $\mathbf{0} \leqslant x_n \uparrow$, $x_n \in \mathbf{X}$, *and* $\sup \|x_n\| < \infty$, *then* $\{x_n\}$ *converges in norm*:

Proof. 1) $\Rightarrow$ 2). As $\mathbf{X}$ satisfies (A), $\mathbf{X}^* = \mathbf{X}_n^\sim$. Since $\mathbf{X}$ is (o)-reflexive, Theorem 2.3 shows that $\pi(\mathbf{X})$ is a band in $\mathbf{X}^{**}$.

2) $\Rightarrow$ 3). By Theorem 2.2, $\mathbf{X}^* = \mathbf{X}_n^\sim$ so $\mathbf{X}$ is (o)-reflexive, by Theorem 2.3. Hence $\mathbf{X}$ is weakly sequentially complete, by Corollary 2 to Theorem 3.4.

3) $\Rightarrow$ 4). Assume the contrary. Then $\mathbf{X}$ contains a closed subspace $\mathbf{Y}$ isomorphic to $\mathbf{c}_0$. As $\mathbf{X}$ is weakly sequentially complete, Corollaries 1 and 3 to Theorem III.3.2 show that $\mathbf{Y}$, and hence also $\mathbf{c}_0$, is also weakly sequentially complete. However, this is not so. For if $x_n = \{\xi_k^{(n)}\}_{k=1}^\infty$, where

$$\xi_k^{(n)} = \begin{cases} 1, & \text{if } k \leqslant n, \\ 0, & \text{if } k > n, \end{cases}$$

then we see easily from the equation $(\mathbf{c}_0)^* = \ell^1$ that $\{x_n\}$ is a weak Cauchy sequence that does not converge in $\mathbf{c}_0$.

4) $\Rightarrow$ 5). Assume the contrary. Then there exists a sequence $\{x_n\}$, $\mathbf{0} \leqslant x_n \uparrow$, which is bounded in norm but does not converge in norm. Write

$$\rho_n = \lim_{m \to \infty} \|x_m - x_n\|, \qquad \rho = \inf \rho_n$$

(the limit in the definition of ρ_n exists since $\{x_n\}$ is monotone). As $\{x_n\}$ is not a Cauchy sequence, $\rho > 0$. By refining $\{x_n\}$, we can arrange that, for $m < n$,

$$(\tfrac{3}{4})\rho \leqslant \|x_n - x_m\| \leqslant (\tfrac{5}{4})\rho.$$

Let $\boldsymbol{\phi}$ be the space of terminating sequences (see p. 103). Define an operator $U : \boldsymbol{\phi} \to \mathbf{X}$ by

$$U(z) = \sum_{i=1}^{m} \lambda_i (x_{i+1} - x_i),$$

where $z = (\lambda_1, \lambda_2, \ldots, \lambda_m, 0, 0, \ldots) \in \boldsymbol{\phi}$. Then

$$U(z) \leqslant \|z\| \, (x_{m+1} - x_1),$$

so that

$$\|U(z)\| \leqslant \|z\| \, \|x_{m+1} - x_1\| \leqslant (\tfrac{5}{4})\rho \, \|z\|. \tag{3}$$

Now we find a lower bond for $\|U(z)\|$. Let

$$\lambda_i^+ = \max(\lambda_i, 0), \qquad \lambda_i^- = \max(-\lambda_i, 0).$$

We set

$$u = \sum_{i=1}^{m} \lambda_i^+ (x_{i+1} - x_i), \qquad v = \sum_{i=1}^{m} \lambda_i^- (x_{i+1} - x_i).$$

Then

$$U(z) = u - v, \qquad U(|z|) = u + v.$$

If $h^+ = \max\{\lambda_i^+ : 1 \leqslant i \leqslant m)$, $h^- = \max\{\lambda_i^- : 1 \leqslant i \leqslant m\}$, then $\|z\| = \max(h^+, h^-)$. We assume, for definiteness, that $\|z\| = h^+$. Since

$$u = \sum_{i=1}^{m} \lambda_i^+ (x_{i+1} - x_i) \geqslant \lambda_j^+ (x_{j+1} - x_j), \qquad 1 \leqslant j \leqslant m,$$

we have

$$\|u\| \geqslant \lambda_j^+ \|x_{j+1} - x_j\| \geqslant \lambda_j^+ (\tfrac{3}{4})\rho, \qquad 1 \leqslant j \leqslant m,$$

and so $\|u\| \geqslant \frac{3}{4}\rho \, \|z\|$. Therefore, using (3), we have

$$\|U(z)\| = \|u - v\| = \|2u - (u + v)\| \geqslant 2\|u\| - \|u + v\| \geqslant (\tfrac{6}{4})\rho\|z\| - \|U(|z|)\| \geqslant (\tfrac{1}{4})\rho \, \|z\|. \tag{4}$$

Comparing (3) and (4), we see finally that

$$(\tfrac{1}{4})\rho \, \|z\| \leqslant \|U(z)\| \leqslant (\tfrac{5}{4})\rho \, \|z\|, \qquad z \in \boldsymbol{\phi}.$$

Since $\boldsymbol{\phi}$ is dense in $\mathbf{c}_0$, the operator U extends to an isomorphism from $\mathbf{c}_0$ onto a subspace of $\mathbf{X}$, contradicting 4).

5) $\Rightarrow$ 1). Let us show that $\mathbf{X}$ satisfies condition (A_σ). Suppose $x_n \downarrow \mathbf{0}$. Then the sequence $\{x_1 - x_n\}$ is bounded in norm and $\mathbf{0} \leqslant x_1 - x_n \uparrow$. By hypothesis, $x_1 - x_n$ converges in norm to some element $\tilde{x}$. Since $x_1 - x_n \uparrow x_1$, it follows that $\tilde{x} = x_1$, and so $x_n = (x_n - x_1) + x_1 \to \mathbf{0}$ in norm.

Condition (B_σ) is obviously satisfied. Thus $\mathbf{X}$ is a Banach K_σ-space satisfying conditions (A_σ) and (B_σ). As we remarked in a footnote on p. 293, $\mathbf{X}$ is a K-space, and therefore also a KB-space.

This completes the proof of the theorem.

Theorem 9 yields an easily verifiable criterion for weak sequential compactness in the class of BLs, as the verification of conditions (A) and (B) in specific spaces does not, as a rule, present any difficulties. First of all, we note that, by Theorem 9, the non-reflexive space $\mathbf{L}^1(T, \Sigma, \mu)$ is weakly sequentially complete. Let us examine further the case of Orlicz spaces on finite continuous measure spaces. By Theorem IV.3.9, the Orlicz space $\mathbf{L}_M$ is a KB-space if and only if the function M satisfies the Δ_2-condition. Hence $\mathbf{L}_M$ is weakly sequentially complete if and only if M satisfies the Δ_2-condition. Similarly, we see that the BFS $\mathbf{E}_M$, which satisfies condition (A), is not weakly sequentially complete unless M satisfies the Δ_2-condition.

The concept of a KB-space enables one to give an easily verifiable criterion for reflexivity (in the sense of the theory of B-spaces).

THEOREM 10 (Ogasavara). *If* $\mathbf{X}$ *is a* BL, *then the following conditions are equivalent*:

1) $\mathbf{X}$ *is reflexive*;

2) $\mathbf{X}$ *and* $\mathbf{X}^*$ *are* KB-*spaces*.

Proof. 1) $\Rightarrow$ 2). Since $\mathbf{X}^*$ is also reflexive, it is sufficient to prove that $\mathbf{X}$ is a KB-space; but this follows from the definition of reflexivity and part 2) of Theorem 9.

2) $\Rightarrow$ 1). Consider the canonical embedding $\pi: \mathbf{X} \to \mathbf{X}^{**}$. Since $\mathbf{X}$ is a KB-space, $\mathbf{X}^* = \mathbf{X}^{\sim}_n$ and $\pi(\mathbf{X}) = (\mathbf{X}^{\sim}_n)^{\sim}_n = (\mathbf{X}^*)^{\sim}_n$. As $\mathbf{X}^*$ is a KB-space, $\mathbf{X}^{**} = (\mathbf{X}^*)^{\sim}_n$, and so $\pi(\mathbf{X}) = \mathbf{X}^{**}$; hence $\mathbf{X}$ is reflexive.

§ 5. Convex sets that are closed with respect to convergence in measure

5.1. In this section it is shown, with the help of the above results from the theory of vector lattices, that the study of bounded convex sets closed with respect to convergence in measure in (o)-reflexive ISs can, in a certain sense, be reduced to the study of compact convex sets in appropriate TVSs. One needs to use non-trivial results from the theory of VLs even for the classical case of the space $\mathbf{L}^1(0, 1)$.

We begin with some preliminary results.

Let $\mathbf{X}$ be a VL, and $\mathbf{Y}$ an ideal in the K-space $\mathbf{X}^{\sim}$ that is total on $\mathbf{X}$. We consider the locally convex topology $|\sigma|(\mathbf{X}, \mathbf{Y})$ on $\mathbf{X}$ generated by the set of seminorms

$$p_f(x) = f(|x|), \quad x \in \mathbf{X},$$

where f ranges over $\mathbf{Y}_+$.

Let us record two properties of the LCS $(\mathbf{X}, |\sigma|(\mathbf{X}, \mathbf{Y}))$.

1) The topology $|\sigma|(\mathbf{X}, \mathbf{Y})$ is the uniform convergence topology on the set of all intervals in $\mathbf{Y}$.

2) The topology $|\sigma|(\mathbf{X}, \mathbf{Y})$ is compatible with the duality $\langle \mathbf{X}, \mathbf{Y} \rangle$.

Statement 1) is obtained from the formula

$$p_f(x) = f(|x|) = \sup\{|g(x)|: |g| \leqslant f\}.$$

By the Corollary to Lemma 3.1, all intervals in $\mathbf{Y}$ are $\sigma(\mathbf{Y}, \mathbf{X})$-compact. Hence 2) follows from 1) and the Mackey–Arens Theorem (see Theorem III.3.8).

Throughout the rest of this section let $\mathbf{X}$ be an FS on (T, Σ, μ), where μ is σ-finite and $\mathbf{X}_n^{\sim}$ is total on $\mathbf{X}$. We recall that $\mathbf{X}_n^{\sim}$ is then isomorphic to $\mathbf{X}'$ and supp $\mathbf{X}' = T$ (Theorem VI.1.1).

LEMMA 1. *Let* $\mathbf{Y}$ *be a foundation in* $\mathbf{X}_n^{\sim}$. *If* $\{x_\alpha\}$ *is a net such that* $x_\alpha \to \mathbf{0}$ *in the topology* $|\sigma|(\mathbf{X}, \mathbf{Y})$, *then* $x_\alpha \to \mathbf{0}(\mu)$.

Proof. We can identify $\mathbf{Y}$ with an FS in $\mathbf{X}'$, in view of the isomorphism between $\mathbf{X}_n^{\sim}$ and $\mathbf{X}'$. By Corollary 2 to Lemma IV.3.1, there exists a partition $\{A_n\}_{n=1}^{\infty}$ of T such that $\chi_{A_n} \in \mathbf{Y}$ and $0 < \mu(A_n) < +\infty$ for all $n \in \mathbb{N}$. Then, for each $n \in \mathbb{N}$, we have

$$\int_{A_n} |x_\alpha| \, d\mu \to 0.$$

Hence $x_\alpha \to \mathbf{0}(\mu)$ on each A_n, and so $x_\alpha \to \mathbf{0}(\mu)$.

LEMMA 2. *Let* $\mathbf{Y}$ *be a foundation in* $\mathbf{X}_n^{\sim}$. *If* $\{x_\alpha\}$ *is a net such that* $x_\alpha \to \mathbf{0}$ $(\sigma(\mathbf{X}, \mathbf{Y}))$, *then there exists a net* $\{y_\beta\}$ *such that* $y_\beta \to \mathbf{0}(\mu)$, *each* y_β *being a convex combination of elements of* $\{x_\alpha\}$.

Proof. Since the dual of $(\mathbf{X}, |\sigma|(\mathbf{X}, \mathbf{Y}))$ is $\mathbf{Y}$, there exists a net $\{y_\beta\}$ such that $y_\beta \to \mathbf{0}$ $(|\sigma|(\mathbf{X}, \mathbf{Y}))$, where y_β is of the type described (see Corollary 2 to Theorem III.3.2). By Lemma 1, $y_\beta \to \mathbf{0}(\mu)$.

LEMMA 3. *Let* $\mathbf{Y}$ *be a foundation in* $\mathbf{X}_n^{\sim}$, *let* $\{x_\alpha\}$ *be a net in* $\mathbf{X}$, *and let* $x, x_0 \in \mathbf{X}$. *If there exists* $y \in \mathbf{S}(T, \Sigma, \mu)$ *such that* $|x_\alpha| \leqslant y$ *for each* α, $x_\alpha \to x$ (μ) *and* $x_\alpha \to x_0$ $(\sigma(\mathbf{X}, \mathbf{Y}))$, *then* $x = x_0$.

Proof. In view of Corollary 2 to Lemma IV.3.1, we may assume that μ is finite and $y \in \mathbf{X}$. If $x' \in \mathbf{Y}$, then there exists a sequence $\{x_{\alpha_n}\}$ such that

$$\left|\int_T (x_{\alpha_n} - x_0) x' \, d\mu\right| < 1/n, \quad x_{\alpha_n} \to x(\mu).$$

Thus, by Lebesgue's Theorem on dominated convergence, we see that $\int_T (x - x_0) \, d\mu = 0$ for each $x' \in \mathbf{Y}$, so that $x_0 = x$.

If M is a subset of $\mathbf{S} = \mathbf{S}(T, \Sigma, \mu)$, then we denote by $\tau(M)$ the topology induced on M by $\mathbf{S}$.

LEMMA 4. *Suppose* $\mathbf{X}$ is *(o)-reflexive, and let* $M \subset \mathbf{X}$ *be bounded in the weak topology* $\sigma(\mathbf{X}, \mathbf{X}_n^{\sim})$. *The following statements are equivalent*:

1) *M is closed in the* TVS $\mathbf{S}(T, \Sigma, \mu)$;

2) *M is closed in* $(\mathbf{X}, \tau(\mathbf{X}))$.

Proof. 1) $\Rightarrow$ 2) is obvious.

2) $\Rightarrow$ 1). Since M is $\sigma(\mathbf{X}, \mathbf{X}_n^{\sim})$-bounded, it is $|\sigma|(\mathbf{X}, \mathbf{X}_n^{\sim})$-bounded (see Theorem III.3.3). Let $\{x_\alpha\} \subset M$, where $x \in \mathbf{S}$ and $x_\alpha \to x$ (μ). We show that $x \in \mathbf{X}$. Since $\mathbf{X} = \mathbf{X}''$, to do this we need only prove that $\int |xx'| \, d\mu < \infty$ for each $x' \in \mathbf{X}'$; but this is clear from the $|\sigma|(\mathbf{X}, \mathbf{X}_n^{\sim})$-boundedness of M and Fatou's Theorem.

5.2. From now on suppose $\mathbf{X}$ is an FS with $\mathbf{X} = \mathbf{X}''$—that is, suppose $\mathbf{X}$ is (o)-reflexive. Denote the canonical embedding of $\mathbf{X}$ into $(\mathbf{X}_n^{\sim})^{\sim}$ by κ. Then $\kappa(\mathbf{X}) = (\mathbf{X}_n^{\sim})_n^{\sim}$ is a band in the K-space $(\mathbf{X}_n^{\sim})^{\sim}$. Denote the projection onto this band $\kappa(\mathbf{X})$ by Pr.

LEMMA 5. *If* $\{x_\alpha\}$ *is a net in* $\mathbf{X}$ *such that* $\kappa(x_\alpha) \to F$ *in the weak topology* $\sigma((\mathbf{X}_n^{\sim})^{\sim}, \mathbf{X}_n^{\sim})$ *and*

$\Pr F = \pi(x)$, $x \in \mathbf{X}$, *then there exists a foundation* $\mathbf{Y}$ *in* $\mathbf{X}_n^\sim$ *such that* $x_\alpha \to x$ $(\sigma(\mathbf{X}, \mathbf{Y}))$.

Proof. By Theorem 3.6, the K-space $(\mathbf{X}_n^\sim)^\sim$ decomposes as a sum of two disjoint bands $(\mathbf{X}_n^\sim)_n^\sim = \kappa(\mathbf{X})$ and $(\mathbf{X}_n^\sim)_s^\sim$. If $F_1 = F - \kappa(x)$, then $F_1 \in (\mathbf{X}_n^\sim)_s^\sim$. Thus the set $\mathbf{Y} = \{f \in \mathbf{X}_n^\sim : |F_1|(|f|) = 0\}$ is a foundation in $\mathbf{X}_n^\sim$. Furthermore, for each $f \in \mathbf{Y}$, we have

$$f(x_\alpha) = [\kappa(x_\alpha)](f) \to F(f) = [\kappa(x)](f) + F_1(f) = f(x),$$

that is, $x_\alpha \to x$ $(\sigma(\mathbf{X}, \mathbf{Y}))$.

Lemmas 2 and 5 are key results in the proof of the following fundamental theorem of this section.

THEOREM 1. *Let* $\mathbf{X}$ *be an* (o)*-reflexive* FS, *let* V *be a non-empty convex subset of* $\mathbf{X}$, *and let* W *be the* $\sigma((\mathbf{X}_n^\sim)^\sim, \mathbf{X}_n^\sim)$*-closure of* $\kappa(V)$. *Then*

a) *if* V *is closed in* $(\mathbf{X}, \tau(\mathbf{X}))$, *then*

$$\Pr W = x(V); \tag{1}$$

b) *if* V *is* $\sigma(\mathbf{X}, \mathbf{X}_n^\sim)$*-bounded and satisfies* (1) *then* V *is closed in* $(\mathbf{X}, \tau(\mathbf{X}))$.

Proof. a) Suppose V is closed in $(\mathbf{X}, \tau(\mathbf{X}))$. Clearly, $\kappa(V) = \Pr W$. Suppose $F \in W$ and $\kappa(x) = \Pr F$. Then there exists a net $\{x_\alpha\} \subset V$ such that $\kappa(x_\alpha) \to F$ in the topology $\sigma((\mathbf{X}_n^\sim)$, $\mathbf{X}_n^\sim)$. By Lemma 5, there exists a foundation $\mathbf{Y}$ in $\mathbf{X}_n^\sim$ such that $x_\alpha \to x$ $(\sigma(\mathbf{X}, \mathbf{Y}))$. Since V is convex and (μ)-closed, Lemma 2 shows that $x \in V$, and hence that $\kappa(x) \in \kappa(V)$.

b) Since $\kappa(V)$ is $\sigma((\mathbf{X}_n^\sim)^\sim, \mathbf{X}_n^\sim)$-bounded, the Corollary to Lemma 3.1 shows that it is relatively compact in this topology, and hence that W is compact.

As μ is σ-finite, to prove that V is (μ)-closed it is sufficient to verify that if $\{x_n\}_{n=1}^\infty \subset V$, $x_n \to x$ a.e., $x \in \mathbf{X}$, then $x \in V$. Since $x_n \to x$ a.e. there exists $y \in \mathbf{S}$ such that $|x_n| \leqslant y$ $(n \in \mathbb{N})$.

As W is compact, there exists a subnet $\{x_\alpha\}$ of $\{x_n\}$ that converges in the topology $\sigma((\mathbf{X}_n^\sim)^\sim, \mathbf{X}_n^\sim)$ to some $F \in W$. Thus, by (1) and Lemma 5 there exist a foundation $\mathbf{Y}$ in $\mathbf{X}_n^\sim$ and an $x_0 \in V$ such that $x_\alpha \to x_0(\sigma(\mathbf{X}, \mathbf{Y}))$. Since $\{x_\alpha\}$ is μ-convergent to x, being a subnet of a (μ)-convergent sequence, it follows from Lemma 3 that $x = x_0 \in V$, which completes the proof of the theorem.

It is Theorem 1 that enables us to reduce the study of convex bounded (μ)-closed sets to that of $\sigma((\mathbf{X}_n^\sim)^\sim, \mathbf{X}^\sim)$-compact sets. As before, let $\mathbf{X}$ be an (o)-reflexive FS.

THEOREM 2. *Let* V_1 *and* V_2 *be non-empty disjoint convex sets in* $\mathbf{X}$, *closed in* $(\mathbf{X}, \tau(\mathbf{X}))$. *If at least one of them is* $\sigma(\mathbf{X}, \mathbf{X}_n^\sim)$*-bounded, then they are strictly separable by a functional in* $\mathbf{X}_n^\sim$: *that is, there exists an* $f \in \mathbf{X}_n^\sim$ *such that* $\sup\{f(x) : x \in V_1\} < \inf\{f(x) : x \in V_2\}$.

Proof. Let W_i be the $\sigma((\mathbf{X}_n^\sim)^\sim, \mathbf{X}_n^\sim)$-closure of $\kappa(V_i)$ $(i = 1, 2)$. Then $W_1 \cap W_2 = \varnothing$, by (1). If, say, V_1 is bounded, then W_1 is $\sigma((\mathbf{X}_n^\sim)^\sim, \mathbf{X}_n^\sim)$-compact (see the beginning of the proof of part b) of Theorem 1). Now we apply the familiar separability theorem (see Theorem III.2.6).

THEOREM 3. *Let* $\{V_\xi\}$ $(\xi \in \Xi)$ *be a system of convex* $\sigma(\mathbf{X}, \mathbf{X}_n^\sim)$*-bounded subsets of* $\mathbf{X}$, *closed in* $(\mathbf{X}, \tau(\mathbf{X}))$, *that has the finite intersection property. Then* $\bigcap_{\xi \in \Xi} V_\xi$ *is non-empty.*

Proof. Let W_ξ be the $\sigma((\mathbf{X}_n^\sim)^\sim, \mathbf{X}_n^\sim)$-closure of $\kappa(V_\xi)$. Since W_ξ is compact in this topology, $\bigcap W_\xi$ is non-empty. Thus $\bigcap_{\xi \in \Xi} V_\xi$ is non-empty, by Theorem 1(a).

THEOREM 4. *Let* V_1 *and* V_2 *be convex* $\sigma(\mathbf{X}, \mathbf{X}_n^\sim)$*-bounded subsets of* $\mathbf{X}$, *closed in* $(\mathbf{X}, \tau(\mathbf{X}))$. *Then* $V = V_1 + V_2$ *is closed in* $(\mathbf{X}, \tau(\mathbf{X}))$.

Proof. Let W_i be the $\sigma((\mathbf{X}_n^\sim)^\sim, \mathbf{X}_n^\sim)$-closure of $\kappa(V_i)$ $(i = 1, 2)$. If $W = W_1 + W_2$, then

$$\Pr W = \Pr W_1 + \Pr W_2 = \kappa(V_1) + \kappa(V_2) = \kappa(V)$$

by Theorem 1(a). Since W is obviously the $\sigma((\mathbf{X}_n^{\sim})^{\check{}}, \mathbf{X}_n^{\sim})$-closure of V, Theorem 1(b) shows that V is (μ)-closed.

THEOREM 5. *Let* $\mathbf{X}$ *be a* BFS *satisfying conditions* (B) *and* (C). *Let* V_1 *and* V_2 *be non-empty convex sets in* $\mathbf{X}$, *closed in* $(\mathbf{X}, \tau(\mathbf{X}))$. *If at least one of these is bounded in norm in* $\mathbf{X}$, *then there exist* $v_1 \in V_1$, $v_2 \in V_2$ *such that*

$$\|v_1 - v_2\|_{\mathbf{X}} = \inf\{\|x_1 - x_2\|_{\mathbf{X}} : x_1 \in V_1, x_2 \in V_2\}.$$

Proof. Assume that V_1 is bounded in norm. Then it is sufficient to look for the closest element $v_2 \in V_2$ to V_1 in the intersection of V_2 with a closed ball of large enough radius. In view of condition (C), this intersection will be closed by Lemma IV.3.4. Therefore we may also assume that V_2 is bounded in norm. Let W_i be the $\sigma((\mathbf{X}_n^{\sim})^*, \mathbf{X}_n^{\sim})$-closure of $\kappa(V_i)$ $(i = 1, 2)$. Since W_i is weak* compact, there exist $\tilde{F}_1 \in W_1$, $\tilde{F}_2 \in W_2$ such that

$$\|\tilde{F}_1 - \tilde{F}_2\|_{(\mathbf{X}_n^{\sim})^*} = \inf\{\|F_1 - F_2\|_{(\mathbf{X}_n^{\sim})^*} : F_1 \in W_1, F_2 \in W_2\}.$$

As $\kappa : \mathbf{X} \to (\mathbf{X}_n^{\sim})^*$ is an isometry, it is easy to see that the elements $v_i = \Pr \tilde{F}_i$, which belong to V_i by Theorem 1(a) $(i = 1, 2)$, are the ones we are seeking.

Theorems 1–5 have applications to convex analysis and the theory of optimal control (see V. L. Levin [2]; we shall present one result from this paper in 5.3). The results of 5.1 and 5.2 were obtained by A. V. Bukhvalov and G. Ya. Lozanovskii (see [1], where there are also other applications and generalizations).

5.3. Using Theorem 3, we now prove a theorem on the attainment of minima of convex functionals on convex sets that are closed with respect to convergence in measure.

A functional p, defined on $\mathbf{L}^1(T, \Sigma, \mu)$ and possibly taking the value $+\infty$, is called *lower semicontinuous* with respect to convergence in measure if

$$x_n \to x(\mu) \text{ implies } p(x) \leqslant \varliminf_{n \to \infty} p(x_n).$$

A functional p is said to be *convex* if

$$p(\lambda x + (1-\lambda)y) \leqslant \lambda p(x) + (1-\lambda)p(y)$$

for all $x, y \in \mathbf{L}^1$, $0 \leqslant \lambda \leqslant 1$.

THEOREM 6. *A convex functional* p *that is lower semicontinuous with respect to convergence in measure attains a minimum on every set* $V \subset \mathbf{L}^1$ *that is closed in* $(\mathbf{L}^1, \tau(\mathbf{L}^1))$ *and bounded in norm.*

Proof. Let

$$m = \inf\{p(x) : x \in V\}.$$

Choose a sequence of numbers $m_n (m_n > m)$ that decreases towards the value m. Consider the sets

$$V_n = \{x \in V : p(x) \leqslant m_n\}.$$

As $m_n > m$, each V_n is non-empty; and as $m_n \downarrow$, the sequence of sets V_n is decreasing and so has the finite intersection property. As p is convex, the V_n are convex; and as p is lower semicontinuous, they are $\tau(\mathbf{L}^1)$-closed. By Theorem 3, we see that $\bigcap_{n=1}^{\infty} V_n \neq \varnothing$. If $x_0 \in \bigcap_{n=1}^{\infty} V_n$, then $x_0 \in V$ and $p(x_0) \leqslant m_n$ for all $n \in \mathbb{N}$, so $p(x_0) = m$.

XI

INTEGRAL OPERATORS

§ 1. Integral representations of operators

1.1. The integral operators constitute an important class of operators which are frequently encountered in applications. We have already had some dealings with integral operators in § 2 of Chapter V. However, there our attention was directed principally to operators on $\mathbf{C}[a, b]$ with continuous kernels. In the present section, we shall consider integral operators with arbitrary measurable kernels, defined on ideal spaces.

Let (S, Σ_s, ν) and (T, Σ_t, μ) be σ-finite measure spaces, and let (R, Σ_R, λ) be the product of these spaces (see I.6.8)—thus $R = S \times T$ and $\lambda = \nu \times \mu$.

Henceforth in this section $\mathbf{X}$ will denote an FS* on (T, Σ_T, μ) and $\mathbf{Y}$ an FS on (S, Σ_S, ν).

An operator $U: \mathbf{X} \to \mathbf{Y}$ is called *integral* if there exists a λ-measurable function $K(s, t)$ $(s \in S, t \in T)$ such that, for every $x \in \mathbf{X}$, the image $y = U(x)$ of x is the function

$$y(s) = \int_T K(s, t) x(t)\, d\mu(t). \tag{1}$$

The function $K(s, t)$ is called the *kernel* of U. It is easy to see that the kernel is finite almost everywhere relative to λ.

An integral operator (1) is called a *regular operator from* $\mathbf{X}$ *into* $\mathbf{Y}$ if the operator $|U|$ defined by

$$z = |U|(x),\ z(s) = \int_T |K(s, t)| x(t)\, d\mu(t) \quad (x \in \mathbf{X}), \tag{2}$$

maps $\mathbf{X}$ into $\mathbf{Y}$. In this situation $|U|$ is called the *modulus*† of U.

Let $U: \mathbf{X} \to \mathbf{S}(S, \Sigma_S, \nu)$ be the integral operator (1). Let us list some of its properties.

I. The operator U is linear.

II. $U(x) \geqslant \mathbf{0}$ for all $x \in \mathbf{X}_+$ if and only if $K(s, t) \geqslant 0$ a.e. (λ).

The sufficiency in II is obvious. Let us prove the necessity. By Corollary 2 to Lemma IV.3.1, we may assume that $\mathbf{L}^\infty(T, \Sigma_T, \mu) \subset \mathbf{X}$. If $\mathbf{1}_T$ is the function that is identically equal to unity on T, then for almost all (λ) s, we have

$$\left|\int_T K(s, t)\, d\mu(t)\right| = |[U(\mathbf{1}_T)]\ (s)| < +\infty,$$

so that, for almost all $(\lambda) s$, we see that

$$y(s) = \int_T |K(s, t)|\, d\mu(t) < +\infty.$$

* In this chapter we shall assume for definiteness that all spaces are real, although the results are valid also in the complex case.

† An integral operator U is regular in the sense of this definition if and only if $U \in L^\sim(\mathbf{X}, \mathbf{Y})$. Then it turns out that $|U|$ is the modulus in the K-space $L^\sim(\mathbf{X}, \mathbf{Y})$ (see p. 281).

Since the function $y(s)$ is a.e. (ν) finite, there exists a partition $\{B_n\}$ of S such that $y\chi_{B_n} \in \mathbf{L}^1(S, \Sigma_S, \nu)$ $(n \in \mathbb{N})$. Thus we may assume that $y \in \mathbf{L}^1(S, \Sigma_S, \nu)$. Then, by Tonelli's Theorem (Theorem I.6.12),

$$\int_R |K(s, t)|\, d\lambda(s, t) = \int_S y(s)\, d\nu(s) < +\infty,$$

that is, $K \in \mathbf{L}^1(R, \Sigma_R, \lambda)$. Hence, by Fubini's Theorem, for any measurable $B \subset S$ and $A \subset T$, we have

$$\int_{B \times A} K(s, t)\, d\lambda(s, t) = \int_B [U(\chi_A)](s)\, d\nu(s) \geqslant 0,$$

so that $K(s, t) \geqslant 0$ a.e. (λ) (see I.6.8).

Applying II to the operators U and $-U$, we have

III. $U = \mathbf{0}$ if and only if $K(s, t) = 0$ a.e. (λ).

IV. The operator $|U|$ defined by (2) maps $\mathbf{X}$ into $\mathbf{S}(S, \Sigma_S, \nu)$.

For it follows from (1) that for every $x \in \mathbf{X}$ we have

$$\int_T |K(s, t)x(t)|\, d\mu(t) < +\infty,$$

for almost all (λ) $s \in S$, and so we deduce that the function $z = |U|(x)$ in (2) is a.e. finite.

V. If $x_n \to \mathbf{0}$ (μ) and $|x_n| \leqslant x \in \mathbf{X}$ $(n \in \mathbb{N})$, then $[U(x_n)](s) \to 0$ a.e. (ν).

Property V follows from IV and Lebesgue's Theorem. As a corollary to V we have:

VI. If $x_n \to \mathbf{0}$ a.e. (μ) and $|x_n| \leqslant x \in \mathbf{X}$ $(n \in \mathbb{N})$, then $[U(x_n)](s) \to 0$ a.e. (ν).

VII. If $\mathbf{X}$ and $\mathbf{Y}$ are BFSs, and the operator U maps $\mathbf{X}$ into $\mathbf{Y}$, then U is continuous—that is, $U \in B(\mathbf{X}, \mathbf{Y})$.

This statement follows from the Closed Graph Theorem (see Theorem XII.1.4 below), Lemma IV.3.2 and V. The problem of expressing, in terms of the kernel $K(s, t)$, the fact that an operator U maps $\mathbf{X}$ into $\mathbf{Y}$ is very difficult. Some results in this direction are presented in §§ 2 and 3.

VIII. If U is the regular integral operator (1) from $\mathbf{X}$ into $\mathbf{Y}$ and an operator $V: \mathbf{Y}' \to \mathbf{X}'$ satisfies the relation

$$\int_S U(x)y'\, d\nu = \int_T xV(y')\, d\mu,$$

for all $x \in \mathbf{X}$, $y' \in \mathbf{Y}'$, then, with $z = V(y')$, we have

$$z(t) = \int_S K(s, t)y'(s)\, d\nu(s) \quad (y' \in \mathbf{Y}'),$$

that is, $V: \mathbf{Y}' \to \mathbf{X}'$ is also a regular integral operator, with kernel $K^*(t, s) = K(s, t)$.

For, as U is regular, the function $\int_T |K(s, t)|x(t)\, d\mu(t)$ belongs to $\mathbf{Y}$ for every $x \in \mathbf{X}$. Thus, for all $x \in \mathbf{X}_+$, $y' \in \mathbf{Y}'_+$, we have, by Tonelli's Theorem,

$$\int_R |K(s, t)|x(t)y'(s)\, d\lambda(s, t) = \int_S \left\{\int_T |K(s, t)|x(t)\, d\mu(t)\right\} y'(s)\, d\nu(s) < +\infty.$$

Hence, by Fubini's Theorem, we see that, for all $x \in \mathbf{X}_+$, $y' \in \mathbf{Y}'_+$,

$$\int_T xV(y')\, d\mu = \int_S U(x)y'\, d\nu = \int_S \left\{\int_T K(s, t)x(t)\, d\mu(t)\right\} y'(s)\, d\nu(s) =$$

$$= \int_T \left\{\int_S K(s, t)y'(s)\, d\nu(s)\right\} x(t)\, d\mu(t) = \int_T zx\, d\mu.$$

From this it follows that the functions $V(y')$ and z determine the same (o)-continuous functional on $\mathbf{X}$. By Theorem VI.1.1, we now see that $z = V(y')$.

1.2. In this section we shall concern ourselves with determining conditions under which a linear operator U from one FS into another is an integral operator—that is, is expressible in the form (1).

We note at once that the situation is very simple in the case of a discrete measure. For suppose the measure μ is discrete. Then every linear operator U from an FS $\mathbf{X}$ on (T, Σ_T, μ) into $\mathbf{S}(S, \Sigma_S, \nu)$ that satisfies condition VI is integral; thus, in this case, condition VI is a necessary and sufficient condition for an operator to be integral. For, as μ is discrete, the set T can be expressed as a union of a countable number of atoms $\{A_n\}$. Thus every $x \in \mathbf{X}$ can be expressed, in a unique way, in the form

$$x = \sum_n \lambda_n \chi_{A_n},$$

where the series on the right-hand side converges almost everywhere. As U satisfies condition VI,

$$U(x) = \sum_n \lambda_n U(\chi_{A_n}),$$

where the series also converges almost everywhere. For $t \in A_n$ $(n \in \mathbb{N})$, we set

$$K(s, t) = \frac{1}{\mu(A_n)}[U(\chi_{A_n})](s).$$

The function $K(s, t)$ is clearly $\nu \times \mu$-measurable. Let us check that $K(s, t)$ is the kernel of U. For every $x \in \mathbf{X}$, we have

$$\int_T K(s, t)x(t)\,d\mu(t) = \sum_n \frac{1}{\mu(A_n)}[U(\chi_{A_n})](s)\lambda_n\mu(A_n) = \sum_n \lambda_n[U(\chi_{A_n})](s) = [U(x)](s),$$

which is what we required to prove.

It is easy to see that, in the case where $\mathbf{X}$ is an FS of sequences, a linear operator $U: \mathbf{X} \to \mathbf{s}$ is integral if and only if it is a matrix operator. The latter means that there exists an infinite matrix $\{a_{ik}\}_{i=1,k=1}^{\infty\;\;\infty}$ such that, if $y = U(x)$, $x = \{\xi_k\} \in \mathbf{X}$, $y = \{\eta_i\} \in \mathbf{s}$, then

$$\eta_i = \sum_{k=1}^{\infty} a_{ik}\xi_k.$$

The situation is significantly more complicated in the case of a continuous measure, which is the most interesting case from the point of view of operator theory. Here the necessary condition VI is no longer sufficient. For if we consider an FS $\mathbf{X}$ on the continuous measure space (T, Σ_T, μ) and denote the identity operator on $\mathbf{X}$ by I, then I obviously satisfies condition VI, but does not satisfy the necessary condition V, since in the case of a continuous measure it is easy to construct a sequence $\{x_n\}$ in $\mathbf{X}$ that is bounded in $\mathbf{X}$ and converges in measure but does not converge almost everywhere (see, for example, Vulikh-III).

However, it turns out that condition V is in fact necessary and sufficient in the general case. Let us give a precise statement of this result, which is non-trivial even for operators on $\mathbf{L}^2(D)$.

THEOREM 1. *Let* $\mathbf{X}$ *be an* FS *on* (T, Σ_T, μ) *and let* $U: \mathbf{X} \to \mathbf{S}(S, \Sigma_S, v)$ *be a linear operator. The following statements are equivalent:*

1) U *is an integral operator;*

2) *if* $x_n \to \mathbf{0}$ (μ) *and* $|x_n| \leqslant x \in \mathbf{X}$ $(n \in \mathbb{N})$, *then* $[U(x_n)](s) \to 0$ *a.e.* (v);

3) U *satisfies condition* VI *and the statement* $\chi_{A_n} \leqslant x \in \mathbf{X}$ $(n \in \mathbb{N})$, $\mu(A_n) \to 0$ *implies that* $[U(\chi_{A_n})](s) \to 0$ *a.e.* (v).

The truth of the implication 1) ⇒ 2) has already been noted (see V). The implication 2) ⇒ 3) is obvious. Theorem 1 was obtained by A. V. Bukhvalov (see Bukhvalov [3], to which we also refer the reader for the proof that 3) ⇒ 1)), and generalizes an earlier result of Nakano.

COROLLARY. *Let* $\Phi(s, t)$ *be a function, not necessarily measurable, such that the almost everywhere* (v) *finite* v*-measurable function* $y(s) = \int_T \Phi(s, t)x(t)\,d\mu(t)$ *is defined for every* $x \in \mathbf{X}$. *Define* $U(x) = y$ $(x \in \mathbf{X})$. *Then there exists a* λ*-measurable function* $K(s, t)$ *such that, for every* $x \in \mathbf{X}$, *we have*

$$[U(x)](s) = \int_T \Phi(s, t)x(t)\,d\mu(t) = \int_T K(s, t)x(t)\,d\mu(t)$$

for almost all (v) s *(the exceptional set will, in general, depend on* x*).*

Proof. For every $x \in \mathbf{X}$, we have

$$\left|\int_T \Phi(s, t)x(t)\,d\mu(t)\right| \leqslant \int_T |\Phi(s, t)x(t)|\,d\mu(t) < +\infty \qquad \text{a.e. } (v)$$

(note that the function on the right may turn out to be non-measurable). Using Lebesgue's Theorem, we can convince ourselves that U satisfies condition 2) of Theorem 1, and thus we obtain the required result.

The corollary enables us to replace the kernels by measurable ones. If the space (T, Σ_T, μ) is separable, then it is easy to deduce from the corollary that, for a.e. $(v)\,s$, we have $K(s, t) = \Phi(s, t)$ for almost all $(\mu)t$ (Gribanov [2]). It can be shown that this is false without the separability condition.

In the remaining part of this section, we shall obtain two theorems about integral representations of operators in certain special classes. We could easily deduce these theorems from Theorem 1, but we shall not do so, both because their proofs, unlike that of Theorem 1, do not require an extensive auxiliary apparatus, and because the proof of Theorem 1 uses a special case of Theorem 6.

1.3. We begin with an auxiliary result, which is also interesting in its own right.

Let E be an arbitrary non-empty set of functions $x \in \mathbf{S}(T, \Sigma_T, \mu)$, and $K(s, t)$ be a λ-measurable, almost everywhere (λ) finite function on $R = S \times T$. We shall say that the set E and the function $K(s, t)$ are *compatible* if the following two conditions are satisfied:

1) there exists a set $S_0' \subset S$, $v(S_0') = 0$, such that, for every $x \in E$, the function $K(s, t)x(t)$ is μ-measurable and $\int_T K(s, t)x(t)\,d\mu(t)$ exists,* for all $s \in S \setminus S_0'$;

2) there exists a set $S_0'' \subset S$, $v(S_0'') = 0$, such that, for every $x \in E$, the functions

$$x_s(t) = \begin{cases} (\operatorname{sign} K(s, t))|x(t)|, & \text{if } K(s, t) \neq 0 \\ |x(t)|, & \text{if } K(s, t) = 0 \end{cases}$$

belong to E, for all $s \in S \setminus S_0''$.

---

* We do not assume that this integral is finite; see I.6.4.

If we write $S_0 = S_0' \cup S_0''$, then $\nu(S_0) = 0$. For every such E and K and for all $s \in S \setminus S_0$, we write

$$d_{K,E}(s) = \sup\{\int_T K(s,t)x(t)\,d\mu(t): x \in E\}. \tag{3}$$

We now concern ourselves with proving that the pointwise supremum $d_{K,E}$ is ν-measurable, which is far from simple to verify in the case where E is uncountable and the measure μ is inseparable. A typical example of a compatible pair E and K is formed by a set E consisting of non-negative μ-measurable functions and a λ-measurable function $K(s, t) \geqslant 0$ a.e. (λ). Condition 2) is obviously satisfied with $S_0'' = \varnothing$. By Fubini's Theorem, the function $K(s, .)$ is μ-measurable for almost all (μ) s, and for almost all (ν) s we have $K(s, t) \geqslant 0$ for almost all $(\mu)t$. Now it is obvious that condition 1) is satisfied. Later we shall become acquainted with other examples.

Note that, for all $s \in S \setminus S_0$, the following equation holds:

$$d_{K,E}(s) = \sup\{\int_T |K(s,t)||x(t)|\,d\mu(t): x \in E\}. \tag{4}$$

In fact, it is obvious that $d_{K,E}(s)$ is less than or equal to the expression on the right-hand side of (4). If $x \in E$, then $x_s \in E$, and so

$$\int_T |K(s,t)||x(t)|\,d\mu(t) = \int_T K(s,t)(\operatorname{sign} K(s,t))|x(t)|\,d\mu(t) = \int_T K(s,t)x_s(t)\,d\mu(t) \leqslant d_{K,E}(s),$$

which gives the opposite inequality. Hence (4) is proved.

THEOREM 2. *Assume that the set E and the function $K(s, t)$ are compatible. Then the function $d_{K,E}(s)$ is ν-measurable. Moreover, there exists a sequence of functions $x_n \in E$ such that*

$$d_{K,E}(s) = \sup_n \int_T K(s,t)x_n(t)\,d\mu(t) \qquad \text{a.e. } (\nu).$$

Before we start the proof of Theorem 2, we note that its non-triviality consists, firstly, in the fact that the pointwise supremum $d_{K,E}$ of a possibly uncountable set of measurable functions is measurable, and secondly in the fact that this coincides with the supremum of the corresponding set in the ordered space $\mathbf{S}(S, \Sigma_s, \nu)$, since by Theorem 2 $d_{K,E}$ is realized on some countable family (cf. I.6.10).

Proof of Theorem 2. Since on S_0 we can set $K(s, t) = 0$ $(t \in T)$, we may assume that $\int_T K(s,t)x(t)\,d\mu(t)$ exists for all $x \in E$ and $s \in S$. We first state a lemma, whose proof we shall present later on in a more general form (see Lemma 2).

LEMMA 1. *Let H denote the set of all functions of the form*

$$\sum_{i=1}^{i_0} \chi_{B_i}(s)g_i(t),$$

where the $B_i \in \Sigma_S(\nu)$ are pairwise disjoint and $g_i \in \mathbf{L}^1(T, \Sigma_T, \mu) \cap \mathbf{L}^\infty(T, \Sigma_T, \mu)(1 \leqslant i \leqslant i_0)$. Then H is everywhere dense in the B-space $\mathbf{L}^1(R, \Sigma_R, \lambda)$.

We recall (see I.6.4) that we use the symbol $[x]_n$ to denote the truncated function associated with a function x. Write $E_n = \{[x]_n : x \in E\}$ $(n \in \mathbb{N})$. Then each E_n is compatible with the function K. By Fatou's Theorem, we see with the help of (4) that, for all $s \in S$,

$$d_{K,E_n}(S) \uparrow d_{K,E}(s). \tag{5}$$

Suppose that

$$K_m(s,t) = \sum_{i=1}^{i_0} \chi_{B_i}(s) g_i(t) \in H.$$

Clearly K_m is compatible with each E_n. Hence, as $B_i \cap B_{i'} = \emptyset\,(i \neq i')$, we have, for each $s \in S$,

$$d_{K_m, E_n}(s) = \sum_{i=1}^{i_0} \left[\sup_{x \in E} \int_T g_i(t)[x]_n(t)\, d\mu(t) \right] \chi_{B_i}(s).$$

Therefore d_{K_m, E_n} is measurable, and there exists an at most countable set $E(n, m) \subset E$ such that, for every $s \in S$,

$$d_{K_m, E_n}(s) = \sup \{ \int_T K_m(s,t)[x]_n(t) d\mu(t) \colon x \in E(n, m) \}. \tag{6}$$

Let K be a function in $\mathbf{L}^1(R, \Sigma_R, \lambda)$. In view of Lemma 1, there exists a sequence of functions $K_m \in H$ such that, for any m, we have $\|K_m - K\|_{\mathbf{L}^1} \leqslant 2^{-m}$. Thus in $\mathbf{L}^1\,(R, \Sigma_R, \lambda)$, and so also almost everywhere (λ), the following series is convergent,

$$L(s,t) = \sum_{m=1}^{\infty} |K_m(s,t) - K(s,t)|.$$

Hence

$$|K_m(s,t) - K(s,t)| \xrightarrow[m \to \infty]{} 0 \qquad \text{a.e. } (\lambda)$$

and

$$|K_m(s,t) - K(s,t)| \leqslant L(s,t) \qquad \text{a.e. } (\lambda).$$

Define a set $S_1 \subset S$ by assigning $s \in S$ to S_1 if and only if the following conditions are satisfied simultaneously:

1) $|K_m(s,t) - K(s,t)| \xrightarrow[m \to \infty]{} 0$ a.e. (μ);
2) $|K_m(s,t) - K(s,t)| \leqslant L(s,t)$ a.e. (μ), for all $m \in \mathbb{N}$;
3) the function $L(s, .)$ is μ-integrable;
4) $d_{K_m, E_n}(s) < +\infty$ and $d_{K, E_n}(s) < +\infty$, for all $m, n \in \mathbb{N}$.

Using Fubini's Theorem, it is easy to prove that each of the conditions 1)–4) is satisfied for almost all $(\nu)s$. Therefore S_1 is measurable and $\nu(S \setminus S_1) = 0$.

For each $s \in S_1$ and for all $m, n \in \mathbb{N}$, we have

$$|d_{K_m, E_n}(s) - d_{K, E_n}(s)| \leqslant d_{|K_m - K|, E_n}(s). \tag{7}$$

By Lebesgue's Theorem on limits under the integral sign, we have, for each $s \in S_1$,

$$d_{|K_m - K|, E_n}(s) \leqslant n \int_T |K_m(s,t) - K(s,t)|\, d\mu(t) \xrightarrow[m \to \infty]{} 0. \tag{8}$$

In view of (7) and (8), we have, for $s \in S_1$,

$$d_{K_m, E_n}(s) \xrightarrow[m \to \infty]{} d_{K, E_n}(s), \quad n \in \mathbb{N}. \tag{9}$$

As we have already proved that d_{K_m, E_n} is ν-measurable, it follows that d_{K_m, E_n} is ν-measurable. Write $E(n) = \bigcup_{m=1}^{\infty} E(n, m)$. Then by (6) and (9), the following equation holds for all $s \in S_1$:

$$d_{K,E_n}(s) = \sup\{\int_T K(s,t)[x]_n(t)\,d\mu(t): x \in E(n)\},\quad n \in \mathbb{N}. \tag{10}$$

Since $E(n)$ is at most countable, d_{K,E_n} is ν-measurable, and so, in view of (5), the function $d_{K,E}$ is also ν-measurable; and, by (5) and (10), we have

$$d_{K,E}(s) = \sup\{\int_T K(s,t)x(t)\,d\mu(t):\ x \in E_0\},$$

for all $s \in S_1$, where the set $E_0 = \bigcup_{n=1}^{\infty} E(n)$ is at most countable. Thus the theorem is proved if $K(s,t)$ is λ-integrable.

In the general case, we choose two fixed sequences $S_n \uparrow S$, $T_n \uparrow T$, where $S_n \in \Sigma_S(\nu)$, $T_n \in \Sigma_T(\mu)$. The functions $K_n(s,t) = [K]_n(s,t)\chi_{S_n \times T_n}(s,t)$ are λ-integrable. Using (4) and Fatou's Theorem, we see that $d_{K_n,E}(s) \uparrow d_{K,E}(s)$ for almost all $(\nu)s$. Since the theorem has already been proved for the $K_n(s,t)$, we conclude that it is true also for $K(s,t)$. This completes the proof.

COROLLARY. *If $U: \mathbf{X} \to \mathbf{S}(S, E_S, \nu)$ is the integral operator* (1), *then for each $x \in \mathbf{X}_+$ there exists a sequence $\{x_n\}$ in $\mathbf{X}$ such that $|x_n| = x$ and*

$$|U|(x)(s) = \int_T |K(s,t)|x(t)\,d\mu(t) = \sup_n \int_T K(s,t)x_n(t)\,d\mu(t)$$

for almost all $(\nu)s$ (the exceptional set will, in general, depend on x).

Proof. We fix an $x \in \mathbf{X}_+$ and consider the set $E = \{\tilde{x} \in X: |\tilde{x}| = x\}$. Write $S_0 = \{s \in S: \int_T |K(s,t)|x(t)\,d\mu(t) = +\infty\}$. By IV, $\nu(S_0) = 0$. Now $\int_T |K(s,t)\tilde{x}(t)|\,d\mu(t) = \int_T |K(s,t)|x(t)\,d\mu(t)$, so for each $\tilde{x} \in E$ the integral $\int_T K(s,t)\tilde{x}(t)\,d\mu(t)$ exists and is finite, for all $s \in S \setminus S_0$. Hence the set E and the function $K(s,t)$ are compatible. Now we need only apply Theorem 2 and the formula (4).

Theorem 2, in a slightly weaker form, was obtained in a paper by Yu. I. Gribanov [1].

1.4. As before, let $\mathbf{X}$ be an FS on (T, Σ_T, μ), and $\mathbf{Y}$ an FS on (S, Σ_S, ν). Assume that $\mathbf{Y}$ is a BFS with condition (C). Denote by $\mathbf{X}[\mathbf{Y}]$ the space of all λ-measurable functions $K(s,t)$ satisfying the two conditions:

1) the function $s \to K(s,t)$ belongs to $\mathbf{Y}$ for almost all (μ) $t \in T$;

2) the function $w_{\mathbf{Y}}(K)(t) = \|K(.,t)\|_{\mathbf{Y}}$ belongs to $\mathbf{X}$.

Even to see that $\mathbf{X}[\mathbf{Y}]$ is a linear manifold, we need to apply Theorem 2. The point is that, *a priori*, the function $w_{\mathbf{Y}}(K)$ might turn out to be non-measurable (this will, in fact, be so in general if $\mathbf{Y}$ does not satisfy condition (C)). Let $K(s,t)$ be a λ-measurable function satisfying condition 1). We verify that $w_{\mathbf{Y}}(K)$ is μ-measurable. As $\mathbf{Y}$ satisfies condition (C), the Nakano–Amemiya–Mori Theorem (see Theorem VI.1.6) shows that, for almost all $(\mu)t$, we have

$$\begin{aligned}\|K(.,t)\|_{\mathbf{Y}} &= \sup\{|\int_S K(s,t)y'(s)d\nu(s)|: y' \in \mathbf{Y}',\ \|y'\| \leq 1\}\\ &= \sup\{\int_S |K(s,t)|y'(s)\,d\nu(s): y' \in \mathbf{Y}'_+,\ \|y'\| \leq 1\}.\end{aligned}$$

It is clear that the set $E = \{y' \in \mathbf{Y}'_+ : \|y'\| \leq 1\}$ and the function $|K|$ are compatible. Thus, by Theorem 2, $w_{\mathbf{Y}}(K)$ is μ-measurable and so we can certainly find a sequence $\{y'_n\} \subset \mathbf{Y}'_+$, $\|y'_n\| \leq 1$, such that

$$\|K(.,t)\|_{\mathbf{Y}} = \sup_n\{\int_S |K(s,t)|\,y'_n(s)\,d\nu(s)\}\ \text{a.e.}\ (\mu).$$

It is now clear that $\mathbf{X}[\mathbf{Y}]$ is a linear manifold in $\mathbf{S}(R, \Sigma_R, \lambda)$ and so also an FS on (R, Σ_R, λ).

If $\mathbf{X}$ is a BFS, then we introduce a norm on $\mathbf{X}[\mathbf{Y}]$ by the formula $\|K\| = \|w_{\mathbf{Y}}(K)\|_{\mathbf{X}}$ $(K \in \mathbf{X}[\mathbf{Y}])$. This is clearly a monotone norm and so $\mathbf{X}[\mathbf{Y}]$ is an NFS. It can be proved that, under these conditions, $\mathbf{X}[\mathbf{Y}]$ is complete in its norm, but we shall not do this, as we shall not need this result (see Bukhvalov [2], [4], where other properties of the spaces $\mathbf{X}[\mathbf{Y}]$ are also given). The spaces $\mathbf{X}[\mathbf{Y}]$ are often called *mixed norm spaces.*

In the case where $\mathbf{X}$ is a BFS on (T, Σ_T, μ) satisfying condition (C) and $\mathbf{Y}$ is an FS on (S, Σ_S, ν), we can, by interchanging $\mathbf{X}$ and $\mathbf{Y}$, define the space $\mathbf{Y}[\mathbf{X}]$ and the function $w_{\mathbf{X}}(K)$, by analogy with the above.

Important examples of mixed norm spaces are provided by the spaces $\mathbf{L}^{p_1, p_2}$, which are generalizations of the spaces $\mathbf{L}^p$. Let $1 \leqslant p_1, p_2 \leqslant \infty$; and let $\mathbf{L}^{p_1} = \mathbf{L}^{p_1}(S, \Sigma_S, \nu)$, $\mathbf{L}^{p_2} = \mathbf{L}^{p_2}(T, \Sigma_T, \mu)$. The space $\mathbf{L}^{p_1, p_2}$ is defined to be the mixed norm space $\mathbf{L}^{p_2}[\mathbf{L}^{p_1}]$; thus the norm in $\mathbf{L}^{p_1, p_2}$ is given by

$$\|K\| = \begin{cases} \left(\int_T \left(\int_S |K(s,t)|^{p_1} d\nu(s) \right)^{p_2/p_1} d\mu(t) \right)^{1/p_2}, & 1 \leqslant p_1, p_2 < \infty, \\ \operatorname*{vrai\,sup}_{t \in T} \left(\int_S |K(s,t)|^{p_1} d\nu(s) \right)^{1/p_1}, & 1 \leqslant p_1 < \infty, p_2 = \infty, \\ \left(\int_T (\operatorname*{vrai\,sup}_{s \in S} |K(s,t)|)^{p_2} d\mu(t) \right)^{1/p_2}, & p_1 = \infty, 1 \leqslant p_2 < \infty, \\ \operatorname*{vrai\,sup}_{(s,t) \in R} |K(s,t)|, & p_1 = p_2 = \infty. \end{cases}$$

Using Fubini's Theorem, it is easy to see that, if $p_1 = p_2 = p$, then $\mathbf{L}^{p_1, p_2} = \mathbf{L}^{p_1}[\mathbf{L}^{p_2}] = \mathbf{L}^p(R, \Sigma_R, \lambda)$. Note that the analogous property may not hold for other BFSs; that is, if $\mathbf{X} \neq \mathbf{L}^p(T, \Sigma_T, \mu)$, and $\mathbf{Y} \neq \mathbf{L}^p(S, \Sigma_S, \nu)$, then the spaces $\mathbf{X}[\mathbf{Y}]$ and $\mathbf{Y}[\mathbf{X}]$ are, as a rule, distinct.

LEMMA 2. *If* $\mathbf{X}$ *and* $\mathbf{Y}$ *are* BFSs *satisfying condition* (A), *then*

1) $\mathbf{X}[\mathbf{Y}]$ *satisfies condition* (A);

2) $\mathbf{X}[\mathbf{Y}]$ *contains an everywhere dense subset* $\mathbf{H}_1$ *of all functions of the form*

$$\sum_{i=1}^{i_0} \chi_{B_i}(s) x_i(t),$$

where the sets $B_i \in \Sigma_S(\nu)$ *are pairwise disjoint,* $\chi_{B_i} \in \mathbf{Y}$, $x_i \in \mathbf{X} \cap \mathbf{L}^\infty(T, \Sigma_T, \mu)$ $(1 \leqslant i \leqslant i_0)$.

Proof. 1) If $K_n \downarrow \mathbf{0}$ in $\mathbf{X}[\mathbf{Y}]$, then for almost all $(\mu)t$ we have $K_n(s,t) \downarrow 0$ for almost all $(\nu)s$. Since $\mathbf{Y}$ satisfies condition (A), $w_{\mathbf{Y}}(K_n) \downarrow \mathbf{0}$, and, since $\mathbf{X}$ also satisfies condition (A), $\|K_n\| = \|w_{\mathbf{Y}}(K_n)\|_{\mathbf{X}} \to 0$.

2) Since $\mathbf{X}[\mathbf{Y}]$ satisfies (A), the set of all functions $K \in \mathbf{X}[\mathbf{Y}]$, each mapping the whole of some set $B \times A$ to zero, where $B \in \Sigma_S(\nu)$, $A \in \Sigma_T(\mu)$, $\chi_B \in \mathbf{Y}$, $\chi_A \in \mathbf{X}$, is everywhere dense in $\mathbf{X}[\mathbf{Y}]$. Therefore, without any loss of generality, we may assume that $\mu(T)$, $\nu(S) < \infty$ and $\mathbf{L}^\infty(T, \Sigma_T, \mu) \subset \mathbf{X}$, $\mathbf{L}^\infty(S, \Sigma_S, \nu) \subset \mathbf{Y}$. By Lemma IV.3.3, the set of all functions of the form

$$\sum_{i=1}^{i_0} \lambda_i \chi_{C_i} \qquad (C_i \in \Sigma_R)$$

is dense in $\mathbf{X}[\mathbf{Y}]$. It is now clear that to prove the Lemma it is enough to approximate an arbitrary function $\chi_C, C \in \Sigma_R$, by functions in H_1. By I.6.8, there exists a sequence of sets $C_n = \bigcup_{k=1}^{k(n)} B_k \times A_k$, where $B_k \in \Sigma_S$, $A_k \in \Sigma_T$, such that $\chi_{C_n} \to \chi_C$ in $\mathbf{L}^1(R, \Sigma_R, \lambda)$. By passing to a subsequence if necessary, we may assume that $\chi_{C_n}(s, t) \to \chi_C(s, t)$ a.e. (λ). Since $\mathbf{L}^\infty(R, \Sigma_R, \lambda) \subset \mathbf{X}[\mathbf{Y}]$ and $\mathbf{X}[\mathbf{Y}]$ satisfies (A), it follows that $\chi_{C_n} \to \chi_C$ in norm in $\mathbf{X}[\mathbf{Y}]$. Hence the set of all functions of the form

$$L(s, t) = \sum_{i=1}^{i_0} \lambda_i \chi_{B_i'}(s) \chi_{A_i}(t)$$

is dense in $\mathbf{X}[\mathbf{Y}]$. Let us show that $L \in H_1$. There exist pairwise disjoint sets $B_j \in \Sigma_S$ $(1 \leqslant j \leqslant j_0)$ such that every B_j is contained in at least one set $B_1', \ldots, B_{i_0}'$, and every B_i' is a union of sets B_j $(1 \leqslant j \leqslant j_0)$. Consider the index set $I_j = \{p\colon B_j \subset B_p',\ 1 \leqslant p \leqslant i_0\}$. Clearly

$$L(s, t) = \sum_{j=1}^{j_0} \chi_{B_j}(s) \left(\sum_{p \in I_j} \lambda_p \chi_{A_p}(t) \right) = \sum_{j=1}^{j_0} \chi_{B_j}(s) x_j(t),$$

and

$$x_j = \sum_{p \in I_j} \lambda_p \chi_{A_p} \in \mathbf{X} \cap \mathbf{L}^\infty(T, \Sigma_T, \mu), \qquad 1 \leqslant j \leqslant j_0.$$

This completes the proof of the lemma.

As we shall see below, mixed norm spaces play an important role for integral representations of operators.

1.5. Let $\mathbf{X}$ be a BFS on (T, Σ_T, μ) and $\mathbf{Y}$ a BFS on (S, Σ_S, ν).

A linear operator $U\colon \mathbf{X} \to \mathbf{S}(S, \Sigma_S, \nu)$ is said to be an *operator with abstract norm* if it maps the unit ball $B_{\mathbf{X}}$ in $\mathbf{X}$ into a set that is bounded under the ordering of $\mathbf{S}(S, \Sigma_S, \nu)$. The ordered space $\mathbf{S}(S, \Sigma_S, \nu)$ contains an element

$$|U| = \sup\{|U(x)|\colon x \in B_{\mathbf{X}}\},$$

which is called the *abstract norm* of U. We denote the set of all operators with abstract norm by $L_A(\mathbf{X}, \mathbf{S})$. Note that $U \in B(\mathbf{X}, \mathbf{L}^\infty)$ if and only if $U \in L_A(\mathbf{X}, \mathbf{S})$ and $|U| \in \mathbf{L}^\infty$; and then $\|U\| = \| |U| \|_{\mathbf{L}^\infty}$. Thus the class of operators with abstract norm is a generalization of the class $B(\mathbf{X}, \mathbf{L}^\infty)$.

We recall that condition (C) is always satisfied in the BFS $\mathbf{X}'$.

THEOREM 3. *If U is an integral operator* (1), *then $U \in L_A(\mathbf{X}, \mathbf{S})$ if and only if $K \in \mathbf{S}(S, \Sigma_S, \nu)[\mathbf{X}']$; and in this case $|U| = w_{\mathbf{X}'}(K)$.*

Proof. If $K \in \mathbf{S}(S, \Sigma_S, \nu)[\mathbf{X}']$, then, for $x \in \mathbf{X}$,

$$|[U(x)](s)| \leqslant \int_T |K(s, t)|\, |x(t)|\, d\mu(t) \leqslant \|K(s, \cdot)\|_{\mathbf{X}'} \|x\|_{\mathbf{X}}$$

for almost all $(\nu)s$, so that $|U| \leqslant w_{\mathbf{X}'}(K)$.

To prove the converse and the opposite inequality, we need the following lemma.

LEMMA 3. *If the integral operator U belongs to $L_A(\mathbf{X}, \mathbf{S})$, then the operator $|U|$ (see (2)) is also an operator with abstract norm; and we then have $|U| = ||U||$.*

Proof. Let $U \in L_A(\mathbf{X}, \mathbf{S})$. By the Corollary to Theorem 2, there exists, for each element

$x \in \mathbf{X}_+$, $\|x\| \leqslant 1$, a sequence $\{x_n\}$ such that $|x_n| = x$ and

$$|U|(x)(s) = \sup_n \left| \int_T K(s,t) x_n(t)\, d\mu(t) \right|.$$

Thus

$$|U|(x) = \sup_n |U(x_n)| \leqslant \mathbf{|}U\mathbf{|},$$

and so $|U| \in L_A(\mathbf{X}, \mathbf{S})$ and $\mathbf{|}\,|U|\,\mathbf{|} \leqslant \mathbf{|}U\mathbf{|}$. The inequality $\mathbf{|}U\mathbf{|} \leqslant \mathbf{|}\,|U|\,\mathbf{|}$ is obvious.

Let us extend the proof of Theorem 3. Let $U \in L_A(\mathbf{X}, \mathbf{S})$. We extend the norm on $\mathbf{X}'$ to all of $\mathbf{S}(T, \Sigma_T, \mu)$, setting $\|x\|_{\mathbf{X}'} = \infty$ if $x \notin \mathbf{X}'$. Then, for every $s \in S$, we have

$$\|K(s,\cdot)\|_{\mathbf{X}'} = \sup \left\{ \int_T |K(s,t)| x(t) d\mu(t) \colon x \in \mathbf{X}_+,\ \|x\| \leqslant 1 \right\}. \tag{11}$$

Clearly $K(s,\cdot) \in \mathbf{X}'$ if and only if the right-hand side of (11) is finite. It is obvious that the set $E = \{x \in \mathbf{X}_+ \colon \|x\| \leqslant 1\}$ and the function $|K(s,t)|$ are compatible. By Theorem 2, there exists a sequence $\{x_n\} \subset \mathbf{X}_+$, $\|x_n\| \leqslant 1$, such that, for almost all $(\nu)s$,

$$\|K(s,\cdot)\|_{\mathbf{X}'} = \sup_n \int_T |K(s,t)| x_n(t)\, d\mu(t).$$

Therefore, using Lemma 3, we see that, for almost all $(\nu)s$,

$$\|K(s,\cdot)\|_{\mathbf{X}'} = \sup_n |U|(x_n)(s) \leqslant \mathbf{|}\,|U|\,\mathbf{|}(s) = \mathbf{|}U\mathbf{|}(s),$$

so that $K \in \mathbf{S}(S, \Sigma_S, \nu)[\mathbf{X}']$ and $w_{\mathbf{X}'}(K) \leqslant \mathbf{|}U\mathbf{|}$, which completes the proof.

It follows from Theorem 3 that an integral operator U is a Hilbert–Schmidt operator if and only if $U \in L_A(\mathbf{L}^2, \mathbf{S})$ and $\mathbf{|}U\mathbf{|} \in \mathbf{L}^2(S, \Sigma_S, \nu)$.

We now consider integral operators from $\mathbf{L}^1 = \mathbf{L}^1(T, \Sigma_T, \mu)$ into a BFS $\mathbf{Y}$. As we have already remarked (see VII in 1.1), every such operator is continuous.

LEMMA 4. *If an integral operator U belongs to $B(\mathbf{L}^1, \mathbf{Y})$, where $\mathbf{Y}$ is a* BFS *satisfying conditions* (B) *and* (C), *then the integral operator $|U|$ also maps $\mathbf{L}^1$ into $\mathbf{Y}$; that is, U is regular. Furthermore,* $\|U\| = \|\,|U|\,\|$.*

Proof. We fix an element $x \in \mathbf{L}^1$, $x \geqslant \mathbf{0}$, and consider the set E_x of all elements of the form

$$y = |U(x\chi_{A_1})| + \ldots + |U(x\chi_{A_n})|,$$

where $\{A_i\}_{i=1}^n$ is a partition of T. If $y_1, y_2 \in E_x$, then, taking a subpartition of the partitions corresponding to y_1 and y_2, we find a $y \in E_x$ such that $y \geqslant y_1 \vee y_2$. Therefore, E_x is a directed set. For each $y \in E_x$, we have

$$\|y\| \leqslant \sum_{i=1}^n \|U(x\chi_{A_i})\| \leqslant \|U\| \sum_{i=1}^n \|x\chi_{A_i}\|_{\mathbf{L}^1} = \|U\|\, \|x\|_{\mathbf{L}^1}. \tag{12}$$

Hence E_x is bounded in the norm of Y. By the Corollary to Theorem 2, there exists a

* In fact, every continuous operator from $\mathbf{L}^1$ into $\mathbf{Y}$ is regular in the sense of the definition in X.2.1. The proof given here ignores this fact.

sequence $\{x_n\}$ such that $|x_n| = x$ and

$$|U|(x)(s) = \sup_n \left| \int_T K(s, t)x_n(t)d\mu(t) \right| \qquad \text{a.e. } (\nu).$$

If $A_1^n = \{t : x_n(t) > 0\}$, $A_2^n = \{t : x_n(t) < 0\}$, $A_3^n = \{t : x(t) = 0\}$, then

$$|U(x_n)| \leqslant |U(x\chi_{A_1^n})| + |U(x\chi_{A_2^n})| + |U(x\chi_{A_3^n})| \in E_x.$$

Write $z_m = \sup_{n=1}^{m} |U(x_n)|$. Then $0 \leqslant z_m \uparrow$, and, as E_x is directed, the sequence $\{z_m\}$ is bounded in norm in $\mathbf{Y}$. As $\mathbf{Y}$ satisfies condition (B) $\sup z_m$ exists in $\mathbf{Y}$. But, by the choice of $\{x_n\}$, we have $z_m \uparrow |U|(x)$, so $|U|(x) \in \mathbf{Y}$. Since x was an arbitrary element, U is regular.

Since $\mathbf{Y}$ satisfies (C), we see that $\|z_m\| \uparrow \| |U|(x)\|$. It is clear from (12) that $\|z_m\| \leqslant \|U\| \, \|x\|_{\mathbf{L}^1}$ $(m \in \mathbb{N})$, so $\| \, |U| \, \| \leqslant \|U\|$. The reverse inequality is obvious. Hence the proof of the lemma is complete.

For brevity, let us write $\mathbf{L}^1 = \mathbf{L}^1(T, \Sigma_T, \mu)$, $\mathbf{L}^\infty = \mathbf{L}^\infty(T, \Sigma_T, \mu)$.

THEOREM 4. *Let* $\mathbf{Y}$ *be a* BFS *satisfying conditions* (B) *and* (C). *If* U *is an integral operator* (1), *then* $U \in B(\mathbf{L}^1, \mathbf{Y})$ *if and only if* $K \in \mathbf{L}^\infty[\mathbf{Y}]$; *and, in that case,* $\|U\| = \|K\|_{\mathbf{L}^\infty[\mathbf{Y}]} = \operatorname{vrai\,sup}_{t \in T} \|K(\cdot, t)\|_{\mathbf{Y}}$.

Proof. If $K \in \mathbf{L}^\infty[\mathbf{Y}]$, then, for all $x \in \mathbf{L}^1$ and $y' \in \mathbf{Y}'$, we have, by Tonelli's Theorem,

$$\int_S |[U(x)](s)y'(s)|d\nu(s) \leqslant \int_S \left\{ \int_T |K(s, t)||x(t)|d\mu(t) \right\} |y'(s)|d\nu(s) =$$

$$= \int_T \left\{ \int_S |K(s, t)||y'(s)|d\nu(s) \right\} |x(t)|d\mu(t) \leqslant \operatorname{vrai\,sup}_{t \in T} \|K(\cdot, t)_{\mathbf{Y}}\|y'\| \|x\|_{\mathbf{L}^1}$$

$$= \|K\|_{\mathbf{L}^\infty[\mathbf{Y}]} \|y'\| \|x\|_{\mathbf{L}^1}.$$

Since $\mathbf{Y}$ satisfies condition (C), Theorem VI.1.6 shows that $\|U(x)\| \leqslant \|K\| \, \|x\|$, so that $U \in B(\mathbf{L}^1, \mathbf{Y})$ and $\|U\| \leqslant \|K\|$.

Conversely, let $U \in B(\mathbf{L}^1, \mathbf{Y})$. Then the adjoint operator U^* belongs to $B(\mathbf{Y}^*, \mathbf{L}^\infty)$. If V is the restriction of U^* to $\mathbf{Y}_n^\sim$, then $V \in B(\mathbf{Y}_n^\sim, \mathbf{L}^\infty)$. Since $\mathbf{Y}_n^\sim$ and $\mathbf{Y}'$ are isomorphic, V can be regarded as an operator from $\mathbf{Y}'$ into $\mathbf{L}^\infty$. Thus, in the notation of VI.1.1, we have, for all $x \in \mathbf{X}$ and $y' \in \mathbf{Y}'$,

$$\int_S U(x)y' \, d\nu = f_{y'}(U(x)) = f_x(U^*(f_{y'})) = f_x(V(f_{y'})) = \int_T xV(y') \, d\mu.$$

Since U is regular by Lemma 4, we see, using property VIII of 1.1, that $V: \mathbf{Y}' \to \mathbf{L}^\infty$ is an integral operator with kernel $K^*(t, s) = K(s, t)$; that is, for each $y' \in \mathbf{Y}'$, we have

$$V(y')(t) = \int_S K^*(t, s)y'(s)d\nu(s).$$

Hence, by Theorem 3, $K^* \in \mathbf{L}^\infty[\mathbf{Y}'']$ and $|V| = w_{\mathbf{Y}''}(K^*)$. But $w_{\mathbf{Y}''}(K^*)(t) = \|K^*(t, \cdot)\|_{\mathbf{Y}''} = \|K(\cdot, t)\|_{\mathbf{Y}''} = w_{\mathbf{Y}''}(K)(t)$, so $K \in \mathbf{L}^\infty[\mathbf{Y}'']$ and $|V| = w_{\mathbf{Y}''}(K)$. By Theorem VI.1.7, $\mathbf{Y}$ and $\mathbf{Y}''$ are equal, both in their elements and in their norms, hence $K \in \mathbf{L}^\infty[\mathbf{Y}]$ and $|V| = w_{\mathbf{Y}}(K)$. Further,

$$\|U\| = \|U^*\| \geqslant \|V\| = \| \, |V| \, \|_{\mathbf{L}^\infty} = \|K\|_{\mathbf{L}^\infty[\mathbf{Y}]},$$

so that $\|U\| = \|K\|$.

1.6. Now we can at last prove some theorems about integral representations of operators. In the following theorem, let

$$\mathbf{S} = \mathbf{S}(S, \Sigma_S, \nu), \qquad \mathbf{L}^1 = \mathbf{L}^1(S, \Sigma_S, \nu), \qquad \mathbf{L}^\infty = \mathbf{L}^\infty(S, \Sigma_S, \nu).$$

THEOREM 5. *Let* $\mathbf{X}$ *be a* BFS *on* (T, Σ_T, μ) *satisfying condition* (A). *The following formula gives a general form for an operator* U *of the class* $L_A(\mathbf{X}, \mathbf{S})$:

$$U(x)(s) = \int_T K(s, t)x(t)\,d\mu(t), \qquad x \in \mathbf{X}, \tag{1}$$

where $K \in \mathbf{S}[\mathbf{X}']$. *Moreover,* $|U| = w_{\mathbf{X}'}(K)$.

Proof. Without loss of generality, we may assume that $|U|(s) > 0$ for all $s \in S$. Consider the operator V on $\mathbf{X}$ defined by

$$V(x)(s) = \frac{1}{|U|(s)}\,U(x)(s).$$

Since $U \in L_A(\mathbf{X}, \mathbf{S})$, it follows that $V \in B(\mathbf{X}, \mathbf{L}^\infty)$. If we show that

$$V(x)(s) = \int_T K_1(s, t)x(t)d\mu(t), \qquad x \in \mathbf{X},$$

then, writing $K(s, t) = K_1(s, t)|U|(s)$, we shall have

$$U(x)(s) = |U|(s)[V(x)](s) = \int_T K_1(s, t)|U|(s)x(t)\,d\mu(t) = \int_T K(s, t)x(t)\,d\mu(t).$$

Hence if we prove the existence of an integral representation for each operator $V \in B(\mathbf{X}, \mathbf{L}^\infty)$, then, since the formula $|U| = w_{\mathbf{X}'}(K)$ follows from Theorem 3, we shall obtain the result of the theorem.

Thus we can assume that $U \in B(\mathbf{X}, \mathbf{L}^\infty)$. Consider the set H_2 of all functions of the form

$$L(s, t) = \sum_{i=1}^{i_0} \chi_{B_i}(s)x_i(t),$$

where the sets $B_i \in \Sigma_S(\nu)$ are pairwise disjoint and $x_i \in \mathbf{X}\ (1 \leqslant i \leqslant i_0)$. Clearly H_2 is a linear manifold. By Lemma 2, H_2 is dense in $\mathbf{L}^1[\mathbf{X}]$. Define a linear functional ϕ on H_2 by

$$\phi(L) = \sum_{i=1}^{i_0} \int_S U(x_i)(s)\chi_{B_i}(s)d\nu(s).$$

We leave it to the reader to check that this is a reasonable definition. As the sets B_i are pairwise disjoint, we have

$$|\phi(L)| \leqslant \sum_{i=1}^{i_0} \int_{B_i} |U(x_i)|d\nu \leqslant \sum_{i=1}^{i_0} \nu(B_i)\|U\|\,\|x_i\| = \|U\|\,\|L\|_{\mathbf{L}^1[\mathbf{X}]},$$

so ϕ is a continuous linear functional on H_2, if we regard H_2 as equipped with norm induced from $\mathbf{L}^1[\mathbf{X}]$. Hence there exists a unique continuous extension of ϕ to the space $\mathbf{L}^1[\mathbf{X}]$, which we shall denote by the same symbol. As $\mathbf{L}^1[\mathbf{X}]$ satisfies condition (A), by Lemma 2, we see from Corollary 2 to Theorem VI.1.4 that there exists a λ-measurable function $K(s, t)$ such that

$$\phi(L) = \int_R L(s, t)K(s, t)d\lambda(s, t), \qquad L \in \mathbf{L}^1[\mathbf{X}].$$

In particular, for $L(s, t) = \chi_B(s)x(t)$, $B \in \Sigma_S(v)$, $x \in \mathbf{X}$, we have

$$\int_B U(x)dv = \phi(L) = \int_R K(s, t)\chi_B(s)x(t)d\lambda(s, t). \tag{13}$$

By Theorem VI.1.1, the function $|K(s, t)|$ also induces a continuous linear functional on $\mathbf{L}^1[\mathbf{X}]$, so, using Fubini's Theorem to rewrite the right-hand side of (13), we deduce that

$$\int_B U(x)dv = \int_B \left\{ \int_T K(s, t)x(t)d\mu(t) \right\} dv(s).$$

As B was arbitrary in $\Sigma_S(v)$, the representation (1) now follows from this. This completes the proof.

COROLLARY. *Let* $\mathbf{X}$ *be a* BFS *on* (T, Σ_T, μ) *satisfying condition* (A), *and let* $\mathbf{L}^\infty = \mathbf{L}^\infty(S, \Sigma_S, v)$. *Formula* (1) *gives the general form for an operator* U *in the class* $B(\mathbf{X}, \mathbf{L}^\infty)$, *where* $K \in \mathbf{L}^\infty[\mathbf{X}']$; *moreover*, $\|U\| = \operatorname{vrai\,sup}_{s \in S} \|K(s, .)\|_{\mathbf{X}'}$.

Theorem 5 enables one to obtain abstract characterizations of Hilbert–Schmidt operators and Carleman operators (see Korotkov [1]).

In the following theorem, let $\mathbf{L}^1 = \mathbf{L}^1(T, \Sigma_T, \mu)$, $\mathbf{L}^\infty = \mathbf{L}^\infty(T, \Sigma_T, \mu)$.

THEOREM 6. *Let* $\mathbf{Y}$ *be a* BFS *on* (S, Σ_S, v) *satisfying condition* (A). *The following formula gives a general form for an operator* U *in the class* $B(\mathbf{L}^1, \mathbf{Y}')$:

$$U(x)(s) = \int_T K(s, t)x(t)\,d\mu(t), \qquad x \in \mathbf{L}^1, \tag{1}$$

where $K \in \mathbf{L}^\infty[\mathbf{Y}']$. *Moreover*, $\|U\| = \operatorname{vrai\,sup}_{t \in T} \|K(., t)\|_{\mathbf{Y}'}$.

Proof. Since $\mathbf{Y}'$ satisfies conditions (B) and (C), Theorem 4 shows that we need only prove that every operator $U \in B(\mathbf{L}^1, \mathbf{Y}')$ has an integral representation (1). Consider the dual operator $U^* \in B((\mathbf{Y}')^*, \mathbf{L}^\infty)$. The space $\mathbf{Y}$ is canonically embedded in $(\mathbf{Y}')^*$, and by Theorem VI.1.6 this embedding preserves norms. Denote the restriction of U^* to $\mathbf{Y}$ by V. Then $V \in B(\mathbf{Y}, \mathbf{L}^\infty)$ and, by Theorem 5, there exists a kernel $K \in \mathbf{L}^\infty[\mathbf{Y}']$ such that

$$V(y)(t) = \int_S K(t, s)y(s)\,dv(s), \qquad y \in \mathbf{Y}.$$

By Lemma 3, V is a regular operator, and so, if we use VIII of 1.1 in a manner analogous to our use of it in the proof of Theorem 4, we see easily that U has a representation (1).

As an important special case, we have

COROLLARY. *Let* $\mathbf{L}^p = \mathbf{L}^p(S, \Sigma_S, v)$, $1 < p \leqslant \infty$. *The following formula gives a general form for an operator* U *in the class* $B(\mathbf{L}^1, \mathbf{L}^p)$:

$$U(x)(s) = \int_T K(s, t)x(t)\,d\mu(t), \qquad x \in \mathbf{L}^1,$$

where $K \in \mathbf{L}^{p, \infty}$. *Moreover*, $\|U\| = \operatorname{vrai\,sup}_{t \in T} \left(\int_S |K(s, t)|^p\,dv(s) \right)^{1/p}$ *if* $1 < p < \infty$, *and* $\|U\| = \operatorname{vrai\,sup}_{(s,t) \in R} |K(s, t)|$, *if* $p = \infty$.

REMARK. This Corollary cannot be extended to the case $p = 1$ since the identity operator on $\mathbf{L}^1(0, 1)$ has no integral representation (see 1.2).

Integral representations for operators attracted the attention of mathematicians long ago. The most important special cases of Theorems 5 and 6 were obtained in the 1930s (see Gel'fand [1], Dunford and Pettis [1], Kantorovich and Vulikh [1], and also

Kantorovich, Vulikh and Pinsker).* In this connection, the interval $[0, 1]$ with Lebesgue measure was used, as a rule, instead of an abstract measure space, and a suitable $\mathbf{L}^p$ space was taken instead of an abstract BFS.

The theorems in 1.5 and 1.6 were obtained, in their present form, in a paper by Bukhvalov [4]; and we have followed this paper in our exposition. Other theorems on integral representations for operators on ideal spaces can be found in the papers by Bukhvalov [3], [4] and Korotkov [2]. Much work has been devoted to analytic representation of operators by means of vector functions (see, for example, Dinculeanu, Diestel, Phillips [1], Grothendieck [2], Korotkov [2], Bukhvalov [5]).

§ 2. Operators on sequence spaces

2.1. We now give a description of the continuous—and, in particular, the compact—linear operators mapping ℓ^p into ℓ^r $(1 < p < \infty, 1 \leqslant r < \infty)$.

Let U be a continuous linear operator from ℓ^p into ℓ^r:

$$y = U(x) \qquad (x = \{\xi_k\} \in \ell^p, \quad y = \{\eta_i\} \in \ell^r).$$

Write $f_i(x) = \eta_i$ $(i = 1, 2, \dots)$. As

$$|f_i(x)| = |\eta_i| \leqslant \|y\| \leqslant \|U\| \, \|x\|,$$

it is clear that f_i is a continuous linear functional on ℓ^p. Also $\|f_i\| \leqslant \|U\|$. By the theorem on the general form of a functional on ℓ^p, the functional f_i can be expressed in the form

$$\eta_i = f_i(x) = \sum_{k=1}^{\infty} a_{ik}\xi_k$$

$$\left(z_i = \{a_{ik}\} \in \ell^q; \ \frac{1}{p} + \frac{1}{q} = 1; \ \|z_i\| = \|f_i\|; \quad i = 1, 2, \dots\right). \tag{1}$$

Thus U is determined, in a natural way, by the matrix

$$A = \begin{pmatrix} a_{11} & a_{12} & \dots & a_{1k} & \dots \\ a_{21} & a_{22} & \dots & a_{2k} & \dots \\ \dots & \dots & \dots & \dots & \dots \\ a_{i1} & a_{i2} & \dots & a_{ik} & \dots \\ \dots & \dots & \dots & \dots & \dots \end{pmatrix}, \tag{2}$$

$$a_{ik} = f_i(x_k) \quad (x_k = \underset{1,\dots,}{0,\dots,} \ \underset{k-1}{0,} \ \underset{k,}{1,} \ \underset{k+1,\dots}{0,\dots;} \ i, k = 1, 2, \dots).$$

Let $x = \{\xi_k\} \in \ell^p$, $y = \{\eta_k\} \in \ell^r$. We recall the notation for "truncated" elements

$$[x]_n = (\xi_1, \xi_2, \dots, \xi_n, 0, \dots), \qquad [y]_n = (\eta_1, \eta_2, \dots, \eta_n, 0, \dots)$$

and consider, along with U, the operators U_n and U_{nm} defined by

$$U_n(x) = [U(x)]_n, \qquad U_{nm}(x) = U_n([x]_m) \qquad (x \in \ell^p).$$

---

* (*Editor's note*). Theorem 5 is often called the Kantorovich–Vulikh Theorem, while Theorem 6 is often called the Dunford–Pettis Theorem.

We also write

$$[a_{ik}]_n = \begin{cases} a_{ik} & (i \leqslant n), \\ 0 & (i > n), \end{cases} \qquad [a_{ik}]_{mn} = \begin{cases} a_{ik} & (i \leqslant n,\ k \leqslant m), \\ 0 & (i > n \text{ or } k > m) \end{cases}$$

and introduce the matrices

$$[A]_n = \begin{pmatrix} [a_{11}]_n & [a_{12}]_n & \dots & [a_{1k}]_n & \dots \\ \dots & \dots & \dots & \dots & \dots \\ [a_{i1}]_n & [a_{i2}]_n & \dots & [a_{ik}]_n & \dots \\ \dots & \dots & \dots & \dots & \dots \end{pmatrix} = \begin{pmatrix} a_{11} & a_{12} & \dots & a_{1k} & \dots \\ \dots & \dots & \dots & \dots & \dots \\ a_{n1} & a_{n2} & \dots & a_{nk} & \dots \\ 0 & 0 & \dots & 0 & \dots \\ \dots & \dots & \dots & \dots & \dots \end{pmatrix},$$

$$[A]_{nm} = \begin{pmatrix} [a_{11}]_{nm} & [a_{12}]_{nm} & \dots & [a_{1k}]_{nm} & \dots \\ \dots & \dots & \dots & \dots & \dots \\ [a_{i1}]_{nm} & [a_{i2}]_{nm} & \dots & [a_{ik}]_{nm} & \dots \\ \dots & \dots & \dots & \dots & \dots \end{pmatrix} = \begin{pmatrix} a_{11} & a_{12} & \dots & a_{1m} & 0 & \dots \\ \dots & \dots & \dots & \dots & \dots & \dots \\ a_{n1} & a_{n2} & \dots & a_{nm} & 0 & \dots \\ \dots & \dots & \dots & \dots & \dots & \dots \\ 0 & 0 & \dots & 0 & 0 & \dots \end{pmatrix}.$$

It is not hard to verify that these matrices correspond to precisely the operators U_n and U_{nm}. In fact, keeping the previous notation for the elements x and $y = U(x)$ and, in addition, setting $U_n(x) = \{[\eta_k]_n\}$, $U_{nm}(x) = \{[\eta_k]_{nm}\}$, we have

$$[\eta_i]_n = \begin{cases} \eta_i = \sum_{k=1}^{\infty} a_{ik}\xi_k & (i \leqslant n), \\ 0 & (i > n), \end{cases}$$

that is,

$$[\eta_i]_n = \sum_{k=1}^{\infty} [a_{ik}]_n \xi_k \qquad (i = 1, 2, \dots).$$

Hence

$$[\eta_i]_{nm} = \sum_{k=1}^{m} [a_{ik}]_n \xi_k = \sum_{k=1}^{\infty} [a_{ik}]_{nm} \xi_k \qquad (i = 1, 2, \dots).$$

Let us prove that the sequences of operators $\{U_n\}$ and $\{U_{nm}\}$ converge to U at every element of ℓ^p.

Since $\lim_{n\to\infty} [y]_n = y$, we have

$$\lim_{n\to\infty} U_n(x) = \lim_{n\to\infty} [U(x)]_n = U(x) \qquad (x \in \ell^p).$$

Notice also that, in view of the obvious inequality $\|[y]_n\| \leqslant \|y\|$, we have $\|U_n(x)\| \leqslant \|U(x)\| \leqslant \|U\|\ \|x\|$; that is,

$$\|U_n\| \leqslant \|U\|. \tag{3}$$

Choose N, M such that

$$\|[x]_m - x\| < \varepsilon, \qquad \|U_n(x) - U(x)\| < \varepsilon \quad \text{when} \quad m \geqslant M,\ n \geqslant N.$$

Then, if $m \geqslant M$, $n \geqslant N$, we have

$$\|U(x)-U_{nm}(x)\| \leqslant \|U(x)-U_n(x)\| + \|U_n(x)-U_n([x]_m)\| < \varepsilon + \|U_n\|\,\|x-[x]_m\| \leqslant \varepsilon + \\ + \|U\|\,\varepsilon.$$

Hence

$$\lim_{n,\, m\to\infty} U_{nm}(x) = U(x) \qquad (x \in \ell^p).$$

In the case where U is a compact operator, both U_n and U_{nm} converge in norm to U in the space of linear operators.

THEOREM 1. *A necessary and sufficient condition for an operator U to be compact is that $U_n \to U$ as $n \to \infty$ or $U_{nm} \to U$ as $m, n \to \infty$; that is, that either of the following relations should hold:*

$$\lim_{n\to\infty} \|U_n - U\| = 0 \qquad \textit{or} \qquad \lim_{n,\, m\to\infty} \|U_{nm} - U\| = 0. \tag{4}$$

Proof. Sufficiency. The operators U_n and U_{nm} are compact, since they map ℓ^p into a finite-dimensional space (of dimension n). And so, by Theorem IX.2.3, the operator $U = \lim U_n = \lim U_{nm}$ is also compact.

Necessity. Denote the set of values $U(x)$, for $\|x\| \leqslant 1$, by Γ. Since U is compact, Γ is a relatively compact set. We now consider the operators V_n, defined by

$$V_n(y) = [y]_n \qquad (y \in \ell^r;\ n = 1, 2, \ldots).$$

We have

$$\lim_{n\to\infty} V_n(y) = y \qquad (y \in \ell^r). \tag{5}$$

Hence Gel'fand's Theorem (Theorem IX.1.3) shows that the convergence in (5) is uniform on the relatively compact set Γ; that is, for all $y \in \Gamma$ and for sufficiently large n $(n \geqslant N)$,

$$\|[y]_n - y\| < \varepsilon$$

or, what is the same thing,

$$\|U_n(x) - U(x)\| < \varepsilon \qquad (\|x\| \leqslant 1,\ n \geqslant N),$$

which means that

$$\|U_n - U\| < \varepsilon \qquad (n \geqslant N). \tag{6}$$

Thus the first of the relations (4) is proved.

To establish the second, we start from the fact that

$$U_N(x) = (f_1(x), f_2(x), \ldots, f_N(x), 0, \ldots).$$

If we set

$$f_{im}(x) = f_i([x]_m) = \sum_{k=1}^{m} a_{ik}\xi_k \qquad (x = \{\xi_k\} \in \ell^p),$$

then, for large enough m $(m \geqslant M)$, we have

$$\|f_i - f_{im}\| = \left\{ \sum_{k=m+1}^{\infty} |a_{ik}|^q \right\}^{1/q} < \frac{\varepsilon}{N} \qquad (i = 1, 2, \ldots, N).$$

In this case,

$$\|U_N(x)-U_{Nm}(x)\| = \|(f_1(x)-f_{1m}(x), f_2(x)-f_{2m}(x), \ldots, f_N(x)-f_{Nm}(x), 0, \ldots)\| \leqslant$$
$$\leqslant \frac{\varepsilon}{N}\|x\|\,\|(1, 1, \ldots, 1, 0, \ldots)\| = \frac{\varepsilon}{N}\|x\|\, N^{1/r} \leqslant \varepsilon\|x\|.$$

Finally, from this and the inequality $\|U_n - U_N\| \leqslant 2\varepsilon$ $(n \geqslant N)$, which follows from (6), we see that

$$\|U_n(x)-U_{nm}(x)\| = \|U_n(x)-U_n([x]_m)\| \leqslant \|U_n(x)-U_N(x)\| + \|U_n([x]_m)-U_N([x]_m)\| +$$
$$+\|U_N(x)-U_N([x]_m)\| \leqslant 2\varepsilon\|x\| + 2\varepsilon\|[x]_m\| + \varepsilon\|x\| \leqslant 5\varepsilon\|x\|.$$

Since, in addition, $\|U_n - U\| \leqslant \varepsilon$, we have

$$\|U_{nm} - U\| \leqslant 6\varepsilon \qquad (n \geqslant N, m \geqslant M),$$

and so the theorem is proved.

REMARK. The statement of this theorem can be expressed in another form. First of all, instead of the operators U_{nm} we can restrict ourselves to operators of the form U_{nn} and replace convergence to U by the Cauchy convergence criterion. Thus we can say that a necessary and sufficient condition for U to be compact is that, for every $\varepsilon > 0$,

$$\|U_{nn} - U_{pp}\| < \varepsilon \qquad (n, p \geqslant N_\varepsilon).$$

Now we note that

$$\|U_{nn}\| = \sup_{\substack{\|x\| \leqslant 1 \\ \|x'\| \leqslant 1}} \Big|\sum_{i,k=1}^{n} a_{ik}\xi_i\xi_k'\Big| \tag{7}$$
$$\left(x = \{\xi_k\} \in \ell^p, \qquad x' = \{\xi_k'\} \in \ell^s, \qquad \frac{1}{r} + \frac{1}{s} = 1\right).$$

For let $\mathbf{X}$ be a B-space. Denote by F_x the functional on $\mathbf{X}^*$ determined by the element $x \in \mathbf{X}$ (see V.7.3). Then

$$\|x\| = \|F_x\| = \sup_{\|f\| \leqslant 1} |F_x(f)| = \sup_{\|f\| \leqslant 1} |f(x)| \qquad (f \in \mathbf{X}^*).$$

If U maps $\mathbf{X}$ into $\mathbf{Y}$, then

$$\|U\| = \sup_{\|x\| \leqslant 1} |U(x)| = \sup_{\|x\| \leqslant 1} \sup_{\|g\| \leqslant 1} |g(U(x))| \qquad (x \in \mathbf{X}, g \in \mathbf{Y}^*). \tag{8}$$

If we take $\mathbf{X}$ to be ℓ^p, $\mathbf{Y}$ to be ℓ^r, and U to be U_{nn}, then we have

$$g(U(x)) = \sum_{i,k=1}^{n} a_{ik}\xi_i\xi_k' \qquad (x = \{\xi_k\} \in \ell^p), \tag{9}$$

where the sequence $x' = \{\xi_k'\} \in \ell^s$ determines the functional g ($\|g\| = \|x'\|$). Comparing (8) and (9), we obtain (7).

By applying (7) to the difference $U_{nn} - U_{pp}$ (assuming $n \geqslant p$), we can write the latter condition in the form

$$\left|\sum_{i,k=1}^{n} a_{ik}\xi_i\xi_k' - \sum_{i,k=1}^{p} a_{ik}\xi_i\xi_k'\right| < \varepsilon \qquad (p \geqslant N)$$

for $\sum_{k=1}^{\infty} |\xi_k|^p \leqslant 1$, $\sum_{k=1}^{\infty} |\xi_k'|^s \leqslant 1$. In other words, we can also formulate the theorem as follows:

A necessary and sufficient condition for U to be compact is that the sequence of bilinear forms corresponding to the "truncated" matrices $[A]_{nn}$ be a uniformly convergent Cauchy sequence on the unit balls in ℓ^p and ℓ^s.

This proposition was in essence established by Hilbert.

2.2. Above we started from a given linear operator U and associated with it the matrix (2). It is of some interest to clarify the conditions under which a preassigned matrix (2) is the matrix corresponding to some continuous or compact operator.

With the aid of the "truncated" matrices we can form operators U_{nm} by setting

$$y = U_{nm}(x) \qquad (x = \{\xi_k\} \in \ell^p, \quad y = \{\eta_k\} \in \ell^r),$$

$$\eta_i = \begin{cases} \sum_{k=1}^{m} a_{ik}\xi_k & (i \leqslant n), \\ 0 & (i > n). \end{cases}$$

THEOREM 2.* *A necessary and sufficient condition for the matrix* (2) *to be the matrix of some continuous linear operator U from ℓ^p into ℓ^r is that the operators U_{nn} (or U_{nm}) determined by the matrix be bounded in norm:*

$$\|U_{nn}\| \leqslant K \qquad (n = 1, 2, \ldots) \qquad [or \qquad \|U_{nm}\| \leqslant K \quad (n, m = 1, 2, \ldots)]. \tag{10}$$

A necessary and sufficient condition for U to be compact is that the operators U_{nn} form a Cauchy sequence:

$$\lim_{n,\, p \to \infty} \|U_{nn} - U_{pp}\| = 0. \tag{11}$$

Proof. Necessity. In the case where the operator U exists, we have $U_{nn} \to U$ on ℓ^p, as we showed above. As ℓ^p and ℓ^r are complete, the norms of the U_{nn} are bounded in aggregate (the Banach–Steinhaus Theorem; VII.1.2).

When the operator is compact, Theorem 1 shows that $\lim_{n \to \infty} U_{nn} = U$, so (11) follows immediately.

Sufficiency. We first consider the behaviour of the operators U_{nn} on truncated elements. Suppose we are given an element $x \in \ell^p$ such that $[x]_m = x$. Taking $n > m$, we have

$$U_{nn}(x) = (\eta_1, \eta_2, \ldots, \eta_n, 0, \ldots)$$

$$\left(\eta_i = \sum_{k=1}^{m} a_{ik}\xi_k;\ x = (\xi_1, \xi_2, \ldots, \xi_m, 0, \ldots);\quad i = 1, 2, \ldots, n\right).$$

Moreover, (10) shows that

$$\|U_{nn}(x)\| = \left[\sum_{i=1}^{n} |\eta_i|^r\right]^{1/r} \leqslant K\|x\|.$$

In this situation, $\{\eta_i\} \in \ell^r$ and this implies the existence of

$$y = \lim_{n \to \infty} U_{nn}(x) = \{\eta_i\}.$$

Thus the sequence $\{U_{nn}\}$ converges on the dense subset of ℓ^p formed by the truncated elements, and since the sequence of norms of these operators is bounded, the convergence

* This theorem was proved for ℓ^2 by Hilbert (see Hilbert).

must take place on the whole of ℓ^p. Let

$$U(x) = \lim_{n\to\infty} U_{nn}(x) \qquad (x\in\ell^p).$$

Since, for any $x\in\ell^p$,

$$U([x]_m) = \lim_{n\to\infty} U_{nn}([x]_m) = (\eta_1^{(m)}, \eta_2^{(m)}, \ldots, \eta_i^{(m)}, \ldots),$$

$$\eta_i^{(m)} = \sum_{k=1}^{m} a_{ik}\xi_k \qquad (x = \{\xi_k\};\quad i = 1, 2, \ldots),$$

we see by letting $m\to\infty$ that

$$U(x) = \lim_{m\to\infty} U([x]_m) = (\eta_1, \eta_2, \ldots, \eta_i, \ldots),$$

where

$$\eta_i = \lim_{m\to\infty} \eta_i^{(m)} = \sum_{k=1}^{\infty} a_{ik}\xi_k \qquad (i = 1, 2, \ldots),$$

that is, U corresponds to the matrix (2).

If the condition (11) is satisfied, then obviously

$$U = \lim_{n\to\infty} U_{nn},$$

and, since the U_{nn} are compact, U will also be compact.

This completes the proof.

Let us present some corollaries.

COROLLARY 1. *A necessary and sufficient condition for a symmetric matrix A to be the matrix corresponding to a continuous linear operator from ℓ^2 into ℓ^2 is that*

$$\sup_n |\lambda_1^{(n)}| < \infty, \tag{12}$$

where $\lambda_1^{(n)}$ is the eigenvalue of the matrix

$$\begin{pmatrix} a_{11} & a_{12} & \ldots & a_{1n} \\ a_{21} & a_{22} & \ldots & a_{2n} \\ \ldots & \ldots & \ldots & \ldots \\ a_{n1} & a_{n2} & \ldots & a_{nn} \end{pmatrix}$$

having the greatest modulus.

For U_{nn} is essentially an operator into the n-dimensional space formed by those elements with all coordinates from the $(n+1)$-st onwards equal to zero. Hence

$$\|U_{nn}\| = |\lambda_1^{(n)}|.$$

Thus (12) implies that $\sup_n \|U_{nn}\| < \infty$, which coincides with (10).

COROLLARY 2. *If the matrix* (2) *satisfies the condition*

$$B = \left\{\sum_{i=1}^{\infty}\left[\sum_{k=1}^{\infty} |a_{ik}|^q\right]^{r/q}\right\}^{1/r} < \infty \qquad \left(\frac{1}{p}+\frac{1}{q}=1\right), \tag{13}$$

then it is the matrix corresponding to a compact operator U, and

$$\|U\| \leqslant B.$$

For let us find a bound for the norm of U_{nn}. We have

$$\|U_{nn}(x)\|^r = \sum_{i=1}^{n} \left| \sum_{k=1}^{n} a_{ik}\xi_k \right|^r \leqslant \sum_{i=1}^{n} \left[\sum_{k=1}^{n} |a_{ik}|^q \right]^{r/q} \left[\sum_{k=1}^{n} |\xi_k|^p \right]^{r/p} \leqslant B^r \|x\|^r.$$

Therefore $\|U_{nn}\| \leqslant B$; that is, (10) is satisfied. Hence A is the matrix corresponding to U, and $\|U\| \leqslant B$.

Taking U_n to be the operator defined in 2.1, we apply the bound just determined for the norm U to $U - U_n$:

$$\|U - U_n\| \leqslant \left\{ \sum_{i=n+1}^{\infty} \left[\sum_{k=1}^{\infty} |a_{ik}|^q \right]^{r/q} \right\}^{1/r},$$

and from this it is clear that $\lim_{n \to \infty} \|U - U_n\| = 0$, so U is compact.

REMARK 1. In the case where $p = r = 2$, the condition (13) takes the form

$$B = \left\{ \sum_{i=1}^{\infty} \sum_{k=1}^{\infty} |a_{ik}|^2 \right\}^{1/2} < \infty.$$

REMARK 2. If $p = 1$, then the condition (13) must be replaced by

$$B = \left\{ \sum_{i=1}^{\infty} \sup_k |a_{ik}|^r \right\}^{1/r} < \infty. \tag{14}$$

The proof is carried out similarly.

§ 3. Integral operators on function spaces

In this section we shall determine conditions under which an integral operator U

$$y = U(x),\ y(s) = \int_D K(s, t)x(t)\,dt \qquad (s \in D') \tag{1}$$

is a continuous (or compact) operator, mapping one function space $\mathbf{X}$ into another one $\mathbf{Y}$. Most of the results here will deal with the case $\mathbf{X} = \mathbf{L}^p(D)$, $\mathbf{Y} = \mathbf{L}^q(D')$ $(1 \leqslant p, q \leqslant \infty)$, where D and D' are bounded regions in μ-dimensional and ν-dimensional Euclidean space, respectively.* Without losing any generality, we shall assume that mes $D = 1$.

The function $K(s, t)$—the kernel of U—will always be assumed measurable in the region $D' \times D$ lying in $(\mu + \nu)$-dimensional space, so that $K(s, t)$ will be measurable with respect to t for almost all $s \in D'$ and measurable with respect to s for almost all $t \in D$.

The facts established in this section find a most important application in § 4, in the presentation of Sobolev's embedding theorems, which in turn play a major role in the theory of differential equations of mathematical physics.

The majority of the results of this section are generalizations of S. L. Sobolev's theorems

* In §§ 3 and 4, owing to the large number of indices, we shall denote finite-dimensional spaces by $\mathbf{R}^\mu$ or $\mathbf{R}^\nu$, and not $\mathbf{R}^m$ or $\mathbf{R}^n$ as in the rest of the book.

on integrals of potential type. The exposition, both here and in §4, follows a paper by L. V. Kantorovich [11].

3.1. THEOREM 1. *Assume that the following conditions are satisfied:*

1)
$$\left[\int_D |K(s,t)^r|\,dt\right]^{1/r} \leqslant C_1 \qquad (r > 0) \tag{2}$$

for almost all $s \in D'$;

2)
$$\left[\int_{D'} |K(s,t)|^\sigma\,ds\right]^{1/\sigma} \leqslant C_2 \qquad (\sigma > 0) \tag{3}$$

for almost all $t \in D'$;

3)
$$q \geqslant p,\ q \geqslant \sigma,\quad \left(1 - \frac{\sigma}{q}\right)p' \leqslant r \qquad (p, q \geqslant 1).* \tag{4}$$

Then the integral operator (1) *is a continuous linear operator from* $\mathbf{L}^p(D)$ *into* $\mathbf{L}^q(D')$, *and*†

$$\|U\| \leqslant C_1^{1-\sigma/q} C_2^{\sigma/q}. \tag{5}$$

Proof. We transform the following obvious bound for $|y(s)|$:

$$|y(s)| \leqslant \int_D |K(s,t)|\,|x(t)|\,dt = \int_D [|K(s,t)|^\sigma |x(t)|^p]^{1/q} |x(t)|^{p(1/p-1/q)} |K(s,t)|^{1-\sigma/q}\,dt.$$

Apply the generalized Hölder inequality (IV.2.4) to the last integral, using the exponents‡

$$\lambda_1 = q,\ \lambda_2 = \frac{1}{\frac{1}{p} - \frac{1}{q}},\ \lambda_3 = p' = \frac{p}{p-1}\ \left(\frac{1}{\lambda_1} + \frac{1}{\lambda_2} + \frac{1}{\lambda_3} = 1\right).$$

This yields

$$|y(s)| \leqslant \left[\int_D |K(s,t)|^\sigma |x(t)|^p\,dt\right]^{1/q} \left[\int_D |x(t)|^p\,dt\right]^{1/p-1/q} \times \left[\int_D |K(s,t)^{p'(1-\sigma/q)}\,dt\right]^{1-1/p}. \tag{6}$$

Writing

$$\lambda = p'\left(1 - \frac{\sigma}{q}\right),$$

we see from (2), remembering that $\operatorname{mes} D = 1$ and $\lambda \leqslant r$, by (4), that

$$\left[\int_D |K(s,t)|^\lambda\,dt\right]^{1-1/p} = \left[\int_D |K(s,t)|^\lambda\,dt\right]^{1/\lambda(1-\sigma/q)} \leqslant \left[\int_D |K(s,t)|^r\,dt\right]^{1/r(1-\sigma/q)} \leqslant C_1^{1-\sigma/q}.$$

Using this in (6), we obtain the following bound for $|y(s)|$:

$$|y(s)| \leqslant \left[\int_D |K(s,t)|^\sigma |x(t)|^p\,dt\right]^{1/q} \|x\|^{1-p/q} C_1^{1-\sigma/q}. \tag{7}$$

---

* Here, and in what follows, the prime denotes the "conjugate" exponent. This is defined by the equation $1/p + 1/p' = 1$.

† This theorem has been generalized by Kh. L. Smolitskii [1].

‡ We assume that $q > p > 1$. The case where $p = q$ is left to the reader. The case $p = 1$ is considered in Remark 3.

This inequality yields the following bound for $\|y\|$:

$$\|y\| = \left[\int_{D'} |y(s)|^q \, ds \right]^{1/q} \leqslant \left\{ \int_{D'} \int_{D} |K(s, t)|^\sigma |x(t)|^p \, dt \, ds \right\}^{1/q} \|x\|^{1 - p/q} C_1^{1 - \sigma/q} =$$

$$= C_1^{1 - \sigma/q} \left\{ \int_{D} |x(t)|^p \left(\int_{D} |K(s, t)|^\sigma \, ds \right) dt \right\}^{1/q} \|x\|^{1 - p/q} \leqslant C_1^{1 - \sigma/q} C_2^{\sigma/q} \|x\|,$$

which proves the theorem.

REMARK 1. The main conclusion of the theorem, that (1) is a continuous operator from $\mathbf{L}^p$ into $\mathbf{L}^q$, is still true in the case where the requirement that $q \geqslant \sigma$ does not hold. For if $q < \sigma$, then, replacing q by $q_1 = \sigma$, we retain the inequality (4). And in this situation U is a continuous operator from $\mathbf{L}^p$ into $\mathbf{L}^{q_1}$, and so *a fortiori* into $\mathbf{L}^q$ (since $q < q_1$).

Also, if the condition $q \geqslant p$ is not satisfied, then U will be continuous from $\mathbf{L}^p$ into $\mathbf{L}^q$, provided we have the inequality

$$p'\left(1 - \frac{\sigma}{p}\right) \leqslant r, \tag{8}$$

obtained by replacing q by $q_1 = p$ in (4).

Note that (8) will be satisfied if $\sigma \geqslant 1$, $r \geqslant 1$.

REMARK 2. The restriction mes $D = 1$ is obviously not essential. We can always arrange for it to be satisfied by applying a similarity transformation to the region D. Then a factor A, depending only on mes D and the exponents, may appear in some of the bounds.

REMARK 3. Let us note the limiting case of the theorem, when $p = 1$: that is, when we are concerned with an operator mapping $\mathbf{L}^1(D)$ into $\mathbf{L}^q(D')$. In this case we can simplify the hypotheses of the theorem. In fact, for U to be continuous from $\mathbf{L}^1(D)$ into $\mathbf{L}^q(D')$ it is sufficient that the following condition be satisfied

$$\left[\int_{D'} |K(s, t)|^q \, ds \right]^{1/q} \leqslant C_2 \tag{3'}$$

for almost all $t \in D$; in this situation, the norm of U is subject to the bound

$$\|U\| \leqslant C_2.$$

Thus conditions 1) and 3) are eliminated, while condition 2) must be satisfied for $\sigma = q$.

The proof of the theorem is also simplified in this case, since the last factor will be missing when we have to establish (6), so we need only use the ordinary Hölder inequality.

REMARK 4. Let us note another important case: when $q = \infty$. Here the theorem can be stated as follows: the integral operator (1) maps $\mathbf{L}^p(D)$ into $\mathbf{L}^\infty(D')$ if condition (2) is satisfied for some $r \geqslant p'$. A bound for the norm of U is given by the inequality

$$\|U\| \leqslant C_1.$$

In this case, a proof is obtained by an immediate application of Hölder's inequality to (1).

REMARK 5. The theorem is still true in the case where the right-hand side of (2) depends on s: that is, when (2) and (3) are replaced by

$$\left[\int_D |K(s, t)|^r \, dt \right]^{1/r} \leqslant C_1 \Psi(s), \tag{2a}$$

$$\left\{ \int_{D'} [|K(s, t)| (\Psi(s))^{q/\sigma - 1}]^\sigma \, ds \right\}^{1/\sigma} \leqslant C_2. \tag{3a}$$

In particular, inequality (5) remains true.

3.2. Let us turn to a description of the compact operators. We first prove a simple theorem.

THEOREM 2. *The integral operator* (1) *is a compact operator from* $\mathbf{L}^p(D)$ *into* $\mathbf{L}^q(D')$ *if its kernel is summable in the region* $D \times D'$ *to the exponent* r', *where* $r = \min(p, q')$, *i.e. if*

$$\left[\int_{D'} \int_D |K(s, t)|^{r'} \, dt \, ds \right]^{1/r'} \leqslant C < \infty. \tag{9}$$

Moreover,

$$\|U\| \leqslant AC, \tag{10}$$

where we can take

$$A = [\operatorname{mes} D']^{\frac{1}{q\left(\frac{r'}{q}\right)'}}.$$

Proof. We first establish the bound for the norm of U. Using Hölder's inequality, we have

$$|y(s)| \leqslant \left[\int_D |K(s, t)|^{r'} \, dt \right]^{1/r'} \left[\int_D |x(t)|^r \, dt \right]^{1/r} \leqslant \left[\int_D |K(s, t)|^{r'} \, dt \right]^{1/r'} \left[\int_D |x(t)|^p \, dt \right]^{1/p}$$

$$= \|x\| \left[\int_D |K(s, t)|^{r'} \, dt \right]^{1/r'}.$$

Hence

$$\|y\| = \left[\int_{D'} |y(s)|^q \, ds \right]^{1/q} \leqslant \|x\| \left\{ \int_{D'} \left[\int_D |K(s, t)|^{r'} \, dt \right]^{q/r'} ds \right\}^{1/q}.$$

Bearing in mind that $r \leqslant q'$, and so $r' \geqslant q$, we apply Hölder's inequality with exponents $\frac{r'}{q}$ and $\left(\frac{r'}{q}\right)'$ to the inner integral; this yields

$$\|y\| \leqslant \|x\| \left\{ \int_{D'} \left[\int_D |K(s, t)|^{r'} \, dt \right] ds \right\}^{1/r'} \left\{ \int_{D'} 1^{\left(\frac{r'}{q}\right)'} ds \right\}^{\left(q\left(\frac{r'}{q}\right)'\right)^{-1}} \leqslant AC \|x\|,$$

from which we obtain (10).

The kernel K is an element of $\mathbf{L}^{r'}(D' \times D)$, so there exists a sequence of continuous kernels K_n such that

$$\left[\int_{D'} \int_D |K(s, t) - K_n(s, t)|^{r'} \, dt \, ds \right]^{1/r'} \leqslant \varepsilon_n \qquad (n = 1, 2, \ldots), \tag{11}$$

where $\varepsilon_n \to 0$. Writing U_n for the integral operator with kernel $K_n(s, t)$, we see that U_n is compact (see IX.2.1) and, by (10) and (11),

$$\|U_n - U\| \leqslant A\varepsilon_n \qquad (n = 1, 2, \dots),$$

that is, the sequence $\{U_n\}$ converges to U. Hence U is compact (Theorem IX.2.3).

REMARK. If the kernel $K(s, t)$ is such that

$$B = \left\{ \int_{D'} \left[\int_D |K(s, t)|^{p'} dt \right]^{q/p'} ds \right\}^{1/q} < \infty, \tag{12}$$

then U is a compact operator from $\mathbf{L}^p(D)$ into $\mathbf{L}^q(D')$, and $\|U\| \leqslant B$.

The proof of this is similar to the proof of the theorem.

THEOREM 3. *Suppose that conditions* 1) *and* 2) *of Theorem* 1 *are satisfied, and that condition* 3) *is satisfied in the stronger form*

$$q \geqslant p, \qquad q > \sigma, \qquad \left(1 - \frac{\sigma}{q}\right) p' < r. \tag{13}$$

Then the integral operator (1) *is a compact operator from* $\mathbf{L}^p(D)$ *into* $\mathbf{L}^q(D')$.

Proof. We choose a number $\rho < r$ such that condition (4) is satisfied,

$$\left(1 - \frac{\sigma}{q}\right) p' < \rho,$$

and introduce kernels $K_n(s, t)$ by setting

$$K_n(s, t) = \begin{cases} -n & (-n > K(s, t)), \\ K(s, t) & (-n \leqslant K(s, t) \leqslant n), \\ n & (K(s, t) > n). \end{cases}$$

Let us find a bound for

$$C_1^{(n)} = \operatorname*{vrai\,sup}_{s \in D'} \left\{ \int_D |K(s, t) - K_n(s, t)|^\rho dt \right\}^{1/\rho} \leqslant \operatorname*{vrai\,sup}_{s \in D'} \left\{ \int_{A_n(s)} |K(s, t)|^\rho dt \right\}^{1/\rho},$$

where $A_n(s)$ denotes the set of those points in D such that $|K(s, t)| > n$. If $t \in A_n(s)$, then, since $|K(s, t)| > n$,

$$|K(s, t)|^\rho \leqslant \frac{1}{n^{r-\rho}} |K(s, t)|^r,$$

and so

$$C_1^{(n)} \leqslant \operatorname*{vrai\,sup}_{s \in D'} \left\{ \int_{A_n(s)} \frac{1}{n^{r-\rho}} |K(s, t)|^r dt \right\}^{\frac{1}{r} \cdot \frac{r}{\rho}} \leqslant n^{1 - r/\rho} C_1^{r/\rho}.$$

If we write U_n, as before, for the operator with kernel $K_n(s, t)$ and apply the bound given by (5) of Theorem 1 to $U - U_n$ (replacing C_1 by $C_1^{(n)} \leqslant n^{1-r/\rho} C_1^{r/\rho}$), we obtain

$$\|U - U_n\| \leqslant (n^{1-r/\rho} C_1^{r/\rho})^{1-\sigma/q} C_2^{\sigma/q},$$

from which it follows that $\lim_{n \to \infty} \|U - U_n\| = 0$.

Since the kernel of U_n is bounded, it is summable to any exponent; hence, by Theorem 2, the operator U_n is compact. Therefore so is U, being the limit of a sequence of compact operators (Theorem IX.2.3).

REMARK 1. The theorem remains true even if we have equality in the third of the conditions (13), provided there exists an increasing function $\Phi(\lambda)$ $(\lambda \geqslant 0)$ such that the ratio $\Phi(\lambda)/\lambda$ tends to infinity as $\lambda \to \infty$, and then the inequality (2) is satisfied in the stronger form:

$$\left\{ \int_D [\Phi(|K(s,t)|)]^r \, dt \right\}^{1/r} \leqslant C_1.$$

REMARK 2. Theorems 1–3 are true also when D and D' are arbitrary finite measure spaces (the proofs then require only insignificant changes). Furthermore they can be generalized to a wide class of ideal spaces.

3.3. Let us introduce the space Lip β of all functions defined in a region D' and satisfying there the Lipschitz condition with exponent β $(0 < \beta \leqslant 1)$; $y \in \operatorname{Lip} \beta$ means, therefore, that there exists a constant B such that

$$|y(s+\Delta s) - y(s)| \leqslant B|\Delta s|^\beta$$

(the interval $[s, s+\Delta s]$ belonging to D'), where $|\Delta s|$ indicates the length of the vector Δs in ν-dimensional space. Clearly, every function $y \in \operatorname{Lip} \beta$ admits of a unique continuous extension to the closure $\bar{D}'$. For if $s_n \to s$, $s_n \in D'$, $s \in \bar{D}'$, then

$$|y(s_n) - y(s_m)| \leqslant B|s_n - s_m|^\beta \xrightarrow[n, m \to \infty]{} 0.$$

Therefore $\{y(s_n)\}$ is a Cauchy sequence, and so has a limit $y(s) = \lim y(s_n)$. Again using the Lipschitz condition, we verify that $y(s)$ does not depend on the choice of the s_n converging to s. The extended function y clearly satisfies the Lipschitz condition on $\bar{D}'$ with the same exponent β, so y is continuous on the compactum $\bar{D}'$. In this way, a natural embedding of Lip β into $\mathbf{C}(D')$ is determined (see IV.4.4). A norm is defined in Lip β by means of the equation

$$\|y\| = \sup_{[s, s+\Delta s] \subset D'} \frac{|y(s+\Delta s) - y(s)|}{|\Delta s|^\beta} + \sup_{s \in D'} |y(s)| = C_\beta(y) + \|y\|_{\mathbf{C}}.$$

We leave it to the reader to check that Lip β is thus turned into a B-space.

Let us consider in more detail the case of the integral operator (1) when the sets D and D' lie in the same space and all the singularities of the kernel $K(s,t)$ are concentrated on the "diagonal" of the set $D' \times D$—that is, where $s = t$. We assume that for $s \neq t$ the kernel is differentiable with respect to the coordinates of s. In what follows we let $|s-t|$ denote the distance between the points s and t—in other words, the length of the vector $s-t$.

In that case we have

THEOREM 4. *Suppose the kernel* $K(s,t)$ *is such that*

1)
$$\left\{ \int_D [|\operatorname{grad}_s K(s,t)| \cdot |s-t|^{1-\beta}]^r \, dt \right\}^{1/r} \leqslant E, \tag{14}$$

where grad_s *denotes the gradient with respect to* s;

2)
$$\left\{ \int_D \left[\frac{|K(s,t)|}{|s-t|^\beta} \right]^r dt \right\}^{1/r} \leqslant F. \tag{15}$$

Then the integral operator (1) *maps* $\ell^{r'}(D)$ *(and any* $\mathbf{L}^p(D)$ *with* $p \geqslant r'$*) into* Lip β, *and*

$$C_\beta(y) \leqslant [E + (2^\beta + 3^\beta)F]\|x\|. \tag{16}$$

Proof. Choose any interval $[s, s+\Delta s]$ lying entirely within D'. The increment of the function $y = U(x)$ along this interval is given by

$$|y(s+\Delta s) - y(s)| = \left| \int_D [K(s+\Delta s, t) - K(s,t)] x(t) \, dt \right| \leqslant \left[\int_D |K(s+\Delta s, t) - K(s,t)|^r \, dt \right]^{1/r} \|x\|_{L^{r'}}. \tag{17}$$

To bound the first factor we divide the region of integration into two parts: the intersection D_1 of D with the ball $|t-s| \leqslant 2|\Delta s|$, and $D_2 = D \setminus D_1$. Taking the integral of $|K(s,t)|^r$ over D_1, we have

$$\left[\int_{D_1} |K(s,t)|^r dt\right]^{1/r} = \left[\int_{D_1} \left(\frac{|K(s,t)|}{|s-t|^\beta}\right)^r |s-t|^{\beta r} dt\right]^{1/r} \leqslant (2|\Delta s|)^\beta F.$$

There is a similar bound for the integral of $|K(s+\Delta s, t)|^r$, with the factor 2^β replaced by 3^β, since

$$|s+\Delta s - t| \leqslant |s-t| + |\Delta s| \leqslant 3|\Delta s|.$$

Taking this into account, we see that

$$I_1 = \left[\int_{D_1} |K(s+\Delta s, t) - K(s,t)|^r dt\right]^{1/r} \leqslant \left[\int_{D_1} |K(s+\Delta s, t)|^r dt\right]^{1/r} + \left[\int_{D_1} |K(s,t)|^r dt\right]^{1/r} \leqslant$$
$$\leqslant (2^\beta + 3^\beta) F\,|\Delta s|^\beta. \tag{18}$$

Also*

$$I_2 = \left[\int_{D_2} |K(s+\Delta s, t) - K(s,t)|^r dt\right]^{1/r} \leqslant \left\{\int_{D_2}\left[\int_s^{s+\Delta s} |\mathrm{grad}_s K(\lambda, t)|\, d\lambda\right]^r dt\right\}^{1/r}$$

Here, since $t \in D_2$, we have $|\lambda - t| \geqslant |s-t| - |\lambda - s| \geqslant |s-t| - |\Delta s| \geqslant \frac{1}{2}|s-t|$, and so

$$I_2 \leqslant 2^{1-\beta}\left\{\int_{D_2}\left[\int_s^{s+\Delta s} |\mathrm{grad}_s K(\lambda,t)|\,|\lambda - t|^{1-\beta} d\lambda\right]^r \frac{dt}{|s-t|^{r(1-\beta)}}\right\}^{1/r} \leqslant$$

$$\leqslant 2^{1-\beta}\left\{\int_{D_2}\left[\int_s^{s+\Delta s} (|\mathrm{grad}_s K(\lambda,t)|\,|\lambda-t|^{1-\beta})^r d\lambda\right] \times\right.$$

$$\left.\times\left[\int_s^{s+\Delta s} 1\cdot d\lambda\right]^{r/r'} \frac{dt}{|s-t|^{r(1-\beta)}}\right\}^{1/r} \leqslant$$

$$\leqslant \frac{2^{1-\beta}|\Delta s|^{1/r'}}{(2|\Delta s|)^{1-\beta}}\left\{\int_s^{s+\Delta s} d\lambda \int_{D_2} [|\mathrm{grad}_s K(\lambda,t)|\,|\lambda-t|^{1-\beta}]^r dt\right\}^{1/r} \leqslant$$

$$\leqslant |\Delta s|^{\frac{1}{r'}+\beta-1+\frac{1}{r}} E = |\Delta s|^\beta E. \tag{19}$$

Comparing (18) and (19), we obtain

$$\left[\int_D |K(s+\Delta s, t) - K(s,t)|^r dt\right]^{1/r} \leqslant I_1 + I_2 \leqslant [E + (2^\beta + 3^\beta)F]\,|\Delta s|^\beta,$$

and thus, by (17),

$$\frac{|y(s+\Delta s) - y(s)|}{|\Delta s|^\beta} \leqslant [E + (2^\beta + 3^\beta)F]\,\|x\|,$$

which gives (16), as required.

A bound for the maximum modulus of $y(s)$ can be obtained using the fact that, by condition (15), U is a continuous operator from $\mathbf{L}^r(D)$ into $\mathbf{L}^\infty(D')$ (see Remark 4 after Theorem (1), and so

$$\max |y(s)| \leqslant A\,\|x\|.$$

Together with (16), this inequality also establishes that U is continuous.

REMARK 1. Since functions having bounded norm in $\mathrm{Lip}\,\beta$ are equicontinuous and uniformly bounded, it follows from this theorem that if $K(s,t)$ satisfies conditions (14) and (15) for some $\beta > 0$, then the operator (1), regarded as an operator from $\mathbf{L}^r(D)$ into $\mathbf{C}(D')$, is compact.

We note that a generalization of Theorem 4 to BFS has been obtained in a paper by Berkolaiko and Rutitskii [1].

---

* The expression $\int_s^{s+\Delta s} \phi(\lambda) d\lambda$ should be understood as an integral over $[s, s+\Delta s]$.

3.4. We now consider the case where the kernel of U depends upon a parameter.

We shall say that a function $K_\tau(s, t)$ is *almost uniformly continuous* in s and t with respect to the parameter τ, at $\tau = \tau_0$ if, given any $\varepsilon > 0$, $h > 0$ and s, there exist a $\delta > 0$ and a set $A(s)$ such that $|K_\tau(s,t) - K_{\tau_0}(s,t)| < \varepsilon$ whenever $|\tau - \tau_0| < \delta$, $t \notin A(s)$, where mes $A(s) < h$.*

The concept of almost uniform continuity is used in the following theorem.

THEOREM 5. *Suppose that the kernel of an integral operator is almost uniformly continuous with respect to τ for $\tau = \tau_0$ and suppose that the conditions of Theorem 3 are satisfied for each τ, where none of the constants (C_1, C_2, σ, etc.) appearing in the statement of that theorem depend on τ. Then the integral operator U_τ, given by*

$$y_\tau = U_\tau(x), \qquad y_\tau(s) = \int_D K_\tau(s,t)x(t)\,dt$$

$$(x \in \mathbf{L}^p(D), \qquad y_\tau \in \mathbf{L}^q(D')), \tag{20}$$

depends continuously on the parameter, in the sense that

$$\lim_{\tau \to \tau_0} \|U_\tau - U_{\tau_0}\| = 0.$$

Proof. We have

$$y_{\tau_0}(s) - y_\tau(s) = \int_D [K_{\tau_0}(s,t) - K_\tau(s,t)]\,x(t)\,dt.$$

Using the notation of the proof of Theorem 3, we find a bound for the constant

$$C_1^{(p)} = \operatorname*{vrai\,sup}_{s \in D'} \left[\int_D |K_{\tau_0}(s,t) - K_\tau(s,t)|^p\,dt\right]^{1/p}$$

Given $\varepsilon > 0$, $h > 0$ and s, we determine the $\delta > 0$ and $A(s) \subset D$ corresponding to these in the definition of almost uniform continuity. Then, assuming for simplicity that $p \geqslant 1$,† we have

$$\left[\int_D |K_{\tau_0}(s,t) - K_\tau(s,t)|^p\,dt\right]^{1/p} \leqslant \left[\int_{D' \setminus A(s)} |K_{\tau_0}(s,t) - K_\tau(s,t)|^p\,dt\right]^{1/p} +$$

$$+ \left[\int_{A(s)} |K_{\tau_0}(s,t) - K_\tau(s,t)|^p\,dt\right]^{1/p} \leqslant$$

$$\leqslant \varepsilon[\text{mes}\, D]^{1/p} + \left[\int_{A(s)} |K_{\tau_0}(s,t) - K_\tau(s,t)|^{p \cdot \frac{r}{p}}\,dt\right]^{1/r} \left[\int_{A(s)} dt\right]^{1/(\frac{r}{p})'p} \leqslant$$

$$\leqslant \varepsilon[\text{mes}\, D]^{1/p} + 2C_1 h^{1/p - 1/r}.$$

Therefore $C_1^{(p)} \leqslant \varepsilon\,[\text{mes}\, D]^{1/p} + 2C_1 h^{1/p - 1/r}$; that is, the value of $C_1^{(p)}$ for the operator $U_{\tau_0} - U_\tau$ can be made as small as we please. Since, by Theorem 1 (see (5)), we have

$$\|U_{\tau_0} - U_\tau\| \leqslant [C_1^{(p)}]^{1-\sigma/q}[2C_2]^{\sigma/q},$$

---

* The values of τ are points in Euclidean space. However, one can also regard τ as an element of an arbitrary metric space.

†If $p < 1$, we need to use the inequality

$$(a+b)^{1/p} \leqslant 2^{1/p - 1}[a^{1/p} + b^{1/p}] \qquad (a, b \geqslant 0).$$

it follows that $\|U_{\tau_0} - U_\tau\|$ is also as small as we please, which proves the theorem.

REMARK. The theorem is immediately applicable to the case where $q = \infty$ (cf. Remark 4 after Theorem 1). More precisely, if $r > p'$ and (2) is satisfied (for every τ), and $K_\tau(s, t)$ is almost uniformly continuous with respect to the parameter at $\tau = \tau_0$, then the operator (20) depends continuously on τ. This now means, clearly, that, if $\eta > 0$, we can choose $\lambda > 0$ such that

$$|y_{\tau_0}(s) - y_\tau(s)| < \eta \quad (|\tau_0 - \tau| < \lambda, \ s \in D').$$

3.5. In conclusion, we consider kernels of a special type. In fact, assuming that D and D' lie in the same space, we consider kernels of the form

$$K(s, t) = \frac{B(s, t)}{|s - t|^m}, \tag{21}$$

where $B(s, t)$ is a bounded function, continuous for $s \neq t$. Such a kernel is called a *kernel of potential type.*

For such a kernel to be summable in t to exponent r, or, more precisely, for (2) to be satisfied, it is enough to require that $mr < \mu$; condition (3) will be satisfied if $m\sigma < \nu$. Thus we may take r and σ to be numbers arbitrarily close to μ/m and ν/m respectively. Then the conditions in (4) may be written in the form

$$q \geqslant p, \ q > \frac{\nu}{m}, \ \left(1 - \frac{\nu}{mq}\right)p' < \frac{\mu}{m}.$$

Bearing in mind Remark 1 after Theorem 1, we can thus say that the conditions

$$q < \frac{\nu p}{\mu - (\mu - m)p}, \quad \nu > \mu - (\mu - m)p \tag{22}$$

guarantee the continuity of the operator (1), with kernel of the form (21), regarded as an operator from $\mathbf{L}^p(D)$ into $\mathbf{L}^q(D')$. Since the conditions of Theorem 3 (and the Remark following it) are then also satisfied, it also follows from (22) that U is compact.* In particular, when $m = \mu - 1$, the conditions (22) take the form

$$q < \frac{\nu p}{\mu - p}, \quad \nu > \mu - p. \tag{23}$$

Theorem 4 is applicable to this case under the following conditions. Since, clearly,

$$|\text{grad}_s K(s, t)| \leqslant \frac{|\text{grad}_s B(s, t)|}{|s - t|^m} + m\frac{|B(s, t)|}{|s - t|^{m+1}},$$

we can guarantee that (14) and (15) are satisfied when

$$\sup_{s \in D'} \left[\int_D \frac{|\text{grad}_s B(s, t)|^r}{|s - t|^{(m + \beta - 1)r}} dt \right]^{1/r} \leqslant E', \tag{24}$$

$$(m + \beta)r < \mu. \tag{25}$$

Under these conditions, the operator (1) is a continuous operator from $\mathbf{L}^p(D)$ into $\text{Lip}\,\beta$.

---

* In fact, the first of the inequalities (22) is equivalent to $\left(1 - \frac{\nu}{mq}\right)p' < \frac{\mu}{m}$. If here $q < p$, then the second guarantees that inequality (8) holds.

If we regard the kernel $K(s+\Delta s, t)$ as a function of the parameter Δs, then, since it becomes uniformly continuous in Δs at $\Delta s = 0$ when we exclude an arbitrarily small neighbourhood of s, we conclude from Theorem 5 that

$$\lim_{\Delta s \to 0} \|U_{\Delta s} - U\| = 0,$$

where $U_{\Delta s}$ is the integral operator with kernel $K(s+\Delta s, t)$. In particular,*

$$\int_{D'} |y(s+\Delta s) - y(s)|^q \, ds \to 0 \qquad \text{as} \qquad \Delta s \to 0.$$

Similarly, suppose D' is a ν-dimensional manifold, continuously and smoothly dependent on a parameter τ, $D' = D'_\tau$; that is, the manifold formed by the points $s + \phi(s, \tau)$, where s ranges over the region D'_0 and $\phi(s, \tau) \to 0$ as $\tau \to 0$, uniformly in s, and where $|\phi(s, \tau) - \phi(s', \tau)| \leqslant \alpha |s - s'|$ $(\alpha < 1)$. Then the function y on the manifold D'_τ, given by

$$y_\tau(s) = y(s + \phi(s, \tau)) = \int_D K(s + \phi(s, \tau), t) x(t) \, dt,$$

is continuous with respect to τ in the metric of $\mathbf{L}^q(D'_0)$: that is,

$$\|y_0 - y_\tau\| = \left\{ \int_{D'_0} |y_0(s) - y_\tau(s)|^q \, ds \right\}^{1/q} \xrightarrow[\tau \to 0]{} 0.$$

Summarizing what has been said, we state the following two theorems.

THEOREM 6. *If the conditions* (22) *are satisfied, then an integral operator with kernel of type* (21) *is a compact operator mapping* $\mathbf{L}^p(D)$ *into* $\mathbf{L}^p(D')$, *where* D *is a* μ-*dimensional region in Euclidean space and* D' *is a* ν-*dimensional manifold in Euclidean space. If, in this situation, the manifold depends continuously on a parameter, then so also does* $y = U(x)$. *In particular,*

$$\lim_{\Delta s \to 0} \left[\int_{D'} |y(s+\Delta s) - y(s)|^q \, ds \right]^{1/q} = 0. \tag{26}$$

REMARK. The theorem is also true for $q = \infty$. Then the conditions (22) are replaced by the following (see Remark 4 following Theorem 1):

$$(\mu - m)p > \mu. \tag{27}$$

In this case, (26) can be written in the form

$$\sup_s |y(s+\Delta s) - y(s)| \xrightarrow[\Delta s \to 0]{} 0,$$

which implies that y is continuous; that is, in this case U actually maps $\mathbf{L}^p(D)$ into $\mathbf{C}(D')$.

We can rewrite condition (27) as $mp' < \mu$. Thus, using the Remark following Theorem 2, we see that U is a compact operator from $\mathbf{L}^p(D)$ into $\mathbf{L}^\infty(D')$ (and hence, as we have shown, into $\mathbf{C}(D')$). We therefore have

THEOREM 7. *If condition* (27) *is satisfied, then an integral operator with kernel of potential type is a compact operator mapping* $\mathbf{L}^p(D)$ *into* $\mathbf{C}(D')$.

---

* Here, and in what follows, y is understood to be the result of applying U to some $x \in \mathbf{L}^p(D)$: thus $y = U(x)$.

§ 4. Sobolev's embedding theorems

In various problems of mathematical physics and other areas of analysis, the mutual relations between the differential properties of functions play an important role. Thus, in some well-known situations, if we know integral estimates for partial derivatives, we can draw conclusions about the boundedness, or even continuity, of the functions. In a number of cases, one can judge the differential properties of a function on a surface from its behaviour on the whole space, and so on.

In Chapter VI we have already made use of the idea that the same function can be regarded as an element of different function spaces. Here this idea will be developed further and more systematically. The spaces that we study in this section are characterized by various differential properties of the functions that belong to them, so that if one such space is contained in another (in the set-theoretical sense), then the group of properties characterizing the first space will entail properties characteristic of the second. Furthermore, if in this situation we associate to a function, regarded as an element of the first space, the same function, but regarded as an element of the second space, then we obtain an *embedding operator*; and the study of this operator makes it possible not only to make more precise the qualitative connections already established between various properties of functions, but also to characterize these connections quantitatively.

The results in this section are principally due to S. L. Sobolev (see Sobolev-I, II) and play a fundamental role for applications in mathematical physics.

4.1. We first prove a lemma on integral representations of differentiable functions (the notation used here and below is the same as in the last section). In what follows D will denote a bounded region in $\mathbf{R}^\mu$ with a sufficiently smooth boundary.

LEMMA 1. *Let $x(s)$ be a continuously differentiable function, defined in a convex region $D \subset \mathbf{R}^\mu$. We have the following identity*

$$x(s) = \frac{1}{\operatorname{mes} D} \int_D x(t)\,dt - \sum_{k=1}^{\mu} \int_D \frac{B_k(s,t)}{|s-t|^{\mu-1}} \frac{\partial x}{\partial t_k}\,dt, \tag{1}$$

where $B_k(s,t)$ $(k = 1, 2, \ldots, \mu)$ are bounded functions,

$$|B_k(s,t)| \leqslant \frac{\delta^\mu}{\mu \operatorname{mes} D} \quad (k = 1, 2, \ldots, \mu), \tag{2}$$

which are continuous for $s \neq t$. Here $t_1, t_2, \ldots, t_\mu$ denote the coordinates of the point t and δ denotes the diameter of D.

Proof. Let s be a fixed point of D and $\bar{l}$ a unit vector. Then every point $t \in D$ lying on the ray going from the point s in the direction of $\bar{l}$ can be expressed in the form $t = s + R\bar{l}$, where $R = |s-t|$ is the distance from t to s. Let $d = d(\bar{l})$ be the length of the portion of this ray lying inside D. We set

$$F(t) = x(t) \int_R^d \rho^{\mu-1}\,d\rho = x(t)\frac{d^\mu - R^\mu}{\mu} \quad (t = s + Rl).$$

Obviously,

$$F(s) = F(t)|_{R=0} = x(s)\frac{d^\mu}{\mu}, \quad F(t)|_{R=d} = 0.$$

Therefore

$$x(s)\frac{d^{\mu}}{\mu} = -\int_0^d \frac{\partial F}{\partial R} dR = \int_0^d x(t)R^{\mu-1} dR - \int_0^d \frac{\partial x}{\partial R}\frac{d^{\mu}-R^{\mu}}{\mu}\frac{R^{\mu-1}}{R^{\mu-1}} dR.$$

We multiply both sides of this equation by an element on the surface of the unit ball in μ-dimensional space and integrate over this surface, using the formula

$$\int_D z(t)\,dt = \int_{\omega} d\omega \int_0^d z(t)R^{\mu-1}\,dR,$$

where ω denotes the surface of the unit ball with centre at s.

After integrating, we obtain

$$x(s)\operatorname{mes} D = \int_D x(t)\,dt - \int_D \frac{\partial x}{\partial R}\frac{d^{\mu}-R^{\mu}}{\mu}\frac{dt}{R^{\mu-1}}.$$

Since $\dfrac{\partial x(t)}{\partial R} = \sum_{k=1}^{\mu} \dfrac{\partial x(t)}{\partial t_k}\cos(\overline{l}, t_k)$, if we divide both sides of the equation by $\operatorname{mes} D$ and write

$$B_k(s, t) = B(s, t)\cos(\overline{e}, t_k) \quad \left(B(s, t) = \frac{d^{\mu}-R^{\mu}}{\mu \operatorname{mes} D}; k = 1, 2, \ldots, \mu \right),$$

we obtain (1). From this we also obtain the bound (2) and the continuity of the function.

REMARK 1. The result of this lemma is also true in a more general form. In fact, the region D need not even be convex; it is sufficient for D to contain a convex subregion D_1 such that D contains the cone D_2 formed by the segments of the rays emerging from an arbitrary point s of D and intersecting with D_1, from s to the closest point of intersection.* Then (1) is replaced by one of the equations

$$x(s) = \frac{1}{\operatorname{mes} D_1}\int_{D_1} x(t)\,dt - \sum_{k=1}^{u} \int_{D_1 \cup D_2} \frac{B_k(s, t)}{|s-t|^{\mu-1}}\frac{\partial x}{\partial t_k} dt, \tag{3}$$

or

$$x(s) = \frac{1}{\operatorname{mes} D_1}\int_{D_1} x(t)\,dt - \sum_{k=1}^{\mu} \int_{D} \frac{B_k(s, t)}{|s-t|^{\mu-1}}\frac{\partial x}{\partial t_k} dt. \tag{3'}$$

The proofs are obtained by arguments similar to those given above.

REMARK 2. In the conditions of the lemma, the convexity of D is used only to ensure that the ray from any interior point of D intersects the boundary in one point only. Hence if we do not assume that the region is convex, the representation (1) is valid for all points $s \in D$ with respect to which D is star-shaped. In particular, it is valid for all $s \in D_1$, where D_1 is as in Remark 1.

REMARK 3. Lemma 1 can be generalized. Namely, we have: if $x(t)$ has continuous derivatives of all orders up to and including l in the convex region D, then the following

* A region D of this kind is said to be *star-shaped* with respect to D_1.

identity is valid:

$$x(s)=\frac{1}{\operatorname{mes} D}\int_D x(t)\mathscr{L}(s,t)\,dt+(-1)^l\sum_{i_1,\ldots,i_l=1}^{\mu}\int_D\frac{\partial^l x(t)}{\partial t_{i_1}\ldots\partial t_{i_l}}\mathscr{L}_{i_1\ldots i_l}(s,t)\frac{dt}{|s-t|^{\mu-l}},\tag{4}$$

where*

$$\mathscr{L}(s,t)=\sum_{i=0}^{l-1}b_i\left(1-\frac{R}{d}\right)^{l-1-i}\left(\frac{R}{d}\right)^i,\tag{5}$$

the b_i being constants, depending only on l and μ,

$$\mathscr{L}_{i_1\ldots i_l}(s,t)=\frac{(\mu+l-1)!}{\mu!\,(l-1)!}\alpha_{i_1}\ldots\alpha_{i_l}\frac{d^\mu}{\operatorname{mes} D}\left(\int_{R/d}^{1}(1-u)^{l-1}u^{\mu-1}du\right),\tag{5'}$$

$$\alpha_i=\cos(\overline{l},t_i)\qquad(i=1,2,\ldots,\mu).$$

The proof of (4) is similar to that of (1). However, in this case the function $F(t)$ is defined as follows:

$$F(t)=x(t)\frac{\partial^{l-1}}{\partial R^{l-1}}\left[\frac{R^{l-1}}{(l-1)!}\psi(R,d)\right]-\frac{\partial x(t)}{\partial R}\frac{\partial^{l-2}}{\partial R^{l-2}}\left[\frac{R^{l-1}}{(l-1)!}\psi(R,d)\right]+$$

$$+\ldots+(-1)^{l-1}\frac{\partial^{l-1}x(t)}{\partial R^{l-1}}\left[\frac{R^{l-1}}{(l-1)!}\psi(R,d)\right],$$

where $\psi(R,d)=\int_R^d(d-\rho)^{l-1}\rho^{\mu-1}d\rho=d^{\mu+l-1}\int_{R/d}^1(1-u)^{l-1}u^{\mu-1}du$. Thus it is easy to see that

$$F(s)=F(t)|_{R=0}=x(s)\psi(0,d)=x(s)\frac{(l-1)!\,(\mu-1)!}{(\mu+l-1)!}d^{\mu+l-1},$$

$$F(t)|_{R=d}=0,$$

$$\frac{\partial F}{\partial R}=x(t)\frac{\partial^l}{\partial R^l}\left[\frac{R^{l-1}}{(l-1)!}\psi(R,d)\right]+(-1)^{l-1}\frac{\partial^l x(t)}{\partial R^l}\left[\frac{R^{l-1}}{(l-1)!}\psi(R,d)\right].$$

Proceeding now as in the proof of Lemma 1 and using the equation

$$\frac{\partial^l x(t)}{\partial R^l}=\sum_{i_1,\ldots,i_l=1}^{\mu}\alpha_{i_1}\ldots\alpha_{i_l}\frac{\partial^l x(t)}{\partial t_{i_1}\ldots\partial t_{i_l}},\quad\alpha_i=\cos(\overline{l},t_i),$$

we obtain the identity (4).

As before, let D be a convex region in $\mathbf{R}^\mu$ bounded by a sufficiently smooth surface S. Let $B_k(s,t)$ be the functions constructed in the proof of Lemma 1.

LEMMA 2. *The functions $B_k(s,t)$ have the property that the integral*

$$\int_D\frac{|\operatorname{grad}_s B_k(s,t)|^r}{|s-t|^{(\mu-2+\beta)r}}dt\tag{6}$$

* We use the notation introduced in the proof of Lemma 1.

is bounded independently of s, provided only that $r < \dfrac{\mu}{\mu-1}$ *and* $\beta > 0$ *is sufficiently small that* $(\mu-1+\beta)r < \mu$.

Proof. Since

$$B_k(s,t) = B(s,t)\cos(\bar{l}, t_k),$$

we have

$$|\operatorname{grad}_s B_k(s,t)| \leqslant |\operatorname{grad}_s B(s,t)|\,|\cos(\bar{l}, t_k)| + |B(s,t)|\,|\operatorname{grad}_s \cos(\bar{l}, t_k)| \leqslant$$
$$\leqslant |\operatorname{grad}_s B(s,t)| + K/R \qquad (R = |s-t|),$$

where K is some constant.

It follows from the condition $(\mu-1+\beta)r < \mu$ that the lemma is true for the term K/R. Let us verify that it is true for the first term. Since $B(s,t) = \dfrac{1}{\mu \operatorname{mes} D}(d^\mu - R^\mu)$ and the lemma is obviously true for R^μ, it only remains to prove it for the function d^μ.

Let u denote the point where the ray determined by the vector $\bar{l}$ intersects the boundary of D, let $\bar{n} = (A_1, \ldots, A_\mu)$ be the unit vector normal to the surface S at u (the A_i are the projections of $\bar{n}$ on the coordinate axes, $1 \leqslant i \leqslant \mu$), and let ϕ be the angle between $\bar{l}$ and $\bar{n}$.

Let $\tilde{s} = s + he_j$, where e_j is a unit vector directed along the j-th coordinate axis, be a point close to s. Let $\tilde{u}$ be the point where the ray emerging from $\tilde{s}$ in the direction of t intersects the boundary of D, and let $\tilde{\phi}$ be the angle between this ray and the vector $\bar{n}$ introduced above. Finally, let v be a point on the tangent hyperplane to S at u situated at the foot of the perpendicular dropped from $\tilde{u}$ to this hyperplane.

We write the coordinates of the points $u = (u_1, \ldots, u_\mu)$ and $\tilde{u} = (\tilde{u}_1, \ldots, \tilde{u}_\mu)$ in the form

$$u_i = s_i + (t_i - s_i)r(s), \qquad \tilde{u}_i = \tilde{s}_i + (t_i - \tilde{s}_i)r(\tilde{s}) \qquad (1 \leqslant i \leqslant \mu).$$

Then

$$d(s) = |u-s| = r(s)|t-s| = r(s)R(s),$$
$$d(\tilde{s}) = |\tilde{u}-\tilde{s}| = r(\tilde{s})|t-\tilde{s}| = r(\tilde{s})R(\tilde{s}).$$

Let us find the derivative of the function $d(s)$ with respect to s_j. We have

$$d'_{s_j}(s) = \lim_{h\to 0}\frac{d(s+he_j)-d(s)}{h} = \lim_{h\to 0}\frac{r(\tilde{s})R(\tilde{s})-r(s)R(s)}{h} =$$
$$= \lim_{h\to 0}\left[\frac{r(\tilde{s})-r(s)}{h}R(\tilde{s}) + \frac{R(\tilde{s})-R(s)}{h}r(s)\right]. \tag{7}$$

Clearly

$$\frac{R(\tilde{s})-R(s)}{h} \xrightarrow[h\to 0]{} R'_{s_j}(s) = \frac{s_j-t_j}{R(s)}. \tag{8}$$

To find the limit of the first quotient on the right-hand side of (7), we express $r(s)$ and $r(\tilde{s})$ in the form:

$$r(s) = \frac{d(s)}{R(s)} = \frac{|u-s|\cos\phi}{|t-s|\cos\phi} = \frac{(u-s,\bar{n})}{(t-s,\bar{n})} = \frac{\sum_{i=1}^{\mu} A_i(u_i-s_i)}{\sum_{i=1}^{\mu} A_i(t_i-s_i)}, \tag{9}$$

$$r(s) = \frac{d(\tilde{s})}{R(\tilde{s})} = \frac{|\tilde{u}-\tilde{s}|\cos\tilde{\phi}}{|t\;\;\tilde{s}|\cos\tilde{\phi}} = \frac{(\tilde{u}-\tilde{s},\bar{n})}{(t\;\;\tilde{s},\bar{n})} =$$

$$= \frac{(\tilde{u}-v, \bar{n}) + (v-u, \bar{n}) + (u-\tilde{s}, \bar{n})}{(t-\tilde{s}, \bar{n})} = \frac{|\tilde{u}-v| + \sum_{i=1}^{\mu} A_i(u_i - s_i) - A_j h}{\sum_{i=1}^{\mu} A_i(t_i - s_i) - A_j h}, \tag{10}$$

since the vector $\tilde{u}-v$ is collinear with $\bar{n}$ and $v-u$ is orthogonal to $\bar{n}$, and also $\tilde{s} = s + he_j = (s_1, \ldots, s_j + h, \ldots, s_n)$.

Notice that, in view of the assumptions about the surface bounding D,

$$\frac{|\tilde{u}-v|}{h} \xrightarrow[h\to 0]{} 0.* \tag{11}$$

Using (9)–(11) and introducing the notation

$$P = \sum_{i=1}^{\mu} A_i(u_i - s_i) = d(s)\cos\phi, \qquad Q = \sum_{i=1}^{\mu} A_i(t_i - s_i) = R(s)\cos\phi,$$

we obtain

$$\frac{r(\tilde{s}) - r(s)}{h} = \frac{|\tilde{u}-v|}{h(Q - A_j h)} + \frac{1}{h}\left[\frac{P - A_j h}{Q - A_j h} - \frac{P}{Q}\right] \xrightarrow[h\to 0]{} \frac{A_j(P-Q)}{Q^2} =$$

$$= \frac{A_j(d(s) - R(s))}{R^2(s)\cos\phi}. \tag{12}$$

Thus, by (7), (8) and (12), we have

$$d'_{s_j}(s) = \frac{A_j(d(s) - R(s))}{R(s)\cos\phi} + \frac{s_j - t_j}{R^2(s)} d(s).$$

Since $|A_j| \leqslant 1$ and $\frac{|s_j - t_j|}{R(s)} \leqslant 1$, this gives the bounds

$$|d'_{s_j}(s)| \leqslant \frac{d-R}{R|\cos\phi|} + \frac{d}{R} \leqslant 2\frac{d}{R|\cos\phi|}$$

and

$$|\text{grad}_s d| \leqslant 2\mu^{1/2} \frac{d}{R|\cos\phi|}.$$

---

* Let us prove this. First of all, in view of the assumption about the surface, we have

$$|\tilde{u}-v| \leqslant A|v-u|^2 \leqslant A|\tilde{u}-u|^2, \tag{*}$$

where A is a constant, depending on the properties of the surface. The segment $\tilde{u}u$ lies in the plane of the lines stu and $\tilde{s}t\tilde{u}$. Choose points w and η on the line $\tilde{u}\tilde{s}$ such that the line uw is parallel to $s\tilde{s}$ and the line $u\eta$ is perpendicular to $\tilde{u}\tilde{s}$. Let $\psi(\tilde{s})$ be the angle between the lines $\tilde{u}u$ and $\tilde{u}\tilde{s}$. Then

$$|\tilde{u}-u| = \frac{|u-\eta|}{\sin\psi(\tilde{s})} \leqslant \frac{|u-w|}{\sin\psi(\tilde{s})} = \frac{d-R}{R\sin\psi(\tilde{s})}|\tilde{s}-s| = \frac{d-R}{R\sin\psi(\tilde{s})}h. \tag{**}$$

Comparing (*) and (**) we see that

$$|\tilde{u}-v| \leqslant A\left(\frac{d-R}{R\sin\psi(\tilde{s})}\right)^2 h^2,$$

and thus (11), follows, because the convexity of D means that $\sin\psi(\tilde{s})$ is different from zero as $\tilde{s} \to s$.

Hence

$$|\text{grad}_s d^\mu| \leqslant 2\mu^{3/2} \frac{d^\mu}{R|\cos\phi|}.$$

Therefore

$$|\text{grad}_s d^\mu|^r R^{(\mu-2+\beta)r} \leqslant 2^r \mu^{(3/2)r} \frac{d^{\mu r}}{R^{(\mu-1+\beta)r}|\cos\phi|^r}.$$

To bound the integral of this expression, note that

$$dt = \frac{R^{\mu-1}}{d^{\mu-1}} \cos\psi \, dS \, dR,$$

where dS is an element of the surface S bounding D. Notice also that, in view of the hypotheses of the lemma,

$$\mu r - (\mu - 1 + \beta)r + 1 > \mu r - \mu + 1 > 0.$$

Taking this into account, we see that

$$\int_D \frac{d^{\mu r}}{R^{(\mu-1+\beta)r}} \frac{dt}{|\cos\phi|^r} = \int_S \frac{\cos\phi}{|\cos\phi|^r} \left(\int_0^d \frac{d^{\mu r-(\mu-1)} R^{\mu-1}}{R^{(\mu-1+\beta)r}} dR \right) dS \leqslant$$

$$\leqslant \frac{\delta^{\mu r-(\mu-1+\beta)r+1}}{\mu - (\mu-1+\beta)^r} \int_S \frac{dS}{|\cos\phi|^{r-1}}.$$

Furthermore, since $r < \dfrac{\mu}{\mu-1} \leqslant 2$ and so $r - 1 < 1$, we have, bearing in mind the conditions imposed on the surface,

$$\int_D \frac{dS}{|\cos\phi|^{r-1}} \leqslant K_1,$$

which, with the above, gives the required result.

The lemma is therefore proved.

REMARK. It follows from what was said in 3.5 that, when the conditions of the lemma are satisfied, the kernel $\dfrac{B_k(s,t)}{|s-t|^{\mu-1}}$ satisfies the requirements of Theorem 3.4.

4.2. Making use of the theorems on integral operators obtained in the last section, we can prove the following fundamental theorem.

THEOREM 1 (Sobolev's inequality). *Suppose the function $x(s)$ is continuously differentiable in the convex region D, and that*

$$q < \frac{\mu p}{\mu - p}. \tag{13}$$

Then we have the inequality

$$\|x - m(x)\|_{L^q(D)} = \left\{ \int_D |x(t) - m(x)|^q \, dt \right\}^{1/q} \leqslant A \| \, |\text{grad}\, x| \, \|_{L^p(D)}$$

$$= A \left\{ \int_D \left[\sum_{k=1}^{\mu} \left(\frac{\partial x}{\partial t_k} \right)^2 \right]^{p/2} dt \right\}^{1/p}, \tag{14}$$

where

$$m(x) = \frac{1}{\text{mes } D} \int_D x(s)\,ds \tag{15}$$

is the mean value of x in the region D, and A is some constant.

Proof. By Lemma 1 (bearing in mind the expression for $B_k(s, t)$), we have

$$|x(s) - m(x)| \leqslant \sum_{k=1}^{\mu} \left| \int_D \frac{B_k(s, t)}{|s-t|^{\mu-1}} \frac{\partial x}{\partial t_k} \right| \leqslant \int_D \frac{|B(s, t)|}{|s-t|^{\mu-1}} |\text{grad } x|\, dt.$$

However, when (13) is true, the integral operator U with kernel $\dfrac{|B(s, t)|}{|s-t|^{\mu-1}}$ is continuous from $\mathbf{L}^p(D)$ into $\mathbf{L}^q(D)$ (see 3.5, condition (23)). Therefore

$$\|x - m(x)\|_{\mathbf{L}^q} \leqslant \|U\| \, \| |\text{grad } x| \|_{\mathbf{L}^p}$$

and (14) is thus established if we take A to be the norm of this integral operator U.

REMARK 1. Inequality (14) is sometimes more conveniently used in another, equivalent, form, namely

$$\|x\|_{\mathbf{L}^q} \leqslant M[\| |\text{grad } x| \|_{\mathbf{L}^p} + |m(x)|]. \tag{16}$$

REMARK 2. If $p = 2$ and $\mu \geqslant 2$, then (13) is satisfied with $q = 2$. In this case, (16) can be written in the form

$$\left[\int_D |x(s)|^2 ds\right]^{1/2} \leqslant M\left\{\left|\int_D x(t)\,dt\right| + \left[\int_D \sum_{k=1}^{\mu} \left(\frac{\partial x}{\partial t_k}\right)^2 dt\right]^{1/2}\right\} \tag{17}$$

or, alternatively,

$$\int_D |x(s)|^2 ds \leqslant M_1 \left\{\left|\int_D x(t)\,dt\right|^2 + \int_D \sum_{k=1}^{\mu} \left(\frac{\partial x}{\partial t_k}\right)^2 dt\right\}. \tag{18}$$

This is known as *Poincaré's inequality.*

REMARK 3. In applications the value of the constant A in (14) can play a role. One can bound the norm of the integral operator U with kernel $\dfrac{|B(s, t)|}{|s-t|^{\mu-1}}$, using Theorem 3.1. Taking $\sigma = r$ and determining the least possible r from the inequality $\left(1 - \dfrac{r}{q}\right)p' \leqslant r$, we find that we can take A to have the value

$$A = C_1 = C_2 \leqslant \max|B(s, t)| \max_s \left[\int_D \frac{dt}{|s-t|^{(\mu-1)r}}\right]^{1/r} \leqslant \frac{\delta^\mu}{\mu \text{ mes } D} \left\{\int_\omega d\omega \int_0^\delta \frac{R^{\mu-1}}{R^{(\mu-1)r}} dR\right\}^{1/r}$$

$$\leqslant \frac{\delta^{1+\mu/r}}{\mu \text{ mes } D} \left\{\frac{\mu \pi^{\mu/2}}{(\mu - (\mu-1)r)\Gamma\left(\dfrac{\mu}{2}+1\right)}\right\}^{1/r}.$$

REMARK 4. The convexity of D was essential in the proof of Sobolev's inequality given above. However, this inequality can also be proved for regions of a significantly more general type. First of all, if for some region there is a continuously differentiable single-valued mapping, with Jacobian equal to unity, into another region for which the inequality holds, then the inequality will also hold in the first region. Apart from this, we can extend Sobolev's inequality to regions of a more general type by means of the following result.

THEOREM 2. *If a region D is the union of two regions D_1 and D_2 whose intersection is of positive measure, and if the inequality* (14) *holds on both D_1 and D_2, then it also holds on the whole of D.*

Proof. Let $D = D_1 \cup D_2$. Write

$$D_3 = D_1 \cap D_2, \qquad D_4 = D_2 \setminus D_3.$$

Denote the mean values of x on each of D, D_1, D_2, D_3, D_4 by $m(x)$, $m_1(x)$, $m_2(x)$, $m_3(x)$, $m_4(x)$. As (14) is applicable on D_1 and D_2, we have

$$\|x - m_1(x)\|_{L^q(D_1)} \leqslant B_1 J, \qquad \|x - m_2(x)\|_{L^q(D_2)} \leqslant B_2 J, \tag{19}$$

where J denotes $\| |\mathrm{grad}\, x| \|_{L^p(D)}$, and B_1 and B_2 are constants.

Furthermore,

$$|m_1(x) - m_2(x)| \leqslant \frac{1}{(\mathrm{mes}\, D_3)^{1/q}} [\|x - m_1(x)\|_{L^q(D_3)} + \|x - m_2(x)\|_{L^q(D_3)}]$$

$$\leqslant \frac{1}{(\mathrm{mes}\, D_3)^{1/q}} [\|x - m_1(x)\|_{L^q(D_1)} + \|x - m_2(x)\|_{L^q(D_2)}] \leqslant B_3 J,$$

$$|m_4(x) - m_2(x)| = \frac{1}{\mathrm{mes}\, D_4} \left| \int_{D_4} [x(s) - m_2(x)]\, ds \right| \leqslant \frac{A}{\mathrm{mes}\, D_4} \|x - m_2(x)\|_{L^q(D_2)} \leqslant B_4 J.$$

However, by the definition of mean value, we have

$$\min[m_1(x), m_4(x)] \leqslant m(x) \leqslant \max[m_1(x), m_4(x)].$$

Therefore

$$|m(x) - m_1(x)| \leqslant |m_4(x) - m_1(x)| \leqslant |m_4(x) - m_2(x)| + |m_2(x) - m_1(x)| \leqslant B_5 J,$$

and thus also

$$|m(x) - m_2(x)| \leqslant |m(x) - m_1(x)| + |m_2(x) - m_1(x)| \leqslant B_6 J.$$

Finally,

$$\|x - m(x)\|_{L^q(D)} \leqslant \|x - m(x)\|_{L^q(D_1)} + \|x - m(x)\|_{L^q(D_2)} \leqslant$$

$$\leqslant \|x - m_1(x)\|_{L^q(D_1)} + \|x - m_2(x)\|_{L^q(D_2)} +$$

$$+ |m(x) - m_1(x)| (\mathrm{mes}\, D_1)^{1/q} + |m(x) - m_2(x)| (\mathrm{mes}\, D_2)^{1/q} \leqslant B_7 J,$$

which establishes that an equality of type (14) is valid for D.

Another extension of the class of regions for which Sobolev's inequality holds can be obtained by using the Remark following Lemma 1. In fact, suppose a region D is star-shaped with respect to some convex region $D_1 \subset D$; and denote the mean value of x over D_1 by $m_1(x)$. Using the representation (3), we obtain

$$\|x - m_1(x)\|_{L^q(D)} \leqslant A_1 J \qquad (J = \| |\mathrm{grad}\, x| \|_{L^p(D)}).$$

But by Remark 2 to Lemma 1, the representation (1) is valid for all $s \in D_1$; thus

$$x(s) = m(x) + \sum_{k=1}^{\mu} \int_D \frac{B_k(s, t)}{|s-t|^{\mu-1}} \frac{\partial x}{\partial t_k} dt \qquad (s \in D_1).$$

Since the right-hand side of this equation clearly defines a continuous operator from $\mathbf{L}^p(D)$ into $\mathbf{L}^q(D_1)$, we have

$$\|x - m(x)\|_{\mathbf{L}^q(D_1)} \leqslant A_2 J.$$

Hence

$$|m_1(x) - m(x)| = \left| \frac{1}{\operatorname{mes} D_1} \int_{D_1} [x(s) - m(x)]\, ds \right| \leqslant$$

$$\leqslant \frac{1}{\operatorname{mes} D_1} \|x - m(x)\|_{\mathbf{L}^q(D_1)} \leqslant \frac{1}{\operatorname{mes} D_1} A_2 J.$$

Comparing this inequality with the one obtained earlier, we conclude finally that

$$\|x - m(x)\| \leqslant A_3 J,$$

which is what we set out to prove.

4.3. We now introduce certain function spaces, each of which has as its elements all possible continuously differentiable functions on a region D. The different spaces will therefore differ only in the norms that we introduce on them.

First we consider the space $\tilde{\mathbf{W}}_p^{(1)} = \tilde{\mathbf{W}}_p^{(1)}(D)$. Set

$$\|x\| = \frac{1}{\operatorname{mes} D} \left| \int_D x(s)\, ds \right| + \left[\int_D |\operatorname{grad} x|^p\, dt \right]^{1/p} = |m(x)| + \|\, |\operatorname{grad} x|\, \|_{\mathbf{L}^p(D)} \tag{20}$$

It is easy to see that the norm thus introduced satisfies all the axioms for a normed space (when linearized in the natural way).

If we regard an element $x \in \tilde{\mathbf{W}}_p^{(1)}$ as a function in $\mathbf{L}^q(D)$, then Sobolev's inequality (in the form (16)) can be expressed as follows:

$$\|x\|_{\mathbf{L}^q(D)} \leqslant M \|x\|_{\tilde{\mathbf{W}}_p^{(1)}}. \tag{21}$$

Hence the embedding operator—that is, the operator assigning to an element $x \in \tilde{\mathbf{W}}_p^{(1)}$ the same function, but viewed as an element of $\mathbf{L}^q(D)$—is a continuous linear operator. Let us show that the embedding operator is compact.

Let $\{x_n\}$ be a bounded sequence in $\tilde{\mathbf{W}}_p^{(1)}$. By Lemma 1, we have

$$x_n(s) = m(x_n) + \sum_{k=1}^{\mu} \int_D \frac{B_k(s, t)}{|s-t|^{\mu-1}} \frac{\partial x_n}{\partial t_k} dt.$$

However, when (16) holds, the operator with kernel $\dfrac{B_k(s, t)}{|s-t|^{\mu-1}}$ is compact (Theorem 3.6).

Since the sequence $\left\{\dfrac{\partial x_n}{\partial t_k}\right\}$ $(k = 1, 2, \ldots, \mu)$ is bounded in $\mathbf{L}^p(D)$, we can choose a sequence of suffices $\{n_i\}$ such that the sequence $\{y_{n_i}^{(k)}\}$, where

$$y_{n_i}^{(k)}(s) = \int\limits_D \frac{B_k(s, t)}{|s-t|^{\mu-1}} \frac{\partial x_{n_i}}{\partial t_k} dt \qquad (k = 1, 2, \ldots, \mu;\ i = 1, 2, \ldots)$$

converges in $\mathbf{L}^q(D)$. We may further assume that the numerical sequence $\{m(x_{n_i})\}$ converges also. In that case, $\{x_{n_i}\}$ converges in $\mathbf{L}^q(D)$ and it follows that the embedding operator is compact. Thus we have

THEOREM 3 (Sobolev–Kondrashev).* *When* (13) *is satisfied, the embedding operator mapping a function of* $\tilde{\mathbf{W}}_p^{(1)}$ *to the same function, regarded as an element of* $\mathbf{L}^q(D)$, *is compact.*

The class of regions D for which Theorem 3 is true coincides with the class of regions for which Sobolev's inequality holds. In particular, the theorem is true for regions that are star-shaped with respect to some convex subregion, or for finite unions of regions of this type (Theorem 2).

One can also prove the following more general result by a method analogous to that for Theorem 3, again using Theorem 3.6.

THEOREM 4. *The embedding operator, associating a function in* $\tilde{\mathbf{W}}_p^{(1)}$ *with the same function, regarded as an element of* $\mathbf{L}^q(D')$ *(where* D' *denotes a* ν*-dimensional surface lying in* D*), is compact, provided*

$$q < \frac{\nu p}{\mu - p}, \qquad \nu > \mu - p. \tag{22}$$

Furthermore, under a translation of the surface D', *the element* x *depends continuously on the amount of translation: that is,*

$$\lim_{\Delta s \to 0} \left\{ \int\limits_{D'} |x(s+\Delta s) - x(s)|^q \, ds \right\}^{1/q} = 0. \tag{23}$$

If we consider the special case where $\mu < p$ and use Theorem 3.7, we have

THEOREM 5. *When* $\mu < p$, *the embedding operator, regarded as an operator from* $\tilde{\mathbf{W}}_p^{(1)}(D)$ *into* $\mathbf{C}(D)$, *is continuous and compact.*

REMARK. Using Theorem 3.4 and the Remarks following Lemma 2, one can sharpen Theorem 5. In fact, the embedding operator from $\tilde{\mathbf{W}}_p^{(1)}$ into Lip β will be continuous if $\beta < \frac{p-\mu}{p}$ (see also Sobolev-I, § 11).

4.4. In the same way as we introduced the spaces $\tilde{\mathbf{W}}_p^{(1)}(D)$, we can also define spaces whose norms are determined by higher derivatives. Namely, we denote by $\tilde{\mathbf{W}}^{(l)} = \tilde{\mathbf{W}}^{(l)}(D)$ the space of all l times continuously differentiable functions on the region D, with the norm defined by

$$\|x\|_{\tilde{\mathbf{W}}_p^{(l)}(D)} = \|x\|_{\mathbf{L}^p(D)} + \sum_{i_1, \ldots, i_l = 1}^{\mu} \left[\int\limits_D \left| \frac{\partial^l x(t)}{\partial t_{i_1} \ldots \partial t_{i_l}} \right|^p dt \right]^{1/p}. \tag{24}$$

Theorem 4 (and hence also Theorem 3, as a special case of Theorem 4) carries over to the case of derivatives of higher order—that is, we have the following theorem.

THEOREM 6. *The embedding operator from* $\tilde{\mathbf{W}}_p^{(l)}(D)$ *(where* D *is a convex region) into*

* S. L. Sobolev proved that the embedding operator is continuous; V. I. Kondrashev established its compactness.

$\mathbf{L}^q(D')$ *(where D' is a v-dimensional surface in D) is compact provided the conditions*

$$q < \frac{vp}{\mu - lp}, \qquad v > \mu - lp \tag{25}$$

are satisfied. In that case, for any deformation of the surface D' depending on a parameter, the operator depends continuously on the parameter.

Proof. In view of the integral representation (4), the following equation holds for every function $x \in \tilde{\mathbf{W}}_p^{(l)}(D)$ and every point $s \in D'$:

$$x(s) = \frac{1}{\text{mes } D} \int_D x(t) \mathscr{L}(s, t)\, dt + (-1)^l \sum_{i_1, \ldots, i_l}^{\mu} \int_D \frac{\partial^l x(t)}{\partial t_{i_1} \ldots \partial t_{i_l}} \frac{\mathscr{L}_{i_1 \ldots i_l}(s, t)}{|s - t|^{\mu - l}}\, dt,$$

where the functions $\mathscr{L}(s, t)$ and $\mathscr{L}_{i_1 \ldots i_l}(s, t)$ are defined by (5) and (5′).

When (25) is satisfied, the integral operator with kernel $\dfrac{\mathscr{L}_{i_1 \ldots i_l}(s, t)}{|s - t|^{\mu - l}}$ is a compact operator from $\mathbf{L}^p(D)$ into $\mathbf{L}^q(D')$, by Theorem 3.6 (condition (28) of that theorem coincides with (25) when $m = \mu - l$); and, by the same theorem, this operator is continuously dependent on the parameter of translation of D'.

The integral operator defined by the first term of the given representation also clearly possesses analogous properties. The theorem thus follows from what has been said.

Finally, the following result is a generalization of Theorem 5 for the case of higher derivatives: when $\mu < lp$, the embedding operator is a compact operator from $\tilde{\mathbf{W}}_p^{(l)}(D)$ into $\mathbf{C}(D)$.

4.5. The spaces $\tilde{\mathbf{W}}_p^{(l)}$ are obviously not complete. However, if we adjoin to them functions that are differentiable in some generalized sense, then they become *B*-spaces.

We shall say that a function $x^{(i)}(s)$, summable in D, is a *generalized derivative* of a function x on D if there is a sequence $\{x_k\}$ of functions, continuously differentiable in D, such that, for every region D_1 whose closure is contained in D, we have

$$\int_{D_1} |x_k(t) - x(t)|\, dt \xrightarrow[k \to \infty]{} 0, \qquad \int_{D_1} \left| \frac{\partial x_k}{\partial t_i} - x^{(i)}(t) \right| dt \xrightarrow[k \to \infty]{} 0. \tag{26}$$

Suppose the function ϕ is continuously differentiable in D and equal to zero on and close to the boundary of D: that is, ϕ is distinct from zero only in a region D_1 whose closure is contained in D. Then we have the equation

$$\int_D x(t) \frac{\partial \phi}{\partial t_i}\, dt = - \int_D \phi(t) x^{(i)}(t)\, dt. \tag{27}$$

For if we substitute x_k for x and $\dfrac{\partial x_k}{\partial t_i}$ for $x^{(i)}$, then (27) is well known as Green's Theorem. The required result is obtained by passing to the limit.

Equation (27) implies, among other things, that the generalized derivative is unique. For if there were two such derivatives, $x^{(i)}$ and $\tilde{x}^{(i)}$, then their difference $\tilde{x}^{(i)} - x^{(i)}$ would satisfy

$$\int_D \phi(t)[\tilde{x}^{(i)}(t) - x^{(i)}(t)]dt = 0, \tag{28}$$

where ϕ is any function of the type mentioned above. Using Luzin's Theorem, we see easily from this that $\tilde{x}^{(i)}(t) = x^{(i)}(t)$ almost everywhere.

In future we shall choose the approximating functions not arbitrarily, but by a special method, namely by means of Steklov averages.

In addition to the requirements imposed on the kernel of the average in IX.1.1, we shall assume that, for every $h > 0$, the function ω_h is continuously differentiable in $[0, \infty)$.

In IX.1.1 we established the following properties of mean functions:

a) $\|x_h\|_{\mathbf{L}^p(D)} \leqslant M\|x\|_{\mathbf{L}^p(D)} \quad (h > 0)$; (29)

b) if $x \in \mathbf{L}^p(D)$ $(p \geqslant 1)$, then $\|x - x_h\|_{\mathbf{L}^p(D)} \to 0$ as $h \to 0$;

c) if $\|x_m - x\|_{\mathbf{L}^p(D)} \to 0$, then $\|(x_m)_h - x_h\|_{\mathbf{L}^p(D)} \to 0$.

In addition, as a result of the special choice of the functions ω_n, we have

d) if x has a generalized derivative $x^{(i)}$ in D, then for every subregion D_1 in the interior of D we have

$$\frac{\partial}{\partial t_i} x_h = [x^{(i)}]_h \tag{30}$$

for all sufficiently small $h > 0$.

Let us prove d). Choose $\{x_k\}$ such that (26) holds. Equation (30) is evidently satisfied for the x_k, since in this case we can differentiate under the integral sign. However, by (c),

$$\left(\frac{\partial x_k}{\partial t_i}\right)_h \to (x^{(i)})_h \text{ in } \mathbf{L}^p(D),$$

$$\frac{\partial}{\partial s_i}[(x_k)_h] = \int \frac{\partial}{\partial s_i}\omega_h(|s-t|)x_k(t)dt \to \int \frac{\partial}{\partial s_i}\omega_h(|s-t|)x(t)dt = \frac{\partial}{\partial s_i}x_h.$$

We can take limits here since $s \in D_1$ and the domain of integration—the ball $K_h(s)$ with radius h and centre at s—is interior to D for small enough h.

Note that it follows from (30) and b) that the mean functions x_h can take the role of the x_k in the definition of the generalized derivatives.

This enables us to extend Sobolev's inequality (16) to functions having only generalized derivatives. For suppose the generalized gradient is summable to exponent p. Then, by b),

$$\int_{D_1} \left|\frac{\partial x_h}{\partial t_i} - x^{(i)}\right|^p dt = \int_{D_1} |[x^{(i)}]_h - x^{(i)}|^p \, dt \xrightarrow[h \to 0]{} 0,$$

and a similar statement holds for the gradient; therefore, applying Sobolev's inequality to the continuously differentiable functions x_h, we obtain

$$\left[\int_{D_1} |x_h|^q \, dt\right]^{1/q} \leqslant M_1 \left\{\frac{1}{\operatorname{mes} D_1}\left|\int_{D_1} x_h(t)\, dt\right| + \left[\int_{D_1} |\operatorname{grad} x_h|^p \, dt\right]^{1/p}\right\}$$

and, letting $h \to 0$, we have Sobolev's inequality for x

$$\left[\int_{D_1} |x|^q \, dt\right]^{1/q} \leqslant M_1 \left\{\frac{1}{\operatorname{mes} D_1}\left|\int_{D_1} x(t)\, dt\right| + \left[\int_{D_1} |\operatorname{grad} x|^p \, dt\right]^{1/p}\right\}.$$

Here we can assume that the constant M_1 is independent of D_1 (Remark 3 following

Theorem 1). If we take an ascending sequence of regions D_1 whose union is D, we obtain the same inequality but now for D. We also see, incidentally, that the function $|x|^q$ is summable—that is, $x \in \mathbf{L}^q(D)$. Thus we have, finally,

$$\left[\int_D |x|^q\,dt\right]^{1/q} \leqslant M_1 \left\{\frac{1}{\operatorname{mes} D_1}\left|\int_D x(t)\,dt\right| + \left[\int_D |\operatorname{grad} x|^p\,dt\right]^{1/p}\right\}.$$

If we adjoin to $\tilde{\mathbf{W}}_p^{(1)}$ the functions having generalized derivatives that are summable to exponent p, then we obtain a complete space,* which we shall call $\mathbf{W}_p^{(1)}$. For suppose that $\|x_k - x_m\|_{\mathbf{W}_p^{(1)}} \to 0$. Then the sequence $\left\{\frac{\partial x_k}{\partial t_i}\right\}$ converges in $\mathbf{L}^p(D)$ to some function $x^{(i)}$, and by Sobolev's inequality the sequence $\{x_k\}$ converges to some function x in $\mathbf{L}^p(D)$. It remains to show that the $x^{(i)}$ are generalized derivatives of x, for if this is so then clearly $x_k \to x$ in $\mathbf{W}_p^{(1)}$. Let $\{D^{(n)}\}$ be a sequence of subregions interior to D whose union is the whole of D. By (26), there exist continuously differentiable functions $\tilde{x}_n$, for $n = 1, 2, \ldots,$ such that

$$\int_{D^{(n)}} |x_n - \tilde{x}_n|\,dt < \frac{1}{n}, \quad \int_{D^{(n)}} \left|x_n^{(i)} - \frac{\partial \tilde{x}_n}{\partial t_i}\right| dt < \frac{1}{n} \quad (i = 1, 2, \ldots, \mu).$$

It is clear from this that the sequences $\{\tilde{x}_n\}$ and $\left\{\frac{\partial \tilde{x}_n}{\partial t_i}\right\}$ converge to x and $x^{(i)}$ respectively in every subregion interior to D, which is what we required to prove.

The space $\mathbf{W}_p^{(1)}$ is separable. For by b) and d) above it is easy to see that we can approximate (in $\mathbf{W}_p^{(1)}$) every function in $\mathbf{W}_p^{(1)}$ by continuously differentiable functions, which in turn can be approximated by polynomials with rational coefficients.

Equation (27) can be used as the basis for the definition of generalized derivatives. In fact, if there exists a function $x^{(i)}$ such that (27) holds for every function ϕ of the relevant type, then $x^{(i)}$ is a generalized derivative of x. For if D_1 is any subregion interior to D then for sufficiently small h we have

$$\frac{\partial x_h}{\partial s_i} = \int \frac{\partial}{\partial s_i}\omega_h(|s-t|)x(t)dt = -\int \frac{\partial}{\partial t_i}[\omega_h(|s-t|)]x(t)dt = \int \omega_h(|s-t|)x^{(i)}(t)dt =$$
$$= (x^{(i)})_h \to x^{(i)},$$

from which it is clear that $x^{(i)}$ is a generalized derivative of x.

Generalized derivatives of higher orders are introduced similarly. Also one can establish, by induction, the completeness and separability of the spaces $\mathbf{W}_p^{(l)}$ obtained from $\tilde{\mathbf{W}}_p^{(l)}$ by adjoining the functions having generalized derivatives of the l-th order that are summable to exponent p. These are called Sobolev spaces.

In the case of the spaces $\mathbf{W}_p^{(l)}$, an embedding of one such space in another ($\mathbf{W}_p^{(l)} \subset \mathbf{W}_q^{(m)}$) not only yields an integral inequality but also tells us that a function that belongs to $\mathbf{W}_p^{(l)}$ is at the same time an element of $\mathbf{W}_q^{(m)}$ (this will be so if $q < \frac{\mu p}{\mu(l-m)p}$).† This is far from trivial, as in the case of the spaces $\tilde{\mathbf{W}}_p^{(l)}$.

* The norm in $\mathbf{W}_p^{(1)}$ is defined by (20), except that we must replace the ordinary derivatives by generalized ones.

† Under these conditions the embedding of $\mathbf{W}_p^{(l)}$ into $\mathbf{W}_q^{(m)}$ will be compact.

We note finally the important fact that the norm can also be introduced in $\mathbf{W}_p^{(l)}$ by other methods equivalent to the preceding one.

Let $f_1, f_2, \ldots, f_k$ be a system of continuous linear functionals on $\mathbf{W}_p^{(l)}$ having the property that the equations

$$f_1(x) = f_2(x) = \ldots = f_k(x) = 0,$$

$$\int_D \left[\sum_{i_1, i_2, \ldots, i_l = 1}^{\mu} \left[\frac{\partial^l(x)}{\partial t_{i_1} \partial t_{i_2} \ldots \partial t_{i_l}} \right]^2 \right]^{p/2} dt = 0 \tag{31}$$

imply that $x = \mathbf{0}$. We shall call such a system of functionals a *defining system.*

Let us establish the inequality

$$\|x\|_{\mathbf{W}_p^{(l)}} \leqslant B\{ |f_1(x)| + |f_2(x)| + \ldots + |f_k(x)| +$$

$$+ \left[\int_D \left[\sum_{i_1, \ldots, i_l} \left(\frac{\partial^l x}{\partial t_{i_1} \partial t_{i_2} \ldots \partial t_{i_l}} \right)^2 \right]^{p/2} dt \right]^{1/p} \Bigg\}, \tag{32}$$

where B is a constant. In fact, if no such constant existed, then there would be a sequence $\{x_n\}$ such that $\|x_n\| = 1$ and

$$|f_1(x_n)| + \ldots + |f_k(x_n)| + \left\{ \int_D \left[\sum_{i_1, \ldots, i_l} \left(\frac{\partial^l x_n}{\partial t_{i_1} \partial t_{i_2} \ldots \partial t_{i_l}} \right)^2 \right]^{p/2} dt \right\}^{1/p} = \varepsilon_n \to 0. \tag{33}$$

Since the mean values of x_n and its generalized derivatives up to the $(l-1)$-st order do not exceed $\|x_n\| = 1$, we can choose a subsequence $\{x_{n_k}\}$ in $\{x_n\}$ such that the mean values of the x_{n_k} and their derivatives form convergent sequences. If we take (33) into account, we see that $\|x_{n_k} - x_{n_j}\| \to 0$. The completeness of $\mathbf{W}_p^{(l)}$ now implies the existence of $y = \lim x_{n_k} \in \mathbf{W}_p^{(l)}$. Applying (33) again, we find that x satisfies the conditions (31), so that $x = \mathbf{0}$. But this is impossible.

The inequality (32) is thus proved. The reverse inequality follows from the continuity of the functionals $f_1, f_2, \ldots, f_k$. Hence we may take the norm in $\mathbf{W}_p^{(l)}$ to be given by the expression

$$\|x\| = |f_1(x)| + |f_2(x)| + \ldots + |f_k(x)| + \left\{ \int_D \left[\sum_{i_1, i_2, \ldots, i_l} \left| \frac{\partial^l x}{\partial t_{i_1} \partial t_{i_2} \ldots \partial t_{i_l}} \right|^2 \right]^{p/2} dt \right\}^{1/p}.$$

Other expressions also can be used for the norm $\mathbf{W}_p^{(l)}$. For example, we can write

$$\|x\| = \left\{ \int_D \left[\sum_{k=0}^{t} \sum_{i_1, \ldots, i_k = 1}^{\mu} \left| \frac{\partial^k x}{\partial t_{i_1} \partial t_{i_2} \ldots \partial t_{i_k}} \right|^2 \right]^{p/2} dt \right\}^{1/p}.$$

Notice that, with this definition $\mathbf{W}_?^{(l)}$ is a Hilbert space, its inner product being defined by

$$(x, y) = \int_D \sum_{k=1}^{l} \sum_{i_1, \ldots, i_k} \frac{\partial^k x}{\partial t_{i_1} \partial t_{i_2} \ldots \partial t_{i_k}} \frac{\partial^k y}{\partial t_{i_1} \partial t_{i_2} \ldots \partial t_{i_k}} dt.$$

Let us also point out the following important fact. Suppose we have an embedding $\mathbf{W}_q^{(m)}(S) \subset \mathbf{W}_p^{(l)}(D)$, where S is a ν-dimensional manifold lying in the region D. Then, for a function in $\mathbf{W}_p^{(l)}(D)$, we can speak of the values of its m-th derivatives on the surface S, which are defined to be functions summable to exponent q on S. When the embedding theorems guarantee that the generalized derivatives of some order are continuous, it is easy to convince oneself that these derivatives exist in the usual sense.*

Finally, it is useful to bear in mind that if certain linearity conditions are imposed on functions in $\mathbf{W}_p^{(l)}(D)$, relating to the values of the m-th derivatives on the surface S, where $\mathbf{W}_p^{(l)}(D) \subset \mathbf{W}_q^{(m)}(S)$, then these conditions determine a closed (linear) manifold in $\mathbf{W}_p^{(l)}(D)$.

4.6. In conclusion, we present, as an application, a proof of the existence of a solution of Dirichlet's problem by variational methods.

Consider the boundary problem

$$\Delta x = 0, \qquad x|_S = \phi. \tag{34}$$

We look for a solution among the functions having a finite Dirichlet integral

$$H(x) = \int_D \sum_{k=1}^{\mu} \left(\frac{\partial x}{\partial t_k}\right)^2 dt < \infty, \tag{35}$$

that is, we assume that $x \in \mathbf{W}_2^{(1)}(D)$. By Theorem 4 it follows from this that $x \in \mathbf{L}^2(S)$. For the dimension ν of the manifold S is $\mu - 1$ and

$$q = 2 < \frac{(\mu-1)2}{\mu-2} = 2 + \frac{2}{\mu-2}, \qquad \nu = \mu - 1 > \mu - 2 = \mu - p.$$

Moreover,

$$\int_S |x(s+\Delta s) - x(s)|^2 \, ds \xrightarrow[\Delta s \to 0]{} 0,$$

that is, each of the values of x on the bounding surface is a limit (in mean) of its values inside (on the translated surface).

Thus a function with finite Dirichlet integral must have boundary values that are square summable on the surface.

However, not every square-summable function ϕ defined on S can be taken as giving the boundary values. We call ϕ *admissible* if there is some $x \in \mathbf{W}_2^{(1)}(D)$ such that $\phi = x|_S$.

In the boundary problem (34) we shall assume ϕ to be admissible.

Denote by $\mathbf{W}_2^{(1)}(\phi)$ the set of functions in $\mathbf{W}_2^{(1)}(D)$ whose boundary values on S are equal to ϕ. We have $0 \leqslant H(x) < \infty$ for $x \in \mathbf{W}_2^{(1)}(\phi)$ so

$$\inf\{H(x): x \in \mathbf{W}_2^{(1)}(\phi)\} = d$$

is finite. We construct a sequence $\{x_n\} \subset \mathbf{W}_2^{(1)}(\phi)$ that minimizes the functional H; that is, one such that

$$d = \lim_{n \to \infty} H(x_n).$$

THEOREM 7. *The minimizing sequence $\{x_n\}$ converges in $\mathbf{W}_2^{(1)}$; the limit function belongs to*

* More precisely, one of the equivalent functions corresponding to x will have this property.

$\mathbf{W}_2^{(1)}(\phi)$ *and the value it assigns to* H *is minimal among all the functions in* $\mathbf{W}_2^{(1)}(\phi)$.

Proof. If we introduce a scalar product in $\mathbf{W}_2^{(1)}$, setting

$$(x, y) = \int_S x(s)y(s)\,ds + \int_D \sum_{k=1}^{\mu} \frac{\partial x}{\partial t_k}\frac{\partial y}{\partial t_k}\,dt \qquad (x, y \in \mathbf{W}_2^{(1)}),$$

then $\mathbf{W}_2^{(1)}$ becomes a Hilbert space. The norm of an arbitrary element $x \in \mathbf{W}_2^{(1)}(\phi)$ and the value of H at this element are connected by the equation

$$\|x\|^2 = \int_S |\phi(s)|^2\,ds + H(x). \tag{36}$$

Denote the set of all $x \in \mathbf{W}_2^{(1)}$ that vanish on S by $\mathring{\mathbf{W}}_2^{(1)}$. As we remarked at the end of 4.5, $\mathring{\mathbf{W}}_2^{(1)}$ is a closed subspace of $\mathbf{W}_2^{(1)}$.

Consider an element $x_0 \in \mathbf{W}_2^{(1)}(\phi)$. We may assume that x_0 is orthogonal to $\mathring{\mathbf{W}}_2^{(1)}$. For otherwise we can express x_0 in the form $x_0 = x_0' + x_0''$, where x_0' is orthogonal to $\mathring{\mathbf{W}}_2^{(1)}$ and $x_0'' \in \mathring{\mathbf{W}}_2^{(1)}$. Since

$$x_0'|_S = x_0|_S - x_0''|_S = \phi,$$

we have $x_0' \in \mathbf{W}_2^{(1)}(\phi)$, and we can take x_0' in place of x_0.

Let us show that $H(x_0) = d$. Let x be an arbitrary element of $\mathbf{W}_2^{(1)}(\phi)$. Since $x - x_0$ belongs to $\mathring{\mathbf{W}}_2^{(1)}$, it is orthogonal to x_0 and so, in view of (36), we have

$$H(x) - H(x_0) = \|x\|^2 - \|x_0\|^2 = \|x - x_0\|^2 \geqslant 0, \tag{37}$$

that is,

$$d \leqslant H(x_0) \leqslant \inf H(x) = d$$

or, in other words, $d = H(x_0)$.

Setting $x = x_n$ in (37), we now obtain

$$H(x_n) - H(x_0) = \|x_n - x_0\|^2 \qquad (n = 1, 2, \ldots).$$

But $H(x_0) = d$, so $H(x_n) - H(x_0) \to 0$. Hence $x_n \to x_0$.

This proves the theorem.

REMARK. It follows from (37) that the element $x_0 \in \mathbf{W}_2^{(1)}(\phi)$ assigning a minimum to H is unique.

We shall not go into the proof of the fact that the function x_0 that minimizes H yields a solution to the boundary problem (34) and is moreover the unique one, but refer the reader for this and also for other applications of the embedding theorems in mathematical physics to the monograph by S. L. Sobolev (see Sobolev-I), where these questions are discussed in detail.

The reader can extend his familiarity with embedding theorems in the following monographs: Sobolev-I, II; Besov, Il'in and Nikol'skii.

PART II

FUNCTIONAL EQUATIONS

XII

THE ADJOINT EQUATION

§ 1. Theorems on inverse operators*

IN THIS section we add to the information on inverse operators given in Part I (see § 4 of Chapter V).

1.1. Let us recall the definitions given in Part I. Let U be a continuous linear operator mapping a normed space $\mathbf{X}$ into a normed space $\mathbf{Y}$. If there exists an operator V mapping $\mathbf{Y}$ into $\mathbf{X}$ such that

$$VU = I_{\mathbf{X}} \qquad (I_{\mathbf{X}}x = x,\ x \in \mathbf{X}), \tag{1}$$

$$UV = I_{\mathbf{Y}} \qquad (I_{\mathbf{Y}}y = y,\ y \in \mathbf{Y}), \tag{2}$$

then V is said to be an inverse of U ($V = U^{-1}$).

The existence of a (not necessarily continuous) inverse operator U^{-1} is equivalent to the condition that U should define a one-to-one mapping from $\mathbf{X}$ onto $\mathbf{Y}$. If, in addition, U^{-1} is continuous, then this mapping will be an isomorphism.

If only one of the equations (1) and (2) is satisfied, then the operator V is called a left inverse or right inverse, respectively (we then write $V = U_l^{-1}$ or $V = U_r^{-1}$, respectively†). It was proved in V.4.4 that a necessary and sufficient condition for the existence of a left inverse is that

$$\|U(x)\| \geqslant m\|x\| \qquad (x \in \mathbf{X}), \tag{3}$$

where $m > 0$ is independent of x. If, moreover, U maps $\mathbf{X}$ onto the whole of $\mathbf{Y}$, then the left inverse operator will also be a right inverse—that is, in this case there exists a continuous (two-sided) inverse operator U^{-1}.

1.2. Let us now prove a theorem.

THEOREM 1. *If a continuous linear operator U, mapping a B-space $\mathbf{X}$ into a normed space $\mathbf{Y}$, has a continuous left inverse, then $\mathbf{Y}' = U(\mathbf{X})$ is a B-space.*

Proof. We need only establish that $\mathbf{Y}'$ is complete. Let $\{y_n\}$ be a Cauchy sequence of elements of $\mathbf{Y}'$. Write $x_n = U^{-1}(y_n)$ ($y_n = U(x_n)$; $n = 1, 2, \ldots$). By our remark in 1.1,

$$\|y_n - y_k\| = \|U(x_n) - U(x_k)\| \geqslant m\|x_n - x_k\|,$$

where m is a positive constant. Hence

$$\lim_{k,\, n \to \infty} \|x_n - x_k\| = 0.$$

Thus the element $x_0 = \lim\limits_{n \to \infty} x_n$ exists in $\mathbf{X}$. Since $\lim y_n = \lim U(x_n) = U(x_0)$, we see,

---

* For the contents of this section, see von Neumann [1], and also Schauder [2].

† The subscripts "l" and "r" will sometimes be omitted.

writing $y_0 = U(x_0)$, that $y_0 \in \mathbf{Y}'$ and $y_n \to y_0$. This proves that $\mathbf{Y}'$ is complete.

COROLLARY. *Under the conditions of the theorem,* $\mathbf{Y}'$ *is a closed set in* $\mathbf{Y}$.

1.3. Suppose we have two normed spaces $\mathbf{X}$ and $\mathbf{Y}$ and a continuous linear operator U mapping $\mathbf{X}$ into $\mathbf{Y}$. The set $\mathbf{X}_0 = U^{-1}(\mathbf{0})$ is obviously a closed subspace of $\mathbf{X}$. Denote the factor-space by $\overline{\mathbf{X}} = \mathbf{X}/\mathbf{X}_0$ (see IV.1.8). Let $\bar{x} \in \overline{\mathbf{X}}$; we consider any $x \in \bar{x}$ and write

$$\bar{U}(\bar{x}) = U(x). \tag{4}$$

The definition of $\bar{U}(\bar{x})$ does not depend on the choice of the element $x \in \bar{x}$, for if $x', x'' \in \bar{x}$, then $x' - x'' \in \mathbf{X}_0$ and therefore $U(x') = U(x'')$. Hence (4) defines an operator $\bar{U}$ mapping $\overline{\mathbf{X}}$ into $\mathbf{Y}$. This operator is homogeneous and additive. It is also continuous, for if we take the infimum of the right-hand side of the inequality

$$\|\bar{U}(\bar{x})\| = \|U(x)\| \leqslant \|U\| \|x\| \qquad (x \in \bar{x}),$$

we can write

$$\|\bar{U}(\bar{x})\| \leqslant \|U\| \, \|\bar{x}\| \qquad (\bar{x} \in \overline{\mathbf{X}}).$$

It follows from the definition of $\bar{U}$ that $U = \bar{U}\phi$, where ϕ is the natural homomorphism of $\mathbf{X}$ onto $\mathbf{X}/\mathbf{X}_0$ (see IV.1.8), and $\|U\| = \|\bar{U}\|$.

Unlike the original operator U, the operator $\bar{U}$ determines a one-to-one mapping (from $\overline{\mathbf{X}}$ into $\mathbf{Y}$). For if $\bar{U}(\bar{x}) = \mathbf{0}$, then $U(x) = \mathbf{0}$ for each $x \in \bar{x}$; that is, $x \in U^{-1}(\mathbf{0}) = \mathbf{X}_0$; and so $\bar{x}$ coincides with the class $\mathbf{X}_0$, which is the zero element of $\overline{\mathbf{X}}$.

In the case where U maps $\mathbf{X}$ onto $\mathbf{Y}$, the operator $\bar{U}$ also maps $\overline{\mathbf{X}}$ onto $\mathbf{Y}$. If, in this situation, there is a continuous inverse operator $\bar{U}^{-1}$, then $\mathbf{X}$ and $\mathbf{Y}$ are said to be *homomorphic* and the operator U is a *homomorphism* from $\mathbf{X}$ onto $\mathbf{Y}$.

The fact that U is a homomorphism from $\mathbf{X}$ onto $\mathbf{Y}$ can be characterized in terms of the original spaces $\mathbf{X}$ and $\mathbf{Y}$ by the following two conditions:

1) $U(\mathbf{X}) = \mathbf{Y}$;
2) there exists $m > 0$ such that for each $y \in \mathbf{Y}$ there is an $x \in \mathbf{X}$ with

$$y = U(x), \ \|y\| \geqslant m\|x\|.$$

For if U is a homomorphism, then the first condition is obviously satisfied. Further, if we choose $x \in \bar{x} = \bar{U}^{-1}(y)$ according to (4) of IV.1.8, we see that

$$\|x\| \leqslant 2\|\bar{x}\| \leqslant 2\|\bar{U}^{-1}\| \, \|y\|$$

and we can take $m = \dfrac{1}{2\|\bar{U}^{-1}\|}$.

Conversely, if both the conditions are satisfied, then from the first we have $\bar{U}(\overline{\mathbf{X}}) = \mathbf{Y}$. Let $\bar{x} \in \overline{\mathbf{X}}$; we associate with $y = \bar{U}(\bar{x})$ an element $x \in \mathbf{X}$ according to the second condition. As

$$\bar{U}(\phi(x)) = U(x) = y = \bar{U}(\bar{x}),$$

the fact that U is one-to-one implies that $\phi(x) = \bar{x}$ and hence

$$\|\bar{U}(\bar{x})\| = \|y\| \geqslant m\|x\| \geqslant m\|\bar{x}\|.$$

As we remarked in 1.1, this, in conjunction with the relation $\bar{U}(\overline{\mathbf{X}}) = \mathbf{Y}$, guarantees the existence of a continuous (two-sided) inverse operator U^{-1}.

1.4. The converse to Theorem 1 plays a fundamental role in the theory of functional equations. It forms part of the following proposition.

LEMMA 1. *Let U be a continuous linear operator mapping a B-space* $\mathbf{X}$ *into a normed space* $\mathbf{Y}$. *If the image $U(B)$ of the unit ball B (with centre at the origin) in* $\mathbf{X}$ *is dense in the ball S_r of radius r (also with centre at the origin) in* $\mathbf{Y}$, *then U is a homomorphism from* $\mathbf{X}$ *onto* $\mathbf{Y}$. *In particular, if U defines a one-to-one mapping, then it has a continuous two-sided inverse U^{-1}.*

Proof. We verify that the two conditions of the preceding subsection are satisfied. We may clearly assume that B and S_r are closed; we prove that

$$U(B) \supset S_{r/2}. \tag{5}$$

Take a sequence $\{\varepsilon_k\}$ of positive numbers such that $\sum_{k=1}^{\infty} \varepsilon_k \leqslant 1$, and consider an element $y \in S_r$. As $\overline{U(B)} \supset S_r$, there exists $y_1 \in U(B)$ such that

$$\|y - y_1\| \leqslant \varepsilon_1 r.$$

Denote by x_1 an element such that $y_1 = U(x_1)$. Let B_h be the closed ball in $\mathbf{X}$ of radius h (with centre at the origin). It follows from the hypotheses of the lemma that $\overline{U(B_h)} \supset S_{hr}$. Therefore, since $y - y_1 \in S_{\varepsilon_1 r}$, there exists an element $x_2 \in B_{\varepsilon_1}$ such that

$$\|y - (y_1 + y_2)\| \leqslant \varepsilon_2 r \qquad (y_2 = U(x_2)).$$

Continuing in this way, we find two sequences $\{y_n\} \subset \mathbf{Y}$ and $\{x_n\} \subset \mathbf{X}$ such that

$$y_n = U(x_n), \ x_n \in B_{\varepsilon_{n-1}}, \left\| y - \sum_{k=1}^{n} y_k \right\| \leqslant \varepsilon_n r \tag{6}$$

$$(n \in \mathbb{N}, \quad \varepsilon_0 = 1).$$

Since $\|x_n\| \leqslant \varepsilon_{n-1}$ and $\mathbf{X}$ is complete, the series $\sum_{k=1}^{\infty} x_k$ converges. If we denote its sum by x, then

$$\|x\| \leqslant \sum_{k=1}^{\infty} \|x_k\| \leqslant \sum_{k=1}^{\infty} \varepsilon_{k-1} \leqslant 2,$$

that is, $x \in B_2$. Further,

$$U(x) = \sum_{k=1}^{\infty} U(x_k) = \sum_{k=1}^{\infty} y_k.$$

But it is clear from (6) that $\sum_{k=1}^{\infty} y_k = y$. Hence $y = U(x)$. We have thus proved that $U(B_2) \supset S_r$, which is equivalent to (5).

Since (5) implies that $U(B_2) \supset S_{nr/2}$, we have

$$U(\mathbf{X}) = \bigcup_{n=1}^{\infty} U(B_n) \supset \bigcup_{n=1}^{\infty} S_{nr/2} = \mathbf{Y}$$

and so the first condition is verified.

Moreover, if $y \neq \mathbf{0}$ is an arbitrary element of $\mathbf{Y}$, then we have

$$y' = \frac{r}{2\|y\|} y \in S_{r/2}$$

and, in view of (5), we can find an element $x' \in B$ such that $y' = U(x')$. Writing $x = \frac{2\|y\|}{r} x'$, we have

$$U(x) = y, \ \|x\| = \frac{2}{r}\|y\|\|x'\| \leqslant \frac{2}{r}\|y\|.$$

Hence the second condition is also satisfied. This proves the lemma.

COROLLARY. *If the conditions of the lemma are satisfied, then* **Y** *is complete.*

For, since U is a homomorphism, the operator $\bar{U}$ mapping the factor-space $\bar{\mathbf{X}} = \mathbf{X}/\mathbf{X}_0$ ($\mathbf{X}_0 = U^{-1}(0)$) onto **Y** has a continuous inverse. As $\mathbf{Y} = \bar{U}(\bar{\mathbf{X}})$, we can apply Theorem 1.

The hypothesis of this lemma is difficult to verify. The hypothesis of the following theorem is more convenient.

THEOREM 2 (Banach). *If the set* $U(\mathbf{X})$ *is of second category in* **Y**, *then the hypothesis of the lemma is satisfied, and hence* U *is a homomorphism from* **X** *onto* **Y**.

Proof. Keeping the notation of the lemma, we prove that if the hypothesis of the lemma is not satisfied, then $U(B)$ is nowhere dense. In fact, if we assume the contrary, then we can find a ball $S(y_0, r)$ in **Y** (with centre at y_0 and radius r) such that

$$\overline{U(B)} \supset S(y_0, r). \tag{7}$$

The set $U(B)$ is symmetric—that is, it contains together with each element y also the element $-y$. Its closure $\overline{U(B)}$ is also clearly symmetric. Thus, by (7), we can write

$$\overline{U(B)} \supset S(-y_0, r).$$

Choose $y \in S_r$. Then $y_0 + y \in S(y_0, r)$ and $-y_0 + y \in S(-y_0, r)$, so these elements belong to $\overline{U(B)}$. But $U(B)$ is convex, and therefore so is $\overline{U(B)}$; hence it must contain half the sum of each pair of its elements. In particular, then,

$$y = \frac{(y_0 + y) + (-y_0 + y)}{2} \in \overline{U(B)}.$$

Hence $\overline{U(B)} \supset S_r$.

Thus if the hypothesis of Lemma 1 is not satisfied, $U(B)$ is nowhere dense. The same is true of the sets $U(B_n)$ $(n \in \mathbb{N})$. However,

$$U(\mathbf{X}) = \bigcup_{n=1}^{\infty} U(B_n),$$

and so we see that $U(\mathbf{X})$ is of first category.

The theorem is therefore proved.

We note one corollary of this theorem, which is a generalization of Theorem 1.

COROLLARY. *If a continuous operator* U *defines a one-to-one mapping from a B-space* **X** *onto a closed subspace of a B-space* **Y**, *then the inverse operator* U^{-1} *is continuous.*

For a closed subspace of a B-space is itself a B-space, and hence is a set of second category in itself (see I.4.7).

1.5. We mention some immediate applications of Theorem 2.

Assume that a norm has been introduced on a vector space **X** in two different ways. We denote the norm of an element $x \in \mathbf{X}$ according to the first method of definition by $\|x\|_1$, and that according to the second by $\|x\|_2$. The set **X** is thus turned into two different normed spaces, which we shall denote by $\mathbf{X}_1$ and $\mathbf{X}_2$ respectively. Although $\mathbf{X}_1$ and $\mathbf{X}_2$

should be regarded as distinct normed spaces, there may in fact be no qualitative difference between them. This will be the case when every sequence $\{x_n\}$ that converges in one space also converges in the other to the same element. In this case the norms in $\mathbf{X}_1$ and $\mathbf{X}_2$ are said to be equivalent; this means that $\mathbf{X}_1$ and $\mathbf{X}_2$ are isomorphic spaces (see IV.1.3).

THEOREM 3. *Let $\mathbf{X}_1$ and $\mathbf{X}_2$ be two B-spaces with $\mathbf{X}_1 \subset \mathbf{X}_2$.* If, whenever $x_n \to x$ in $\mathbf{X}_1$, we have also $x_n \to x$ in $\mathbf{X}_2$, then either $\mathbf{X}_1 = \mathbf{X}_2$ and the norms in $\mathbf{X}_1$ and $\mathbf{X}_2$ are equivalent, or $\mathbf{X}_1$ is a set of the first category in $\mathbf{X}_2$.*

Proof. Write U for the embedding operator from $\mathbf{X}_1$ into $\mathbf{X}_2$: that is, the operator associating to an element $x \in \mathbf{X}_1$ the same element x, but viewed as an element of $\mathbf{X}_2$. Under the conditions of the theorem, U is a continuous linear operator. If the set $\mathbf{X}_1 = U(\mathbf{X}_1)$ is of second category in $\mathbf{X}_2$, then, by Theorem 2, $\mathbf{X}_1 = \mathbf{X}_2$ and U has a continuous inverse. Hence if $x_n \to x$ in $\mathbf{X}_2$, then $x_n = U^{-1}(x_n) \to U^{-1}(x) = x$ in $\mathbf{X}_1$; that is, the norms in the spaces are equivalent.

Taking $\mathbf{X}_1 = \mathbf{C}^{(1)}(D)$, $\mathbf{X}_2 = \mathbf{C}(D)$ in the theorem, we see that the set of all continuously differentiable functions is of first category in the space $\mathbf{C}$ of all continuous functions. In precisely the same way, we can verify that the set of measurable almost everywhere bounded functions is of first category in the space $\mathbf{L}^1$, and so on.

1.6. A (not necessarily linear) mapping T from a set Ω in a normed space $\mathbf{X}$ into a normed space $\mathbf{Y}$ is said to be *closed* if, whenever

$$x_n \in \Omega \quad (n = 1, 2, \ldots), \qquad x_n \to x_0, \qquad T(x_n) \to y_0,$$

we have $x_0 \in \Omega$ and $T(x_0) = y_0$.

A continuous linear operator defined on a closed set is obviously closed. The converse statement is also true. Namely, we have

THEOREM 4. *Let T be a closed linear operator, mapping a closed linear manifold Ω in a B-space $\mathbf{X}$ into a B-space $\mathbf{Y}$. Then $\mathbf{Y}$ is a continuous operator.*

Proof. We may assume that $\Omega = \mathbf{X}$ (since Ω is itself a B-space). We introduce a new norm in $\mathbf{X}$ by setting

$$\|x\|_1 = \|x\| + \|T(x)\| \quad (x \in \mathbf{X}). \tag{8}$$

It is not hard to verify that the axioms for a normed space are satisfied for this norm. Let us verify that $\mathbf{X}$ is also complete in the new norm. Suppose

$$\lim_{k, n \to \infty} \|x_n - x_k\|_1 = 0.$$

This means that

$$\lim_{k, n \to \infty} \|x_n - x_k\| = 0 \quad \text{and} \quad \lim_{k, n \to \infty} \|T(x_n) - T(x_k)\| = 0.$$

Since the spaces $\mathbf{X}$ (with the given norm) and $\mathbf{Y}$ are complete, we conclude from this that the limits

$$\lim_{n \to \infty} x_n = x \text{ and } \lim_{n \to \infty} T(x_n) = y_0$$

both exist.

As T is a closed operator, $y_0 = T(x_0)$. But then we have

$$\lim_{n \to \infty} \|x_n - x_0\|_1 = \lim_{n \to \infty} \|x_n - x_0\| + \lim_{n \to \infty} \|T(x_n) - T(x_0)\| = 0$$

* Here we assume that the embedding $\mathbf{X}_1 \subset \mathbf{X}_2$ preserves the algebraic operations—that is, that $\mathbf{X}_1$ can be regarded as a linear manifold in $\mathbf{X}_2$.

and this proves that $\mathbf{X}$ is complete in the new norm.

Since

$$\|x\| \leqslant \|x\|_1,$$

we have $\|x_n\| \to 0$ whenever $\|x_n\|_1 \to 0$. Applying the preceding theorem, we see that

$$\|x\|_1 \leqslant M\|x\|.$$

Hence we certainly have

$$\|Tx\| \leqslant M\|x\|,$$

which means that T is a continuous operator.

REMARK. The class of closed linear operators defined on the whole space (or on a closed linear manifold) coincides with the class of continuous linear operators. However, if we consider closed linear operators on a (non-closed) linear manifold, then these form a substantially wider class than that of continuous operators. For example, in $\mathbf{L}^2(a, b)$ one can easily verify that the operator T given by

$$y = T(x), \qquad y(t) = \frac{dx(t)}{dt},$$

defined on the set Ω of all absolutely continuous functions whose derivatives belong to $\mathbf{L}^2(a, b)$, is closed but not continuous.

Operators that are closed but not continuous are studied mainly in the case where $\mathbf{X} = \mathbf{Y}$ is a Hilbert space. The study of the general case is greatly complicated by the complexity of the structure of an arbitrary B-space.

§ 2. The connection between an equation and its adjoint

In this section we shall consider, in addition to the equation

$$U(x) = y, \tag{1}$$

also the equation

$$U^*(g) = f, \tag{2}$$

which we call the equation *adjoint* to (1). Here we assume, as usual, that U is a continuous linear operator mapping a space $\mathbf{X}$ into a space $\mathbf{Y}$. We shall assume, without any explicit statement in what follows, that these spaces are complete, although some of the theorems of the present section are also true without assuming the spaces complete.

The theorems in this section were proved for the space $\mathbf{L}^2(a, b)$ by Hellinger and Toeplitz [1], [2], and for the spaces l^p and $\mathbf{L}^p(a, b)$ by Riesz [3]; for the general case, see Banach; Zaanen-I.

2.1. The so-called zero sets will play a large role in what follows. Let Γ be a set of linear functionals on a space $\mathbf{X}$. We denote by $\mathbf{N}(\Gamma)$ the set of all $x \in \mathbf{X}$ such that $f(x) = 0$ for every $f \in \Gamma$. If $E \subset \mathbf{X}$, then we denote by $\mathbf{N}^*(E)$ the set of all functionals $f \in \mathbf{X}^*$ that vanish for each element of E. The sets $\mathbf{N}(\Gamma)$ and $\mathbf{N}^*(E)$ are called the *zero sets* of Γ and E respectively.

If we consider the dual pairing $\langle \mathbf{X}, \mathbf{X}^* \rangle$, then clearly $\mathbf{N}(\Gamma)$ is the annihilator $\Gamma^\perp$ of Γ, while $\mathbf{N}^*(E)$ is the annihilator of E, introduced in a general situation in III.3.2. Bearing in mind the properties of annihilators and polars set out in III.3.2, and also the fact that closure in the norm and weak closure coincide in $\mathbf{X}$, we see that: 1) $\mathbf{N}(\Gamma)$ is a closed linear manifold in $\mathbf{X}$, 2) $\mathbf{N}(\mathbf{N}^*(E))$ is the closed linear hull of E, and 3) if $\mathbf{X}_0$ is a closed subspace of $\mathbf{X}$, then

$$\mathbf{X}_0 = \mathbf{N}(\mathbf{N}^*(\mathbf{X}_0)).$$

Similarly we see that: 1) $\mathbf{N}^*(E)$ is a weak* closed linear manifold in $\mathbf{X}^*$, 2) $\mathbf{N}^*(\mathbf{N}(\Gamma))$ is the weak* closed linear hull of Γ, and 3) if $\mathbf{Z}$ is a weak* closed subspace of $\mathbf{X}^*$, then

$$\mathbf{Z} = \mathbf{N}^*(\mathbf{N}(\mathbf{Z})).$$

Since not every subspace of $\mathbf{X}^*$ that is closed in the norm is weak* closed when $\mathbf{X}$ is non-reflexive, it follows that not every closed subspace has the form $\mathbf{N}^*(E)$ $(E \subset \mathbf{X})$.

Let $\pi : \mathbf{X} \to \mathbf{X}^{**}$ be the canonical embedding. If E is a subset of $\mathbf{X}$, then we write E_0^{**} for $\pi(E)$: that is, E_0^{**} is the set of all functionals on $\mathbf{X}^*$ having the form F_x $(x \in \mathbf{X})$, where as usual

$$F_x(f) = \overline{f(x)} \qquad (f \in \mathbf{X}^*).$$

Similarly, for $\mathbf{Y}$ we use the notation G_y $(y \in \mathbf{Y})$.

Obviously we have the equation

$$\mathbf{N}^*(E) = \mathbf{N}(E_0^{**}).$$

For a set $\Gamma \subset \mathbf{X}^*$, we can consider two zero sets: $\mathbf{N}(\Gamma) \subset \mathbf{X}$ and $\mathbf{N}^*(\Gamma) \subset \mathbf{X}^{**}$. It is easy to see that

$$[\mathbf{N}(\Gamma)]_0^{**} = \mathbf{N}^*(\Gamma) \cap \mathbf{X}_0^{**},$$

where, in accordance with the notation we are using, $\mathbf{X}_0^{**}$ means $\pi(\mathbf{X})$.

Finally, we introduce the set $\mathbf{N}(U)$, *the zero set of the operator* U, by which we mean the set of all $x \in \mathbf{X}$ mapped to zero by U. In other words, we have $\mathbf{N}(U) = U^{-1}(\mathbf{0})$.

2.2. The following theorem gives some information about which elements belong to the image $U(\mathbf{X})$ of the space $\mathbf{X}$.

THEOREM 1. *Write* $\mathbf{Y}' = U(\mathbf{X})$ *and let* $\mathbf{Y}_1$ *be the closure of* $\mathbf{Y}'$. *Then*

$$\mathbf{Y}_1 = \mathbf{N}(\mathbf{N}(U^*)),$$

that is, $\mathbf{Y}_1$ *is the set of common zeros of those functionals* $g \in \mathbf{Y}^*$ *at which the operator* U^* *vanishes.*

Proof. Let us first prove that

$$\mathbf{N}^*(\mathbf{Y}') = \mathbf{N}(U^*). \tag{3}$$

Let $g \in \mathbf{N}^*(\mathbf{Y}')$. Then, as $U(x) \in \mathbf{Y}'$ for each $x \in \mathbf{X}$, we have

$$U^*(g)(x) = g(U(x)) = 0 \qquad (x \in \mathbf{X}), \tag{4}$$

that is, $U^*(g) = \mathbf{0}$ and $g \in \mathbf{N}(U^*)$. Equation (4) also shows that if $U^*(g) = \mathbf{0}$, then $g \in \mathbf{N}^*(\mathbf{Y}')$. Hence (3) is established. Taking the zero set of each side of this equation, we obtain

$$\mathbf{N}(\mathbf{N}^*(\mathbf{Y}')) = \mathbf{N}(\mathbf{N}(U^*)).$$

But as $\mathbf{Y}'$ is a linear manifold its closure coincides with $\mathbf{N}(\mathbf{N}^*(\mathbf{Y}'))$, as we remarked in 2.1. Thus the theorem is proved.

The dual theorem is also true.

THEOREM 1*. *Let $\tilde{\mathbf{X}}^*$ denote the weak* closure of the set $U^*(\mathbf{Y}^*)$. Then*

$$\tilde{\mathbf{X}}^* = \mathbf{N}^*(\mathbf{N}(U)).$$

Proof. We rewrite (3), replacing U by U^*:

$$\mathbf{N}^*(U^*(\mathbf{Y}^*)) = \mathbf{N}(U^{**}),$$

and consider the intersections of both sides with $\mathbf{X}_0^{**}$. Since we have

$$U^{**}(F_x)(g) = F_x(U^*(g)) = g(U(x)) = G_{U(x)}(g)$$
$$(g \in \mathbf{Y}^*, \ x \in \mathbf{X}),$$

that is,

$$U^{**}(F_x) = G_{U(x)} \quad (x \in \mathbf{X}),$$

the intersection $\mathbf{N}(U^{**}) \cap \mathbf{X}_0^{**}$ consists of those functionals $F_x \in \mathbf{X}_0^{**}$ for which $x \in \mathbf{N}(U)$. In other words,

$$\mathbf{N}(U^{**}) \cap \mathbf{X}_0^{**} = [\mathbf{N}(U(\mathbf{Y}^*))]_0^{**}.$$

On the other hand, we have

$$\mathbf{N}^*(U^*(\mathbf{Y}^*)) \cap \mathbf{X}_0^{**} = [\mathbf{N}(U^*(\mathbf{Y}^*))]_0^{**}.$$

Comparing the equations thus obtained, we find that

$$\mathbf{N}(U^*(\mathbf{Y}^*)) = \mathbf{N}(U),$$

and therefore

$$\mathbf{N}^*(\mathbf{N}(U^*(\mathbf{Y}^*))) = \mathbf{N}^*(\mathbf{N}(U)).$$

But $\tilde{\mathbf{X}}^* = \mathbf{N}^*(\mathbf{N}(U^*(\mathbf{Y}^*)))$, and this proves the theorem.

2.3. The following two theorems show how the solubility of one of the equations (1) and (2) for every right-hand side is related to the unique solubility of the other.

THEOREM 2. *A necessary and sufficient condition for* (2) *to have a solution for every $f \in \mathbf{X}^*$ is that the operator U have a continuous left inverse U^{-1}.*

Proof. Necessity. Suppose (2) has a solution for every $f \in \mathbf{X}^*$. In other words, we suppose that $U^*(\mathbf{Y}^*) = \mathbf{X}^*$. Consider an arbitrary element $x \neq \mathbf{0}$ in $\mathbf{X}$. By Theorem V.7.2, there exists a functional $f \in \mathbf{X}^*$ such that

$$\|f\| = 1, \qquad f(x) = \|x\|.$$

Also, by Theorem 1.2, U^* is a homomorphism, so there exists a $g \in \mathbf{Y}^*$ such that

$$U^*(g) = f, \qquad \|g\| \leqslant m\|f\|$$

where m is independent of f, and hence of x. Taking into account what has been said, we can write

$$\|x\| = f(x) = g(U(x)) \leqslant \|g\|\,\|U(x)\| \leqslant m\|U(x)\|.$$

Thus

$$\|U(x)\| \geqslant \frac{1}{m}\|x\| \qquad (x \in \mathbf{X}),$$

and so we deduce the existence of a continuous left inverse operator U^{-1}.

Sufficiency. Let $f_0 \in \mathbf{X}^*$. Assuming the existence of a continuous $U_l^{-1} = U^{-1}$, write

$$g'(y) = f_0(U^{-1}(y)) \qquad (y \in U(\mathbf{X})).$$

The functional g' defined on $\mathbf{Y}' = U(\mathbf{X})$ is continuous and linear. Extending it to the whole of $\mathbf{Y}$, we obtain a functional $g_0 \in \mathbf{Y}^*$. As $U(x) \in \mathbf{Y}'$, for each $x \in \mathbf{X}$, we have

$$U^*(g_0)(x) = g_0(U(x)) = g'(U(x)) = f_0(U^{-1}U(x)) = f_0(x).$$

Therefore $U^*g = f_0$; that is, (2) is soluble for any right-hand side.

THEOREM 2*. *A necessary and sufficient condition for* (1) *to have a solution for every* $y \in \mathbf{Y}$ *is that* U^* *have a continuous left inverse* U^{*-1}.

Proof. Necessity. This is verified in an analogous manner to the necessity of the condition in Theorem 2. Let $y \in \mathbf{Y}$ be an arbitrary element with $\|y\| \leqslant 1$. By Theorem 1.2, there exists an element x in $\mathbf{X}$ such that

$$U(x) = y, \qquad \|x\| \leqslant m\|y\| \leqslant m.$$

If, further, $g \in \mathbf{Y}^*$ and $f \in U^*(g)$, then

$$|g(y)| = |g(U(x))| = |f(x)| \leqslant \|f\|\,\|x\| \leqslant m\|f\|.$$

Taking the supremum of the left-hand side over y (with $\|y\| \leqslant 1$) and bearing in mind that m is here independent of y, we obtain

$$\|g\| = \sup_{\|y\| \leqslant 1} |g(y)| \leqslant m\|f\| = m\|U^*(g)\|,$$

and we have a condition that guarantees the existence of a continuous left inverse U^{*-1}.

Sufficiency. We prove that U satisfies the conditions of Lemma 1 of the preceding section, and hence, by that lemma, that $U(\mathbf{X}) = \mathbf{Y}$.

Let B be the unit ball in $\mathbf{X}$ (with centre at the origin). The set $U(B)$ is absolutely convex, so its closure in $\mathbf{Y}$ is $U(B)^{\circ\circ}$ (see III.3.2). The polar $U(B)^\circ$ is the set of all functionals $g \in \mathbf{Y}^*$ such that $|g(y)| \leqslant 1$ for all $y \in U(B)$. In other words,

$$|g(U(x))| \leqslant 1 \text{ for all } x \in B.$$

Setting $f = U^*g$, we can write the latter inequality as

$$|f(x)| \leqslant 1 \qquad (x \in B).$$

Hence, as g ranges over the set $U(B)^\circ$, the functional $f = U^*(g)$ does not go out of the unit ball B° in $\mathbf{X}^*$: that is,

$$U^*(U(B)^\circ) \subset B^\circ \cap U^*(\mathbf{Y}^*).$$

We rewrite this as follows:

$$U(B)^\circ \subset U^{*-1}(B^\circ \cap U^*(\mathbf{Y}^*)).$$

In view of the continuity of U^{*-1}, the set $S = U^{*-1}(B^\circ \cap U^*(\mathbf{Y}^*))$ is bounded in $\mathbf{Y}^*$. Hence the polar S° is a neighbourhood of zero in $\mathbf{Y}$. But

$$\overline{U(B)} = (U(B))^{\circ\circ} \supset S^\circ,$$

and this is what we wish to prove.

From Theorems 2 and 2* we obtain the

COROLLARY. *A necessary and sufficient condition for the existence of a continuous two-sided inverse* U^{-1} *is that there exist a continuous two-sided inverse operator* U^{*-1}. *Then we have*

$$U^{*-1} = (U^{-1})^*.$$

Let us prove the latter equation. Let $f \in \mathbf{X}^*$ and let $g = (U^{-1})^*(f)$. We have

$$g(y) = f(U^{-1}(y)) = f(x) \qquad (y = U(x) \in \mathbf{Y}).$$

On the other hand, if we write $g' = U^{*-1}(f)$, so that $f = U^*(g')$, then

$$f(x) = g'(U(x)) = g'(y) \qquad (x = U^{-1}(y) \in \mathbf{X}).$$

For every $y \in \mathbf{Y}$ we thus have $g(y) = g'(y)$; or, in other words, $(U^{-1})^*(f) = U^{*-1}(f)$, as we set out to prove.

2.4. The following two theorems are generalizations of Theorems 2 and 2* in which conditions are established for the solubility of (1) and (2).

THEOREM 3. *If the set $U^*(\mathbf{Y}^*)$ is closed, then*

$$U(\mathbf{X}) = \mathbf{N}(\mathbf{N}(U^*)),$$

that is, equation (1) *is soluble if and only if $U^*(g) = \mathbf{0}$ implies $g(y) = 0$.*

Proof. Consider the subspace $\mathbf{Y}_1 = \overline{U(\mathbf{X})}$ of $\mathbf{Y}$ and the operator U_1, mapping $\mathbf{X}$ into $\mathbf{Y}_1$, given by*

$$U_1(x) = U(x) \qquad (x \in \mathbf{X}).$$

We prove that

$$U_1^*(\mathbf{Y}_1^*) = U^*(\mathbf{Y}^*). \tag{5}$$

To do this, let ω denote the embedding operator from $\mathbf{Y}_1$ into $\mathbf{Y}$. Since this clearly has a continuous left inverse ω^{-1}, we conclude from Theorem 2 that $\omega^*(\mathbf{Y}^*) = \mathbf{Y}_1^*$. But $U = \omega U_1$, so $U^* = U_1^* \omega^*$ (see IX.3.1) and therefore

$$U^*(\mathbf{Y}^*) = U_1^*(\omega^*(\mathbf{Y}^*)) = U_1^*(\mathbf{Y}_1^*).$$

As $\overline{U_1(\mathbf{X})} = \mathbf{Y}_1$, Theorem 1 shows that $\mathbf{Y}_1 = \mathbf{N}(\mathbf{N}(U_1^*))$, which is only possible if $\mathbf{N}(U_1^*) = \{\mathbf{0}\}$, so U_1^* determines a one-to-one mapping from the B-space $\mathbf{Y}_1^*$ onto $U_1^*(\mathbf{Y}^*)$, which by (5) is also a B-space. By the Corollary to Theorem 1.2, we conclude that there is a continuous left inverse U_1^{*-1}, and thus, by Theorem 2,

$$U(\mathbf{X}) = U_1(\mathbf{X}) = \mathbf{Y}_1,$$

from which the required result follows by an application of Theorem 1.

THEOREM 3*. *If the set $U(\mathbf{X})$ is closed, then*

$$U^*(\mathbf{Y}^*) = \mathbf{N}^*(\mathbf{N}(U)), \tag{6}$$

that is, (2) *has a solution if and only if f vanishes on the set of zeros of U.*

Proof. We first consider the case where $U(\mathbf{X}) = \mathbf{Y}$. We write $\mathbf{X}_0 = \mathbf{N}(U) = U^{-1}(\mathbf{0})$ and consider the factor-space $\overline{\mathbf{X}} = \mathbf{X}/\mathbf{X}_0$. Next we construct the operator $\overline{U}$ mapping $\overline{\mathbf{X}}$ onto $\mathbf{Y}$ as described in 1.3. Then $U = \overline{U}\phi$, where ϕ is the natural homomorphism from $\mathbf{X}$ onto $\overline{\mathbf{X}}$. As $\overline{\mathbf{X}}$ is complete and $\overline{U}$ is one-to-one, the Corollary to Theorem 1.2 shows that $\overline{U}$ has a continuous inverse $\overline{U}^{-1}$ and hence, by Theorem 2, we have $U^*(\mathbf{Y}^*) = \overline{\mathbf{X}}^*$. But $U^* = \phi^* \overline{U}^*$, and so

$$U^*(\mathbf{Y}^*) = \phi^*(\overline{U}^*(\mathbf{Y}^*)) = \phi^*(\overline{\mathbf{X}}^*).$$

* The difference between the operators U and U_1 is that the spaces in which the images $U(\mathbf{X})$ and $U_1(\mathbf{X})$ lie are different, so that the adjoint operators U^* and U_1^* act on distinct spaces.

Thus it remains to prove that $\phi^*(\overline{\mathbf{X}}^*) = \mathbf{N}^*(\mathbf{X}_0)$. Take any functional $f \in \phi^*(\overline{\mathbf{X}}^*)$. Let $\bar{f} \in \overline{\mathbf{X}}^*$, such that $f = \phi^*(\bar{f})$. As

$$f(x) = \bar{f}(\phi(x)) \qquad (x \in \mathbf{X}),$$

and $\phi(x) = 0$ for $x \in \mathbf{X}_0$, we have $f \in \mathbf{N}^*(\mathbf{X}_0)$.

Now assume that f is an arbitrary functional in $\mathbf{N}^*(\mathbf{X}_0)$. Using literally the same arguments that were used in 1.3 to construct $\overline{U}$, we see that the functional

$$\bar{f}(\bar{x}) = \bar{f}(x) \qquad (x \in \bar{x} = \phi(x))$$

is well defined and is a continuous linear functional on $\overline{\mathbf{X}}$. But then we have $f = \phi^*(\bar{f}) \in \phi^*(\overline{\mathbf{X}}^*)$.

In the general case, when we may have $U(\mathbf{X}) = \mathbf{Y}_1 \neq \mathbf{Y}$, we introduce, as in the proof of Theorem 3, an operator U_1 mapping $\mathbf{X}$ onto $\mathbf{Y}_1$. The operator U_1 comes under the special case already considered, and, by this case, bearing in mind (5), we can write

$$U^*(\mathbf{Y}^*) = U_1^*(\mathbf{Y}^*) = \mathbf{N}^*(\mathbf{N}(U_1)) = \mathbf{N}^*(\mathbf{N}(U)).$$

The theorem is thus completely proved.

COROLLARY. *If the set $U^*(\mathbf{Y}^*)$ is closed, then it is also weak* closed.*

For $U(\mathbf{X})$ is closed by Theorem 3 and we can thus apply Theorem 3 to deduce that $U^*(\mathbf{Y}^*)$ is weak* closed.

We note a special case of this Corollary—the case where U^* has a continuous left inverse. In that case, $U^*(\mathbf{Y}^*)$ is closed, by the Corollary to Theorem 1.1.

2.5. We now mention some consequences of the above theorems.

THEOREM 4. *A necessary and sufficient condition for* (1) *to have a unique solution for each $y \in \mathbf{Y}$ is that* (2) *have a unique solution for each $f \in \mathbf{X}^*$.*

THEOREM 5. *Assume that* (1) *and* (2) *are soluble, for every choice of $y \in \mathbf{Y}$ and $f \in \mathbf{X}^*$. Then U and U^* have continuous* (*two-sided*) *inverses U^{-1} and U^{*-1}. Hence the solutions of* (1) *and* (2) *are unique.*

THEOREM 6. *Assume that* (1) *and* (2) *have only the zero solution when $y = \mathbf{0}$ and $f = \mathbf{0}$. If one of the sets $U(\mathbf{X})$ and $U^*(\mathbf{Y}^*)$ is closed, then* (1) *and* (2) *are soluble for all $y \in \mathbf{Y}$ and $f \in \mathbf{X}^*$, and the solutions are unique.*

REMARK. The condition that $U(\mathbf{X})$ or $U^*(\mathbf{Y}^*)$ be closed is equivalent (under the hypotheses of the theorem) to the existence of a continuous left inverse U^{-1} or U^{*-1}.

2.6. If $\mathbf{X} = \mathbf{H}$ is a Hilbert space and $\mathbf{Y} = \mathbf{X}$, then Theorem 6 yields a result that was obtained in Chapter IX by different arguments, and even in a somewhat more general form (cf. Theorem IX.5.3): a necessary and sufficient condition for a self-adjoint operator U to have a continuous (two-sided) inverse is that it have a continuous left inverse U^{-1}—that is, for each $x \in \mathbf{H}$, we should have

$$\|U(x)\| \geqslant m\|x\| \qquad (m > 0).$$

From Theorem 5 we obtain

THEOREM 7. *If a self-adjoint operator U maps $\mathbf{H}$ onto itself, then it has a continuous inverse U^{-1}.*

2.7. Let us illustrate the theorems proved in this section with the example of a system of linear algebraic equations

$$\begin{aligned} a_{11}\xi_1 + a_{12}\xi_2 + \ldots + a_{1m}\xi_m &= \eta_1, \\ a_{21}\xi_1 + a_{22}\xi_2 + \ldots + a_{2m}\xi_m &= \eta_2, \\ \ldots\ldots\ldots\ldots\ldots\ldots\ldots\ldots \\ a_{n1}\xi_1 + a_{n2}\xi_2 + \ldots + a_{nm}\xi_m &= \eta_n. \end{aligned} \tag{7}$$

We can interpret this system as a single equation

$$U(x) = y,$$

where U is the operator from m-dimensional normed space $\mathbf{X}_m$ into n-dimensional normed space* $\mathbf{Y}_n$, defined by the matrix

$$A = \begin{pmatrix} a_{11} & a_{12} & \ldots & a_{1m} \\ a_{21} & a_{22} & \ldots & a_{2m} \\ \ldots & \ldots & \ldots & \ldots \\ a_{n1} & a_{n2} & \ldots & a_{nm} \end{pmatrix}.$$

The adjoint operator U^* from $\mathbf{Y}_n^*$ into $\mathbf{X}_m^*$ is defined by the matrix

$$A^* = \begin{pmatrix} a_{11} & a_{21} & \ldots & a_{n1} \\ a_{21} & a_{22} & \ldots & a_{n2} \\ \ldots & \ldots & \ldots & \ldots \\ a_{1m} & a_{2m} & \ldots & a_{nm} \end{pmatrix},$$

so that

$$f = U^*(g) \quad (f = (\phi_1, \phi_2, \ldots, \phi_m) \in \mathbf{X}_m^*, \quad g = (\psi_1, \psi_2, \ldots, \psi_n) \in \mathbf{Y}_n^*)$$

means that

$$\phi_k = \sum_{j=1}^{n} a_{jk}\psi_j \qquad (k = 1, 2, \ldots, m).$$

Since $U(\mathbf{X}_m)$ is automatically closed, being a finite-dimensional linear manifold (IV.4.6), Theorem 1 yields

$$U(\mathbf{X}_m) = \mathbf{N}(\mathbf{N}(U^*)). \tag{8}$$

But $\mathbf{N}(U^*)$ consists of those functionals $g = (\psi_1, \psi_2, \ldots, \psi_n) \in \mathbf{Y}_n^*$ for which $U^*(g) = \mathbf{0}$; that is, for which

$$\sum_{j=1}^{n} a_{jk}\psi_j = 0 \qquad (k = 1, 2, \ldots, m).$$

Hence $y = (\eta_1, \eta_2, \ldots, \eta_n) \in U(\mathbf{X}_m)$ if and only if (8) implies

$$g(y) = \sum_{j=1}^{n} \psi_j\eta_j = 0.$$

In other words, a necessary and sufficient condition for the system (7) to have a solution is that the right-hand sides of the system form a vector orthogonal to some solution of the homogeneous system with the adjoint matrix.

* Precisely how the metrics on $\mathbf{X}_m$ and $\mathbf{Y}_n$ are defined is insignificant. For definiteness we can assume, say, that $\mathbf{X} = \ell_m^2$, $\mathbf{Y} = \ell_n^2$. The spaces are assumed to be real spaces.

Now assume that $m = n$. If the system (17) is soluble for all right-hand sides, then U^* has a left inverse (Theorem 2*). It is not difficult to see that, in such a case, $U^*(\mathbf{Y}_n^*) = \mathbf{X}_n^*$, for otherwise U^* would map an n-dimensional space into a space of smaller dimension, which would contradict* the existence of the left inverse U^{*-1}. Applying Theorem 4, we find that the solution of the system (7) is unique.

Using Theorem 5 and applying similar arguments, one can establish the converse result: if the system (7) has at most one solution, then the solution exists for all right-hand sides.

For systems with symmetric matrices, both these results follow immediately from Theorem 7.

* For suppose the set $U^*(\mathbf{Y}_n^*)$ had dimension $n_0 < n$; that is, suppose there existed elements $f_1, f_2, \ldots, f_{n_0} \in U^*(\mathbf{Y}_n^*)$ such that, for every $f \in U^*(\mathbf{Y}_n^*)$,

$$f = \sum_{k=1}^{n_0} c_k f_k.$$

Then, for every $g \in \mathbf{Y}_n^*$, we would have $g = \sum_{k=1}^{n_0} c_k g_k$, where $g_k = U^{*-1}(f_k)$, that is, the dimension of $\mathbf{Y}_n^*$ would be less than n.

XIII

FUNCTIONAL EQUATIONS OF THE SECOND KIND

IN THIS chapter we study the equation of the form

$$x - \lambda U(x) = y, \tag{$*$}$$

where U is a continuous linear operator mapping a B-space $\mathbf{X}$ into itself. Such an equation will be called an *equation of the second kind*, and the operator U will be called the *kernel of the equation*. This terminology is taken from the theory of integral equations, where the equation

$$x(s) - \lambda \int_a^b K(s, t)x(t)dt = y(s) \qquad (s \in [a, b])$$

is called an equation of the second kind, to distinguish it from the equation of the first kind:

$$\int_a^b K(s, t)x(t)dt = y(s) \qquad (s \in [a, b]).$$

Although one can formally write the functional equation ($*$) as an equation "of the first kind"

$$T(x) = y \qquad (T = I - \lambda U),$$

it turns out to be advantageous to separate out the identity operator, as U may have more desirable properties than T which allow one to make a more complete study of the equation ($*$).

§ 1. Equations with compact kernels

In this section we study the equation

$$x - U(x) = y \qquad (x, y \in \mathbf{X}) \tag{1}$$

and its adjoint

$$g - U^*(g) = f \qquad (f, g \in \mathbf{X}^*), \tag{2}$$

assuming that U (and hence also U^*, by Theorem IX.3.3) is a compact operator on the B-space $\mathbf{X}$. Let us write $T = I - U$, where as usual I denotes the identity operator on $\mathbf{X}$. Then (1) can be written more succinctly as

$$T(x) = y, \tag{1'}$$

and (2) as

$$T^*(g) = f, \tag{2'}$$

since $T^* = I^* - U^*$ (IX.3.1) and I^* is the identity operator on $\mathbf{X}^*$.

1.1. Let us prove three preliminary lemmas.

LEMMA 1. *The set $T(\mathbf{X})$ is closed.*

Proof. We write $\mathbf{X}_0 = \mathbf{N}(T) = T^{-1}(\mathbf{0})$, and consider the factor-space $\overline{\mathbf{X}} = \mathbf{X}/\mathbf{X}_0$ and the operator $\overline{T}$ mapping $\overline{\mathbf{X}}$ into $\mathbf{X}$ (see XII.1.3). As in XII.1.3, we denote the natural homomorphism from $\mathbf{X}$ onto $\overline{\mathbf{X}}$ by ϕ. Let $\{y_n\}$ be a sequence in $T(\mathbf{X})$ converging to an element $y_0 \in \mathbf{X}$. Since $\overline{T}(\overline{\mathbf{X}}) = T(\mathbf{X})$, there exist elements $\bar{x}_n \in \overline{\mathbf{X}}$ such that $y_n = \overline{T}(\bar{x}_n)$ ($n = 1, 2, \ldots$). Also we can find elements $x_n \in \mathbf{X}$ in accordance with (1) in IV.1.8—that is, elements such that

$$\bar{x}_n = \phi(x_n) \qquad \|\bar{x}_n\| \geqslant \tfrac{1}{2}\|x_n\| \qquad (n = 1, 2, \ldots). \tag{3}$$

Let us prove that the sequence $\{\bar{x}_n\}$ is bounded. For otherwise we could arrange, by passing to a subsequence if necessary, that $c_n = \|\bar{x}_n\| \to \infty$. By (3), the sequence $\left\{\dfrac{x_n}{c_n}\right\}$ is bounded, so by passing once more to a subsequence we can assume that $\left\{U\left(\dfrac{x_n}{c_n}\right)\right\}$ converges. Assume, say, that $U\left(\dfrac{x_n}{c_n}\right) \to z$. Since $T(x_n) = \overline{T}(\bar{x}_n) = y_n$, we can write

$$\frac{x_n}{c_n} = U\left(\frac{x_n}{c_n}\right) + T\left(\frac{x_n}{c_n}\right) = U\left(\frac{x_n}{c_n}\right) + \frac{y_n}{c_n} \to z,$$

and hence

$$T(z) = \lim_{n\to\infty} T\left(\frac{x_n}{c_n}\right) = \lim_{n\to\infty}\left(\frac{y_n}{c_n}\right) = \mathbf{0},$$

that is, $z \in \mathbf{X}_0$. But then we have

$$\frac{\bar{x}_n}{c_n} = \phi\left(\frac{x_n}{c_n}\right) \to \phi(z) = \mathbf{0},$$

which is impossible since $\left\|\dfrac{\bar{x}_n}{c_n}\right\| = 1$ for every $n = 1, 2, \ldots$.

Thus the sequence $\{\bar{x}_n\}$, and by (3) also the sequence $\{x_n\}$, is bounded. We may therefore assume that $\{U(x_n)\}$ is a convergent sequence. If $U(x_n) \to x$, say, then

$$x_n = T(x_n) + U(x_n) = y_n + U(x_n) \to y_0 + x = x_0.$$

We thus obtain

$$y_0 = \lim_{n\to\infty} y_n = \lim_{n\to\infty} T(x_n) = T(x_0) \in T(\mathbf{X}),$$

as we required to prove.

LEMMA 2. *The sequence of sets*

$$\mathbf{N}(T), \qquad \mathbf{N}(T^2), \ldots, \mathbf{N}(T^n), \ldots$$

is increasing and contains only a finite number of distinct sets.

Proof. The first half of the statement of the lemma is almost obvious, for if $x \in \mathbf{N}(T^n)$ then $T^n(x) = \mathbf{0}$ and so *a fortiori* $T^{n+1}(x) = \mathbf{0}$; that is, $x \in \mathbf{N}(T^{n+1})$, and hence $\mathbf{N}(T^n) \subset \mathbf{N}(T^{n+1})$.

To prove the second half, we write $\mathbf{X}_n = \mathbf{N}(T^n)$ and show that if $\mathbf{X}_n = \mathbf{X}_{n+1}$ for some n, then also $\mathbf{X}_{n+1} = \mathbf{X}_{n+2}$. Choose an $x \in \mathbf{X}_{n+2}$. Then $T^{n+2}(x) = T^{n+1}(T(x)) = \mathbf{0}$, and therefore $T(x) \in \mathbf{X}_{n+1} = \mathbf{X}_n$. Consequently $T^{n+1}(x) = T^n(T(x)) = \mathbf{0}$; that is, $x \in \mathbf{X}_{n+1}$.

Thus $\mathbf{X}_{n+2} \subset \mathbf{X}_{n+1}$. The reverse inclusion is always true, so we conclude that $\mathbf{X}_n = \mathbf{X}_{n+1} = \mathbf{X}_{n+2} = \ldots$.

Now assume that, for every $n = 1, 2, \ldots$, we have

$$\mathbf{X}_n \neq \mathbf{X}_{n+1}.$$

Each $\mathbf{X}_n$ is a subspace of $\mathbf{X}_{n+1}$ so, by the Lemma on Almost Perpendicularity (IV.1.7), we can find an element x_{n+1} of unit norm in $\mathbf{X}_{n+1}$ such that

$$\rho(x_{n+1}, \mathbf{X}_n) > \tfrac{1}{2} \qquad (n = 0, 1, \ldots). \tag{4}$$

Let $m > n$. We consider the element

$$U(x_m) - U(x_n) = x_m - T(x_m) - [x_n - T(x_n)] = x_m - \tilde{x},$$

where we have written $\tilde{x} = T(x_m) + x_n - T(x_n)$. Let us prove that $\tilde{x} \in \mathbf{X}_{m-1}$. We have

$$T^{m-1}(\tilde{x}) = T^m(x_m) + T^{m-1}(x_n) - T^m(x_n) = \mathbf{0},$$

as $x_n \in \mathbf{X}_n \subset \mathbf{X}_{m-1}$, and $x_m \in \mathbf{X}_m$.

Bearing in mind the inequality (4), we have

$$\|U(x_m) - U(x_n)\| = \|x_m - \tilde{x}\| > \tfrac{1}{2} \qquad (m > n;\ m, n = 1, 2, \ldots). \tag{5}$$

However, $\{x_n\}$ is a bounded sequence and, as U is compact, we can select a convergent subsequence from the sequence $\{U(x_n)\}$; however, this contradicts (5).

LEMMA 3. *The sequence*

$$T(\mathbf{X}), T^2(\mathbf{X}), \ldots, T^n(\mathbf{X}), \ldots \tag{6}$$

contains only a finite number of distinct sets.

Proof. The proof is similar, generally speaking, to that of the preceding lemma, so we give it without going into any details.

Note that the sets (6) are closed, by Lemma 1, and furthermore form a decreasing sequence. Clearly, if we have $T^n(\mathbf{X}) = T^{n+1}(\mathbf{X})$ for some n, then it follows that

$$T^n(\mathbf{X}) = T^{n+1}(\mathbf{X}) = T^{n+2}(\mathbf{X}) = \ldots,$$

and thus the proof is complete in this case.

Assuming that $T^n(\mathbf{X}) \neq T^{n+1}(\mathbf{X})$ $(n = 0, 1, \ldots)$, we use the Lemma on Almost Perpendicularity (IV.1.7) to construct a sequence $\{x_n\}$ such that

$$\|x_n\| = 1, \quad x_n \in T^n(\mathbf{X}), \quad \rho(x_n, T^{n+1}(\mathbf{X})) > \tfrac{1}{2} \qquad (n = 1, 2, \ldots). \tag{7}$$

Let $m > n$. As in Lemma 2, we have

$$U(x_n) - U(x_m) = x_n - T(x_n) - [x_m - T(x_m)] = x_n - \tilde{x}.$$

However,

$$T(x_n) \in T^{n+1}(\mathbf{X}), \qquad x_m \in T^m(\mathbf{X}) \subset T^{n+1}(\mathbf{X}),$$
$$T(x_m) \in T^{m+1}(\mathbf{X}) \subset T^{n+1}(\mathbf{X}),$$

and hence $\tilde{x} = T(x_n) + x_m - T(x_m) \in T^{m+1}(\mathbf{X})$. It therefore follows from (7) that

$$\|U(x_n) - U(x_m)\| = \|x_n - \tilde{x}\| > \tfrac{1}{2} \qquad (m > n;\ m, n = 1, 2, \ldots),$$

which contradicts the assumption that U is compact.

1.2. Let r denote the least non-negative number n for which $T^n(\mathbf{X}) = T^{n+1}(\mathbf{X})$. In the special case where $T(\mathbf{X}) = \mathbf{X} = T^{\circ}(\mathbf{X})$, we set $r = 0$.

Also, let $\mathbf{X}' = T^r(\mathbf{X}), \mathbf{X}'' = \mathbf{N}(T^r)$.

The characteristic properties of T, and hence those of equation (1), are described in the following theorem.

THEOREM 1. a) *The operator T is a one-to-one mapping from $\mathbf{X}'$ onto itself.*

b) *The subspace $\mathbf{X}''$ is finite-dimensional. The operator T maps $\mathbf{X}''$ into itself.*

c) *Each element $x \in \mathbf{X}$ can be expressed, in a unique way, in the form*

$$x = x' + x'' \qquad (x' \in \mathbf{X}',\ x'' \in \mathbf{X}''); \tag{8}$$

moreover, there exists a constant $M > 0$ such that

$$\|x'\| \leqslant M\|x\|, \qquad \|x''\| \leqslant M\|x\|. \tag{9}$$

d) *The operator U is expressible in the form*

$$U = U' + U'', \tag{10}$$

where U' and U'' are compact operators mapping $\mathbf{X}$ into $\mathbf{X}'$ and $\mathbf{X}''$ respectively. Furthermore, $T' = I - U'$ has a continuous two-sided inverse, and the following equation is satisfied:

$$U'U'' = U''U' = \mathbf{0}. \tag{11}$$

Proof. a) Since $\mathbf{X}' = T^r(\mathbf{X})$, we have

$$T(\mathbf{X}') = T^{r+1}(\mathbf{X}) = T^r(\mathbf{X}) = \mathbf{X}'.$$

If $T(x) = \mathbf{0}$, where $x \in \mathbf{X}'$, then choosing $n \geqslant r$ so that $\mathbf{N}(T^n) = \mathbf{N}(T^{n+1})$, in accordance with Lemma 2, we have $x \in T^n(\mathbf{X})$; hence there exists $\tilde{x} \in \mathbf{X}$ such that $x = T^n(\tilde{x})$. But then $\mathbf{0} = T(x) = T^{n+1}(\tilde{x})$ and so $\tilde{x} \in \mathbf{N}(T^{n+1}) = \mathbf{N}(T^n)$; that is, $x = T^n(\tilde{x}) = \mathbf{0}$.

b) We have

$$T^r = (1 - U)^r = I - U_1,$$

where the operator U_1 is a linear combination of positive powers of U. Hence, by Theorem IX.2.2, U_1 is compact. Since $U_1(x) = x$ for $x \in \mathbf{X}''$, every bounded set in $\mathbf{X}''$ is compact. By Theorem IV.1.3, $\mathbf{X}''$ is finite-dimensional.

In the case where $r > 0$, the set $T(\mathbf{X}'')$ is clearly equal to $\mathbf{N}(T^{r-1})$, and so is contained in $\mathbf{N}(T^r) = \mathbf{X}''$. But if $r = 0$, then $\mathbf{X}'' = \{\mathbf{0}\}$ and the inclusion $T(\mathbf{X}'') \subset \mathbf{X}''$ is trivial.

c) Let T_0 denote the operator T, acting on $\mathbf{X}'$ only. Applying Lemma 1 to $T^r = I - U_1$, we conclude that $\mathbf{X}'$ is closed and so is a B-space. Hence, by Theorem XII.1.2, T_0, being a one-to-one mapping of $\mathbf{X}'$ onto itself, has a continuous inverse T_0^{-1}.

Let x be an arbitrary element of $\mathbf{X}$; we set

$$x' = T_0^{-r}T^r(x), \qquad x'' = x - x' = x - T_0^{-r}T^r(x). \tag{12}$$

Clearly $x' \in \mathbf{X}'$, and as

$$T^r(x'') = T^r(x) - T^rT_0^{-r}T^r(x) = T^r(x) - T^r(x) = \mathbf{0},$$

we have $x'' \in \mathbf{X}''$, showing that x can be expressed in the form (8).

If $x = x_1' + x_1''$ is some other expression for x in the form (8), with $x_1' \in \mathbf{X}'$, $x_1'' \in \mathbf{X}''$, then

$$T^r(x) = T^r(x_1') + T^r(x_1'') = T^r(x_1').$$

But, since $x_1' \in \mathbf{X}'$, we have $T^r(x_1') = T_0^r(x_1')$, so

$$x_1' = T_0^{-r}T^r(x) = x'$$

and this proves the uniqueness of the expression (8) for x.

In view of (12), the inequalities (9) follow from the continuity of T_0^{-1}.

d) Since $U = I - T$, we have

$$U(x) = x - T(x) \in \mathbf{X}'$$

for $x \in \mathbf{X}'$, so U maps $\mathbf{X}'$ into itself. Similarly we can verify that $U(\mathbf{X}'') \subset \mathbf{X}''$.

For any $x \in \mathbf{X}$, we set

$$U'(x) = U(x'), \qquad U''(x) = U(x''), \tag{13}$$

where $x' \in \mathbf{X}'$ and $x'' \in \mathbf{X}''$ are from the expression for x in the form (8). Using the inequalities (9), we can easily convince ourselves that U' and U'' are continuous linear operators. Furthermore, it is clear that $U = U' + U''$, and $U'(\mathbf{X}) \subset \mathbf{X}'$, $U''(\mathbf{X}) \subset \mathbf{X}''$. Also, we evidently have

$$U'(\mathbf{X}'') = U''(\mathbf{X}') = \{\mathbf{0}\}. \tag{14}$$

From these relations we see that $U'U'' = U''U' = \mathbf{0}$, so (11) holds.

The operator U'' maps $\mathbf{X}$ into the finite-dimensional space $\mathbf{X}''$, in which every bounded set is compact. Hence U'' is a compact operator. But $U' = U - U''$, and we therefore conclude from Theorem IX.2.2 that U' is compact.

Finally we show that the operator $T' = I - U'$ has a continuous two-sided inverse. For this it is enough to prove, firstly, that $T'(x) = \mathbf{0}$ implies $x = \mathbf{0}$, and, secondly, that $T'(\mathbf{X}) = \mathbf{X}$. Suppose $T'(x) = \mathbf{0}$. Expressing x in the form (8), we see that

$$\mathbf{0} = T'(x) = x - U'(x) = x' - U(x') + x'' = T(x') + x''.$$

Since $T(x') \in \mathbf{X}'$, it follows from the uniqueness of the expression for $\mathbf{0}$ in the form (8) that

$$T(x') = x'' = \mathbf{0}$$

and so, by (a), $x' = \mathbf{0}$. Thus also $x = x' + x'' = \mathbf{0}$.

Now consider an arbitrary element $y \in \mathbf{X}$. Write it in the form (8) $y = y' + y''$ ($y' \in \mathbf{X}'$, $y'' \in \mathbf{X}''$) and set

$$x = T_0^{-1}(y') + y''.$$

As $T_0^{-1}(y') \in \mathbf{X}'$, we have

$$U'(x) = U(T_0^{-1}(y'))$$

and

$$T'(x) = x - U'(x) = T_0^{-1}(y') - U(T_0^{-1}(y')) + y'' = TT_0^{-1}(y') + y'' = y' + y'' = y.$$

Thus $T'(\mathbf{X}) = \mathbf{X}$.

This completes the proof of the theorem.

REMARK. Let m be the smallest of the non-negative integers n such that $\mathbf{N}(T^n) = \mathbf{N}(T^{n+1})$. Then we have $m = r$.

For if we take an $x \in \mathbf{N}(T^{r+1})$ and express it in the form (8), we have

$$\mathbf{0} = T^{r+1}(x) = T^{r+1}(x') + T^{r+1}(x'') = T^{r+1}(x'),$$

which, by (a), is possible only for $x' = \mathbf{0}$. Hence $x = x'' \in \mathbf{N}(T^r)$, and therefore $m \leqslant r$.

Also if $y = T^m(x)$ ($x \in \mathbf{X}$), then, writing x in the form (8), we have

$$y = T^m(x) = T^m(x') + T^m(x'') = T^m(x') = T^{m+1}(T_0^{-1}(x'))$$

and so $y \in T^{m+1}(\mathbf{X})$; hence we must also have $r \leqslant m$.

THEOREM 2. *A necessary and sufficient condition for* (1) *to be soluble for every* $y \in \mathbf{X}$ *is that the homogeneous equation*

$$T(x) = \mathbf{0} \tag{15}$$

have a unique solution (which obviously will be $x = \mathbf{0}$*).*

For the solubility of (1) for every $y \in \mathbf{X}$ amounts to the requirement that $T(\mathbf{X}) = \mathbf{X}$: that is, that $r = 0$. The uniqueness of the solution of (15) is equivalent to the condition that $m = 0$.

REMARK. If one takes into account the results of § 2 of the preceding chapter, one can obtain a proof of the theorem that does not depend on Theorem 1, and uses only the fact that $T(\mathbf{X})$ is closed. We leave it to the reader to supply the argument for himself.

1.3. In the following theorem we establish the connections between equations (1) and (2).

THEOREM 3. *The spaces* $\mathbf{N}(T)$ *and* $\mathbf{N}(T^*)$ *have the same finite dimension.*

Proof. Since $\mathbf{N}(T) \subset \mathbf{N}(T^r) = \mathbf{X}''$, and since $\mathbf{X}''$ is finite-dimensional by Theorem 1, it follows that $\mathbf{N}(T)$ is also finite-dimensional. Moreover, as U^* is a compact operator, the same is true of $\mathbf{N}(T^*)$ as well.

Suppose the dimension of $\mathbf{N}(T)$ is n and that of $\mathbf{N}(T^*)$ is m. Let $x_1, x_2, \ldots, x_n$ be a system of linearly independent elements in $\mathbf{N}(T)$, and let $g_1, g_2, \ldots, g_m$ be linearly independent elements in $\mathbf{N}(T^*)$.

As $x_1, x_2, \ldots, x_n$ are linearly independent, we can apply the biorthogonalization theorem (Theorem V.7.4) to them, and we conclude that there is a biorthogonal system of functionals $f_1, f_2, \ldots, f_n$:

$$f_j(x_k) = \begin{cases} 1, & j = k, \\ 0, & j \neq k \end{cases} \qquad (j, k = 1, 2, \ldots, n). \tag{16}$$

In the same way, using Lemma III.3.1, we can find elements $y_1, y_2, \ldots, y_m$ such that

$$g_j(y_k) = \begin{cases} 1, & j = k, \\ 0, & j \neq k \end{cases} \qquad (j, k = 1, 2, \ldots, m). \tag{17}$$

Assume first that $n < m$. Consider the operator $V = U + W$ on $\mathbf{X}$, where

$$W(x) = \sum_{k=1}^{n} f_k(x) y_k \qquad (x \in \mathbf{X}).$$

Since W is a linear operator mapping $\mathbf{X}$ into a finite-dimensional space, it is compact. Hence V is also compact. Now consider the equation

$$\tilde{T}(x) = x - V(x) = T(x) - \sum_{k=1}^{n} f_k(x) y_k = \mathbf{0}. \tag{18}$$

Let x_0 be a solution:

$$\tilde{T}(x_0) = T(x_0) - \sum_{k=1}^{n} f_k(x_0) y_k = \mathbf{0}. \tag{19}$$

It follows from this equation that

$$g_s(T(x_0)) - \sum_{k=1}^{n} f_k(x_0)g_s(y_k) = 0 \qquad (s = 1, 2, \ldots, n), \tag{20}$$

that is, in view of (17),

$$g_s(T(x_0)) - f_s(x_0) = 0$$

and so, as $T^*(g_s) = \mathbf{0}$, we have

$$f_s(x_0) = 0 \qquad (s = 1, 2, \ldots, n). \tag{21}$$

Together with (19), this yields $T(x_0) = \mathbf{0}$—that is, $x_0 \in \mathbf{N}(T)$—and consequently x_0 can be expressed in the form

$$x_0 = \sum_{k=1}^{n} \alpha_k x_k.$$

Now (16) shows that $\alpha_s = f_s(x_0)$, so (21) implies that $\alpha_s = 0$, and hence that $x_0 = \mathbf{0}$. Hence (18) has a unique solution. By Theorem 2, the corresponding inhomogeneous equation is soluble, whatever its right-hand side is. In particular, the following equation has a solution:

$$\tilde{T}(x) = T(x) - \sum_{k=1}^{n} f_k(x)y_k = y_{n+1}.$$

Let x^* be a solution of this equation. We have also

$$g_{n+1}\left(T(x^*) - \sum_{k=1}^{n} f_k(x^*)y_k\right) = T^*(g_{n+1})\ (x^*) - \sum_{k=1}^{n} f_k(x^*)g_{n+1}(y_k) = 0.$$

Also $g_{n+1}(y_{n+1}) = 1$.

Hence we must have $m \leqslant n$.

A similar argument rules out the possibility that $m < n$. Namely, we consider instead of (13) the following equation in $\mathbf{X}^*$:

$$T^*(g) - \sum_{k=1}^{m} g(y_k)f_k = \mathbf{0}.$$

1.4. Putting together the theorems proved above, we obtain the following result.

THEOREM 4. *Either equations* (1) *and* (2) *are soluble for all right-hand sides, in which case their solutions are unique, or the homogeneous equations*

$$T(x) = \mathbf{0} \qquad \text{and} \qquad T^*(g) = \mathbf{0}$$

have the same number of linearly independent solutions $x_1, x_2, \ldots, x_n$ *and* $g_1, g_2, \ldots, g_n$, *respectively. In the latter case, the following are necessary and sufficient conditions for the existence of solutions of* (1) *and* (2), *respectively:*

$$g_k(y) = 0 \qquad (k = 1, 2, \ldots, n)$$

and

$$f(x_k) = 0 \qquad (k = 1, 2, \ldots, n).$$

When these conditions are satisfied, the general solution of (1) *has the form*

$$x = x^* + \sum_{k=1}^{n} c_k x_k,$$

while that of (2) *has the form*

$$g = g^* + \sum_{k=1}^{n} c_k g_k,$$

where x^* *and* g^* *are arbitrary solutions of* (1) *and* (2), *respectively, and* $c_1, c_2, \ldots, c_n$ *are arbitrary constants.*

The second half of this theorem is obtained by applying Theorems XII.2.3 and XII.2.3* to equations (1) and (2); the hypotheses are satisfied in view of Lemma 1 (however, one can also prove this part of the theorem directly).

In view of the analogy with the well-known theorem in the theory of integral equations, this theorem is known as the *Fredholm alternative.*

§ 2. Complex normed spaces

As will become clear below (see § 3), it is natural to consider the equation

$$x - \lambda U(x) = y$$

in a complex space, in particular when we assign complex values to λ. In this connection, we introduce below some auxiliary concepts relating to complex spaces, which will enable us to include the real case under the complex case.

2.1. Let $\mathbf{Z}$ be a complex normed space. We shall say that $\mathbf{Z}$ has a *real kernel* if there is an operator C, called an *involution,* defined on $\mathbf{Z}$ that maps $\mathbf{Z}$ into itself and has the properties:

1. $C(\lambda_1 z_1 + \lambda_2 z_2) = \overline{\lambda_1} C(z_1) + \overline{\lambda_2} C(z_2) \qquad (z_1, z_2 \in \mathbf{Z}).$
2. $C^2(z) = z \qquad (z \in \mathbf{Z}).$
3. $\|C(z)\| = \|z\|.$

The set of elements for which $C(z) = z$ is called the *real kernel* of $\mathbf{Z}$ and is denoted by $\operatorname{Re} \mathbf{Z}$; elements of this set are called *real elements.*

Let $z \in \mathbf{Z}$; the element $x = \frac{1}{2}[z + C(z)]$ is called the *real part* of z and is denoted by $x = \operatorname{Re} z$.

The element $y = \frac{1}{2i}[z - C(z)]$ is called the *imaginary part* of z and is denoted by $y = \operatorname{Im} z$.

Since

$$C(x) = \frac{1}{2}[C(z) + C^2(z)] = x, \qquad C(y) = -\frac{1}{2i}[C(z) - C^2(z)] = y,$$

both $x = \operatorname{Re} z$ and $y = \operatorname{Im} z$ are real elements. Clearly

$$z = x + yi \tag{1}$$

and

$$C(z) = x - yi.$$

The last equation justifies us in calling $C(z)$ the *conjugate* of z and introducing the usual notation $C(z) = \bar{z}$.

If an element z is expressed in any way in the form (1), for real x and y, then we necessarily have $x = \text{Re } z$ and $y = \text{Im } z$. For $\bar{z} = \overline{\bar{x} - yi} = x - yi$, and so $x = \frac{1}{2}(z + \bar{z}) = \text{Re } z$, $y = \frac{1}{2i}(z - \bar{z}) = \text{Im } z$. Hence the representation (1) is unique; the elements x and y are uniquely determined by z.

The real kernel **X** of the space **Z** is a real normed space, and it is complete if the original space **Z** is complete.

For a linear combination of elements of **X** with real coefficients is itself an element of **X**. The axioms for a normed space are satisfied in **X** because they are satisfied in **Z**. Let us verify that **X** is complete (assuming that **Z** is). Let $\{x_n\}$ be a Cauchy sequence in **X**. Regarding this as a sequence in **Z** and bearing in mind that **Z** is complete, we see that $\lim_{n \to \infty} x_n = z$ exists in **Z**. Since $\overline{x_n - z} = \bar{x}_n - \bar{z}$ by condition 1, $\|x_n - z\| = \|\overline{x_n - z}\| = \|x_n - \bar{z}\|$, and so $x_n \to \bar{z}$. By the uniqueness of the limit, $\bar{z} = z$; that is, $z \in \mathbf{X}$.

All the concrete examples of real spaces considered above are the real kernels of the corresponding complex spaces. For example, the real space $\mathbf{C}(K)$ is the real kernel of the complex space $\mathbf{C}_{\mathbb{C}}(K)$. The involution here is the complex conjugation operator $x(t) \to \overline{x(t)}$. The same is true of the spaces $\mathbf{L}^p$, **V**, $\mathbf{c}_0$, and so on.

In general, one can consider an arbitrary real space **X** as the real kernel of a certain complex space, namely the space **Z** whose elements are ordered pairs $z = (x, y)$ $(x, y \in \mathbf{X})$ of elements of **X**, where the operations are introduced for the pairs according to the following rules: $(x_1, y_1) + (x_2, y_2) = (x_1 + x_2, y_1 + y_2)$, $\lambda(x, y) = (\alpha x - \beta y, \beta x + \alpha y)$ $(\lambda = \alpha + \beta i$, α, β real),

$$\|(x, y)\| = \max_{\theta} \|x \cos\theta + y \sin\theta\|.^* \tag{2}$$

To convince oneself that **X** is the real kernel of **Z**, one need only write

$$C((x, y)) = (x, -y).$$

Then the real elements in **Z** are all the elements $(x, \mathbf{0})$, and only these. Since

$$\alpha(x_1, \mathbf{0}) + \beta(x_2, \mathbf{0}) = (\alpha x_1 + \beta x_2, \mathbf{0}),$$

$$\|(x, \mathbf{0})\| = \max_{\theta} \|x \cos\theta\| = \|x\|,$$

we obtain the required result if we identify the element $(x, \mathbf{0})$ with x. This identification allows one to use the familiar notation $x + iy$ instead of (x, y). The space **Z** is called the *complexification* of **X**.

2.2. Let **Z** and **W** be complex spaces, having real kernels $\mathbf{X} = \text{Re } \mathbf{Z}$ and $\mathbf{Y} = \text{Re } \mathbf{W}$. A continuous linear operator $\tilde{U}$ mapping **Z** into **W** is said to be real if it maps real elements in **Z** to real elements in **W**: that is, if $\tilde{U}(\mathbf{X}) \subset \mathbf{Y}$. A real operator thus induces a continuous linear operator from the real space **X** into the real space **Y**.

Conversely, if U is a continuous linear operator mapping **X** into **Y**, and if $\mathbf{X} = \text{Re } \mathbf{Z}$ and $\mathbf{Y} = \text{Re } \mathbf{W}$, then, by setting

$$\tilde{U}(z) = \tilde{U}(x + yi) = U(x) + iU(y) \qquad (z = x + yi),$$

* Of course, for a specific space one may also be able to introduce a norm in another way. For instance, if $\mathbf{X} = \mathbf{L}^p$, then the usual norm in the complex space $\mathbf{L}^p_{\mathbb{C}}$ is equivalent to, but not the same as, that defined by (2).

we obtain a continuous linear operator $\tilde{U}$ from the complex space $\mathbf{Z}$ into the complex space $\mathbf{W}$. Obviously, $\tilde{U}$ and U coincide on $\mathbf{X}$. The operator $\tilde{U}$ is said to be a *complex extension* of U.

Let us note the inequality

$$\|U\| \leqslant \|\tilde{U}\| \leqslant 2\|U\|. \tag{3}$$

The first part of this inequality is obvious. As for the second, it follows from the chain of inequalities

$$\|\tilde{U}(z)\| = \|U(x)+iU(y)\| \leqslant \|U(x)\| + \|U(y)\| \leqslant \|U\|(\|x\| + \|y\|) \leqslant 2\|U\|\|z\|.$$

(Here we have used the inequality $\|\operatorname{Re} z\| \leqslant \|z\|$, which is proved as follows: $\|\operatorname{Re} z\| = \frac{1}{2}\|z+\bar{z}\| \leqslant \frac{1}{2}[\|z\| + \|\bar{z}\|] = \|z\|$.)

In a number of cases it is possible to show that $\|\tilde{U}\| = \|U\|$. For instance, if a norm is defined on $\mathbf{Z}$ and $\mathbf{W}$ by (2), then

$$\|\tilde{U}(z)\| = \max_{\theta}\|U(x)\cos\theta + U(y)\sin\theta\| \leqslant \|U\| \max_{\theta}\|x\cos\theta + y\sin\theta\| = \|U\|\|z\|$$

and hence $\|\tilde{U}\| \leqslant \|U\|$, which, together with (3), yields $\|\tilde{U}\| = \|U\|$.

Further, if for every $\varepsilon > 0$ there exists a real element x of unit norm such that $\|\tilde{U}(x)\| > \|\tilde{U}\| - \varepsilon$, then we also have $\|\tilde{U}\| = \|U\|$, since in this case

$$\|\tilde{U}\| \leqslant \|\tilde{U}(x)\| + \varepsilon = \|U(x)\| + \varepsilon \leqslant \|U\| + \varepsilon.$$

This latter condition is satisfied, for example, when $\mathbf{Z}$ and $\mathbf{W}$ are any of the function spaces mentioned above and $\tilde{U}$ is the integral operator

$$w = \tilde{U}(z), \qquad w(s) = \int_a^b K(s,t)x(t)\,dt$$

with real kernel $K(s, t)$.

In exactly the same way, the condition is satisfied if $\mathbf{Z}$ and $\mathbf{W}$ are the space of sequences and $\tilde{U}$ is determined by a real matrix.

2.3. If a real operator $\tilde{U}$ is compact, then clearly so is the operator U from $\mathbf{X} = \operatorname{Re}\mathbf{Z}$ into $\mathbf{Y} = \operatorname{Re}\mathbf{W}$ that it induces. The converse of this is also true: that is, if U is a compact operator from $\mathbf{X}$ into $\mathbf{Y}$, then its complex extension—namely, the operator $\tilde{U}$—is also compact.

To prove this, we need only use the fact that, if a sequence $\{z_n\}$ converges, then so do the sequences $\{\operatorname{Re} z_n\}$ and $\{\operatorname{Im} z_n\}$, and conversely.

2.4. If $\mathbf{Z}$ is a space with real kernel, then its dual space $\mathbf{Z}^*$ also has a real kernel.

For let $f \in \mathbf{Z}^*$; define the involution by setting $C(f) = \bar{f}$:

$$\bar{f}(z) = \overline{f(\bar{z})} \qquad (z \in \mathbf{Z}). \tag{4}$$

Let us first verify that $\bar{f}$ is a continuous linear functional. In fact, we have

$$\bar{f}(\lambda z_1 + \mu z_2) = \overline{f(\overline{\lambda z_1 + \mu z_2})} = \overline{f(\bar{\lambda}\bar{z}_1 + \bar{\mu}\bar{z}_2)} = \overline{\bar{\lambda} f(\bar{z}_1) + \bar{\mu} f(\bar{z}_2)} = \lambda\bar{f}(z_1) + \mu\bar{f}(z_2)$$

and

$$|\bar{f}(z)| = |f(\bar{z})| \leqslant \|f\|\|\bar{z}\| = \|f\|\|z\|. \tag{5}$$

Now let us verify that conditions 1–3 in the definition of an involution are satisfied.

Bearing in mind the rule for multiplying a functional by a complex number, we have

$$\overline{(\lambda f_1+\mu f_2)}(z) = \overline{(\lambda f_1+\mu f_2)(\bar{z})} = \overline{\lambda f_1(\bar{z})+\mu f_2(\bar{z})} = \lambda\overline{f_1(\bar{z})}+\mu\overline{f_2(\bar{z})} = \lambda\bar{f}_1(z)+\mu\bar{f}_2(z)$$
$$= (\lambda\bar{f}_1+\mu\bar{f}_2)\ (z),$$
$$\bar{\bar{f}}(z) = \overline{\bar{f}(\bar{z})} = \overline{\overline{f(\bar{\bar{z}})}} = f(z).$$

Finally, we have the inequality $\|\bar{f}\| \leqslant \|f\|$ by (5). On the other hand, $\|f\| = \|\bar{\bar{f}}\| \leqslant \|\bar{f}\|$, and thus $\|\bar{f}\| = \|f\|$.

Denote the real kernel of **Z** by **X**. We show that, if ϕ is a linear functional on a real space **X**, then the complex extension f of ϕ belongs to the real kernel of **Z***, and that the latter is composed solely of such functionals. Both these assertions follow without difficulty from the definition of a dual functional. For if f is the complex extension of a functional $\phi \in$ **X***, then

$$\bar{f}(z) = \overline{f(\bar{z})} = \overline{\phi(x)-\phi(y)i} = \phi(x)+\phi(y)i = f(z)$$
$$(z = x+yi \in \mathbf{Z}).$$

Conversely, if f is a real element of **Z***—that is, if $\bar{f} = f$—then, by (4) we have $\overline{f(\bar{z})} = f(z)$. In particular, if $z = x \in$ **X**, then $\overline{f(x)} = f(x)$; that is, $f(x)$ is real.

The proposition we have just proved can be expressed as follows: the real kernel of a dual space is the dual space of the real kernel of the given space.

2.5. It is natural to expect that an operator adjoint to a real operator is itself real. Let $\tilde{U}$ be a real operator mapping a space **Z** into a space **W**, and let U be the induced operator from **X** into **Y** (where **X** = Re **Z**, **Y** = Re **W**). We show that $\tilde{U}^*$ is a real operator, and that the operator it induces from **Y*** into **X*** is just U^*. The last assertion must be understood as meaning that, if ϕ and ψ are functionals on **X** and **Y** respectively, where

$$\phi = U^*(\psi), \tag{6}$$

then

$$f = \tilde{U}^*(g), \tag{7}$$

where f is the complex extension of ϕ and g is that of ψ; conversely, if real functionals $f \in$ **Z*** and $g \in$ **W*** are connected by (7) then the functionals $\phi \in$ **X*** and $\psi \in$ **Y*** induced by them are connected by (6).

The proof that $\tilde{U}^*$ is real is very simple: if $g \in$ **W*** is a real functional—that is, if $\bar{g} = g$ and $f = \tilde{U}^*(g)$—then

$$\bar{f}(z) = \overline{f(\bar{z})} = \overline{g(\tilde{U}(\bar{z}))} = \overline{g(\overline{\tilde{U}(z)})} = g(\tilde{U}(z)) = f(z).$$

Next suppose that $\phi = U^*\psi$; then (for $x = \operatorname{Re} z$, $y = \operatorname{Im} z$) we have

$$f(z) = \phi(x)+i\phi(y) = \psi(U(x))+i\psi(U(y)) = g(U(x)+iU(y)) = g(\tilde{U}(z)),$$

so that $f = \tilde{U}^*(g)$. It is just as simple to show that (6) follows from (7).

§ 3. The spectrum

3.1. In this and the following section we study the behaviour of the equation

$$T_\lambda(x) = x-\lambda U(x) = y, \tag{1}$$

or, what is the same, the equation

$$T'_\mu(x) = \mu x - U(x) = y \tag{1'}$$

depending on the complex parameter λ(or μ). Here, and in the sequel, U is understood to be a continuous linear operator on a complex* B-space $\mathbf{X}$. We shall consider both (1) and (1′), on account of the fact that (1) is used for investigations in the theory of integral equations, some applications to which we shall be giving in § 6, while (1′) is used for investigations in abstract functional analysis, for the study of spectral properties of U.

In studying properties related to the solubility of (1) the complex plane is divided into two sets: the set $\pi(U)$ of those λ for which (1) has a unique solution whatever $y \in \mathbf{X}$ is taken on the right-hand side (so that T_λ has a continuous inverse (see XII.1.3)), and the set $\chi(U)$ of all other λ. The points of $\pi(U)$ are called the *non-singular* values of U; the set $\chi(U)$ is called the *characteristic set* of U.

Similarly we introduce the set $\rho(U)$ of those μ for which (1′) has a unique solution for every right-hand side, and the complementary set $\sigma(U)$. The points of $\rho(U)$ are called the *regular values* of U and the set $\rho(U)$ itself is called the *resolvent set* of U; the set $\sigma(U)$ is called the *spectrum* of U.

If, for some λ, the homogeneous equation

$$T_\lambda(x) = x - \lambda U(x) = \mathbf{0} \tag{2}$$

has a non-zero solution, then this λ is called a *characteristic value* of U. The set $\chi_0(U)$ of all characteristic values is obviously contained in $\chi(U)$. Every solution of (2) is called an *eigenvector* corresponding to the given characteristic value. The set $\bigcup_{n=1}^{\infty} \mathbf{N}(T_\lambda^n)$ is called the *root subspace*, and its dimension (finite or infinite) is the *multiplicity* of λ. The number r of distinct sets $\mathbf{N}(T_\lambda^n)$ $(n = 1, 2, \ldots)$ is called the *rank* of λ (cf. Lemma 2 of XIII.1.1).

If instead of (2) we consider the homogeneous equation corresponding to (1′):

$$T'_\mu(x) = \mu x - U(x) = \mathbf{0}, \tag{2'}$$

then we obtain the concepts of *eigenvalue*, *eigenvector* and *root subspace* corresponding to a given eigenvalue which we have already defined for operators in Hilbert space (IX.4.1).

Note that if U is a self-adjoint operator on Hilbert space and μ is an eigenvalue of U, then the rank of μ is $r = 1$, that is

$$\mathbf{N}(T'_\mu) = \mathbf{N}(T'^2_\mu) = \ldots = \mathbf{N}(T'^n_\mu) = \ldots, \tag{3}$$

and hence in this case the root subspace is $\mathbf{N}(T'_\mu)$: that is, it consists of all eigenvalues of U.

Let us prove (3). Since the eigenvalues of a self-adjoint operator are real, T'_μ and all its powers are self-adjoint operators. If we omit the subscript μ for simplicity and take $n = 2^k$, then for $x \in \mathbf{N}(T'^{2^k})$ we have

$$(T'^{2^{k-1}}x, T'^{2^{k-1}}x) = (T'^{2^k}x, x) = 0,$$

and so

$$T'^{2^{k-1}}x = \mathbf{0}.$$

Proceeding in this way, we eventually obtain the equation $T'x = \mathbf{0}$, that is, $x \in \mathbf{N}(T')$, and so $\mathbf{N}(T') \supset \mathbf{N}(T'^{2^k})$. The reverse inclusion holds for any operator. Hence $\mathbf{N}(T') = \mathbf{N}(T'^{2^k})$ $(k = 1, 2, \ldots)$. If now we take an arbitrary $n = 2, 3, \ldots$, then by choosing k such that $n \leqslant 2^k$, we have the obvious inclusions $\mathbf{N}(T') \subset \mathbf{N}(T'^n) \subset \mathbf{N}(T'^{2^k})$. Hence $\mathbf{N}(T') = \mathbf{N}(T'^n)$.

Let us note a simple connection between the spectrum and the characteristic set of a given operator U. It is easy to see that if $\lambda \in \chi(U)$, then $\mu = 1/\lambda \in \sigma(U)$, and conversely.

---

* If the original space $\mathbf{X}$ is real, then instead of U we consider its complex extension.

Obviously the characteristic values and eigenvalues of U are related in the same way. Here it is important to keep in mind that an eigenvector corresponding to a characteristic value λ will also be an eigenvector corresponding to the eigenvalue $\mu = 1/\lambda$, and conversely. Further, since we have $\mathbf{N}(T_\lambda^n) = \mathbf{N}(T_\mu'^n)$ $(n = 1, 2, \ldots)$ when $\lambda\mu = 1$, the last remark also extends to eigenspaces. Because of this, it is not necessary to distinguish between the concept of an eigenvector (or characteristic vector) corresponding to a characteristic value and that of an eigenvector corresponding to an eigenvalue. This fact is reflected in the terminology used above.

This connection between the characteristic set and the spectrum allows us, when it is convenient, to consider only one of these two parallel, and essentially equivalent, concepts, and to provide statements for both only in exceptional cases.

Let us now list some fairly simple propositions relating to the concepts introduced above.

I. The statement that $\lambda \in \pi(U)$ is equivalent to asserting the existence of the two-sided inverse operator $B_\lambda = T_\lambda^{-1} = (1 - \lambda U)^{-1}$.

See XII.1.3, Corollary to Theorem XII.1.2.

II. The set of non-singular values is open, and consequently the characteristic set is closed.

This follows from the theorem which says that, if an operator has a continuous inverse, then so does any operator close to it in norm (V.4.6). In our case,

$$\|T_{\lambda'} - T_\lambda\| = |\lambda - \lambda'| \|U\|,$$

so that if T_λ^{-1} exists then so does $T_{\lambda'}^{-1}$, provided the difference $\lambda - \lambda'$ is sufficiently small.

III. The disc $|\lambda| < 1/\|U\|$ is contained in $\pi(U)$; consequently the spectrum $\sigma(U)$ lies entirely within the disc $|\mu| \leqslant \|U\|$.

To see the truth of this, it is sufficient to apply Banach's theorem on inverse operators (V.4.5).

IV. The sets $\chi(U)$ and $\chi(U^*)$ are situated symmetrically with respect to the real axis.

For we have (IX.3.1)

$$T_{\bar{\lambda}}^* = [I - \lambda U]^* = I^* - \bar{\lambda} U^*.$$

And so, by Theorem XII.2.4, the operators* T_λ^{-1} and $(T_{\bar{\lambda}}^*)^{-1}$ will exist simultaneously.

V. If $\mathbf{X}$ is a space with real kernel and U is a real operator, then $\chi_0(U)$ is symmetrical about the real axis. Moreover, if $\lambda \in \chi_0(U)$ and z is the corresponding eigenvector, then the characteristic value $\bar{\lambda}$ will have $\bar{z}$ as a corresponding eigenvector.

For

$$\bar{z} - \bar{\lambda} U(\bar{z}) = \overline{z - \lambda U(z)},$$

so that the equations $z - \lambda U(z) = \mathbf{0}$ and $\bar{z} - \bar{\lambda} U(\bar{z}) = \mathbf{0}$ are equivalent.

Remark. In IX.5.3, we gave the definition of the spectrum for the case of a self-adjoint operator in Hilbert space. Theorem IX.5.3, which we proved in that section, demonstrates the equivalence of the two definitions (that of Chapter IX and the one given above).

3.2. In the case where U is a compact operator, one can determine the structure of the characteristic set fairly completely.

Theorem 1. *If U is a compact operator, then*

a) *the characteristic set contains only characteristic values, that is, $\chi(U) = \chi_0(U)$; also, every characteristic value has finite multiplicity;*

---

* T_λ^* is understood to mean $(T_\lambda)^*$.

b) *for every $r > 0$, the disc $|\lambda| \leqslant r$ contains at most a finite number of characteristic values;*

c) *if $\lambda_1 \in \chi(U)$ and $\lambda_2 \in \chi(U^*)$, where $\lambda_1 \neq \overline{\lambda_2}$, and if $x_1 \in \mathbf{X}$ is an eigenvector corresponding to λ_1 and $g_2 \in \mathbf{X}^*$ is an eigenvector corresponding to λ_2, then*

$$g_2(x_1) = 0.$$

Proof. a) By Theorem XII.1.4, if $\lambda \notin \pi(U)$, then the homogeneous equation (2) has a non-zero solution. The fact that the eigenspace has finite dimension follows from Lemma 2 (XIII.1.1). For according to that lemma, there exists an n_0 such that $\mathbf{N}(T_\lambda^n) = \mathbf{N}(T_\lambda^{n_0})$ ($n \geqslant n_0$). Hence in the present case the eigenspace is $\mathbf{N}(T_\lambda^{n_0})$. However,

$$T_\lambda^{n_0} = (I - \lambda U)^{n_0} = I - \tilde{U},$$

where $\tilde{U} = \sum_{k=1}^{n_0} (-1)^{k-1} C_{n_0}^k \lambda^k U^k$ is clearly a compact operator. Therefore, by Theorem XII.1.4, which we just referred to, the set of solutions of the homogeneous equation

$$T_\lambda^{n_0}(x) = \mathbf{0}$$

is a finite-dimensional subspace, and $\mathbf{N}(T_\lambda) \subset \mathbf{N}(T_\lambda^{n_0})$.

b) Assume the contrary: that is, assume that some disc $|\lambda| \leqslant r$ contains an infinite number of characteristic values. We select a sequence $\{\lambda_n\}$ of distinct characteristic values belonging to this set: $|\lambda_n| \leqslant r$ ($n = 1, 2, \ldots$). Let $\{x_n\}$ be a sequence of corresponding eigenvectors:

$$x_n - \lambda_n U(x_n) = \mathbf{0} \quad (n = 1, 2, \ldots). \tag{4}$$

We show (by induction) that, for each $n = 1, 2, \ldots$, the elements $x_1, x_2, \ldots, x_n$ are linearly independent. This is true for $n = 1$. Assume it is true for some $n \geqslant 1$. We verify that the elements $x_1, x_2, \ldots, x_n, x_{n+1}$ are linearly independent. Otherwise,

$$x_{n+1} = \sum_{k=1}^{n} \alpha_k x_k,$$

so that, by (4),

$$\frac{x_{n+1}}{\lambda_{n+1}} = \sum_{k=1}^{n} \frac{\alpha_k}{\lambda_k} x_k.$$

Comparing this with the previous equation, we find that

$$\sum_{k=1}^{n} \left(1 - \frac{\lambda_{n+1}}{\lambda_k}\right) \alpha_k x_k = \mathbf{0}.$$

Since $1 - \frac{\lambda_{n+1}}{\lambda_k} \neq 0$ ($k = 1, 2, \ldots, n$), it follows that $x_1, x_2, \ldots, x_n$ are linearly dependent, contrary to the inductive hypothesis.

Now form the set $\mathbf{X}_n = \mathscr{L}(\{x_1, x_2, \ldots, x_n\})$. Since, by what we have proved, $\mathbf{X}_{n+1} \neq \mathbf{X}_n$, the Lemma on Almost Perpendicularity (IV.1.7) shows that there exist elements $y_1, y_2, \ldots, y_n$ such that

$$y_n \in \mathbf{X}_n, \quad \|y_n\| = 1, \quad \rho(y_{n+1}, \mathbf{X}_n) > \frac{1}{2} \qquad (n = 1, 2, \ldots). \tag{5}$$

If $x \in \mathbf{X}_n$, that is, if $x = \sum_{k=1}^{n} \beta_k x_k$, then $U(x) = \sum_{k=1}^{n} \frac{\beta_k}{\lambda_k} x_k \in \mathbf{X}_n$. Moreover, $T_{\lambda_n}(x) = x - \lambda_n U(x) = \sum_{k=1}^{n} \beta_k \left(1 - \frac{\lambda_n}{\lambda_k}\right) x_k \in \mathbf{X}_{n-1}$ (as the coefficient of x_n is zero).

Let $m > n$. Consider the expression

$$U(\lambda_m y_m) - U(\lambda_n y_n) = y_m - [T_{\lambda_m}(y_m) + U(\lambda_n y_n)] = y_m - \tilde{y}.$$

By what we have proved, $T_{\lambda_m}(y_m) \in \mathbf{X}_{m-1}$ and $U(\lambda_n y_n) \in \mathbf{X}_n \subset \mathbf{X}_{m-1}$. Therefore

$$\tilde{y} = T_{\lambda_m}(y_m) + U(\lambda_n y_n) \in \mathbf{X}_{m-1}.$$

In view of (5), we have

$$\|U(\lambda_m y_m) - U(\lambda_n y_n)\| = \|y_m - \tilde{y}\| > \frac{1}{2}.$$

However, this contradicts the compactness of U, as the sequence $\{\lambda_n y_n\}$ is bounded ($\|\lambda_n y_n\| \leqslant r$).

c) We have $\lambda_1 U(x_1) = x_1$ and $\lambda_2 U^*(g_2) = g_2$. Therefore

$$g_2(x_1) = g_2(\lambda_1 U(x_1)) = \lambda_1 U^*(g_2)\ (x_1) = \lambda_1 \left(\frac{g_2}{\lambda_2}\right)(x_1) = \frac{\lambda_1}{\lambda_2} g_2(x_1),$$

and, since $\lambda_1 \neq \lambda_2$, this is possible only if $g_2(x_1) = 0$.

Note, in conclusion, that if U is a compact operator on an infinite-dimensional space $\mathbf{X}$, then the point zero belongs to the spectrum of U.

§ 4. Resolvents

4.1. Here we continue our investigation of the equation

$$x - \lambda U(x) = y, \tag{1}$$

but, in contrast to the preceding section, we shall now be interested in the case where the solution is unique.

Let $\lambda \neq 0$ be a non-singular value of U. The operator B_λ defined by

$$I + \lambda B_\lambda = (I - \lambda U)^{-1} \tag{2}$$

is called the *resolvent* of U. When $\lambda = 0$, we set $B_0 = U$.

If one is studying the spectrum and the set of regular values, it is more convenient to use instead of B_λ the operator

$$R_\mu = (\mu I - U)^{-1}, \tag{3}$$

which is meaningful for all regular values of U. We shall call R_μ the resolvent as well. There is no danger of confusing the two, as it will always be clear from the context which resolvent is being discussed; moreover, to distinguish the two resolvents we shall retain the different notations. We remark that the resolvent B_λ is often encountered in the theory of integral equations, where it is called the Fredholm resolvent, while in functional analysis it is R_μ that is generally called the resolvent (cf. 3.1).

If $\mu \neq 0$, then clearly

$$R_\mu = \frac{1}{\mu} I + \frac{1}{\mu^2} B_{1/\mu}. \tag{4}$$

Conversely, the equation

$$(I+\lambda B_\lambda)(I-\lambda U)=(I-\lambda U)(I+\lambda B_\lambda)=I$$

yields

$$B_\lambda = U(I-\lambda U)^{-1}=(I-\lambda U)^{-1}U. \tag{5}$$

Therefore, if $\lambda \neq 0$,

$$B_\lambda=\frac{1}{\lambda}UR_{1/\lambda}=\frac{1}{\lambda}R_{1/\lambda}U. \tag{6}$$

Equations (4) and (6) enable every proposition stated for B_λ to be reformulated for R_μ, and conversely.

4.2. Let us investigate the behaviour of B_λ for small λ. Consider the series

$$I+\lambda U+\lambda^2U^2+\ldots+\lambda^nU^n+\ldots. \tag{7}$$

If this converges in the space of operators $B(\mathbf{X},\mathbf{X})$, then, in accordance with the Remark following Banach's Theorem (V.4.5), its sum will be $(I-\lambda U)^{-1}$: thus

$$(I-\lambda U)^{-1}=I+\lambda U+\ldots+\lambda^nU^n+\ldots,$$

and so, by (5),

$$B_\lambda=U+\lambda U^2+\ldots+\lambda^nU^{n+1}+\ldots. \tag{8}$$

This formula is valid for precisely those λ for which (7) is convergent. However, as we proved in V.4.2, (7) converges if

$$\lim_{n\to\infty}\sqrt[n]{\|\lambda^nU^n\|}<1,$$

and diverges if

$$\lim_{n\to\infty}\sqrt[n]{\|\lambda^nU^n\|}>1.$$

We have thus arrived at the following theorem.

THEOREM 1. *The resolvent B_λ can be expanded as a series* (8) *in powers of λ, whose radius of convergence is*

$$r=\frac{1}{\lim\limits_{n\to\infty}\sqrt[n]{\|U^n\|}}=\frac{1}{\lim\limits_{n\to\infty}\sqrt[n]{\|B_0^n\|}}.$$

If we use (4) to pass from B_λ to R_μ, we obtain

COROLLARY. *The resolvent R_μ can be expanded as a series in powers of μ^{-1}:*

$$R_\mu=\frac{1}{\mu}I+\frac{1}{\mu^2}U+\ldots+\frac{1}{\mu^n}U^{n-1}+\ldots\left(|\mu|>\frac{1}{r}=\lim_{n\to\infty}\sqrt[n]{\|U^n\|}\right).$$

4.3. One can determine another expression for the radius of convergence of the series (8), related to the location of the characteristic set in the complex plane.

First we prove two auxiliary lemmas.

LEMMA 1. *For any $\lambda,\mu\in\pi(U)$, we have*

$$B_\lambda-B_\mu=(\lambda-\mu)B_\mu B_\lambda. \tag{9}$$

Proof. From (5) we obtain

$$B_\lambda-B_\mu=U(I-\lambda U)^{-1}-(I-\mu U)^{-1}U.$$

Multiplying this on the right by $I - \lambda U$ and on the left by $I - \mu U$, we obtain

$$(I - \mu U)(B_\lambda - B_\mu)(I - \lambda U) = (I - \mu U)U - U(I - \lambda U) = (\lambda - \mu)U^2$$

and hence

$$B_\lambda - B_\mu = (\lambda - \mu)\ (I - \mu U)^{-1}U \cdot U(I - \lambda U)^{-1} = (\lambda - \mu)B_\mu B_\lambda,$$

which is the required result.

COROLLARY. *The operators B_λ and B_μ are permutable: that is, $B_\lambda B_\mu = B_\mu B_\lambda$.*

Similarly it can be proved that, for all $\lambda, \mu \in \rho(U)$,

$$R_\lambda - R_\mu = -(\lambda - \mu)R_\lambda R_\mu.$$

LEMMA 2. *The resolvent B_λ is a continuous function of the parameter λ at every point of $\pi(U)$—that is, if $\lambda_n \to \lambda_0$ ($\lambda_n, \lambda_0 \in \pi(U)$), then $B_{\lambda_n} \to B_{\lambda_0}$.*

Proof. We first prove that the real function $\|B_\lambda\|$ is continuous on $\pi(U)$. If $U = \mathbf{0}$, then $B_\lambda = \mathbf{0}$ and the assertion is proved. But if $U \neq \mathbf{0}$, then $B_\lambda \neq \mathbf{0}$, and so we can prove the continuity of the function $\dfrac{1}{\|B_\lambda\|}$. From (9) we obtain

$$|\,\|B_\lambda\| - \|B_\mu\|\,| \leqslant \|B_\lambda - B_\mu\| = |\lambda - \mu|\ \|B_\mu B_\lambda\| \leqslant |\lambda - \mu|\ \|B_\mu\|\ \|B_\lambda\|.$$

Hence

$$\left|\frac{1}{\|B_\mu\|} - \frac{1}{\|B_\lambda\|}\right| \leqslant |\lambda - \mu|,$$

and thus we obtain the required result.

Now let us prove that B_λ is continuous. Since $\pi(U)$ is an open set and $\lambda_0 \in \pi(U)$, there exists a disc $|\lambda - \lambda_0| \leqslant \varepsilon$ that lies wholly within $\pi(U)$. The continuous function $\|B_\lambda\|$ is bounded in this disc; let us assume, say, that

$$\|B_\lambda\| \leqslant M \qquad (|\lambda - \lambda_0| \leqslant \varepsilon). \tag{10}$$

By (9) and (10), we have

$$\|B_\lambda - B_{\lambda_0}\| \leqslant |\lambda - \lambda_0|\ \|B_{\lambda_0}\|\ \|B_\lambda\| \leqslant M^2|\lambda - \lambda_0| \qquad (|\lambda - \lambda_0| \leqslant \varepsilon).$$

The lemma is thus proved.

THEOREM 2. *The radius of convergence r of the series* (8) *is equal to the distance r_0 from the point $\lambda = 0$ to the characteristic set $\chi(U)$.*

Proof. Firstly, the series (8) converges in the disc $|\lambda| < r$ and so the resolvent exists for such λ; hence this disc is contained in the set of non-singular values. Therefore $r \leqslant r_0$.

Now choose any element $x \in \mathbf{X}$ and any functional $f \in \mathbf{X}^*$ and consider the function ϕ of the complex variable λ defined by

$$\phi(\lambda) = f(B_\lambda(x)).$$

Let us prove that ϕ is regular on $\pi(U)$. In fact, if $\lambda, \mu \in \pi(U)$, then by (9)

$$\frac{\phi(\mu) - \phi(\lambda)}{\mu - \lambda} = \frac{f(B_\mu(x)) - f(B_\lambda(x))}{\mu - \lambda} = f(B_\mu B_\lambda(x)).$$

As $\mu \to \lambda$, the right-hand side tends to the limit $f(B_\lambda^2(x))$ (Lemma 2). Thus ϕ has a continuous derivative

$$\phi'(\lambda) = f(B_\lambda^2(x)).$$

Now expand ϕ in a Taylor series in a neighbourhood of the point $\lambda_0 = 0$:

$$\phi(\lambda) = \phi(0) + \frac{\phi'(0)}{1!}\lambda + \ldots + \frac{\phi^{(n)}(0)}{n!}\lambda^n + \ldots . \tag{11}$$

This expansion is valid in a disc not containing singularities of ϕ, and so certainly in the disc $|\lambda| < r_0$. But, by (8), we have

$$\phi(\lambda) = \sum_{n=0}^{\infty} f(U^{n+1}(x))\lambda^n \qquad (|\lambda| < r). \tag{12}$$

By a well-known theorem in complex function theory, the two series (11) and (12) are identical, so (12) converges for $|\lambda| < r_0$.

Choose any λ_1, with $0 < \lambda_1 < r_0$. As (12) is convergent for $\lambda = \lambda_1$, it follows that

$$\lambda_1^n f(U^{n+1}(x)) \underset{n\to\infty}{\longrightarrow} 0$$

and so, as f was arbitrary,

$$\lambda_1^n U^{n+1}(x) \to \mathbf{0}(\sigma(\mathbf{X}, \mathbf{X}^*)).$$

However, a weakly convergent sequence of elements is bounded (VIII.1.1):

$$\sup_n \|\lambda_1^n U^{n+1}(x)\| < \infty.$$

Since this inequality is satisfied for every $x \in \mathbf{X}$ and $\mathbf{X}$ is complete, Theorem VII.1.1 shows that

$$\sup \|\lambda_1^n U^{n+1}\| = M < \infty. \tag{13}$$

Therefore

$$\lambda_1 \lim_{n\to\infty} \sqrt[n]{\|U^n\|} \leqslant \lim_{n\to\infty} M^{1/n} = 1$$

and

$$\lambda_1 \leqslant \frac{1}{\lim\limits_{n\to\infty} \sqrt[n]{\|U^n\|}} = r.$$

Since λ_1 can be chosen arbitrarily close to r_0, we have $r_0 \leqslant r$. Bearing in mind the inequality $r \leqslant r_0$ that we proved above, we deduce that $r = r_0$, as we set out to prove.

REMARK 1. Let λ_0 be a non-singular value of U. As above, the expansion

$$B_\lambda = B_{\lambda_0} + (\lambda - \lambda_0)B_{\lambda_0}^2 + \ldots + (\lambda - \lambda_0)^n B_{\lambda_0}^{n+1} + \ldots$$

is justifiable, and holds in the disc $|\lambda - \lambda_0| < \rho_0$, where ρ_0 is the distance from λ_0 to the characteristic set, or, as in Theorem 1,

$$\rho_0 = \frac{1}{\lim\limits_{n\to\infty} \sqrt[n]{\|B_{\lambda_0}^{n+1}\|}}.$$

Replacing B_λ by R_μ, we obtain the following result.

COROLLARY 1. *The expansion*

$$R_\mu = \frac{1}{\mu}I + \frac{1}{\mu^2}U + \ldots + \frac{1}{\mu^{n+1}}U^n + \ldots \tag{14}$$

is valid for $|\mu| > 1/r$, *where* $1/r$ *is the radius of the smallest circle centred at the origin that contains the spectrum.*

The number $1/r$ is called the *spectral radius* of U.

REMARK 2. If U is a self-adjoint operator in Hilbert space, then, by IX.5.3, $1/r = \|U\|$. Comparing this with the Corollary to Theorem 1, we obtain the interesting equation*

$$\|U\| = \lim_{n\to\infty} \sqrt[n]{\|U^n\|}.$$

COROLLARY 2. *The spectrum* $\sigma(U)$ *of a continuous linear operator* U *on a complex B-space is non-empty.*

Proof. If $\sigma(U) = \phi$, then, on account of the relationship between $\sigma(U)$ and $\chi(U)$ and the fact that $0 \notin \chi(U)$, we see that the set $\pi(U)$ of non-singular values is the whole complex plane. Since we may assume that $U \neq \mathbf{0}$, we have $B_\lambda \neq \mathbf{0}$ for every λ. Suppose $y_0 = B_{\lambda_0}(x_0) \neq \mathbf{0}$. Choose $f \in \mathbf{X}^*$, $f(y_0) \neq 0$. An argument similar to that in the proof of Theorem 2 shows that the function

$$\phi(\lambda) = f(B_\lambda(x_0))$$

is regular in the whole complex plane. Further, we have

$$|\phi(\lambda)| = |f(B_\lambda(x_0))| \leqslant \|f\| \, \|B_\lambda\| \, \|x_0\|.$$

Since $\sigma(U) = \phi$, there exists a continuous inverse operator U^{-1}, and so an argument similar to that of Lemma 2 shows that $\|R_{1/\lambda}\| \to \|U^{-1}\|$ as $\lambda \to \infty$. Hence, in view of (6),

$$\|B_\lambda\| \leqslant 1/|\lambda| \, \|R_{1/\lambda}\| \, \|U\| \xrightarrow[\lambda\to\infty]{} 0.$$

Thus $\phi(\lambda)$ is bounded, and so, by Liouville's Theorem, it is a constant function; and clearly this can only be the zero function. However, $\phi(\lambda_0) = f(y_0) \neq 0$, so we have a contradiction.

Notice that if we attempted to define the spectrum of an operator U on a real space by a method analogous to that of 3.1, the spectrum could turn out to be empty. This is the basic reason for considering the complex case.

4.4. Let us apply the result proved above to the study of the convergence of the method of successive approximations for the equation

$$x - U(x) = y. \tag{15}$$

As we showed in V.5.1, the convergence of the series

$$I + U + U^2 + \ldots + U^n + \ldots \tag{16}$$

* This equation is also a consequence of the stronger relation:

$$\|U\| = \sqrt[n]{\|U^n\|} \qquad (n = 1, 2, \ldots),$$

which can be obtained by applying the theorem on the spectrum of an operator $\phi(U)$ (Theorem IX.5.4) to the functions $\phi(t) = t^n$ $(n = 1, 2, \ldots)$.

ensures that the method of successive approximations converges, whatever initial value x_0 we take to start the successive approximation process. Setting $\mu = 1$ in (14), we obtain the following convergence criterion for the method of successive approximations for equation (15).

THEOREM 3. *If the spectrum of the operator U lies in the disc $|\mu| < 1$, then the method of successive approximations for* (15) *converges, whatever values we take for $y \in \mathbf{X}$ and the initial approximation $x_0 \in \mathbf{X}$. If, however, there are points of the spectrum outside the disc $|\mu| \leqslant 1$, then there exists a set $E \subset \mathbf{X}$, which is residual* in $\mathbf{X}$, such that the successive approximation process for* (15), *starting with $x_0 = \mathbf{0}$, diverges when $y \in E$.*

Proof. Only the second part of the theorem needs to be proved. Note that, in this case, we have

$$\lim_{n\to\infty} \|U^n\| = \infty$$

and, in view of the Remark after Theorem VII.1.1,

$$\sup_n \|U^n(y)\| = \infty \tag{17}$$

for all $y \in \mathbf{X}$, except possibly for a set G of first category in $\mathbf{X}$. However, convergence of the successive approximation process starting with $x_0 = \mathbf{0}$ is equivalent to convergence of the series

$$y + U(y) + U^2(y) + \ldots + U^n(y) + \ldots,$$

and, in view of (17), this series diverges if $y \notin G$.

REMARK 1. If the operator U is such that all non-zero points of the spectrum are eigenvalues, then the theorem may be stated in a more precise form. Namely, in this case a necessary and sufficient condition for convergence of the method of successive approximations is that all the eigenvalues of U lie in the disc $|\mu| < 1$.

For, according to the theorem proved above, if the method of successive approximations is convergent, then the spectrum of U lies in the disc $|\mu| \leqslant 1$. If we assume that there exists an eigenvalue μ_0 lying on the circumference $|\mu| = 1$, then, setting $y = y'$ in (15), where y' is an eigenvector corresponding to μ_0, and taking $x_0 = \mathbf{0}$, we obtain for the n-th approximation x_n the expression

$$x_n = (1 + \mu_0 + \ldots + \mu_0^{n-1})y',$$

which does not tend to any limit.

REMARK 2. The results of the theorem and Remark 1 are greatly simplified if the operator U in (15) is a self-adjoint operator in Hilbert space. In this case, bearing in mind Remark 2 in 4.3, we have: a sufficient condition for the method of successive approximations to be convergent is that $\|U\| < 1$. If all the non-zero points of the spectrum are eigenvalues, then this condition is also necessary.

Finally, let us formulate Theorem 3 in terms of characteristic sets.

THEOREM 3′. *If the characteristic set of the operator U lies outside the disc $|\lambda| \leqslant 1$, then the method of successive approximations for* (15) *is convergent, whatever values we take for $y \in \mathbf{X}$ and the initial approximation $x_0 \in \mathbf{X}$. If, however, there are points of the characteristic set in the disc $|\lambda| < 1$, then there exists a set $E \subset \mathbf{X}$, which is residual in $\mathbf{X}$, such that the successive approximation process, starting from $x_0 = \mathbf{0}$, is divergent when $y \in E$.*

One can also make remarks analogous to Remarks 1 and 2 in connection with this theorem. We leave this for the reader to do.

4.5. If U is a compact operator, then we can add to the results of Theorems 1 and 2 some

* Recall that a set E in a space $\mathbf{X}$ is said to be residual if its complement is of first category in $\mathbf{X}$ (see I.4.7).

more subtle facts relating to the behaviour of a resolvent in the neighbourhood of a characteristic value.

We shall call a continuous linear operator V *finite-dimensional* if it maps a space $\mathbf{X}$ into a finite-dimensional subspace $\tilde{\mathbf{X}} \subset \mathbf{X}$. Let us choose a complete system of linearly independent elements $x_1, x_2, \ldots, x_n$ in $\tilde{\mathbf{X}}$. The definition implies that, for any $x \in \mathbf{X}$,

$$V(x) = \sum_{k=1}^{n} \alpha_k x_k.$$

The coefficients $\alpha_1, \alpha_2, \ldots, \alpha_n$ obviously depend on x. If we set $\alpha_k = f_k(x)$ $(k = 1, 2, \ldots, n)$, we can verify without any difficulty that the functionals f_k $(k = 1, 2, \ldots, n)$ are linear and continuous: their linearity is not in doubt, and their continuity follows from the fact that, if a sequence of elements in a finite-dimensional space tends to zero, then so do all the sequences of coordinates of these elements (see IV.1.6). Hence we have

$$V(x) = f_1(x)x_1 + f_2(x)x_2 + \ldots + f_n(x)x_n. \tag{18}$$

Conversely, every operator V expressible in the form (18) is evidently finite-dimensional. We remark that a finite-dimensional operator is necessarily compact.

Now consider a compact operator U, and let λ_0 be a characteristic value of U. We have

THEOREM 4. *The operator U is expressible in the form*

$$U = U' + U'',$$

where U'' is a finite-dimensional operator and U' is compact; also the characteristic set of U'' consists of the single point λ_0, while that of U' is obtained from the characteristic set of U by removing the point λ_0.

Proof. We may assume that $\lambda_0 = 1$ (otherwise we would consider the operator $(1/\lambda_0)U$ instead of U). With this assumption, we shall show that the decomposition of U into the sum $U' + U''$ of Theorem 1.1 satisfies the requirement of the theorem.

We shall use the notation of Theorem 1.1. Let us verify that U'' has the unique characteristic value $\lambda_0 = 1$. In fact, if $x_0 \in \mathbf{N}(T)$, then also $x_0 \in \mathbf{N}(T^r) = \mathbf{X}''$; hence, by the definition of U'',

$$U''(x_0) = U(x_0) = x_0$$

and $\lambda_0 = 1$ is a characteristic value of U''.

If

$$\lambda U''(x) = x \tag{19}$$

for some $x \in \mathbf{X}$ $(x \neq \mathbf{0})$, then, since $U''(x) \in \mathbf{X}''$, we have $x \in \mathbf{X}''$ and therefore $U''(x) = U(x)$. Let $m \geqslant 0$ be such that $T^{m+1}(x) = \mathbf{0}$, but $T^m(x) \neq \mathbf{0}$. On account of (19), we have

$$\mathbf{0} = T^m(x - \lambda U(x)) = T^m(x - \lambda[x - T(x)]) = (1-\lambda)T^m(x) + \lambda T^{m+1}(x) =$$
$$= (1-\lambda)T^m(x),$$

which is possible only if $\lambda = 1$.

Thus the unique characteristic value of U'' is $\lambda_0 = 1$.

Next we prove the part of the theorem relating to the characteristic set of U'.

Since, by Theorem 1.1, the operator $T' = I - U'$ has an inverse, $\lambda_0 = 1$ is not a characteristic value of U'. Let $\lambda \neq 1$ be any characteristic value of U and let $x_0 \neq \mathbf{0}$ be a corresponding eigenvector. If it were the case that $x_0 \in \mathbf{X}''$, then, arguing as above, we would obtain $\lambda = 1$. Hence $x_0 \notin \mathbf{X}''$. Therefore, in the decomposition

$$x_0 = x'_0 + x''_0 \qquad (x'_0 \in \mathbf{X}',\ x''_0 \in \mathbf{X}'') \tag{20}$$

(see part (c) of Theorem 1.1) we must have $x'_0 \neq \mathbf{0}$. In view of the uniqueness of the representation (20), the equation

$$x_0 = \lambda U(x'_0) + \lambda U(x''_0)$$

implies that $x'_0 = \lambda U(x'_0) = \lambda U'(x'_0)$, that is, λ is a characteristic value of U'.

Conversely, suppose λ is a characteristic value of U' and $x \neq \mathbf{0}$ is a corresponding eigenvector. As

$$x = \lambda U'(x) \in \mathbf{X}',$$

we have $U'(x) = U(x)$ and so $x = \lambda U(x)$—that is, λ is a characteristic value of U.

The remaining assertions of the theorem form part of Theorem 1.1.

The theorem is thus proved.

4.6. One can form a more complete idea of the behaviour of the resolvent in the neighbourhood of a characteristic value on the basis of the following theorem.

THEOREM 5. *Let λ_0 be a characteristic value of a compact operator U. Then, in a sufficiently small neighbourhood of λ_0, we have the expansion*

$$B_\lambda = \frac{U_{-r}}{(\lambda-\lambda_0)^r} + \dots + \frac{U_{-1}}{\lambda-\lambda_0} + U_0 + U_1(\lambda-\lambda_0) + \\ + \dots + U_n(\lambda-\lambda_0)^n + \dots . \tag{21}$$

Here r is the rank of the characteristic value λ_0; the operators $U_{-r}, \dots, U_{-1}$ are finite-dimensional; and $U_{-r} \neq \mathbf{0}$.

The series on the right-hand side of (21) *is convergent in the space of operators $B(\mathbf{X}, \mathbf{X})$.*

Proof. As in the proof of the preceding theorem, we shall assume that $\lambda_0 = 1$. We note at once that, by Lemma 2 of 1.1, the rank of the characteristic value λ_0 is finite. Using again the notation of Theorem 1.1, we find, from the Remark following that theorem, that $\mathbf{X}' = T^r(\mathbf{X})$, $\mathbf{X}'' = \mathbf{N}(T^r)$.

By expressing an element $x \in \mathbf{X}$ in the form

$$x = x' + x''(x' \in \mathbf{X}', x'' \in \mathbf{X}'')$$

(part (c) of Theorem 1.1) and associating with x the elements $x' = P'(x)$ and $x'' = P''(x)$, we construct two operators P' and P''—the projections of $\mathbf{X}$ onto $\mathbf{X}'$ and $\mathbf{X}''$ respectively. In view of inequality (9) of § 1, these operators are continuous. Notice also that

$$P'(\mathbf{X}) = \mathbf{X}', \quad P''(\mathbf{X}) = \mathbf{X}'', \quad P' + P'' = I.$$

Consider an arbitrary element $y \in \mathbf{X}$. The element $x = B_\lambda(y)$ is a solution of the equation

$$x - \lambda U(x) = U(y). \tag{22}$$

If here we substitute $x = P'(x) + P''(x)$, $y = P'(y) + P''(y)$ and bear in mind that

$U(\mathbf{X}') \subset \mathbf{X}'$ and $U(\mathbf{X}'') \subset \mathbf{X}''$, then we can rewrite (22) as a system of two equations

$$\left.\begin{aligned} P'(x) - \lambda U P'(x) &= U P'(y) \\ P''(x) - \lambda U P''(x) &= U P''(y) \end{aligned}\right\}. \tag{23}$$

If we observe that $UP' = U'P'$, we can rewrite the first of the equations in the form

$$T'_\lambda(P'(x)) = U'P'(y),$$

where we have written $T'_\lambda = I - \lambda U'$. By Theorem 4, $\lambda_0 = 1$ is a regular value of U'. Therefore, by the Remark following Theorem 2, provided the difference $\lambda - 1$ is small enough, the resolvent B'_λ of U' has an expansion

$$B'_\lambda = U'_0 + (\lambda - 1)U'_1 + \ldots + (\lambda - 1)^n U'_n + \ldots,$$

and the series on the right-hand side converges in $B(\mathbf{X}, \mathbf{X})$. Taking what has been said into account, we can write

$$P'(x) = B'_\lambda P'(y) = [U_0 + (\lambda - 1)U_1 + \ldots + (\lambda - 1)^n U_n + \ldots]\ (y), \tag{24}$$

where $U_n = U'_n P'$ $(n = 1, 2, \ldots)$ and the series on the right-hand side converges, as before, in $B(\mathbf{X}, \mathbf{X})$.

Let us now concern ourselves with the second of the equations (23).

Form the factor-spaces

$$\mathbf{X}^{(k)} = \mathbf{N}(T^k)/\mathbf{N}(T^{k-1}) \qquad (k = 1, 2, \ldots, r)$$

and denote by ϕ_k the unique homomorphism from $\mathbf{N}(T^k)$ onto $\mathbf{X}^{(k)}$. The space $\mathbf{X}^{(r)}$ is clearly finite-dimensional. Choose a complete system of linearly independent elements $\bar{x}_1^{(r)}$, $\bar{x}_2^{(r)}, \ldots, \bar{x}_{n_r}^{(r)}$ in this space, and let $x_1^{(r)}, x_2^{(r)}, \ldots, x_{n_r}^{(r)}$ be elements in $\mathbf{N}(T^r)$ such that $\phi_r(x_j^{(r)}) = \bar{x}_j^{(r)}$. The elements $x_j^{(r-1)} = T(x_j^{(r)})$ $(j = 1, 2, \ldots, n_r)$ belong to $\mathbf{N}(T^{r-1})$. Further, their images $\bar{x}_j^{(r-1)} = \phi_{r-1}(x_j^{(r-1)})$ are linearly independent, for if $\sum_{j=1}^{n_r} \alpha_j \bar{x}_j^{(r-1)} = \mathbf{0}$, then

$$\sum_{j=1}^{n_r} \alpha_j x_j^{(r-1)} = T\left(\sum_{j=1}^{n_r} \alpha_j x_j^{(r)}\right) \in \mathbf{N}(T^{r-2});$$

in other words, we have $\sum_{j=1}^{n_r} \alpha_j x_j^{(r)} \in \mathbf{N}(T^{r-1})$, and therefore

$$\sum_{j=1}^{n_r} \alpha_j \bar{x}_j^{(r)} = \phi_r\left(\sum_{j=1}^{n_r} \alpha_j x_j^{(r)}\right) = \mathbf{0},$$

which is possible only if $\alpha_1 = \alpha_2 = \ldots = \alpha_{n_r} = 0$.

Now adjoin to the elements $\bar{x}_1^{(r-1)}, \bar{x}_2^{(r-1)}, \ldots, x_{n_r}^{(r-1)}$ additional elements $\bar{x}_{n_r+1}^{(r-1)}, \ldots, \bar{x}_{n_{r-1}}^{(r-1)} \in \mathbf{X}^{(r-1)}$ so that we obtain a basis for $\mathbf{X}^{(r-1)}$. Also choose $x_{n_r+1}^{(r-1)}, \ldots, x_{n_{r-1}}^{(r-1)} \in \mathbf{N}(T^{r-1})$ so that $\bar{x}_j^{(r-1)} = \phi_{r-1}(x_j^{(r-1)})$.

Continuing the argument this way, we construct, for each $k = 1, 2, \ldots, r$, elements $x_1^{(k)}, x_2^{(k)}, \ldots, x_{n_k}^{(k)}$ such that

$$x_j^{(k)} \in \mathbf{N}(T^k), \ T(x_j^{(k)}) = x_j^{(k-1)}$$

$$(j = 1, 2, \ldots, n_k;\ k = 1, 2, \ldots, r;\ x_j^{(0)} = \mathbf{0}). \tag{25}$$

Then, for each $k = 1, 2, \ldots, r$, the elements

$$\bar{x}_1^{(k)} = \phi_k(x_1^{(k)}), \quad \bar{x}_2^{(k)} = \phi_k(x_2^{(k)}), \ldots, \quad \bar{x}_{n_k}^{(k)} = \phi_k(x_{n_k}^{(k)})$$

form a basis for $\mathbf{X}^{(k)}$.

Write

$$\mathbf{X}_k = \mathscr{L}(\{x_1^{(k)}, x_2^{(k)}, \ldots, x_{n_k}^{(k)}\}) \quad (k = 1, 2, \ldots, r).$$

By (25), we have

$$T(\mathbf{X}_k) \subset \mathbf{X}_{k-1} \quad (k = 1, 2, \ldots, r;\ \mathbf{X}_0 = \{\mathbf{0}\}). \tag{26}$$

We now show that the elements $\{x_j^{(k)}\}$ $(j = 1, 2, \ldots, n_k;\ k = 1, 2, \ldots, r)$ form a complete system of linearly independent elements in $\mathbf{X}''$. Suppose

$$\sum_{k=1}^{r} \sum_{j=1}^{n_k} \lambda_j^{(k)} x_j^{(k)} = \mathbf{0}.$$

Since $x_j^{(k)} \in \mathbf{N}(T^{r-1})$ for $k < r$, when we apply the operator ϕ, we get

$$\sum_{k=1}^{r} \sum_{j=1}^{n_k} \lambda_j^{(k)} \phi_r(x_j^{(k)}) = \sum_{j=1}^{n_r} \lambda_j^{(r)} \bar{x}_j^{(r)} = \mathbf{0}$$

and hence $\lambda_j^{(r)} = 0$ $(j = 1, 2, \ldots, n_r)$. Similarly one can verify that all the other coefficients $\lambda_j^{(k)}$ are zero. Now consider an arbitrary element $x'' \in \mathbf{X}''$. We have $\phi_r(x'') \in \mathbf{X}^{(r)}$, so we can find coefficients $\alpha_1^{(r)}, \alpha_2^{(r)}, \ldots, \alpha_{n_r}^{(r)}$ such that

$$\phi_r(x'') = \sum_{j=1}^{n_r} \alpha_j^{(r)} \bar{x}_j^{(r)} = \phi_r\left(\sum_{j=1}^{n_r} \alpha_j^{(r)} x_j^{(r)} \right).$$

Hence

$$x'' - \sum_{j=1}^{n_r} \alpha_j^{(r)} x_j^{(r)} \in \mathbf{N}(T^{r-1}).$$

Continuing in a similar way, we eventually find that, for some $\alpha_j^{(k)}$,

$$x'' - \sum_{k=1}^{r} \sum_{j=1}^{n_k} \alpha_j^{(k)} x_j^{(k)} \in \mathbf{N}(T^0) = \{\mathbf{0}\},$$

and so $x'' = \sum_{k=1}^{r} \sum_{j=1}^{n_k} \alpha_j^{(k)} x_j^{(k)}$.

Let x be an arbitrary element of $\mathbf{X}$. Then $P''(x) \in \mathbf{X}''$, so

$$P''(x) = \sum_{k=1}^{r} \sum_{j=1}^{n_k} \alpha_j^{(k)}(x) x_j^{(k)},$$

and here the coefficients $\alpha_j^{(k)}$ are, as we mentioned in 4.5, linear functionals. If we write

$$P_k(x) = \sum_{j=1}^{n_k} \alpha_j^{(k)}(x) x_j^{(k)} \qquad (k = 1, 2, \ldots, r),$$

then, this implies that P_k is a continuous operator mapping $\mathbf{X}$ onto $\mathbf{X}_k$, and that

$$P'' = P_1 + P_2 + \ldots + P_r, \tag{27}$$

$$P_m P_k = \begin{cases} \mathbf{0}, & m \neq k, \\ P_m, & m = k \end{cases} \qquad (k, m = 1, 2, \ldots, r). \tag{28}$$

Using the equation $UP'' = P''U$, we now write the right-hand side of the second equation in (23) in the form $P''U(y)$ and then apply the operator P'' to the sum $P_1 + P_2 + \ldots + P_r$:

$$\sum_{k=1}^{r} P_k(x) - \lambda \sum_{k=1}^{r} UP_k(x) = \sum_{k=1}^{r} P_k U(y).$$

Applying the operator P_m to both sides of this equation, we obtain, in view of (28),

$$P_m(x) - \lambda \sum_{k=1}^{r} P_m U P_k(x) = P_m U(y) \quad (m = 1, 2, \ldots, r). \tag{29}$$

However,

$$UP_k(x) = P_k(x) - TP_k(x),$$

so, by (26),

$$P_m U P_k(x) = \begin{cases} P_m(x), & k = m, \\ -TP_{m+1}(x), & k = m+1, \\ \mathbf{0}, & k \neq m, m+1 \end{cases} \qquad (k, m = 1, 2, \ldots, r;\ P_{r+1} = \mathbf{0}).$$

Using this, we can rewrite (29) in the simpler form

$$P_r(x) - \lambda P_r(x) = P_r U(y), \tag{30}$$

$$P_m(x) - \lambda[P_m(x) - TP_{m+1}(x)] = P_m U(y) \qquad (m = 1, 2, \ldots, r-1). \tag{31}$$

From (30), we find that

$$P_r(x) = \frac{P_r U(y)}{1-\lambda},$$

so that, by (31),

$$P_{r-1}(x) = \frac{\lambda}{1-\lambda} TP_r(x) + \frac{1}{1-\lambda} P_{r-1}U(y) = \frac{\lambda}{(1-\lambda)^2} TP_r U(y) + $$

$$+ \frac{1}{1-\lambda} P_{r-1} U(y)$$

and, in general, for $m = 1, 2, \ldots, r$,

$$P_m(x) = \sum_{k=0}^{r-m} \frac{\lambda^k}{(1-\lambda)^{k+1}} T^k P_{m+k} U(y).$$

Adding together all the resulting equations, we have

$$P''(x) = \sum_{m=1}^{r} P_m(x) = \sum_{m=1}^{r} \sum_{k=0}^{r-m} \frac{\lambda^k}{(1-\lambda)^{k+1}} T^k P_{m+k} U(y) =$$

$$= \sum_{k=0}^{r-1} \frac{\lambda^k}{(1-\lambda)^{k+1}} \sum_{m=1}^{r-k} T^k P_{m+k} U(y). \tag{32}$$

Now express $\dfrac{\lambda^k}{(1-\lambda)^{k+1}}$ in partial fractions:

$$\frac{\lambda^k}{(1-\lambda)^{k+1}} = \sum_{j=0}^{k} \frac{c_j^{(k)}}{(\lambda-1)^{j+1}}.$$

Here the $c_j^{(k)}$ are constants, and $c_k^{(k)} = (-1)^{k+1}$. Substituting this in (32), we obtain

$$P''(x) = \sum_{s-1}^{r} \frac{U_{-s}(y)}{(\lambda-1)^s}, \tag{33}$$

where the operators U_{-s} $(s = 1, 2, \ldots, r)$ are linear combinations of operators of the form $T^k P_{m+k} U$. Since $P_{m+k}(\mathbf{X}) = \mathbf{X}_{m+k} \subset \mathbf{X}''$, and since by Theorem 1.1(b) we have $T(\mathbf{X}'') \subset \mathbf{X}''$, it follows that the operators U_{-s} map $\mathbf{X}$ into $\mathbf{X}''$, and so are finite-dimensional. Furthermore, it is clear from (32) that

$$U_{-r} = (-1)^r T^{r-1} P_r U.$$

Hence if, for example, we had $y = x_1^{(r)}$, then by (25),

$$U_{-r}(y) = (-1)^r T^{r-1} P_r U(x_1^{(r)}) = (-1)^r T^{r-1} P_r(x_1^r - x_1^{(r-1)}) = (-1)^r T^{r-1}(x_1^{(r)}) =$$
$$= (-1)^r x_1^{(1)} \neq \mathbf{0},$$

so that $U_{-r} \neq \mathbf{0}$.

Adding equations (24) and (33), we obtain the required decomposition of the resolvent B_λ.

This completes the proof of the theorem.

REMARK. If U is a self-adjoint operator in Hilbert space, then the theorem can be made somewhat more precise, since in this case $r = 1$ (see 3.1) and therefore the decomposition (21) contains only one term for which the exponent of $\lambda - \lambda_0$ is negative, namely $(\lambda - \lambda_0)^{-1} U_{-1}$. We do not give a detailed statement of the relevant result, as it has already been formulated (in fact in a stronger form) in IX.4.5.

§ 5. The Fredholm alternative

In this section we give conditions that a continuous linear operator T, mapping a B-space $\mathbf{X}$ into itself, must satisfy for the Fredholm alternative to be valid for it (see 1.4). It turns out, in particular, that if some power U^m of a linear operator U is compact, then the Fredholm alternative is valid for $T = I - U$. The results presented below are due to S. M. Nikol'skii [1].

5.1. We consider the equation

$$T(x) = y \tag{1}$$

and its adjoint

$$T^*(g) = f. \tag{2}$$

We also consider the corresponding homogeneous equations

$$T(x) = \mathbf{0}, \tag{3}$$

$$T^*(g) = \mathbf{0}. \tag{4}$$

We recall that for the Fredholm alternative to hold for an operator T means that either

1) equations (1) and (2) are soluble for all right-hand sides, and the solutions are then unique, or

2) the homogeneous equations (3) and (4) have the same finite number of linearly independent solutions, $x_1, x_2, \ldots, x_n$ and $g_1, g_2, \ldots, g_n$, respectively; in this case, necessary and sufficient conditions for (1) and (2), respectively, to hold are

$$g_k(y) = 0 \qquad (k = 1, 2, \ldots, n),$$

and

$$f(x_k) = 0 \qquad (k = 1, 2, \ldots, n);$$

then the general solution of (1) is given by

$$x = x^* + \sum_{k=1}^{n} c_k x_k,$$

and that of (2) by

$$g = g^* + \sum_{k=1}^{n} c_k g_k,$$

where x^* and g^* are arbitrary solutions of (1) and (2), respectively, and $c_1, c_2, \ldots, c_n$ are arbitrary constants.

The following theorem shows that the class of operators T for which the Fredholm alternative holds is essentially only slightly different from the class of operators of the form $T = I - U$, where U is a compact operator.

THEOREM 1. *Each of the following two conditions is both necessary and sufficient for the Fredholm alternative to hold for an operator T:*

1) *T is expressible in the form*

$$T = W + V,$$

where W is an operator having a continuous two-sided inverse and V is a compact operator.

2) *T is expressible in the form*

$$T = W_1 + V_1,$$

where W_1 is an operator having a continuous two-sided inverse and V_1 is a finite-dimensional operator.

Proof. We may clearly restrict ourselves to proving the sufficiency of condition 1) and the necessity of condition 2).

Sufficiency of 1). Suppose

$$T = W + V,$$

where W has a continuous two-sided inverse W^{-1} and V is compact. In that case, (1) is equivalent to

$$W^{-1}T(x) = W^{-1}(y). \tag{5}$$

Furthermore, W^* has a continuous two-sided inverse $(W^*)^{-1} = (W^{-1})^*$ (XII.2.3), so (2) is equivalent to

$$T^*W^{*-1}(g) = f \tag{6}$$

in the sense that, if g_0 is a solution of (6), then $W^{*-1}(g_0)$ is a solution of (2), while if g_0' is a solution of (2) then $W^*(g_0')$ is a solution of (6).

Let us now introduce the notation $U = -W^{-1}V$. Bearing in mind that $U^* =$

$-V^*(W^{-1})^* = -V^* W^{*-1}$ (see IX.3.1), we can rewrite (5) and (6) in the forms

$$x - U(x) = W^{-1}(y), \tag{7}$$

$$g - U^*(g) = f. \tag{8}$$

As U is compact, Theorem 1.4 applies to (7) and (8). Hence the homogeneous equations

$$x - U(x) = \mathbf{0}, \tag{9}$$

$$g - U^*(g) = \mathbf{0} \tag{10}$$

have the same finite number of linearly independent solutions $x_1, x_2, \ldots, x_n$ and $g'_1, g'_2, \ldots, g'_n$. The homogeneous equation (3) will evidently have the same complete system of linearly independent solutions as (9), namely $x_1, x_2, \ldots, x_n$. Let us show that the functionals

$$g_k = W^{*-1}(g'_k) \qquad (k = 1, 2, \ldots, n) \tag{11}$$

form a complete system of linearly independent solutions of (4). The fact that every functional (11) is a solution of (4) follows from the equivalence of (2) and (6) referred to above. The functionals (11) are linearly independent, since an equation $\sum_{k=1}^{n} \alpha_k g_k = \mathbf{0}$ implies $\sum_{k=1}^{n} \alpha_k g'_k = \sum_{k=1}^{n} \alpha_k W^*(g_k) = \mathbf{0}$, and this is only possible if $\alpha_1 = \alpha_2 = \ldots = \alpha_n = 0$. Finally, if (4) had a a solution g_0 that was not a linear combination of the functionals (11), then the functional $W^*(g_0)$ would be a solution of (10) that was not a linear combination of $g'_1, g'_2, \ldots, g'_n$, and this is impossible.

Thus (3) and (4) have the same finite number of linearly independent solutions.

Moreover, by Theorem 1.4 again, (5), and hence also (1), is soluble if and only if

$$g'_k(W^{-1}(y)) = 0 \qquad (k = 1, 2, \ldots, n). \tag{12}$$

In view of (11), this condition is equivalent to the following:

$$(W^{-1})^*(g'_k)(y) = W^{*-1}(g_k)(y) = g_k(y) = 0 \qquad (k = 1, 2, \ldots, n).$$

Similarly, one can check that the condition

$$f(x_k) = 0 \qquad (k = 1, 2, \ \ldots, n)$$

is necessary and sufficient for (2) to be soluble.

Necessity of 2). Let $x_1, x_2, \ldots, x_n$ and $g_1, g_2, \ldots, g_n$ be complete systems of linearly independent solutions of (3) and (4) respectively. Using Theorem V.7.4 and Lemma III.3.1, we can find functionals $f_1, f_2, \ldots, f_n \in \mathbf{X}^*$ and elements $y_1, y_2, \ldots, y_n \in \mathbf{X}$ such that

$$f_j(x_k) = \begin{cases} 0, & j \neq k, \\ 1, & j = k \end{cases} \qquad (j, k = 1, 2, \ldots, n), \tag{13}$$

$$g_k(y_j) = \begin{cases} 0, & j \neq k, \\ 1, & j = k \end{cases} \qquad (j, k = 1, 2, \ldots, n). \tag{14}$$

Write $\mathbf{Y}' = T(\mathbf{X})$, $\mathbf{Y}'' = \mathscr{L}(\{y_1, y_2, \ldots, y_n\})$. Each element $y \in \mathbf{X}$ can be expressed in a unique way in the form

$$y = y' + y'' \qquad (y' \in \mathbf{Y}', y'' \in \mathbf{Y}''). \tag{15}$$

For if we set

$$y'' = \sum_{k=1}^{n} g_k(y) y_k, \qquad y' = y - y'',$$

then, by (14),

$$g_j(y') = g_j(y) - \sum_{k=1}^{n} g_k(y) g_j(y_k) = 0 \qquad (j = 1, 2, \ldots, n),$$

so that the equation $T(x) = y'$ is soluble, and hence $y' \in \mathbf{Y}'$. The uniqueness of (15) follows from the fact that, if $y'' = \sum_{k=1}^{n} \alpha_k y_k \in \mathbf{Y}'$, then the equation $T(x) = y''$ must have a solution and so $g_j(y'') = \alpha_j = 0$ $(j = 1, 2, \ldots, n)$.

Let us also write $\mathbf{X}' = \mathbf{N}(f_1, f_2, \ldots, f_n)$ and $\mathbf{X}'' = \mathscr{L}(\{x_1, \ldots, x_n\})$. As above, we can show that every $x \in \mathbf{X}$ can be expressed uniquely in the form

$$x = x' + x'' \qquad (x' \in \mathbf{X}',\ x'' \in \mathbf{X}''). \tag{16}$$

We construct the operator W_1 by setting

$$W_1(x) = T(x) + \sum_{k=1}^{n} f_k(x) y_k,$$

and show that W_1 determines a one-to-one mapping of $\mathbf{X}$ onto itself, and hence has a continuous two-sided inverse (Theorem XII.1.2). In fact, let y be any element of $\mathbf{X}$. We express y in the form (15): $y = y' + y''$. Then $y' \in \mathbf{Y}' = T(\mathbf{X})$, $y'' = \sum_{k=1}^{n} \alpha_k y_k \in \mathbf{Y}''$, that is, the equation $T(x) = y'$ has a solution x', which, we may assume, belongs to $\mathbf{X}'$.*

If we now set

$$x'' = \sum_{k=1}^{n} \alpha_k x_k, \qquad x^* = x' + x''$$

and bear in mind that $T(x'') = \mathbf{0}$, $f_j(x') = 0$, and also that we have (13), we obtain

$$W_1(x^*) = T(x') + T(x'') + \sum_{k=1}^{n} f_k(x'') y_k = T(x') + \sum_{k=1}^{n} f_k\Big(\sum_{j=1}^{n} \alpha_j x_j\Big) y_k = y' + \sum_{k=1}^{n} \alpha_k y_k$$
$$= y.$$

We now show that there are no other solutions of the equation $W_1(x) = y$, besides the element x^*. For otherwise there would exist an element $x_0 \neq \mathbf{0}$ such that

$$W_1(x_0) = \mathbf{0},$$

that is,

$$T(x_0) + \sum_{k=1}^{n} f_k(x_0) y_k = \mathbf{0}.$$

Here $T(x_0) \in \mathbf{Y}$, and $\sum_{k=1}^{n} f_k(x_0) y_k \in \mathbf{Y}''$. Since the expression for an element in the form

* For if we write a solution x in the form (16): $x = x' + x''$ and bear in mind that $T(x'') = \mathbf{0}$, then we have $y' = T(x) = T(x')$.

(15) is unique, we obtain the equations

$$T(x_0) = \mathbf{0}, \quad \sum_{k=1}^{n} f_k(x_0) y_k = \mathbf{0}, \quad f_k(x_0) = 0 \qquad (k = 1, 2, \ldots, n),$$

and hence $x_0 = \mathbf{0}$ (since we have both $x_0 \in \mathbf{X}''$ and $x_0 \in \mathbf{X}'$).

To complete the proof of the theorem, we need only write

$$V_1(x) = -\sum_{k=1}^{n} f_k(x) y_k.$$

REMARK. We leave it to the reader to prove that, if T is replaced by T^* in condition 1) or 2), then the resulting two conditions are also necessary and sufficient for the Fredholm alternative to hold for T.

5.2. Our subsequent discussion is based on two simple lemmas.

LEMMA 1. *Let A and B be continuous linear operators mapping a normed space $\mathbf{X}$ into itself. If these operators are permutable and the operator $C = AB$ has a (two-sided) inverse, then A and B also have inverses.*

Proof. We first show that A and C^{-1} are permutable. In fact,

$$A = C^{-1}CA = C^{-1}ABA = C^{-1}A(AB) = C^{-1}AC.$$

Multiplying this equation on the right by C^{-1}, we see that $AC^{-1} = C^{-1}A$.

Now, using the permutability of A and C^{-1}, we can write

$$B(AC^{-1}) = BAC^{-1} = CC^{-1} = I$$

and

$$(AC^{-1})B = C^{-1}AB = C^{-1}C = I,$$

and so it follows that B has the inverse $B^{-1} = AC^{-1}$. Similarly, one can show that A has the inverse $A^{-1} = BC^{-1}$.

REMARK. If C^{-1} is continuous, then so are A^{-1} and B^{-1}.

LEMMA 2. *Let U be a continuous operator on a space $\mathbf{X}$. The characteristic set $\chi(U)$ of U and that of U^m are related by*

$$[\chi(U)]^m \subset \chi(U^m),$$

that is, if $\lambda \in \chi(U)$, then $\lambda^m \in \chi(U^m)$.

Proof. Write $\varepsilon = e^{2\pi i/m}$. We have

$$I - \lambda^m U^m = (I - \lambda U)(I - \lambda\varepsilon U) \ldots (I - \lambda\varepsilon^{m-1} U).$$

If $\lambda^m \notin \chi(U^m)$, then, writing

$$A = I - \lambda U, \quad B = (I - \lambda\varepsilon U) \ldots (I - \lambda\varepsilon^{m-1} U),$$

$$C = I - \lambda^m U^m,$$

we see that C has a continuous inverse C^{-1}. Hence, by the Remark following Lemma 1, A has a continuous inverse A^{-1}, and thus $\lambda \notin \chi(U)$.

5.3. Taking $\mathbf{X}$ to be a B-space, as in 5.1, we now consider a continuous linear operator U on $\mathbf{X}$.

THEOREM 2. *Suppose that there exists a natural number m such that U^m is a compact operator. Then the Fredholm alternative applies to $T = I - U$.*

Proof. By Lemma 2, the characteristic set $\chi(U)$ consists of isolated points, so the unit circle in the complex plane contains only a finite number of points $\lambda_1, \lambda_2, \ldots, \lambda_\nu \in \chi(U)$.

Let p range over the set of all prime numbers; then the numbers

$$e^{2k\pi i/p} \qquad (k = 1, 2, \ldots, p-1)$$

are all distinct, so for large enough p_0

$$\lambda_j \neq e^{2k\pi i/p} \qquad (p \geqslant p_0;\ k = 1, 2, \ldots, p-1;\ j = 1, 2, \ldots, \nu). \tag{17}$$

We may assume that m is a prime number, and $m \geqslant p_0$. We have the factorization

$$I - U^m = (I - U)(I - \varepsilon U) \ldots (I - \varepsilon^{m-1} U) = (I - U)V, \tag{18}$$

where

$$\varepsilon = e^{2\pi i/m}, \qquad V = (I - \varepsilon U) \ldots (I - \varepsilon^{m-1} U).$$

By (17), the operators $I - \varepsilon^k U$ $(k = 1, 2, \ldots, m-1)$ are invertible, and so V has a continuous inverse V^{-1}. But then we have

$$T = I - U = (I - U^m)V^{-1} = V^{-1} - U^m V^{-1}.$$

Since V^{-1} is invertible and $U^m V^{-1}$ is compact, we can now apply Theorem 1.

This completes the proof of the theorem.

5.4. The theorem on characteristic sets of compact operators (Theorem 2) can be extended to operators of the type considered in the last section. Namely, we have

THEOREM 3. *If U^m is compact for some m, then:* 1) *the characteristic set $\chi(U)$ of U consists just of the characteristic values, and each characteristic value has finite rank, while the corresponding eigenspace is finite-dimensional;* 2) *each disc $|\lambda| \leqslant R$ in the complex plane contains only a finite number of characteristic values.*

Proof. In view of Lemma 2, we need only prove the first statement. Moreover, the first part of this follows in an obvious manner from Theorem 2. Hence it remains only to prove that each characteristic value has finite rank and that the corresponding eigenspace is finite-dimensional.

Without losing any generality, we can assume that the characteristic value under consideration is $\lambda_0 = 1$. Write $T_m = I - U^m$. By (18), we can write

$$T_m = VT, \qquad T = V^{-1} T^m.$$

Since the operators T, T_m, V and V^{-1} are permutable with one another, we have

$$T_m^n = V^n T^n, \qquad T^n = V^{-n} T_m^n \ (n = 1, 2, \ldots).$$

From this it follows at once that

$$\mathbf{N}(T_m^n) = \mathbf{N}(T^n). \tag{19}$$

Hence, as U^m is a compact operator and the required result has already been established for such operators, we conclude from (19) that the result is also true in the present case.

Finally, we give an example of a continuous linear operator U whose square U^2 is compact, but which is not compact itself.

Let $\mathbf{X}$ be one of the spaces ℓ^p $(1 \leqslant p \leqslant \infty)$, $\mathbf{c}$, $\mathbf{c}_0$. If $x = \{\xi_n\} \in \mathbf{X}$, we set $y = U(x)$, where

$$y = \{\eta_n\} \qquad \left(\eta_n = \begin{cases} 0, & n \text{ odd}, \\ \xi_{n-1}, & n \text{ even}, \end{cases} \quad n = 1, 2, \ldots \right).$$

Obviously $U^2 = \mathbf{0}$.

REMARK. Since the theorems of § 4, which were proved for compact operators, used only those properties of compact operators involved in Theorem 1.1 and Theorem 3.1, and since these theorems also carry over to operators of the type considered above without any change in their formulation, it follows that these results of § 4 are also true when we assume only that some power of U is compact.

§ 6. Applications to integral equations

6.1. We consider the integral equation

$$x(s) - \lambda \int_0^1 K(s, t) x(t)\, dt = y(s), \tag{1}$$

where the kernel $K(s, t)$ is assumed to be a continuous function in the square $[0, 1; 0, 1]$. If we regard the integral term as a linear operator on the space $\mathbf{C}[0, 1]$, then (1) turns out to be an equation of the type considered in the preceding sections.

We could also consider an integral equation of a more general form than (1), namely

$$x(s) - \lambda \int_T K(s, t) x(t)\, dt = y(s), \tag{1'}$$

where T is any closed bounded set in n-dimensional Euclidean space (in this case, s and t denote points of n-dimensional space). However, everything that we prove about (1) will carry over, without any essential changes in the proofs, to (1′), so we shall in fact consider the simpler case.

The integral operator U, given by

$$z = U(x), \qquad z(s) = \int_0^1 K(s, t) x(t)\, dt, \tag{2}$$

which we regard as an operator from $\mathbf{C}[0, 1]$ to $\mathbf{C}[0, 1]$, has the norm (see III.2.4)

$$\|U\| = \max_s \int_0^1 |K(s, t)|\, dt$$

and is compact (IX.2.1).

We write (1) in the form

$$x - \lambda U(x) = y. \tag{3}$$

A solution x^* of this equation, expressed in terms of y by means of the formula

$$x^* = y + \lambda B_\lambda(y),$$

can, in accordance with Theorem 4.1, be expanded as a power series

$$x^* = y + \lambda U(y) + \dots + \lambda^n U^n(y) + \dots, \tag{4}$$

which converges for all λ with

$$|\lambda| < \frac{1}{d} = r,$$

where $d = \lim_{n \to \infty} \sqrt[n]{\|U^n\|}$ and r is the distance from the point $\lambda = 0$ to the characteristic set of U (Theorem 4.2). The series (4) will always converge provided

$$|\lambda| < \frac{1}{\|U\|} = \frac{1}{\max\limits_{s} \int\limits_0^1 |K(s, t)|\, dt}.$$

As we showed in V.3.8, the powers of U are also integral operators. In fact,

$$z = U^n(x),\ z(s) = \int\limits_0^1 K_n(s, t)x(t)dt \qquad (n = 1, 2, \ldots), \tag{5}$$

where $K_n(s, t)$ is the iterated kernel.

Substituting (5) in (4), we obtain an expansion for a solution of (1) as a power series in the parameter:

$$x^*(s) = y(s) + \lambda \int\limits_0^1 K(s, t)y(dt) + \ldots + \lambda^n \int\limits_0^1 K_n(s, t)y(t)dt + \ldots.$$

This series converges uniformly in $s \in [0, 1]$.

Since the series

$$B_\lambda = U + \lambda U^2 + \ldots + \lambda^{n-1} U^n + \ldots \tag{6}$$

is convergent in the space of operators from $\mathbf{C}[0, 1]$ to $\mathbf{C}[0, 1]$, we have

$$\left\| \sum_{j=m+1}^{m+p} \lambda^{j-1} U^j \right\| = \max_s \int\limits_0^1 \left| \sum_{j=m+1}^{m+p} \lambda^{j-1} K_j(s, t) \right| dt \xrightarrow[m \to \infty]{} 0.$$

Therefore, for any fixed $s \in [0, 1]$, the series

$$\sum_{j=1}^{\infty} \lambda^{j-1} K_j(s, t) \tag{7}$$

converges uniformly in $s \in [0, 1]$, in the space $\mathbf{L}^1$. The sum of this series is a function $\Gamma(s, t; \lambda)$ called the *resolvent* of the integral equation (1). Clearly,

$$B_\lambda(y)(s) = \int\limits_0^1 \Gamma(s, t; \lambda) y(t)\, dt,$$

so that (4) can also be written in the form

$$x^*(s) = y(s) + \lambda \int\limits_0^1 \Gamma(s, t; \lambda) y(t)\, dt.$$

If $|\lambda| < r$ then the successive approximation process for (3) is convergent, by Theorem 1.3. Applying this to the integral equation (1) we obtain the following result: for the values of λ mentioned, a solution of (1) is obtainable as a limit of the uniformly convergent sequence of continuous functions $\{x_n(s)\}$ defined recursively by

$$x_{n+1}(s) = \lambda \int\limits_0^1 K(s, t)x_n(t)dt + y(s) \qquad (n = 0, 1, \ldots),$$

where $x_0(t)$ is an arbitrary continuous function.

6.2. If the kernel $K(s,t)$ vanishes when $s < t$, then (1) can be written in the form

$$x(s) - \lambda \int_0^s K(s, t)x(t)\,dt = y(s). \tag{8}$$

An equation of this type is called a *Volterra integral equation.*

It is not hard to verify that the iterated kernels of a Volterra equation also vanish when $s < t$.

Assuming that the kernel $K(s,t)$ is continuous when $0 \leqslant t \leqslant s \leqslant 1$, we show that the expansion (4) is valid for all complex numbers—that is, $r = \infty$.

Let us put $|K(s,t)| \leqslant M$. The following inequalities for the kernels $K_n(s,t)$ are obtained without difficulty:

$$|K_n(s, t)| \leqslant \frac{s^{n-1}}{(n-1)!} M^n \qquad (n = 1, 2, \ldots). \tag{9}$$

For, when $n = 1$, the inequality is trivial, and if it is true for some $n > 1$, then

$$|K_{n+1}(s, t)| \leqslant \int_0^s |K(s, u)K_n(u, t)|\,dt \leqslant M \int_0^s \frac{u^{n-1}}{(n-1)!} M^n du = \frac{s^n}{n!} M^{n+1}.$$

From (9) we see that

$$\|U^n\| = \max_s \int_0^s |K_n(s, t)|\,dt \leqslant M^n \max_s \int_0^s \frac{s^{n-1}}{(n-1)!}\,dt = \frac{M^n}{(n-1)!},$$

so that

$$\sqrt[n]{\|U^n\|} \leqslant M \sqrt[n]{\frac{1}{(n-1)!}} \xrightarrow[n\to\infty]{} 0.$$

Hence $r = \infty$, and a Volterra type integral operator has no characteristic values.

6.3. Let us now consider equation (1) again. Since the operator (2) is compact, the Fredholm alternative is valid for equation (3). This leads to the following result for the integral equation (1).

THEOREM 1. *Either* (1) *has a unique continuous solution, whatever continuous function we take for* $y(s)$, *or the equation*

$$x(s) - \lambda \int_0^1 K(s, t)x(t)\,dt = 0 \tag{10}$$

has a finite number of linearly independent solutions $x_1(s), x_2(s), \ldots, x_n(s)$. *In the latter case, the equation*

$$\psi(t) - \bar{\lambda} \int_0^1 \overline{K(s, t)}\, \psi(s)\,ds = 0$$

also has n linearly independent continuous solutions $\psi_1(t), \psi_2(t), \ldots, \psi_n(t)$, *and* (1) *is soluble if and only if*

$$\int_0^1 \overline{\psi_k(s)}\, y(s)\,ds = 0.$$

The non-zero values of λ for which (10) is soluble are called the *characteristic values* of (1), or of the kernel $K(s,t)$. In other words, the characteristic values of (1) are just the characteristic values of the operator U. Hence we have the following inequality for the characteristic value λ_1 of smallest absolute value:

$$|\lambda_1| = R = \frac{1}{d} \geqslant \frac{1}{\|U\|} = \frac{1}{\max \int_0^1 |K(s, t)|\, dt} \geqslant \frac{1}{\max\limits_{s,\, t} |K(s, t)|}.$$

If we make use of the dependence of the solutions of (3) on λ, Theorem 4.5 yields the following result.

THEOREM 2. *In a neighbourhood of a characteristic value λ_0, a solution of* (1) *can be expressed in the form*

$$x^*(s) = \frac{y_{-r}(s)}{(\lambda - \lambda_0)^r} + \ldots + \frac{y_{-1}(s)}{\lambda - \lambda_0} + y_0(s) + \ldots + y_n(s)(\lambda - \lambda_0)^n + \ldots,$$

where $y_k(s)$ $(k = -r, \ldots, 0, \ldots)$ are continuous functions depending only on y. The series on the right-hand side converges uniformly in $s \in [0, 1]$.

Hence, for each fixed $s \in [0, 1]$, $x^*(s)$ is a meromorphic function of λ having poles at the characteristic values.

6.4. Now we consider the following integral equation in the space $\mathbf{L}^p(D)$, where D is a bounded region in n-dimensional space:

$$x(s) - \lambda \int_D K(s, t) x(t) dt = y(s) \qquad (s \in D). \tag{11}$$

Assume that the kernel $K(s, t)$ satisfies the conditions of Theorem XI.3.3 (with $q = p$)—that is, assume that

1) $\left\{ \int_D |K(s,t)|^r dt \right\}^{1/r} \leqslant C_1 \qquad (s \in D);$
2) $\left\{ \int_D |K(s, t)|^\sigma ds \right\} \leqslant C_2 \qquad (t \in D);$
3) $p - r(p-1) < \sigma < p.$

Then the theorem in question shows that the integral operator U with kernel $K(s, t)$ is a compact operator from $\mathbf{L}^p(D)$ into $\mathbf{L}^p(D)$, with norm $\|U\| \leqslant C_1^{1-\sigma/p} C_2^{\sigma/p}$; hence everything we have said above about (1) is applicable to (11).

In particular, if the kernel $K(s, t)$ is of potential type—that is, if

$$K(s, t) = \frac{B(s, t)}{|s - t|^m},$$

where $B(s, t)$ is a bounded function, continuous for $s \neq t$, and where $m < n$, then the above conditions are satisfied (Theorem XI.3.6).

If, in addition, $B(s, t)$ is a continuous function and

$$p > \frac{n}{n - m},$$

then, by Theorem XI.3.7, U will map $\mathbf{L}^p(D)$ into $\mathbf{C}(D)$, so that the power series representing a solution $x^*(s)$ in a neighbourhood of a characteristic value, or of zero, will be uniformly convergent.

We leave it to the reader to make the precise statements.

§ 7. Invariant subspaces of compact operators. The approximation problem

In the late nineteen-sixties and early nineteen-seventies, a number of important discoveries were made in the theory of B-spaces. One of these was the characterization of Hilbert spaces by Lindenstrauss and Tzafriri (see Theorem V.3.3). Here we devote some space to two other important results. (The reader can learn about the present state of the theory of B-spaces in the monograph by Lindenstrauss and Tzafriri.)

7.1. Let $\mathbf{X}$ be a B-space, and U a continuous linear operator on $\mathbf{X}$. A closed linear manifold $E \subset \mathbf{X}$ is called an *invariant* subspace for U if $U(E) \subset E$. Obviously, $\{\mathbf{0}\}$ and $\mathbf{X}$ are invariant subspaces. An invariant subspace E is called *non-trivial* if $E \neq \mathbf{X}, \{\mathbf{0}\}$. If an operator has an eigenvector, then it will have a one-dimensional invariant subspace. However, even compact operators may have no eigenvectors.

An example of this is a Volterra type integral operator (see 6.2). However, this operator has many invariant subspaces. For example, for each $a \in (0, 1)$, the set of all functions $x \in \mathbf{C}[0, 1]$ that are equal to zero in the interval $[0, a]$ is such a subspace.

In 1935 J. von Neumann proved that every non-zero compact operator on a Hilbert space has a non-trivial invariant subspace.* Recently V. I. Lomonosov [1] obtained a simple proof of a more general result (we shall present a variant of the proof, put forward by Khilden).

Let U be a compact operator on an infinite-dimensional B-space $\mathbf{X}$. Write $\mathfrak{A}$ for the set of all operators that commute with U, that is, the set of all $T \in B(\mathbf{X}, \mathbf{X})$ such that $TU = UT$. Obviously, $\mathfrak{A}$ is a linear manifold in $B(\mathbf{X}, \mathbf{X})$, and if $T_1, T_2 \in \mathfrak{A}$, then $T_1 T_2 \in \mathfrak{A}$.

THEOREM 1 (Lomonosov). *There is a subspace E in $\mathbf{X}$ that is a non-trivial invariant subspace for every $T \in \mathfrak{A}$.*

Proof. Since $U \neq \mathbf{0}$, we may assume that $\|U\| = 1$. Furthermore, there exists an $x_0 \in \mathbf{X}$ such that

$$\|x_0\| > 1 \quad \text{and} \quad \|U(x_0)\| > 1. \tag{1}$$

Let B be the open unit ball with centre at x_0. As U is compact, $K = \overline{U(B)}$ is a compactum. Every $x \in B$ is expressible in the form $x = x_0 + z$, where $\|z\| < 1$. If $y = U(x)$, then $y = U(x_0) + U(z)$ and $\|U(z)\| \leqslant 1$. Hence

$$\|y\| \geqslant \|U(x_0)\| - \|U(z)\| \geqslant \|U(x_0)\| - 1 = \delta > 0,$$

so that for every $y \in K$ we have

$$\|y\| \geqslant \delta > 0, \tag{2}$$

where δ is independent of y. We consider two cases.

a) Assume U has a non-zero eigenvalue λ. Denote the eigenspace corresponding to λ by N_λ. We show that N_λ is an invariant subspace for every $T \in \mathfrak{A}$ (clearly $N_\lambda \neq \mathbf{X}$, as U is compact). If $x \in N_\lambda$, then

$$U(T(x)) = T(U(x)) = T(\lambda x) = \lambda T(x),$$

so that $T(x) \in N_\lambda$.

b) Now assume that U has no eigenvalues, so that its spectrum is $\sigma(U) = \{\mathbf{0}\}$. Then the

---

* This result of von Neumann was published in a paper by Aronszain and Smith [1], where it was generalized to B-spaces.

spectral radius of U is zero, and 4.1 shows that, for each λ,

$$\|\lambda^n U^{n+1}\| \underset{n\to\infty}{\longrightarrow} 0. \tag{3}$$

Assume that no non-trivial subspace is invariant for all $T\in\mathfrak{A}$. For an arbitrary $x \neq \mathbf{0}$ in $\mathbf{X}$ consider the closure $\mathbf{L}_x$ of the set $\mathbf{L}'_x = \{T(x): T\in\mathfrak{A}\}$. Since $\mathfrak{A}$ is a linear manifold, so is $\mathbf{L}'_x$. If $y\in\mathbf{L}'_x$, then $y = T_0(x)$, $T_0\in\mathfrak{A}$. For all $T\in\mathfrak{A}$, we have $TT_0\in\mathfrak{A}$, so $T(y) = (TT_0)(y)\in\mathbf{L}'_x$. Hence, by the continuity of the operators in $\mathfrak{A}$, it follows that $\mathbf{L}_x$ is also invariant for each operator in $\mathfrak{A}$.

As the identity operator I belongs to $\mathfrak{A}$, we have $x\in\mathbf{L}'_x$, and so $\mathbf{L}_x \neq \{\mathbf{0}\}$. As we assumed that there was no non-trivial invariant subspace, $\mathbf{L}_x = \mathbf{X}$ for every $x \neq \mathbf{0}$, and so $\mathbf{L}'_x \cap B$ is non-empty.

Hence, for each $x \neq \mathbf{0}$, there exists $T\in\mathfrak{A}$ such that $T(x)\in B$, and so

$$\bigcup_{T\in\mathfrak{A}} T^{-1}(B) = \mathbf{X}\setminus\{\mathbf{0}\}.$$

By (2), $K \subset \mathbf{X}\setminus\{\mathbf{0}\}$, so that $\{T^{-1}(B): T\in\mathfrak{A}\}$ is an open cover for the compactum K. Choose a finite subcover:

$$K \subset \bigcup_{i=1}^{m} T_i^{-1}(B).$$

Now we consider again the point x_0 that was chosen at the beginning of the proof (recall that x_0 satisfies (1)). As $U(x_0)\in U(B)\subset K$, there exists an i_1 with $1 \leqslant i_1 \leqslant m$, such that $U(x_0)\in T_{i_1}^{-1}(B)$. Then we have

$$x_1 = T_{i_1}(U(x_0))\in B.$$

Similarly, there exists an i_2, with $1 \leqslant i_2 \leqslant m$, such that

$$x_2 = T_{i_2}(U(x_1))\in B,$$

and so on. We obtain a sequence $\{x_n\} \subset B$ such that

$$x_n = T_{i_n} U T_{i_{n-1}} U T_{i_{n-2}}, \ldots, T_{i_1} U(x_0).$$

Since the T_i are permutable with U, we have

$$x_n = T_{i_n} T_{i_{n-1}} T_{i_{n-2}} \ldots T_{i_1} U^n(x_0).$$

Now apply U to x_n:

$$z_n = U(x_n) = T_{i_n} T_{i_{n-1}} T_{i_{n-2}} \ldots T_{i_1} U^{n+1}(x_0)\in K.$$

By (2), we have

$$\|z_n\| \geqslant \delta > 0 \qquad (n\in\mathbb{N}). \tag{4}$$

Next we estimate $\|z_n\|$ in another way. Write $\lambda = \max\{\|T_i\|: 1 \leqslant i \leqslant n\}$. We have

$$\|z_n\| \leqslant \prod_{j=1}^{n} \|T_{i_j}\|\,\|U^{n+1}\|\,\|x_0\| \leqslant \|\lambda^n U^{n+1}\|\,\|x_0\|. \tag{5}$$

Using (3), we see that the right-hand side of (5) tends to zero as $n\to\infty$, contradicting (4).

This contradiction proves the theorem. The theorem shows, in particular, that U itself has a non-trivial invariant subspace—thus we have the original result of von Neumann.

As we have seen, the question of whether the operators in a given class have a non-trivial invariant subspace is highly complicated. However, the inverse problem of finding a continuous linear operator acting on a separable Banach space with no non-trivial invariant subspace is also difficult.

7.2. As we have seen in this chapter, compact operators are very close in their properties to finite-dimensional operators. This prompted A. Grothendieck to make the following definition.

A B-space $\mathbf{X}$ is said to have the *approximation property* if every compact operator U form a B-space $\mathbf{Y}$ into $\mathbf{X}$ is the limit of a sequence of finite-dimensional operators in the norm of $B(\mathbf{Y}, \mathbf{X})$. This is equivalent to requiring that, for every compactum K in $\mathbf{X}$ and every $\varepsilon > 0$, there exist a finite-dimensional operator $T \in B(\mathbf{X}, \mathbf{X})$ such that

$$\|Tx - x\| < \varepsilon \qquad (x \in K).$$

Grothendieck posed the question whether every B-space has the approximation property. He proved that a positive answer to this question, which is called the approximation problem, is equivalent to a number of very specific assumptions being true. For example, the following one. If $K(s, t)$ is a continuous function on the square $[0, 1; 0, 1]$ and

$$\int_0^1 K(s, t)K(t, u)\,dt = 0 \quad \text{for all} \quad s, u \in [0, 1],$$

then

$$\int_0^1 K(t, t)\,dt = 0.$$

Even earlier, the basis problem (see Banach) had been posed. This is contained in the following problem.

A sequence $\{x_n\}$ of elements of a B-space $\mathbf{X}$ is said to form a *basis** in $\mathbf{X}$ if every $x \in \mathbf{X}$ is uniquely decomposable as a series

$$x = \sum_{n=1}^{\infty} \lambda_n x_n,$$

which converges in norm in $\mathbf{X}$.

Clearly, a space with a basis is necessarily separable. In a separable Hilbert space a complete orthonormal system is a basis (see IV.5.8). The spaces $\mathbf{L}^p(0, 1)$ and ℓ^p, $1 \leqslant p < \infty$, and $\mathbf{C}[0, 1]$ have bases. For example, in ℓ^p the sequence of unit vectors—that is, the elements $x_n = \{\xi_k\}_{k=1}^{\infty}$ with $\xi_n = 1$, $\xi_k = 0$ $(k \neq n)$—is clearly a basis.

The basis problem asks whether every separable B-space has a basis.

It is not difficult to prove that every space with a basis has the approximation property.

Both the above problems remained unsolved for a long time. It was only in 1972 that the Swedish mathematician P. Enflo constructed a remarkable example solving both the basis problem and the approximation problem (see Enflo [1]). In fact, he constructed an example of a separable reflexive B-space not having the approximation property, and hence also having no basis. For more details on these problems, see the paper by Pelchinskii and Figel' [1].

In 1975 A. Shankovskii constructed an example of a separable reflexive BFS not having the approximation property.

---

* The concept of a basis was introduced by J. Schauder [1]. For a more detailed account of bases see Singer.

XIV

A GENERAL THEORY OF APPROXIMATION METHODS

In response to the need to obtain approximate solutions to problems of mathematical analysis—such as differential and integral equations, boundary value problems of mathematical physics, conformal mapping, and so on—mathematicians have proposed, and are currently using in practice, a large number of methods, based on various ideas. For instance, in the case of boundary value problems, there are variational methods, and methods similar to these (the methods of Ritz and Galerkin, the method of moments, reduction to ordinary differential equations), and also difference methods and interpolation methods. To judge the effectiveness of these methods, and the extent to which they can be justified, the methods need to be investigated theoretically. Such an investigation gives rise to three problems. In order of increasing specificity and difficulty, these are:

a) to establish that the algorithm is feasible, and that it converges;

b) to investigate its speed of convergence;

c) to give an effective estimate of the error.

Different techniques for tackling these problems have been used for each class of equations and each approximation method, and they have frequently encountered considerable difficulties, which in a number of cases have remained insuperable up to the present day.

The problem of combining these investigations and constructing a unified theory of approximation methods is an important one. A natural approach to its solution is afforded by the ideas of functional analysis.

In the present chapter we shall be concerned with constructing a theory of approximation methods for a class of linear equations. In this theory we consider an exact equation in one space and an approximate equation in a different space—these spaces being, of course, related in some definite manner. This is a situation that occurs in many specific methods—for example, when one approximates to an integral equation by replacing it by a finite system of linear algebraic equations. As we shall show, moreover, one can always confine the discussion to the case where the approximation space is a subspace of the space in which the exact equation is defined.

Below we shall prove theorems of two kinds: those that enable one, on the basis of data relating to the exact equation, to establish whether the "approximate" equations are soluble and whether their solutions converge to exact solutions; and, conversely, those that make use of results concerning the approximate solutions to obtain information about the solubility of the exact equation and the closeness of its solutions to those of the approximate equations.

The results of this chapter were obtained, under somewhat different hypotheses, by L. V. Kantorovich (see Kantorovich [9]). Some theorems in § 1 have been made more precise through the efforts of G. P. Akilov. Alternative theories of approximation methods using the techniques of functional analysis may be found in papers by S. G. Mikhlin, M. A. Krasnosel'skii, N. I. Pol'skii, G. M. Vainikko, M. K. Gavurin, and others. See Vainikko; Gavurin; Krasnosel'skii *et al.*; Krasnosel'skii [1]; Mikhlin-II.

§ 1. A general theory for equations of the second kind

1.1. Let $\mathbf{X}$ be a normed space and $\tilde{\mathbf{X}}$ a complete subspace* of $\mathbf{X}$. Assume that there exists a continuous linear operator P that projects $\mathbf{X}$ onto $\tilde{\mathbf{X}}$, i.e. satisfies

$$P(\mathbf{X}) = \tilde{\mathbf{X}}, \qquad P^2 = P.$$

Clearly P fixes every element of $\tilde{\mathbf{X}}$.

For example, let $\mathbf{X} = \mathbf{C}[a, b]$ and let $\tilde{\mathbf{X}}$ be the set of polynomials of degree at most $n-1$. The operator P assigns to a function $x \in \mathbf{C}[a, b]$ its interpolation polynomial, constructed with respect to a preassigned system of points $t_1, t_2, \ldots, t_n$.

Next, consider two equations; the first one

$$Kx \equiv x - \lambda Hx = y \tag{1}$$

in $\mathbf{X}$, and the second in $\tilde{\mathbf{X}}$:

$$\tilde{K}\tilde{x} \equiv \tilde{x} - \lambda \tilde{H}\tilde{x} = Py. \tag{2}$$

Here H is a continuous linear operator on $\mathbf{X}$ and $\tilde{H}$ is a continuous linear operator on $\tilde{\mathbf{X}}$. We shall call (1) the *exact* equation and (2) an *approximate* equation (corresponding to the exact equation (1)).

In what follows, the spaces $\mathbf{X}$ and $\tilde{\mathbf{X}}$ and the operators H and $\tilde{H}$ will be related by the following conditions.

I. (Closeness condition for H and $\tilde{H}$.) For every $\tilde{x} \in \tilde{\mathbf{X}}$,

$$\|PH\tilde{x} - \tilde{H}\tilde{x}\| \leqslant \eta\|\tilde{x}\|.$$

This last condition is equivalent to:

$$\|PK\tilde{x} - \tilde{K}\tilde{x}\| \leqslant |\lambda|\eta\|\tilde{x}\| \qquad (\tilde{x} \in \tilde{\mathbf{X}}).$$

II. (Close approximability of elements Hx by elements of $\tilde{\mathbf{X}}$.) For each $x \in \mathbf{X}$, there exists $\tilde{x} \in \tilde{\mathbf{X}}$, such that

$$\|Hx - \tilde{x}\| \leqslant \eta_1\|x\|.$$

III. (Close approximability of the free term of the exact equation.) There exists $\tilde{y} \in \tilde{\mathbf{X}}$ such that

$$\|y - \tilde{y}\| \leqslant \eta_2\|y\|.$$

Here the η_2, unlike the constants in the previous conditions, will in general depend on y.

1.2. We now prove some theorems relating the exact and approximate equations when one or more of the above conditions is satisfied.

THEOREM 1 (Solubility of the approximate equation). *Suppose conditions* I *and* II are *satisfied, and assume that the operator K has a continuous* (*two-sided*) *inverse. If*

$$q = |\lambda|[\eta + \|I - P\|\eta_1]\|K^{-1}\| < 1, \tag{3}$$

then $\tilde{K}$ also has a continuous inverse $\tilde{K}^{-1}$. Moreover, we then have

$$\|\tilde{K}^{-1}\| \leqslant \frac{\|K^{-1}\|}{1-q}. \tag{4}$$

* We do not assume that $\mathbf{X}$ itself is complete.

Proof. Assume first that $\mathbf{X}$ is complete. Consider the operator $K_1 = I - \lambda PH$. For an arbitrary $x \in \mathbf{X}$, choose $\tilde{x} \in \tilde{\mathbf{X}}$ according to condition II, so that

$$\|Hx - \tilde{x}\| \leqslant \eta_1 \|x\|.$$

Then

$$\|(K - K_1)x\| = |\lambda|\|Hx - PHx\| = |\lambda|\|Hx - \tilde{x} + P\tilde{x} - PH\tilde{x}\| = |\lambda|\|(I-P)(Hx - \tilde{x})\|$$
$$\leqslant |\lambda|\|I - P\|\eta_1 \|x\|.$$

Since x was arbitrary, this shows that

$$\|K - K_1\| \leqslant |\lambda|\|I - P\|\eta_1.$$

We can express K_1 in the form

$$K_1 = K(I - K^{-1}(K - K_1)), \tag{5}$$

and the operator $K^{-1}(K - K_1)$ satisfies the inequality

$$\|K^{-1}(K - K_1)\| \leqslant |\lambda|\|I - P\|\eta_1\|K^{-1}\| \leqslant q < 1.$$

By Banach's Theorem (Theorem V.4.3), the inverse operator $(I - K^{-1}(K - K_1))^{-1}$ exists, and

$$\|(I - K^{-1}(K - K_1))^{-1}\| \leqslant \frac{1}{1 - |\lambda|\|I - P\|\eta_1\|K^{-1}\|}.$$

It now follows from (5) that K_1 is invertible, with $K_1^{-1} = (I - K^{-1}(K - K_1))^{-1}K^{-1}$, and so

$$\|K_1^{-1}\| \leqslant \frac{\|K^{-1}\|}{1 - |\lambda|\|I - P\|\eta_1\|K^{-1}\|}. \tag{6}$$

Now consider the operator $\tilde{K}_1$ on $\tilde{\mathbf{X}}$ defined by $\tilde{K}_1\tilde{x} = \tilde{x} - \lambda PH\tilde{x}$. We clearly have $\tilde{K}_1\tilde{x} = K_1\tilde{x}$ for all $\tilde{x} \in \tilde{\mathbf{X}}$. The operator K_1 also has the property that, if $\tilde{y} \in \tilde{\mathbf{X}}$, then $K_1^{-1}\tilde{y} \in \tilde{\mathbf{X}}$. For if $x' = K_1^{-1}\tilde{y}$, then $x' - PHx' = \tilde{y}$, $x' = \tilde{y} + PHx' \in \tilde{\mathbf{X}}$. Therefore $\tilde{K}_1$ has a continuous inverse which coincides with K_1^{-1} on $\tilde{\mathbf{X}}$, and

$$\|\tilde{K}_1^{-1}\| \leqslant \|K_1^{-1}\|. \tag{7}$$

We now find a bound for $\tilde{K} - \tilde{K}_1$. By condition 1 we have

$$\|\tilde{K}\tilde{x} - \tilde{K}_1\tilde{x}\| = |\lambda|\|PH\tilde{x} - \tilde{H}\tilde{x}\| \leqslant |\lambda|\eta\|\tilde{x}\| \tag{8}$$

for every $\tilde{x} \in \tilde{\mathbf{X}}$ and thus

$$\|\tilde{K} - \tilde{K}_1\| \leqslant |\lambda|\eta. \tag{9}$$

Hence, by (7), (6) and (9),

$$\|\tilde{K}_1^{-1}\|\|\tilde{K} - \tilde{K}_1\| \leqslant \frac{\|K^{-1}\||\lambda|\eta}{1 - |\lambda|\|I - P\|\eta_1\|K^{-1}\|} = 1 - \frac{1 - q}{1 - |\lambda|\|I - P\|\eta_1\|K^{-1}\|} < 1. \tag{10}$$

Expressing $\tilde{K}$ in the form $\tilde{K} = \tilde{K}_1(I - \tilde{K}_1^{-1}(\tilde{K}_1 - \tilde{K}))$ and applying Banach's Theorem again, we obtain a proof of the existence of the inverse operator $\tilde{K}_1^{-1}$ and we also find the bound

$$\|\tilde{K}^{-1}\| \leqslant \frac{\|\tilde{K}_1^{-1}\|}{1-\|\tilde{K}_1^{-1}\|\,\|\tilde{K}-\tilde{K}_1\|} \leqslant \frac{(1-|\lambda|\|I-P\|\eta_1\|K^{-1}\|)\|K_1^{-1}\|}{1-q} \leqslant \frac{\|K^{-1}\|}{1-q}. \tag{11}$$

Thus, if **X** is complete, the theorem is proved.

The case where **X** is not complete can easily be reduced to the case just considered. For if we introduce the completion of **X**, then the closures (see Theorem V.8.2) of K and K^{-1} will be mutual inverses, and the closure of H will also be related to $\tilde{H}$ by conditions I and II, with the same constant η and a new constant η_1, as close to the old one as we please.

1.3. A solution of the approximate equation (2) is naturally regarded as an approximate solution to the exact equation (1). Bounds for the error of this approximate solution are given in the following theorem.

THEOREM 2 (Bounds for the error of the approximate solution). *If conditions* I, II *and* III *are satisfied and* $\tilde{K}$ *has a continuous inverse* (in particular, if the hypotheses of Theorem 1 are satisfied), *and if* (1) *has a solution* x^*, *then we have the inequality*

$$\|x^*-\tilde{x}^*\| \leqslant p\|x^*\|, \tag{12}$$

where $\tilde{x}^*$ *is a solution of* (2) *and*

$$p = 2|\lambda|\eta\|\tilde{K}^{-1}\| + (\eta_1|\lambda| + \eta_2\|K\|)(1+\|\tilde{K}^{-1}PK\|). \tag{13}$$

Proof. We first show that the conditions of the theorem enable us to approximate to x^* by an element of $\tilde{\mathbf{X}}$ to within an accuracy of the order of $\eta_1+\eta_2$. More precisely, we shall prove that there exists an $\tilde{x}\in\tilde{\mathbf{X}}$ such that

$$\|x^*-\tilde{x}\| \leqslant \varepsilon\|x^*\|, \tag{14}$$

where

$$\varepsilon = \min\,[1, \eta_1|\lambda| + \eta_2\|K\|]. \tag{15}$$

In fact, if we choose $\tilde{y}\in\tilde{\mathbf{X}}$ and $\tilde{z}\in\tilde{\mathbf{X}}$ according to II and III so that

$$\|Hx^*-\tilde{z}\| \leqslant \eta_1\|x^*\|, \qquad \|y-\tilde{y}\| \leqslant \eta_2\|y\| \leqslant \eta_2\|K\|\|x^*\|,$$

and write $\tilde{x} = \lambda\tilde{z}+\tilde{y}$, then we have

$$\|x^*-\tilde{x}\| = \|\lambda Hx^* + y - (\lambda\tilde{z}+\tilde{y})\| \leqslant (\eta_1|\lambda| + \eta_2\|K\|)\|x^*\|.$$

On the other hand, (14) is satisfied with $\varepsilon = 1$ if we set $\tilde{x} = \mathbf{0}$. Hence we can take ε to be given by (15).

Now we turn to the proof of the inequality (12).

Writing $\tilde{x}_0 = \tilde{K}^{-1}PK\tilde{x}$, we have

$$\|x^*-\tilde{x}^*\| \leqslant \|x^*-\tilde{x}\| + \|\tilde{x}-\tilde{x}_0\| + \|\tilde{x}_0-\tilde{x}^*\|. \tag{16}$$

We estimate each summand separately. We already have a bound for the first one, namely (14).

A bound for the second is obtained as follows: by condition I,

$$\|\tilde{x}-\tilde{x}_0\| = \|\tilde{K}^{-1}\tilde{K}\tilde{x} - \tilde{K}^{-1}PK\tilde{x}\| \leqslant \|\tilde{K}^{-1}\|\|\tilde{K}\tilde{x}-PK\tilde{x}\| \leqslant |\lambda|\eta\|\tilde{K}^{-1}\|\|\tilde{x}\|. \tag{17}$$

However, by (14),

$$\|\tilde{x}\| \leqslant \|x^*\| + \|x^*-\tilde{x}\| \leqslant (1+\varepsilon)\|x^*\|.$$

Substituting this in (17), we finally have

$$\|\tilde{x}-\tilde{x}_0\| \leqslant |\lambda|\eta(1+\varepsilon)\|\tilde{K}^{-1}\|\|x^*\|. \tag{18}$$

Bearing in mind that $\tilde{x}^*$ is a solution of (2), and hence that

$$\tilde{x}^* = \tilde{K}^{-1}PKx^*,$$

we can bound the last summand in (16). By (14), we have

$$\|\tilde{x}_0-\tilde{x}^*\| = \|\tilde{K}^{-1}PK\tilde{x}-\tilde{K}^{-1}PKx^*\| \leqslant \|\tilde{K}^{-1}PK\|\|\tilde{x}-x^*\| \leqslant \varepsilon\|\tilde{K}^{-1}PK\|\|x^*\|.$$

Comparing this with (14) and (18), we obtain the bound

$$\|x^*-\tilde{x}^*\| \leqslant [|\lambda|\eta(1+\varepsilon)\|\tilde{K}^{-1}\|+\varepsilon(1+\|\tilde{K}^{-1}PK\|)]\|x^*\|. \tag{19}$$

It remains to note that $1+\varepsilon \leqslant 2$ and $\varepsilon \leqslant \eta_1|\lambda|+\eta_2\|K\|$.

This completes the proof.

REMARK 1. It may turn out that one can manage to show directly that x^* can be approximated by an element of $\tilde{\mathbf{X}}$, without using II and III. In this case, one can use (19) as before. It is then, of course, unnecessary to assume that conditions II and III are satisfied.

It is also useful to note that in this case we are concerned with approximating a specific element x^*, and ε may depend on this element, whereas in condition II η_1 is assumed to be independent of x.

REMARK 2. On the basis of Theorem 2 one can easily obtain a bound for the closeness of the approximate solution to the exact one, without using any data relating to the exact solution. In fact, if under the hypotheses of Theorem 2 we assume that $p < 1$, then we have

$$\|x^*-\tilde{x}^*\| \leqslant \frac{p}{1-p}\|\tilde{x}^*\|. \tag{20}$$

For

$$\|x^*\| \leqslant \|\tilde{x}^*\|+\|x^*-\tilde{x}^*\|.$$

Substituting this in (12), we obtain

$$\|x^*-\tilde{x}^*\| \leqslant p\|\tilde{x}^*\|+p\|x^*-\tilde{x}^*\|,$$

and, bringing the terms involving $\|x^*-\tilde{x}^*\|$ together, we obtain (20).

1.4. Now assume that we have a sequence of approximate equations and a sequence of approximate solutions obtained from these. In this case, the space $\tilde{\mathbf{X}}$, the operators $\tilde{H}$ ($\tilde{K}$), P and the constants $\eta, \eta_1, \eta_2, q, p, \varepsilon, \ldots$ all depend on an index n; however, for simplicity we shall omit this from our notation.

The theorem which follows gives conditions for the sequence $\{\tilde{x}_n^*\}$ of approximate solutions to converge to an exact solution.

THEOREM 3. *If the following conditions are satisfied:*

1) *K has a continuous inverse;*

2) *conditions* I, II, III *are satisfied for each n, and*

$$\lim_{n\to\infty}\eta = 0, \qquad \lim_{n\to\infty}\eta_1\|P\| = 0, \qquad \lim_{n\to\infty}\eta_2\|P\| = 0, \tag{21}$$

then, for sufficiently large n, the approximate equations are soluble and the sequence of approximate solutions converges to an exact solution:

$$\lim_{n\to\infty}\|x^*-\tilde{x}_n^*\| = 0.$$

More precisely,

$$\|x^* - \tilde{x}_n^*\| \leqslant Q_0\eta + Q_1\eta_1\|P\| + Q_2\eta_2\|P\|, \tag{22}$$

where Q_0, Q_1, Q_2 are constants.

Proof. Since $\|P\| \geqslant 1$ it follows from (21) that $\lim_{n\to\infty} \eta_1 = 0$, and so, for large enough n, we have $q < \frac{1}{2}$ in Theorem 1. Hence, for such values of n, there exists a continuous operator $\tilde{K}^{-1}$, and

$$\|\tilde{K}^{-1}\| \leqslant \frac{\|K^{-1}\|}{1-q} < 2\|K^{-1}\|:$$

From this it is clear that $\|\tilde{K}^{-1}\|$ is bounded independently of n. Taking this fact into account, we obtain (22) by using the inequality (12) in Theorem 2.

REMARK. If we use (19) instead of (12), then we obtain

$$\|x^* - \tilde{x}_n^*\| \leqslant Q'\eta + Q''\varepsilon\|P\| \tag{23}$$

and the convergence conditions (21) are replaced by

$$\lim_{n\to\infty} \eta = 0, \qquad \lim_{n\to\infty} \varepsilon\|P\| = 0. \tag{24}$$

Condition III then turns out to be superfluous (condition II is necessary, as it is used in Theorem 1; however, the condition that $\eta_1\|P\| \to 0$ can be replaced by the requirement that $q \leqslant q_0 < 1$ for sufficiently large n).

1.5. Theorem 1 makes it possible to determine the solubility of an approximate equation from that of the exact one. The following theorem enables us to do the opposite.

THEOREM 4. *If the operator $\tilde{K}$ has a continuous inverse, and conditions* I *and* II *are satisfied, and if*

$$r = |\lambda|\eta(1 + |\lambda|\eta_1)\|\tilde{K}^{-1}\| + |\lambda|\eta_1(1 + \|\tilde{K}^{-1}PK\|) < 1, \tag{25}$$

then K has a continuous left inverse whose norm satisfies the inequality

$$\|K^{-1}\| \leqslant \frac{1 + \|\tilde{K}^{-1}P\| + |\lambda|\eta\|\tilde{K}^{-1}\| + \|\tilde{K}^{-1}PK\|}{1-r}. \tag{26}$$

Proof. Let x^* be an arbitrary non-zero element of $\mathbf{X}$. This element is obviously a solution of the equation

$$Kx = Kx^*.$$

By condition II, there is an element $\tilde{x} \in \mathbf{X}$ such that

$$\|Hx^* - \tilde{x}\| \leqslant \eta_1\|x^*\|.$$

Since $x^* = \lambda Hx^* + Kx^*$, we have

$$\|x^* - \lambda\tilde{x}\| \leqslant |\lambda|\eta_1\|x^*\| + \|Kx^*\| = \left(|\lambda|\eta_1 + \frac{\|Kx^*\|}{\|x^*\|}\right)\|x^*\|.$$

This inequality allows us to apply Theorem 2 (in the form given in Remark 1 after the theorem) with

$$\varepsilon = |\lambda|\eta_1 + \frac{\|Kx^*\|}{\|x^*\|}. \tag{27}$$

Since the approximate equation corresponding to this exact equation is

$$\tilde{K}\tilde{x} = PKx^*,$$

which has the solution $\tilde{x}^* = \tilde{K}^{-1}PKx^*$, we see, on substituting for ε in (19) the value given by (27), that

$$\|x^* - \tilde{K}^{-1}PKx^*\| \leqslant \left[|\lambda|\eta\left(1+|\lambda|\eta_1+\frac{\|Kx^*\|}{\|x^*\|}\right)\|\tilde{K}^{-1}\| + \right.$$

$$\left. + \left(|\lambda|\eta_1 + \frac{\|Kx^*\|}{\|x^*\|}\right)(1+\|\tilde{K}^{-1}PK\|)\right]\|x^*\| =$$

$$= [|\lambda|\eta(1+|\lambda|\eta_1)\|\tilde{K}^{-1}\| + |\lambda|\eta_1(1+\|\tilde{K}^{-1}PK\|)]\|x^*\| +$$

$$+ [|\lambda|\eta\|\tilde{K}^{-1}\| + 1 + \|\tilde{K}^{-1}PK\|]\|Kx^*\|.$$

Hence

$$\|x^*\| \leqslant \|x^* - \tilde{K}^{-1}PKx^*\| + \|\tilde{K}^{-1}P\|\|Kx^*\| \leqslant$$

$$\leqslant r\|x^*\| + [1 + \|\tilde{K}^{-1}P\| + |\lambda|\eta\|\tilde{K}^{-1}\| + \|\tilde{K}^{-1}PK\|]\|Kx^*\|$$

or

$$\|Kx^*\| \geqslant \frac{1-r}{1+\|K^{-1}P\|+|\lambda|\eta\|\tilde{K}^{-1}\|+\|\tilde{K}^{-1}PK\|}\|x^*\| = Q\|x^*\|.$$

Since $r < 1$, we have $Q > 0$, and so, in view of what was said in V.4.4, the theorem is proved.

COROLLARY. *Assume that K satisfies the following condition:*

(A) *If* (1) *has a unique solution, then it is soluble for all right-hand sides.*

Then, under the hypotheses of the theorem, K has a continuous two-sided inverse K^{-1}.

1.6. If condition III is satisfied only for large η_2—that is, if the right-hand side of (1) does not have a good approximation by elements of $\tilde{\mathbf{X}}$—then the solution x^* will also have no good approximation and the approximate solution $\tilde{x}^*$ will automatically be a poor one.

In this case it turns out to be advantageous to "regularize" the free term. Let us substitute $x = y + z$ in (1). The new equation (in z) will then be

$$Kz = y' \qquad (y' = \lambda Hy). \tag{28}$$

The right-hand side of this equation is already "regular", by condition II.

The approximate equation corresponding to (28) is

$$\tilde{K}\tilde{z} = \lambda PHy. \tag{29}$$

Writing $\tilde{z}^*$ for a solution of this equation, we now naturally take as an approximate solution to (1) the element

$$x' = y + \tilde{z}^*. \tag{30}$$

Since

$$\|x^* - x'\| = \|z^* - \tilde{z}^*\|,$$

where z^* denotes a solution to (28), we can use Theorem 2 to estimate the error in the approximate solution to (30)—here we must take η_2 to be $|\lambda|\eta_1$.* However, we can obtain a more precise estimate by a direct approach in this case.

* Or even the smaller quantity $|\lambda|\eta_1^*$, provided that for some $\tilde{y} \in \tilde{\mathbf{X}}$ we have

$$\|Hy - \tilde{y}\| \leqslant \eta_1^*\|y\|.$$

Thus, if the second of the conditions (21) is satisfied, we can always arrange that the third is too.

THEOREM 5. *If conditions* I *and* II *are satisfied*, $\tilde{K}$ *has a continuous inverse, and* (1) *has a solution* x^*, *then we have*

$$\|x^* - x'\| \leqslant p'\|x^*\|, \tag{31}$$

where

$$p' = |\lambda|[\eta_1(1 + \|\tilde{K}^{-1}PK\|) + |\lambda|\eta\|\tilde{K}^{-1}\|(\|H\| + \eta_1)]. \tag{32}$$

Proof. Since

$$z^* = \lambda H z^* + \lambda H y = \lambda H x^*, \tag{33}$$

condition II enables us to find an element $\tilde{z} \in \tilde{\mathbf{X}}$ such that

$$\|z^* - \tilde{z}\| \leqslant |\lambda|\eta_1\|x^*\|. \tag{34}$$

Write

$$\tilde{z}_0 = \tilde{K}^{-1}PK\tilde{z}.$$

Then since

$$\tilde{K}\tilde{z}^* = PKz^*,$$

we have

$$\tilde{z}^* = \tilde{K}^{-1}PKz^*,$$

so that

$$\|\tilde{z}^* - \tilde{z}_0\| \leqslant \|\tilde{K}^{-1}PK\|\|z^* - \tilde{z}\| \leqslant |\lambda|\eta_1\|\tilde{K}^{-1}PK\|\|x^*\|. \tag{35}$$

Also, by condition I and (34) and (33),

$$\|\tilde{z} - \tilde{z}_0\| \leqslant \|\tilde{K}^{-1}\|\|\tilde{K}\tilde{z} - PK\tilde{z}\| \leqslant |\lambda|\eta\|\tilde{K}^{-1}\|\|\tilde{z}\| \leqslant |\lambda|\eta\|\tilde{K}^{-1}\|[\|z^*\| + \|\tilde{z} - z^*\|] \leqslant$$
$$\leqslant |\lambda|\eta\|\tilde{K}^{-1}\|(|\lambda|\|H\| + |\lambda|\eta_1)\|x^*\|.$$

However, (34) and (35) give

$$\|x^* - x'\| \leqslant \|z^* - \tilde{z}\| + \|\tilde{z} - \tilde{z}_0\| + \|\tilde{z}_0 - \tilde{z}^*\| \leqslant$$
$$\leqslant [|\lambda|\eta_1 + |\lambda|^2\eta\|\tilde{K}^{-1}\|(\|H\| + \eta_1) + |\lambda|\eta_1\|\tilde{K}^{-1}PK\|]\|x^*\| = p'\|x^*\|,$$

which is what we wished to prove.

Theorem 4 can also be made more precise in this case.

1.7. Let us now consider the question of whether the characteristic values of $\tilde{H}$ converge to those of H, restricting ourselves to the case where both operators are compact, so that the characteristic set is discrete.

First of all, under the hypotheses of Theorem 3, we can say that the characteristic values of $\tilde{H}$ can converge only to characteristic values of H. For let C_0 be a disc in the λ-plane with radius R and centre at the origin, and let $\lambda_1, \lambda_2, \ldots, \lambda_m$ be the characteristic values of H that lie in C_0. Let D denote the closed region obtained from C_0 by removing δ-neighbourhoods of $\lambda_1, \lambda_2, \ldots, \lambda_m$. In this region, the operator $K_\lambda = I - \lambda H$ has an inverse K_λ^{-1}, and $\|K_\lambda^{-1}\|$ is bounded, as a function of λ. Then, by Theorem 3, $\|\tilde{K}_\lambda^{-1}\|$ is also bounded for sufficiently large n by a number independent of n (here $\tilde{K}_\lambda = \tilde{I} - \lambda\tilde{H}$). Hence, under the above conditions, the characteristic values of $\tilde{H}$ can only lie in the deleted δ-neighbourhoods or outside C_0, and it is thus clear that if they tend to a finite limit as $n \to \infty$, then this limit must be one of the characteristic values of H.

Furthermore, each characteristic value of H is a limit of characteristic values of $\tilde{H}$. For if some λ_k were not a limit, then some neighbourhood of λ_k bounded by the circumference of C_k would contain no characteristic values of $\tilde{H}$ (for sufficiently large n). The function $\|K_\lambda^{-1}\|$ is bounded on the circumference of C_k, and hence, by (11), so is $\|\tilde{K}_\lambda^{-1}\|$. In this case, $\|\tilde{K}_\lambda^{-1}\|$ would also be bounded inside the circumference by the same number. This allows us to apply the maximum modulus principle to the analytic function $\psi(\lambda) = f(\tilde{K}_\lambda^{-1}\tilde{x})$ (f being any functional on $\tilde{\mathbf{X}}$ and $\tilde{x}$ any element of $\tilde{\mathbf{X}}$). Using Theorem 4, we would then find that K_λ^{-1} existed in every neighbourhood, which is a contradiction.

The question of convergence of eigenvalues and eigenvectors is discussed in the context of the general theory of approximation methods by Troitskaya [1] and Vainikko (see also Mikhlin-II).

1.8. We now note an important case where one can simplify the statements of the theorems proved above.

Often the approximate equation (2) is constructed in a special way. Namely $\tilde{H}$ is taken to be the operator PH;* then the approximate equation takes the form:

$$\tilde{x} - \lambda PH\tilde{x} = Py \qquad (\tilde{K} = PK). \tag{36}$$

* Here H must be regarded as an operator from $\tilde{\mathbf{X}}$ into $\mathbf{X}$.

With this choice, condition I is obviously satisfied with $\eta = 0$. This simplifies the statement of the theorems. Thus, in Theorem 1 we must take

$$q = |\lambda|\eta_1 \|I - P\| \, \|K^{-1}\|. \tag{37}$$

In Theorem 2,

$$p = (\eta_1 |\lambda| + \eta_2 \|K\|)(1 + \|\tilde{K}^{-1}PK\|). \tag{38}$$

Condition (21) in Theorem 3 is replaced by the conditions

$$\lim_{n\to\infty} \eta_1 \|P\| = \lim_{n\to\infty} \eta_2 \|P\| = 0. \tag{39}$$

In Theorem 4,

$$r = |\lambda|\eta_1(1 + \|K^{-1}PK\|), \quad \|K^{-1}\| \leqslant \frac{1 + \|\tilde{K}^{-1}P\| + \|\tilde{K}^{-1}PK\|}{1 - r}. \tag{40}$$

Let us determine conditions under which we can ensure that (39) is satisfied in the present case, and hence that the sequence of approximate solutions converges to an exact one. Assume that we have a sequence of operators P_n, projecting $\mathbf{X}$ onto spaces $\tilde{\mathbf{X}}_n$. To each P_n there corresponds an approximate equation (36).

THEOREM 6. *Suppose the following conditions are satisfied:*

1) $\mathbf{X}$ *is a complete space;*

2) $P_n \to I$ *on* $\mathbf{X}$; *that is,* $\lim_{n\to\infty} P_n x = x (x \in \mathbf{X})$.

3) *the operator* H *is compact.*

Then the constants $\eta_1^{(n)}$ *and* $\eta_2^{(n)}$ *occurring in conditions* II *and* III *can be chosen so that* $\eta_1^{(n)} \to 0$ *and* $\eta_2^{(n)} \to 0$ *as* $n \to \infty$.

Proof. Let B be the unit ball in $\mathbf{X}$. The set $H(B)$ is compact, and so by Gel'fand's Theorem (IX.1.4), $P_n \to I$ uniformly on this set. Writing

$$\eta_1^{(n)} = \sup_{z \in H(B)} \|P_n z - z\| \qquad (n = 1, 2, \ldots),$$

we have

$$\eta_1^{(n)} \to 0.$$

On the other hand, for each $x \in \mathbf{X}$,

$$\|P_n Hx - Hx\| \leqslant \eta_1^{(n)} \|x\|.$$

Since $P_n Hx \in \tilde{\mathbf{X}}_n$, it follows from this inequality that $\eta_1^{(n)}$ satisfies condition II.

We can take $\eta_2^{(n)}$ to be given by

$$\eta_2^{(n)} = \frac{\|P_n y - y\|}{\|y\|},$$

which obviously tends to zero.

COROLLARY 1. *Under the conditions of the theorem, if* λ *is not a characteristic value of* H, *then the conditions of Theorem* 3 *are satisfied, and hence the sequence* $\{\tilde{x}_n^*\}$ *of approximate solutions converges to an exact solution.*

In fact, (39) is satisfied, since, by condition 2) of the theorem, we have $\sup_n \|P_n\| < \infty$ (VII.1.2).

COROLLARY 2. *The characteristic values of H are limits of the characteristic values of the $\tilde{H}_n$.*

Let us consider in more detail the case where $\tilde{\mathbf{X}}$ is a finite-dimensional space (of dimension n). Each $\tilde{x} \in \tilde{\mathbf{X}}$ is uniquely expressible in the form

$$\tilde{x} = c_1\omega_1 + c_2\omega_2 + \ldots + c_n\omega_n, \tag{41}$$

where the elements $\omega_1, \omega_2, \ldots, \omega_n$ form a basis in $\tilde{\mathbf{X}}$.

Let $f_1, f_2, \ldots, f_n$ be a complete system of linear functionals on $\tilde{\mathbf{X}}$—that is, a system for which the relations

$$f_j(\tilde{x}) = 0 \qquad (j = 1, 2, \ldots, n)$$

imply that $\tilde{x} = 0$.

The approximate equation (36) is equivalent to the system of equations

$$f_j(PK\tilde{x}) = f_j(Py) \qquad (j = 1, 2, \ldots, n).$$

If we seek a solution to (36) in the form (41), we obtain a system of linear algebraic equations for the coefficients c_k:

$$\sum_{k=1}^{n} \alpha_{jk}c_k - \lambda \sum_{k=1}^{n} a_{jk}c_k = b_j \qquad (j = 1, 2, \ldots, n), \tag{42}$$

where

$$\alpha_{jk} = f_j(\omega_k), \qquad a_{jk} = f_j(PH\omega_k), \qquad b_j = f_j(Py) \qquad (j, k = 1, 2, \ldots, n).$$

If the system of functionals $f_1, f_2, \ldots, f_n$ is biorthogonal to the basis $\omega_1, \omega_2, \ldots, \omega_n$, then the system (42) simplifies, and takes the form

$$c_j - \lambda \sum_{k=1}^{n} a_{jk}c_k = b_j \qquad (j = 1, 2, \ldots, n). \tag{43}$$

In particular, if $\mathbf{X}$ is a Hilbert space and P is a projector—that is, an orthogonal projection operator—then, assuming that the system $\omega_1, \omega_2, \ldots, \omega_n$ is orthonormal, we find that

$$a_{jk} = f_j(PH\omega_k) = (PH\omega_k, \omega_j) = (H\omega_k, P\omega_j) = (H\omega_k, \omega_j)$$
$$(j, k = 1, 2, \ldots, n),$$
$$b_j = f_j(Py) = (Py, \omega_j) = (y, P\omega_j) = (y, \omega_j) \qquad (j = 1, 2, \ldots, n),$$

and the system (42) is expressible in the form

$$c_j - \lambda \sum_{k=1}^{n} c_k(H\omega_k, \omega_j) = (y, \omega_j) \qquad (j = 1, 2, \ldots, n). \tag{44}$$

The system (42) will be called the system for *Galerkin's method* in abstract form.*

We note another special case of the general theory, where even greater simplifications are possible.

If $\mathbf{X}$ coincides with $\tilde{\mathbf{X}}$, then conditions II and III are obviously satisfied with $\eta_1 = \eta_2 = 0$. Condition I now becomes a closeness condition for the operators H and $\tilde{H}$ (which in this case act on the same space):

$$\|H - \tilde{H}\| \leqslant \eta.$$

* Galerkin's method for non-linear equations is studied in Krasnosel'skii *et al.* and Vainikko.

However, it should be noted that this case does not present any significant interest from the point of view of the general theory, as all the theorems can in this case be proved by elementary methods (cf. V.4.6).

1.9. It is not uncommon for the approximate equation for some method to be viewed as an equation, not in $\mathbf{X}$, but in another space $\overline{\mathbf{X}}$, which as a rule is isomorphic to a subspace of $\mathbf{X}$.

We shall assume that a continuous linear operator ϕ_0 is defined on a *complete* subspace $\tilde{\mathbf{X}} \subset \mathbf{X}$, and maps $\tilde{\mathbf{X}}$ in a one-to-one fashion onto a *complete* space $\overline{\mathbf{X}}$. These hypotheses ensure that the inverse operator ϕ_0^{-1} is continuous. We assume further that ϕ is a continuous linear operator which is an extension of ϕ_0 to all of $\mathbf{X}$; thus ϕ denotes a continuous linear operator mapping $\mathbf{X}$ onto $\overline{\mathbf{X}}$ and coinciding with ϕ_0 on $\tilde{\mathbf{X}}$.

If we take P to be the projection operator from $\mathbf{X}$ onto $\tilde{\mathbf{X}}$ introduced in 1.1, then we can take ϕ to be

$$\phi = \phi_0 P. \tag{45}$$

Multiplying both sides of this equation by ϕ_0^{-1}, we obtain

$$P = \phi_0^{-1}\phi. \tag{46}$$

In view of the one-to-one correspondence between the elements of the spaces $\tilde{\mathbf{X}}$ and $\overline{\mathbf{X}}$, the approximate equation (2) can be transformed into an equivalent equation in $\overline{\mathbf{X}}$. To do this, we substitute $\tilde{x} = \phi_0^{-1}\bar{x}$ in (2), and then apply ϕ_0 to both sides of the equation. After this transformation, the equation takes the form

$$\bar{x} - \phi_0 \tilde{H} \phi_0^{-1} \bar{x} = \phi_0 P y.$$

Writing $\overline{H}$ for the operator from $\overline{\mathbf{X}}$ into itself given by

$$\overline{H} = \phi_0 \tilde{H} \phi_0^{-1}, \tag{47}$$

and bearing in mind equation (45), we can rewrite the approximate equation in the form:

$$\overline{K}\bar{x} \equiv \bar{x} - \lambda \overline{H}\bar{x} = \phi y \qquad (\overline{K} = \phi_0 \tilde{K} \phi_0^{-1}). \tag{48}$$

Since cases where the approximate equation is given in the form (48) often occur in applications, we now determine the form that the basic theorems of the general theory take when they are expressed in terms of the operators $\overline{H}$ and ϕ (and $\overline{K}$). To do this, P must be replaced everywhere by the expression (46), and we find the following expression for $\tilde{H}$ (and $\tilde{K}$):

$$\tilde{H} = \phi_0^{-1} \overline{H} \phi_0 \qquad (\tilde{K} = \phi_0^{-1} \overline{K} \phi_0). \tag{49}$$

Let us now look at condition I. In the new terminology, it takes the form:

$$\|\phi_0^{-1} \overline{H} \phi_0 \tilde{x} - \phi_0^{-1} \phi H \tilde{x}\| \leqslant \eta \|\tilde{x}\| \qquad (\tilde{x} \in \tilde{\mathbf{X}}).$$

This inequality will be satisfied if the following condition holds:

Ia. For each $\tilde{x} \in \tilde{\mathbf{X}}$,

$$\|\overline{H}\phi_0 \tilde{x} - \phi H \tilde{x}\| \leqslant \bar{\eta} \|\tilde{x}\|.$$

Here we must take η in condition I to be

$$\eta = \bar{\eta} \|\phi_0^{-1}\|. \tag{50}$$

Conditions II and III do not involve $\tilde{H}$ and P, so they are unchanged in the new situation.

We now present the new formulation of Theorem 1.

THEOREM 1a. *Suppose that conditions* Ia *and* II *are satisfied, and that K has a continuous inverse. If*

$$\bar{q} = |\lambda|[\bar{\eta}\|\phi_0^{-1}\| + \|I - \phi_0^{-1}\phi\|\eta_1]\|K^{-1}\| < 1, \tag{51}$$

then $\bar{K}$ also has a continuous inverse $\bar{K}^{-1}$. Moreover,

$$\|\bar{K}^{-1}\| \leqslant \frac{\bar{N}}{1-\bar{q}}, \tag{52}$$

where

$$\bar{N} = \|K^{-1}\|\,\|\phi_0\|\,\|\phi_0^{-1}\|. \tag{53}$$

For $\bar{K} = \phi_0 \tilde{K} \phi_0^{-1}$, and $\tilde{K}$ has an inverse, by Theorem 1, and satisfies the inequality (4).

If $\bar{x}^*$ denotes a solution of (48), then an approximate solution $\tilde{x}^*$ of (1) is given by $\tilde{x}^* = \phi_0^{-1}\bar{x}^*$. An estimate for the error of the approximate equation can be obtained using Theorem 2, if we make the necessary changes in the wording.

THEOREM 2a. *If conditions* Ia, II, *and* III *are satisfied, $\bar{K}$ has a continuous inverse and* (1) *has a solution x^*, then we have*

$$\|x^* - \phi_0^{-1}\bar{x}^*\| \leqslant \bar{p}\|x^*\|, \tag{54}$$

where

$$\bar{p} = (1+\varepsilon)|\lambda|\bar{\eta}\|\phi_0^{-1}\bar{K}^{-1}\| + \varepsilon(1 + \|\phi_0^{-1}\bar{K}^{-1}\phi K\|), \tag{55}$$

and

$$\varepsilon \leqslant \eta_1|\lambda| + \eta_2\|K\|.$$

Here $\bar{p}$ may be taken to have the value

$$\bar{p} = 2|\lambda|\bar{\eta}\|\phi_0^{-1}\|\,\|\phi_0^{-1}\bar{K}^{-1}\phi_0\| + \varepsilon(1 + \|\phi_0^{-1}\bar{K}^{-1}\phi K\|). \tag{56}$$

The fact that (56) is a value for $\bar{p}$ comes directly from (13). Equation (55) requires a little further elucidation. Retaining the notation of the proof of Theorem 2, we have, by condition Ia,

$$\|\tilde{x} - \tilde{x}_0\| = \|\phi_0^{-1}\bar{K}^{-1}\bar{K}\phi_0\tilde{x} - \phi_0^{-1}\bar{K}^{-1}\phi_0\phi_0^{-1}\phi K\tilde{x}\| \leqslant$$
$$\leqslant \|\phi_0^{-1}\bar{K}^{-1}\|\,\|\bar{K}\phi_0\tilde{x} - \phi K\tilde{x}\| \leqslant |\lambda|\bar{\eta}\|\phi_0^{-1}\bar{K}^{-1}\|\,\|\tilde{x}\|.$$

If we use this inequality in place of (17) we obtain (55).

Note that the facts referred to in the Remarks following Theorem 2 remain true, with the appropriate modifications.

Finally, let us reformulate the theorem about convergence of sequences of approximate solutions.

THEOREM 3a. *Suppose the following conditions are satisfied:*

1) *K has a continuous inverse;*

2) *for each $n = 1, 2, \ldots$, conditions* Ia, II *and* III *are satisfied, and*

$$\lim_{n\to\infty} \bar{\eta}\|\phi_0^{-1}\| = \lim_{n\to\infty} \eta_1\|\phi_0^{-1}\phi\| = \lim_{n\to\infty} \eta_2\|\phi_0^{-1}\phi\| = 0. \tag{57}$$

Then, for sufficiently large n, the approximate equation (48) *is soluble, and the sequence of approximate solutions converges to an exact solution. Moreover,*

$$\|x^* - \tilde{x}_n^*\| \leqslant \bar{Q}\bar{\eta}\,\|\phi_0^{-1}\| + \bar{Q}_1\eta_1\,\|\phi_0^{-1}\phi\| + \bar{Q}_2\eta_2\,\|\phi_0^{-1}\phi\|, \tag{58}$$

where $\tilde{x}_n^* = \phi_0^{-1}\tilde{x}_n^*$, *and* $\bar{Q}, \bar{Q}_1, \bar{Q}_2$ *are constants.*

In fact, if we bear in mind (50) and (46), we conclude that all the conditions of Theorem 3 are fulfilled, and so the required result is a consequence.

We do not give a complete statement of Theorem 4, but note that condition (25), in the new terminology, takes the form

$$\bar{r} = |\lambda|\,[(1 + |\lambda|\,\eta_1)\bar{\eta}\,\|\phi_0^{-1}\bar{K}^{-1}\| + \eta_1\,\|\phi_0^{-1}\bar{K}^{-1}\phi K\|\,] < 1, \tag{59}$$

while (26) can be rewritten as

$$\|K^{-1}\| \leqslant \frac{1 + \|\phi_0^{-1}\bar{K}^{-1}\phi\| + |\lambda|\,\bar{\eta}\,\|\phi_0^{-1}\bar{K}^{-1}\| + \|\phi_0^{-1}\bar{K}^{-1}\phi K\|}{1 - \bar{r}}. \tag{60}$$

The proof of these use the same arguments as in the proof of Theorem 4, together with inequality (54) of Theorem 2a and the $\bar{p}$ of (55).

Direct substitution in (25) and (26) gives the following values for $\bar{r}$:

$$\bar{r} = |\lambda|\,[(1 + |\lambda|\eta_1)\bar{\eta}\,\|\phi_0^{-1}\|\,\|\phi_0^{-1}\bar{K}^{-1}\phi_0\| + \eta_1(1 + \|\phi_0^{-1}\bar{K}^{-1}\phi K\|)], \tag{61}$$

and, in place of (60), we have the inequality

$$\|K^{-1}\| \leqslant \frac{1 + \|\phi_0^{-1}\bar{K}^{-1}\phi\| + |\lambda|\bar{\eta}\,\|\phi_0^{-1}\|\,\|\phi_0^{-1}\bar{K}^{-1}\phi_0\| + \|\phi_0^{-1}\bar{K}^{-1}\phi K\|}{1 - \bar{r}}, \tag{62}$$

where $\bar{r}$ is given by (61).

§ 2. Equations reducible to equations of the second kind

2.1. In the equations that we consider next, the identity operator does not appear as a separate term on the left-hand side; in fact, the operator on the left-hand side does not even map the given space into itself, but into another space. Nevertheless, these equations can be reduced to equations of the second kind, like those we considered in § 1, owing to the fact that the operator has a principal part of a simple type.

Thus we assume that we have two normed spaces $\mathbf{X}$ and $\mathbf{Y}$, and that complete subspaces $\tilde{\mathbf{X}} \subset \mathbf{X}$ and $\tilde{\mathbf{Y}} \subset \mathbf{Y}$ have been distinguished in these spaces. We further assume that we have a continuous linear operator $\mathbf{\Phi}$ that projects $\mathbf{Y}$ onto $\tilde{\mathbf{Y}}$.

We consider two equations: the exact equation,

$$K_1 x \equiv Gx - \lambda Tx = y_1, \tag{1}$$

and the approximate equation corresponding to it,

$$\tilde{K}_1\tilde{x} \equiv G\tilde{x} - \lambda\tilde{T}\tilde{x} = \Phi y_1. \tag{2}$$

Here G and T (and K_1) are continuous linear operators mapping $\mathbf{X}$ into $\mathbf{Y}$, and $\tilde{T}$ (and $\tilde{K}$) is a continuous linear operator from $\tilde{\mathbf{X}}$ into $\tilde{\mathbf{Y}}$. In addition, we make the following assumptions about these operators:

1) G has a continuous inverse;

2) G determines a one-to-one correspondence between $\tilde{\mathbf{X}}$ and $\tilde{\mathbf{Y}}$: that is, $G(\tilde{\mathbf{X}}) = \tilde{\mathbf{Y}}$, and hence $G^{-1}(\tilde{\mathbf{Y}}) = \tilde{\mathbf{X}}$.

Now we introduce conditions similar to conditions I–III of § 1.

Ib. For each $\tilde{x} \in \tilde{\mathbf{X}}$,

$$\|\Phi T\tilde{x} - \tilde{T}\tilde{x}\| \leqslant \mu \|\tilde{x}\|.$$

IIb. For each $x \in \mathbf{X}$, there exists $\tilde{y} \in \tilde{\mathbf{Y}}$ such that

$$\|Tx - \tilde{y}\| \leqslant \mu_1 \|x\|.$$

IIIb. There exists $\tilde{y}_1 \in \tilde{\mathbf{Y}}$ such that

$$\|y_1 - \tilde{y}_1\| \leqslant \mu_2 \|y_1\|.$$

Let us show that, with these assumptions, the study of equations (1) and (2) is reduced to the study of equations of the second kind.

In fact, let us apply G^{-1} to both sides of (1) and (2). This yields

$$Kx \equiv G^{-1}K_1x \equiv x - \lambda G^{-1}Tx = G^{-1}y_1, \tag{3}$$

$$\tilde{K}\tilde{x} \equiv G^{-1}\tilde{K}_1\tilde{x} \equiv \tilde{x} - \lambda G^{-1}\tilde{T}\tilde{x} = G^{-1}\Phi y_1. \tag{4}$$

Equations (3) and (4) have the same form as equations (1) and (2) of § 1, the roles of the operators H, $\tilde{H}$, $P(K, \tilde{K})$ now being played by $G^{-1}T$, $G^{-1}\tilde{T}$, $G^{-1}\Phi G(G^{-1}K_1, G^{-1}\tilde{K}_1)$, and the role of y being played by $G^{-1}y_1$.

We now verify that conditions I, II and III are satisfied.

I. In view of Ib, we have

$$\|PH\tilde{x} - \tilde{H}\tilde{x}\| = \|G^{-1}\Phi GG^{-1}T\tilde{x} - G^{-1}\tilde{T}\tilde{x}\| \leqslant \|G^{-1}\|\mu\|\tilde{x}\|.$$

II. Given $x \in \mathbf{X}$, we find $\tilde{y} \in \tilde{\mathbf{Y}}$ according to IIb.
Then, if we set $\tilde{x} = G^{-1}\tilde{y}$, we have

$$\|Hx - \tilde{x}\| = \|G^{-1}Tx - G^{-1}\tilde{y}\| \leqslant \|G^{-1}\|\mu_1 \|x\|.$$

III. Let $\tilde{y}_1$ be determined as in IIIb. Write $\tilde{y} = G^{-1}\tilde{y}_1$. Then

$$\|y - \tilde{y}\| = \|G^{-1}y_1 - G^{-1}\tilde{y}_1\| \leqslant \|G^{-1}\|\mu_2\|y_1\| \leqslant \|G^{-1}\| \, \|G\| \, \mu_2 \|y\|.$$

Thus conditions I, II and III of § 1 are satisfied; also the constants η, η_1, η_2 may be expressed in terms of μ, μ_1, μ_2 as follows:

$$\eta = \mu\|G^{-1}\|, \quad \eta_1 = \mu_1\|G^{-1}\|, \quad \eta_2 = \mu_2\|G^{-1}\| \, \|G\|. \tag{5}$$

REMARK. Instead of IIb and IIIb, we could have stipulated the existence of $\tilde{x} \in \tilde{\mathbf{X}}$ and $\tilde{y} \in \tilde{\mathbf{X}}$ such that the following inequalities hold:

$$\|G^{-1}Tx - \tilde{x}\| \leqslant \mu_1' \|x\|, \qquad \|G^{-1}y_1 - \tilde{y}\| \leqslant \mu_2' \|G^{-1}y_1\|. \tag{6}$$

Then conditions II and III would be satisfied with $\eta_1 = \mu_1'$ and $\eta_2 = \mu_2'$.

Since conditions I, II and III are satisfied for equations (3) and (4), we can obtain from the theorems of the last section corresponding theorems for equations (3) and (4), or for the equations (1) and (2) equivalent to these. We state these theorems explicitly below, but first make a further simplifying assumption.

Namely, we assume that the norms in $\mathbf{X}$ and $\mathbf{Y}$ are compatible with G:

$$\|x\|_{\mathbf{X}} = \|Gx\|_{\mathbf{Y}}, \quad \|y\|_{\mathbf{Y}} = \|G^{-1}y\|_{\mathbf{X}} \qquad (x \in \mathbf{X}, \, y \in \mathbf{Y}). \tag{7}$$

Under these conditions, G is obviously an isometry; $\|G\| = \|G^{-1}\| = 1$. Consequently,

$$\|P\| = \|\Phi\|, \quad \|K\| = \|K_1\|, \quad \|\tilde{K}\| = \|\tilde{K}_1\|, \quad \|H\| = \|T\|, \quad \|\tilde{H}\| = \|\tilde{T}\|. \tag{8}$$

2.2. Now we restate the basic theorems of § 1 for equations (1) and (2).

THEOREM 1b. *If conditions* Ib *and* IIb *are satisfied,* K_1 *has a continuous inverse; and if*

$$q = |\lambda|[\mu + \|I - \Phi\|\mu_1]\|K_1^{-1}\| < 1, \tag{9}$$

then $\tilde{K}_1$ *also has a continuous inverse, and*

$$\|\tilde{K}_1^{-1}\| \leqslant \frac{\|K_1^{-1}\|}{1-q}. \tag{10}$$

THEOREM 2b. *If conditions* Ib, IIb *and* IIIb *are satisfied, if* $\tilde{K}_1$ *has a continuous inverse (in particular, if the conditions of Theorem* 1b *are satisfied) and if there exists a solution* x^* *to* (1), *then*

$$\|x^* - \tilde{x}^*\| \leqslant p\|x^*\|, \tag{11}$$

where $\tilde{x}^*$ *is a solution of* (2) *and*

$$p = 2|\lambda|\mu\|\tilde{K}_1^{-1}\| + (\mu_1|\lambda| + \mu_2\|K_1\|)\ (1 + \|\tilde{K}_1^{-1}\Phi K_1\|). \tag{12}$$

REMARK. If there is an $\tilde{x} \in \tilde{\mathbf{X}}$ such that

$$\|x^* - \tilde{x}\| \leqslant \varepsilon\|x^*\|, \tag{13}$$

then the inequality (11) is valid even with conditions IIb and IIIb; in this case, p is taken to be

$$p = 2|\lambda|\mu\|\tilde{K}_1^{-1}\| + \varepsilon(1 + \|\tilde{K}_1^{-1}\Phi K_1\|). \tag{14}$$

THEOREM 3b. *If conditions* Ib, IIb *and* IIIb *are satisfied for each* $n = 1, 2, \ldots$, *if* K_1 *has a continuous inverse, and if*

$$\lim_{n\to\infty} \mu = \lim_{n\to\infty} \mu_1\|\Phi\| = \lim_{n\to\infty} \mu_2\|\Phi\| = 0, \tag{15}$$

then the sequence of approximate solutions (that is, solutions of (2)) *converges to a solution* x^* *of* (1). *Moreover, we then have*

$$\|x^* - \tilde{x}_n^*\| \leqslant Q\mu + Q_1\mu_1\|\Phi\| + Q_2\mu_2\|\Phi\|. \tag{16}$$

THEOREM 4b. *If* $\tilde{K}_1$ *has a continuous inverse, if conditions* Ib *and* IIb *are satisfied, and if*

$$r = |\lambda|\mu(1 + |\lambda|\mu_1)\|\tilde{K}_1^{-1}\| + |\lambda|\mu_1(1 + \|\tilde{K}_1^{-1}\Phi K_1\|) < 1, \tag{17}$$

then K_1 *has a continuous left inverse, and*

$$\|K_1^{-1}\| \leqslant \frac{1 + \|\tilde{K}_1^{-1}\Phi\| + |\lambda|\mu\|\tilde{K}_1^{-1}\| + \|\tilde{K}_1^{-1}\Phi K_1\|}{1-r}. \tag{18}$$

2.3. Finally, we note a special case where the statements of the theorems take a simpler form. Assume that the approximate equation (2) is constructed in a special way—namely, by projecting the exact equation. In other words, the approximate equation is obtained by applying the operator Φ to both sides of (1):

$$\tilde{K}_1\tilde{x} \equiv \Phi K_1\tilde{x} \equiv G\tilde{x} - \lambda\Phi T\tilde{x} = \Phi y_1, \tag{19}$$

so that, in this case, $\tilde{T} = \Phi T$.

It is easy to see that, under these conditions, condition Ib is satisfied with $\mu = 0$. For

$$\|\tilde{T}\tilde{x} - \Phi T\tilde{x}\| = \|\Phi T\tilde{x} - \Phi T\tilde{x}\| = 0.$$

Hence we can set $\mu = 0$ throughout the statements of the theorems.

We now add a further theorem, which is obtained from Theorem 1.6 and relates specifically to equations of the type we are now considering.

THEOREM 6b. *If we have a sequence of approximate equations of the form* (19) *and corresponding mappings* Φ_n, *if*

1) $\mathbf{Y}$ *is complete,*
2) $\Phi_n \to I$ *on* $\mathbf{Y}$,
3) *the operator* $G^{-1}T$ *is compact,*

and if, furthermore, K_1^{-1} *exists, then the approximate equations are soluble, for sufficiently large n, and the approximate solutions converge to an exact solution.*

§ 3. Applications to infinite systems of equations*

3.1. Suppose we are given an infinite system of equations

$$\xi_j - \lambda \sum_{k=1}^{\infty} a_{jk}\xi_k = b_j \qquad (j = 1, 2, \ldots), \tag{1}$$

with the assumption that

$$\sum_{j,\,k=1}^{\infty} |a_{jk}|^2 < \infty, \quad \sum_{j=1}^{\infty} |b_j|^2 < \infty. \tag{2}$$

We are required to find a solution of this system satisfying the condition

$$\sum_{k=1}^{\infty} |\xi_k|^2 < \infty. \tag{3}$$

One common method of solving this system is the so-called *reduction method*, which consists in replacing the system (1) by the system of n equations in n unknowns,

$$\xi_j - \lambda \sum_{k=1}^{n} a_{jk}\xi_k = b_j \qquad (j = 1, 2, \ldots, n). \tag{4}$$

A solution of this system is regarded as an approximate solution of the original system (1).

We shall be interested in the problem of bounding the error for the approximate solution, and in whether the approximate solutions converge to an exact solution as $n \to \infty$.

To apply the general theory, we take $\mathbf{X} = \ell^2$. The system (1) is then expressible as a single equation in $\mathbf{X}$:

$$Kx \equiv x - \lambda Hx = y \qquad (x = \{\xi_n\},\ y = \{b_n\}). \tag{5}$$

Here H is understood to be the continuous linear (and compact—see XI.2.2) operator in ℓ^2 defined by the matrix of the system:

$$z = Hx,\ \zeta_j = \sum_{k=1}^{\infty} a_{jk}\xi_k \qquad (j = 1, 2, \ldots;\ x = \{\xi_n\},\ z = \{\zeta_n\}).$$

* For infinite systems, see Riesz [4], Hellinger and Toeplitz [1], and also Kantorovich and Krylov.

We take $\bar{\mathbf{X}}$ to be the finite-dimensional Euclidean space ℓ_n^2. We then naturally take $\tilde{\mathbf{X}}$ to be the set of elements in ℓ^2 all of whose coordinates from the $(n+1)$-st onwards are zero. The meaning of the operators ϕ_0 and ϕ_0^{-1} needs no explanation, and ϕ is the operator associating to an element $x = \{\xi_m\} \in \ell^2$ the element

$$\bar{x} = \phi x = (\xi_1, \xi_2, \ldots, \xi_n) \in \ell_n^2.$$

Clearly

$$\|\phi\| = \|\phi_0\| = \|\phi_0^{-1}\| = 1.$$

We can now express the system (4) as a single equation in $\bar{\mathbf{X}}$, namely

$$\bar{K}\bar{x} \equiv \bar{x} - \lambda\bar{H}\bar{x} = \phi y, \tag{6}$$

where H is defined by the "truncated" matrix

$$A_n = \begin{pmatrix} a_{11} & a_{12} & \ldots & a_{1n} \\ a_{21} & a_{22} & \ldots & a_{2n} \\ \ldots & \ldots & \ldots & \ldots \\ a_{n1} & a_{n2} & \ldots & a_{nn} \end{pmatrix}.$$

Let us verify that conditions Ia, II and III are satisfied. For any $\tilde{x} = (\xi_1, \xi_2, \ldots, \xi_n, 0, \ldots) \in \tilde{\mathbf{X}}$, we have

$$\bar{z} = \phi H\tilde{x} - \bar{H}\phi_0\tilde{x} \qquad (\bar{z} = (\zeta_1, \zeta_2, \ldots, \zeta_n)),$$

$$\zeta_j = \sum_{k=1}^{n} a_{jk}\xi_k - \sum_{k=1}^{n} a_{jk}\xi_k = 0 \qquad (j = 1, 2, \ldots, n).$$

Hence condition Ia is satisfied with $\bar{\eta} = 0$.

To verify condition II, take any $x = \{\xi_m\} \in \ell^2$ and set* $\tilde{x} = [Hx]_n = (\zeta_1, \zeta_2, \ldots, \zeta_n, 0, \ldots)$. Then

$$\|Hx - \tilde{x}\| = \left[\sum_{j=n+1}^{\infty} \left|\sum_{k=1}^{\infty} a_{jk}\xi_k\right|^2\right]^{1/2} \leqslant \left[\sum_{j=n+1}^{\infty}\sum_{k=1}^{\infty} |a_{jk}|^2 \sum_{k=1}^{\infty} |\xi_k|^2\right]^{1/2} = \eta_1 \|x\|,$$

where

$$\eta_1 = \left[\sum_{j=n+1}^{\infty}\sum_{k=1}^{\infty} |a_{jk}|^2\right]^{1/2}.$$

It is clear from (2) that $\eta_1 \to 0$ as $n \to \infty$. Finally, if we set $\tilde{y} = [y]_n$, then we have

$$\|y - \tilde{y}\| = \left[\sum_{j=n+1}^{\infty} |b_j|^2\right]^{1/2} = \left[\frac{\sum_{j=n+1}^{\infty} |b_j|^2}{\sum_{j=1}^{\infty} |b_j|^2}\right]^{1/2} \|y\|,$$

* We use the symbol $[x]_n$ to denote a "truncated" element, i.e. the element obtained from $x \in \ell^2$ by replacing all the coordinates from the $(n+1)$-st onwards by zeros. Obviously, $[x]_n = \phi_0^{-1}\phi x = Px$.

so that in condition III we can take

$$\eta_2 = \left[\frac{\sum_{j=n+1}^{\infty} |b_j|^2}{\sum_{j=1}^{\infty} |b_j|^2}\right]^{1/2}$$

and $\eta_2 \to 0$ as $n \to \infty$.

The theory put forward in 1.9 is therefore applicable to our present case. In particular, we conclude from Theorem 3a that, if λ is not a characteristic value of the system (1) (that is, if it is not a characteristic value of H), then for sufficiently large n the system (4) is soluble, and the approximate solutions converge to an exact solution. The rate of convergence is determined from the inequality

$$\|x^* - \phi_0^{-1}\bar{x}_n^*\| \leqslant Q_1\left[\sum_{j=n+1}^{\infty}\sum_{k=1}^{\infty} |a_{jk}|^2\right]^{1/2} + Q_2\left[\frac{\sum_{j=n+1}^{\infty} |b_j|^2}{\sum_{j=1}^{\infty} |b_j|^2}\right]^{1/2},$$

where $x^* = (\xi_1^*, \xi_2^*, \ldots, \xi_n^*, \ldots)$ denotes a solution of the system (1) and $\bar{x}_n^* = (\xi_1^{(n)}, \xi_2^{(n)}, \ldots, \xi_n^{(n)})$ denotes a solution of the system (4).

From this it is clear that ξ_k^* differs only slightly from $\xi_k^{(n)}$ for $k = 1, 2, \ldots, n$, while ξ_k^* is small for $k > n$. Another obvious consequence is that

$$\lim_{n \to \infty} \xi_k^{(n)} = \xi_k^* \qquad (k = 1, 2, \ldots).$$

Using Theorem 1.4 (with condition (59) of § 1), we obtain the following theorem.

THEOREM 1. *If the system* (4) *has a unique solution and*

$$\bar{r} = |\lambda|\eta_1\|\bar{K}^{-1}\|\|\phi K\| < 1, \tag{7}$$

then λ is not a characteristic value of the system (1); *furthermore*

$$\|K^{-1}\| \leqslant \frac{1 + \|\bar{K}^{-1}\|(1 + \|\phi K\|)}{1 - \bar{r}}.$$

This theorem merits attention in so far as a conclusion relating to the solubility of an infinite system is drawn from data relating to the solution of a finite system; furthermore, all the values occurring in the inequality (7) can be found without difficulty, so the solubility criterion provided by this theorem is a completely effective one.

The value of η_1 is given above. For $\|\phi K\|$ we have the inequality

$$\|\phi K\| \leqslant 1 + \|\phi H\| \leqslant 1 + \left[\sum_{j=1}^{n}\sum_{k=1}^{\infty} |a_{jk}|^2\right]^{1/2}.$$

Finally, if the matrix A_n is symmetric, then $\|\bar{K}^{-1}\|$ is given by the equation

$$\|\bar{K}^{-1}\| = \max_{j=1,2,\ldots,n} \frac{1}{|1 - \lambda\lambda_j^{(n)}|},$$

where the $\lambda_j^{(n)}$ are the eigenvalues of A_n. In the general case, the determination of $\|\bar{K}^{-1}\|$ is more complicated (see V.2.8).

REMARK 1. Instead of systems satisfying the condition (2), we could consider the more extensive class of systems for which H is a compact operator from ℓ^2 into ℓ^2. Then η_1 would be given by the formula

$$\eta_1 = \|H - \phi_0^{-1}\phi H \phi_0^{-1}\phi\|,$$

and, as before, $\eta_1 \to 0$ as $n \to \infty$ (XI.2.2), so the theorems of the general theory are applicable here also.

REMARK 2. The general theory can also be applied to other classes of infinite systems, which can be regarded as functional equations in other spaces of sequences. These include, for example, the Riesz type systems, which satisfy the condition

$$\sum_{j=1}^{\infty} \left[\sum_{k=1}^{\infty} |a_{jk}|^{p/(p-1)} \right]^{p-1} < \infty \qquad (1 < p < \infty)$$

(here H must be regarded as an operator on ℓ^p). Also regular and completely regular systems belong in this scheme (in the space ℓ^∞).

§ 4. Applications to integral equations*

4.1. One of the most effective methods for the numerical solution of integral equations is the method of replacing the integral equation by an algebraic system of linear equations, using a quadrature formula.

Suppose we are given the equation

$$x(s) - \lambda \int_0^1 h(s, t) x(t)\, dt = y(s). \tag{1}$$

If we replace the integral, using a numerical interpolation formula of the form

$$\int_0^1 x(t)\, dt = \sum_{k=1}^{n} A_k x(t_k), \tag{2}$$

based on the points $t_1, t_2, \ldots, t_n$, and require (1) to be satisfied only at these points, then we obtain a system

$$x(t_j) - \lambda \sum_{k=1}^{n} A_k h(t_j, t_k) x(t_k) = y(t_j) \qquad (j = 1, 2, \ldots, n), \tag{3}$$

any solution of which determines an approximate value for the required solution at the points $t_1, t_2, \ldots, t_n$.

Let us see how the general theory is applied to estimate the error of this method. For definiteness we assume that the right-hand side of (1) is a continuous periodic function (of period 1) and that the kernel $h(s, t)$ is a continuous periodic function both in s and t (also of period 1). Under these conditions the solution will also be a continuous periodic function. Equation (1) will be regarded as a functional equation in the space $\mathbf{X} = \tilde{\mathbf{C}}$ of continuous periodic functions. The algebraic system (3) will be regarded as the approximate functional equation in the space $\tilde{\mathbf{X}} = \ell^\infty$,

$$\bar{K}\bar{x} \equiv \bar{x} - \lambda \bar{H}\bar{x} = \phi y. \tag{4}$$

* For the approximate solution of integral equations, and literature relating to this problem, see Kantorovich and Krylov.

We choose the points $t_1, t_2, \ldots, t_n$ and the coefficients in the quadrature formula (2) as follows:

$$A_k = \frac{1}{n}, \quad t_k = \frac{2k-1}{2n} \qquad (k = 1, 2, \ldots, n),$$

that is, we use as our starting-point the mean rectangle formula, which is one of the most convenient for periodic functions. With this choice, the operator $\bar{H}$ in (4) is defined by the matrix

$$\begin{pmatrix} \frac{1}{n}h\left(\frac{1}{2n}, \frac{1}{2n}\right) & \frac{1}{n}h\left(\frac{1}{2n}, \frac{3}{2n}\right) & \cdots & \frac{1}{n}h\left(\frac{1}{2n}, \frac{2n-1}{2n}\right) \\ \frac{1}{n}h\left(\frac{3}{2n}, \frac{1}{2n}\right) & \frac{1}{n}h\left(\frac{3}{2n}, \frac{3}{2n}\right) & \cdots & \frac{1}{n}h\left(\frac{3}{2n}, \frac{2n-1}{2n}\right) \\ \cdots & \cdots & \cdots & \cdots \\ \frac{1}{n}h\left(\frac{2n-1}{2n}, \frac{1}{2n}\right) & \frac{1}{n}h\left(\frac{2n-1}{2n}, \frac{3}{2n}\right) & \cdots & \frac{1}{n}h\left(\frac{2n-1}{2n}, \frac{2n-1}{2n}\right) \end{pmatrix}. \tag{5}$$

The operator ϕ is defined as follows:

$$\phi x = (x(t_1), x(t_2), \ldots, x(t_n)), \qquad \|\phi\| = 1 \qquad (x \in \check{\mathbf{C}}).$$

We define the space $\tilde{\mathbf{X}}$ and the associated mapping ϕ_0^{-1} by two different methods, thus obtaining two estimates for the error.

First we take $\tilde{\mathbf{X}}$ to be the set of all continuous periodic functions which are linear in each of the intervals $[t_k, t_{k+1}]$, where $t_k = \dfrac{2k-1}{2n}$ $(k = 0, \pm 1, \pm 2, \ldots)$. Every such function is determined by its values at the points $t_1, t_2, \ldots, t_n$; hence ϕ_0^{-1} is defined by

$$\tilde{x} = \phi_0^{-1} \bar{x},$$

$$(\bar{x} = (\xi_1, \ldots, \xi_n) \in \ell_n^\infty; \quad \tilde{x}(t_k) = \xi_k; \quad k = 1, 2, \ldots, n; \quad \tilde{x} \in \tilde{\mathbf{X}}).$$

We then have $\|\phi_0^{-1}\| = 1$.

Now write $\omega_s(\delta)$ for the modulus of continuity of $h(s, t)$ as a function of s: thus

$$\omega_s(\delta) = \sup |h(s+\sigma, t) - h(s, t)| \qquad (0 \leqslant s, t \leqslant 1, |\sigma| \leqslant \delta);$$

and define $\omega_t(\delta)$ analogously.

To verify condition Ia, we estimate the error of the quadrature formula (2) for functions of the form $z(t)\tilde{x}(t)$, where z is a periodic function whose modulus of continuity is at most $\omega(\delta)$ and $\tilde{x} \in \tilde{\mathbf{X}}$. In view of the periodicity, we have

$$\left| \int_0^1 z(t)\tilde{x}(t)\,dt - \sum_{k=1}^{n} \frac{1}{n} z(t_k)\tilde{x}(t_k) \right| =$$

$$= \left| \int_{t_0}^{t_n} z(t)\tilde{x}(t)\,dt - \sum_{k=0}^{n-1} \frac{1}{2n}(z(t_k)\tilde{x}(t_k) + z(t_{k+1})\tilde{x}(t_{k+1})) \right| =$$

$$= \left| \sum_{k=0}^{n-1} \int_{t_k}^{t_{k+1}} \left[z(t) - z\left(\frac{t_k + t_{k+1}}{2}\right) \right] \tilde{x}(t)\,dt + \sum_{k=0}^{n} \frac{1}{2n} \left[\tilde{x}(t_k)\left(z\left(\frac{t_k + t_{k+1}}{2}\right) - z(t_k) \right) + \right.\right.$$

$$\left.\left. + \tilde{x}(t_{k+1})\left(z\left(\frac{t_k + t_{k+1}}{2}\right) - z(t_{k+1}) \right) \right] \right| \leqslant 2\omega\left(\frac{1}{2n}\right)\|\tilde{x}\|,$$

since the linearity of $\tilde{x}(t)$ in each of the intervals implies that

$$\int_{t_k}^{t_{k+1}} \tilde{x}(t)\,dt = \frac{1}{2n}(\tilde{x}(t_k) + \tilde{x}(t_{k+1})) \qquad (k = 0, 1, \ldots, n-1).$$

Setting $z(t) = h(t_j, t)$ in the above inequality, we obtain

$$\|\phi H\tilde{x} - \bar{H}\phi_0\tilde{x}\| = \max_{j=1,2,\ldots,n} \left| \int_0^1 h(t_j, t)\tilde{x}(t)\,dt - \frac{1}{n}\sum_{k=1}^{n} h(t_j, t_k)\tilde{x}(t_k) \right| \leqslant 2\omega_t\left(\frac{1}{2n}\right)\|\tilde{x}\|.$$

Hence condition Ia is satisfied with

$$\bar{\eta} = 2\omega_t\left(\frac{1}{2n}\right).$$

Now we turn to verifying condition II. We show first that if $z \in \tilde{\mathbf{C}}$ is any function whose modulus of continuity does not exceed $\omega(\delta)$, then

$$\|z - \tilde{z}\| \leqslant \omega\left(\frac{1}{n}\right),$$

where $\tilde{z} = \phi_0^{-1}\phi z$; that is, $\tilde{z}$ is a piecewise-linear function whose values at $t_1, t_2, \ldots, t_n$ coincide with the corresponding values of z. In fact, if $t_j \leqslant s \leqslant t_{j+1}$, then

$$|z(s) - \tilde{z}(s)| = |z(s) - [(t_{j+1} - s)z(t_j) + (s - t_j)z(t_{j+1})]n| \leqslant n[|(t_{j+1} - s)(z(s) -$$

$$- z(t_j))| + |(s - t_j)(z(s) - z(t_{j+1}))|] \leqslant n\omega\left(\frac{1}{n}\right)(t_{j+1} - t_j) = \omega\left(\frac{1}{n}\right),$$

and so, as s was arbitrary, we obtain the required result. Now let us estimate the modulus of continuity of the element $z = Hx\,(x \in \tilde{\mathbf{C}})$:

$$|z(s) - z(s')| \leqslant \int_0^1 |h(s,t) - h(s',t)|\,|x(t)|\,dt \leqslant \omega_s(\delta)\|x\|$$

$$(|s - s'| \leqslant \delta).$$

Hence we can take $\omega(\delta)$ to be

$$\omega(\delta) = \omega_s(\delta)\|x\|$$

and, if we write $\tilde{z} = \phi_0^{-1}\phi Hx$, the above shows that

$$\|Hx - \tilde{z}\| \leqslant \omega_s\left(\frac{1}{n}\right)\|x\|.$$

Therefore condition II is satisfied with

$$\eta_1 = \omega_s\left(\frac{1}{n}\right).$$

Finally, applying what we have said to the element y, we find a $\tilde{y} \in \tilde{\mathbf{X}}$ such that

$$\|y - \tilde{y}\| \leqslant \bar{\omega}\left(\frac{1}{n}\right),$$

where $\bar{\omega}(\delta)$ is the modulus of continuity of y. It is thus clear that condition III is satisfied with

$$\eta_2 = \frac{1}{\|y\|}\,\bar{\omega}\left(\frac{1}{n}\right).$$

As the kernel $h(s, t)$ and the right-hand side $y(s)$ of (1) are continuous, $\bar{\eta}, \eta_1, \eta_2$ all tend to zero as $n \to \infty$, and $\|\phi\| = \|\phi_0^{-1}\| = 1$, so the theorems of the general theory are applicable. In particular, if λ is not a characteristic value of (1), then we conclude from Theorem 1.3a that the system (3) is soluble (for sufficiently large n) and that the approximate solutions $\phi_0^{-1}\bar{x}_n^*$ (the piecewise-linear functions constructed using the values $\{\xi_k\}$ found from (3)) converge to an exact solution.

For later use, it is important to note that by Theorem 1.1a the norm of the inverse operator $\|\bar{K}^{-1}\|$ is bounded independently of n.

When $h(s, t)$ and $y(s)$ satisfy a Lipschitz condition of order α, then Theorem 1.3a enables us to estimate the speed of convergence of the sequence of approximate solutions, as follows:

$$\|x^* - \phi_0^{-1}\bar{x}_n^*\| = O\left(\frac{1}{n^\alpha}\right).$$

For in this case $\omega_s(\delta)$, $\omega_t(\delta)$, $\bar{\omega}(\delta)$ all have the form $O(\delta^\alpha)$.

4.2. Let us consider another way of estimating the error of the same method, obtained using a different choice for the subspace $\tilde{\mathbf{X}}$.

For convenience we assume that n is odd: $n = 2m+1$. We take $\tilde{\mathbf{X}}$ to be the set of all trigonometric polynomials of degree at most m: that is, $\tilde{x} \in \tilde{\mathbf{X}}$ means that

$$\tilde{x}(t) = \frac{a_0}{2} + \sum_{k=1}^{m} (a_k \cos 2k\pi t + b_k \sin 2k\pi t).$$

The operator ϕ_0^{-1} clearly associates to an element $\bar{x} = (\xi_1, \xi_2, \ldots, \xi_n) \in \ell_n^\infty$ the interpolation polynomial with values $\xi_1, \xi_2, \ldots, \xi_n$ at the base-points $t_1, t_2, \ldots, t_n$, respectively. By Bernstein's Lemma,* if the trigonometric polynomial $\tilde{x}$ satisfies

$$|\tilde{x}(t_j)| \leqslant 1 \qquad (j = 1, 2, \ldots, n)$$

at the points $t_1, t_2, \ldots, t_n$, then

$$|\tilde{x}(t)| \leqslant A \ln m + B \qquad (0 \leqslant t \leqslant 1).$$

Hence we obtain a bound for the norm of ϕ_0^{-1}:

$$\|\phi_0^{-1}\| \leqslant A \ln m + B. \tag{6}$$

* See Natanson-I.

Now write E_m^t and E_m^s for the error of the best approximations to $h(s, t)$ as a function of t, and as a function of s, respectively, by trigonometric polynomials of degree m in the corresponding variables. In other words, we are assuming that there exist functions of the form

$$h_1(s, t) = \frac{a_0'(s)}{2} + \sum_{k=1}^{m} [a_k'(s) \cos 2k\pi t + b_k'(s) \sin 2k\pi t],$$

$$h_2(s, t) = \frac{a_0''(t)}{2} + \sum_{k=1}^{m} [a_k''(t) \cos 2k\pi s + b_k''(t) \sin 2k\pi s]$$

such that

$$|h(s, t) - h_1(s, t)| \leqslant E_m^t, \quad |h(s, t) - h_2(s, t)| \leqslant E_m^s \quad (0 \leqslant s,\ t \leqslant 1).$$

To verify condition Ia we note,* first of all, that the quadrature formula (2) is exact if the integrand is a trigonometric polynomial of degree at most $n-1 = 2m$.

Since $h_1(s, t)\tilde{x}(t)$ is such a polynomial in t, we have

$$\int_0^1 h_1(s, t)\tilde{x}(t)dt = \frac{1}{n}\sum_{k=1}^{n} h_1(s, t_k)\tilde{x}(t_k) \qquad (0 \leqslant s \leqslant 1).$$

Hence

$$\|\phi H\tilde{x} - \bar{H}\phi_0\tilde{x}\| = \max_{j=1, 2, \ldots, n} \left| \int_0^1 h(t_j, t)\tilde{x}(t)dt - \frac{1}{n}\sum_{k=1} h(t_j, t_k)\tilde{x}(t_k) \right| \leqslant$$

$$\leqslant \max_{j=1, 2, \ldots, n} \left| \int_0^1 [h(t_j, t) - h_1(t_j, t)]\tilde{x}(t)dt \right| +$$

$$+ \max_{j=1, 2, \ldots, n} \frac{1}{n} \left| \sum_{k=1}^{n} [h(t_j, t_k) - h_1(t_j, t_k)]\tilde{x}(t_k) \right| \leqslant 2E_m^t\|\tilde{x}\|.$$

Thus condition Ia is satisfied with

$$\bar{\eta} = 2E_m^t.$$

To verify condition II we look for an approximation to the element $z = Hx$ by constructing the function†

$$\tilde{x}(s) = \int_0^1 h_2(s, t)x(t)dt.$$

From the form of $h_2(s, t)$ it is clear that $\tilde{x}(s)$ is a trigonometric polynomial of degree m: that is, $\tilde{x} \in \tilde{\mathbf{X}}$. At the same time, we have

$$\|Hx - \tilde{x}\| = \max_s \left| \int_0^1 [h(s, t) - h_2(s, t)]x(t)dt \right| \leqslant E_m^s\|x\|.$$

* See Natanson-I.

† It is not hard to show that the function $h_2(s, t)$ can always be chosen to be continuous in t, so that the given integral exists. However, this proviso does not need to be made if the integral is understood in the sense of Banach (see II.4.2), since it is then defined for an arbitrary bounded function.

Condition II is therefore satisfied with

$$\eta_1 = E_m^s.$$

Finally, condition III is satisfied if we set

$$\eta_2 = \frac{1}{\|y\|} E_m(y),$$

where $E_m(y)$ is the error of the best approximation to y by a trigonometric polynomial of degree m.

On the basis of these inequalities, we can use Theorem 1.2a to obtain an estimate of how close the approximate solution is to the exact solution:

$$\|x^* - \phi_0^{-1}\bar{x}^*\| \leqslant \bar{p}\,\|x^*\|,$$

$$\bar{p} = 4|\lambda|E_m^t\|\phi_0^{-1}\|\,\|\bar{K}^{-1}\| + \left(|\lambda|E_m^s + \frac{E_m(y)\|K\|}{\|y\|}\right) \times$$

$$\times\,(1 + \|\phi_0^{-1}\|\,\|\bar{K}^{-1}\|\,\|K\|).$$

Notice that, as we showed in 4.1, $\|\bar{K}^{-1}\|$ is bounded independently of n.

Assume that the kernel $h(s, t)$ has derivatives up to order v in s and t, and that $\frac{\partial^v h}{\partial s^v}$ and $\frac{\partial^v h}{\partial t^v}$ satisfy a Lipschitz condition of order α; and make the same assumptions for the function $y(s)$. Then, by a well-known theorem of Jackson (see Natanson-I),

$$E_m^t = O\left(\frac{1}{m^{v+\alpha}}\right) = O\left(\frac{1}{n^{v+\alpha}}\right), \qquad E_m^s = O\left(\frac{1}{m^{v+\alpha}}\right) = O\left(\frac{1}{n^{v+\alpha}}\right),$$

$$E_m(y) = O\left(\frac{1}{m^{v+\alpha}}\right) = O\left(\frac{1}{n^{v+\alpha}}\right),$$

and, as (6) implies that

$$\|\phi_0^{-1}\| = O(\ln m) = O(\ln n),$$

we obtain the following estimate for the rate of decrease in the error of the approximate equation, by Theorem 1.3a:

$$\|x^* - \phi_0^{-1}\bar{x}_n^*\| = \left(\frac{\ln n}{n^{v+\alpha}}\right).$$

By means of Theorem 1.4, we can use the results of our approximate solution to determine a possible region for the characteristic values of (1) to lie in.

Finally, we note that similar arguments could be used to estimate the error of the same method for different quadrature formulae. In particular, in the non-periodic case, if we were to use Gauss's formula, we should have to rely on interpolation by algebraic polynomials, and on theorems concerning orders of approximations of functions by algebraic polynomials.

4.3. Let us now consider some other approximation methods for the integral equation (1). Firstly, there is the *method of moments*. This consists in finding an approximate solution to (1) of the form

$$\tilde{x}(t) = \sum_{k=1}^{n} c_k\omega_k(t) \qquad (0 \leqslant t \leqslant 1), \tag{7}$$

where ω_k ($k = 1, 2, \ldots$) are the functions of a complete orthonormal system. Here the coefficients c_k are determined from the system

$$c_j - \lambda \sum_{k=1}^{n} c_k \int_0^1 \int_0^1 h(s, t)\omega_j(s)\omega_k(t) ds\, dt = \int_0^1 y(s)\omega_j(s)\, ds$$

$$(j = 1, 2, \ldots, n), \tag{8}$$

which replaces the condition that $Kx - y$ should vanish by the condition that it should be orthogonal to the first n functions in the given system.

Assuming that the kernel $h(s, t)$ of (1) satisfies the condition

$$\int_0^1 \int_0^1 |h(s, t)|^2\, ds\, dt < \infty,$$

we take $\mathbf{X}$ to be the space $\mathbf{L}^2(0, 1)$ and $\tilde{\mathbf{X}}$ to be the subspace of elements of the form (7). We define P to be the orthogonal projection operator onto $\tilde{\mathbf{X}}$: thus $\tilde{x} = Px$ means that

$$\tilde{x} = \sum_{k=1}^{n} (x, \omega_k)\omega_k.$$

We can express (8) as a single equation in $\tilde{\mathbf{X}}$, namely

$$\tilde{x} - PH\tilde{x} = Py.$$

From this it is clear that we are back to the conditions we were discussing in 1.8. Moreover, as the system $\{\omega_k\}$ is complete, the conditions of Theorem 1.6 are satisfied, and thus the approximate solutions converge to an exact solution. The rate of convergence is determined by the properties of the kernel $h(s, t)$ and of the term $y(s)$ on the right-hand side; it depends on the rate of convergence of the orthogonal expansions for these functions in terms of the functions of the system. Let us look at the case where the kernel is periodic and satisfies a Lipschitz condition of order $\alpha + \frac{1}{2}$ ($\alpha > 0$) in s. We impose similar conditions on $y(s)$. We take the orthonormal system $\{\omega_k\}$ to be the system of trigonometric functions. Using Jackson's Theorem, it is easy in this case to obtain the estimates

$$\eta_1 = O(n^{-1/2-\alpha}), \qquad \eta_2 = O(n^{-1/2-\alpha}),$$

and hence we have

$$\|x^* - \tilde{x}_n^*\|_{\mathbf{L}^2} = O(\eta_1 + \eta_2) = O(n^{-1/2-\alpha}). \tag{9}$$

Let us show that in fact in this case the approximate solutions converge uniformly to an exact solution.

Suppose $\tilde{x} = \sum_{k=1}^{n} c_j\omega_j$. We find a bound for $\|\tilde{x}\|_{\mathbf{C}}$ in terms of $\|\tilde{x}\|_{\mathbf{L}^2}$. We have $\|\omega_j\|_{\mathbf{C}} \leqslant M$ ($j = 1, 2, \ldots$), so using the Cauchy–Buniakowski inequality, we have

$$\|\tilde{x}\|_{\mathbf{C}} \leqslant M \sum_{k=1}^{n} |c_j| \leqslant M\sqrt{n}\left[\sum_{j=1}^{n} |c_j|^2\right]^{1/2} = M\sqrt{n}\,\|x\|_{\mathbf{L}^2}.$$

Now apply this inequality to the difference between two approximations $\tilde{x}_n^*$ and $\tilde{x}_{2^k}^*$, where $2^{k-1} \leqslant n \leqslant 2^k$. We obtain

$$\|\tilde{x}_{2^k}^* - \tilde{x}_n^*\|_{\mathbf{C}} \leqslant M2^{k/2}\|\tilde{x}_{2^k}^* - \tilde{x}_n^*\|_{\mathbf{L}^2}.$$

Also, using (9), we have

$$\|\tilde{x}_{2^k}^* - \tilde{x}_n^*\|_{L^2} \leqslant \|\tilde{x}_{2^k}^* - x^*\|_{L^2} + \|x^* - \tilde{x}_n^*\|_{L^2} \leqslant$$

$$\leqslant \frac{C}{2^{\frac{k}{2}+\alpha k}} + \frac{C}{n^{\frac{1}{2}+\alpha}} \leqslant \frac{C_2}{2^{\frac{k}{2}+\alpha k}},$$

and thus

$$\|\tilde{x}_{2^k}^* - \tilde{x}_n^*\|_C \leqslant \frac{C_2}{2^{\alpha k}}. \tag{10}$$

In particular, the last inequality is also true for $n = 2^{k-1}$; that is,

$$\|\tilde{x}_{2^k}^* - \tilde{x}_{2^{k-1}}^*\|_C \leqslant \frac{C_2}{2^{\alpha k}}.$$

We conclude from this that the series

$$\tilde{x}_1^*(t) + \sum_{k=1}^{\infty} [\tilde{x}_{2^k}^*(t) - \tilde{x}_{2^{k-1}}^*(t)] = \lim_{k\to\infty} \tilde{x}_{2^k}^*(t) = x^*(t)$$

is uniformly convergent. Remembering (10), we see that the convergence

$$\lim_{n\to\infty} \tilde{x}_n^*(t) = x^*(t)$$

is also uniform. Further, we clearly have

$$\|x^* - \tilde{x}_n^*\|_C = O(n^{-\alpha}).$$

Note that if we regard the integral equation (1) as an equation in $\mathbf{C}[0, 1]$, then Theorem 1.3 yields a stronger result than that just given. In fact, if we assume a Lipschitz condition of order $\beta > 0$ for both the kernel (as a function of s) and the right-hand term $y(s)$, then the approximate solutions converge to an exact solution and

$$\|x^* - \tilde{x}_n^*\|_C = O(n^{-\beta} \ln n).$$

4.4. Now we consider the method of replacing the kernel by a neighbouring one (particularly a degenerate one). The method consists, as is clear from its name, in replacing (1) by an equation

$$\tilde{x}(s) - \lambda \int_0^1 \tilde{h}(s, t)\tilde{x}(t)dt = y(s), \tag{11}$$

whose kernel $\tilde{h}(s, t)$ is close to $h(s, t)$, and whose solution is already known.

If we write (11) as a functional equation in the same space as (1), namely

$$\tilde{K}\tilde{x} \equiv \tilde{x} - \lambda\tilde{H}\tilde{x} = y,$$

we find ourselves in the same situation that we described at the end of section 1.8. As we have already mentioned, condition I reduces to the inequality

$$\|H - \tilde{H}\| \leqslant \eta, \tag{12}$$

and conditions II and III are satisfied with $\eta_1 = \eta_2 = 0$.

Taking $\mathbf{X} = \tilde{\mathbf{X}} = \mathbf{L}^{\infty}(0, 1)$, the space of bounded measurable functions, we rewrite (12) in the form

$$\int_0^1 |h(s,t)-\tilde{h}(s,t)|\,dt \leqslant \eta$$

$$(0 \leqslant s \leqslant 1).$$

The norm of $\tilde{K}^{-1}$ can be estimated with no difficulty if we know the resolvent $\tilde{\Gamma}(\lambda; s, t)$ of the approximate equation (11). As the inverse operator $\tilde{K}^{-1}$ has the form

$$z = \tilde{K}^{-1}y,$$

$$z(s) = y(s) + \lambda \int_0^1 \tilde{\Gamma}(\lambda; s, t)y(t)dt,$$

we have

$$\|\tilde{K}^{-1}\| \leqslant 1 + |\lambda| B,$$

where B is found from the inequality

$$\int_0^1 |\tilde{\Gamma}(\lambda; s, t)|\,dt \leqslant B \qquad (0 \leqslant s \leqslant 1).$$

Using the estimate given by (19) in § 1, we obtain the following estimate for the closeness of the approximate solution to the exact solution

$$|x^*(t) - \tilde{x}^*(t)| \leqslant |\lambda|\eta(1 + |\lambda|B)\|x^*\|. \tag{13}$$

A similar estimate is obtained by interchanging the roles of the two equations.

It follows from (13) that the approximate solutions converge uniformly to an exact solution when the kernels converge uniformly.

The general theory can also be applied to the method in which one seeks a solution of the form (7), requiring the terms on the left-hand and right-hand sides to be equal only at specified points. We do not describe or investigate this method (see Kantorovich [9]), as we shall be considering a similar method for differential equations in the next section.

§ 5. Applications to ordinary differential equations

5.1. We first consider the so-called *interpolation method*, otherwise known as the *method of coincidence* (or collocation),* in the case of the boundary-value problem for the equation

$$\frac{d^{2m}x}{dt^{2m}} - \lambda\left[p_1 \frac{d^{2m-1}x}{dt^{2m-1}} + \ldots + p_{2m}x\right] = y \tag{1}$$

subject to the conditions

$$x(a) = x'(a) = \ldots = x^{(m-1)}(a) = 0,$$

$$x(b) = x'(b) = \ldots = x^{(m-1)}(b) = 0. \tag{2}$$

We seek an approximate solution of the form

$$\tilde{x}(t) = (t-a)^m(b-t)^m \sum_{k=1}^{n} c_k t^{k-1}, \tag{3}$$

* The interpolation method was apparently first put forward in Kantorovich [1]. A proof of convergence based on the theorems of §§ 1–2 was obtained in Karpilovskaya [1].

so that the boundary conditions are satisfied for $\tilde{x}$; the coefficients $c_1, c_2, \ldots, c_n$ are found by requiring (1) to be satisfied for some system of points (nodes) $t_1, t_2, \ldots, t_n$:

$$\left\{\frac{d^{2m}\tilde{x}}{dt^{2m}} - \lambda\left[p_1 \frac{d^{2m-1}\tilde{x}}{dt^{2m-1}} + \ldots + p_{2m}\tilde{x}\right]\right\}_{t=t_j} = y(t_j). \tag{4}$$

To apply the general theory of § 2 we regard (1) as a functional equation in the space $\mathbf{X} = \mathbf{C}^{(2m)}[a, b]$ of $2m$ times continuously differentiable functions satisfying the conditions (2). The definition of the norm in $\mathbf{X}$ will be given below.

We take $\tilde{\mathbf{X}}$ to be the set of functions of the form (3).

Next, we take $\mathbf{Y}$ to be the space $\mathbf{C}[a, b]$ of functions continuous in $[a, b]$, with the usual definition of norm.

Finally, we take $\tilde{\mathbf{Y}}$ to be the set of polynomials of degree $n-1$. We define the mapping Φ from $\mathbf{Y}$ onto $\tilde{\mathbf{Y}}$ by taking $\tilde{y} = \Phi(y)$ to be the interpolation polynomial of degree $n-1$ having the same values at the points $t_1, t_2, \ldots, t_n$ as y. As we know, in the case of Chebyshev base points, we have

$$\|\Phi\| \leqslant A \ln n + B. \tag{5}$$

It is also well known that, for Gaussian base points,

$$\|\Phi\| \leqslant A\sqrt{n}, \tag{6}$$

while if the base points are roots of the n-th orthogonal polynomial with bounded weights $\omega(t) \geqslant c > 0$, then*

$$\|\Phi\| \leqslant An. \tag{7}$$

The boundary value problem we have been discussing is equivalent to the functional equation

$$K_1 x \equiv Gx - \lambda Tx = y, \tag{8}$$

where

$$Gx = \frac{d^{2m}x}{dt^{2m}}, \quad z = Tx, \quad z(t) = \sum_{s=1}^{2m} p_s(t) x^{(2m-s)}(t). \tag{9}$$

The operator inverse to G is an integral operator whose kernel is the Green's function of the differential operator $\dfrac{d^{2m}x}{dt^{2m}}$, under the conditions (2).

The existence of the Green's function follows from the fact that the differential equation

$$\frac{d^{2m}x}{dt^{2m}} = 0 \tag{10}$$

subject to the conditions (2) has only the zero solution, since a general solution of (10) is an arbitrary polynomial of degree $2m-1$, and a non-zero polynomial satisfying (2) must have a factor $(t-a)^m(b-t)^m$ of degree $2m$.

Hence

$$x = G^{-1}y, \quad x(s) = \int_a^b g(s, t) y(t) dt \quad (x \in \mathbf{X},\ y \in \mathbf{Y}).$$

The function $g(s, t)$ has continuous derivatives of orders up to $2m-2$, while its $(2m-1)$-st

* For (5) and (7) see Natanson-I. The estimate (6) is in the book by Szegö.

derivative has a jump when $s = t$. Thus

$$x^{(k)}(s) = \int_a^b \frac{\partial^k}{\partial s^k} g(s, t) y(t) dt \qquad (k = 0, 1, \ldots, 2m-1),$$

and so

$$\max_s |x^{(k)}(s)| \leqslant A_k \|y\|_{\mathbf{C}} \qquad (k = 0, 1, \ldots, 2m-1). \tag{11}$$

If we take $A_{2m} = 1$, then (11) is also true for $k = 2m$.

If we wish to make the norms of $\mathbf{X}$ and $\mathbf{Y}$ compatible, so that G induces an isometry between the spaces, then we must set

$$\|x\|_{\mathbf{X}} = \|Gx\|_{\mathbf{Y}} = \max_t |x^{(2m)}(t)|.$$

Then (11) makes it possible to find a bound for any derivative of x in terms of the norm of x in $\mathbf{X}$:

$$\max_t |x^{(k)}(t)| \leqslant A_k \|x\| \qquad (k = 0, 1, \ldots, 2m). \tag{12}$$

Note that G maps elements of $\tilde{\mathbf{X}}$ into elements of $\tilde{\mathbf{Y}}$, the latter being polynomials of degree at most $n-1$.

Let us find a bound for the norm of T, as an operator from $\mathbf{C}^{(2m)}[a, b]$ into $\mathbf{C}[a, b]$. Using (12), we have

$$\|Tx\| = \max_t \left| \sum_{s=1}^{2m} p_s(t) x^{(2m-s)}(t) \right| \leqslant \left[\sum_{s=1}^{2m} B_s A_{2m-s} \right] \|x\|,$$

where $B_s = \max_t |p_s(t)|$ $(s = 1, 2, \ldots, 2m)$.

Hence

$$\|T\| \leqslant \sum_{s=1}^{2m} B_s A_{2m-s}. \tag{13}$$

Similarly, if we know the Green's function for (1), subject to the conditions (2), then we can find a bound for the norm of K_1^{-1}.

The approximate equation—namely, the system (4)—can clearly be written in the form

$$\tilde{K}_1 \tilde{x} \equiv G\tilde{x} - \lambda \Phi T \tilde{x} = \Phi y, \tag{14}$$

as the system (4) is obtained by substituting the expression (3) for $\tilde{x}$ in (1) and equating the two sides at each of the points $t_1, t_2, \ldots, t_n$, and this is equivalent to equating the interpolation polynomials constructed from the values at these points.

If we compare (14) with equation (22) of 2.3, we see that it is constructed in the special way for which condition Ib is satisfied with $\mu = 0$.

To verify condition IIb—that is, to determine the order of the approximation of elements of the form Tx by elements of $\tilde{\mathbf{Y}}$—we find a bound for $\frac{d}{dt} Tx$. Using (12), we have

$$\left\| \frac{d}{dt} Tx \right\|_{\mathbf{C}} = \max_t \left| \frac{d}{dt} \sum_{s=1}^{2m} p_s(t) x^{(2m-s)}(t) \right| =$$

$$= \max_t \left| \sum_{s=0}^{2m} (p_s'(t) + p_{s+1}(t))x^{(2m-s)}(t) \right| \leqslant M \|x\|$$

$$(p_0(t) = p_{2m+1}(t) = 0).$$

By Jackson's Theorem, there exists a polynomial $\tilde{y} \in \tilde{\mathbf{Y}}$ such that

$$\|Tx - \tilde{y}\| \leqslant \frac{AM\|x\|}{n}.$$

Hence condition IIb is satisfied with $\mu_1 = O(1/n)$.

As for condition IIIb, if the right-hand term of equation (1)—namely, the function y—has a continuous derivative, then by applying Jackson's Theorem again we obtain the value $\mu_2 = O(1/n)$.

Notice also that, if $p_1 = 0$ in (1) (this can always be arranged by an appropriate change of variables) and the coefficients p_s are twice continuously differentiable, then by applying the above arguments to $\dfrac{d^2}{dt^2}Tx$, we obtain $\mu_1 = O(1/n^2)$. If y is a twice differentiable function as well, then $\mu_2 = O(1/n^2)$.

Using the results of § 2, the estimates we have found for μ_1, and μ_2, and the estimate for $\|\Phi\|$, we see from Theorem 2.1b that the system (4) is soluble for sufficiently large n.

If we use Theorem 2.3b, we see that the approximate solutions converge to exact solutions, for either Chebyshev or Gaussian base points. We then have

$$\|x^* - \tilde{x}_n^*\| = O\left(\frac{\ln n}{n}\right), \qquad \|x^* - \tilde{x}_n^*\| = O\left(\frac{1}{\sqrt{n}}\right)$$

for Chebyshev base points and Gaussian base points respectively.

In the case mentioned above, where $p_1 = 0$, we find that the system (4) is also soluble when we take the base points to be the roots of the n-th orthogonal polynomial with weight $\omega(t) \geqslant c > 0$. The estimate for the error of the approximate solution in this case has the form

$$\|x^* - \tilde{x}_n^*\| = O\left(\frac{1}{n}\right).$$

It can be shown that the rate of convergence increases when the coefficients are smoother. Assume that $p_1, p_2, \ldots, p_{2m}$ and y are r times differentiable and that their r-th derivatives satisfy a Lipschitz condition of order α. We use the Remark following Theorem 1.3, which shows that the rate of convergence is given by

$$\|x^* - \tilde{x}_n^*\| = O(\varepsilon \|P\|) \qquad (P = G^{-1}\Phi G, \ \|P\| = \|\Phi\|), \tag{15}$$

where ε characterizes the possibility of approximating the solution x^* by an element $\tilde{x} \in \tilde{\mathbf{X}}$. Since

$$\|x^* - \tilde{x}\|_{\mathbf{X}} = \|G(x^* - \tilde{x})\|_{\mathbf{Y}} = \left\| \frac{d^{2m}x^*}{dt^{2m}} - \tilde{y} \right\|_{\mathbf{Y}} \qquad (\tilde{y} = Gx \in \tilde{\mathbf{Y}}),$$

ε is determined by the order of approximation of the $2m$-th derivative of the solution by a polynomial of degree $n-1$. However, it is easy to see, by induction on r, that $\dfrac{d^{2m}x^*}{dt^{2m}}$ has

r-th derivative satisfying a Lipschitz condition of order α. Hence, by Jackson's Theorem, there exists a polynomial $\tilde{y} \in \tilde{\mathbf{Y}}$ such that

$$\left\| \frac{d^{2m}x^*}{dt^{2m}} - \tilde{y} \right\| \leqslant \frac{D}{n^{r+\alpha}}.$$

Hence ε is of the order $n^{-r-\alpha}$, and so, by (15), we have

$$\|x^* - \tilde{x}_n^*\| = O\left(\frac{\ln n}{n^{r+\alpha}}\right), \qquad \|x^* - \tilde{x}_n^*\| = O\left(\frac{1}{n^{r+\alpha-\frac{1}{2}}}\right),$$

$$\|x^* - \tilde{x}_n^*\| = O\left(\frac{1}{n^{r+\alpha-1}}\right)$$

for Chebyshev and Gaussian base points and for the roots of orthogonal polynomials with weights bounded below, respectively.

REMARK 1. Without essentially changing the arguments, one can show that this method converges not only for the boundary-value problem but also for the Cauchy problem. The character of the arguments remains unchanged if we consider a system of differential equations rather than a single one.

REMARK 2. The convergence theorem for this approximation method can be regarded as a convergence theorem for an interpolation process—$\tilde{x}_n^*(t)$ can be viewed as an interpolation polynomial for the solution $x^*(t)$, constructed in accordance with the boundary conditions and the prescribed values of the differential operator Kx^* at the points $t_1, t_2, \ldots, t_n$.

REMARK 3. It is curious that this process does not converge for equally spaced base points even in the case of the simplest equation $\frac{d^2x}{dt^2} = y$, as one can easily see if one recalls that the ordinary interpolation process with equally spaced base points can diverge (see Natanson-I).

5.2. Now we consider an application of the general theory to the study of the convergence of the Galerkin method and the method of moments.*

Suppose we are given the differential equation (1) and homogeneous boundary conditions

$$M(x) = 0 \tag{16}$$

at the ends of the interval. The *method of moments* consists in taking a system of functions $\{\omega_k\}$ satisfying the boundary conditions (16) and seeking an approximate solution of the form

$$\tilde{x}(t) = \sum_{k=1}^{n} c_k \omega_k(t), \tag{17}$$

where the c_k are determined by the condition that the result of substituting $\tilde{x}$ in (1) (or, more precisely, the discrepancy arising from this) be orthogonal to the first n functions of some system $\{\zeta_j\}$:

$$\int_a^b L\left(\sum_{k=1}^{n} c_k\omega_k\right)\zeta_j(t)dt = \int_a^b y(t)\zeta_j(t)dt \qquad (j = 1, 2, \ldots, n),$$

* The convergence of the Galerkin method was first established in its general form by M. V. Keldysh [1]. A special case of the Galerkin method—namely, the Ritz method—and the method of moments were considered earlier in papers by N. M. Krylov and N. N. Bogolyubov (see Krylov, and also Mikhlin-II).

where $L(x)$ denotes the differential expression

$$L(x) = \frac{d^{2m}x}{dt^{2m}} - \lambda \sum_{s=1}^{2m} p_s \frac{d^{2m-s}x}{dt^{2m-s}}.$$

If here $\zeta_j = \omega_j$, then we obtain *Galerkin's method.*

5.3. Let us consider, in particular, the equation

$$x''(t) - \lambda[p_1(t)x'(t) + p_2(t)x(t)] = y(t) \qquad (|t| \leqslant 1) \tag{18}$$

subject to the conditions

$$x(-1) = x(1) = 0. \tag{19}$$

Here we take the functions ω_k to be

$$\omega_k(t) = t^{k-1}(1-t^2) \qquad (k = 1, 2, \ldots).$$

Equation (18) can be written in the form

$$Gx - \lambda Tx = y, \tag{20}$$

where $Gx = x''$. We take $\mathbf{X}$ to be the (incomplete) unitary space of twice continuously differentiable functions satisfying the boundary conditions (19), in which

$$\|x\|_{\mathbf{X}} = \left[\int_{-1}^{1} |x''(t)|^2\, dt\right]^{1/2},$$

$$(x_1, x_2)_{\mathbf{X}} = \int_{-1}^{1} x_1''(t)x_2''(t)\, dt.$$

We introduce a metric in $\mathbf{Y}$—which in this case consists of all the continuous functions—in such a way that the metric agrees, via G, with that of $\mathbf{X}$:

$$\|y\|_{\mathbf{Y}} = \|G^{-1}y\|_{\mathbf{X}} = \left[\int_{-1}^{1} |y(t)|^2 dt\right]^{1/2}, \qquad (y_1, y_2)_{\mathbf{Y}} = \int_{-1}^{1} y_1(t)y_2(t)\, dt.$$

We take $\tilde{\mathbf{X}}$ to be the subspace of functions of the form (17), and define $\tilde{\mathbf{Y}}$ to be $G(\tilde{\mathbf{X}})$, which, as one can easily verify, is the set of all polynomials of degree $n-1$. Finally, we take Φ to be the orthogonal projection operator from $\mathbf{Y}$ onto $\tilde{\mathbf{Y}}$. Thus, if $\zeta_1, \zeta_2, \ldots, \zeta_n$ is a system of linearly independent elements of $\tilde{\mathbf{Y}}$, the equation $\Phi y = \mathbf{0}$ is equivalent to the system of equations $(y, \zeta_j) = 0$ $(j = 1, 2, \ldots, n)$.

Under these hypotheses, the equation of the method of moments is equivalent to the functional equation

$$G\tilde{x} - \lambda\Phi T\tilde{x} = \Phi y, \tag{21}$$

so that the approximate equation arises in the special way described in 2.3. Hence condition Ib is satisfied with $\mu = 0$.

To verify condition IIb, we need to show that we can approximate an element of the form Tx by elements of $\tilde{\mathbf{Y}}$ We note first that the norm of a function in $\mathbf{X}$ yields a bound for both the maximum value of the function itself and that of its first derivative:

$$|x'(t)| = \frac{1}{2}\left|\int_{-1}^{1} [x'(t) - x'(\tau)]d\tau\right| = \frac{1}{2}\left|\int_{-1}^{1}\left[\int_{\tau}^{t} x''(u)du\right]d\tau\right| \leqslant$$

$$\leqslant \frac{1}{2} \int_{-1}^{1} \left[\int_{\tau}^{t} |x''(u)|^2 du \right]^{1/2} \left[\int_{\tau}^{t} du \right]^{1/2} d\tau \leqslant A\|x\|,$$

$$|x(t)| \leqslant \int_{-1}^{t} |x'(\tau)| d\tau \leqslant B\|x\|.$$

In this case, if we assume that the coefficients of (18) are continuously differentiable, we can write

$$\left\| \frac{d}{dt} Tx \right\|_{\mathbf{Y}} = \|p_1 x'' + (p_1 + p_2')x' + p_2 x\|_{\mathbf{Y}} \leqslant C\|x\|.$$

From this we easily see that Tx satisfies a Lipschitz condition of order 1/2:

$$|(Tx)(t_1) - (Tx)(t_2)| = \left| \int_{t_1}^{t_2} \frac{d}{dt} Tx\, dt \right| \leqslant \left\| \frac{d}{dt} Tx \right\|_{\mathbf{Y}} |t_2 - t_1|^{1/2},$$

and it therefore can be approximated by a polynomial $y \in \mathbf{Y}$ such that

$$\|Tx - \tilde{y}\| \leqslant \frac{D\|x\|}{\sqrt{n}}.$$

Thus we may assume that $\mu_1 = O(1/\sqrt{n})$.

Similar arguments show that $\mu_2 = O(1/\sqrt{n})$ (assuming that the derivative of the right-hand side is square summable).

If $p_1 = \mathbf{0}$, then we can estimate $\frac{d^2}{dt^2} Tx$ and derive the result that $\mu_1 = O(1/n\sqrt{n})$ in the same way.

An application of Theorem 2.3b gives us the order of magnitude of the rate at which the sequence of approximate solutions converge to the exact solution:

$$\|x^* - \tilde{x}_n^*\| = O\left(\frac{1}{\sqrt{n}}\right).$$

If, moreover, $p_1 = \mathbf{0}$ and y'' is square summable, then

$$\|x^* - \tilde{x}_n^*\| = O\left(\frac{1}{n\sqrt{n}}\right).$$

If p_1, p_2 and y have derivatives of higher orders, then one can establish a higher order of magnitude for the decrease in the error of approximate solutions, since in this case x^* is several times differentiable and so can be approximated by polynomials with more precision.

In every case, the approximate solutions converge to an exact solution in the space $\mathbf{X}$—that is, the approximate solutions themselves and their first derivatives converge *uniformly* to an exact solution and its first derivative, respectively. The second derivatives of the approximate solutions converge in mean to second derivative of the solution.

5.4. Now we consider equation (18) again, but with different definitions for the norms in $\mathbf{X}$ and $\mathbf{Y}$. Specifically, we set

$$\|x\|_{\mathbf{X}} = \max_{|t| \leqslant 1} |x''(t)|, \quad \|y\|_{\mathbf{Y}} = \max_{|t| \leqslant 1} |y(t)| \qquad (x \in \mathbf{X},\ y \in \mathbf{Y}).$$

These norms are evidently compatible, so $\|G\| = \|G^{-1}\| = 1$. We retain our previous definition of Φ, which enables us to regard the approximate equation as being of the special type for which $\mu = 0$. We now find a bound for the norm of Φ. Let $\zeta_1, \zeta_2, \ldots, \zeta_n$ be orthogonal polynomials with unit weights (Legendre polynomials). Then we can express the projection Φ from $\mathbf{Y}$ onto $\tilde{\mathbf{Y}}$ in the form

$$\Phi y = \sum_{k=1}^{n} (y, \zeta_k)\zeta_k.$$

However, using a well-known estimate for the partial sums of the expansion of a continuous function with respect to Legendre polynomials (see, e.g., Agakhanov and Natanson [1]), we have

$$\max_{|t| \leqslant 1} \left| \sum_{k=1}^{n} (y, \zeta_k)\zeta_k(t) \right| \leqslant A\sqrt{n},$$

so that

$$\|\Phi\| \leqslant A\sqrt{n}. \tag{22}$$

To approximate Tx by a polynomial—that is, by an element of $\tilde{\mathbf{Y}}$—we make use of the fact that

$$\left\| \frac{d}{dt} Tx \right\|_{\mathbf{Y}} \leqslant C\|x\|.$$

Hence we can find $\tilde{y} \in \tilde{\mathbf{Y}}$ such that

$$\|Tx - \tilde{y}\| \leqslant \frac{D\|x\|}{n}.$$

Therefore $\mu_1 = O(1/n)$ (if $p_1 = 0$, we can take $\mu_1 = O(1/n^2)$). Finally, we have $\mu_2 = \dfrac{E_n(y)}{\|y\|}$.

If we take (22) into account, Theorem 2.2b shows that

$$\|x^* - \tilde{x}_n^*\| = (\mu_1 + \mu_2)O(\sqrt{n}).$$

If $E_n(y)\sqrt{n} \to 0$, then both the approximating functions and their first and second derivatives converge uniformly to the values for the exact solution. If, in addition, $p_1 = 0$ and y is twice differentiable, then

$$\|x^* - \tilde{x}_n^*\| = O\left(\frac{1}{n\sqrt{n}}\right).$$

5.5. For our next application, we consider the equation

$$L(x) \equiv \frac{d}{dt}\left[p\frac{dx}{dt}\right] - \lambda\left\{\frac{d}{dt}[qx] + rx\right\} = y \tag{23}$$

subject to the conditions

$$x(0) = x(1) = 0. \tag{24}$$

We assume, moreover, that p and q are continuously differentiable, and that $p(t) > 0$. In addition, we suppose λ chosen so that the boundary value problem has a unique solution.

In Galerkin's method, we seek a solution of the form

$$\tilde{x}(t) = \sum_{k=1}^{n} c_k \omega_k(t) \qquad (0 \leqslant t \leqslant 1), \tag{25}$$

where ω_k $(k = 1, 2, \ldots)$ are continuously differentiable functions satisfying (24):

$$\omega_k(0) = \omega_k(1) = 0 \qquad (k = 1, 2, \ldots)$$

such that their derivatives, together with the function identically equal to 1, form a complete system. We assume this system $\{\omega_k'\}$ to be orthonormal with weight p:

$$\int_0^1 p(t)\,\omega_j'(t)\,\omega_k'(t)\,dt = \begin{cases} 0, & j \neq k, \\ 1, & j = k \end{cases} \qquad (j, k = 1, 2, \ldots). \tag{26}$$

We adjoin to this system the function $\omega_0'(t) = \dfrac{1}{p(t)} \left[\int_0^1 \dfrac{ds}{p(s)} \right]^{-\frac{1}{2}}$; then (26) will also be true for $j, k = 0$, and $\{\omega_k'\}$ $(k = 0, 1, \ldots)$ will be a complete orthonormal system.

We regard (23) and the conditions (24) as a single functional equation

$$Gx - \lambda Tx = y \tag{27}$$

in the space $\mathbf{X}$ consisting of twice continuously differentiable functions satisfying the boundary conditions (24). We define a norm on $\mathbf{X}$ by setting

$$\|x\| = \left[\int_0^1 p(t)|x'(t)|^2\,dt \right]^{1/2}, \qquad (x_1, x_2) = \int_0^1 p(t)x_1'(t)x_2'(t)\,dt.$$

We take $\mathbf{Y}$ to be the space of continuous functions and introduce a metric in it in such a way that it is compatible with that of $\mathbf{X}$:*

$$\|y\|_{\mathbf{Y}} = \|G^{-1}y\|_{\mathbf{X}}, \qquad (y_1, y_2)_{\mathbf{Y}} = (G^{-1}y_1, G^{-1}y_2)_{\mathbf{X}}.$$

We take $\tilde{\mathbf{X}}$ to be the set of elements of the form (25). We take $\tilde{\mathbf{Y}}$ to be the set of elements of the form

$$y = \sum_{k=1}^{n} c_k G\omega_k. \tag{28}$$

Finally, $\mathbf{\Phi}$ is defined to be the orthogonal projection operator from $\mathbf{Y}$ onto $\tilde{\mathbf{Y}}$, so that $\|\mathbf{\Phi}\| = 1$. The equation $\mathbf{\Phi}y = \mathbf{0}$ is equivalent to the condition that $(y, G\omega_k) = 0$ $(k = 1, 2, \ldots, n)$; and, in view of the definition of the inner product in $\mathbf{Y}$, this is equivalent to

$$(G^{-1}y, \omega_k)_{\mathbf{X}} = \int_0^1 p(t)\frac{d}{dt}(G^{-1}y)\omega_k'(t)\,dt = -\int_0^1 y(t)\omega_k(t)\,dt = 0.$$

The equation of the Galerkin method

$$\int_0^1 [L(x)(t) - y(t)]\omega_j(t)\,dt = 0 \qquad (j = 1, 2, \ldots, n)$$

* The element $x = G^{-1}y$ is the solution of the equation $\dfrac{d}{dt}\left(p\dfrac{dx}{dt}\right) = y$, satisfying the conditions (24).

thus amounts simply to

$$\Phi[Gx - \lambda Tx] = \Phi y,$$

in other words, the approximate equation constructed by the special method of (2.3).

In applying the results of § 2, we have $\mu = 0$. We now estimate μ_1, using the remark in 2.1 (inequality (6)); that is, we approximate $z = G^{-1}Tx$, where $\|x\| < 1$, by an element of $\tilde{\mathbf{X}}$, in the metric of $\mathbf{X}$. We have $Gz = Tx$, or

$$\frac{d}{dt}\left[p\frac{dz}{dt}\right] = \frac{d}{dt}[qx] + rx.$$

Integrating this equation from 0 to $u \leqslant 1$, we find that

$$p(u)z'(u) = q(u)x(u) + \int_0^u r(t)x(t)\,dt + C = q(u)\int_0^u x'(t)\,dt + \int_0^u \left(\int_t^u r(\tau)\,d\tau\right)x'(t)\,dt + C$$

$$= \int_0^1 h(u, t)x'(t)\,dt + C,$$

where

$$h(u, t) = \begin{cases} q(u) + \int_t^u r(\tau)\,d\tau & (t \leqslant u), \\ 0 & (t > u). \end{cases}$$

To eliminate C, we integrate the resulting equation, after first dividing both sides by $p(u)$:

$$0 = \int_0^1 \left[\int_0^1 \frac{1}{p(u)} h(t, u)\,du\right] x'(t)\,dt + C\int_0^1 \frac{du}{p(u)}.$$

Substituting the value thus obtained for C in the preceding equation, we obtain

$$z'(u) = \int_0^1 K(u, t)x'(t)\,dt,$$

where

$$K(u, t) = \frac{1}{p(u)} h(u, t) - \frac{1}{p(u)\int_0^1 \frac{ds}{p(s)}} \int_0^1 \frac{1}{p(s)} h(s, t)\,ds.$$

Notice that we clearly have

$$\int_0^1 K(u, t)\,du = 0 \qquad (0 \leqslant t \leqslant 1).$$

We can approximate $K(u, t)$ by a partial sum of its Fourier series relative to $\{\omega_j'(u)\}$ ($j = 1, 2, \ldots$):

$$K_n(u, t) = \sum_{j=1}^n \alpha_j(t)\omega_j'(u) \qquad \left(\alpha_j(t) = \int_0^1 p(u)K(u, t)\omega_j'(u)\,du\right),$$

since the kernel is orthogonal to $\dfrac{1}{p(u)}$ and the system $\{\omega_j'\}$ is complete. Hence

$$\int_0^1 p(u)[K(u,t)-K_n(u,t)]^2\,du \underset{n\to\infty}{\longrightarrow} 0 \qquad (0 \leqslant t \leqslant 1), \tag{29}$$

the left-hand side decreasing monotonically as n increases.

Determining $\tilde{x}$ from the equation

$$\tilde{x}'(u) = \int_0^1 K_n(u,t)x'(t)\,dt = \sum_{j=1}^n c_j\omega_j'(u) \qquad \left(c_j = \int_0^1 \alpha_j(t)x'(t)\,dt\right),$$

we obtain the required approximation. For we have

$$\|z-\tilde{x}\|^2 = \int_0^1 p(u)|z'(u)-\tilde{x}'(u)|^2\,du = \int_0^1 p(u)\left|\int_0^1 [K(u,t)-K_n(u,t)]x'(t)\,dt\right|^2 du \leqslant$$

$$\leqslant \int_0^1 p(t)|x'(t)|^2\,dt \int_0^1\left[\int_0^1 |K(u,t)-K_n(u,t)|^2 p(u)\,du\right]\frac{dt}{p(t)} = \eta_n^2\|x\|^2,$$

where

$$\eta_n^2 = \int_0^1 \frac{dt}{p(t)} \int_0^1 [K(u,t)-K_n(u,t)]^2 p(u)\,du \underset{n\to\infty}{\longrightarrow} 0.$$

In a similar way, we can estimate how small μ_2 is. We need to show that the element $z = G^{-1}y$ can be approximated by an element of $\tilde{\mathbf{X}}$. We have

$$\frac{d}{dt}\left(p\frac{dz}{dt}\right) = y, \qquad p\frac{dz}{dt} = \int y\,dt = F,$$

where F is chosen so that $\displaystyle\int_0^1 \frac{F(t)}{p(t)}\,dt = 0$, and we can now approximate $\dfrac{dz}{dt} = \dfrac{F}{p}$ by a sum of the form $\sum_{k=1}^n c_k\omega_k'$.

In the case where the ω_k are trigonometric functions, we can estimate the order of approximation. For suppose

$$\omega_k(t) = \sin k\pi t \qquad (k = 1, 2, \ldots).$$

We now need to approximate $K(u,t)$ by sums of the form

$$\sum_{k=1}^n \alpha_k(t)\cos k\pi u,$$

where the functions

$$\alpha_k(t) = 2\int_0^1 K(u,t)\cos k\pi u\,du,$$

being Fourier coefficients of a function of bounded variation, are of the order $1/k$, and the deviation satisfies

$$\int_0^1 [K(u,t) - K_n(u,t)]^2\,du = 2\sum_{k=n+1}^{\infty} |\alpha_k(t)|^2 = O\left(\frac{1}{n}\right).$$

Hence $\mu_1 = O(1/\sqrt{n})$. In the same way, one can easily show that $\mu_2 = O(1/\sqrt{n})$.

Thus Theorem 2.3b yields, in the present case, the estimate

$$\|x^* - \tilde{x}_n^*\| = O\left(\frac{1}{\sqrt{n}}\right),$$

that is, the approximate solutions converge to an exact solution in the metric of **X**, which ensures, in particular, that we have uniform convergence of the same rate.

If the differential expression $L(x)$ is self-adjoint—that is, if $q = 0$—then Galerkin's method reduces to Ritz's method. If we assume, in this case, that $p(u)K(u,t)$ has a first derivative in u, and that this derivative is a function of bounded variation (this will be so if p and r have the same property), then we find that the Fourier coefficients are of the order $1/k^2$, which gives the estimate

$$\mu_1 = O\left(\frac{1}{n\sqrt{n}}\right)$$

and a corresponding estimate for the rate of convergence of the approximate solutions to the exact solution.

§ 6. Applications to boundary-value problems for equations of elliptic type

6.1. We now consider some applications of the general theory of approximation methods to the approximate solution of boundary-value problems for certain equations of elliptic type. In particular, we consider the boundary-value problem for the equation

$$\Delta u + \lambda a u = v \tag{1}$$

in the region D bounded by a curve Γ, with the boundary condition

$$u|_\Gamma = 0 \tag{2}$$

for the case of m (= 2, 3) dimensions. For this it will be necessary to use, in addition to propositions from the general theory, some facts from mathematical physics and approximation theory, which we shall state without proof.

GYUNTER–KORN THEOREM.* *If u is a solution of the equation*

$$\Delta u = f, \qquad u|_\Gamma = 0, \tag{3}$$

where Γ is a smooth contour and the function f satisfies a Lipschitz condition of order β, with constant M (i.e. $f \in \mathrm{Lip}_M \beta$), then $u(x, y)$ has second partial derivatives satisfying a Lipschitz

* See Gyunter [1].

condition of order $\beta' < \beta$; *in fact*,

$$\frac{\partial^2 u}{\partial x^2}, \frac{\partial^2 u}{\partial x \partial y}, \frac{\partial^2 u}{\partial y^2} \in \mathrm{Lip}_{M'}\beta' \qquad (M' \leqslant CM, \qquad C = C(\beta, \beta')).$$

It is well known that, in the case of a function of one variable, various differential properties of the function are sufficient for the function and its derivatives to be simultaneously approximable by arbitrary algebraic or trigonometric polynomials. If, in addition, the function vanishes at the ends of an interval $[a, b]$, then one can arrange that the approximating functions also vanish at the ends of $[a, b]$; in particular, in the case of algebraic polynomials, one can arrange that they contain the factor $(t-a)(b-t)$.

Theorems of a similar type hold also for functions of several variables. We now state two such theorems: one establishes that the relevant system of functions is complete, and the other gives a characterization of the degree of precision of the approximation.

COMPLETENESS THEOREM.* *Suppose we are given a region* D, *bounded by the contour* Γ *defined by the equation*

$$\omega(x, y) = 0,$$

where ω *is a piecewise continuously differentiable function, and where*

$$\omega(x, y) > 0 \qquad ((x, y) \in D), \qquad \operatorname{grad} \omega(x, y) \neq 0 \qquad ((x, y) \in \Gamma). \tag{4}$$

Then the system of functions of the form

$$\tilde{u}(x, y) = \omega(x, y) P(x, y) \qquad (P \text{ being a polynomial})$$

is complete in the space $\mathring{\mathbf{W}}_2^{(1)}$ (*that is, in the space of functions belonging to* $\mathbf{W}_2^{(1)}$ *and vanishing on* Γ).

KHARRIK'S THEOREM.† *Suppose that the function* ω *satisfies the conditions of the Completeness Theorem, and is, in addition,* k *times differentiable, and that its* k-th *derivatives satisfy a Lipschitz condition. If* u *is* k *times differentiable and its* k-th *derivatives satisfy a Lipschitz condition of order* α, *then there exists a sequence of polynomials* P_n, *where the degree of* P_n *is at most* n, *such that the functions* ωP_n *approximate both* u *and its derivatives. More precisely,*

$$\|u - \omega P_n\|_{C^{(r)}} = O\left(\frac{1}{n^{k+\alpha-r}}\right), \tag{5}$$

where

$$\|u\|_{C^{(r)}} = \sum_{i=0}^{r} \sum_{\nu_1+\nu_2=i} \max_D \left|\frac{\partial^i u}{\partial x^{\nu_1} \partial y^{\nu_2}}\right|. \tag{6}$$

6.2. Now we consider the boundary-value problem posed at the beginning of the section. We seek an approximate solution of the form

$$\tilde{u}(x, y) = \omega(x, y) P(x, y), \tag{7}$$

where P is a polynomial of degree at most n, and ω satisfies the conditions of Kharrik's Theorem for $k = 1$.

---

* See Kantorovich and Krylov.

† See Kharrik [2].

The N coefficients of P are determined from the system of equations

$$\int_D\int (\Delta\tilde{u} + \lambda a\tilde{u})\zeta_k\,dx\,dy = \int_D\int v\zeta_k\,dx\,dy \qquad (k = 1, 2, \ldots, N) \tag{8}$$

(we shall be saying something about the choice of the ζ_k below).

We introduce the space $\mathbf{U}$ of twice differentiable functions u vanishing on Γ, with metric and inner product defined by

$$\|u\| = \left[\int_D\int |\Delta u|^2\,dx\,dy\right]^{1/2}, \qquad (u_1, u_2) = \int_D\int \Delta u_1\,\Delta u_2\,dx\,dy.$$

We take the subspace $\tilde{\mathbf{U}}$ to be the set of functions of the form $\tilde{u}$.

We take the space $\mathbf{V}$ to consist of functions of the form $v = \Delta u$ ($u \in \mathbf{U}$)—in other words, those v for which the equation $\Delta u = v$ has a solution belonging to $\mathbf{U}$. We take the metric in $\mathbf{V}$ to be that of the Hilbert space $\mathbf{L}^2(D)$. Finally, $\tilde{\mathbf{V}}$ is the set of functions of the form $\tilde{v} = \Delta\tilde{u}$ ($\tilde{u} \in \tilde{\mathbf{U}}$).

With these definitions, our boundary-value problem can be regarded as a functional equation of the form

$$Gu + \lambda Tu = v \qquad (Gu = \Delta u, Tu = au). \tag{9}$$

Here the metrics of $\mathbf{U}$ and $\mathbf{V}$ are compatible, so G determines an isometric mapping from $\mathbf{U}$ onto $\mathbf{V}$.

For $\zeta_1, \zeta_2, \ldots, \zeta_N$ we can take any linearly independent system of functions in $\tilde{\mathbf{V}}$, for example,

$$\zeta_{ij} = \Delta[\omega(x, y)x^i y^j] \qquad (i + j \leqslant n).$$

If Φ is taken to be the orthogonal projection of $\mathbf{V}$ onto $\tilde{\mathbf{V}}$, then the equation $\Phi v = \mathbf{0}$ is equivalent to the system of equations

$$(v, \zeta_k) = \int_D\int v\zeta_k\,dx\,dy = 0 \qquad (k = 1, 2, \ldots, n).$$

Hence the system (8) can be expressed, in the new notation, as

$$\Phi G\tilde{u} - \Phi T\tilde{u} = \Phi v, \tag{10}$$

so that the approximate equation is of the special type considered in 2.3, and therefore condition Ib is satisfied with $\mu = 0$.

To verify condition IIb, we show that, if $u \in \mathbf{U}$, then $Tu = au \in \operatorname{Lip}\beta$, where $\beta < 1$ in the case $m = 2$ and $\beta < \frac{1}{2}$ if $m = 3$.

In fact, since $u \in \mathbf{U}$, we have $u \in \mathbf{W}_2^{(2)}$. By the Remark following the Embedding Theorem (XI.4.4), we have $u \in \operatorname{Lip}\beta$, where

$$\beta < 1 \qquad (m = 2), \qquad \beta < \frac{1}{2} \qquad (m = 3).$$

Hence it follows from the Gyunter–Korn Theorem that $z = G^{-1}Tu = \Delta^{-1}au$ has second derivatives belonging to $\operatorname{Lip}\beta'$, where $\beta' < \beta$; that is, we again have $\beta' < 1$ for $m = 2$ and $\beta' < \frac{1}{2}$ for $m = 3$. Moreover,

$$\left\|\frac{\partial^2 z}{\partial x^2}\right\|_{\operatorname{Lip}\beta'} \leqslant C\,\|u\|.$$

Now, by Kharrik's Theorem, z and its second derivatives can be approximated by a function of the form $\tilde{u}$, of the order $1/n^{\beta'}$; and this gives us the bound

$$\|G^{-1}Tu-\tilde{u}\| \leqslant \frac{C_1}{n^{\beta'}}.$$

Thus condition IIb is satisfied with $\mu = 0(1/n^{\beta'})$.

In precisely the same way, if v belongs to Lip β—in particular, if $v \in \mathbf{W}_2^{(2)}$—then the same argument shows that $\mu_2 = O(1/n^{\beta'})$. By what we have said, we conclude that the method of moments is convergent for base functions of the given form, the convergence being in the metric of $\mathbf{U}$:

$$\|u^*-\tilde{u}_n^*\|_{\mathrm{U}} = O\left(\frac{1}{n^{\beta'}}\right).$$

From this it is easy to see that the functions are uniformly convergent. For if we express u in terms of a Green's function, we have

$$|u^*(x,y)-\tilde{u}_n^*(x,y)| = \left|\int_D\int g(x,y)\Delta(u^*-\tilde{u}_n^*)\,dx\,dy\right|$$

$$\leqslant \left[\int_D\int |g(x,y)|^2\,dx\,dy\right]^{1/2}\|u^*-\tilde{u}_n^*\| = O\left(\frac{1}{n^{\beta'}}\right).$$

We note that a better rate of convergence can be established if we know further data on the differential properties of the solution.

6.3. One can establish the actual convergence of Galerkin's method, without any information on the precise rate of convergence, under significantly more general conditions than those postulated in 6.2 for the method of moments.

Consider the boundary-value problem for the equation

$$\Delta u + \lambda\left(au + b\frac{\partial u}{\partial x} + c\frac{\partial u}{\partial y}\right) = v \tag{11}$$

subject to the condition

$$u|_\Gamma = 0. \tag{12}$$

We assume that the coefficients a, b, c are continuously differentiable functions.

We seek a solution of the same form as in 6.2 above. The coefficients are determined from the system for the Galerkin method,

$$\int_D\int\left[\Delta\tilde{u} + \lambda\left(a\tilde{u} + b\frac{\partial\tilde{u}}{\partial x} + c\frac{\partial\tilde{u}}{\partial y}\right)\right]\zeta_k\,dx\,dy = \int_D\int v\zeta_k\,dx\,dy$$

$$(k = 1, 2, \ldots, N).$$

This time the functions ζ_k have the form $\zeta_k = \omega(x,y)\,P_k(x,y)$, where the $P_k(x,y)$ are polynomials of degree at most n. The metric in $\mathbf{U}$ is now introduced in a different way:

$$(u_1, u_2) = -\int_D\int u_1\Delta u_2\,dx\,dy = \int_D\int\left[\frac{\partial u_1}{\partial x}\frac{\partial u_2}{\partial x} + \frac{\partial u_1}{\partial y}\frac{\partial u_2}{\partial y}\right]dx\,dy,$$

$$\|u\|^2 = -\iint_D u\Delta u\,dx\,dy = \iint_D \left[\left(\frac{\partial u}{\partial x}\right)^2 + \left(\frac{\partial u}{\partial y}\right)^2\right] dx\,dy.$$

We introduce a metric in $\mathbf{V}$ by a corresponding method:

$$(v_1, v_2)_{\mathbf{V}} = (G^{-1}v_1, G^{-1}v_2)_{\mathbf{U}}, \qquad \|v\|_{\mathbf{V}} = \|G^{-1}v\|_{\mathbf{U}},$$

thus if u_1 and u_2 are solutions of $\Delta u = v_1$ and $\Delta u = v_2$ belonging to $\mathbf{U}$ then we set

$$(v_1, v_2) = -\iint_D u_1 v_2\,dx\,dy = -\iint_D u_2 v_1\,dx\,dy,$$

$$\|v_1\|^2 = -\iint_D u_1 v_1\,dx\,dy.$$

As before we can write our boundary-value problem as a single functional equation in $\mathbf{U}$:

$$Gu + \lambda Tu = v \qquad \left(Gu = \Delta u, \quad Tu = au + b\frac{\partial u}{\partial x} + c\frac{\partial u}{\partial y}\right)$$

and the approximate equation takes the form

$$G\tilde{u} + \lambda\Phi T\tilde{u} = \Phi v.$$

With a view to using Theorem 1.6, we verify that the operator $G^{-1}T$ is compact.

It is easy to see that $\mathbf{U}$ is linearly isometric to a dense subset of $\mathring{\mathbf{W}}_2^{(1)}$. To see this, one need only take as one of the possible metrics for $\mathbf{W}_2^{(1)}$ the metric determined by the functional $\int_\Gamma u(x, y)\,d\Gamma$,

$$\|u\|_{\mathbf{W}_2^{(1)}} = \left|\int_\Gamma u(x, y)\,d\Gamma\right| + \left\{\iint_D |\operatorname{grad} u|^2\,dx\,dy\right\}^{1/2} =$$

$$= \left\{\iint_D \left[\left(\frac{\partial u}{\partial x}\right)^2 + \left(\frac{\partial u}{\partial y}\right)^2\right] dx\,dy\right\}^{1/2} = \|u\|_{\mathbf{U}},$$

so that T can be viewed as a continuous linear operator mapping $\mathring{\mathbf{W}}_2^{(1)}$ into $\mathbf{L}^2$. By the Bernstein–Ladyzhenskaya inequality, the inverse operator Δ^{-1} is continuous, as an operator from $\mathbf{L}^2$ into $\mathring{\mathbf{W}}_2^{(2)}$ (see Ladyzhenskaya, Ch. II, § 6). Hence $\Delta^{-1}T \in B(\mathring{\mathbf{W}}_2^{(1)}, \mathring{\mathbf{W}}_2^{(2)})$. The operator embedding $\mathring{\mathbf{W}}_2^{(2)}$ in $\mathring{\mathbf{W}}_2^{(1)}$ is compact (XI.3.5), so $\Delta^{-1}T$ is also compact, as an operator from $\mathring{\mathbf{W}}_2^{(1)}$ into $\mathring{\mathbf{W}}_2^{(1)}$.

By the Completeness Theorem, the sequence of projections $P_n = G^{-1}\Phi G$ converges to the identity operator on $\mathbf{U}$. Hence we can apply Theorem 1.6, which shows that Galerkin's method is convergent for the space $\mathbf{U}$. We can also deduce from this that the approximate solutions converge uniformly to an exact solution.

Notice that the arguments we have given can be used, without any additional considerations, for cases where the number of independent variables for the problem is more than two.

In conclusion, we mention that other interesting applications of the methods developed in this chapter may be found in a number of works: see, in particular, Kalandiya [1], Vladimirov [1], Filippov and Ryaben'kii.

XV

THE METHOD OF STEEPEST DESCENT

THE METHOD of steepest descent is now one of the most widely used methods of solving unconditional extremal problems. It was originally regarded as a variational method for solving linear functional equations and determining eigenvalues of linear operators. As with every variational method, the problem of solving an equation (or determining an eigenvalue) is reduced to a problem of determining an extremal value of a functional of a special type, defined on the whole space. It turned out, however, that the method was suitable for minimizing functionals of a far more general type. After a few steps it leads one into a reasonably small neighbourhood of a stationary point (of a minimum, if the functional is convex).

The plan behind the method is as follows: one constructs a sequence of approximations to the minimum, in such a way that the route from each approximation to the next is along the direction in which the given functional decreases most rapidly. This, of course, is the origin of the method's name.

The idea of the method of steepest descent goes back as far as Cauchy, who considered it for a finite-dimensional space. The method was developed in the general case for quadratic functionals by L. V. Kantorovich (see Kantorovich [7], [9]). The results of M. Sh. Birman [1], [2] were used in proving the general theorems that appear below (§§ 1 and 2). For some variations to the method of steepest descent and applications of the method to various problems, see Krasnosel'skii *et al.*; Lyubich and Maistrovskii [1]. The application of the method of steepest descent to the solution of elliptic differential equations (§ 3) was developed in L. Tsakh's thesis (1955). In §4 we use an approach that was earlier applied to the study of the conditional gradient method: see Dem'yanov and Rubinov.

§ 1. The solution of linear equations

1.1. Let Φ be a (non-linear) real functional, defined on a (real or complex) normed space $\mathbf{X}$. We assume that Φ is bounded below on $\mathbf{X}$ and pose the problem of finding an element $x^* \in \mathbf{X}$ (if one exists) for which Φ attains a least value:

$$\Phi(x) \geqslant \Phi(x^*) \qquad (x \in \mathbf{X}).$$

To solve this problem one usually proceeds as follows: somehow or other, one constructs a sequence $\{x_n\}$ that "minimizes" Φ, in the sense that

$$\lim_{n\to\infty} \Phi(x_n) = \inf_{x\in\mathbf{X}} \Phi(x).$$

In certain cases, one can construct this sequence $\{x_n\}$ so that it converges to an element x^*. If Φ is assumed continuous, then this element will be a solution to the problem.

The method of steepest descent is a method for constructing a sequence $\{x_n\}$ for minimizing a fairly wide class of functionals. It can be formulated for functionals having a directional derivative. Let us now define this concept.

Let x be a fixed element of $\mathbf{X}$. Consider the ray emanating from x in the direction z: that is, the set of elements of the form $x + \alpha z$, where α is a non-negative real number and $z \neq \mathbf{0}$ is an element of $\mathbf{X}$ that specifies the direction of the ray.* The restriction of Φ to this ray is a function of a real variable; we shall denote it by $\phi(\alpha; x, z)$ in what follows. Thus, to be more precise, we have

$$\phi(\alpha; x, z) = \Phi(x + \alpha z) \qquad (\alpha \geqslant 0).$$

The *derivative* of the functional Φ in the direction of z (at the point x) is the name naturally given to the expression

$$\frac{\partial \Phi}{\partial z}(x) = \frac{1}{\|z\|}\,\phi'(\alpha; x, z)\Big|_{\alpha = 0} = \frac{1}{\|z\|}\lim_{h \to 0^+}\frac{\Phi(x + hz) - \Phi(x)}{h}.$$

We assume that the derivative $\dfrac{\partial \Phi}{\partial z}(x)$ exists in every direction, and, in addition, that there exists a direction for which this derivative takes a minimum value. We call such a direction the direction of steepest descent of Φ at the point x.

Now we describe the method of steepest descent. Suppose Φ is differentiable in every direction, at each point x, and that a direction of steepest descent exists at every $x \in \mathbf{X}$. Choose an element $x_0 \in \mathbf{X}$ arbitrarily (the zeroth approximation to the minimum). Assume that we have already found x_k. For the next approximation x_{k+1} we naturally seek a point of the form

$$x_{k+1} = x_k + \varepsilon_{k+1} z_{k+1}.$$

Here z_{k+1} is the direction of steepest descent at x_k. The numerical parameter ε_{k+1} is called the descent value. It can be found in various ways. We give just three of these.

1) Assume that $\dfrac{\partial \Phi}{\partial z_{k+1}}(x) < 0$ and that $\dfrac{\partial \Phi}{\partial z}(x_k + \alpha z_{k+1})$ is continuous as a function of α. Then the function $\phi(\alpha; x_k, z_{k+1})$ is decreasing at least in the interval $[0, \alpha'_k]$, where α'_k is the smallest positive root of the equation $\phi'(\alpha; x_k, z_{k+1}) = 0$. This root α'_k is taken as the descent value ε_{k+1}.

2) Assume that $\phi(\alpha; x_k, z_{k+1})$ attains a minimum on the positive semi-axis at some point. Then this point is taken as the descent value.

When Φ is a strictly convex functional, these methods coincide (we consider this in more detail in 4.4).

3) We specify a sequence of positive numbers $\{\varepsilon_k\}$ such that $\varepsilon_k \to 0$ and $\sum\limits_{k=0}^{\infty} \varepsilon_k = \infty$ (assume that the z_k are normalized: that is, $\|z_k\| = 1$). We take the descent value at the k-th step to be ε_k.

Having chosen a definite method for determining the descent value, we obtain a sequence $\{x_k\}$, which in many cases turns out to be a minimizing sequence. This construction in fact constitutes the main element in the method of steepest descent.

An analogous plan can be used when one is determining the least value of a function that is bounded above. Here, instead of the direction of steepest descent, one has to use the direction of steepest ascent.

---

* Two non-zero elements z_1 and z_2 specify the same direction if and only if $z_1 = \lambda z_2$, for some $\lambda > 0$, or—what is the same thing—z_1 and z_2 lie on a ray emanating from the origin. Usually a direction is defined by specifying some vector on such a ray.

1.2. The connection between the problem of minimizing a functional and that of solving a linear functional equation is established as follows.

Let U be a self-adjoint operator on Hilbert space $\mathbf{H}$. Denoting the bounds of U, as always, by m and M (see IX.4.3), we assume that $m > 0$ and consider the equation

$$Ux = y. \tag{1}$$

Since $\lambda = 0$ does not belong to the spectrum of U, the inverse operator U^{-1} exists (Theorem IX.5.3), so (1) has a unique solution $x^* = U^{-1}y$, whatever element $y \in \mathbf{H}$ we choose.

Now we form the functional

$$F(x) = (Ux, x) - [(x, y) + (y, x)] \qquad (x \in \mathbf{H}). \tag{2}$$

THEOREM 1. *A solution x^* of (1) yields a minimum of (2). Conversely, if (2) attains a minimum at $\tilde{x}$, then $\tilde{x}$ is a solution of (1): that is, $\tilde{x} = x^*$.*

Proof. Since $y = Ux^*$, we can express F in the form

$$F(x) = (Ux, x) - (x, Ux^*) - (Ux^*, x) = (U(x - x^*), x - x^*) - (Ux^*, x^*). \tag{3}$$

Hence

$$F(x) \geqslant m(x - x^*, x - x^*) - (Ux^*, x^*) \geqslant -(Ux^*, x^*) = F(x^*),$$

that is, x^* minimizes the functional F.

The second part of the theorem follows from the same inequality, for

$$0 = F(\tilde{x}) - F(x^*) = (U(\tilde{x} - x^*), \tilde{x} - x^*) \geqslant m(\tilde{x} - x^*, \tilde{x} - x^*)$$

and therefore $\tilde{x} = x^*$.

REMARK. If the existence of the solution x^* is known, then it is clear from the proof of the theorem that the condition $m > 0$ can be replaced by a weaker condition, in which we require only that $U(x, x) > 0$ for all $x \neq \mathbf{0}$.

1.3. The application of the method of steepest descent to the functional (2) leads to a sequence converging to x^*, a solution to (1), so one can regard it as a method for the approximate solution of (1).

Let us clarify the plan of the method of steepest descent as it applies to the functional (2). If $x, z \in \mathbf{H}$, then

$$\begin{aligned} F(x+z) &= (U(x+z), x+z) - [(x+z, y) + (y, x+z)] = \\ &= (Ux, x) - [(x, y) + (y, x)] + [(Ux - y, z) + (z, Ux - y)] + (Uz, z) = \\ &= F(x) + [(Ux - y, z) + (z, Ux - y)] + (Uz, z). \end{aligned} \tag{4}$$

Hence

$$\frac{\partial F}{\partial z}(x) = \frac{1}{\|z\|}[(Ux - y, z) + (z, Ux - y)] = \frac{2\,\mathrm{Re}(Ux - y, z)}{\|z\|}.$$

Thus the direction of steepest descent at x_0 is given by the element $(-z_1)$, where

$$z_1 = Ux_0 - y.$$

To determine ε, we form the equation

$$\phi'(\alpha; x_0, -z_1) = 0,$$

where, in view of (4), we have

$$\phi(\alpha; x_0, -z_1) = F(x_0 - \alpha z_1) = F(x_0) - 2\alpha(z_1, z_1) + \alpha^2(Uz_1, z_1).$$

Hence

$$\varepsilon_1 = \frac{(z_1, z_1)}{(Uz_1, z_1)}.$$

Finally,

$$x_1 = x_0 - \varepsilon_1 z_1.$$

In exactly the same way, we have

$$x_2 = x_1 - \varepsilon_2 z_2 \quad \left(z_2 = Ux_1 - y,\ \varepsilon_2 = \frac{(z_2, z_2)}{(Uz_2, z_2)}\right)$$

and, in general,

$$x_n = x_{n-1} - \varepsilon_n z_n \qquad (n = 1, 2, \ldots),$$

where

$$z_n = Ux_{n-1} - y, \quad \varepsilon_n = \frac{(z_n, z_n)}{(Uz_n, z_n)} \qquad (n = 1, 2, \ldots).$$

The sequence $\{x_n\}$ thus constructed is a minimizing sequence. Furthermore, we have

THEOREM 2. *The sequence $\{x_n\}$ converges to x^*. Its speed of convergence is given by the inequality*

$$\|x_n - x^*\| \leq C\left(\frac{M-m}{M+m}\right)^n \quad \left(n = 0, 1, \ldots;\ C = \frac{\|z_1\|}{m}\right). \tag{5}$$

Proof. We transform equation (1), writing it in the form

$$x = x - kUx + ky. \tag{6}$$

We choose the numerical factor $k > 0$ so that the operator

$$T = I - kU$$

has the smallest possible norm. Since T has upper bound $1 - km$ and lower bound $1 - km$, the minimum norm will occur if

$$1 - km = -(1 - kM),$$

so that

$$k = \frac{2}{M+m}.$$

Then

$$\|T\| = 1 - km = kM - 1 = \frac{M-m}{M+m}. \tag{7}$$

Let us write

$$\tilde{x}_1 = Tx_0 + ky = x_0 - k(Ux_0 - y) = x_0 - kz_1. \tag{8}$$

Now we introduce the operator $V = U^{1/2}$ (see Theorem V.6.2). Using (3), we can rewrite F in the form

$$\begin{aligned} F(x) &= (U(x - x^*), x - x^*) - (Ux^*, x^*) \\ &= (V(x - x^*), V(x - x^*)) - (Vx^*, Vx^*) \\ &= \|V(x - x^*)\|^2 - \|Vx^*\|^2. \end{aligned} \tag{9}$$

However, in view of the obvious inequality $F(\tilde{x}_1) \geqslant F(x_1)$, we have

$$F(x_1) - F(x^*) \leqslant F(\tilde{x}_1) - F(x^*),$$

which we can rewrite, using (9), in the form

$$\| V(x_1 - x^*)\| \leqslant \| V(\tilde{x}_1 - x^*)\|. \tag{10}$$

Equation (6) is equivalent to equation (1). Hence

$$x^* = Tx^* + ky.$$

Subtracting this from (8), we obtain

$$\tilde{x}_1 - x^* = T(x_0 - x^*)$$

and hence

$$V(\tilde{x}_1 - x^*) = TV(x_0 - x^*),$$

so that

$$\| V(\tilde{x}_1 - x^*)\| \leqslant \|T\| \, \| V(x_0 - x^*)\| = \frac{M-m}{M+m} \| V(x_0 - x^*)\|.$$

Comparing this with (10), we have

$$\| V(x_1 - x^*)\| \leqslant \frac{M-m}{M+m} \| V(x_0 - x^*)\|. \tag{11}$$

Similar arguments for $n = 1, 2, \ldots$ lead to the inequalities

$$\| V(x_n - x^*)\| \leqslant \frac{M-m}{M+m} \| V(x_{n-1} - x^*)\| \qquad (n = 1, 2, \ldots).$$

Finally, we obtain

$$\| V(x_n - x^*)\| \leqslant \left(\frac{M-m}{M+m}\right)^n \| V(x_0 - x^*)\| \qquad (n = 1, 2, \ldots). \tag{12}$$

As the function $t^{-1/2}$ is continuous in $[m, M]$, and so certainly on the spectrum of U, the inverse operator V^{-1} exists. Moreover (see Theorem IX.5.2),

$$\|V^{-1}\| = \max_{t \in S_U} \frac{1}{\sqrt{t}} = \frac{1}{\sqrt{m}}.$$

Notice also that

$$\|V\| = \max_{t \in S_U} \sqrt{t} = \sqrt{M}.$$

By what has been said, we find, using (12), that

$$\|x_n - x^*\| = \|V^{-1}V(x_n - x^*)\| \leqslant \|V^{-1}\| \left(\frac{M-m}{M+m}\right)^n \| V(x_0 - x^*)\| \leqslant$$

$$\leqslant \|V^{-1}\| \, \|V\| \left(\frac{M-m}{M+m}\right)^n \|x_0 - x^*\| =$$

$$= \sqrt{\frac{M}{m}} \left(\frac{M-m}{M+m}\right)^n \|x_0 - x^*\|.$$

This bound is awkward, since the unknown element x^* appears on the right-hand side. To get round this we note that

$$\|V(x_0-x^*)\| = \|V^{-1}U(x_0-x^*)\| \leqslant \|V^{-1}\|\,\|Ux_0-y\| = \frac{\|z_1\|}{\sqrt{m}}.$$

This enables us to write

$$\|x_n-x^*\| \leqslant \frac{\|z_1\|}{m}\left(\frac{M-m}{M+m}\right)^n \qquad (n=1, 2, \ldots),$$

which is the same as (5).

This proves the theorem.

1.4. Let us express the p-th approximation x_p in terms of x_0 and z_1. We have

$$x_2 = x_1-\varepsilon_2 z_2 = x_1-\varepsilon_2(Ux_1-y) = x_0-\varepsilon_1 z_1-\varepsilon_2(Ux_0-y-\varepsilon_1 Uz_1) =$$
$$= x_0-(\varepsilon_1+\varepsilon_2)z_1+\varepsilon_1\varepsilon_2 Uz_1 = x_0+\lambda_1^{(2)}z_1+\lambda_2^{(2)}Uz_1.$$

It can be shown without difficulty by induction that

$$x_p = x_0+\lambda_1^{(p)}z_1+\ldots+\lambda_p^{(p)}U^{p-1}z_1. \tag{13}$$

We now consider the element

$$x_{(1)} = x_0+\lambda_1 z_1+\ldots+\lambda_p U^{p-1}z_1$$

and determine the complex coefficients $\lambda_1, \lambda_2, \ldots, \lambda_p$ from the condition that $F(x_{(1)})$ be minimal. Since

$$F(x_{(1)}) = \left(Ux_0+\sum_{j=1}^{p}\lambda_j U^j z_1,\ x_0+\sum_{j=1}^{p}\lambda_j U^{j-1}z_1\right)-$$
$$-[(x_{(1)}, y)+(y, x_{(1)})] = F(x_0)+\sum_{j=1}^{p}(\lambda_j+\overline{\lambda}_j)(U^{j-1}z_1, z_2)+$$
$$+\sum_{j=1}^{p}\sum_{k=1}^{p}\lambda_j\overline{\lambda}_k(U^j z_1, U^{k-1}z_1),$$

we have, writing $\lambda_k = \sigma_k+\tau_k i$ $(k=1, 2, \ldots, p)$, the following two systems of equations for defining the λ_k:

$$\frac{\partial F(x_{(1)})}{\partial\sigma_i} = 0, \quad \frac{\partial F(x_{(1)})}{\partial\tau_j} = 0 \qquad (j=1, 2, \ldots, p),$$

that is, in explicit form,

$$(U^{j-1}z_1, z_1)+\sum_{k=1}^{p}\sigma_k(U^j z_1, U^{k-1}z_1) = 0, \quad \sum_{k=1}^{p}\tau_k(U^j z_1, U^{k-1}z_1) = 0.$$

If we multiply the second equation by i, add it to the first and bear in mind that $(U^j z_1, U^{k-1}z_1) = (U^{j+k-1}z_1, z_1)$, we can combine these systems into one:

$$(U^{j-1}z_1, z_2)+\sum_{k=1}^{p}\lambda_k(U^{j+k-1}z_1, z_1) = 0 \qquad (j=1, 2, \ldots, p). \tag{14}$$

Since the minimum we are seeking obviously exists, this system has a solution.*

---

* However, this solution need not be unique. It should be noted, however, that the value of $F(x_{(1)})$ does not depend on the choice of the solution.

Starting from $x_{(1)}$, we construct in exactly the same way an element $x_{(2)}$, and so on, yielding a sequence $x_{(0)} = x_0, x_{(1)}, \ldots$, corresponding to the *p-step variant* of the method of steepest descent.

It is clear from the way the sequence in the p-step process was constructed that it converges to x^* at least p times faster than $\{x_n\}$. For it is easy to verify, using the proof of Theorem 2, that

$$\|x_{(n)} - x^*\| \leqslant \frac{\|z_1\|}{m} \left[\left(\frac{M-m}{M+m}\right)^p\right]^n. \tag{15}$$

However, as one should expect, the convergence is in fact even faster.

THEOREM 3 (Birman). *We have the inequality*

$$\|x_{(n)} - x^*\| \leqslant \frac{\|z_1\|}{m} \left[\frac{1}{\vartheta_p\left(\dfrac{M+m}{M-m}\right)}\right]^n \qquad (n = 1, 2, \ldots), \tag{16}$$

where $\vartheta_p(t)$ *is the Chebyshev polynomial of degree p:*

$$\vartheta_p(t) = \cos p \arccos t \qquad (\textit{for } t \in [-1, 1]).$$

Proof. Consider the equation

$$x = Tx + \tilde{y} \qquad (T = I - U\phi(U), \tilde{y} = \phi(U)y), \tag{17}$$

where $\phi(t)$ is a polynomial of degree $p-1$.

Write

$$\tilde{x}_1 = Tx_0 + \tilde{y} = x_0 - \phi(U)(Ux_0 - y) = x_0 - \phi(U)z_1 = x_0 + \sum_{k=1}^{p} c_k U^{k-1} z_1.$$

Clearly $F(\tilde{x}_1) \geqslant F(x_{(1)})$, so if we repeat the arguments of Theorem 2, we find that

$$\|x_{(n)} - x^*\| \leqslant \frac{\|z_1\|}{m} \|T\|^n \qquad (n = 1, 2, \ldots). \tag{18}$$

Since

$$\|T\| \leqslant \max_{t \in [m, M]} |1 - t\phi(t)|,$$

it is natural to define the polynomial $\phi(t)$ by the condition that $\max\limits_{t \in [m, M]} |1 - t\phi(t)|$ be minimal. If we write $\psi(t) = 1 - t\phi(t)$, then $\psi(t)$ is the polynomial of degree p deviating least from zero in the interval $[m, M]$, with $\psi(0) = 1$. As is well known,*

$$\psi(t) = \frac{\vartheta_p\left(\dfrac{2t - M - m}{M - m}\right)}{\vartheta_p\left(\dfrac{M+m}{M-m}\right)}.$$

Since we have

$$\max_{t \in [m, M]} \left|\vartheta_p\left(\frac{2t - M - m}{M - m}\right)\right| = \max_{t \in [-1, 1]} |\vartheta_p(t)| = 1,$$

* See Goncharov [1], p. 230.

$\|T\|$ satisfies the inequality

$$\|T\| \leqslant \max_{t \in [m, M]} |\psi(t)| = \frac{1}{\vartheta_p\left(\dfrac{M+m}{M-m}\right)},$$

and, by (18), this proves the theorem.

REMARK 1. Let $\rho(t)$ be an arbitrary polynomial of degree p. Then we have the inequality (Natanson-I)

$$|\rho(t_0)| \leqslant |\vartheta_p(t_0)| \max_{t \in [-1, 1]} |\rho(t)| \qquad (|t_0| > 1).$$

If here we take $\rho(t) = t^p$, $t_0 = \dfrac{M+m}{M-m}$, we have

$$\frac{1}{\vartheta_p\left(\dfrac{M+m}{M-m}\right)} \leqslant \left(\frac{M-m}{M+m}\right)^p.$$

Hence (15) is a consequence of the inequality (16).

REMARK 2. Since

$$\frac{\vartheta_p(t)}{t^p} \underset{t \to \infty}{\to} \frac{1}{2^{p-1}},$$

it follows that, if $\dfrac{M+m}{M-m}$ is sufficiently large, the bound in (16) differs from that in (15) by a factor close to $\left(\dfrac{1}{2^{p-1}}\right)^n$.

In a paper by Samokish [1], a more precise bound than that in (16) is obtained, under certain additional assumptions concerning the spectrum of U.

REMARK 3. When $m = 0$, it can be shown that, if a solution (not necessarily unique) to (1) exists, then the method of steepest descent converges to the solution x^* closest to x_0 (Fridman [1]). Moreover, we have $F(x_n) - F(x^*) = O(1/n)$; however, the rate of convergence of x_n to x^* can be arbitrarily slow.

§ 2. Determination of the eigenvalues of compact operators

2.1. We now consider a compact self-adjoint operator U. We assume, without losing any generality, that $M > |m| \geqslant 0$ (see IX.4.3). Then

$$M = \sup_{\|x\| = 1} (Ux, x) = \sup_{x \neq 0} \frac{(Ux, x)}{(x, x)} = \frac{(Ux^*, x^*)}{(x^*, x^*)},$$

where x^* is an eigenvector corresponding to the largest eigenvalue $\lambda_1 = M$. Hence λ_1 is the largest value of the functional

$$L(x) = \frac{(Ux, x)}{(x, x)},$$

and, by IX.4.3, every element for which the maximum is attained is an eigenvector.

To determine the maximum of L, we apply the method of steepest descent.* Choose any

* Strictly speaking, the method of "steepest ascent".

normalized element $x_0 \in \mathbf{H}$. As

$$L(x_0 + \alpha z) = \frac{(Ux_0, x_0) + \alpha(Ux_0, z) + \alpha(z, Ux_0) + \alpha^2(Uz, z)}{1 + \alpha(x_0, z) + \alpha(z, x_0) + \alpha^2(z, z)}, \tag{1}$$

we have

$$\frac{\partial L}{\partial z}(x_0) = \frac{[(Ux_0, z) + (z, Ux_0)] - [(x_0, z) + (z, x_0)](Ux_0, x_0)}{\|z\|} =$$

$$= \frac{(Ux_0 - \mu_0 x_0, z) + (z, Ux_0 - \mu_0 x_0)}{\|z\|},$$

where μ_0 denotes the quantity

$$\mu_0 = L(x_0) = (Ux_0, x_0).$$

It is clear from the expression for $\dfrac{\partial L}{\partial z}(x_0)$ that the direction of the "gradient" is given by the element $z_1 = Ux_0 - \mu_0 x_0$ (we assume that $z_1 \neq \mathbf{0}$).

Hence ε_1 is found from the equation

$$\phi'(\alpha; x_0, z_1) = 0, \tag{2}$$

where

$$\phi(\alpha; x_0, z_1) = L(x_0 + \alpha z_1).$$

Remembering that

$$(x_0, z_1) = (x_0, Ux_0 - \mu x_0) = (x_0, Ux_0) - \mu_0(x_0, x_0) = 0,$$

$$(Ux_0, z_1) = (\mu_0 x_0 + z_1, z_1) = \mu_0(x_0, z_1) + (z_1, z_1) = (z_1, z_1),$$

we can rewrite (2), after some elementary manipulations, in the form

$$\frac{2\{(z_1, z_1) + [(Uz_1, z_1) - \mu_0(z_1, z_1)]\alpha - (z_1, z_1)^2\alpha^2\}}{[1 + \alpha^2(z_1, z_1)]^2} = 0.$$

Thus ε_1 is the positive root of the equation

$$(z_1, z_1)^2\alpha^2 - [(Uz_1, z_1) - \mu_0(z_1, z_1)]\alpha - (z_1, z_1) = 0,$$

that is,

$$\varepsilon_1 = \frac{[(Uz_1, z_1) - \mu_0(z_1, z_1)] + \sqrt{[(Uz_1, z_1) - \mu_0(z_1, z_1)]^2 + 4(z_1, z_1)^3}}{2(z_1, z_1)^2}.$$

Setting

$$x^{(1)} = x_0 + \varepsilon_1 z_1,$$

we take our next approximation to be the element

$$x_1 = \frac{x^{(1)}}{\|x^{(1)}\|} = \frac{x_0 + \varepsilon_1 z_1}{\|x_0 + \varepsilon_1 z_1\|}.$$

Carrying out the same operations for x_1 and proceeding in the same way, we obtain a sequence $x_0, x_1, \ldots, x_n, \ldots$

$$x_n = \frac{x^{(n)}}{\|x^n\|}, \quad x^{(n)} = x_{n-1} + \varepsilon_n z_n \qquad (n = 1, 2, \ldots),$$

where, for $n = 1, 2, \ldots$, we have

$$z_n = Ux_{n-1} - \mu_{n-1}x_{n-1}, \quad \mu_{n-1} = L(x_{n-1}) = (Ux_{n-1}, x_{n-1}),$$

$$\varepsilon_n = \frac{[(Uz_n, z_n) - \mu_{n-1}(z_n, z_n)] + \sqrt{[(Uz_n, z_n) - \mu_{n-1}(z_n, z_n)]^2 + 4(z_n, z_n)^3}}{2(z_n, z_n)^2}.$$

2.2. The following theorem is concerned with the convergence of the sequence $\{\mu_n\}$ to λ_1 and of $\{x_n\}$ to the corresponding eigenvector.

THEOREM 1. *If x_0 is not orthogonal to the eigenspace corresponding to the largest eigenvalue $\lambda_1 = M$, then $\mu_n \to \lambda_1$ and $x_n \to x^*$, where x^* is an eigenvector belonging to the eigenvalue λ_1.*

Proof. Write $\lambda_k (k = 1, 2, \ldots)$ for the distinct eigenvalues of U, write $\mathbf{H}_{\lambda_k} = \mathbf{H}_k$ for the corresponding eigenspaces, and write P_k for the projections onto these eigenspaces. Also put*

$$x_k^* = \frac{P_k x_0}{\|P_k x_0\|}.$$

By IX.4.5, we have

$$x_0 = \sum_k \|P_k x_0\| = \sum_k \|P_k x_0\| x_k^* = \sum_k c_{0k} x_k^* \quad (c_{0k} = \|P_k x_0\|, \ k = 1, 2, \ldots). \tag{3}$$

Clearly $c_{0k} \geqslant 0$ $(k = 1, 2, \ldots)$, and, by the hypotheses of the theorem, $c_{01} > 0$. Since the terms of the right-hand side of (3) are pairwise orthogonal and $\|x_k^*\| = 1$, we have

$$\sum_k c_{0k}^2 = \|x_0\|^2 = 1$$

and so finally

$$\mu_0 = (Ux_0, x_0) = \sum_k \lambda_k c_{0k}^2 \qquad (\mu_0 \leqslant \lambda_1).$$

We can also express x_1 in terms of $x_1^*, x_2^*, \ldots$. In fact,

$$x_1 = \frac{x^{(1)}}{\|x^{(1)}\|} = \frac{x_0 + \varepsilon_1 z_1}{\|x^{(1)}\|} = \frac{x_0 + \varepsilon_1(Ux_0 - \mu_0 x_0)}{\|x^{(1)}\|}.$$

Also, since

$$Ux_0 = \sum \lambda_k c_{0k} x_k^*,$$

we have

$$x_1 = \sum_k c_{1k} x_k^* \quad \left(c_{1k} = \frac{[1 + \varepsilon_1(\lambda_k - \mu_0)]c_{0k}}{\|x^{(1)}\|}; \ k = 1, 2, \ldots\right).$$

As above, we have

$$\sum_k c_{1k}^2 = 1, \qquad \mu_1 = \sum_k \lambda_k c_{1k}^2, \quad c_{11} > 0.$$

In general,

$$x_n = \sum_k c_{nk} x_k^*$$

$$\left(c_{nk} = \frac{[1 + \varepsilon_n(\lambda_k - \mu_{n-1})]c_{n-1k}}{\|x^{(n)}\|}; \quad k, n = 1, 2, \ldots\right), \tag{4}$$

* If $P_k x_0 = \mathbf{0}$, then $x_k^* = \mathbf{0}$.

where

$$\sum_k c_{nk}^2 = 1, \ \mu_n = \sum_k \lambda_k c_{nk}^2, \ c_{n1} > 0 \quad (n = 0, 1, \ldots). \tag{5}$$

Let us express μ_n in terms of μ_{n-1}. By (1), we have

$$\mu_n = L(x_n) = L(x^{(n)}) = L(x_{n-1} + \varepsilon_n z_n) =$$

$$= \frac{(Ux_{n-1}, x_{n-1}) + 2\varepsilon_n(Ux_{n-1}, z_n) + \varepsilon_n^2(Uz_n, z_n)}{1 + 2\varepsilon_n(x_{n-1}, z_n) + \varepsilon_n^2(z_n, z_n)}.$$

However,

$$(x_{n-1}, z_n) = (x_{n-1}, Ux_{n-1} - \mu_{n-1}x_{n-1}) = 0,$$

$$(Ux_{n-1}, z_n) = (\mu_{n-1}x_{n-1} + z_n, z_n) = (z_n, z_n).$$

Therefore

$$\mu_n = \frac{\mu_{n-1} + 2\varepsilon_n(z_n, z_n) + \varepsilon_n^2(Uz_n, z_n)}{1 + \varepsilon_n^2(z_n, z_n)} =$$

$$= \mu_{n-1} + \frac{2\varepsilon_n(z_n, z_n) + \varepsilon_n^2[(Uz_n, z_n) - \mu_{n-1}(z_n, z_n)]}{1 + \varepsilon_n^2(z_n, z_n)}.$$

Now ε_n satisfies the equation

$$(z_n, z_n)^2\varepsilon_n^2 - [(Uz_n, z_n) - \mu_{n-1}(z_n, z_n)]\varepsilon_n - (z_n, z_n) = 0, \tag{6}$$

so that

$$\varepsilon_n^2[(Uz_n, z_n) - \mu_{n-1}(z_n, z_n)] = \varepsilon_n^3(z_n, z_n)^2 - \varepsilon_n(z_n, z_n).$$

Taking this into account, we obtain

$$\mu_n = \mu_{n-1} + \frac{2\varepsilon_n(z_n, z_n) + \varepsilon_n^3(z_n, z_n)^2 - \varepsilon_n(z_n, z_n)}{1 + \varepsilon_n^2(z_n, z_n)}$$

so we finally have

$$\mu_n = \mu_{n-1} + \varepsilon_n(z_n, z_n). \tag{7}$$

As $\varepsilon_n > 0$, we have $\mu_n > \mu_{n-1}$ and so, as the sequence $\{\mu_n\}$ is bounded ($\mu_n \leqslant \lambda_1$), it has a limit, which we denote by μ:

$$\mu = \lim_{n\to\infty} \mu_n.$$

It follows from (7) that

$$\lim_{n\to\infty} \varepsilon_n(z_n, z_n) = 0, \tag{8}$$

and, in view of the fact that

$$|\varepsilon_n(Uz_n, z_n)| \leqslant \|U\|\varepsilon_n(z_n, z_n),$$

we also have

$$\lim_{n\to\infty} \varepsilon_n(Uz_n, z_n) = 0. \tag{9}$$

Taking (8) and (9) into account, and passing to the limit in (6), we obtain

$$\lim_{n\to\infty} (z_n, z_n) = 0.$$

Hence

$$\lim_{n\to\infty} [Ux_n - \mu_n x_n] = \lim_{n\to\infty} z_{n+1} = \mathbf{0}. \tag{10}$$

Since U is compact and the sequence $\{x_n\}$ is bounded, there is a subsequence $\{x_{n_j}\}$ such that $\{Ux_{n_j}\}$ is convergent. It follows from (10) that $\{x_{n_j}\}$ is convergent. Suppose $x_{n_j} \to \tilde{x}$. Taking limits in the equation

$$Ux_{n_j} - \mu_{n_j} x_{n_j} = z_{n_{j+1}},$$

we find that

$$U\tilde{x} - \mu\tilde{x} = \mathbf{0}.$$

Hence μ is an eigenvalue, and $\tilde{x}$ is a corresponding eigenvector. We now show that $\mu = \lambda$ and $\tilde{x} = x_1^*$. Assume that $\mu = \lambda_s$ $(s > 1)$. Bearing (4) in mind, we have

$$\tilde{x} = \sum_k c_k x_k^* \qquad (c_k = \lim_{j\to\infty} c_{n_j k};\ k = 1, 2, \ldots).$$

But $(\tilde{x}, x_k^*) = 0$ $(k \neq s)$, so $\tilde{x} = c_s x_s^*$, where $|c_s| = 1$. As $c_{ns} \geqslant 0$, we find that $c_s = 1$ and $\tilde{x} = x_s^*$.

Now consider the ratio c_{ns}/c_{n1}. By (4), we have

$$\frac{c_{ns}}{c_{n1}} = \frac{1 + \varepsilon_n(\lambda_s - \mu_{n-1})}{1 + \varepsilon_n(\lambda_1 - \mu_{n-1})} \frac{c_{n-1,\,s}}{c_{n-1,\,1}} < \frac{c_{n-1,\,s}}{c_{n-1,\,1}},$$

so the ratio decreases with increasing n. But $\lim_{j\to\infty} c_{n_j s} = c_s = 1$ and so $\lim_{j\to\infty} c_{n_j 1} = c_1 = 0$, showing that

$$\lim_{j\to\infty} \frac{c_{n_j s}}{c_{n_j 1}} = \infty,$$

which contradicts what we established above. Thus $\mu = \lambda_1$, $\tilde{x} = x_1^*$.

Now we prove that the whole sequence $\{x_n\}$ converges to x_1^*. Let d be the distance from the point λ_1 to the remainder of the spectrum of U. We have

$$\|x_n - x_1^*\|^2 = \|x_n\|^2 + \|x_1^*\|^2 - (x_n, x_1^*) - (x_1^*, x_n) =$$

$$= 2(1 - c_{n1}) \leqslant 2(1 - c_{n1}^2) = 2 \sum_{k \geqslant 2} c_{nk}^2 \leqslant \frac{2}{d} \sum_k (\lambda_1 - \lambda_k) c_{nk}^2.$$

Since, by (23), we have

$$\sum_k (\lambda_1 - \lambda_k) c_{nk}^2 = \lambda_1 \sum_k c_{nk}^2 - \sum_k \lambda_k c_{nk}^2 = \lambda_1 - \mu_n,$$

it follows that

$$\|x_n - x_1^*\| \leqslant \frac{2}{d}(\lambda_1 - \mu_n) \qquad (n = 0, 1, \ldots). \tag{11}$$

If we bear in mind that $\mu_n \to \lambda_1$, then it also follows from this that $x_n \to x_1^*$.

The theorem is thus proved.

REMARK. Concerning the rate of convergence of $\{\mu_n\}$ to λ_1, one can establish the following fact:

$$\lambda_1 - \mu_n \leqslant q^2(1+\alpha_n)\ (\lambda_1 - \mu_{n-1}) \quad (n = 1, 2, \ldots),$$

where

$$q = \frac{M-m-d}{M-m+d},$$

and $\{\alpha_n\}$ is a monotonically decreasing sequence that tends to zero.

Hence, if n is so large that $q(1+\alpha_n) < 1$, the rate of convergence is that of a geometric progression.

Using (11), one can also describe the rate of convergence of $\{x_n\}$ to x_1^*.

See Samokish [2] for applications of the method of steepest descent to the determination of eigenvalues.

§ 3. Applications to elliptic differential equations

3.1. Let us consider how the method of steepest descent is applied to the solution of an elliptic differential equation. For simplicity, we consider the boundary-value problem for a self-adjoint equation of second order, with two independent variables, in the case of a closed disc D (with centre at the origin) bounded by a circle S:

$$Lx \equiv -\frac{\partial}{\partial s}\left(a\frac{\partial x}{\partial s}\right) - \frac{\partial}{\partial t}\left(b\frac{\partial x}{\partial t}\right) + cx = \phi, \qquad x|_S = 0; \tag{1}$$

the coefficients a and b of the equation are assumed to be continuously differentiable functions, c and ϕ are continuous functions, and we assume, furthermore, that $a, b > 0$; $c \geqslant 0$ in the disc D.

Let us apply the operator $-\Delta^{-1}$ to both sides of this equation. We obtain a new equation

$$Ux \equiv -\Delta^{-1}Lx = \psi \qquad (\psi = -\Delta^{-1}\phi), \tag{2}$$

which we shall solve in the subspace $\mathring{\mathbf{W}}_2^{(1)}$ consisting of functions in $\mathbf{W}_2^{(1)}$ that vanish on S. Notice that, as we remarked in XIV.6.3, Δ^{-1} is a continuous linear operator mapping $\mathbf{L}^2$ onto $\mathring{\mathbf{W}}_2^{(2)}$, so that $\psi \in \mathring{\mathbf{W}}_2^{(2)}$.

We verify that the operator $U = -\Delta^{-1}L$ satisfies the conditions of § 1. Let x be a twice continuously differentiable function in $\mathring{\mathbf{W}}_2^{(1)}$. As Lx is a continuous function, we have $y = Ux = -\Delta^{-1}Lx \in \mathring{\mathbf{W}}_2^{(2)} \subset \mathring{\mathbf{W}}_2^{(1)}$. Using Green's formula twice, we have

$$(Ux, x) = (-\Delta^{-1}Lx, x) = (y, x) = \iint_D \left(\frac{\partial y}{\partial s}\frac{\partial x}{\partial s} + \frac{\partial y}{\partial t}\frac{\partial x}{\partial t}\right) ds\,dt =$$

$$= -\iint_D x\left[\frac{\partial^2 y}{\partial s^2} + \frac{\partial^2 y}{\partial t^2}\right] ds\,dt = -\iint_D x\,\Delta y\,ds\,dt =$$

$$= \iint_D x\left[-\frac{\partial}{\partial s}\left(a\frac{\partial x}{\partial s}\right) - \frac{\partial}{\partial t}\left(b\frac{\partial x}{\partial t}\right) + cx\right] ds\,dt =$$

$$= \iint_D \left[a\left(\frac{\partial x}{\partial s}\right)^2 + b\left(\frac{\partial x}{\partial t}\right)^2 + cx^2\right] ds\,dt. \tag{3}$$

This yields

$$(Ux, x) \geqslant \alpha \int_D\int\left[\left(\frac{\partial x}{\partial s}\right)^2 + \left(\frac{\partial x}{\partial t}\right)^2\right] ds\,dt = \alpha(x, x), \tag{4}$$

where

$$\alpha = \min[\min a(s, t), \quad \min b(s, t)] \qquad ((s, t) \in D).$$

Since the Embedding Theorem (Theorem XI.4.3) shows that

$$\int_D\int |x(s, t)|^2\, ds\,dt \leqslant A\,\|x\|^2_{\mathring{\mathbf{W}}_2^{(1)}},$$

we have, on applying this to (3),

$$(Ux, x) \leqslant \beta \int_D\int\left[\left(\frac{\partial x}{\partial s}\right)^2 + \left(\frac{\partial x}{\partial t}\right)^2\right] ds\,dt = \beta(x, x), \tag{5}$$

where

$$\beta = 2\max\left[\max a(s, t), \quad \max b(s, t), \quad A \max c(s, t)\right] \quad ((s, t) \in D).$$

Moreover, if we argue exactly as we did in deducing (3), we obtain the equation

$$(Ux, y) = (x, Uy), \tag{6}$$

which is valid for any twice continuously differentiable functions $x, y \in \mathring{\mathbf{W}}_2^{(1)}$. Comparing (4), (5) and (6), we see that U, regarded as an operator on the set of twice continuously differentiable functions in $\mathring{\mathbf{W}}_2^{(1)}$, is bounded, and so, since this set is dense in $\mathring{\mathbf{W}}_2^{(1)}$, it follows that U is bounded also on $\mathring{\mathbf{W}}_2^{(1)}$. We can now extend the relations (4), (5) and (6) throughout the whole of the space $\mathring{\mathbf{W}}_2^{(1)}$. Thus U is a self-adjoint operator, and its bounds m and M satisfy

$$m \geqslant \alpha > 0, \qquad M \leqslant \beta. \tag{7}$$

Let us apply the method of steepest descent to equation (2). If $x_0 \in \mathring{\mathbf{W}}_2^{(1)}$ is the first approximation, then the successive approximations x_n are found from the formulae of XV.1:3 to be

$$x_n = x_{n-1} - \varepsilon_n z_n \qquad (n = 1, 2, \ldots),$$

where

$$z_n = Ux_{n-1} - \psi = \Delta^{-1}(Lx_{n-1} - \phi),$$

that is, z_n is found from the equation

$$\Delta z = Lx_{n-1} - \phi, \qquad z|_S = 0.$$

For ε_n we find, from (3), the expression

$$\varepsilon_n = \frac{(z_n, z_n)}{(Uz_n, z_n)} = \frac{\displaystyle\int_D\int\left[\left(\frac{\partial z_n}{\partial s}\right)^2 + \left(\frac{\partial z_n}{\partial t}\right)^2\right] ds\,dt}{\displaystyle\int_D\int\left[a\left(\frac{\partial z_n}{\partial s}\right)^2 + b\left(\frac{\partial z_n}{\partial t}\right)^2 + cz_n^2\right] ds\,dt}.$$

To see that the process converges, we note that, by Theorem 1.2 and (7), we have

$$\|x_n - x^*\| \leqslant C\left(\frac{M-m}{M+m}\right)^n \leqslant C\left(\frac{\beta-\alpha}{\beta+\alpha}\right)^n \qquad (n = 0, 1, \ldots), \tag{8}$$

where x^* is a solution of (2), or—what is the same—a solution of the boundary-value problem (1).

3.2. The process of successive approximations described above is particularly effective if the coefficients a, b, c of the given equation and its right-hand side ϕ are polynomials. In that case, taking x_0 also to be a polynomial, we have for the definition of z_1 the equation

$$\Delta z = p, \tag{9}$$

where p is a polynomial. It is not hard to see that a solution of this equation (in $\mathring{\mathbf{W}}_2^{(1)}$) is also a polynomial, whose degree exceeds that of p by 2. For let us write z in the form

$$z(s, t) = (s^2 + t^2 - 1)\pi(s, t), \tag{10}$$

where π is a polynomial with undetermined coefficients, of the same degree as p. Substituting the expression (10) in equation (9) and equating the coefficients of the two sides, we obtain a system of linear algebraic equations for determining the coefficients of π, in which the number of unknowns is equal to the number of equations. Further, the matrix of this system is clearly independent of p. Since a polynomial solution for (9), if one exists, is unique, the determinant of the system is non-zero, which ensures that a polynomial solution of (9) does exist,* whatever polynomial we take for p.

Thus the function z_1 is a polynomial, and consequently $x_1 = x_0 - \varepsilon_1 z_1$ is also a polynomial; hence, in this case, all the approximations x_n turn out to be polynomials.

This situation makes it possible to establish, assuming certain regularity properties for equation (1), the existence not only of a generalized solution, but also of a solution in the classical sense. For this we shall need certain facts from the constructive theory of functions.

3.3. First we record the two-dimensional analogues of the inequalities of Bernstein and A. A. Markov. Let p be a polynomial of degree n, where

$$|p(s, t)| \leqslant M \qquad (s, t \in D),$$

D being an arbitrary bounded region having a smooth contour. Then in D we have

$$\left|\frac{\partial p(s, t)}{\partial s}\right| \leqslant An^2 M, \qquad \left|\frac{\partial p(s, t)}{\partial t}\right| \leqslant An^2 M, \tag{11}$$

while, in any interior subregion D_1 of D, we have

$$\left|\frac{\partial p(s, t)}{\partial s}\right| \leqslant BnM, \qquad \left|\frac{\partial p(s, t)}{\partial t}\right| \leqslant BnM, \tag{12}$$

where A and B are constants depending only on D and D_1.

---

* When solving (9) in practice, it is simpler first to find a solution disregarding the boundary condition, and then to obtain a solution with zero boundary values by subtracting a solution of Laplace's equation having the same boundary values as the polynomial first found. The first solution can be found particularly simply using a Fourier series expansion.

The inequalities (11) and (12) can be derived in an obvious way from the corresponding inequalities in the one-dimensional case (see Natanson-I).

We now prove the following lemma.

LEMMA 1. *If u is a polynomial of degree at most n, then*

$$\|u\|_{\mathbf{C}(D)} \leqslant A_1 n^2 \|u\|_{\mathbf{L}^2(D)}, \qquad \|u\|_{\mathbf{C}(D_1)} \leqslant B_1 n \|u\|_{\mathbf{L}^2(D)}, \tag{13}$$

$$\|u\|_{\mathbf{C}^{(1)}(D)} \leqslant A_2 n^2 \|u\|_{\mathbf{W}_2^{(1)}(D)}, \qquad \|u\|_{\mathbf{C}^{(1)}(D_1)} \leqslant B_1 n \|u\|_{\mathbf{W}_2^{(1)}(D)}, \tag{14}$$

where D_1 denotes an interior subregion of D and $\mathbf{C}^{(1)}$ denotes the space of continuously differentiable functions.

Proof. We restrict ourselves to the case where D is bounded by a circle centred at the origin (not many additional considerations are needed for the general case). Writing

$$\phi(s, t) = \int_0^s \int_0^t p^2(s, t)\, ds dt,$$

we have

$$|\phi(s, t) \leqslant \|p\|_{\mathbf{L}^2}^2 \qquad ((s, t) \in D),$$

and, applying inequality (11) twice, we have

$$|p(s, t)|^2 = \left| \frac{\partial^2 \phi(s, t)}{\partial s \partial t} \right| \leqslant A^2 (2n+2)^2 (2n+1)^2 \max|\phi(s, t)| \leqslant A_1^2 n^4 \|p\|_{\mathbf{L}^2}^2,$$

which gives the first of the inequalities (13). The second inequality is obtained similarly, but using (12) instead of (11).

Applying the inequality just derived to the derivative, we have

$$\left| \frac{\partial p(s, t)}{\partial s} \right| \leqslant A_1 n^2 \left\| \frac{\partial p}{\partial s} \right\|_{\mathbf{L}^2} \leqslant A_1 n^2 \|p\|_{\mathbf{W}_2^{(1)}};$$

and a similar inequality holds for $\partial p/\partial t$. Finally, by (11) and the Embedding Theorem, we have

$$|p(s, t)| \leqslant A n^2 \|p\|_{\mathbf{L}^2} \leqslant C A n^2 \|p\|_{\mathbf{W}_2^{(1)}}.$$

Combining all these, we obtain the first of the inequalities (14):

$$\|p\|_{\mathbf{C}^{(1)}(D)} = \max_D |p(s, t)| + \max_D \left| \frac{\partial p(s, t)}{\partial s} \right| + \max_D \left| \frac{\partial p(s, t)}{\partial t} \right| \leqslant A_2 n^2 \|p\|_{\mathbf{W}_2^{(1)}}.$$

The second inequality is obtained similarly.

3.4. We now return to equation (1). Assume first that the coefficients a, b, c and the right-hand side ϕ are polynomials. As we showed in 3.2, if the initial approximation x_0 is a polynomial, then all the subsequent approximations x_n $(n = 1, 2, \ldots)$ will also be polynomials. Moreover, if the degrees of a, b, c, ϕ are at most m and x_0 is a polynomial of degree N, then Lx_0 will be a polynomial of degree at most $m + N$, while z_1 will have degree at most $m + N + 2$. In particular, if $x_0 = \mathbf{0}$, then z_1, and hence also x_1, will be a polynomial of degree at most $m + 2$. Repeating this argument, we find that x_n will be a polynomial of degree at most $(m + 2)n$.

Using (8), we obtain

$$\|x_k - x_{k-1}\|_{\mathring{\mathbf{W}}_2^{(1)}} \leqslant \|x_k - x^*\| + \|x_{k-1} - x^*\| \leqslant 2C\left(\frac{\beta - \alpha}{\beta + \alpha}\right)^{k-1}$$

$$(k = 1, 2, \ldots).$$

Since $x_k - x_{k-1}$ is a polynomial of degree at most $(m+2)k$, the lemma shows that

$$\|x_k - x_{k-1}\|_{C^{(1)}(D)} \leqslant A_3 (m+2)^2 k^2 q \qquad (k = 1, 2, \ldots),$$

where

$$q = \frac{\beta - \alpha}{\beta + \alpha}.$$

Hence applying the inequality (11) repeatedly, we obtain

$$\|x_k - x_{k-1}\|_{C^{(p)}(D)} \leqslant A_4 [(m+2)k]^{2p} q^k \qquad (k, p = 1, 2, \ldots). \tag{15}$$

Thus the series

$$x^*(s, t) = \sum_{k=1}^{\infty} [x_k(s, t) - x_{k-1}(s, t)] \tag{16}$$

can be differentiated term-by-term p times. In other words, the solution of (1) has derivatives of all orders, and

$$\|x^* - x_n\|_{C^{(p)}} = \sum_{k=n+1}^{\infty} \|x_k - x_{k-1}\| \leqslant A_5 [(m+2)n]^{2p} q^n \tag{17}$$

$$(n = 0, 1, \ldots),$$

that is, the sequence $\{x_n\}$ of approximations obtained through the method of steepest descent converges uniformly to a solution x^*, together with all derivatives.

In this connection, it is worth noticing that if we use Bernstein's Theorem on approximating analytic functions by polynomials (Natanson-I), we can conclude that the solution x^* is also analytic.

Now we turn to the case where the coefficients of (1) and the right-hand side are arbitrary several times differentiable functions. In fact, we shall assume that the functions a and b have derivatives of orders up to $v+1$, that c and ϕ have derivatives of orders up to v, and that the $(v+1)$-st derivative of a and b and the v-th derivative of c and ϕ satisfy a Lipschitz condition of order $\alpha > 0$. By Jackson's Theorem (see e.g. Kharrik [1]), we can find polynomials a_n, b_n, c_n and ϕ_n of degree at most n such that

$$|a(s,t) - a_n(s,t)| \leqslant \frac{K}{n^{v+1+\alpha}},$$

$$\left|\frac{\partial}{\partial s}[a(s, t) - a_n(s, t)]\right| \leqslant \frac{K}{n^{v+\alpha}},$$

$$\left|\frac{\partial}{\partial t}[a(s, t) - a_n(s, t)]\right| \leqslant \frac{K}{n^{v+\alpha}},$$

$$|b(s, t) - b_n(s, t)| \leqslant \frac{K}{n^{v+1+\alpha}},$$

$$\left|\frac{\partial}{\partial s}[b(s, t) - b_n(s, t)]\right| \leqslant \frac{K}{n^{v+\alpha}},$$

$$\left|\frac{\partial}{\partial t}[b(s, t) - b_n(s, t)]\right| \leqslant \frac{K}{n^{v+\alpha}},$$

$$|c(s, t) - c_n(s, t)| \leqslant \frac{K}{n^{\nu+\alpha}},$$

$$|\phi(s, t) - \phi_n(s, t)| \leqslant \frac{K}{n^{\nu+\alpha}}.$$

Write L_n for the operator obtained from L by replacing the coefficients a, b, c by a_n, b_n, c_n respectively. If n is sufficiently large, then clearly the method of steepest descent is applicable to L_n, and we may assume that the values α and β also serve for L_n. We write $x^{(n)}$ for a solution of the equation

$$L_n x = \phi_n, \qquad x|_S = 0, \tag{18}$$

and look for a bound for the difference between two such solutions:

$$\|Lx^{(n)} - L_n x^{(n)}\|_{\mathbf{L}^2} \leqslant \left\| \frac{\partial}{\partial s}(a - a_n)\frac{\partial x^{(n)}}{\partial s} \right\| + \left\| (a - a_n)\frac{\partial^2 x^n}{\partial s^2} \right\| + \left\| \frac{\partial}{\partial t}(b - b_n)\frac{\partial x^n}{\partial t} \right\| +$$

$$+ \left\| (b - b_n)\frac{\partial^2 x^{(n)}}{\partial t^2} \right\| + \|(c - c_n)x^{(n)}\| \leqslant \frac{K_1}{n^{\nu+\alpha}}\left[\left\| \frac{\partial x^{(n)}}{\partial s} \right\|_{\mathbf{L}^2} + \left\| \frac{\partial^2 x^{(n)}}{\partial s^2} \right\|_{\mathbf{L}^2} + \right.$$

$$\left. + \left\| \frac{\partial x^{(n)}}{\partial t} \right\|_{\mathbf{L}^2} + \left\| \frac{\partial^2 x^{(n)}}{\partial t^2} \right\|_{\mathbf{L}^2} + \|x^{(n)}\|_{\mathbf{L}^2} \right] \leqslant \frac{K_2}{n^{\nu+\alpha}}\|x^{(n)}\|_{\mathring{\mathbf{W}}_2^{(2)}}.$$

We also have

$$\|L(x^{(n)} - x^{(m)})\|_{\mathbf{L}^2} \leqslant \|(L - L_n)x^{(n)}\| + \|(L - L_m)x^{(m)}\| + \|\phi_n - \phi_m\| \leqslant$$

$$\leqslant \frac{K_2}{n^{\nu+\alpha}}\|x^{(n)}\|_{\mathring{\mathbf{W}}_2^{(2)}} + \frac{K_2}{m^{\nu+\alpha}}\|x^{(m)}\|_{\mathring{\mathbf{W}}_2^{(2)}} + \frac{K}{n^{\nu+\alpha}} + \frac{K}{m^{\nu+\alpha}}.$$

By the Bernstein–Ladyzhenskaya inequality (see Ladyzhenskaya, Ch. II, §6) we thus obtain

$$\|x^{(n)} - x^{(m)}\|_{\mathring{\mathbf{W}}_2^{(2)}} \leqslant K_3\|L(x^{(n)} - x^{(m)})\|_{\mathbf{L}^2} \leqslant$$

$$\leqslant K_4\left[\frac{1}{n^{\nu+\alpha}}\|x^{(n)}\|_{\mathring{\mathbf{W}}_2^{(2)}} + \frac{1}{m^{\nu+\alpha}}\|x^{(m)}\|_{\mathring{\mathbf{W}}_2^{(2)}} + \frac{1}{n^{\nu+\alpha}} + \frac{1}{m^{\nu+\alpha}} \right]. \tag{19}$$

Applying this inequality to the pair $x^{(n_0)}, x^{(m)}$ (where n_0 is fixed and sufficiently large), and replacing $\|x^{(m)}\|$ on the right-hand side by the larger sum $\|x^{(n_0)}\| + \|x^{(n_0)} - x^{(m)}\|$, we find that $\|x^{(n_0)} - x^{(m)}\|_{\mathring{\mathbf{W}}_2^{(2)}}$ is bounded independently of m and hence that $\|x^{(m)}\|$ is bounded. Taking $n < m$ in (19), we use this to rewrite the inequality in the form

$$\|x^{(n)} - x^{(m)}\|_{\mathring{\mathbf{W}}_2^{(2)}} \leqslant \frac{K_5}{n^{\nu+\alpha}}. \tag{20}$$

Now we choose a number $h > 1$, to be specified more precisely below, and consider the sequence $\{n_k\}$ of natural numbers $n_k = [h^k]$ (where $[\gamma]$ is the integral part of γ, and $k = 1, 2, \ldots$). For each of the functions $x^{(n_k)}$ we construct the k-th approximation $x_k^{(n_k)}$ according to the method of steepest descent (taking the initial approximations to be zero). As we showed above, $x_k^{(n_k)}$ is a polynomial of degree at most $(n_k + 2)k$. Let us find a bound for the difference between two such polynomials. In the first place,

$$\|x_k^{(n_k)} - x_{k-1}^{(n_{k-1})}\| \leqslant \|x_k^{(n_k)} - x^{(n_k)}\| + \|x_{k-1}^{(n_{k-1})} - x^{(n_{k-1})}\| + \|x^{(n_k)} - x^{(n_{k-1})}\|.$$

By (20), we have

$$\|x^{(n_k)} - x^{(n_{k-1})}\|_{\mathring{\mathbf{W}}_2^{(2)}} \leqslant \frac{K_5}{n_{k-1}^{\nu+\alpha}} \leqslant \frac{K_6}{n_k^{\nu+\alpha}}. \tag{21}$$

Furthermore, using (17), we have

$$\|x^{(n_k)} - x_k^{(n_k)}\|_{\mathbf{C}^{(2)}} \leqslant A_5[(n_k+2)k]^4 q^k \qquad (k = 1, 2, \ldots). \tag{22}$$

Hence

$$\|x_k^{(n_k)} - x_{k-1}^{(n_{k-1})}\|_{\mathring{\mathbf{W}}_2^{(2)}} \leqslant K_7[(n_k+2)k]^4 q^k + \frac{K_6}{n_k^{\nu+\alpha}}.$$

Since the norm symbol contains a polynomial of degree at most $(n_k+2)k$, we obtain, using the lemma and inequality (11),

$$\|x_k^{(n_k)} - x_{k-1}^{(n_{k-1})}\|_{\mathbf{C}^{(p)}} \leqslant K_8[(n_k+2)k]^{2p-2}\left\{K_7[(n_k+2)k]^4 q^k + \frac{K_6}{n_k^{\nu+\alpha}}\right\} = \varepsilon_k.$$

As $n_k \leqslant h^k$, it is clear that, if $\nu+\alpha > 2p-2$ and h is chosen so that $h^{2p+2}q < 1$, then the series $\sum_{k=1}^{\infty} \varepsilon_k$ is convergent, and so the sequence $\{x_k^{(n_k)}\}$ is convergent in the space $\mathbf{C}^{(p)}$ and its limit belongs to $\mathbf{C}^{(p)}$: that is, it is the p times continuously differentiable function

$$x(s, t) = \lim_{k\to\infty} x_k^{(n_k)}(s, t).$$

If $p \geqslant 2$, then using the same inequalities, it is not hard to show that x satisfies equation (1), that is, is a classical solution of the boundary-value problem. Hence a classical solution exists if $\nu+\alpha > 2$, and in particular if $\nu = 2$: that is, if the coefficients a and b have third derivatives, and c and ϕ have second derivatives, satisfying a Lipschitz condition of some order $\alpha > 0$.

Similarly, if we use the inequalities for interior subregions, we find that the solution is twice continuously differentiable inside the region, even for $\nu = 1$.

In conclusion, we note that the same method can be used for more intricate problems. The case where the region is not circular can be reduced to the case just studied, provided the region can be smoothly transformed into a circular region. The method can also be used, with insignificant changes, for higher order equations having more than two independent variables.

§ 4. Minimization of convex differentiable functionals

4.1. Now we study applications of the method of steepest descent to the minimization of differentiable functionals. A functional Φ defined on a B-space $\mathbf{X}$ is said to be differentiable at a point $x \in \mathbf{X}$ if there exists a linear functional f such that, for $h \in \mathbf{X}$, we have

$$\Phi(x+h) = \Phi(x) + f(h) + o(h),$$

where $o(h)/\|h\| \to 0$ as $\|h\| \to 0$. The functional f is called the (Fréchet) derivative of Φ at x and is denoted by $\Phi'(x)$. (For further details on differentiation, see Ch. XVII.)

Let $\mathbf{X}$ be a B-space. From the point of view of extremal problems it is more convenient to assume (in contrast to the preceding sections) that $\mathbf{X}$ is real. We consider a functional Φ, defined and differentiable on $\mathbf{X}$, and show how to determine the direction of steepest descent at a point $x \in \mathbf{X}$. Since Φ is differentiable, the derivative of the functional at x in the direction of z can be expressed in the form

$$\frac{\partial \Phi}{\partial z}(x) = \frac{1}{\|z\|} \lim_{\alpha \to 0+} \frac{1}{\alpha}(\Phi(x+\alpha z) - \Phi(x)) = \frac{1}{\|z\|}\Phi'(x)(z)$$
$$= \Phi'(x)\left(\frac{z}{\|z\|}\right).$$

It follows from what has been said that a vector y of unit norm determining the direction of steepest descent of Φ at x coincides with a minimum point of the linear functional $\Phi'(x)$ on the unit sphere, and so is found from the condition

$$\Phi'(x)(y) = \inf_{\|z\|=1} \Phi'(x)(z). \tag{1}$$

Since a linear functional need not attain a minimum on the unit sphere, a direction of steepest descent may, in fact, not exist. However, it is not hard to find conditions for the minimum in (1) to be attained. We can guarantee this, for example, when $\mathbf{X}$ is a space dual to some Banach space $\mathbf{Y}$ and the functionals $\Phi'(x)$ ($x \in \mathbf{X}$) belong to $\mathbf{Y}$. (More precisely, to the image of $\mathbf{Y}$ under the canonical embedding of $\mathbf{Y}$ in $\mathbf{Y}^{**} = \mathbf{X}^*$.) For the unit ball in $\mathbf{X} = \mathbf{Y}^*$ is compact in the weak* topology, and so $\Phi'(x)$ attains a minimum on this ball, and hence also on the sphere. In particular, if $\mathbf{X}$ is reflexive, then the direction of steepest descent automatically exists.

In what follows, we shall assume, without any specific mention of the fact, that the space $\mathbf{X}$ and the functional Φ are such that the direction of steepest descent exists.

If $\mathbf{X}$ is strictly convex,* then the direction of steepest descent is unique, as follows from (1). In general, of course, this is not so.

The formula (1) shows that the direction of steepest descent (and hence also the use of the method) depends essentially not only on the functional but also on the norm of the space. If we change to another norm, even one equivalent to the original, we may change the direction of steepest descent substantially. It follows from (1) and the equation

$$\inf_{\|z\|=1} \Phi'(x)(z) = -\sup_{\|z\|=1}(-\Phi'(x)(z)) = -\|\Phi'(x)\|$$

that a direction y ($\|y\| = 1$) is the direction of steepest descent if and only if

$$\Phi'(x)(y) = -\|\Phi'(x)\|. \tag{2}$$

For all $z \in \mathbf{X}$ we have the inequality

$$-\Phi'(x)(z) \leqslant \|\Phi'(x)\| \, \|z\|.$$

The direction of steepest descent is characterized by the presence of equality in this relation. This enables one to find the direction fairly simply in certain cases. Let us give some examples.

1) Let $\mathbf{X}$ be a Hilbert space. The equation $-\Phi'(x)(z) = \|\Phi'(x)\| \, \|z\|$, or, what is the same, $(-\Phi'(x), z) = \|-\Phi'(x)\| \, \|z\|$, shows that we have equality in the Cauchy–Buniakowski

* A B-space $\mathbf{X}$ is said to be *strictly convex* if the triangle inequality holds with equality only for elements that are proportional by a positive scalar: that is, if $\|x+y\| = \|x\| + \|y\|$ implies that $x = \lambda y$ for some $\lambda > 0$. If $\mathbf{X}$ is strictly convex, $x, y \in \mathbf{X}$ and $\|x\| = \|y\| = \frac{1}{2}\|x+y\| = 1$, then $x = y$. An example of a strictly convex space is the space $\mathbf{L}^p(a, b)$ $(1 < p < +\infty)$.

inequality for the vectors $-\Phi'(x)$ and z. This means that there exists $\lambda > 0$ such that $-\Phi'(x) = \lambda z$. Hence the direction of steepest descent is determined by the vector $-\Phi'(x)$.

REMARK. If $\mathbf{X}$ is a Hilbert space, then the derivative $\Phi'(x)$ can be regarded as an element of $\mathbf{X}$; this element is usually called the *gradient* of Φ at x. In this connection, the direction of steepest descent is, in this situation, called the direction of the *antigradient*.

2) Let $\mathbf{X} = \mathbf{L}^p(a, b)$ $(1 < p < \infty)$. By examining the conditions for equality to be attained in Hölder's inequality, it is not hard to verify that the direction of steepest descent z at a point x is given by

$$z(t) = -[\text{sign } y(t)]\,|y(t)|^{q-1},$$

where $1/p + 1/q = 1$ and y is an element of $\mathbf{L}^q(a, b)$ such that $\Phi'(x)(u) = \int_a^b y(t)u(t)\,dt$ for all $u \in \mathbf{L}^p(a, b)$. As in the preceding case, the direction of steepest descent is unique.

3) Let $\mathbf{X} = \mathbf{L}^\infty(a, b)$, and suppose the functionals $\Phi'(x)$ $(x \in \mathbf{L}^\infty(a, b))$ belong to $\mathbf{L}^1(a, b)$ (i.e. $\Phi'(x)(u) = \int_a^b y(t)u(t)\,dt$), where $y \in \mathbf{L}^1(a, b)$. Here the direction of steepest descent is, generally speaking, not unique. Any such direction is determined by a measurable function z such that vrai sup $|z(t)| = 1$ and $z(t) = -\text{sign } y(t)$, if $y(t) \neq 0$. (If the set $\{t : y(t) = 0\}$ has measure zero, then the direction is unique.)

4.2. Before we turn to the study of the method of steepest descent we introduce some definitions.

A point $x \in \mathbf{X}$ is called a local minimum of a functional Φ if $\Phi(z) \geqslant \Phi(x)$ for elements z in some ball centred at this point. If $\Phi(z) \geqslant \Phi(x)$ for all $z \in \mathbf{X}$, then x is called a global minimum. A point $x \in \mathbf{X}$ is called a stationary point of a differentiable functional Φ if $\Phi'(x) = 0$.

In the study of minima, an important role is played by the function of a real variable $\phi(\alpha; x, z) = \Phi(x + \alpha z)$ that we considered in 1.1.

It is not hard to verify that if Φ is differentiable then $\phi(\alpha; x, z)$ is differentiable (with respect to α) for all x and z, and $\phi'(\alpha; x, z) = \Phi'(x + \alpha z)(z)$. Using this, let us show that a local minimum is a stationary point. If a local minimum is attained at x, then the function $\phi(\alpha; x, z)$ attains a local minimum at zero on the real line for each $z \in \mathbf{X}$; therefore, $\phi(0; x, z) = \Phi'(x)(z) = 0$, which means that x is a stationary point.

The converse statement is, of course, false. However, if Φ is a convex functional, then the stationary points coincide with global minima.

A functional Φ defined on a linear space $\mathbf{X}$ is said to be convex if, for all elements $z_1, z_2 \in \mathbf{X}$ and all numbers $t_1, t_2 \geqslant 0$ with $t_1 + t_2 = 1$, we have

$$\Phi(t_1 x_1 + t_2 x_2) \leqslant t_1 \Phi(x_1) + t_2 \Phi(x_2).$$

If Φ is a convex functional, then for all x and z the function $\phi(\alpha; x, z)$ is convex on the real line, that is

$$\phi(t_1\alpha_1 + t_2\alpha_2; x, z) \leqslant t_1\phi(\alpha_1; x, z) + t_2\phi(\alpha_2; x, z);$$

here $t_1, t_2, \alpha_1, \alpha_2$ are real numbers; $t_1, t_2 \geqslant 0$, $t_1 + t_2 = 1$. The convexity of $\phi(\alpha; x, z)$ implies that its derivative is non-decreasing.

LEMMA 1. *Let Φ be a convex differentiable functional. Then every stationary point of Φ is a global minimum of Φ.*

Proof. Let x be a stationary point, z an arbitrary element of $\mathbf{X}$. Since the derivative of

$\phi(\alpha;\, x, z)$ is non-decreasing, Lagrange's formula shows that, for some $\theta \in (0, 1)$,

$$\Phi(x+z)-\Phi(z) = \phi(1;\, x, z)-\phi(0;\, x, z) =$$
$$= \phi'(\theta;\, x, z) \geqslant \phi'(0;\, x, z) = \Phi'(x)(z) = 0.$$

Hence $\Phi(x+z) \geqslant \Phi(x)$ for all $z \in \mathbf{X}$, which proves the lemma.

In what follows we shall be concerned with the convergence of the method of steepest descent to a stationary point of the functional Φ (and thus, in the case of a convex functional, to a global minimum of the functional).

The basic results on convergence hold only when the derivative of Φ satisfies a Lipschitz condition in some ball, that is, when there exists a number L such that, for all x, z in the ball, we have

$$\|\Phi'(x)-\Phi'(z)\| \leqslant L\,\|x-z\|. \tag{3}$$

Here we make use of the following lemma.

LEMMA 2. *Suppose Φ' satisfies the Lipschitz condition* (3) *in the ball $B_{a+b}(\mathbf{0})$ of radius $a+b$ centred at the origin. If $\|x\| \leqslant a$, then*

$$\Phi(x+z) \leqslant \Phi(z)+\Phi'(x)(z)+\frac{L}{2}\|z\|^2.$$

Proof. Using the differentiability of $\phi(\alpha;\, x, z)$ and the expression for its derivative, we have

$$\Phi(x+z) = \phi(1;\, x, z) = \phi(0;\, x, z)+\int_0^1 \phi'(\alpha;\, x, z)\,d\alpha =$$
$$= \Phi(x)+\int_0^1 \Phi'(x+\alpha z)(z)\,d\alpha =$$
$$= \Phi(x)+\Phi'(x)(z)+\int_0^1 (\Phi'(x+\alpha z)-\Phi'(x))(z)\,d\alpha \leqslant$$
$$\leqslant \Phi(x)+\Phi'(x)(z)+\int_0^1 \|\Phi'(x+\alpha z)-\Phi'(x)\|\,\|z\|\,d\alpha \leqslant$$
$$\leqslant \Phi(x)+\Phi'(x)(z)+L\|z\|^2\int_0^1 \alpha\,d\alpha =$$
$$= \Phi(x)+\Phi'(x)(z)+(L/2)\|z\|^2.$$

This proves the lemma.

4.3. We now apply the method of steepest descent to the minimization of Φ; more precisely, we investigate the behaviour of the sequence $x_0, x_1, \ldots, x_n, \ldots$, where

$$x_n = x_{n-1}+\varepsilon_n z_n \qquad (n = 1, 2, \ldots), \tag{4}$$

z_n is the direction of steepest descent at x_{n-1} (if there are several such directions, then z_n is any one of them); ε_n is found from the condition

$$\Phi(x_{n-1}+\varepsilon_n z_n) = \min_{\alpha \geqslant 0} \Phi(x_{n-1}+\alpha z_n).$$

Thus the descent value is chosen by the second of the methods listed in 1.1. Notice, however, that if Φ is a strictly convex functional, then the first two of these methods coincide. (Strict convexity of Φ means that $\Phi(t_1 x_1 + t_2 x_2) < t_1 \Phi(x_1) + t_2 \Phi(x_2)$ for $x_1, x_2 \in \mathbf{X}, x_1 \neq x_2; t_1, t_2 > 0, t_1 + t_2 = 1$; it follows from this that the derivative $\phi'(\alpha; x, z)$ is strictly increasing, so it can vanish only at one point; hence the equivalence of the two conditions just mentioned.)

We assume below, without any special stipulations, that the initial point $x_0 \in \mathbf{X}$ is such that the Lebesgue set $\Omega_0 = \{x \in \mathbf{X} : \Phi(x) \leqslant \Phi(x_0)\}$ is bounded. Note that $x_n \in \Omega_0$ ($n = 0, 1, 2, \ldots$). We also stipulate that $\|z_n\| = 1$ for all n.

THEOREM 1. *Suppose the derivative of Φ satisfies the Lipschitz condition* (3) *in the ball with centre at the origin and radius R', where $R' > R = \sup_{x \in \Omega_0} \|x\|$. Then the sequence $\{x_n\}$ constructed according to* (4) *satisfies $\Phi'(x_n) \to 0$.*

Proof. Suppose $0 < \alpha \leqslant R' - R$. Then, by the definition of the descent value ε_n, we have $\Phi(x_{n-1} + \alpha z_n) \geqslant \Phi(x_n)$. Using Lemma 2, we obtain

$$\Phi(x_n) \leqslant \Phi(x_{n-1} + \alpha z_n) \leqslant \Phi(x_{n-1}) - \alpha \Phi'(x_{n-1})(z_n) + \tfrac{1}{2} L \alpha^2$$

(where L is the constant appearing in the Lipschitz condition). It follows from this inequality and (2) that

$$\|\Phi'(x_n)\| \leqslant \frac{\Phi(x_{n-1}) - \Phi(x_n)}{\alpha} + \frac{1}{2} L\alpha.$$

Let ε be any positive number and suppose $\alpha < \min(\varepsilon, R' - R)$. The boundedness of Ω_0 and the Lipschitz condition imply that Φ is bounded below. Since the sequence $\Phi(x_n)$ is decreasing (by construction) and bounded, it converges, so for sufficiently large n we have $\frac{1}{\alpha}(\Phi(x_{n-1}) - \Phi(x_n)) < \varepsilon$. For such n, we have

$$\|\Phi'(x_n)\| \leqslant \varepsilon(1 + \tfrac{1}{2} L),$$

and this proves the theorem.

COROLLARY. *The cluster points of the sequence* (4) (*if they exist*) *are stationary points.*

In certain cases, by using compactness considerations, convergence to a stationary point can be proved for functionals whose derivatives are only continuous. Let us give one result of this type (Curry [1]).

THEOREM 2. *Suppose the space* $\mathbf{X}$ *is finite-dimensional, and the functional Φ is continuously differentiable. Then the cluster points of the sequence $\{x_n\}$ are stationary points.*

Proof. Notice first that the finite-dimensionality of $\mathbf{X}$ and the boundedness of Ω_0 imply that the sequence x_n has cluster points. Let y be one of these: $y = \lim x_{n_k}$. Without loss of generality, we may assume that $\lim z_{n_k+1} = \tilde{y}$ exists. For all positive α, we set

$$w_{n_k}(\alpha) = \frac{1}{\alpha}(\Phi(x_{n_k} + \alpha z_{n_k+1}) - \Phi(x_{n_k})) - \Phi'(x_{n_k})(z_{n_k+1}). \tag{5}$$

It follows from the hypotheses of the theorem that $\lim_{k \to \infty} w_{n_k}(\alpha) = w(\alpha)$, and

$$w(\alpha) = \frac{1}{\alpha}(\Phi(y + \alpha\tilde{y}) - \Phi(y)) - \Phi'(y)(\tilde{y}).$$

Since Φ is differentiable at y, we have $\lim_{\alpha \to 0} w(\alpha) = 0$. By our choice of descent value,

$$\Phi(x_{n_k+1}) = \Phi(x_{n_k} + \varepsilon_{n_k+1} z_{n_k+1}) \leqslant \Phi(x_{n_k} + \alpha z_{n_k+1}),$$

and hence, by (5), we have

$$\Phi(x_{n_k+1}) \leqslant \Phi(x_{n_k}) + \alpha\Phi'(x_{n_k})(z_{n_k+1}) + \alpha w_{n_k}(\alpha).$$

As $\alpha > 0$, we have

$$-\Phi'(x_{n_k})(z_{n_k+1}) \leqslant \frac{1}{\alpha}(\Phi(x_{n_k}) - \Phi(x_{n_k+1})) + w_{n_k}(\alpha). \tag{6}$$

Letting $k \to \infty$ in (6), remembering that $\lim \Phi(x_{n_k}) - \Phi(x_{n_k+1}) = 0$ and using (2), we have

$$\|\Phi'(y)\| = -\Phi'(y)\tilde{y} \leqslant w(\alpha).$$

The theorem now follows from the fact that $w(\alpha) \xrightarrow[\alpha\to 0]{} 0$.

4.4. Substantially more can be said about the convergence of the method of steepest descent if Φ is a convex functional.

LEMMA 3. *Let Φ be a convex functional bounded below on $\mathbf{X}$ and let $Q = \inf_{x\in\mathbf{X}} \Phi(x)$. Then there exists a number $c > 0$ such that*

$$\Phi(x_n) - Q \leqslant c\,\|\Phi'(x_n)\|.$$

Proof. As $\Omega_0 = \{z \in \mathbf{X} : \Phi(z) \leqslant \Phi(x_0)\}$ is bounded by hypothesis, the set $\Omega_0 - \Omega_0 = \{z : z = z_1 - z_2, z_1, z_2 \in \Omega_0\}$ is also bounded. We show that we can take c to be the radius of a ball B centred at the origin containing $\Omega_0 - \Omega_0$. Bearing in mind that $\phi'(\alpha: x_n, z)$ is non-increasing (as Φ is convex) and using Lagrange's formula, we have

$$\Phi(x_n + z) - \Phi(x_n) = \phi(1; x_n, z) - \phi(0; x_n, z) =$$
$$= \phi'(\theta; x_n, z) \geqslant \phi'(0; x_n, z) = \Phi'(x_n)(z)$$

(here $\theta \in (0, 1)$). Hence

$$\min_{\|z\|\leqslant c} \Phi(x_n + z) - \Phi(x_n) \geqslant \min_{\|z\|\leqslant c} \Phi'(x_n)(z). \tag{7}$$

As $x_n \in \Omega_0$, we have $B \supset \Omega_0 - \Omega_0 \supset \Omega_0 - x_n$, so that $x_n + B \supset \Omega_0$. Therefore

$$Q = \inf_{x\in\mathbf{X}} \Phi(x) \leqslant \inf_{\|z\|\leqslant c} \Phi(x_n + z) \leqslant \inf_{x\in\Omega_0} \Phi(x).$$

It follows easily from the definition of Ω_0 that $Q = \inf_{x\in\Omega_0} \Phi(x)$. Furthermore, $\inf_{\|z\|\leqslant c} \Phi'(x_n)(z) = c \inf_{\|z\|\leqslant 1} \Phi'(x_n)(z) = -c\|\Phi'(x_n)\|$. From the relations obtained above and (7) it follows that

$$Q - \Phi(x_n) \geqslant -c\,\|\Phi'(x_n)\|.$$

The lemma is thus proved.

COROLLARY. *If the hypotheses of Lemma 3 and those of either Theorem 1 or Theorem 2 are satisfied, then the sequence $\{x_n\}$ is a minimizing sequence.*

If we assume that $\Phi'(x)$ satisfies a Lipschitz condition, then we can give a bound for the rate of convergence of the sequence $\{\Phi(x_n)\}$.

THEOREM 3. *Let Φ be a convex functional, bounded below on $\mathbf{X}$, and suppose $\Phi'(x)$ satisfies the Lipschitz condition (3) on $\mathbf{X}$. Then $\Phi(x_n) - Q = O(1/n)$ (here $Q = \inf_{x\in\mathbf{X}} \Phi(x)$).*

The proof depends on the following lemma.

LEMMA 4. *Let $\lambda_n > 0$ $(n = 1, 2, \ldots)$ be numbers such that, for some $\mu > 0$, we have $\lambda_n - \lambda_{n+1} \geqslant \mu\lambda_n^2$ for all n. Then $\lambda_n = O(1/n)$.*

Proof. Since the sequence λ_n is decreasing,

$$\lambda_n - \lambda_{n+1} = \lambda_{n+1}\left(\frac{\lambda_n}{\lambda_{n+1}} - 1\right) \geqslant \mu\lambda_{n+1}^2,$$

and so it follows that $\dfrac{\lambda_n}{\lambda_{n+1}} - 1 \geqslant \mu\lambda_{n+1}$.

Setting $v_n = \lambda_n n$, we conclude from the last inequality, after some simple rearrangements,

$$\frac{v_n}{v_{n+1}} \geqslant \frac{n}{n+1}\left(1 + \mu\frac{v_{n+1}}{n+1}\right). \tag{8}$$

If $v_{n+1} \geqslant 2/\mu$ for some n, then it follows from (8) that $v_{n+1} \leqslant v_n$. Hence $v_{n+1} \leqslant \max(2/\mu, v_n)$, and it thus follows that $v_n \leqslant \max(2/\mu, v_1)$ for all n. The lemma is thus proved.

Proof of Theorem 3. Write $\lambda_n = \Phi(x_n) - Q$. Assume that $\lambda_n > 0$ for all n. We show that the sequence λ_n satisfies the hypotheses of Lemma 4 (for some $\mu > 0$). To this end we first seek a bound for $\lambda_n - \lambda_{n+1} = \Phi(x_n) - \Phi(x_{n+1})$. It follows from Lemma 2 that, for all real α,

$$\Phi(x_n + \alpha z_{n+1}) \leqslant \Phi(x_n) + \alpha\Phi'(x_n)(z_{n+1}) + \frac{\alpha^2}{2} L\|z_{n+1}\|^2.$$

Since $\|z_{n+1}\| = 1$ and $\Phi'(x_n)(z_{n+1}) = -\|\Phi'(x_n)\|$, we have

$$\Phi(x_n + \alpha z_{n+1}) \leqslant \Phi(x_n) - \alpha\|\Phi' x_n\| + \frac{\alpha^2}{2} L. \tag{9}$$

Let $\tilde{\alpha} = \dfrac{1}{L}\|\Phi'(x_n)\|$. (The quadratic trinomial on the right-hand side of (9) attains a minimum at this point.) Using (9), we have

$$\lambda_n - \lambda_{n+1} = \Phi(x_n) - \Phi(x_{n+1}) \geqslant \Phi(x_n) - \Phi(x_n + \tilde{\alpha} z_{n+1}) \geqslant$$

$$\geqslant \tilde{\alpha}\|\Phi'(x_n)\| + \frac{\tilde{\alpha}^2}{2} L = \frac{L}{2}\|\Phi'(x_n)\|^2.$$

On the other hand, in view of Lemma 3, we have, for some $c > 0$,

$$\lambda_n = \Phi(x_n) - Q \leqslant c\|\Phi'(x_n)\| \qquad (n = 1, 2, \ldots).$$

Hence, for all n,

$$\lambda_n - \lambda_{n+1} \geqslant \frac{L}{2c^2}\lambda_n^2.$$

To complete the proof, we refer to Lemma 4.

For the wide class of functionals considered in Theorem 3, the bound for the convergence can hardly be improved. However, for "good" functionals the method converges much more rapidly. A twice differentiable functional Φ defined on Hilbert space **H** is said to be strongly convex if there exist positive numbers M and m such that

$$m\|z\|^2 \leqslant (\Phi''(x)(z), z) \leqslant M\|z\|^2 \tag{10}$$

for all $x, z \in \mathbf{H}$. Here* $\Phi''(x)$ is the second derivative of the functional Φ at x. It is not hard to verify that, if Φ is strongly convex, then it is convex. The inequality (10) means, in fact, that the operators $\Phi''(x)$ are positive definite, for all $x \in \mathbf{H}$ and the bounds of these operators are uniformly bounded by positive constants. In particular, the quadratic functional F considered in § 1 is strongly convex (for this functional $F''(x) = U$ for all x).

The results relating to the method of steepest descent for a strongly convex functional are in many respects analogous to the results of §1 for a quadratic functional. It can be shown that for every $x_0 \in \mathbf{H}$ the set $\Omega_0 = \{z \in \mathbf{H}\, \Phi(z) \leqslant \Phi(x_0)\}$ is bounded and the functional Φ attains a minimum on this set (and therefore also on the whole space); moreover, the minimum x^* is unique. As for quadratic functionals, we have the following bound for the convergence,

$$\|x_n - x^*\| \leqslant C\left(\frac{M-m}{M+m}\right)^n,$$

where C is some constant depending on the initial point x and the functional Φ.

4.5. The idea of steepest descent turns out to be highly fruitful also for solving problems on conditional extrema. One use of the idea in this situation leads to the so-called *conditional gradient method.*

Let Ω be a convex weakly compact set in a Banach space $\mathbf{X}$. We assume that the structure of Ω is sufficiently simple, in the sense that a solution of the minimization problem on Ω is known for a linear functional. Suppose further that Φ is a (non-linear) differentiable functional defined on some open region containing Ω. We are required to find the minimum of Φ on Ω. The conditional gradient method enables one to construct a sequence $x_0, x_1, \ldots, x_n, \ldots$, which in many cases turns out to be a minimizing sequence. We take x_0 to be an arbitrary element of Ω. If x_n is already known, then x_{n+1} is defined by

$$x_{n+1} = x_n + \varepsilon_n(z_n - x_n),$$

where $z_n \in \Omega$ is chosen from the condition $\Phi'(x_n)(z_n) = \min\limits_{z \in \Omega} \Phi'(x_n)(z)$, and the descent value ε_n is found from the equation $\Phi(x_{n+1}) = \min\limits_{\alpha \in [0,1]} \Phi(x_n + \alpha(z_n - x_n))$. (Other methods of defining the descent value are also possible, but we shall not dwell on them.) Thus, whereas in the method of steepest descent the direction of descent is found by minimizing the derivative $\Phi'(x)$ on the unit sphere (the direction sphere) and the descent value by minimizing the functional on a ray, in the conditional gradient method the direction of descent is found by minimizing the derivative on a specified set and the descent value by minimizing the functional on an interval. Roughly speaking, the conditional gradient method is a modification of the method of steepest descent for determining extrema subject to restrictions.

Analogues of Theorems 1, 2 and 3 hold for the conditional gradient method. Here a point $y \in \Omega$ is called a stationary point of Φ on Ω if $\Phi'(y)(y) = \min\limits_{z \in \Omega} \Phi'(y)(z)$. It is easy to verify that a local minimum of Φ on Ω is stationary, and that, if Φ is a convex functional, then the stationary values of Φ on Ω are the global minima. If $\mathbf{X}$ is a Hilbert space, Φ is a

---

* $\Phi''(x)$ is a linear operator from $\mathbf{X} = \mathbf{H}$ into $\mathbf{X}^*$ defined by $\Phi'(x+h) = \Phi'(x) + \Phi''(x)(h) + o(h)$, where $\dfrac{o(h)}{\|h\|} \to 0$ as $h \to 0$. For more details, see Ch. XVII.

convex functional, $\inf_{x \in \Omega} \|\Phi'(x)\| > 0$, and Ω is a strongly convex set (i.e. there exists $\gamma > 0$ such that the set contains with each pair of elements z, x also the elements $\frac{x+z}{2} + \gamma u$, where $\|u\| \leqslant \|x-z\|^2$), then the sequence constructed using the conditional gradient method converges to a unique minimum of Φ on Ω at the rate of a geometrical progression. For a detailed account of the conditional gradient method, see Dem'yanov and Rubinov.

§5. Minimization of convex functionals on finite-dimensional spaces

5.1. In §4 we considered the method of steepest descent in the problem of minimizing a continuously differentiable functional. Below we shall be studying the problem of minimizing a convex functional on finite-dimensional Euclidean space $\mathbf{R}^n$.

Let Φ be an arbitrary convex functional.* It is well known (see, for example, Rockafellar) that a convex functional defined on $\mathbf{R}^n$ is continuous.

A vector $v \in \mathbf{R}^n$ is called a *subgradient* of Φ at x if

$$\Phi(z) \geqslant \Phi(x) + (v, z-x) \tag{1}$$

for all $z \in \mathbf{R}^n$.

The set of subgradients is non-empty, convex, closed and bounded. It is called the *subdifferential* of Φ at x and is denoted by $\partial\Phi(x)$. The subdifferential is a many-valued mapping; we shall be concerned with these in more detail in Ch. XVI, §5. Here we merely note that the mapping $\partial\Phi(x)$ is semicontinuous from above: that is, if $x_k \to x$, $v_k \to v$, where $v_k \in \partial\Phi(x_k)$, then $v \in \partial\Phi(x)$.

The convex functional Φ has directional derivatives: that is, for every $x \in \mathbf{R}^n$ and $y \in \mathbf{R}^n$ the following limit exists

$$\frac{1}{\|y\|} \lim_{\alpha \to +0} \frac{1}{\alpha}[\Phi(x+\alpha y) - \Phi(x)] \equiv \frac{\partial\Phi}{\partial y}(x),$$

and

$$\|y\| \frac{\partial\Phi}{\partial y}(x) = \max_{v \in \partial\Phi(x)} (v, y). \tag{2}$$

A direction $y(x)$, $\|y(x)\| = 1$, is called a *direction of steepest descent* for Φ at x if (cf. (4.1))

$$\frac{\partial\Phi}{\partial y(x)}(x) = \inf_{\|y\|=1} \frac{\partial\Phi}{\partial y}(x).$$

One can see from (2) that if $\inf_{\|y\|=1} \frac{\partial\Phi}{\partial y}(x) < 0$, then a direction of steepest descent exists, is unique, and can be found from the following formula (cf. example 1) (in 4.1):

$$y(x) = -\frac{z(x)}{\|z(x)\|}, \tag{3}$$

where

$$\|z(x)\| = \min_{z \in \partial\Phi(x)} \|z\| \equiv \rho(x).$$

* For the definition of convex functional, see subsection 4.2.

LEMMA 1. *A necessary and sufficient condition for a convex functional* Φ *to attain a minimum on* $\mathbf{R}^n$ *at a point* x *is that*

$$\inf_{\|y\|=1} \frac{\partial \Phi}{\partial y}(x) \geqslant 0. \tag{4}$$

Proof. The necessity of (4) is obvious. Let us prove its sufficiency. Assume that (4) is satisfied. We shall show that x is then a minimum. Assume the contrary. Suppose that a point z exists for which $\Phi(z) < \Phi(x)$. Consider the direction $y = \dfrac{z-x}{\|z-x\|}$. Then by the definition of a convex functional, we have, for small α,

$$\Phi(x+\alpha y) = \Phi\left(\left(1 - \frac{\alpha}{\|z-x\|}\right)x + \frac{\alpha}{\|z-x\|}z\right) \leqslant$$

$$\leqslant \left(1 - \frac{\alpha}{\|z-x\|}\right)\Phi(x) + \frac{\alpha}{\|z-x\|}\Phi(z),$$

and so

$$\frac{\partial \Phi}{\partial y}(x) \equiv \lim_{\alpha \to +0} \frac{1}{\alpha}[\Phi(x+\alpha y) - \Phi(x)] \leqslant \frac{1}{\|z-x\|}(\Phi(z) - \Phi(x)) < 0,$$

contradicting (4).

Geometrically, the condition (4) is equivalent to

$$\mathbf{0} \in \partial \Phi(x). \tag{5}$$

Condition (5) is a generalization of the necessary condition for a minimum for continuously differentiable functions: $\Phi'_x = \mathbf{0}$.

Let us give an example. Let

$$\Phi(x) = \max_{z \in G} F(x, z), \tag{6}$$

where G is a compactum, the functionals $F(x, z)$ and $F'_x(x, z)$ are continuous with respect to the variables in $\mathbf{R}^n \times G$, and $F(x, z)$ is convex in x for each fixed $z \in G$. Then Φ is a convex functional, and

$$\partial \Phi(x) = \operatorname{co}\{F'_x(x, z):\ z \in R(x)\},$$

where

$$R(x) = \{z \in G:\ F(x, z) = \Phi(x)\},$$

and co A is the convex hull of the set A.

REMARK. The functional (6), generally speaking, also has directional derivatives when F is not convex in x. What is more, formula (2) holds, although $\partial \Phi(x)$ is no longer a subdifferential.

Let $\varepsilon > 0$. A vector v is called an ε-*subgradient* of a functional Φ at a point x if

$$\Phi(z) \geqslant \Phi(x) + (v, z - x) - \varepsilon \tag{7}$$

for all $z \in \mathbf{R}^n$.

We denote the set of all ε-subgradients by $\partial_\varepsilon \Phi(x)$. This set is closed, convex and bounded. A point x is called an ε-*stationary point* of Φ if

$$\mathbf{0} \in \partial_\varepsilon \Phi(x). \tag{8}$$

One sees easily from (7) that

$$0 \leqslant \Phi(x) - \min_{z \in \mathbf{R}^n} \Phi(z) \leqslant \varepsilon. \tag{9}$$

If $\mathbf{0} \notin \partial_\varepsilon \Phi(x)$, then the direction $y_\varepsilon(x) = -\dfrac{z_\varepsilon(x)}{\|z_\varepsilon(x)\|}$, where $\|z_\varepsilon(x)\| = \min_{z \in \partial_\varepsilon \Phi(x)} \|z\| \equiv \rho_\varepsilon(x)$, is called the *direction of ε-steepest descent* for Φ at the point x. The mapping $\partial_\varepsilon \Phi(x)$ is also semicontinuous from above.

Let $v \in \partial \Phi(x)$. Then, from (1), we have

$$\Phi(z) \geqslant \Phi(x) + (z - x, v) = \Phi(x_0) + (v, z - x_0) + [\Phi(x) - \Phi(x_0) + (v, x_0 - x)].$$

It is thus clear that, if x is sufficiently close to x_0,

$$\partial \Phi(x) \subset \partial_\varepsilon \Phi(x_0). \tag{10}$$

5.2. In this situation it is natural to consider the method of steepest descent.

Choose any $x_0 \in \mathbf{R}^n$. Assume that we have already determined $x_k \in \mathbf{R}^n$. If $\mathbf{0} \in \partial \Phi(x_k)$, then the point x_k is a minimum, and the process terminates. However, if $\mathbf{0} \notin \partial \Phi(x_n)$, then we find $y_{k+1} = y(x_k)$ (see (3)). The direction y_{k+1} is the direction of steepest descent for Φ at x_k. We now consider the ray

$$x_{k\alpha} = x_k + \alpha y_{k+1} \qquad (\alpha \geqslant 0)$$

and find

$$\min_{\alpha \geqslant 0} \Phi(x_{k\alpha}) = \Phi(x_{k\alpha_k}).$$

Set $x_{k+1} = x_{k\alpha_k}$. Obviously, $\Phi(x_{k+1}) < \Phi(x_k)$.

However, the method we have just described may, in fact, not converge to a minimum (the "jamming" effect), in which case $\partial \Phi(x)$ is said to be discontinuous (for an example of this, see the book by Dem'yanov and Malozemov).

5.3. Let $\varepsilon > 0$. To determine the ε-stationary points we can use the following method. Take any $x_0 \in \mathbf{R}^n$. Assume that the set $\Omega_0 = \{x : \Phi(x) \leqslant \Phi(x_0)\}$ is bounded. Suppose we have already found $x_k \in \mathbf{R}^n$. If $\mathbf{0} \in \partial_\varepsilon \Phi(x_k)$, then x_k is an ε-stationary point, and the process terminates. However, if $\mathbf{0} \notin \partial_\varepsilon \Phi(x_k)$, then we find $y_{k+1} = y_\varepsilon(x_k)$ and consider the ray $x_{k\alpha} = x_k + \alpha y_{k+1}$ $(\alpha \geqslant 0)$. We find $\min_{\alpha \geqslant 0} \Phi(x_{k\alpha}) = \Phi(x_{k\alpha_k})$. Set $x_{k+1} = x_{k\alpha_k}$. Obviously, $\Phi(x_{k+1}) < \Phi(x_k)$.

Now we proceed in a similar manner. As a result, we obtain a sequence of points $\{x_k\}$. If this sequence contains a finite number of points, then the point constructed last is ε-stationary. However, if the sequence is infinite, we have

THEOREM 1. *Every cluster point of the sequence $\{x_k\}$ is an ε-stationary point of Φ.*

Proof. Suppose that $x_{k_s} \to \tilde{x}$. We need to show that $\mathbf{0} \in \partial_\varepsilon \Phi(\tilde{x})$. Assume the contrary. Suppose $\rho_\varepsilon(\tilde{x}) = a > 0$. As the mapping $\partial_\varepsilon \Phi(x)$ is semicontinuous from above, there exists $\delta > 0$ such that $\rho_\varepsilon(x) \geqslant a/2$ for all

$$x \in S_\delta(\tilde{x}) = \{x : \|x - \tilde{x}\| \leqslant \delta\}.$$

This means that, for large enough k_s, we have

$$\rho_\varepsilon(x_{k_s}) \geqslant \frac{a}{2}. \tag{11}$$

We conclude from (10) that there exists $\delta_1 > 0$ such that $\partial\Phi(x) \subset \partial_\varepsilon\Phi(x_{k_s})$ for all $x \in S_{\delta_1}(x_{k_s})$ (and for sufficiently large k_s). Thus the functional Φ is decreasing on the ray $x_{k_s a}$, at a rate of absolute value at least $a/2$ and at a distance of at least δ_1. Hence

$$\Phi(x_{k_s+1}) \leqslant \Phi(x_{k_s}) - \delta_1 \frac{a}{2}.$$

Thus $\Phi(x_k) \to -\infty$, which contradicts the hypothesis that Ω_0 is a bounded set.

5.4. To determine minima of Φ we can now proceed as follows: choose $\varepsilon_0 > 0$, $\rho_0 > 0$, $x_0 \in \mathbf{R}^n$ arbitrarily. Suppose that $\varepsilon_k > 0$, $\rho_k > 0$, $x_k \in \mathbf{R}^n$ have already been found. If $\mathbf{0} \in \partial\Phi(x_k)$, then x_k is a minimum. Otherwise, by applying the method described in § 5.3 (and taking x_k as initial point), we reach, after a finite number of steps, a point x_{k+1} such that $\rho_{\varepsilon_k}(x_{k+1}) \leqslant \rho_k$. Now set $\varepsilon_{k+1} = \beta\varepsilon_k$, $\rho_{k+1} = \beta\rho_k$, where $\beta \in (0, 1)$ is independent of k. We thus obtain a sequence $\{x_k\}$. One can then prove the following theorem without any difficulty (see, for example, the proof of Theorem III.7.2 in Dem'yanov and Malozemov):

THEOREM 2. *Every cluster point of the sequence $\{x_k\}$ is a minimum of the functional Φ.*

REMARK. Above we have described the "principal" steps of the algorithms. In applying them in practice, a number of auxiliary problems (determination of the direction of descent, minimization on a ray) are solved by approximate methods.

5.5. Now we consider the generalized gradient method (see Ermol'ev and Shor [1]). Let Φ be a convex functional on $\mathbf{R}^n$, attaining a minimum, let M be the set of minima, and let $\rho(x) = \min_{z \in M} \|z - x\|$. We write M_ε for an ε-neighbourhood of M:

$$M_\varepsilon = \{x : \rho(x) \leqslant \varepsilon\}, \qquad \varepsilon > 0.$$

Let $\partial\Phi(x)$ be the subdifferential of Φ at x:

$$\partial\Phi(x) = \{z \in \mathbf{R}^n : \Phi(y) - \Phi(x) \geqslant (z, y - x) \text{ for all } y \in \mathbf{R}^n\}.$$

We choose an $x_0 \in \mathbf{R}^n$ and a sequence of positive numbers $\{\lambda_k\}$ such that $\lambda_k \to 0$, $\sum_{k=0}^{\infty} \lambda_k = \infty$.

We construct the sequence $\{x_k\}$ according to the formula

$$x_{k+1} = x_k - \lambda_k y_{k+1}, \qquad y_{k+1} = z_{k+1}/\|z_{k+1}\|,$$

where z_{k+1} is an arbitrary vector in $\partial\Phi(x_k)$. If, at some stage, $z_k = \mathbf{0}$, then $x_k \in M$ and the process terminates.

Suppose now that the sequence $\{x_k\}$ is infinite.

THEOREM 3. *If the set M is bounded, then*

$$\rho(x_k) \to 0, \quad \Phi(x_k) \to \Phi \equiv \min_{x \in \mathbf{R}^n} \Phi(x).$$

Proof. Let $\Omega(x) = \{y : \Phi(y) \leqslant \Phi(x)\}$. As M is bounded, $\Omega(x)$ is also bounded, for each $x \in \mathbf{R}^n$ (see Rockafellar). Write

$$T(x) = \{y : \Phi(y) = \Phi(x)\}, \qquad b(x) = \min_{y \in T(x)} \rho(y).$$

It is not hard to show that

$$M_{b(x)} \subset \Omega(x). \tag{12}$$

Suppose $x \notin M$. Then $b(x) > 0$. Choose an arbitrary direction $y = z/\|z\|$, $z \in \partial\Phi(x)$. For all $v \in \Omega(x)$, we have

$$(y, v - x) \leqslant \frac{1}{\|z\|}(\Phi(v) - \Phi(x)) \leqslant 0.$$

In particular, by (12), we have, for all $z \in M$,

$$0 \geqslant (y, z + b(x)y - x).$$

From this one sees without difficulty that the functional ρ is decreasing in the direction of $-y$, and that

$$\lim_{k \to \infty} b(x_k) = 0. \tag{13}$$

It follows from (13) that there is a subsequence $\{x_{k_s}\}$ such that $b(x_{k_s}) \to 0$. Then also $\rho(x_{k_s}) \to 0$. In fact it can be shown that the whole sequence $\rho(x_k)$ also converges to zero: thus $\Phi(x_k) \to \tilde{\Phi}$.

REMARK. In the case where Φ is continuously differentiable, its subdifferential consists of a single point (its gradient). Then the method of steepest descent (in subsection 5.4) coincides with the method of steepest descent considered in §4, provided the descent value is chosen by the second method mentioned in 1.1. The generalized gradient method (subsection 5.5) coincides with the method of steepest descent, provided the descent value is chosen by the third method of subsection 1.1.

The method of steepest descent is described in detail in Dem'yanov and Malozemov, while the generalized gradient method can be found, for example, in the paper by Ermol'ev and Shor [1].

The methods described in this section have also been extended to the case of minimization subject to restrictions. The extension is based on the necessary conditions for minima (see, for example, Dubovitskii and Milyutin [1], and Pshenichnii).

XVI

THE FIXED-POINT PRINCIPLE

§ 1. The Caccioppoli–Banach principle

IN THIS chapter and the one following we shall mainly be considering non-linear operators and functionals.

1.1. We begin our study of non-linear equations with the simplest case—a generalization of Banach's Theorem on inverse operators (V.4.5).

Consider a complete metric space $\mathbf{X}$ (not necessarily linear) and a closed subset Ω of $\mathbf{X}$. Assume that there is a mapping P defined on Ω that maps Ω into itself. A point $x^* \in \Omega$ is a *fixed point* of P if

$$x^* = P(x^*).$$

Thus the fixed points of P are the solutions of the equation

$$x = P(x). \tag{1}$$

The mapping P need not have fixed points; this will be the case, for instance, if $\mathbf{X} = \Omega$ is a metric vector space and

$$P(x) = x + x_0 \qquad (x_0 \neq \mathbf{0}).$$

We shall say that a mapping P is a *contraction operator* if there exists $\alpha < 1$ such that

$$\rho(P(x), P(x')) \leqslant \alpha\rho(x, x') \tag{2}$$

for all $x, x' \in \Omega$. It turns out that, if P is a contraction operator, one can guarantee the existence, and even uniqueness, of a fixed point. In fact, we have

THEOREM 1. *If P is a contraction operator, then there is a unique solution x^* in Ω for equation* (1).

Moreover, x^* can be obtained as the limit of the sequence $\{x_n\}$, where

$$x_{n+1} = P(x_n) \qquad (n = 0, 1, \ldots),$$

and where x_0 is an arbitrary element of Ω.

The rate of convergence of $\{x_n\}$ to the solution is given by the inequality

$$\rho(x_n, x^*) \leqslant \frac{\alpha^n}{1-\alpha}\rho(x_1, x_0) \qquad (n = 0, 1, \ldots). \tag{3}$$

Proof. Since

$$x_{n+1} = P(x_n), \qquad x_n = P(x_{n-1}),$$

we have, from (2),

$$\rho(x_{n+1}, x_n) \leqslant \alpha\rho(x_n, x_{n-1}).$$

Using inequalities analogous to this in succession, we obtain

$$\rho(x_{n+1}, x_n) \leqslant \alpha^n \rho(x_1, x_0).$$

Therefore

$$\rho(x_{n+p}, x_n) \leqslant \rho(x_{n+p}, x_{n+p-1}) + \ldots + \rho(x_{n+1}, x_n) \leqslant$$

$$\leqslant (\alpha^{n+p-1} + \ldots + \alpha^n)\rho(x_1, x_0) \leqslant \frac{\alpha^n}{1-\alpha}\rho(x_1, x_0). \quad (4)$$

Since $\alpha^n \to 0$, this bound shows that $\{x_n\}$ is a Cauchy sequence, and so, as $\mathbf{X}$ is complete, it converges to an element $x^* \in \mathbf{X}$. As every x_n belongs to Ω and Ω is closed, we have $x^* \in \Omega$, so $P(x^*)$ makes sense. Again, using (2), we have

$$\rho(x_{n+1}, P(x^*)) = \rho(P(x_n), \ P(x^*)) \leqslant \alpha\rho(x_n, x^*) \qquad (n = 0, 1, \ldots)$$

and, since the right-hand side tends to zero,

$$x_{n+1} \to P(x^*),$$

so that

$$x^* = P(x^*).$$

The uniqueness of the solution also follows from (2). For if there were another solution $\tilde{x} \in \Omega$, then we would have

$$\rho(\tilde{x}, x^*) = \rho(P(\tilde{x}), \ P(x^*)) \leqslant \alpha\rho(\tilde{x}, x^*),$$

but this is only possible if $\rho(\tilde{x}, x^*) = 0$, i.e. if $\tilde{x} = x^*$.

We obtain the inequality (3) by letting $p \to \infty$ in (4).

REMARK. The inequality (3) yields a possible domain of solution of (1). In particular, with $n = 0$,

$$\rho(x^*, x_0) \leqslant \frac{1}{1-\alpha}\rho(x_1, x_0). \quad (5)$$

1.2. In general, we cannot replace (2) by the weaker condition

$$\rho(P(x), \ P(x')) < \rho(x, x')(x, x' \in \Omega, \ x \neq x'). \quad (6)$$

For suppose $\mathbf{X} = \mathbf{R}^1$, the set of real numbers, $\Omega = \mathbf{X}$, and suppose P is defined as follows:

$$P(x) = \frac{\pi}{2} + x - \operatorname{arctg} x.$$

It is not hard to see that this P has no fixed points, although

$$\rho(P(x), \ P(x')) = |P(x) - P(x')| = |P'(\xi)| \, |x - x'| = \left| \frac{\xi^2}{1+\xi^2} \right| |x - x'| < \rho(x, x')$$

(ξ being a point between x and x'). However, we have

THEOREM 2. *If P maps the closed set Ω into a relatively compact set $\Delta \subset \Omega$, and if condition (6) is satisfied, then P has a unique fixed point.*

Proof. We consider P on the compactum $\Delta \subset \Omega$. As P is obviously continuous, so is the function $\phi(x) = \rho(x, P(x))$. The function ϕ assumes its least value on the compactum $\overline{\Delta}$ at a point $x_0 \in \bar{\Delta}$:

$$\rho(x_0, P(x_0)) = \min_{x \in \bar{\Delta}} \rho(x, P(x)).$$

Assume that $\rho(x_0, P(x_0)) > 0$. By (6), we have

$$\rho(P(x_0), P^2(x_0)) < \rho(x_0, P(x_0)) = \min_{x \in \bar{\Delta}} \rho(x, P(x)). \tag{7}$$

However, $P(x_0) \in \bar{\Delta}$ and so we have a contradiction to (7). Therefore $\rho(x_0, P(x_0)) = 0$ and $x_0 = P(x_0)$.

If $\tilde{x} \in \Omega$ were some other fixed point of P then we should have

$$\rho(\tilde{x}, x_0) = \rho(P(\tilde{x}), \; P(x_0)) < \rho(\tilde{x}, x_0),$$

which leads to a contradiction. This proves the theorem.

1.3. It often happens that the mapping P depends on a parameter, either numerical or of some other type. Then the solution of (1) will also depend on this parameter. It can be shown that if P depends continuously on the parameter then the solution also depends continuously on the parameter. Let us state this more precisely.

Suppose we have another metric space $\mathbf{Y}$, as well as the space $\mathbf{X}$. Assume that with each $y \in \mathbf{Y}$ there is associated a mapping P_y of $\Omega \subset \mathbf{X}$ into itself. We say that P_y is *continuous in y at a point $y_0 \in \mathbf{Y}_0$* if for every sequence $\{y_n\} \subset \mathbf{Y}_0$ with $y_n \to y_0$, we have

$$P_{y_n}(x) \to P_{y_0}(x), \qquad \text{for each } x \in \Omega. \tag{8}$$

Consider the following equation (or, more precisely, family of equations):

$$x = P_y(x). \tag{9}$$

Assume that this has a unique solution for each $y \in \mathbf{Y}$, which will obviously depend on y, so we naturally denote it by x_y^*. We shall say that the solution of (9) *depends continuously on y at $y = y_0$* if, for every sequence $\{y_n\} \subset \mathbf{Y}$ with $y_n \to y_0$, we have $x_{y_n}^* \to x_{y_0}^*$.

In the case where P_y is a contraction operator for each $y \in \mathbf{Y}_0$ the continuity of P_y implies that of the solution of (9). More precisely, we have

THEOREM 3. *If P_y satisfies condition* (2) *for each $y \in \mathbf{Y}_0$, with α independent of y, and if P_y is continuous in y at a point $y_0 \in \mathbf{Y}_0$, then the solution of* (9) *depends continuously on y at $y = y_0$.*

Proof. Let y be an arbitrary element of $\mathbf{Y}_0$. We construct a solution x_y^* of (9) as the limit of $\{x_n\}$:

$$x_{n+1} = P_y(x_n) \qquad (n = 0, 1, \ldots; \; x_0 = x_{y_0}^*).$$

Since $x_{y_0}^* = P_{y_0}(x_{y_0}^*)$, we find from (5) that

$$\rho(x_y^*, x_{y_0}^*) \leqslant \frac{1}{1-\alpha} \rho(x_1, x_0) = \frac{1}{1-\alpha} \rho(P_y(x_{y_0}^*), P_{y_0}(x_{y_0}^*)).$$

The continuity of x_y^* at $y = y_0$ is now easily obtained from this using (8).

REMARK. We can regard the set of mappings P_y as a single mapping P associating to each pair (x, y) of elements $x \in \Omega$, $y \in \mathbf{Y}_0$, an element $P(x, y) = P_y(x)$. In exactly the same way, a solution x_y^* of (9) can be regarded as a mapping F associating to the element $y \in \mathbf{Y}_0$ an element $F(y) = x_y^*$.

The theorem above can thus be formulated as follows:

If, for each $y \in \mathbf{Y}_0$,

$$\rho(P(x, y),\ P(x', y)) \leqslant \alpha\rho(x, x') \qquad (x, x' \in \Omega)$$

($\alpha < 1$ being independent of y), and if, for each $x \in \Omega$, the mapping P is continuous in y at a point $y_0 \in \mathbf{Y}$, then the mapping F is also continuous at y_0.

The fixed point principle, in the form it is presented here, was established by Banach [1] and Caccioppoli. For generalizations to many-valued mappings, see Ioffe and Tikhomirov.

§ 2. Auxiliary propositions

In the next section we shall establish a second fixed point principle. Its proof involves certain difficulties, as it relies on deep results concerning the topological structure of finite-dimensional spaces. We present this auxiliary material from topology in the present section.

2.1. Let us first introduce some new concepts that will be needed in what follows. Consider a B-space $\mathbf{X}$ and a set of elements in $\mathbf{X}$:

$$x_0, x_1, x_2, \ldots, x_n. \tag{1}$$

Assume that the differences

$$x_1 - x_0, x_2 - x_0, \ldots, x_n - x_0 \tag{2}$$

are linearly independent, and form the convex hull $S(x_0, x_1, \ldots, x_n)$ of the elements (1). The set $S(x_0, x_1, \ldots, x_n)$ is called the *simplex* spanned by the *vertices* $x_0, x_1, \ldots, x_n$. The number n is called the *dimension* of the simplex.

REMARK. In the definition of simplex given here, the vertices do not all occur on the same footing. However, it can be shown (we leave it to the reader to do so) that if the differences (2) are linearly independent, then so are the differences in each of the n sets

$$x_0 - x_k, \ldots, x_{k-1} - x_k, x_{k+1} - x_k, \ldots, x_n - x_k \quad (k = 1, 2, \ldots, n).$$

In this way one can restore the symmetry in the roles of the vertices.

Suppose we have a simplex $S(x_0, x_1, \ldots, x_n)$; let us single out k distinct vertices x_{i_1}, $x_{i_2}, \ldots, x_{i_k}$ and form the simplex spanned by the remaining ones.* This simplex is called the ($(n-k)$-dimensional) *face* of the original simplex $S(x_0, x_1, \ldots, x_n)$ opposite the vertices $x_{i_1}, x_{i_2}, \ldots, x_{i_k}$.

Now again let $S = S(x_0, x_1, \ldots, x_n)$ be a simplex. Each element $x \in S$ can be expressed in the form

$$x = \alpha_0 x_0 + \alpha_1 x_1 + \ldots + \alpha_n x_n, \tag{3}$$

where the coefficients α_k are connected by the relation

$$\alpha_0 + \alpha_1 + \ldots + \alpha_n = 1, \quad \alpha_k \geqslant 0 \qquad (k = 0, 1, \ldots, n).$$

Let us prove that the representation (3) is unique. In fact, since $\alpha_0 = 1 - \sum_{k=1}^{n} \alpha_k$, we find from (3) that

$$x = x_0 + \sum_{k=1}^{n} \alpha_k (x_k - x_0). \tag{4}$$

---

* It follows from the remark made above that this can be done.

If, in addition to (3), we had another representation of the same form

$$x = \alpha'_0 x_0 + \alpha'_1 x_1 + \ldots + \alpha'_n x_n,$$

then, as above, we would find that

$$x = x_0 + \sum_{k=1}^{n} \alpha'_k (x_k - x_0),$$

which, with (4), yields

$$\sum_{k=1}^{n} \alpha_k (x_k - x_0) = \sum_{k=1}^{n} \alpha'_n (x_k - x_0),$$

but, as the differences (2) are linearly independent, this is only possible if

$$\alpha_1 = \alpha'_1,\ \alpha'_2 = \alpha_2, \ldots, \alpha'_n = \alpha_n,\ \alpha'_0 = \alpha_0.$$

We call the numbers $\alpha_0, \alpha_1, \ldots, \alpha_n$ determined uniquely by the element the *simplicial coordinates** of the point $x \in S$.

The *centre* of a simplex is the point for which all the coordinates are the same.

We now introduce the concept of a subdivision of a complex $S = S(x_0, x_1, \ldots, x_n)$. By a *subdivision* of a simplex we mean a representation of the simplex as a sum of a finite number of simplexes of the same dimension as the given one, these simplexes being determined inductively in the following way.

For a one-dimensional simplex $S(x_0, x_1)$, a subdivision is the set of two simplexes $S(x_0, \frac{1}{2}(x_0 + x_1))$ and $S(\frac{1}{2}(x_0 + x_1), x_1)$. Assume that we have already defined a subdivision for all simplexes of dimension less than n, and consider an n-dimensional simplex $S = S(x_0, x_1, \ldots, x_n)$. Choose any $(n-1)$-dimensional face of S. Since the face is an $(n-1)$-dimensional simplex, a subdivision is already defined for it. Suppose this consists of the ($(n-1)$-dimensional) simplexes $S_1, S_2, \ldots, S_m$. Adjoining the centre of S to the vertices of $S_k (k = 1, 2, \ldots, m)$, we form the n-dimensional simplex spanned by these $n+1$ points.† Carrying out a similar construction for each $(n-1)$-dimensional face of the given simplex, we obtain the required subdivision.

It is not hard to see that the simplexes constituting a subdivision have no interior points in common. Their intersection is a common face of all of them.

Now consider a simplex S and a subdivision of S. By forming a subdivision of each partial simplex, we obtain a double subdivision of S, and, if we continue the process, we obtain subdivisions of higher multiplicities.

Obviously the diameters of the partial simplexes tend to zero as the multiplicity of the subdivision increases.

2.2. Now we record a number of lemmas on which the proof of the second fixed point principle depends.

Suppose we have an n-dimensional simplex $S = S(x_0, x_1, \ldots, x_n)$ and some subdivision of S (of arbitrary multiplicity), consisting of the simplexes

$$S_1, S_2, \ldots, S_m. \tag{5}$$

* Sometimes $\alpha_0, \alpha_1, \ldots, \alpha_n$ are called the *barycentric coordinates* of $x \in S$, since if one places a mass α_k at each vertex x_k then the centre of gravity of the resulting system is situated at the point x.

† We leave it to the reader to convince himself that this is possible.

Assume that with each vertex z of each of the simplexes (5) there is associated one of the numbers $0, 1, \ldots, n$: that is, that a function v is defined on the set of vertices of the simplexes (5), taking the values $0, 1, \ldots, n$. Thus with each of the simplexes (5) there is associated a set $v(S_j)$ of $n+1$ numbers, corresponding to the vertices of this simplex. We shall call the simplex S_j normal if $v(S_j) = \{0, 1, \ldots, n\}$, the order of the numbers here being taken as insignificant.

If the function v is arbitrary, then, generally speaking, there may not be any normal simplexes. However, we have

LEMMA 1. *Suppose the function v satisfies the following condition: if a vertex z of a simplex in* (5) *belongs to a k-dimensional face $S(x_{i_0}, x_{i_1}, \ldots, x_{i_k})$ of S, then $v(z)$ can only be one of the numbers $i_0, i_1, \ldots, i_k$.*

Then there is at least one normal simplex among the simplexes (5).

Proof. We shall prove that the number of normal simplexes is odd. If the dimension of S is zero—that is, if S reduces to a single point—then the assertion is trivial. We assume that the assertion has been proved for simplexes of dimension $n-1$, and verify that it is true for a simplex of dimension n. Consider an $(n-1)$-dimensional face $S_k^{(i)}$ of S_k. We call this face distinguished if $v(S_k^{(i)}) = \{0, 1, \ldots, n\}$. If S_k is normal then one can easily verify that the number ρ_k of its distinguished faces is one; otherwise $\rho_k = 2$. Hence the sum $\rho = \sum_{k=1}^{m} \rho_k$ has the same parity as the number of normal simplexes. We now divide the distinguished faces of the simplexes (5) into two groups. In the first we put those faces that are not contained in any $(n-1)$-dimensional face of the original simplex S. We denote the number of such faces by ρ'. Since every distinguished face of this type is a common face of precisely two simplexes (5), it must be counted twice in calculating the sum ρ. In the second group we put those distinguished faces that are wholly contained in one of the $(n-1)$-dimensional faces of S. Since such a face is a face of only one of the simplexes (5), the number of these is $\rho'' = \rho - 2\rho'$. Using the hypothesis of the lemma, one can easily see that all the distinguished faces in the second group are contained in the face $S' = S(x_0, x_1, \ldots, x_{n-1})$ of S. But the $(n-1)$-dimensional faces of the simplexes (5) lying in S' form a subdivision of S', and the function v, regarded as pertaining only to this subdivision, satisfies the condition of the lemma. Hence, by the inductive assumption, the number of normal simplexes in this subdivision is odd. If we note that the normal simplexes in this case are just the distinguished faces of the second group, then we conclude that ρ'' is odd, and hence so is $\rho = 2\rho' + \rho''$.

Thus the lemma is proved.

LEMMA 2. *Suppose we have an n-dimensional simplex $S = S(x_0, x_1, \ldots, x_n)$ in a B-space of dimension n, and suppose there exist closed sets $F_0, F_1, \ldots, F_n$ such that*

$$S(x_{i_0}, x_{i_1}, \ldots, x_{i_k}) \subset \bigcup_{m=0}^{k} F_{i_m},$$

for every set $i_0, i_1, \ldots, i_k$ $(k = 1, 2, \ldots, n)$.

Then the intersection $\bigcap_{k=0}^{n} F_k$ is non-empty.

Proof. Consider a subdivision of S of arbitrary multiplicity. Let z be a vertex of one of the simplexes forming the subdivision, and let $S(x_i, x_{i_1}, \ldots, x_{i_k})$ be a face containing z. By the hypotheses of the lemma, $z \in \bigcup_{m=0}^{k} F_{i_m}$. Let us assume that $z \in F_{i_s}$, and write $v(z) = i_s$. This function v clearly satisfies the condition of Lemma 1, and so, by that lemma, there exists a normal partial simplex, which in this case is characterized by the fact that its intersection with each F_j is non-empty (containing one of the vertices).

Now consider a sequence of subdivisions of multiplicities $p = 1, 2, \ldots$. The p-th subdivision contains points $z_0^{(p)}, z_1^{(p)}, \ldots, z_n^{(p)}$—the vertices of a normal partial simplex—such that

$$z_k^{(p)} \in F_k \qquad (k = 0, 1, \ldots, n;\ p = 1, 2, \ldots). \tag{6}$$

As S is compact, we can find an increasing sequence $\{p_j\}$ of natural numbers such that the sequence $\{z_0^{(p_j)}\}$ is convergent. As S is closed, the limit z^* of this sequence belongs to S. Since the diameters of the partial simplexes tend to zero as $p \to \infty$, we have

$$z_k^{(p_j)} \to z^* \qquad (k = 0, 1, 2, \ldots, n).$$

Hence, if we bear in mind that the sets F_k are closed, we see from (6) that

$$z^* \in F_k \qquad (k = 0, 1, \ldots, n),$$

as we set out to prove.

2.3. The following lemma in fact yields a fixed point principle (Brouwer's principle), though its hypotheses are excessively restricted.

LEMMA 3. *A continuous mapping P from an n-dimensional simplex S in a* B*-space* $\mathbf{X}$ *into itself has a fixed point.*

Proof. Let $\alpha_0, \alpha_1, \ldots, \alpha_n$ be the simplicial coordinates of an arbitrary point $x \in S$. Write

$$\beta_j = f_j(\alpha_0, \alpha_1, \ldots, \alpha_n) \qquad (j = 0, 1, \ldots, n)$$

for the simplicial coordinates of the point $P(x)$. This notation is meaningful since $P(x) \in S$. Obviously, we have

$$\beta_0 + \beta_1 + \ldots + \beta_n = 1, \qquad \beta_j \geqslant 0 \qquad (j = 0, 1, \ldots, n).$$

It is not hard to verify, using the continuity of P, that the functions f_j are continuous. Write F_j $(j = 0, 1, \ldots, n)$ for the set of those points $x \in S$ for which

$$\alpha_j \geqslant \beta_j.$$

As the f_j are continuous, the sets F_j are closed. Let us also check that they satisfy the hypotheses of Lemma 2.

In fact, let $x \in S(x_{i_0}, x_{i_1}, \ldots, x_{i_k})$. Assume that $x \notin \bigcup_{m=0}^{k} F_{i_m}$. This means that the simplicial coordinates $\alpha_0, \alpha_1, \ldots, \alpha_n$ and $\beta_0, \beta_1, \ldots, \beta_n$ of x and $P(x)$, respectively, satisfy

$$\alpha_{i_m} < \beta_{i_m} \qquad (m = 0, 1, \ldots, k).$$

Since we must have $\alpha_i = 0$ for $i = i_m\,(m = 0, 1, \ldots, k)$, and hence $\alpha_i \leqslant \beta_i$ for such i, we see that

$$\sum_{i=0}^{n} \alpha_i < \sum_{i=0}^{n} \beta_i,$$

which is impossible, as both these sums must be equal to 1.

By Lemma 2, there exists a point $x^* \in \bigcap_{k=0}^{n} F_k$. If $\alpha_0^*, \alpha_1^*, \ldots, \alpha_n^*$ and $\beta_0^*, \beta_1^*, \ldots, \beta_n^*$ are the simplicial coordinates of x^* and $P(x^*)$ respectively, then

$$\alpha_j^* \geqslant \beta_j^* \qquad (j = 0, 1, \ldots, n),$$

so that

$$\sum_{i=0}^{n} \alpha_i^* \geqslant \sum_{i=0}^{n} \beta_i^*. \tag{7}$$

Moreover, we have equality here only if

$$\alpha_i^* = \beta_i^* \qquad (i = 0, 1, \ldots, n). \tag{8}$$

However, both the sums in (7) are equal to 1, so (8) must be true: that is, we have $x^* = P(x^*)$.

The lemma is thus proved.

2.4. We now extend Lemma 3 to arbitrary closed bounded convex sets in a finite-dimensional space. For this we first prove another lemma.

Recall (III.2.3) that a convex body in a B-space $\mathbf{X}$ is a convex closed set Ω having non-empty interior.

LEMMA 4. *Let Ω be a convex body in a* B*-space $\mathbf{X}$ which is bounded in norm, and let $B_{\mathbf{X}}$ be a closed unit ball in $\mathbf{X}$. Then Ω and $B_{\mathbf{X}}$ are homeomorphic.*

Proof. Without any loss of generality we may assume that $\mathbf{0}$ is an interior point of Ω and that $B_{\mathbf{X}} \subset \Omega$. Then Ω is a neighbourhood of $\mathbf{0}$ and we can consider its Minkowski functional p (II.3.2).

By the general properties of Minkowski functionals proved in Lemma III.2.1, we see that p is continuous, and that $\Omega = \{x \in \mathbf{X} : p(x) \leqslant 1\}$. If $x \neq \mathbf{0}$ is an arbitrary element of $\mathbf{X}$, then $x/\|x\| \in B_{\mathbf{X}} \subset \Omega$, and so $p(x/\|x\|) \leqslant 1$, leading to the inequality

$$p(x) \leqslant \|x\|, \tag{9}$$

which, of course, is also true when $x = \mathbf{0}$.

Furthermore, as Ω is bounded, there exists an open ball K_r, with centre at the origin and radius $r > 0$, containing Ω. Since $rx/\|x\| \notin K_r$, for all $x \neq \mathbf{0}$, we have also $rx/\|x\| \notin \Omega$, so that $p(rx/\|x\|) > 1$. Therefore

$$p(x) \geqslant \frac{1}{r}\|x\| \qquad (x \in \mathbf{X}). \tag{10}$$

We now define a mapping T by setting

$$T(x) = x\frac{p(x)}{\|x\|} \qquad (x \neq \mathbf{0}),\ T(\mathbf{0}) = \mathbf{0} \tag{11}$$

for all $x \in \Omega$.

As Ω coincides with the set of x for which $p(x) \leqslant 1$, we clearly have $T(\Omega) = B_{\mathbf{X}}$. Furthermore, it is immediately clear that T is one-to-one. Its inverse T^{-1} is given by the formula

$$T^{-1}(x) = x\frac{\|x\|}{p(x)} \qquad (x \neq \mathbf{0}), \quad T^{-1}(\mathbf{0}) = \mathbf{0}. \tag{12}$$

The continuity of T at a point $x_0 \neq \mathbf{0}$ follows from that of the functional p. Also if $x_n \to \mathbf{0}$, then, since $\|x_n/\|x_n\|\| = 1$ and $p(x_n) \to \mathbf{0}$, we have $T(x_n) \to \mathbf{0} = T(\mathbf{0})$.

In exactly the same way it can be checked that T^{-1} is continuous. If $x_n \to x_0 \neq \mathbf{0}$, then, since $p(x_0) \geqslant \frac{1}{r}\|x\| > 0$ (see (10)), the continuity of p implies that $T^{-1}(x_n) \to T^{-1}(x_0)$. For $x_0 = \mathbf{0}$, we have, using (10) again,

$$\|T^{-1}(x_n)\| = \frac{\|x_n\|}{p(x_n)} \|x_n\| \leqslant \frac{r\|x_n\|}{\|x_n\|} \|x_n\| = r\|x_n\| \to 0,$$

that is, $T^{-1}(x_n) \to \mathbf{0} = T^{-1}(\mathbf{0})$.

This completes the proof of the lemma.

COROLLARY. *A convex bounded body Ω in an n-dimensional* B*-space is homeomorphic to an n-dimensional simplex.*

For an n-dimensional simplex is a convex body in n-dimensional space, and so like Ω it is homeomorphic to the ball B.

Now we are in a position to prove the fundamental lemma.

LEMMA 5. *Let Ω be a convex bounded body in an n-dimensional* B*-space* $\mathbf{X}$. *A continuous mapping P from Ω into itself has a fixed point.*

Proof. Choose linearly independent elements $x_1, x_2, \ldots, x_n$ in $\mathbf{X}$. Taking $x_0 = \mathbf{0}$, we form a simplex S with vertices $x_0, x_1, \ldots, x_n$. By the Corollary to Lemma 4 there is a one-to-one bicontinuous mapping T from S onto Ω. Consider the mapping

$$P_0 = T^{-1}PT.$$

This mapping P_0 is continuous and maps S into itself. Hence by Lemma 3 it has a fixed point x_0:

$$x_0 = P_0(x_0).$$

But then $x^* = T(x_0)$ is a fixed point of P, since

$$P(x^*) = PT(x_0) = TT^{-1}PT(x_0) = TP_0(x_0) = T(x_0) = x^*.$$

The fixed point theorem for a continuous transformation of a simplex into itself was proved by Brouwer [1] in connection with establishing the invariance of dimension. The proof above is due to Kuratowski.

§ 3. Schauder's principle

3.1. Using the results of the last section we can establish a second fixed-point principle (Schauder's principle).

THEOREM 1. *A continuous mapping P that transforms a compact convex set Ω in a* B*-space* $\mathbf{X}$ *into itself has a fixed point.*

Proof. Take any $\varepsilon > 0$. As Ω is a compact set, it has a finite ε-net. Suppose this consists of the elements

$$x_1, x_2, \ldots, x_m. \tag{1}$$

Form the convex hull Ω_0 of the elements (1). Clearly $\Omega_0 \subset \Omega$. Furthermore, Ω has finite dimension* $n \leqslant m-1$.

---

* We say that the *dimension* of a set $E \subset \mathbf{X}$ is n if there exist elements $\bar{x}_0, \bar{x}_1, \ldots, \bar{x}_n$ such that (1) the differences $\bar{x}_k - \bar{x}_0$ ($k = 1, 2, \ldots, n$) are linearly independent and (2) each $x \in E$ is expressible in the form $x = \bar{x}_0 + \sum_{k=1}^{n} \alpha_k(\bar{x}_k - \bar{x}_0)$; moreover, for each $k = 1, 2, \ldots, n$, there exists an $x \in E$ such that the corresponding coefficient α_k is non-zero.

Using induction on the dimension, it is not hard to prove that Ω_0 can be expressed as a sum of n-dimensional simplexes such that

a) all the points (1) appear as vertices of the simplexes,

b) two simplexes either have no common points or intersect in a common (k-dimensional) face ($k = 0, 1, \ldots, n-1$).

Now we form a subdivision of each of these simplexes, of large enough multiplicity that the diameters of all the partial simplexes resulting from the subdivision are less than ε. Denote these simplexes by

$$S_1, S_2, \ldots, S_p. \tag{2}$$

Obviously the set of all vertices of the simplexes (2) also forms an ε-net. Notice also that the simplexes (2) satisfy both conditions a) and b).

Now consider the mapping P. As Ω is compact, continuity implies uniform continuity for P; so for any $\varepsilon > 0$ we can find a $\delta > 0$ such that $\|x - x'\| < \delta$ implies

$$\|P(x) - P(x')\| < \varepsilon \qquad (x, x' \in \Omega). \tag{3}$$

By subdividing the simplexes (2) if necessary, we can arrange that their diameters are less than δ. We shall assume that such a subdivision has already been carried out, so that the diameters of the simplexes (2) are not only less than ε but also less than δ.

Now we construct a mapping P_ε, a simplicial approximation to P, mapping Ω_0 into itself. First we define P_ε at the vertices of the simplexes (2). Let z be a vertex of one of the simplexes (2). Since $P(z) \in \Omega$ and the set of vertices of these simplexes forms an ε-net, there exists a vertex $\bar{z}$ such that $\|\bar{z} - P(z)\| < \varepsilon$. We set $\bar{z} = P_\varepsilon(z)$.

Now suppose $x \in \Omega_0$ is not a vertex of any of the simplexes (2). Assume that $x \in S_k$. We denote the vertices of S_k by $x_0^{(k)}, x_1^{(k)}, \ldots, x_n^{(k)}$ and express x in the form

$$x = \sum_{i=0}^{n} \alpha_i^{(k)} x_i^{(k)} \quad \left(\sum_{i=0}^{n} \alpha_i^{(k)} = 1;\ \alpha_i^{(k)} \geqslant 0;\ i = 0, 1, \ldots, n \right).$$

Write

$$P_\varepsilon(x) = \sum_{i=0}^{n} \alpha_i^{(k)} P_\varepsilon(x_i^{(k)}). \tag{4}$$

This definition requires no further explanation if the simplex containing x is unique. If, however, we have also $x \in S_r$ ($r \neq k$), then we need to show that $P_\varepsilon(x)$ does not depend on which of these simplexes we choose. According to condition b), the intersection of S_k and S_r is a common face of both. Assume that this face is spanned by the vertices $x_{i_0}^{(k)}$, $x_{i_1}^{(k)}, \ldots, x_{i_s}^{(k)}$ of S_k; if we denote the vertices of S_r by $x_0^{(r)}, x_1^{(r)}, \ldots, x_n^{(r)}$, then we can assume that

$$x_{i_j}^{(k)} = x_{i_j}^{(r)} \qquad (j = 1, 2, \ldots, s). \tag{5}$$

Let us express x in the form

$$x = \sum_{i=1}^{n} \alpha_i^{(r)} x_i^{(r)} \quad \left(\sum_{i=1}^{n} \alpha_i^{(r)} = 1;\ \alpha_i^{(r)} \geqslant 0,\ i = 0, 1, \ldots, n \right).$$

Bearing in mind the remark made in 2.1, we have

$$\alpha_i^{(k)} = \alpha_i^{(r)} = 0 \qquad (i \neq i_j;\ j = 1, 2, \ldots, s). \tag{6}$$

But then, by (5), we have

$$\sum_{j=0}^{s} \alpha_{ij}^{(k)} x_{ij}^{(k)} = \sum_{j=0}^{s} \alpha_{ij}^{(r)} x_{ij}^{(k)},$$

from which we obtain

$$\alpha_{ij}^{(k)} = \alpha_{ij}^{(r)} \qquad (j = 0, 1, \ldots, s). \tag{7}$$

Thus if we define $P_\varepsilon(x)$ starting with the simplex S_r, then, by (6), (5) and (7),

$$P_\varepsilon(x) = \sum_{i=0}^{n} \alpha_i^{(r)} P_\varepsilon(x_i^{(r)}) = \sum_{j=0} \alpha_{ij}^{(r)} P_\varepsilon(x_{ij}^{(r)}) = \sum_{j=0} \alpha_i^{(k)} P_\varepsilon(x_{ij}^{(k)})$$

$$= \sum_{i=0}^{n} \alpha_i^{(k)} P_\varepsilon(x_i^{(k)}),$$

as we wished to prove.

We can now convince ourselves by quite straightforward arguments that the mapping P_ε thus constructed is continuous on Ω_0 and maps Ω_0 into itself.

Thus the mapping P_ε satisfies the hypotheses of Lemma 5 of the last section. Hence we conclude from that lemma that P_ε has a fixed point $x_\varepsilon \in \Omega_0$:

$$x_\varepsilon = P_\varepsilon(x_\varepsilon).$$

Let $z_0, z_1, \ldots, z_n$ be the vertices of that one of the simplexes (2) to which x_ε belongs. Since $\|z_i - z_j\| < \delta$, it follows from (3) that

$$\|P(z_i) - P(z_j)\| < \varepsilon \qquad (i, j = 0, 1, \ldots, n).$$

Hence, if we bear in mind that, by definition, we have

$$\|P_\varepsilon(z_i) - P(z_i)\| < \varepsilon \qquad (i = 0, 1, \ldots, n), \tag{8}$$

then we find that

$$\|P_\varepsilon(z_i) - P_\varepsilon(z_j)\| < 3\varepsilon \qquad (i, j = 0, 1, \ldots, n). \tag{9}$$

Next, if x_ε is expressed in the form

$$x_\varepsilon = \sum_{i=0}^{n} \alpha_i^{(\varepsilon)} z_i \qquad \left(\sum_{i=0}^{n} \alpha_i^{(\varepsilon)} = 1;\ \alpha_i^{(\varepsilon)} \geqslant 0;\ i = 0, 1, \ldots, n \right),$$

then, by the definition of P_ε, we have

$$x_\varepsilon = P_\varepsilon(x_\varepsilon) = \sum_{i=0}^{n} \alpha_i^{(\varepsilon)} P_\varepsilon(z_i).$$

Using (9), we see from this that

$$\|x_\varepsilon - P_\varepsilon(z_j)\| = \left\| \sum_{i=0}^{n} \alpha_i^{(\varepsilon)} [P_\varepsilon(z_i) - P_\varepsilon(z_j)] \right\| \leqslant 3\varepsilon \sum_{i=0} \alpha_i^{(\varepsilon)} = 3\varepsilon \tag{10}$$

$$(j = 0, 1, \ldots, n).$$

Finally, using the inequalities

$$\|x_\varepsilon - z_i\| < \delta \qquad (i = 0, 1, \ldots, n),$$

we deduce from (3) that

$$\|P(x_\varepsilon) - P(z_i)\| < \varepsilon \qquad (i = 0, 1, \ldots, n). \tag{11}$$

Now using (10), (8) and (11), we finally conclude that

$$\|x_\varepsilon - P(x_\varepsilon)\| \leqslant \|x_\varepsilon - P_\varepsilon(z_i)\| + \|P_\varepsilon(z_i) - P(z_i)\| + \|P(z_i) - P(x_\varepsilon)\| < 5\varepsilon.$$

Thus for each $\varepsilon > 0$ there exists a point $x_\varepsilon \in \Omega$ such that

$$\|x_\varepsilon - P(x_\varepsilon)\| < 5\varepsilon.$$

Take a sequence $\varepsilon_k \to 0$ and construct the point $x_k = x_{\varepsilon_k}$ for each $k = 1, 2, \ldots$. As Ω is compact and closed, we may assume without loss of generality that

$$x_k \underset{k \to \infty}{\to} x^* \in \Omega.$$

But then we have

$$\|x^* - P(x^*)\| \leqslant \|x^* - x_k\| + \|x_k - P(x_k)\| + \|P(x_k) - P(x^*)\|,$$

and all the terms on the right-hand side tend to zero (the last one does so because P is continuous). Hence

$$x^* = P(x^*),$$

and the existence of a fixed point has thus been proved.

REMARK. We leave it to the reader to convince himself that both hypotheses of the theorem—the convexity and compactness of Ω—are essential. In the same way, one can easily construct an example to show that we cannot replace the compactness of Ω by the weaker requirement that it be bounded.

3.2. Theorem 1 can be stated in a rather more general form if we make use of the fact that the closed convex hull of a relatively compact set in a B-space is compact (Corollary to Theorem III.2.3).

THEOREM 2. *A continuous mapping P that transforms a closed convex subset Ω of a* B-*space* **X** *into a compact subset $\Delta \subset \Omega$ has a fixed point.*

Proof. Consider the closed convex hull $\Omega_0 = \overline{\text{co}}(\Delta)$ of the set Δ. This is a convex compact set, and it is contained in Ω.

The mapping P takes Ω_0 into Δ, and hence, as $\Delta \subset \Omega_0$, it maps Ω_0 into itself. Hence, if we regard P as a mapping on Ω_0, we find ourselves in the situation of Theorem 1, and we complete the proof by applying that theorem.

By making appropriate modifications to the proofs of Theorems 1 and 2 one can prove similar fixed-point theorems for the case where **X** is a locally convex space (see Dunford and Schwartz-I).

The fixed-point principle (Theorems 1 and 2) was established in a paper by Schauder [1]. Deeper fixed-point theorems have been proved by Leray (see Leray and Schauder [1]), in particular for locally convex spaces. The study of these problems has recently been significantly advanced by M. A. Krasnosel'skii (see Krasnosel'skii-I, Krasnosel'skii [1]–[3]).

§ 4. Applications of the fixed-point principle

We now consider some applications of the theorems proved above.

4.1. Suppose we have a function $\Phi(s, u)$ of two real variables defined in the strip $a \leqslant s \leqslant b$, $-\infty < u < \infty$; we assume that Φ is continuous and has a continuous derivative in u

satisfying the condition

$$0 < m \leqslant \Phi'_u(s, u) \leqslant M \qquad (a \leqslant s \leqslant b, \ -\infty < u < \infty). \tag{1}$$

Under these conditions, there exists a unique function $u = x^*(s)$ continuous in $[a, b]$ and satisfying

$$\Phi(s, x^*(s)) = 0. \tag{2}$$

This result can be obtained using Theorem 1.1. To this end we consider the operator P on the space $\mathbf{X} = \mathbf{C}[a, b]$, defined by

$$y = P(x), \ y(s) = x(s) - \frac{2}{M+m}\Phi(s, x(s)) \qquad (s \in [a, b]).$$

It can be verified without any difficulty that P is a contraction operator. In fact, if $y = p(x)$, $\tilde{y} = P(\tilde{x})$, then, for any $s \in [a, b]$, we have

$$|y(s) - \tilde{y}(s)| = \left| x(s) - \tilde{x}(s) - \frac{2}{M+m}[\Phi(s, x(s)) - \Phi(s, x(s))] \right| =$$

$$= |x(s) - \tilde{x}(s)| \left| 1 - \frac{2}{M+m}\Phi'_u(s, \theta(s)) \right| \leqslant \alpha \|x - \tilde{x}\|,$$

where

$$\alpha = \frac{M-m}{M+m},$$

and thus, by (1), we have $\alpha < 1$.

The operator P has a unique fixed point in $\mathbf{C}[a, b]$. It remains to note that the equation $x^* = P(x^*)$ is equivalent to equation (2).

4.2. Next we consider the system of ordinary differential equations

$$\frac{dx_i}{dt} = \phi_i(x_1, x_2, \ldots, x_n; t) \qquad (i = 1, 2, \ldots, n). \tag{3}$$

If we introduce n-dimensional vector functions—in other words, ordered systems of n real functions—we can write (3) as a single equation

$$\frac{dx}{dt} = \phi(x, t), \tag{4}$$

where $x(t)$ is the vector having components $x_1(t), x_2(t), \ldots, x_n(t)$, and $\phi(x, t)$ has components $\phi_1(x_1, \ldots, x_n;\ t), \ldots, \phi_n(x_1, x_2, \ldots, x_n;\ t)$. Here the derivative is understood to be the vector function having components $\frac{dx_1}{dt}, \frac{dx_2}{dt}, \ldots, \frac{dx_n}{dt}$.

Suppose that we wish to find a solution of the system (1) in the interval $[a, b]$, satisfying the initial conditions

$$x_i(a) = y_0^{(i)} \qquad (i = 1, 2, \ldots, n)$$

or, in vector form,

$$x(a) = y_0 \tag{5}$$

(y_0 being the element of $\mathbf{R}^n$ with coordinates $y_0^{(1)}, y_0^{(2)}, \ldots, y_0^{(n)}$). Solving this problem is equivalent to finding a solution of the integral equation

$$x(t) = y_0 + \int_a^t \phi(x(\tau), \tau)\, d\tau. \tag{6}$$

THEOREM 1. *Suppose the vector function* $\phi(y, t)$ *is defined for**

$$y \in \mathbf{R}^n, |y - y_0| < \delta, \qquad t \in [a, b'],$$

that it is bounded for these values of y, t, *and measurable in* t *for each value of* y, *and that it satisfies a Lipschitz condition in* y:

$$|\phi(y, t) - \phi(\tilde{y}, t)| \leqslant K|y - \tilde{y}|, \tag{7}$$

where K *is independent of* t.

Write

$$M = \sup|\phi(y, t)| \qquad (|y - y_0| \leqslant \delta, \qquad t \in [a, b']).$$

Then, provided the interval $[a, b] \subset [a, b']$ *is sufficiently small, provided, in fact, that*

$$b - a < \min\left[\frac{\delta}{M}, \frac{1}{K}\right], \tag{8}$$

the integral equation (6), *and hence also the system of differential equations* (3), *has a unique solution satisfying the given initial conditions. Moreover, the solution depends continuously on the initial vector* y_0.

Proof. We introduce the space $\mathbf{C}_n = \mathbf{C}([a, b], \mathbf{R}^n)$ of continuous n-dimensional real vector functions defined on $[a, b]$. If it is linearized in the natural way, $\mathbf{C}_n$ is turned into a B-space by setting

$$\|x\| = \max_{t \in [a, b]} |x(t)|$$

for each $x \in \mathbf{C}_n$.

With a view to applying Theorem 1.3 to equation (6), we set $\mathbf{X} = \mathbf{C}_n$, $\mathbf{Y} = \mathbf{R}^n$. Equation (6) can be written as a functional equation

$$x = P_{y_0}(x); \tag{9}$$

here P_y is the integral operator

$$z = P_y(x) \qquad (y \in \mathbf{R}^n, x \in \mathbf{C}_n),$$
$$z(t) = y + \int_a^t \phi(x(\tau), \tau)\, d\tau, \tag{10}$$

defined on the set Ω of all $x \in \mathbf{C}_n$ satisfying the condition

$$|x(t) - y_0| \leqslant \delta \qquad (t \in [a, b]).$$

We must now take $\mathbf{Y}_0$ to be the set of vectors $y \in \mathbf{R}^n$ for which

$$|y - y_0| < \delta_1 = \delta - M(b - a).$$

For each $y \in \mathbf{Y}_0$, the operator P_y maps Ω into itself. For

$$|z(t) - y_0| \leqslant |y - y_0| + \left|\int_a^t \phi(x(\tau), \tau)\, d\tau\right| \leqslant \delta_1 + M(b - a) = \delta.$$

Let us verify also that condition (2) of § 1 is satisfied. Let $x, \tilde{x} \in \Omega$ and $y \in \mathbf{Y}_0$; then

$$\|P_y(x) - P_y(\tilde{x})\| = \max_{t \in [a, b]} \left|\int_a^t [\phi(x(\tau), \tau) - \phi(\tilde{x}(\tau), \tau)]\, d\tau\right| \leqslant K(b - a)\|x - \tilde{x}\|.$$

* Here and below we denote the Euclidean length of a vector y by $|y|$.

Taking

$$\alpha = K(b-a),$$

we have $\alpha < 1$, in view of (8).

The continuity of P_y in y is obvious.

Thus, by Theorem 1.3, equation (9), and hence also the integral equation (6), has a unique solution depending continuously on y.

REMARK. In the present case, the fact that the solution x_y^* of system (3) is continuously dependent on the initial vector y at $y = y_0$ means this: for each $\varepsilon > 0$, there exists $\eta > 0$ such that $\|x_y^* - x_{y_0}^*\|_{\mathbf{C}} < \varepsilon$ whenever $|y - y_0| < \eta$.

4.3. By applying Schauder's principle to the operator (9) one can establish the existence of a solution to equation (6), and hence to the system (1), under weaker assumptions on the function $\phi(y, t)$. However, in this case one cannot prove the uniqueness of the solution, still less that it depends continuously on the initial vector.

THEOREM 2. *Suppose the function ϕ satisfies all the conditions of Theorem 1 except for condition (7), which is replaced by the weaker requirement: for each $t \in [a, b']$, the function $\phi(y, t)$ is continuous in y, uniformly for all $t \in [a, b]$.*

Then, provided

$$b - a \leqslant \frac{\delta}{M},$$

the system (3) has at least one solution satisfying the initial condition (5).

Proof. We use the notation of Theorem 1 and set $P = P_{y_0}$. It is easy to see that P maps Ω into itself. For when we proved this in Theorem 1, we used only the inequality $M(b-a) \leqslant \delta$, which is also satisfied in the present case.

It is also easy to verify that P is a continuous operator. For suppose $x_n \to x_0$, $x_n \in \Omega$, $z_n = P(x_n)$ $(n = 0, 1, \ldots)$. Since $\phi(y, t)$ is continuous, for any $\varepsilon > 0$ there exists an $\eta > 0$ such that, whenever $|y - y'| < \eta$, we have

$$|\phi(y, t) - \phi(y', t)| < \varepsilon \qquad (t \in [a, b]). \tag{11}$$

Since $\|x_n - x_0\| \to 0$, we have $\|x_n - x_0\| < \eta$ for sufficiently large n $(n \geqslant n_0)$, and so certainly

$$|x_n(t) - x_0(t)| < \eta \qquad (t \in [a, b]).$$

If we take (11) into account, we can write, for these values of n,

$$|\phi(x_n(t), t) - \phi(x_0(t), t)| < \varepsilon \qquad (t \in [a, b]).$$

Hence, for $n \geqslant n_0$,

$$\|z_n - z_0\| = \max_{t \in [a, b]} |z_n(t) - z_0(t)| \leqslant$$

$$\leqslant \max_{t \in [a, b]} \int_a^t |\phi(x_n(\tau), \tau) - \phi(x_0(\tau), \tau)| d\tau \leqslant \varepsilon(b-a).$$

Therefore $z_n = P(x_n) \to z_0 = P(x_0)$.

Since Ω is clearly a closed convex set, in order to apply Theorem 3.2 we need only establish that $P(\Omega)$ is a relatively compact set. This will complete the proof of the theorem.

The compactness criteria for sets in the space $\mathbf{C}$ (I.5.4) can be carried over without difficulty to the space $\mathbf{C}_n$. Hence we need to prove that the functions in the family $P(\Omega)$ are equicontinuous (that they are uniformly bounded follows from the fact that $P(\Omega) \subset \Omega$, and that Ω is a bounded set). However, this follows from the inequality

$$|z(t')-z(t)| = \left|\int_t^{t'} \phi(x(\tau), \tau)\,d\tau\right| \leqslant M|t'-t| \qquad (x\in\Omega,\ z = P(x)).$$

The theorem is therefore proved.

4.4. Consider the integral equation

$$x(s) = \lambda \int_a^b \phi(s, t, x(t))\,dt. \tag{12}$$

By applying Theorem 1.1 we can prove that (12) has a unique solution for sufficiently small λ.

THEOREM 3. *Suppose the function $\phi(s, t, u)$ is defined and continuous in the parallelepiped*

$$a \leqslant s,\ t \leqslant b,\ |u| \leqslant h$$

and satisfies a Lipschitz condition in u (for each fixed s and t):

$$|\phi(s, t, u) - \phi(s, t, u')| \leqslant K|u-u'|,$$

where K is independent of s and t.

If we write

$$M = \max|\phi(s, t, u)| \quad (a \leqslant s,\ t \leqslant b,\ |u| \leqslant h),$$

then, provided λ satisfies

$$|\lambda| < \min\left[\frac{h}{M(b-a)}, \frac{1}{K(b-a)}\right], \tag{13}$$

equation (12) *has a unique continuous solution.*

Proof. Take $\mathbf{X} = \mathbf{C}[a, b]$ in Theorem 1.1, and take Ω to be the set of all $x\in\mathbf{C}$ such that $\|x\| \leqslant h$. Define the operator P by:

$$z = P(x), \qquad z(s) = \lambda\int_a^b \phi(s, t, x(t))\,dt.$$

The condition (13) ensures that $P(x)\in\Omega$. For

$$\|P(x)\| \leqslant |\lambda|M(b-a) < h \qquad (x\in\Omega).$$

Moreover,

$$\begin{aligned}\|P(x)-P(x')\| &\leqslant |\lambda| \max_{s\in[a,b]} \int_a^b |\phi(s, t, x(t)) - \phi(s, t, x'(t))|\,dt \\ &\leqslant |\lambda|K(b-a)\|x-x'\|.\end{aligned}$$

Hence, if we set $\alpha = |\lambda|K(b-a)$, then we see from (13) that $\alpha < 1$; that is, P is a contraction operator.

Since we can write (12) in the form

$$x = P(x), \tag{14}$$

the existence and uniqueness of a solution of (12) follow from Theorem 1.1, all of whose hypotheses are fulfilled, in view of what we have just shown.

REMARK 1. Take an arbitrary continuous function x_0 ($\|x_0\| \leqslant h$) and construct from it a sequence $\{x_n\}$ of functions

$$x_{n+1}(s) = \lambda \int_a^b \phi(s, t, x_n(t))\,dt \qquad (n = 0, 1, \ldots).$$

Then this sequence will converge uniformly to a solution of (12), as Theorem 1.1 shows.

REMARK 2. By applying Theorem 1.3, we can show that the solution of (12) depends continuously on λ, for values of λ satisfying (13).

4.5. Using Schauder's principle, we can in certain cases extend the values of λ for which we can claim that (12) is soluble.

Assume that the function ϕ has the form

$$\phi(s, t, u) = K(s, t)\psi(t, u), \tag{15}$$

where $K(s, t)$ is continuous in the square $[a, b; a, b]$, and $\psi(t, u)$ is continuous for $t \in [a, b]$, $|u| \leqslant h$. Then we have

THEOREM 4. *If the function in the integral equation* (12) *has the form* (15) *and*

$$|\lambda| \leqslant \frac{h}{M_1 M_2 (b-a)}, \tag{16}$$

where $M_1 = \max\limits_{a \leqslant b,\, t \leqslant b} |K(s, t)|$, $M_2 = \max\limits_{\substack{a \leqslant t \leqslant b \\ |u| \leqslant h}} |\psi(t, u)|$, *then* (12) *has a continuous solution.*

Proof. Keeping the previous notation, we prove that, as above, P maps Ω into itself. Since Ω is convex and closed, and P is continuous (cf. the proof of Theorem 2), it is sufficient, as in Theorem 2, to verify that $P(\Omega)$ is relatively compact; then we can apply Theorem 3.2. As $P(\Omega)$ is a subset of $\mathbf{C}[a, b]$, we need to prove that the functions in $P(\Omega)$ are uniformly bounded and equicontinuous. The first is obvious, while the second is a consequence of the following inequality ($z = P(x)$, $x \in \Omega$):

$$|z(s) - z(s')| = |\lambda| \left| \int_a^b [K(s, t) - K(s', t)]\psi(t, x(t))\,dt \right| \leqslant$$

$$\leqslant |\lambda| M_2 \int_a^b |K(s, t) - K(s', t)|\,dt.$$

By regarding the integral operator as an operator on $\mathbf{L}^2[a, b]$, we can use Schauder's principle to obtain a result analogous to Theorem 4.

THEOREM 5. *Suppose that the function* $\phi(s, t, u)$ *in equation* (12) *has the form* (15), *where* $K(s, t)$ *is measurable in the square* $[a, b; a, b]$ *and satisfies the condition*

$$A^2 = \sup_{s \in [a, b]} \int_a^b |K(s, t)|^2\,dt < \infty,$$

and $\psi(t, u)$ *is defined and continuous in the rectangle* $[a, b; -h, h]$.

Then, provided

$$|\lambda| \leqslant \frac{h}{AM_2\sqrt{b-a}}, \tag{17}$$

equation (12) *has a square-summable solution on* $[a, b]$.

Proof. As above, the proof relies on Theorem 3.2. But this time we take $\mathbf{X} = \mathbf{L}^2[a, b]$.

We take Ω to be the set of all $x \in \mathbf{L}^2$ for which

$$|x(t)| \leqslant h \qquad \text{for almost all} \qquad t \in [a, b],$$

and we let P have the same meaning as before.

Let us check that (17) ensures that $P(\Omega) \subset \Omega$. If $x \in \Omega$ and $z = P(x)$, then by Buniakowski's inequality, we have, for $s \in [a, b]$,

$$|z(s)| = |\lambda| \left| \int_a^b K(s, t)\psi(t, x(t))\, dt \right| \leqslant$$

$$\leqslant |\lambda| \left[\int_a^b |K(s, t)|^2 \right]^{1/2} \left[\int_a^b |\psi(t, x(t))|^2\, dt \right]^{1/2} \leqslant |\lambda| A M_2 \sqrt{b-a} \leqslant h.$$

We now show that P is continuous. Let $x, x' \in \Omega$ and let $z = P(x)$, $z' = P(x')$. We have

$$|z(s) - z'(s)|^2 \leqslant |\lambda|^2 \int_a^b |K(s, t)|^2\, dt \int_a^b |\psi(t, x(t)) - \psi(t, x'(t))|^2\, dt$$

$$\leqslant |\lambda|^2 A^2 \int_a^b |\psi(t, x(t)) - \psi(t, x'(t))|^2\, dt. \tag{18}$$

Let any $\varepsilon > 0$ be given. As ψ is continuous, there exists $\eta > 0$ such that, whenever $|u - u'| < \eta$, we have

$$|\psi(t, u) - \psi(t, u')| < \varepsilon \qquad (t \in [a, b]).$$

We next divide the interval $[a, b]$ into two sets e_1 and e_2, putting in e_1 those points for which $|x(t) - x'(t)| < \eta$ and in e_2 the remaining points. Writing $\|x - x'\| = \gamma$ we have

$$\gamma^2 = \int_a^b |x(t) - x'(t)|^2\, dt \geqslant \int_{e_2} |x(t) - x'(t)|^2\, dt \geqslant \eta^2 \operatorname{mes} e_2,$$

so that

$$\operatorname{mes} e_2 \leqslant \gamma^2/\eta^2.$$

We now find a bound for the integral $I = \int_a^b |\psi(t, x(t)) - \psi(t, x'(t))|^2 dt$:

$$I = \int_{e_1} + \int_{e_2} |\psi(t, x(t)) - \psi(t, x'(t))|^2\, dt \leqslant$$

$$\leqslant \varepsilon^2 \operatorname{mes} e_1 + 4M_2^2 \operatorname{mes} e_2 \leqslant \varepsilon^2 (b-a) + \frac{4M_2^2 \gamma^2}{\eta^2}.$$

Using this bound, we find from (18) that

$$\|z - z'\|^2 = \int_a^b |z(s) - z'(s)|^2\, ds \leqslant$$

$$\leqslant |\lambda|^2 A^2 (b-a) \left[\varepsilon^2 (b-a) + \frac{4M_2^2 \|x - x'\|^2}{\eta^2} \right].$$

Hence z and z' can be as close as we want, if x and x' are close enough.

Since Ω is clearly convex and closed, it remains to prove that $P(\Omega)$ is relatively compact. To do this, we use M. Riesz's Theorem (IX.1.3), according to which we need only verify that

$$\int_a^b |z(t+\eta)-z(t)|^2\,dt \underset{\eta\to 0}{\longrightarrow} 0$$

uniformly for $z\in P(\Omega)$.*

Let $x\in\Omega$, $z=P(x)$. Then

$$\int_a^b |z(s+\eta)-z(s)|^2\,ds =$$

$$=\int_a^b \left|\lambda\int_a^b [K(s+\eta,t)-K(s,t)]\psi(t,x(t))\,dt\right|^2 ds \leqslant$$

$$\leqslant M_2^2|\lambda|^2\int_a^b\int_a^b |K(s+\eta,t)-K(s,t)|^2\,ds\,dt.$$

We show that the last integral tends to zero.

This is obvious if $K(s,t)$ is continuous. If, however, $K(s,t)$ is of a more general type, then we can choose a sequence $\{K_n(s,t)\}$ of continuous functions such that

$$\int_a^b\int_a^b |K(s,t)-K_n(s,t)|^2\,ds\,dt \underset{n\to\infty}{\longrightarrow} 0.$$

We have

$$\left[\int_a^b\int_a^b |K(s+\eta,t)-K(s,t)|^2\,ds\,dt\right]^{1/2} \leqslant$$

$$\leqslant \left[\int_a^b\int_a^b |K_n(s+\eta,t)-K_n(s,t)|^2\,ds\,dt\right]^{1/2} +$$

$$+\left[\int_a^b\int_a^b |K(s+\eta,t)-K_n(s+\eta,t)|^2\,ds\,dt\right]^{1/2} +$$

$$+\left[\int_a^b\int_a^b |K(s,t)-K_n(s,t)|^2\,ds\,dt\right]^{1/2}.$$

Choosing n large enough for the last two integrals on the right-hand side to be less than an arbitrary $\varepsilon>0$, and then choosing η small enough for the first integral to be less than ε as well, we obtain the required result in the general case.

The theorem is thus completely proved.

REMARK 1. If the function $\psi(t,u)$ is defined in the strip $a\leqslant t\leqslant b$, $-\infty<u<\infty$, and, in addition,

$$\frac{\psi(t,h)}{h}\underset{h\to\infty}{\longrightarrow} 0, \tag{19}$$

then, for sufficiently large h, condition (16) of Theorem 4 and condition (17) of Theorem 5 will be satisfied, whatever values we choose for λ, M_1, M_2 and $b-a$. Hence (19) ensures that (12) is soluble for all λ.

* Riesz's Theorem contains a further condition, that the norm be bounded; this is not mentioned in the text since it is obviously satisfied.

REMARK 2. Theorems 4 and 5 can be proved with somewhat different conditions on the functions appearing in the equations. It is also not difficult to formulate theorems in which (12) is regarded as an equation in a function space other than $\mathbf{C}[a, b]$ or $\mathbf{L}^2(a, b)$ (for example, in $\mathbf{L}^1(a, b)$).

See Krasnosel'skii [1] for applications of the fixed-point principle to the proof of existence of solutions for non-linear equations.

The most important applications of Schauder's principle are put forward in his papers on elliptic differential equations, while those of the Leray–Schauder theorems are in papers on hyperbolic equations (see Schauder [3]).

§ 5. Kakutani's Theorem

5.1. The theorem of Kakutani (see Kakutani [2]) to be presented in this section is a generalization of Schauder's Theorem (3.1). Kakutani's Theorem has numerous applications in mathematical economics (see Nikaido). Below we shall present an application of the theorem to the theory of games.

To state Kakutani's Theorem we shall require certain facts about many-valued mappings. Let $\mathbf{X}$ and $\mathbf{Y}$ be metric spaces; we denote by $2^{\mathbf{Y}}$ the set of all subsets of $\mathbf{Y}$. We call every mapping $f: \mathbf{X} \to 2^{\mathbf{Y}}$ a *many-valued function*. In this situation there is associated with each point $x \in \mathbf{X}$ a subset $f(x)$ of $\mathbf{Y}$, which from now on will be assumed non-empty.

The *graph* of a many-valued mapping $f: \mathbf{X} \to 2^{\mathbf{Y}}$ is the subset G_f of $\mathbf{X} \times \mathbf{Y}$ defined by

$$G_f = \{(x, y): \ y \in f(x)\}.$$

A many-valued mapping $f: \mathbf{X} \to 2^{\mathbf{Y}}$ is said to be *upper semicontinuous at a point* $x \in \mathbf{X}$ if, sequences $x_n \to x$, $y_n \to y$, with $y_n \in f(x_n)$, we have $y \in f(x)$. The mapping f is said to be *closed* if it is closed at each point $x \in \mathbf{X}$. Clearly, f is closed if and only if its graph G_f is closed in $\mathbf{X} \times \mathbf{Y}$.

A many-valued mapping $f: \mathbf{X} \to 2^{\mathbf{Y}}$ is said to be *upper semicontinuous at a point* $x \in \mathbf{X}$ if, for every open set U with $f(x) \subset U$, there exists a neighbourhood V of x satisfying $f(V) \subset U$, where $f(V) = \bigcup_{z \in V} f(z)$. The mapping f is said to be *upper semicontinuous* if it is upper semicontinuous at each point $x \in \mathbf{X}$.

If f is an ordinary single-valued mapping, then upper semicontinuity for f is equivalent to continuity, whereas a closed mapping need not be continuous.

LEMMA 1. *If the set* $\mathbf{Y}$ *is compact, then a closed mapping* f *is upper semicontinuous.*

Proof. Choose $x \in \mathbf{X}$ and an open set $U \supset f(x)$. Consider $V = \{z \in \mathbf{X}: f(z) \subset U\}$. Since $x \in V$ and $f(V) \subset U$, we need only prove that V is open, or that $\mathbf{X} \setminus V = \{z \in \mathbf{X}: f(z) \cap (\mathbf{Y} \setminus U) \neq \varnothing\}$ is closed. If a sequence $\{z_n\}$ converges to $z \in \mathbf{X}$ and $z_n \in \mathbf{X} \setminus V$, then there exist points $y_n \in f(z_n) \cap (\mathbf{Y} \setminus U)$. As $\mathbf{Y}$ is compact, there is a subsequence $\{y_{n_k}\}$ such that $y_{n_k} \to y \in \mathbf{Y} \setminus U$. Since f is a closed mapping, it follows that $y \in f(z)$. Therefore $y \in f(z) \cap (\mathbf{Y} \setminus U)$, and so $z \in \mathbf{X} \setminus V$.

5.2. A point $x^* \in \mathbf{X}$ is called a *fixed point* of a mapping $f: \mathbf{X} \to 2^{\mathbf{X}}$ if $x^* \in f(x^*)$. Now we can state Kakutani's Theorem.

THEOREM 1. *Let* $\mathbf{X}$ *be a* B*-space,* $\boldsymbol{K}$ *a non-empty compact convex set in* $\mathbf{X}$*, and let* $f: \boldsymbol{K} \to 2^{\boldsymbol{K}}$ *be a many-valued mapping satisfying the conditions:*

1) *for each* $x \in \boldsymbol{K}$*, the set* $f(x)$ *is a non-empty convex subset of* $\boldsymbol{K}$*;*

2) f *is a closed mapping.*

Then f *has a fixed point.*

*Proof.** As K is compact, Hausdorff's Theorem shows that for each $\varepsilon > 0$ there exists a finite ε-net for K, consisting of $x_{\varepsilon 1}, x_{\varepsilon 2}, \ldots, x_{\varepsilon n(\varepsilon)}$, say. For every $x \in K$, we set

$$\phi_{\varepsilon i}(x) = \max(\varepsilon - \|x - x_{\varepsilon i}\|, 0) \qquad (i = 1, \ldots, n(\varepsilon)).$$

Obviously, the $\phi_{\varepsilon i}$ are continuous non-negative functions on K. As $\{x_{\varepsilon i}\}_{i=1}^{n(\varepsilon)}$ is an ε-net, for each $x \in K$ there is at least one i with $\|x - x_{\varepsilon i}\| < \varepsilon$. Therefore we have $\phi_{\varepsilon i}(x) > 0$ for this i. This argument shows that the following definition is meaningful:

$$w_{\varepsilon i}(x) = \phi_{\varepsilon i}(x) \Big/ \sum_{j=1}^{n(\varepsilon)} \phi_{\varepsilon j}(x) \qquad (i = 1, \ldots, n(\varepsilon)).$$

Now we fix an arbitrary point $y_{\varepsilon i} \in f(x_{\varepsilon i})$ $(i = 1, \ldots, n(\varepsilon))$ and define a continuous single-valued mapping $f_\varepsilon: K \to K$, which acts according to the formula

$$f_\varepsilon(x) = \sum_{i=1}^{n(\varepsilon)} w_{\varepsilon i}(x) y_{\varepsilon i}.$$

From the conditions $y_{\varepsilon i} \in K$, $w_{\varepsilon i}(x) \geqslant 0$, $\Sigma w_{\varepsilon i}(x) = 1$, and the convexity of K it follows that $f_\varepsilon(x) \in K$. Hence, for any $\varepsilon > 0$, we have a continuous single-valued mapping $f_\varepsilon: K \to K$. By Schauder's Theorem (3.1), this mapping has a fixed point x_ε: thus $f_\varepsilon(x_\varepsilon) = x_\varepsilon$.

As K is compact, there exist a sequence $\{\varepsilon_n\}$ of positive numbers and a point $x^* \in K$ such that the following conditions are satisfied:

$$\text{a)} \lim_{n \to \infty} \varepsilon_n = 0; \qquad \text{b)} x_{\varepsilon_n} \to x^*; \qquad \text{c)} f_{\varepsilon_n}(x_{\varepsilon_n}) = x_n.$$

We show that x^* is the fixed point of f we are seeking. Set $U_\delta = f(x^*) + B_\delta$, where $B_\delta = \{y: \|y\| < \delta\}$, $\delta > 0$. We shall show that $x^* \in U_\delta$ for every $\delta > 0$, and so, as $f(x^*)$ is closed, we obtain the required result, that $x^* \in f(x^*)$ (the fact that $f(x^*)$ is closed follows from the fact that f is a closed mapping).

It is clear that U_δ is an open convex set and that $f(x^*) \subset U_\delta$. Given x^* and U_δ, we can choose, by Lemma 1, a ball $K_\varepsilon = \{x: \|x - x^*\| < \varepsilon\}$ such that $f(K_\varepsilon) \subset U_\delta$. By conditions a) and b), there exists a number N such that $\varepsilon_n < \varepsilon/2$ and $x_{\varepsilon_n} \in K_{\varepsilon/2}$ whenever $n \geqslant N$. If $w_{\varepsilon_n i}(x_{\varepsilon_n}) > 0$, then

$$\|x_{\varepsilon_n i} - x_{\varepsilon_n}\| < \varepsilon_n < \varepsilon/2$$

and

$$\|x_{\varepsilon_n i} - x^*\| \leqslant \|x_{\varepsilon_n i} - x_{\varepsilon_n}\| + \|x_{\varepsilon_n} - x^*\| < \varepsilon/2 + \varepsilon/2 = \varepsilon.$$

Hence, for $n > N$ we have $x_{\varepsilon_n i} \in K_\varepsilon$, for all i such that $w_{\varepsilon_n i}(x_{\varepsilon_n}) > 0$. For these i, we have

$$y_{\varepsilon_n i} \in f(x_{\varepsilon_n i}) \subset f(K_\varepsilon) \subset U_\delta. \tag{1}$$

By condition c), we deduce that

$$x_{\varepsilon_n} = \Sigma \omega_{\varepsilon_n i}(x_{\varepsilon_n}) y_{\varepsilon_n i}. \tag{2}$$

From (1) and (2) we see that, for $n \geqslant N$, the point x_{ε_n} is a convex combination of just those points $y_{\varepsilon_n i}$ that lie in U_δ, and so, as U_δ is convex, we have $x_{\varepsilon_n} \in U_\delta$. Taking limits as n

* For the proof of Theorem 1 we follow Nikaido.

$\to \infty$, we see that $x^* \in \overline{U_\delta} \subset U_{2\delta}$. As we have already remarked, it follows from this that $x^* \in f(x^*)$. This completes the proof of Kakutani's Theorem.

5.3. Now we present an application of Kakutani's Theorem to the theory of games.

We consider a game between two people (player I and player II) having zero sum: this means that one player wins by whatever amount the other loses. A player's *strategy* is a complete enumeration of all moves that this player will make in each of the possible positions in the course of the game. We denote the set of all strategies (or the space of strategies) of player I by X, and the set of strategies of player II by Y.

A *pay-off function* is a real-valued function $K(x, y)$ defined on $X \times Y$, the number $K(x, y)$ being interpreted as the winnings of player I if player I chooses strategy x and player II strategy y. The number $-K(x, y)$ is interpreted as the winnings of player II in the same situation. A triple (X, Y, K), where X and Y are sets and $K(x, y)$ is a function on $X \times Y$, is called a *game*.

When player I chooses his strategy $x \in X$, he must assume, knowing nothing about player II's strategy, that his opponent will choose the worst strategy from player I's point of view—that is, that player I's winnings will be $\inf_{y \in Y} K(x, y)$. Hence it is natural for player *I* to seek a strategy $x_0 \in X$ such that

$$\inf_{y \in Y} K(x_0, y) = \max_{x \in X} \inf_{y \in Y} K(x, y). \tag{3}$$

Bearing in mind that player II's winnings are $-K(x, y)$, we see from (3) that player II must look for a strategy $y_0 \in Y$ such that

$$\inf_{x \in X} \{-K(x, y_0)\} = \max_{y \in Y} \inf_{x \in X} [-K(x, y)],$$

or, what is the same thing,

$$\sup_{x \in X} K(x, y_0) = \min_{y \in Y} \sup_{x \in X} K(x, y). \tag{4}$$

Comparison of (3) and (4) shows that a necessary and sufficient condition for the existence of strategies $x_0 \in X$ and $y_0 \in Y$ satisfying (3) and (4) is that for all $x \in X$, $y \in Y$ the following inequality should hold:

$$K(x, y_0) \leqslant K(x_0, y_0) \leqslant K(x_0, y), \tag{5}$$

and a necessary and sufficient condition for this is that

$$\min_{y \in Y} \sup_{x \in X} K(x, y) = \max_{x \in X} \inf_{y \in Y} K(x, y). \tag{6}$$

The common value of the two sides of (6) is called the *value* of the game. Each pair of strategies x_0, y_0 satisfying (5) is called a *solution* of the game, and the strategies x_0, y_0 themselves are called *optimal strategies.** To prove the existence of a solution for a game one is required to prove a *minimax theorem*, which consists in verifying equation (6) for the given game. With the help of Kakutani's Theorem, we now prove one very general theorem of this type. First we state the following lemma, whose proof is left to the reader.

* For more details, see the monograph by Karlin, which we follow in part for the exposition of this subsection.

LEMMA 2. 1) *For all real-valued functions* $K(x, y)$, *we have*

$$\inf_{y \in Y} \sup_{x \in X} K(x, y) \geqslant \sup_{x \in X} \inf_{y \in Y} K(x, y). \tag{7}$$

2) *If* X *and* Y *are metric compacta, and the function* $K(x, y)$ *is continuous on* $X \times Y$, *then the functions* $\phi(x) = \max_{y \in Y} K(x, y)$ *and* $\psi(x) = \min_{y \in Y} K(x, y)$ *are continuous on* X. *Furthermore, both* $\min_{y \in Y} \max_{x \in X} K(x, y)$ *and* $\max_{x \in X} \min_{y \in Y} K(x, y)$ *exist.*

THEOREM 2. *Suppose the following conditions are satisfied:*

a) X *is a compact convex set in a* B-*space* $\mathbf{X}$;

b) Y *is a compact convex set in a* B-*space* $\mathbf{Y}$;

c) *the function* $K(x, y)$ *is continuous on* $X \times Y$, *convex in* y *for each* $x \in X$ *and concave in* x *for each* $y \in Y$, *i.e.*

$$K(x, \lambda y_1 + (1-\lambda)y_2) \leqslant \lambda K(x, y_1) + (1-\lambda)K(x_2, y),$$

$$K(\lambda x_1 + (1-\lambda)x_2, y) \geqslant \lambda K(x_1, y) + (1-\lambda)K(x_2, y),$$

where $x, x_1, x_2 \in X$; $y, y_1, y_2 \in Y$; $0 \leqslant \lambda \leqslant 1$.

Then the game (X, Y, K) *has a solution. In addition,*

$$\min_{y \in Y} \max_{x \in X} K(x, y) = \max_{x \in X} \min_{y \in Y} K(x, y). \tag{8}$$

Proof. In view of Lemma 2 (2), condition (6), which expresses the fact that the game has a solution, coincides with (8). Hence, we prove (8).

For each $x \in X$, set

$$B_x = \{y \in Y\colon\ K(x, y) = \min_{z \in Y} K(x, z)\}.$$

As Y is compact, $B_x \neq \varnothing$. Clearly B_x is closed. Let us show that it is convex. If $y_1, y_2 \in B_x$, $0 \leqslant \lambda \leqslant 1$, then

$$K(x, \lambda y_1 + (1-\lambda)y_2) \leqslant \lambda K(x, y_1) + (1-\lambda)K(x, y_2) =$$

$$= \lambda \min_{z \in Y} K(x, z) + (1-\lambda) \min_{z \in Y} K(x, z) = \min_{z \in Y} K(x, z). \tag{9}$$

On the other hand, as Y is convex, we have $\lambda y_1 + (1-\lambda)y_2 \in Y$, and so

$$K(x, \lambda y_1 + (1-\lambda)y_2) \geqslant \min_{z \in Y} K(x, z). \tag{10}$$

Comparing (9) and (10), we obtain

$$K(x, \lambda y_1 + (1-\lambda)y_2) = \min_{z \in Y} K(x, z).$$

Hence $\lambda y_1 + (1-\lambda)y_2 \in B_x$; that is, B_x is convex.

Similarly we conclude that, for each $y \in Y$, the set

$$A_y = \{x \in X\colon\ K(x, y) = \max_{w \in X} K(w, y)\}$$

is non-empty, closed and convex.

Consider the convex set $C = X \times Y$ in the B-space $\mathbf{X} \times \mathbf{Y}$. By Tychonoff's Theorem (I.2.8), C is compact in $\mathbf{X} \times \mathbf{Y}$. It is clear that the sets $A_y \times B_x$ $(x \in X, y \in Y)$ are non-empty convex closed subsets of C. Now consider the many-valued mapping $f: C \to 2^C$ taking a point $(x, y) \in C$ into $A_y \times B_x$. For Kakutani's Theorem to be applicable, it remains to show that f is a closed mapping.

Suppose $(x_n, y_n) \to (x, y)$, $(u_n, v_n) \to (u, v)$ in C and that $(u_n, v_n) \in f(x_n, y_n)$. We show that $(u, v) \in f(x, y) = A_y \times B_x$. Since $u_n \in A_{y_n}$, $v_n \in B_{x_n}$, we have

$$K(u_n, y_n) = \max_{w \in X} K(w, y_n)$$

and

$$K(x_n, v_n) = \min_{z \in Y} K(x_n, z).$$

Since $u_n \to u$, it follows from Lemma 2 (2) that

$$K(u, y) = \lim_{n \to \infty} K(u_n, y_n) = \lim_{n \to \infty} \max_{w \in X} K(w, y_n) = \max_{w \in X} K(w, y).$$

Hence $u \in A_y$. Similarly $v \in B_x$.

Thus f is closed and Kakutani's Theorem is applicable. This shows that there exists a point $(x_0, y_0) \in C$ such that $(x_0, y_0) \in f(x_0, y_0) \in A_{y_0} \times B_{x_0}$. Therefore

$$K(x_0, y_0) = \max_{x \in X} K(x, y_0) \qquad \text{and} \qquad K(x_0, y_0) = \min_{y \in Y} K(x_0, y).$$

Hence

$$\min_{y \in Y} \max_{x \in X} K(x, y) \leqslant K(x_0, y_0) \leqslant \max_{x \in X} \min_{y \in Y} K(x, y).$$

Comparing this inequality with (7), we obtain (8). This completes the proof of the theorem.

XVII

DIFFERENTIATION OF NON-LINEAR OPERATORS

NON-LINEAR operators can be investigated further by establishing a connection between them and linear operators—more precisely, by the technique of local approximation to a non-linear operator by a linear one.

In this chapter we develop the techniques of differential calculus for non-linear operators that are needed for this purpose, but we do not investigate them in full detail until the following chapter.

§ 1. The first derivative

1.1. Suppose we have an operator P mapping an open set Ω of one B-space $\mathbf{X}$ into a set Δ of another B-space $\mathbf{Y}$. We take a fixed point $x_0 \in \Omega$ and assume that there exists a continuous linear operator* $U \in B(\mathbf{X}, \mathbf{Y})$ such that, for every $x \in \mathbf{X}$,

$$\lim_{t \to 0} \frac{P(x_0 + tx) - P(x_0)}{t} = U(x). \tag{1}$$

In this situation we say that the linear operator U is the *derivative* of P at the point x_0. We then write

$$U = P'(x_0).$$

The derivative just defined is often called the *Gateaux derivative*, or the *weak derivative*, and the element $U(x)$ is called the *Gateaux differential.*

Denote by $\bar{K}$ the set of all $x \in \mathbf{X}$ with $\|x\| = 1$. If the limit in equation (1) is uniform for $x \in \bar{K}$, then P is said to be *differentiable* at x_0 and in this case $P'(x_0)$ is called the *Fréchet derivative*, or *strong derivative*.†

The differentiability of an operator P at a point x_0 means, in other words, that there exists a linear operator $U \in B(\mathbf{X}, \mathbf{Y})$ such that, for every $\varepsilon > 0$, we can find a $\delta > 0$ such that, whenever $\|\Delta x\| < \delta$ ($\Delta x \in \mathbf{X}$), we have

$$\|P(x_0 + \Delta x) - P(x_0) - U(\Delta x)\| \leqslant \varepsilon \|\Delta x\|. \tag{2}$$

We leave the simple proof of this statement to the reader.

1.2. Let us record some of the simplest properties of the derivative.

I. If an operator P is differentiable at a point x_0, then it is continuous at that point.‡ This follows immediately from (2).

* Recall that we use the symbol $B(\mathbf{X}, \mathbf{Y})$ for the space of all continuous linear operators mapping $\mathbf{X}$ into $\mathbf{Y}$.

† The definition of the Fréchet derivative was given by Fréchet [3], and that of the weak derivative by Gateaux [1].

‡ It is insufficient to assume only the existence of a derivative.

II. Let $P = \alpha_1 P_1 + \alpha_2 P_2$. If the derivatives $P_1'(x_0)$ and $P_2'(x_0)$ exist, then so does

$$P'(x_0) = \alpha_1 P_1'(x_0) + \alpha_2 P_2'(x_0).$$

Moreover, if P_1 and P_2 are differentiable at x_0, then so is P.

This follows immediately from the definition.

III. If $P = U \in B(\mathbf{X}, \mathbf{Y})$, then P is differentiable at each point $x_0 \in \mathbf{X}$ and

$$P'(x_0) = U.$$

For we have

$$\frac{P(x_0 + tx) - P(x_0)}{t} = U(x).$$

IV. Let P be an operator mapping $\Omega \subset \mathbf{X}$ into an open set $\Delta \subset \mathbf{Y}$, and let Q be an operator mapping Δ into some B-space $\mathbf{Z}$. Set $R = QP$, and suppose that Q is differentiable at $y_0 = P(x_0)$ $(x_0 \in \Omega)$ and that P has a derivative at x_0. Then R has a derivative at x_0 and

$$R'(x_0) = Q'(P(x_0))P'(x_0) = Q'(y_0)P'(x_0).$$

For write $U = P'(x_0)$, $V = Q'(y_0)$, and choose any $x \in \mathbf{X}$. Set $\Delta y = P(x_0 + tx) - P(x_0)$. Then

$$\frac{R(x_0 + tx) - R(x_0)}{t} = \frac{QP(x_0 + tx) - QP(x_0)}{t} = \frac{Q(y_0 + \Delta y) - Q(y_0)}{t} =$$

$$= \frac{V(\Delta y) + \zeta(\Delta y)}{t} = V\left(\frac{P(x_0 + tx) - P(x_0)}{t}\right) + \frac{\zeta(\Delta y)}{\|\Delta y\|} \left\| \frac{P(x_0 + tx) - P(x_0)}{t} \right\|, \tag{3}$$

where $\zeta(\Delta y) = Q(y_0 + \Delta y) - Q(y_0) - V(\Delta y)$, $\|\xi(\Delta y)\| = o(\|\Delta y\|)$. As $t \to 0$, the first term in (3) tends to the limit

$$[VP'(x_0)](x) = VU(x),$$

and the second term tends to zero, since the ratio $\dfrac{\xi(\Delta y)}{\|\Delta y\|}$ tends to zero and the scalar multiple is bounded. Hence

$$\lim_{t \to 0} \frac{R(x_0 + tx) - R(x_0)}{t} = VU(x),$$

as we set out to prove.

REMARK. If P is differentiable at x_0 then so is R.

V. If, with the conditions of IV, the operator $Q = V$ is continuous and linear, then

$$R'(x_0) = VP'(x_0).$$

1.3. We now note an inequality that, in the general case, replaces the Mean Value Theorem for ordinary real-valued functions.

Let x_0 and x belong to Ω; assume also that all points of the linear interval $[x_0, x]$ joining x_0 and x belong to Ω, and that the derivative of the operator P exists at each point of $[x_0, x]$. Write $\Delta x = x - x_0$, and for any functional $g \in \mathbf{Y}^*$, set

$$\phi(t) = g(P(x_0 + t\Delta x)). \tag{4}$$

It is not hard to see that the real-valued function ϕ has a derivative in $[0, 1]$. In fact,

$$\frac{\phi(t+\Delta t)-\phi(t)}{\Delta t} = g\left(\frac{P(x_0+t\Delta x+\Delta t\,\Delta x)-P(x_0+t\,\Delta x)}{\Delta t}\right). \tag{5}$$

As $\Delta t \to 0$, the expression appearing as the argument of g tends to the limit $P'(x_0 + t\Delta x)(\Delta x)$, and hence, as g is continuous, the right-hand side of (5) tends to

$$\phi'(t) = g(P'(x_0+t\,\Delta x)(\Delta x)). \tag{6}$$

We now apply the Mean Value Theorem to ϕ:

$$\phi(1)-\phi(0) = \phi'(\theta) \qquad (0<\theta<1).$$

Taking (4) and (6) into account, we see from this that

$$g(P(x_0+\Delta x)-P(x_0)) = g(P'(x_0+\theta\,\Delta x)(\Delta x))$$

and hence that

$$|g(P(x_0+\Delta x)-P(x_0))| \leqslant \|g\| \sup_{0<\theta<1} \|P'(x_0+\theta\,\Delta x)\|\,\|\Delta x\|. \tag{7}$$

Now choose the functional $g \neq \mathbf{0}$ so that

$$g(P(x_0+\Delta x)-P(x_0)) = \|g\|\,\|P(x_0+\Delta x)-P(x_0)\|.$$

Then from (7) we see that

$$\|P(x)-P(x_0)\| \leqslant \sup_{0<\theta<1} \|P'(x_0+\theta\,\Delta x)\|\,\|\Delta x\| \tag{8}$$

$$(\Delta x = x - x_0).$$

Applying the inequality (8) to the operator $P-U$ ($U \in B(\mathbf{X}, \mathbf{Y})$) and then using II and III in 1.2, we obtain the inequality

$$\|P(x)-P(x_0)-U(\Delta x)\| \leqslant \sup_{0<\theta<1} \|P'(x_0+\theta\Delta x)-U\|\,\|\Delta x\|.$$

In particular, if we set $U = P'(x_0)$ here, we obtain

$$\|P(x_0+\Delta x)-P(x_0)-P'(x_0)\Delta x\| \leqslant$$

$$\leqslant \|\Delta x\| \sup_{0<\theta<1} \|P'(x_0+\theta\,\Delta x)-P'(x_0)\|. \tag{9}$$

We call (8) the *Mean Value Theorem* and (9) the *Mean Value Theorem with remainder term.*

1.4. We now mention some applications of the Mean Value Theorem.

Assume that the derivative of the operator P exists at every point of some open set $\Omega_0 \subset \Omega$. Then there is an operator P' defined on Ω_0, associating to an element $x \in \Omega_0$ the element $P'(x) \in B(\mathbf{X}, \mathbf{Y})$. Let us prove that, if P' is continuous at a point $x_0 \in \Omega_0$, then P is differentiable at that point.

In fact, by the assumption of continuity, for every $\varepsilon > 0$, there exists $\delta > 0$ such that $\|P'(x)-P'(x_0)\| < \varepsilon$, provided $\|x-x_0\| < \delta$; that is, if $\|\Delta x\| < \delta$, then $\sup_{0<\theta<1} \|P'(x_0$

$+\theta\Delta x)-P'(x_0)\| \leqslant \varepsilon$. Applying this to (9), we arrive at the inequality

$$\|P(x_0+\Delta x)-P(x_0)-P'(x_0)(\Delta x)\| \leqslant \varepsilon\|\Delta x\|,$$

which means that P is differentiable at x_0.

The Mean Value Theorem can be useful for proving that an operator P has a fixed point.

THEOREM 1. *Suppose the operator P maps a set $\Omega \subset \mathbf{X}$ into $\mathbf{X}$, and that it has a derivative at every point of a convex closed subset $\Omega_0 \subset \Omega$. If*

1) $P(\Omega_0) \subset \Omega$;
2) $\sup\limits_{x\in\Omega_0} \|P'(x)\| = \alpha < 1$,

then P has a unique fixed point in Ω_0.

Proof. It is sufficient to verify that, in this situation, the hypotheses of Theorem XVI.1.1 are satisfied—that is, that P is a contraction operator.

Let $x_1, x_2 \in \Omega$. By the Mean Value Theorem, we have

$$\|P(x_2)-P(x_1)\| \leqslant \|x_2-x_1\| \sup_{0<\theta<1} \|P'(x_1+\theta(x_2-x_1))\| \leqslant \alpha\|x_2-x_1\|,$$

as we required to prove.

REMARK. If we assume that P' is continuous in Ω_0, then condition 2) turns out also to be a necessary condition for P to be a contraction operator.

For if

$$\sup_{x\in\Omega_0} \|P'(x)\| \geqslant 1,$$

then there exist a numerical sequence $\alpha_n \to 1$ and a sequence $\{x_n\}$ of elements of Ω_0 such that

$$\|P'(x_n)\| > \alpha_n \qquad (n = 1, 2, \dots);$$

and we may assume that the points x_n are interior to Ω_0.

By the definition of derivative, there exists an element x_n', for each $n = 1, 2, \dots$, such that, for small enough t,

$$\left\|\frac{P(x_n+tx_n')-P(x_n)}{t}\right\| > \alpha_n \|x_n'\|$$

$$(|t| \leqslant t_n; \qquad n = 1, 2, \dots).$$

Setting $y_n = x_n + t_n x_n'$, we obtain a sequence of pairs of elements x_n, y_n such that

$$\frac{\|P(y_n)-P(x_n)\|}{\|y_n-x_n\|} > \alpha_n, \tag{10}$$

so P is not a contraction operator.

1.5. Let us clarify the meaning of the concepts introduced above in the case where one of the spaces $\mathbf{X}$ and $\mathbf{Y}$ is finite-dimensional.

Assume first that $\mathbf{X}$ is a one-dimensional real space, whose elements we shall identify with numbers.

Let $U \in B(\mathbf{X}, \mathbf{Y})$. It is not difficult to see that in this case U has the form

$$U(t) = ty_0 \qquad (t \in \mathbf{X}), \tag{11}$$

where y_0 is some element of $\mathbf{Y}$. Further, since $\|U(t)\| = |t|\,\|y_0\|$,

$$\|U\| = \|y_0\|. \tag{12}$$

Conversely, for any element $y_0 \in \mathbf{Y}$, the formula (11) yields an operator $U \in B(\mathbf{X}, \mathbf{Y})$. Clearly the correspondence between elements of $\mathbf{Y}$ and those of $B(\mathbf{X}, \mathbf{Y})$ is one-to-one and linear, and, by (12), it preserves norms. Hence we may assume that $B(\mathbf{X}, \mathbf{Y}) = \mathbf{Y}$.

Now consider an operator F mapping $\Omega \subset \mathbf{X}$ into $\mathbf{Y}$. Thus F is a function of a numerical argument, with values in $\mathbf{Y}$.

The existence of a derivative $F'(t_0) = y_0 \in \mathbf{Y}$ means that, for each $t \in \mathbf{X}$,

$$\lim_{\tau \to 0} \frac{F(t_0 + \tau t) - F(t_0)}{\tau} = t y_0. \tag{13}$$

Putting $\Delta t = \tau t$, we can rewrite (13) in the form

$$\lim_{\Delta t \to 0} \frac{F(t_0 + \Delta t) - F(t_0)}{\Delta t} = y_0.$$

Thus in the present case the definition of $F'(t_0)$ is in no way different from that of the usual derivative of a real-valued function of a real argument. Note that, as $F'(t_0) \in \mathbf{Y}$, the operator F', like F, maps some numerical set into $\mathbf{Y}$.

As in the case of real-valued functions, it is not hard to prove that the existence of the derivative $F'(t_0)$ ensures that F is differentiable at t_0.

If $\Omega = [a, b]$, then it is natural to regard $F(\Omega)$ as a *curve* in $\mathbf{Y}$. Then the element $F'(t_0)$ gives the direction of the tangent to this curve at the point $y_0 = F(t_0)$.

1.6. Now suppose $\mathbf{X}$ and $\mathbf{Y}$ are finite-dimensional spaces. Denote the dimension of $\mathbf{X}$ by m, and that of $\mathbf{Y}$ by n. As we noted in V.2.8, each operator $U \in B(\mathbf{X}, \mathbf{Y})$ is determined by a rectangular matrix

$$\begin{pmatrix} a_{11} & a_{12} & \dots & a_{1m} \\ a_{21} & a_{22} & \dots & a_{2m} \\ \dots & \dots & \dots & \dots \\ a_{n1} & a_{n2} & \dots & a_{nm} \end{pmatrix} \tag{14}$$

by means of the formulae

$$y = U(x) \quad (y = (\eta_1, \eta_2, \dots, \eta_n) \in \mathbf{Y}, \quad x = (\xi_1, \xi_2, \dots, \xi_m) \in \mathbf{X}),$$

$$\eta_i = \sum_{k=1}^{m} a_{ik} \xi_k \qquad (i = 1, 2, \dots, n). \tag{15}$$

Consider an operator P mapping $\Omega \subset \mathbf{X}$ into $\mathbf{Y}$. Defining P amounts to prescribing n numerical functions $\phi_1, \phi_2, \dots, \phi_n$ of m variables, so that, if

$$y = P(x) \quad (y = (\eta_1, \eta_2, \dots, \eta_n) \in \mathbf{Y}, \quad x = (\xi_1, \xi_2, \dots, \xi_m) \in \Omega),$$

then

$$\eta_i = \phi_i(\xi_1, \xi_2, \dots, \xi_m) \qquad (i = 1, 2, \dots, n). \tag{16}$$

Assume that the derivative $P'(x_0) = U$ exists $(x_0 = (\xi_1^{(0)}, \xi_2^{(0)}, \dots, \xi_m^{(0)}))$. Let U be determined by the matrix (14). If we write out the equation

$$\lim_{t \to 0} \frac{P(x_0 + tx) - P(x_0)}{t} = U(x)$$

in full and bear in mind (15) and (16), we obtain n equations

$$\lim_{t\to 0}\frac{\phi_i(\xi_1^{(0)}+t\xi_1,\ldots,\xi_m^{(0)}+t\xi_m)-\phi_i(\xi_1^{(0)},\ldots,\xi_m^{(0)})}{t}=\sum_{k=1}^{m}a_{ik}\xi_k$$

$$(i=1,2,\ldots,n). \tag{17}$$

Since these equations must be true for all $x\in\mathbf{X}$, if we take successively values of x, y for which all coordinates are zero except one, equal to unity, we find that $\phi_1,\phi_2,\ldots,\phi_n$ have partial derivatives in $\xi_1,\xi_2,\ldots,\xi_m$, namely

$$\frac{\partial\phi_i(\xi_1^{(0)},\ldots,\xi_m^{(0)})}{\partial\xi_k}=a_{ik}\qquad (i=1,2,\ldots,n;\quad k=1,2,\ldots,m).$$

Hence the derivative $P'(x_0)$ is the linear operator determined by the matrix of partial derivatives of the functions $\phi_1,\phi_2,\ldots,\phi_n$.

Notice, however, that the existence of the partial derivatives of $\phi_1,\phi_2,\ldots,\phi_n$ does not guarantee that the derivative $P'(x_0)$ will exist, as the following example shows. Let $n=1$, $m=2$, and let

$$\phi_1(\xi_1,\xi_2)=\frac{\xi_1\xi_2}{(\xi_1^2+\xi_2^2)^2},\qquad \phi_1(0,0)=0;\quad x_0=(0,0).$$

Clearly $\dfrac{\partial\phi_1(0,0)}{\partial\xi_1}=\dfrac{\partial\phi_1(0,0)}{\partial\xi_2}=0$. Thus, if the derivative of $P'(x_0)$ existed, it would be the zero operator and so (17) would give

$$\lim_{t\to 0}\frac{\phi_1(t\xi_1,t\xi_2)}{t}=0.$$

However, in fact this limit is equal to infinity, provided $\xi_1\neq 0$ and $\xi_2\neq 0$.

We leave it to the reader to prove that a necessary and sufficient condition for an operator P to be differentiable is that the functions $\phi_1,\phi_2,\ldots,\phi_n$ have differentials. Note that, in the general case, the differentiability of P is not a consequence of the existence of the derivative, as is true in the case of elementary analysis.

1.7. Consider a function F of a real argument whose values are elements of a B-space $\mathbf{X}$. If F is defined on an interval $[a,b]$, then it makes sense to speak of the integral of F, where we mean by this the limit of the sums

$$\sum_{k=0}^{n-1}F(\tau_k)(t_{k-1}-t_k)$$

$$(a=t_0<t_1<\ldots<t_n=b;\quad \tau_k\in[t_k,t_{k+1}];\, k=0,1,\ldots,n-1)$$

as $\lambda=\max\limits_k\,[t_{k+1}-t_k]\to 0$. When this limit exists, it is called the *integral* of F, and is denoted by $\int_a^b F(t)\,dt$. As for functions with real values, it can be shown that a continuous function is uniformly continuous and so integrable—that is, possesses an integral. The properties of the familiar Riemann integral, and their proofs, carry over to this abstract integral. Among these we note the following three.

I. If $U \in B(\mathbf{X}, \mathbf{Y})$, then

$$\int_a^b U(F(t))\,dt = U\left(\int_a^b F(t)\,dt\right).$$

II. If $F(t) = \phi(t)x_0$ ($t \in [a, b]$), where $x_0 \in \mathbf{X}$ is a fixed element and $\phi(t)$ is a real-valued integrable function, then

$$\int_a^b F(t)\,dt = x_0 \int_a^b \phi(t)\,dt.$$

III.

$$\left\| \int_a^b F(t)\,dt \right\| \leqslant \int_a^b \|F(t)\|\,dt.$$

Now we turn to the study of non-linear operators.

Suppose we have an operator R, defined on the interval $[x_0, x_0 + \Delta x]$ ($x_0, x_0 + \Delta x \in \mathbf{X}$), with values in the space $B(\mathbf{X}, \mathbf{Y})$. By definition, we set

$$\int_{x_0}^{x_0 + \Delta x} R(x)\,dx = \int_0^1 R(x_0 + t\Delta x)\Delta x\,dt = \lim \sum_{k=0}^{n-1} R(x_0 + \tau_k \Delta x)\Delta x(t_{k+1} - t_k).$$

Obviously, if R is continuous, then the integral exists and is an element of $\mathbf{Y}$. Integrals of this type were first introduced by M. K. Gavurin [1].

In particular, if $R = P'$, where P is an operator mapping $\Omega \subset \mathbf{X}$ into $\mathbf{Y}$ which has a continuous derivative in the interval $[x_0, x_0 + \Delta x] \subset \Omega$, then the integral of $P'(x)$ exists, and, as we shall prove, the following is true:

$$\int_{x_0}^{x_0 + \Delta x} P'(x)\,dx = P(x_0 + \Delta x) - P(x_0), \tag{18}$$

a generalization of the fundamental Newton—Leibniz formula of the integral calculus.

In fact,

$$\int_{x_0}^{x_0 + \Delta x} P'(x)\,dx = \lim_{\lambda \to 0} \sum_{k=0}^{n-1} P'(x_0 + \tau_k \Delta x)\Delta x\,(t_{k+1} - t_k) = \lim_{\lambda \to 0} \sum_{k=0}^{n-1} P'(\bar{x}_k)\Delta x_k$$

$$(\bar{x}_k = x_0 + \tau_k \Delta x; \qquad \Delta x_k = (t_{k+1} - t_k)\Delta x; \qquad k = 0, \ldots, n-1).$$

On the other hand,

$$P(x_0 + \Delta x) - P(x_0) = \sum_{k=0}^{n-1} [P(x_0 + t_{k+1}\Delta x) - P(x_0 + t_k \Delta x)] = \sum_{k=0}^{n-1} [P(x_{k+1}) - P(x_k)].$$

Using the Mean Value Theorem, we have

$$\left\| \sum_{k=0}^{n-1} [P(x_{k+1}) - P(x_k) - P'(\bar{x}_k)\Delta x_k] \right\| \leqslant \|\Delta x\| \sum_{k=0}^{n-1} (t_{k+1} - t_k) \sup_{0 < \theta < 1} \|P'(x_k + \theta \Delta x_k) - P'(\bar{x}_k)\|,$$

so that, as P' is continuous and therefore also uniformly continuous on $[x_0, x_0 + \Delta x]$, we obtain the required result.

REMARK. It follows from property III of the integral that, if

$$\|R(x_0 + \tau \Delta x)\| \leqslant \phi(t_0 + \tau \Delta t) \qquad (\tau \in [0, 1]) \tag{19}$$

and

$$\|\Delta x\| \leqslant \Delta t, \tag{20}$$

then

$$\left\| \int_{x_0}^{x_0+\Delta x} R(x)\,dx \right\| \leqslant \int_{t_0}^{t_0+\Delta t} \phi(t)\,dt. \tag{21}$$

For we have

$$\left\| \int_{x_0}^{x_0+\Delta x} R(x)\,dx \right\| = \left\| \int_0^1 R(x_0+\tau\Delta x)\ (\Delta x) d\tau \right\| \leqslant \|\Delta x\| \int_0^1 \|R(x_0+\tau\Delta x)\|\, d\tau \leqslant$$

$$\leqslant \Delta t \int_0^1 \phi(t_0+\tau\Delta t)\,d\tau = \int_{t_0}^{t_0+\Delta t} \phi(t)\,dt.$$

It follows, in particular, from (21) that, if the inequality

$$\|R(x)\| \leqslant \phi(t) \tag{22}$$

holds for all x and t such that

$$\|x-x_0\| \leqslant t-t_0, \tag{23}$$

then

$$\left\| \int_{x_0}^{x_1} R(x)dx \right\| \leqslant \left\| \int_{t_0}^{t_1} \phi(t)\,dt, \right. \tag{24}$$

where x_1 is any element satisfying $\|x_1-x_0\| \leqslant t_1-t_0$.

For if we set $\Delta x = x_1-x_0$, $\Delta t = t_1-t_0$, $x = x_0+\tau\Delta x$, $t = t_0+\tau\Delta t$, then we find that $\|\Delta x\| \leqslant \Delta t$ and

$$\|x-x_0\| = \tau\|\Delta x\| \leqslant \tau\Delta t = t-t_0 \qquad (\tau\in[0, 1]),$$

that is, (23) is satisfied, and so, by (22),

$$\|R(x_0+\tau\Delta x)\| \leqslant \phi(t_0+\tau\Delta t) \qquad (\tau\in[0, 1]).$$

Applying (21), we obtain (24).

§ 2. Second derivatives and bilinear operators

2.1. Suppose an operator P, mapping an open set Ω in a B-space $\mathbf{X}$ into a B-space $\mathbf{Y}$, has a derivative throughout Ω. As we noted above, P' can be regarded as an operator mapping Ω into $B(\mathbf{X}, \mathbf{Y})$. Hence it makes sense to refer to the derivative of this operator at a point $x_0\in\Omega$, which, when it exists, is called the *second derivative* of the given operator P, and is denoted by $P''(x_0)$. If P' is differentiable, then P is said to be *twice differentiable.*

If the second derivative exists at every point of Ω, then an operator P'' is defined, whose derivative is called the *third derivative* of the given operator P; and, in general, the *derivative* $P^{(n)}(x_0)$ *of order* n at the point x_0 is, by definition, the derivative of $P^{(n-1)}$. Clearly,

$$P^{(n)}(x_0)\in \overbrace{B(\mathbf{X}, B(\mathbf{X}, \ldots, B(\mathbf{X}}^{n}, \mathbf{Y})\ldots)).$$

2.2. The elements of the space $B(\mathbf{X}, B(\mathbf{X}, \mathbf{Y}))$ are not very readily interpreted on an intuitive level. To clarify the essential nature of these mappings it is convenient to introduce a new concept.

Suppose we are given two B-spaces $\mathbf{X}$ and $\mathbf{Y}$, and suppose that with each pair of elements $x, x' \in \mathbf{X}$ there is associated an element

$$y = B(x, x') \in \mathbf{Y}.$$

The binary operator B is said to be *bilinear* if two further conditions are satisfied.

I. For all $x_1, x_2, x'_1, x'_2 \in \mathbf{X}$ and all numbers α, β,

$$\left.\begin{aligned} B(\alpha x_1 + \beta x_2, x') &= \alpha B(x_1, x') + \beta B(x_2, x'),\\ B(x, \alpha x'_1 + \beta x'_2) &= \alpha B(x, x'_1) + \beta B(x, x'_2) \end{aligned}\right\} \quad (x, x' \in \mathbf{X}). \tag{1}$$

II. There exists a real $M > 0$ such that, for all $x, x' \in \mathbf{X}$,

$$\|B(x, x')\| \leqslant M\|x\|\|x'\|. \tag{2}$$

As for linear operators, the least value of M in (2) is called the *norm* of B, and is written in the usual way: $\|B\|$. It is clear that

$$\|B\| = \sup_{\|x\|, \|x'\| \leqslant 1} \|B(x, x')\|. \tag{3}$$

The set of all bilinear operators, with the norm (3) and linearized in the natural way, is, as one can easily verify, a B-space, which we shall denote by $B(\mathbf{X}^2, \mathbf{Y})$.

Without dwelling on the details, we remark that one can similarly introduce the B-spaces $B(\mathbf{X}^n, \mathbf{Y})$ of n-linear operators.*

Let us consider some examples of bilinear operators.

In the simplest case, $\mathbf{X} = \mathbf{Y} = \mathbf{R}^1$, the general form for a bilinear operator is

$$B(x, x') = \alpha x x'$$

(α being a real number).

Now let $\mathbf{X}$ and $\mathbf{Y}$ be finite-dimensional spaces: suppose $\mathbf{X}$ is m-dimensional and $\mathbf{Y}$ n-dimensional. Denote by x_i $(i = 1, 2, \ldots, m)$ the element of $\mathbf{X}$ having zero for each coordinate except the i-th, and unity for the i-th coordinate. Let $y_j \in \mathbf{Y}$ $(j = 1, 2, \ldots, n)$ have a similar meaning. Consider an operator $B \in B(\mathbf{X}^2, \mathbf{Y})$. Setting

$$B(x_i, x_j) = \left(a_{ij}^{(1)}, a_{ij}^{(2)}, \ldots, a_{ij}^{(n)}\right) \qquad (i, j = 1, 2, \ldots, m),$$

we see from (1) that, for arbitrary elements $x, x' \in \mathbf{X}$ $(x = (\xi_1, \xi_2, \ldots, \xi_m), x' = (\xi'_1, \xi'_2, \ldots, \xi'_m))$, we have

$$y = B(x, x') = B\left(\sum_{i=1}^{m} \xi_i x_i, \sum_{j=1}^{m} \xi'_j x_j\right) = \sum_{i,j=1}^{m} \xi_i \xi'_j B(x_i, x_j).$$

Hence, if $y = (\eta_1, \eta_2, \ldots, \eta_n)$, then

$$\eta_k = \sum_{i,j=1}^{m} a_{ij}^{(k)} \xi_i \xi'_j \qquad (k = 1, 2, \ldots, n), \tag{4}$$

that is, the coordinates of y are bilinear forms in the coordinates of x and x'.

---

* For a detailed account of the theory of multilinear operators, see Gavurin [2], Hille and Phillips.

Clearly, however we choose the three-dimensional array

$$\left(a_{ij}^{(k)}\right) \qquad (i, j = 1, 2, \ldots, m;\ k = 1, 2, \ldots, n) \tag{5}$$

the operator B defined by (4) will be bilinear.

Thus, in this case, every bilinear operator determines, and is determined by, a three-dimensional array (5).

The norm of B depends, of course, on the norms of **X** and **Y**. For instance, if $\mathbf{X} = \ell_m^2$, $\mathbf{Y} = \ell_n^2$, then we see from (4), using the Cauchy–Buniakowski inequality, that

$$|\eta_k| = \left|\sum_{i,j=1}^{m} a_{ij}^{(k)} \xi_i \xi_j'\right| \leqslant \left[\sum_{i=1}^{m} |\xi|^2\right]^{1/2} \left[\sum_{i=1}^{m} \left|\sum_{j=1}^{m} a_{ij}^{(k)} \xi_j'\right|^2\right]^{1/2} \leqslant \Lambda_k \|x\| \|x'\|,$$

where Λ_k^2 denotes the least eigenvalue of the matrix $A_k A_k^*$ obtained by multiplying the matrix

$$A_k = \begin{pmatrix} a_{11}^{(k)} & a_{12}^{(k)} & \ldots & a_{1m}^{(k)} \\ a_{21}^{(k)} & a_{22}^{(k)} & \ldots & a_{2m}^{(k)} \\ \ldots & \ldots & \ldots & \ldots \\ a_{m1}^{(k)} & a_{m2}^{(k)} & \ldots & a_{mm}^{(k)} \end{pmatrix} \qquad (k = 1, 2, \ldots, n)$$

by its transpose A_k^* (cf. V.2.8). Hence

$$\|y\| = \left[\sum_{k=1}^{n} |\eta_k|^2\right]^{1/2} \leqslant \left[\sum_{k=1}^{n} \Lambda_k^2\right]^{1/2} \|x\| \|x'\|.$$

This yields a bound for $\|B\|$:

$$\|B\| \leqslant \left[\sum_{k=1}^{n} \Lambda_k^2\right]^{1/2}. \tag{6}$$

Since

$$\Lambda_k^2 \leqslant \sum_{i,j=1}^{m} |a_{ij}^{(k)}|^2 \qquad (k = 1, 2, \ldots, n),$$

(6) yields the cruder, but simpler, bound

$$\|B\| \leqslant \left[\sum_{k=1}^{n} \sum_{i=1}^{m} \sum_{j=1}^{m} |a_{ij}^{(k)}|^2\right]^{1/2}.$$

If we set $\mathbf{X} = \ell_m^\infty$, $\mathbf{Y} = \ell_n^\infty$, then, using obvious arguments, one can obtain the following bounds for $\|B\|$:

$$\|B\| \leqslant \max_k \sum_{i=1}^{m} \sum_{j=1}^{m} |a_{ij}^{(k)}|.$$

Examples of bilinear operators on infinite-dimensional spaces—particularly spaces of functions—are of considerable interest.

The operator B defined by

$$y = B(x, x'), \qquad y(s) = \int_0^1 \int_0^1 K(s, t, u) x(t) x'(u)\, dt\, du \tag{7}$$

is bilinear on various spaces of functions, depending on what conditions we impose on K, the kernel of the operator. For instance, if K is assumed continuous, B will be bilinear for any choice of $\mathbf{X}$ and $\mathbf{Y}$ from among the spaces $\mathbf{C}$, $\mathbf{L}^p\,(1 \leqslant p \leqslant \infty)$.

Another case of great interest for applications is where the operator is of the form

$$y = B(x, x'), \tag{8}$$

$$y(s) = \int_0^1 K(s, t)x(t)x'(t)\,dt. \tag{9}$$

Suppose first that $\mathbf{X} = \mathbf{Y} = \mathbf{L}^p(0, 1)$ $(p \geqslant 2)$. Then, provided $M < \infty$, where

$$M^p = \begin{cases} \int_0^1 \left[\int_0^1 |K(s, t)|^{\frac{p}{p-2}}\,dt\right]^{p-2} ds & (p > 2), \\ \int_0^1 \sup_s |K(s, t)|\,dt & (p = 2), \end{cases}$$

we have $B \in B(\mathbf{X}^2, \mathbf{Y})$.

In fact, the integrand in (9) is summable for almost all $s \in [0, 1]$. To see this for $p > 2$, we need only apply the generalized Hölder inequality (II.3.4) to the integral (9), with exponents $\dfrac{p}{p-2}$, p, p. We then obtain

$$|y(s)|^p = \left|\int_0^1 K(s, t)x(t)x'(t)\,dt\right|^p \leqslant \left[\int_0^1 |K(s, t)|^{\frac{p}{p-2}}\,dt\right]^{p-2} \int_0^1 |x(t)|^p\,dt \int_0^1 |x'(t)|^p\,dt$$

and so

$$\int_0^1 |y(s)|^p\,ds \leqslant M^p \|x\|^p\,\|x'\|^p < \infty. \tag{10}$$

Thus $y \in \mathbf{L}^p(0, 1)$. We have also proved that B satisfies the inequality (2). As it is obviously additive (condition (1)), it follows that B is bilinear. At the same time, (10) yields a bound for the norm:

$$\|B\| \leqslant M.$$

The above arguments applied to the case $p > 2$. The case $p = 2$ presents no difficulties.

Now suppose $\mathbf{X} = \mathbf{Y} = \mathbf{C}[0, 1]$. In this case, we can ensure that the operator (8) is bilinear by requiring that $K(s, t)$ be continuous in s and that

$$M = \max_s \int_0^1 |K(s, t)|\,dt < \infty.$$

For the summability of the integrand in (9) is not in doubt. The continuity of y is a consequence of the fact that we can take limits under the integral sign in (9). Equations (1) are obvious. Equation (2) is also easy to verify, as

$$|y(s)| \leqslant \max_t |x(t)| \max_t |x'(t)| \int_0^1 |K(s, t)|\,dt \qquad (0 \leqslant s \leqslant 1),$$

and so

$$\|y\| \leqslant M\,\|x\|\,\|x'\|.$$

Hence we obtain the bound

$$\|B\| \leqslant M = \sup_s \int_0^1 |K(s, t)|\,dt,$$

which actually holds with equality:

$$\|B\| = \sup_s \int_0^1 |K(s, t)|\, dt$$

(this is established in exactly the same way as the corresponding equation for linear operators in V.2.4).

2.3. The space $B(\mathbf{X}^2, \mathbf{Y})$ introduced above turns out to be closely related to the space $B(\mathbf{X}, B(\mathbf{X}, \mathbf{Y}))$, which we encountered in our discussion of the second derivative. In fact, we shall now prove that these spaces are linearly isometric.

Let $W \in B(\mathbf{X}, B(\mathbf{X}, \mathbf{Y}))$. Take an arbitrary $x' \in \mathbf{X}$ and write $U_{x'} = W(x')$. Clearly $U_{x'} \in B(\mathbf{X}, \mathbf{Y})$, so $y = B(x, x') = U_{x'}(x)$ is an element of $\mathbf{Y}$. The binary operator B thus defined obviously satisfies condition I of 2.2; and condition II is also satisfied, since

$$\|y\| \leqslant \|U_{x'}\|\, \|x\| \leqslant \|W\|\, \|x\|\, \|x'\|. \tag{11}$$

Thus $B \in B(\mathbf{X}^2, \mathbf{Y})$, and from (11) we obtain

$$\|B\| \leqslant \|W\|. \tag{12}$$

Hence in this way we can associate with each operator $W \in B(\mathbf{X}, B(\mathbf{X}, \mathbf{Y}))$ a bilinear operator $B \in B(\mathbf{X}^2, \mathbf{Y})$. Let us show that, in this correspondence, every bilinear operator $B \in B(\mathbf{X}^2, \mathbf{Y})$ is the image of some $W \in B(\mathbf{X}, B(\mathbf{X}, \mathbf{Y}))$.

For this it is sufficient to note that, if B is given, then for any fixed $x' \in \mathbf{X}$ we have*

$$W(x') = B(\cdot, x') \in B(\mathbf{X}, \mathbf{Y}).$$

Moreover,

$$\|W(x')\| = \sup_{\|x\| \leqslant 1} \|W(x')(x)\| = \sup_{\|x\| \leqslant 1} \|B(x, x')\| \leqslant \|B\|\, \|x'\|,$$

and so it follows that $W \in B(\mathbf{X}, B(\mathbf{X}, \mathbf{Y}))$, and we have

$$\|W\| \leqslant \|B\|.$$

Comparing this with (12) we obtain

$$\|W\| = \|B\|. \tag{13}$$

Since the correspondence between the elements of $B(\mathbf{X}, B(\mathbf{X}, \mathbf{Y}))$ and $B(\mathbf{X}^2, \mathbf{Y})$ is additive and homogeneous, it follows from (13) that it is one-to-one, and hence determines a linear isometry between these spaces.

This linear isometry provides grounds for identifying the corresponding elements of $B(\mathbf{X}, B(\mathbf{X}, \mathbf{Y}))$ and $B(\mathbf{X}^2, \mathbf{Y})$, and we shall do this in future.

We note without proof (the reader can supply one without difficulty) that, in general, the space $B(\mathbf{X}, \ldots, B(\mathbf{X}, \mathbf{Y}), \ldots)$ is linearly isometric to the space $B(\mathbf{X}^n, \mathbf{Y})$ of n-linear operators (see Gavurin [2]).

2.4. As we showed in 2.1, the second derivative of an operator P mapping $\mathbf{X}$ into $\mathbf{Y}$ is an element of the space $\mathbf{B}(\mathbf{X}, B(\mathbf{X}, \mathbf{Y}))$, and so, by the results just proved, $P''(x_0)$ may be regarded as a bilinear operator. If we do this, then, by 2.3,

$$\lim_{t \to 0} \frac{P'(x_0 + tx') - P'(x_0)}{t} = P''(x_0)(\cdot, x'). \tag{14}$$

* $B(\cdot, x')$ denotes the operator U such that $U(x) = B(x, x')$. We shall use similar notation in what follows.

Consequently, for any $x \in \mathbf{X}$, we have

$$P''(x_0)(x, x') = \lim_{t \to 0} \frac{P'(x_0 + tx')x - P'(x_0)x}{t}. \tag{15}$$

Note that (14) and (15) are not equivalent. To be more precise, it may happen that there exists a bilinear operator $B \in B(\mathbf{X}^2, \mathbf{Y})$ such that, for all $x, x' \in \mathbf{X}$, we have

$$\lim_{t \to 0} \frac{P'(x_0 + tx')x - P'(x_0)x}{t} = B(x, x') \tag{16}$$

while $P''(x_0)$ does not exist. It is not hard to see that a necessary and sufficient condition for $P''(x_0)$ to exist is that, for each $x' \in \mathbf{X}$, the limit in equation (16) should be uniform in all $x \in \mathbf{X}$ with $\|x\| = 1$. This situation enables us to extend the concept of the second derivative somewhat. We shall call a bilinear operator B the *weak second derivative* of P at x_0 if (16) holds for all $x, x' \in \mathbf{X}$. We can clearly retain the previous notation for the weak derivative as well, since if the second derivative in the usual sense exists then it coincides with the weak second derivative.

REMARK. It is easy to see that the fact that P is twice differentiable at a point x_0 means that (16) holds uniformly for all $x, x' \in \mathbf{X}$ with $\|x\| = \|x'\| = 1$.

Let us see what the second derivative is like when $\mathbf{X}$ and $\mathbf{Y}$ are finite-dimensional spaces. Suppose that $P''(x_0)$ exists and is determined, as a bilinear operator, by the array

$$(a_{ij}^{(k)}) \qquad (i, j = 1, 2, \ldots, m; \quad k = 1, 2, \ldots, n).$$

If we equate corresponding coordinates on the left and right of (15), we see that, in the notation of 1.6,*

$$\sum_{i,j=1}^{m} a_{ij}^{(k)} \xi_i \xi_j' = \lim_{t \to 0} \frac{\sum_{i=1}^{m} \frac{\partial \phi_k}{\partial \xi_i}(\xi_j^{(0)} + t\xi_j')\xi_i - \sum_{i=1}^{m} \frac{\partial \phi_k}{\partial \xi_i}(\xi_j^{(0)})\xi_i}{t}$$

$$(k = 1, 2, \ldots, n).$$

Taking x and x' to be the elements with ones in the i-th and j-th coordinates and zeros elsewhere, we find that the functions ϕ_k have second derivatives and that

$$a_{ij}^{(k)} = \frac{\partial^2 \phi_k(\xi_1^{(0)}, \xi_2^{(0)}, \ldots, \xi_m^{(0)})}{\partial \xi_i \partial \xi_j} \qquad (i, j = 1, 2, \ldots, m;\ k = 1, 2, \ldots, n).$$

2.5. To conclude this section, we establish a generalization of Taylor's formula for real-valued functions.

Suppose, as above, that the operator P maps a subset Ω of $\mathbf{X}$ into $\mathbf{Y}$ and has continuous second derivative in the interval $[x_0, \bar{x}] \subset \Omega$. Then we have the formula

$$P(\bar{x}) - P(x_0) - P'(x_0)\ (\bar{x} - x_0) = \int_{x_0}^{\bar{x}} P''(x)\ (\bar{x} - x, \cdot)dx. \tag{17}$$

To prove this, we consider the operator Q defined by

$$Q(x) = P(x) + P'(x)\ (\bar{x} - x) = P(x) + Q_0(x) \qquad (x \in \Omega)$$

* Here we use the abbreviated notation $\phi(\xi_j) = \phi(\xi_1, \xi_2, \ldots, \xi_m)$.

and verify that, for each $\tilde{x} \in [x_0, \bar{x}]$,

$$Q'(\tilde{x})\ (x) = P''(\tilde{x})\ (\bar{x} - \tilde{x}, x) \qquad (x \in \mathbf{X}). \tag{18}$$

If we establish this, then (17) is an obvious consequence of the Newton–Leibniz formula (see (18), 1.7).

It is sufficient to consider just the operator Q_0. We have

$$\frac{Q_0(\tilde{x}+tx) - Q_0(\tilde{x})}{t} = \frac{P'(\tilde{x}+tx)(\bar{x}-\tilde{x}-tx) - P'(\tilde{x})(\bar{x}-\tilde{x})}{t} =$$

$$= \frac{P'(\tilde{x}+tx)(\bar{x}-\tilde{x}) - P'(\tilde{x})(\bar{x}-\tilde{x})}{t} - P'(\tilde{x}+tx)x.$$

As $t \to 0$, the first term tends to the limit $P''(\tilde{x})\,(\bar{x}-\tilde{x}, x)$ and the second—since P' is obviously continuous—tends to $P'(\tilde{x})\,(x)$. Hence

$$Q_0'(\tilde{x})(x) = P''(\tilde{x})\ (\bar{x}-\tilde{x}, x) - P'(\tilde{x})x,$$

which gives the required expression (18) for $Q'(\tilde{x})\,(x)$.

REMARK. It is clear from the above argument that (17) remains true if $P''(x)$ is taken to be a weak second derivative.

§ 3. Examples

In the preceding sections we have already met examples in which we determined derivatives of non-linear operators in the simplest case—where the spaces are finite-dimensional. The examples we shall consider in this section are more complex and richer in content, and will find application in the next section and, more particularly, in the next chapter.

3.1. Consider the integral operator P defined by

$$y = P(x), \qquad y(s) = \int_0^1 K(s, t, x(t))\,dt. \tag{1}$$

THEOREM 1. *Suppose that $K(s, t, u)$ is continuous and twice continuously differentiable in u for $0 \leqslant s, t \leqslant 1$, $-\infty < u < \infty$, and that, for these values of s, t, u,*

$$|K''_{u^2}(s, t, u)| \leqslant M|u|^{p-2} + N. \tag{2}$$

Then, for $p \geqslant 2$, the operator (1) *maps $\mathbf{L}^p(0, 1)$ into $\mathbf{L}^q(0, 1)$ $(1 \leqslant q < \infty)$, is differentiable and has a second derivative at each point $x_0 \in \mathbf{L}^p(0, 1)$. Furthermore, $P'(x_0) = U$, $P''(x_0) = B$, where U and B are defined by*

$$z = U(x), \qquad z(s) = \int_0^1 K'_u(s, t, x_0(t))x(t)\,dt,$$

$$v = B(x, x'), \qquad v(s) = \int_0^1 K''_{u^2}(s, t, x_0(t))x(t)x'(t)\,dt.$$

Proof. We first check that the operator (1) maps $\mathbf{L}^p(0, 1)$ into $\mathbf{L}^\infty(0, 1)$ (and so, *a fortiori*, into $\mathbf{L}^q(0, 1)$ $(q < \infty)$). With this aim, we note that, since $K(s, t, u)$ is a continuous function, the compound function $K(s, t, x(t))$ is measurable.

Now, by Taylor's formula, we have

$$K(s, t, x(t)) = K(s, t, 0) + K'_u(s, t, 0)x(t) + \frac{1}{2}K''_{u^2}(s, t, \theta x(t))x^2(t) \qquad (0 < \theta < 1).$$

Let us now estimate the integrals of the first two terms:

$$\left|\int_0^1 K(s, t, 0)\,dt\right| \leqslant \max_{s,t}|K(s, t, 0)| = A_1,$$

$$\left|\int_0^1 K'_u(s, t, 0)x(t)\,dt\right| \leqslant \max_{s,t}|K'_u(s, t, 0)|\int_0^1 |x(t)|\,dt \leqslant A_2\|x\|_{\mathbf{L}^p}.$$

To estimate the integral of the third term, we use (2):

$$\left|\frac{1}{2}\int_0^1 K''_{u^2}(s, t, \theta x(t))x^2(t)dt\right| \leqslant \frac{1}{2}\int_0^1 [M|x(t)|^{p-2} + N]|x(t)|^2\,dt \leqslant A_3\|x\|^p_{\mathbf{L}^p}.$$

It thus follows that $y = P(x)$ is well defined and, moreover, that

$$|y(s)| \leqslant A_1 + A_2\|x\|_{\mathbf{L}^p} + A_3\|x\|^p_{\mathbf{L}^p} \qquad (s \in [0, 1]),$$

so that $y \in \mathbf{L}^\infty(0, 1)$.

We now prove that P is differentiable, and that $P'(x_0) = U$. Let $z = U(x)$ and set

$$z_\tau = \frac{P(x_0 + \tau x) - P(x_0)}{\tau}.$$

We have

$$|z_\tau(s) - z(s)| = \left|\int_0^1 \left[\frac{K(s, t, x_0(t) + \tau x(t)) - K(s, t, x_0(t))}{\tau} - K'_u(s, t, x_0(t))x(t)\right]dt\right|. \tag{3}$$

Since by Taylor's formula we have

$$K(s, t, x_0(t) + \tau x(t)) = K(s, t, x_0(t)) + \tau K'_u(s, t, x_0(t))x(t) + \frac{\tau^2}{2}K''_{u^2}(s, t, x_0(t) + \theta\tau x(t))x^2(t)$$

$$(0 < \theta < 1),$$

we can rewrite (3):

$$|z_\tau(s) - z(s)| = \frac{|\tau|}{2}\left|\int_0^1 K''_{u^2}(s, t, x_0(t) + \theta\tau x(t))\,dt\right|.$$

By similar arguments we obtain

$$|z_\tau(s) - z(s)| \leqslant A_4|\tau| \qquad (0 \leqslant s \leqslant 1),$$

where the constant A_4 depends only on $\|x\|$, and not on x itself. Hence $z_\tau \to z$ as $\tau \to 0$, uniformly for all x with $\|x\| = 1$.

We now turn to the second derivative. Write

$$v = B(x, x'); \qquad v_\tau = \frac{P'(x_0 + \tau x')(x) - P'(x_0)(x)}{\tau}.$$

As we remarked in 2.4, it is sufficient to prove that, as $\tau \to 0$,

$$v_\tau \to v$$

uniformly for $x \in \mathbf{L}^p(0, 1)$ with $\|x\| = 1$.

Using the Mean Value Theorem and then Hölder's inequality, we obtain

$$|v_\tau(s) - v(s)| = \left| \int_0^1 \left[\frac{K'_u(s, t, x_0(t) + \tau x'(t)) - K'_u(s, t, x_0(t))}{\tau} - K''_{u^2}(s, t, x_0(t))x'(t) \right] x(t)\, dt \right| =$$

$$= \left| \int_0^1 [K''_{u^2}(s, t, x_0(t) + \tau\theta x'(t)) - K''_{u^2}(s, t, x_0(t))] x(t) x'(t)\, dt \right| \leqslant$$

$$\leqslant \|x\| \left\{ \int_0^1 |[K''_{u^2}(s, t, x_0(t) + \theta\tau x'(t)) - \right.$$

$$\left. - K''_{u^2}(s, t, x_0(t))] x'(t)|^{\frac{p}{p-1}}\, dt \right\}^{\frac{p-1}{p}}. \tag{4}$$

If $x_0(t)$ and $x'(t)$ are finite, then the integrand in the last integral tends to zero as $\tau \to 0$. Further, by (2), we have*

$$|[K''_{u^2}(s, t, x_0(t) + \theta\tau x'(t)) - K''_{u^2}(s, t, x_0(t))] x'(t)|^{\frac{p}{p-1}} \leqslant$$

$$\leqslant \{[M|x_0(t) + \theta\tau x'(t)|^{p-2} + M|x_0(t)|^{p-2} + 2N] \times$$

$$\times |x'(t)|\}^{\frac{p}{p-1}} \leqslant A_5 |x_0(t)|^{\frac{(p-2)p}{p-1}} |x'(t)|^{\frac{p}{p-1}} + A_6 |x'(t)|^p.$$

Since $|x_0(t)|^{p - \frac{p}{p-1}}$ and $|x'(t)|^p$ are summable, we can take limits under the integral sign in (4), which yields

$$\lim_{\tau \to 0} v_\tau(s) = v(s) \qquad (0 \leqslant s \leqslant 1);$$

moreover, this obviously holds uniformly for all x with $\|x\| = 1$.

It also follows from the arguments just given that

$$|v_\tau(s)| \leqslant \int_0^1 |K''_{u^2}(s, t, x_0(t) + \theta\tau x'(t)) x(t) x'(t)|\, dt \leqslant A_7,$$

so we can take limits under the integral sign in the integral $\|v_\tau - v\|_{\mathbf{L}^q} = \left\{ \int_0^1 |v_\tau(s) - v(s)|^q ds \right\}^{1/q}$; that is, $\|v_\tau - v\| \to 0$, uniformly for all x with $\|x\| = 1$.

This proves the theorem.

Remark. If we replace condition (2) by the condition

$$|K''_{u^2}(s, t, u)| \leqslant M|u|^\lambda + N \qquad (\lambda < p - 2) \tag{5}$$

---

* Here we use the inequality $|a + b|^\alpha \leqslant 2^\alpha(|a|^\alpha + |b|^\alpha)$ for $\alpha > 0$.

and if $p > 2$, then it can be shown that P, regarded as an operator from $\mathbf{L}^p$ into $\mathbf{L}^\infty$, is twice differentiable at each point $x_0 \in \mathbf{L}^p$ for which x_0 is a bounded function.

3.2. The operator (1) can also be regarded as an operator from $\mathbf{C}[0, 1]$ into $\mathbf{C}[0, 1]$.

Let $x_0 \in \mathbf{C}[0, 1]$ be a fixed element. Write Ω for the ball in $\mathbf{C}[0, 1]$ with centre at x_0 and radius r, and G for the set of points (s, t, u) in three-dimensional space defined by the inequalities $0 \leqslant s, t \leqslant 1$, $|u - x_0(t)| \leqslant r$.

THEOREM 2. *Assume that the function $K(s, t, u)$ is defined and continuous, and has continuous derivatives $K'_u(s,t,u)$ and $K''_{u^2}(s,t,u)$, throughout the region G. Then the operator P defined by* (1) *maps Ω into $\mathbf{C}[0,1]$ and is twice differentiable at each interior point $\bar{x} \in \Omega$; and $P'(\bar{x}) = U$, $P''(x) = B$ are given by*

$$z = U(x), \qquad z(s) = \int_0^1 K'_u(s, t, \bar{x}(t))x(t)\,dt,$$

$$v = B(x, x'), \qquad v(s) = \int_0^1 K''_{u^2}(s, t, \bar{x}(t))x(t)x'(t)\,dt.$$

Proof. Choose any x in Ω. It follows immediately from elementary facts in the theory of integrals depending on a parameter that the function $y = P(x)$ defined by (1) is continuous; hence $y \in \mathbf{C}[0, 1]$.

Now let $\bar{x} \in \Omega$ and $x \in \mathbf{C}[0, 1]$. Consider the element

$$z_\tau = \frac{P(\bar{x} + \tau x) - P(\bar{x})}{\tau} - U(x).$$

We have

$$z_\tau(s) = \int_0^1 \left[\frac{K(s, t, \bar{x}(t) + \tau x(t)) - K(s, t, \bar{x}(t))}{\tau} - K'_u(s, t, \bar{x}(t))x(t)\right]dt.$$

As $\tau \to 0$ the integrand tends to zero and is uniformly bounded for all x with $\|x\| = 1$ and all $s \in [0, 1]$ (this can be established using the Mean Value Theorem). Hence $z_\tau \to \mathbf{0}$, uniformly for all $x \in \mathbf{C}[0, 1]$ with $\|x\| = 1$, and it follows that P is differentiable at $\bar{x}$.

Now we have

$$P'(\bar{x}) = U.$$

Similar arguments show that $P''(\bar{x}) = B$, and also that P is twice differentiable.

3.3. We now consider a non-linear operator associated with the following system of ordinary differential equations of first order:

$$\frac{dy_i}{ds} = f_i(s, y_k(s)) \qquad (i, k = 1, 2, \ldots, n),$$

which we shall find it convenient to rewrite as a system of integral equation

$$y_i(s) = y_i^{(0)} + \int_0^s f_i(t, y_k(t))\,dt \qquad (i, k = 1, 2, \ldots, n)$$

or, more concisely, in vector form

$$y(s) = y^{(0)} + \int_0^s f(t, y(t))\,dt,$$

where $y(s)$, $y^{(0)}$, $f(t, u)$, u are the vectors having components

$$(y_1(s), y_2(s), \ldots, y_n(s)), \qquad (y_1^{(0)}, y_2^{(0)}, \ldots, y_n^{(0)}),$$

$$(f_1(t, u_k), f_2(t, u_k), \ldots, f_n(t, u_k)), \qquad (u_1, u_2, \ldots, u_n).$$

We introduce the space $\mathbf{C}_n^\alpha$ $(\alpha > 0)$, whose elements are absolutely continuous vector-functions on $[0, \infty)$ such that*

$$\sup_t |y(t)e^{\alpha t}| < \infty, \qquad \sup_t |y'(t)e^{\alpha t}| < \infty \qquad (t \geqslant 0).$$

We introduce a norm in $\mathbf{C}_n^\alpha$ by means of the equation

$$\|y\| = \sup_t |y(t)e^{\alpha t}| + \sup_t |y'(t)e^{\alpha t}| \qquad (y \in \mathbf{C}_n^\alpha).$$

In addition to $\mathbf{C}_n^\alpha$ we consider the space $\mathbf{D}_n^\alpha$ consisting of absolutely continuous vector-functions satisfying the conditions

$$\sup_t |z(t)| < \infty, \qquad \sup_t |z'(t)e^{\alpha t}| < \infty \qquad (z \in \mathbf{D}_n^\alpha).$$

We define a norm in $\mathbf{D}_n^\alpha$ by

$$\|z\| = \sup_t |z(t)| + \sup_t |z'(t)e^{\alpha t}| \qquad (z \in \mathbf{D}_n^\alpha).$$

It is not hard to verify that $\mathbf{C}_n^\alpha$ and $\mathbf{D}_n^\alpha$ are B-spaces.

Consider the operator P defined by

$$z = P(y), \tag{6}$$

$$z(s) = y(s) - \int_0^s f(t, y(t))\, dt. \tag{7}$$

THEOREM 3. *Let $f(t, u)$ be a vector function defined on the subset G of $(n+1)$-dimensional space specified by*

$$(t, u) \in G \textit{ if and only if } t \geqslant 0,\ |u| \leqslant re^{-\alpha t}.$$

Suppose that

1) $f(t, 0) = 0$;

2) $f(t, u)$ has a bounded derivative† $f_u'(t, u)$ in G, which is continuous in u, uniformly for all $t \geqslant 0$.

If Ω denotes the ball in $\mathbf{C}_n^\alpha$ with centre at the origin and radius r, then the operator (6) maps Ω into $\mathbf{D}_n^\alpha$ and is differentiable at each point of Ω; moreover, $P'(y_0) = U$, where U is defined by

$$z = U(y), \qquad z(s) = y(s) - \int_0^1 f_u'(t, y_0(t))y(t)\, dt. \tag{8}$$

Proof. First we prove that the function defined by (7) belongs to $\mathbf{D}_n^\alpha$ (assuming that $y \in \Omega$). In fact, since $f(t, 0) = 0$, we have

$$z(s) = y(s) - \int_0^s [f(t, y(t)) - f(t, 0)]\, dt.$$

* As above (in Chapter XVI), we denote the length of a vector u by the symbol $|u|$ and the norm of a matrix A, regarded as an operator from ℓ_n^2 into ℓ_n^2, by $|A|$.

† Recall that the derivative $f_u'(t, u)$ of a vector function $f(t, \cdot)$ is the $n \times n$ matrix of derivatives $\left(\dfrac{\partial f_i}{\partial u_k}\right)$ (see 1.6).

Using the Mean Value Theorem, we obtain the following bound for the integrand:

$$|f(t, y(t)) - f(t, 0)| \leqslant \sup_{0<\theta<1} |f'_u(t, \theta y(t))| |y(t)| \leqslant M\|y\| e^{-\alpha t}$$

$$(M = \sup |f'_u(t, u)|, \qquad (t, u) \in G).$$

Using this, we have

$$\sup_s |z(s)| \leqslant \sup_s |y(s)| + M\|y\| \sup_s \int_0^s e^{-\alpha t}\, dt \leqslant \|y\| + \frac{M}{\alpha}\|y\| < \infty.$$

Furthermore,

$$z'(s) = y'(s) - f(s, y(s)),$$

and, by similar arguments, we find that

$$\sup_s |z'(s) e^{\alpha s}| < \infty.$$

Hence z is well defined and is an element of $\mathbf{D}_n^\alpha$.

Now we prove that U, defined by (8), is a continuous linear operator mapping $\mathbf{C}_n^\alpha$ into $\mathbf{D}_n^\alpha$.

Repeating the above arguments, we find that $z = U(y)$ satisfies

$$\sup_s |z(s)| + \sup_s |z'(s) e^{\alpha s}| = \sup_s \left| y(s) - \int_0^s f'_u(t, y_0(t)) y(t)\, dt \right| +$$

$$+ \sup_s |[y'(s) - f'_u(s, y_0(s)) y(s)] e^{\alpha s}| \leqslant$$

$$\leqslant \|y\| + \frac{M}{\alpha}\|y\| + (M+1)\|y\| = L\|y\|,$$

so that $z = U(y) \in \mathbf{D}_n^\alpha$ and $\|z\| \leqslant L\|y\|$.

As in the preceding theorems, we now consider the element

$$v_\tau = \frac{P(y_0 + \tau y) - P(y_0)}{\tau} - U(y), \qquad v_\tau(s) = \int_0^s J_\tau(t)\, dt$$

$$\left(J_\tau(t) = \frac{f(t, y_0(t) + \tau y(t)) - f(t, y_0(t))}{\tau} - f'_u(t, y_0(t)) y(t) \right).$$

Applying the Mean Value Theorem to the integrand, we find that

$$|J_\tau(t)| \leqslant \sup_{0<\theta<1} |f'_u(t, y_0(t) + \theta\tau y(t)) - f'_u(t, y_0(t))| \, |y(t)|.$$

Hence, as $f'_u(t, u)$ is uniformly continuous, for each $\varepsilon > 0$, we have, for sufficiently small τ,

$$|J_\tau(t)| \leqslant \varepsilon e^{-\alpha t}$$

uniformly for all y with $\|y\| = 1$; and so

$$\sup_s |v_\tau(s)| \leqslant \frac{\varepsilon}{\alpha}.$$

Similarly, for these values of τ, we have

$$\sup_s |v'_\tau(s)e^{\alpha s}| \leqslant \varepsilon,$$

so that

$$\|v_\tau\| \leqslant \varepsilon\left(\frac{1}{\alpha}+1\right).$$

Thus $v_\tau \to 0$ as $\tau \to 0$, uniformly in $y \in \mathbf{C}_n^\alpha$ with $\|y\| = 1$, as we set out to prove.

§ 4. The implicit function theorem

4.1. Let $\mathbf{X}$ and $\mathbf{Y}$ be two B-spaces. Form the set $\mathbf{X} \times \mathbf{Y}$ of all pairs (x, y) $(x \in \mathbf{X}, y \in \mathbf{Y})$. The set $\mathbf{X} \times \mathbf{Y}$, linearized in the natural way, becomes a B-space if we introduce a norm in it—for example, by setting

$$\|(x, y)\| = \|x\| + \|y\|.$$

This B-space is called the *direct* (or *Cartesian*) *product* of the spaces $\mathbf{X}$ and $\mathbf{Y}$ (cf. IV.1.8).

If we identify pairs of the form $(x, \mathbf{0_Y})$ $(x \in \mathbf{X}, \mathbf{0_Y}$ the zero element of $\mathbf{Y})$ with elements of $\mathbf{X}$, we may assume that $\mathbf{X}$—and similarly $\mathbf{Y}$— is a subspace of the direct product $\mathbf{X} \times \mathbf{Y}$. Furthermore, each element $u = (x, y) \in \mathbf{X} \times \mathbf{Y}$ can be expressed in a unique way in the form

$$u = (x, y) = (x, \mathbf{0_Y}) + (\mathbf{0_X}, y) = x + y.$$

Every linear operator $U \in B(\mathbf{X} \times \mathbf{Y}, \mathbf{Z})$ induces a pair $(U_\mathbf{X}, U_\mathbf{Y})$ of linear operators $U_\mathbf{X} \in B(\mathbf{X}, \mathbf{Z})$, $U_\mathbf{Y} \in B(\mathbf{Y}, \mathbf{Z})$:

$$U_\mathbf{X}(x) = U((x, \mathbf{0_Y})), \qquad U_\mathbf{Y}(y) = U((\mathbf{0_X}, y)).$$

Conversely, if the pair of operators $(U_\mathbf{X}, U_\mathbf{Y})$ is given, then the operator U, defined by

$$U((x, y)) = U_\mathbf{X}(x) + U_\mathbf{Y}(y)$$

is a continuous linear operator from $\mathbf{X} \times \mathbf{Y}$ into $\mathbf{Z}$.

Now let P be a non-linear operator mapping a set $\Omega \subset \mathbf{X} \times \mathbf{Y}$ into a B-space $\mathbf{Z}$. Fixing $y_0 \in \mathbf{Y}$, we consider the operator

$$P^{(y_0)} = P(\cdot, y_0),$$

which maps the set $\Omega^{(y_0)}$ of all x such that $(x, y_0) \in \Omega$ (a "section" of Ω) into $\mathbf{Z}$. Similarly we define the operator $P^{(x_0)}$ $(x_0 \in \mathbf{X})$ mapping $\Omega^{(x_0)} \subset \mathbf{Y}$ into $\mathbf{Z}$.

If $x_0 \in \mathbf{X}$ is an interior point of $\Omega^{(y_0)}$ then we can speak of the derivative $P^{(y_0)\prime}(x_0)$. This derivative is called the *partial derivative with respect to* x of P at (x_0, y_0), and is denoted by $P'_x(x_0, y_0)$. Similarly, we introduce the partial derivative with respect to y, denoted by $P'_y(x_0, y_0)$. Obviously,

$$P'_x(x_0, y_0) \in B(\mathbf{X}, \mathbf{Z}), \qquad P'_y(x_0, y_0) \in B(\mathbf{Y}, \mathbf{Z}).$$

If P has a derivative $P'(x_0, y_0) = U$ at (x_0, y_0), then $P'_x(x_0, y_0) = U_\mathbf{X}$ and $P'_y(x_0, y_0) = U_\mathbf{Y}$; this justifies the notation we introduced for partial derivatives.

We can regard P as an abstract "function of two variables", which we shall reflect in our notation by writing $P(x, y)$ instead of $P((x, y))$. The fundamental propositions of the elementary theory of functions of several variables extend without difficulty to such "functions". Let us note one of these specifically.

Assume that the derivatives P'_x and P'_y exist in some neighbourhood of a point $(x_0, y_0) \in \Omega$. Then we have the inequality

$$\|P(x_0 + \Delta x, y_0 + \Delta y) - P(x_0, y_0) - P'_x(x_0, y_0)\ (\Delta x) - P'_y(x_0, y_0)\ (\Delta y)\| \leqslant$$

$$\leqslant \sup_{0 < \theta, \theta_1 < 1} \|P'_x(x_0 + \theta\Delta x, y_0 + \theta_1 \Delta y) - P'_x(x_0, y_0)\|\ \|\Delta x\| +$$

$$+ \sup_{0 < \theta, \theta_1 < 1} \|P'_y(x_0 + \theta\Delta x, y_0 + \theta_1 \Delta y) - P'_y(x_0, y_0)\|\ \|\Delta y\|. \tag{1}$$

For it is clear that

$$A = \|P(x_0 + \Delta x, y_0 + \Delta y) - P(x_0, y_0) - P'_x(x_0, y_0)\ (\Delta x) - P'_y(x_0, y_0)\ (\Delta y)\| \leqslant$$
$$\leqslant \|P(x_0 + \Delta x, y_0 + \Delta y) - P(x_0, y_0 + \Delta y) - P'_x\ (x_0, y_0)\ (\Delta x)\| +$$
$$+ \|P(x_0, y_0 + \Delta y) - P(x_0, y_0) - P'_y(x_0, y_0)\ (\Delta y)\|.$$

Applying the Mean Value Theorem to each term (see (8), § 1), we obtain

$$A \leqslant \sup_{0 < \theta < 1} \|P'_x(x_0 + \theta\Delta x, y_0 + \Delta y) - P'_x(x_0, y_0)\|\ \|\Delta x\| +$$

$$+ \sup_{0 < \theta_1 < 1} \|P'_y(x_0, y_0 + \theta_1 \Delta y) - P'_y(x_0, y_0)\|\ \|\Delta y\| \leqslant$$

$$\leqslant \sup_{0 < \theta, \theta_1 < 1} \|P'_x(x_0 + \theta\Delta x, y_0 + \theta_1 \Delta y) - P'_x(x_0, y_0)\|\ \|\Delta x\| +$$

$$+ \sup_{0 < \theta, \theta_1 < 1} \|P'_y(x_0 + \theta\Delta x, y_0 + \theta_1 \Delta y) - P'_y(x_0, y_0)\|\ \|\Delta y\|$$

as required.

4.2. The basic result of this section is concerned with the question of whether there exists a solution to the equation

$$P(x, y) = \mathbf{0},$$

that is, whether the equation defines an implicit function. As in elementary analysis, we have

THEOREM 1. *Let P be an operator, defined in a neighbourhood Ω of a point $(x_0, y_0) \in \mathbf{X} \times \mathbf{Y}$ and mapping Ω into a space $\mathbf{Z}$, and suppose P is continuous at (x_0, y_0). If, in addition, the following conditions are satisfied:*

1) $P(x_0, y_0) = \mathbf{0}$;
2) *P'_y exists in Ω and is continuous at (x_0, y_0);*
3) *the operator $P'_y(x_0, y_0) \in B(\mathbf{Y}, \mathbf{Z})$ has a continuous inverse*

$$\Gamma = [P'_y(x_0, y_0)]^{-1} \in B(\mathbf{Z}, \mathbf{Y}),$$

then there exists an operator F, defined in a neighbourhood $G \subset \mathbf{X}$ of x_0 and mapping G into

Y, *with the following properties*:

a) $P(x, F(x)) = \mathbf{0} \quad (x \in G)$;

b) $F(x_0) = y_0$;

c) F *is continuous at* x_0.

The operator F determined by properties a)–c) *is unique in the sense that, if* F_1 *is another operator with these properties, then there exists an* $\eta > 0$ *such that, for* $\|x - x_0\| < \eta$,

$$F_1(x) = F(x).$$

Proof. Without any loss of generality, we may assume that $x_0 = \mathbf{0}_{\mathbf{X}}$ and $y_0 = \mathbf{0}_{\mathbf{Y}}$. Choose $x \in \mathbf{X}$ so small that $\mathbf{0}_{\mathbf{Y}} \in \Omega^{(x)}$, where, as above, $\Omega^{(x)}$ denotes the set of elements $y \in \mathbf{Y}$ such that $(x, y) \in \Omega$, and consider the operator defined on $\Omega^{(x)}$ by

$$Q^{(x)}(y) = y - \Gamma P(x, y).$$

We prove that, for all sufficiently small $\varepsilon > 0$, we can find a $\delta > 0$ such that, when $\|x\| < \delta$, the operator $Q^{(x)}$ maps the ball $\|y\| \leqslant \varepsilon$ into itself. To do this we determine, and give a bound for, the derivative of $Q^{(x)}$. By results III and V of 1.2, we have

$$Q^{(x)\prime}(\bar{y})\ (y) = y - \Gamma P_y'(x, \bar{y})\ (y) = -\Gamma[P_y'(x, \bar{y}) - P_y'(\mathbf{0}, \mathbf{0})](y)$$

and so

$$\|Q^{(x)\prime}(\bar{y})\| \leqslant \|\Gamma\|\ \|P_y'(x, \bar{y}) - P_y'(\mathbf{0}, \mathbf{0})\|.$$

As $P_{\mathbf{Y}}'$ was assumed continuous at $(\mathbf{0}, \mathbf{0})$, the second factor here can be made as small as we wish, so by choosing ε and δ small enough we can arrange that

$$\|Q^{(x)\prime}(\bar{y})\| \leqslant \alpha < 1 \qquad (\|x\| < \delta, \quad \|\bar{y}\| \leqslant \varepsilon). \tag{2}$$

Next we find a bound for $Q^{(x)}(\mathbf{0})$. Bearing in mind condition 1), we have

$$\|Q^{(x)}(\mathbf{0})\| = \|\Gamma P(x, \mathbf{0})\| \leqslant \|\Gamma\|\ \|P(x, \mathbf{0}) - P(\mathbf{0}, \mathbf{0})\|$$

and therefore, as P is continuous at $(\mathbf{0}, \mathbf{0})$, we can make the right-hand side as small as we wish by decreasing δ. We shall assume that δ has already been decreased, if necessary, by an amount sufficient to make

$$\|Q^{(x)}(\mathbf{0})\| \leqslant \varepsilon(1 - \alpha) \qquad (\|x\| < \delta). \tag{3}$$

Our assertion can now be proved easily using the Mean Value Theorem. For, if $\|x\| < \delta$, $\|y\| < \varepsilon$, then, by (2) and (3),

$$\|Q^{(x)}(y)\| \leqslant \|Q^{(x)}(\mathbf{0})\| + \|Q^{(x)}(y) - Q^{(x)}(\mathbf{0})\| \leqslant$$

$$\leqslant \varepsilon(1 - \alpha) + \sup_{0 < \theta < 1} \|Q^{(x)\prime}(\theta y)\|\ \|y\| \leqslant \varepsilon(1 - \alpha) + \alpha\varepsilon = \varepsilon.$$

Thus the operator $Q^{(x)}$ maps the closed ball $\|y\| \leqslant \varepsilon$ into itself, and the derivative $Q^{(x)\prime}$ satisfies (2). Hence the conditions of Theorem 1.1 are satisfied, so $Q^{(x)}$ has a unique fixed point $y^* = F(x)$ in the above ball:

$$y^* = y^* - \Gamma P(x, y^*),$$

that is,

$$P(x, y^*) = \mathbf{0}.$$

The operator F is the one we require. For we have already verified a). The truth of b) follows from condition 1)—which we can rewrite in the form $Q_0(\mathbf{0}) = \mathbf{0}$—and the uniqueness of the fixed point of Q_0. Finally, c) is true since the value of ε is unrestricted from below, and so can be made as small as we require.

Let us prove the uniqueness of F. In the ball $\|y\| \leqslant \varepsilon$ there is a unique fixed point for $Q^{(x)}$ ($\|x\| < \delta$), and, since F_1 is continuous, we have, for small enough η,

$$\|F_1(x)\| = \|F_1(x) - F_1(\mathbf{0})\| \leqslant \varepsilon \qquad (\|x\| < \eta),$$

so that $F_1(x) = F(x)$ for such values of η.

THEOREM 2. *If, in Theorem* 1, *the operator P is assumed continuous at each point of Ω, then F will be continuous in some neighbourhood of x_0.*

Proof. In this case we can apply Theorem XVI.1.3 to $Q^{(x)}$. For the operator $Q^{(x)}$ ($\|x\| < \delta$) maps the ball $\|y\| \leqslant \varepsilon$ into itself and (2) ensures that $Q^{(x)}$ is a contraction operator (uniformly, for all $x \in \mathbf{X}$ with $\|x\| < \delta$). The continuity of $Q^{(x)}$ in its parameter (that is, in x, in the present instance) follows from the continuity of P.

Applying the theorem just referred to, we see that $y^* = F(x)$ depends continuously on x, for $\|x\| < \delta$.

REMARK. Theorems 1 and 2 are of a local type. Moreover, they provide no effective way of estimating ε as a function of δ. This deficiency will be removed in the next chapter, where more precise results will be established by enlisting the help of the second derivative.

We can, in fact, make some progress in this direction with the help of Theorem 3 below.

4.3. THEOREM 3. *Suppose that the conditions of Theorem* 1 *are satisfied, and suppose further that the partial derivative P'_x exists in Ω and is continuous at (x_0, y_0). Then the operator F is differentiable at x_0, and*

$$F'(x_0) = U,$$

where

$$U = -\Gamma P'_x(x_0, y_0) = -[P'_y(x_0, y_0)]^{-1} P'_x(x_0, y_0). \tag{4}$$

Proof. As in the proof of Theorem 1, we shall assume that $x_0 = \mathbf{0}$, $y_0 = \mathbf{0}$.

By 1.1, we need to prove that, for any $\varepsilon > 0$, we can find a $\delta > 0$ such that, for all $x \in \mathbf{X}$ with $\|x\| < \delta$,

$$\|F(x) - F(\mathbf{0}) - U(x)\| \leqslant \varepsilon \|x\|,$$

or, in other words, since $F(\mathbf{0}) = \mathbf{0}$,

$$\|F(x) - U(x)\| \leqslant \varepsilon \|x\|. \tag{5}$$

Setting $F(x) = y$ and replacing U by the expression (4), we rewrite the expression inside the norm symbol on the left-hand side of (5) in the form

$$F(x) - U(x) = y + \Gamma P'_x(\mathbf{0}, \mathbf{0})x = \Gamma[P'_x(\mathbf{0}, \mathbf{0})x + P'_y(\mathbf{0}, \mathbf{0})y].$$

But $P(x, y) = P(\mathbf{0}, \mathbf{0}) = \mathbf{0}$, so by (1)

$$\|F(x) - U(x)\| \leqslant \|\Gamma\| \, \|P(x, y) - P(\mathbf{0}, \mathbf{0}) - P'_x(\mathbf{0}, \mathbf{0})x - P'_y(\mathbf{0}, \mathbf{0})y\| \leqslant$$

$$\leqslant \|\Gamma\| \Big[\sup_{0 < \theta, \theta_1 < 1} \|P'_x(\theta x, \theta_1 y) - P'_x(\mathbf{0}, \mathbf{0})\| \, \|x\| +$$

$$+ \sup_{0 < \theta, \theta_1 < 1} \|P'_y(\theta x, \theta_1 y) - P'_y(\mathbf{0}, \mathbf{0})\| \, \|y\| \Big] \leqslant \eta[\|x\| + \|y\|],$$

and, as the partial derivatives P'_x and P'_y are continuous, we can make η as small as we wish by choosing δ small enough. Hence

$$\|F(x)-U(x)\| \leqslant \eta[\|x\|+\|F(x)\|] \leqslant \eta[\|x\|+\|U(x)\|+\|F(x)-U(x)\|].$$

Thus, if η is sufficiently small,

$$\|F(x)-U(x)\| \leqslant \eta \frac{1+\|U\|}{1-\eta}\|x\|,$$

and so, if η is chosen so that

$$\eta \frac{1+\|U\|}{1-\eta} \leqslant \varepsilon,$$

then inequality (5) will hold for all sufficiently small x, as we required.

4.4. To conclude the section, we present an example of how Theorem 1 is applied. Consider the system of differential equations

$$y_i'(s) = f_i(s, y_k(s)) \qquad (i, k = 1, 2, \ldots, n), \tag{6}$$

which, as in XVII.3.3, can be expressed also in the form

$$y(s) = y(0) + \int_0^s f(t, y(t))\,dt. \tag{7}$$

Assume that $f(t, 0) = 0$ $(t \geqslant 0)$. Then the system (6), or equivalently equation (7), has a trivial solution, identically equal to zero. This solution corresponds to the initial condition $y(0) = 0$.

There arises a question of great practical significance: under what conditions does (7) have a solution with sufficiently small initial data, defined on $[0, \infty)$, and more importantly when does it follow from the fact that the initial data are close to zero that the solution is close to zero on the whole semi-axis $[0, \infty)$? Let us give some precise definitions.

Suppose that, for each $\varepsilon < 0$, we can find a $\delta < 0$ such that, whenever a vector x has norm less than δ, the equation

$$y(s) = x + \int_0^s f(t, y(t))\,dt \tag{8}$$

has a unique bounded solution y^*, and

$$\sup_{s \geqslant 0} |y^*(s)| \leqslant \varepsilon.$$

Then the zero solution of (7) is said to be *stable*. If, in addition, $y^*(s) \to 0$ as $s \to \infty$, then the zero solution is said to be *asymptotically stable.*

Before investigating sufficient conditions for stability, we introduce another useful concept.

Let T be an $n \times n$ matrix. Form the series

$$E + \frac{1}{1!}T + \frac{1}{2!}T^2 + \ldots + \frac{1}{n!}T^n + \ldots,$$

where E is the identity matrix. Since $|T^k| \leqslant |T|^k$ $(k = 0, 1, \ldots)$, this series converges, whatever matrix we take for T. Its sum is denoted by e^T.

If x_0 is a fixed vector, then the vector-function

$$u(s) = e^{sT}x_0 \qquad (s \geqslant 0)$$

satisfies the differential equation

$$u'(s) = Tu(s), \tag{9}$$

as one can verify exactly as for ordinary real-valued functions. We make use of this fact in the proof of the following lemma.

LEMMA 1. *Let $\mu_1, \mu_2, \ldots$ be the eigenvalues of the matrix T. Then the elements of the matrix e^{sT} have the form $\sum_k p_k(s)e^{s\mu_k}$, where the p_k are polynomials of degree at most n.*

Hence, if $\mu = \max_k \operatorname{Re}\mu_k$, then the norm of e^{sT} satisfies

$$|e^{sT}| \leqslant K_\eta e^{(\mu+\eta)s}, \tag{10}$$

where $\eta > 0$ is arbitrary.

Proof. The first assertion follows immediately from elementary facts in the theory of systems of linear differential equations with constant coefficients, if one takes into account that the columns of the matrix e^{sT} form, by the remark made above, a complete system of linearly independent solutions for (9), the latter being, in essence, a system of linear differential equations with constant coefficients.*

Since the norm of a matrix is bounded by the square root of the sum of the squares of its entries, (10) follows immediately from what was said above.

Now let us return to equation (7). Suppose, as in 3.3, that the vector-function $f(t, u)$ is defined in a region G in $(n+1)$-dimensional space ($t \geqslant 0$, $|u| \leqslant re^{-\alpha t}$, $\alpha > 0$), that it is continuous in u at $u = 0$ and has a derivative $f'_u(t, u)$ which is continuous in u at $u = 0$. Assume that the matrix $A = f'_u(t, 0)$ is independent of t, and denote its eigenvalues by $\lambda_1, \lambda_2, \ldots$.

THEOREM 4. *If*

$$\operatorname{Re}\lambda_k < 0 \qquad (k = 1, 2, \ldots),$$

then the zero solution of (7) is asymptotically stable. More precisely, we have

$$|y^*(s)| \leqslant L_x e^{-\alpha s} \qquad (s \geqslant 0),$$

where the constant L_x depends only on x and tends to zero as $x \to 0$, and where $\alpha > 0$ is such that $\operatorname{Re}\lambda_k < -\alpha$ $(k = 1, 2, \ldots)$.

Proof. The proof is based on Theorem 1: take $\mathbf{X} = \ell^2_n$, $\mathbf{Y} = \mathbf{C}^\alpha_n$, $\mathbf{Z} = \mathbf{D}^\alpha_n$, and define the operator P as follows:

$$z = P(x, y), \qquad z(s) = y(s) - x - \int_0^s f(t, y(t))\,dt;$$

also take Ω to be the set of points $(x, y) \in \mathbf{X} \times \mathbf{Y}$, where $x \in \mathbf{X}$ is arbitrary and $\|y\| < r$.

Using the result of Theorem 3.3, we see at once that the first two conditions of Theorem 1 are satisfied. The verification of condition 3) is more complicated. First of all, by Theorem 3.3, the derivative $P'_{\mathbf{Y}}(\mathbf{0}, \mathbf{0})$ exists and has the form

$$z = U(y), \qquad z(s) = y(s) - \int_0^s Ay(t)\,dt,$$

* See Stepanov [1], p. 214.

and so we need to prove that U has a continuous inverse. Since $\mathbf{C}_n^\alpha$ and $\mathbf{D}_n^\alpha$ are complete spaces, it is sufficient to show that U defines a one-to-one correspondence between them (see XII.1.3)—that is, that, for each $\tilde{z} \in \mathbf{D}_n^\alpha$, there exists a unique element $\tilde{y} \in \mathbf{C}_n^\alpha$ such that $\tilde{z} = U(\tilde{y})$. In other words, we need to show that the equation

$$\tilde{z}(s) = y(s) - \int_0^s Ay(t)\,dt \tag{11}$$

has a unique solution in $\mathbf{C}_n^\alpha$.

Elementary results from the theory of differential equations show that a solution $\tilde{y}$ of (11) always exists, and that it is unique and has the form

$$\tilde{y}(s) = \int_0^s e^{(s-t)A}\,\tilde{z}'(t)\,dt + e^{sA}z(0).$$

Consequently, if we choose a number σ such that

$$\alpha < \sigma < -\max_k \operatorname{Re}\lambda_k,$$

then, by (10), we have

$$|\tilde{y}(s)| \leqslant K\left[\int_0^s e^{-(s-t)\sigma}|\tilde{z}'(t)|\,dt + e^{-s\sigma}|\tilde{z}(0)|\right] \leqslant$$

$$\leqslant K\left[e^{-s\sigma}\int_0^s e^{(\sigma-\alpha)t}\|\tilde{z}\|\,dt + e^{-s\sigma}\|\tilde{z}\|\right] =$$

$$= K\|\tilde{z}\|e^{-s\sigma}\left[\frac{e^{(\sigma-\alpha)s}-1}{\sigma-\alpha}+1\right] =$$

$$= K\|\tilde{z}\|e^{-\alpha s}\left[\frac{1-e^{(\alpha-\sigma)s}}{\sigma-\alpha}+e^{(\alpha-\sigma)s}\right] \leqslant K_1\|\tilde{z}\|e^{-\alpha s},$$

so that

$$\sup|\tilde{y}(s)e^{\alpha s}| \leqslant K_1\|\tilde{z}\| < \infty. \tag{12}$$

Since

$$\tilde{y}'(s) = \tilde{z}'(x) - A\tilde{y}(s),$$

it follows that

$$\sup_s|\tilde{y}'(s)e^{\alpha s}| \leqslant \sup_s|\tilde{z}'(s)e^{\alpha s}| + |A|\sup_s|\tilde{y}(s)e^{\alpha s}| \leqslant (1+K_1|A|)\|\tilde{z}\| < \infty. \tag{13}$$

Thus, comparing (12) and (13), we see that $\tilde{y} \in \mathbf{C}_n^\alpha$, and the conditions of Theorem 1 are all verified.

Thus, by what we have proved, we can now assert that the equation $P(x, y) = \mathbf{0}$ (that is, equation (8)) has a unique solution $y^* = F(x) \in \mathbf{C}_n^\alpha$, provided that the initial vector x is sufficiently small. Furthermore, we have

$$|y^*(s)| \leqslant \|F(x)\|e^{-\alpha s},$$

and, since F is continuous, $\|F(x)\| \to 0$ as $x \to \mathbf{0}$.

This completes the proof of the theorem.

XVIII

NEWTON'S METHOD

IN THIS chapter we develop in detail the theory of a method of solving functional equations known as Newton's method, or the method of tangents. At present, this method and a certain modification of it are two out of a small number of methods used in practice for finding an actual solution of a functional equation.

The theoretical significance of the method should also be noted, as it can be used to draw conclusions about the existence and uniqueness of a solution, and about the region in which it is located, without finding the solution itself; and this is sometimes more important than the actual knowledge of the solution.

The results of this chapter are mainly due to L. V. Kantorovich (see Kantorovich [9]). Our exposition follows a paper by Kantorovich [12].

§ 1. Equations of the form $P(x) = \mathbf{0}$

1.1. Suppose we have an operator P mapping an open subset Ω of one B-space $\mathbf{X}$ into another B-space $\mathbf{Y}$. Assume that there is a zero of P in Ω—that is, an element x^* such that

$$P(x^*) = \mathbf{0}.$$

Choose an arbitrary element $x_0 \in \Omega$. If we assume that P has a continuous derivative in Ω, we can replace the element $P(x_0) = P(x_0) - P(x^*)$ by the approximation $P'(x_0)(x_0 - x^*)$. We therefore have reason to suppose that a solution of the equation

$$P'(x_0)(x_0 - x) = P(x_0)$$

will be close to x^*. But the latter equation is linear, so its solution is easy to find. This solution is

$$x_1 = x_0 - [P'(x_0)]^{-1}(P(x_0))$$

(assuming, of course, that $[P'(x_0)]^{-1}$ exists).

If we continue this process, we obtain, starting from the initial approximation x_0, a sequence $\{x_n\}$, where

$$x_{n+1} = x_n - [P'(x_n)]^{-1}(P(x_n)) \qquad (n = 0, 1, \ldots). \tag{1}$$

Each x_n is an approximate solution to the equation

$$P(x) = \mathbf{0}, \tag{2}$$

and, in general, the larger n is, the more accurate this solution will be.

It is this method of generating the sequence $\{x_n\}$ that is known as *Newton's method.**

Newton's method is clearly not always feasible. Firstly, x_n may pass beyond the set Ω for some value of n, and secondly, even if this does not happen, $[P'(x_n)]^{-1}$ may not exist.

If the sequence $\{x_n\}$ converges to a root x^* and x_0 is chosen close enough to x^*, then, by the continuity of P', the operators $P'(x_n)$ and $P'(x_0)$ will only differ by a small amount. This is a justification for replacing formulae (1) by the simplified formulae

$$x'_{n+1} = x'_n - [P'(x_0)]^{-1}(P(x'_n)) \qquad (n = 0, 1, \ldots; \; x'_0 = x_0), \tag{3}$$

which are significantly simpler than formulae (1), although in general they yield poorer approximations.

We shall call this method of generating the sequence $\{x'_n\}$ *the modified Newton's method.*†

Below we shall study in detail the conditions under which Newton's method (both the original and the modified version) is feasible and converges—that is, the sequences $\{x_n\}$ and $\{x'_n\}$ converge to a solution of (2).

We observe that Newton's method for equation (2) coincides with the familiar method of successive approximations, applied to the equation

$$x = x - \Gamma(x)(P(x)) \qquad (\Gamma(x) = [P'(x)]^{-1}), \tag{4}$$

which is obviously equivalent to (2).

In exactly the same way, the modified Newton's method amounts to the method of successive approximations applied to the equation

$$x = x - \Gamma(x_0)(P(x)). \tag{5}$$

In view of this connection, we now occupy ourselves with an investigation of the usual method of successive approximations.

1.2. Consider the equation

$$x = S(x), \tag{6}$$

where S is an operator defined on the ball $\|x - x_0\| < R$ in some B-space $\mathbf{X}$ ($x_0 \in \mathbf{X}$). As well as (6), we consider the real equation

$$t = \phi(t), \tag{7}$$

where the function ϕ is defined in the interval $[t_0, t']$ ($t' = t_0 + r < t_0 + R$). We shall say that equation (7) (or the function ϕ) *majorizes* equation (6) (or the operator S) if

1) $\|S(x_0) - x_0\| \leqslant \phi(t_0) - t_0$; (8)

2) $\|S'(x)\| \leqslant \phi'(t)$ when $\|x - x_0\| \leqslant t - t_0$.

THEOREM 1. *Suppose that the operator S has a continuous derivative in a closed ball Ω_0 ($\|x - x_0\| \leqslant r$) and that the function ϕ is differentiable in the interval $[t_0, t']$. If equation (7) majorizes equation (6) and if (7) has a root in the interval $[t_0, t']$, then (6) also has a solution*

---

* If P is a real-valued function of a real argument, then we can write (1) in the form

$$x_{n+1} = x_n - \frac{P(x_n)}{P'(x_n)} \qquad (n = 0, 1, \ldots),$$

which reduces to Newton's method in the strict sense of the phrase.

† (*Editor's note*) These two methods are often called the Newton–Kantorovich methods.

x^*, and the sequence $\{x_n\}$ of successive approximations

$$x_{n+1} = S(x_n) \qquad (n = 0, 1, \ldots), \tag{9}$$

starting from x_0, converges to x^*.

Furthermore, we have

$$\|x^* - x_0\| \leqslant t^* - t_0, \tag{10}$$

where t^* denotes the smallest root of (7) in $[t_0, t']$.

Proof. We prove first that the successive approximations for equation (7), namely

$$t_{n+1} = \phi(t_n) \qquad (n = 0, 1, \ldots), \tag{11}$$

form a convergent sequence. To this end we note that condition 2) implies that

$$\phi'(t) \geqslant 0 \qquad (t \in [t_0, t']),$$

so that ϕ is increasing in the interval $[t_0, t']$. It therefore follows that t_n is well defined for each n, and that

$$t_n \leqslant \bar{t} \qquad (n = 0, 1, \ldots), \tag{12}$$

where $\bar{t}$ denotes the root of (7) whose existence is postulated in the theorem.

In fact, when $n = 0$ equation (12) is obvious; and if it has already been proved for $n = k$, then since ϕ is monotone $t_k \leqslant \bar{t}$ implies $\phi(t_k) \leqslant \phi(\bar{t})$ —that is, $t_{k+1} \leqslant \bar{t}$. By induction, the inequality holds for all n.

Using the fact that ϕ is monotone, we can also show by induction that the sequence $\{t_n\}$ is monotone.

For $t_n \leqslant t_{n+1}$ implies $t_{n+1} \leqslant \phi(t_n) \leqslant \phi(t_{n+1}) = t_{n+2}$, and the inequality $t_0 \leqslant t_1$ is a consequence of condition 1).

Thus we have shown that $t^* = \lim_{n\to\infty} t_n$ exists, and, by (11) and the continuity of ϕ, this is a root of (7); moreover, it follows from (12) that this is the smallest root in $[t_0, t']$.

Next we prove that the elements (9) are well defined and form a convergent sequence.

Firstly, writing (8) in the form

$$\|x_1 - x_0\| \leqslant t_1 - t_0,$$

we see that $x_1 \in \Omega_0$. Assume it has already been shown that $x_1, x_2, \ldots, x_n \in \Omega_0$ and that

$$\|x_{k+1} - x_k\| \leqslant t_{k+1} - t_k \qquad (k = 0, 1, \ldots, n-1). \tag{13}$$

Then, by the results of XVII.1.7, we have

$$x_{n+1} - x_n = S(x_n) - S(x_{n-1}) = \int_{x_{n-1}}^{x_n} S'(x)\,dx.$$

If we write x and t for corresponding points in $[x_{n-1}, x_n]$ and $[t_{n-1}, t_n]$, that is, if

$$x = x_{n-1} + \tau(x_n - x_{n-1}), \quad t = t_{n-1} + \tau(t_n - t_{n-1}) \qquad (0 \leqslant \tau \leqslant 1),$$

then, in view of (13), we have

$$\begin{aligned}\|x - x_0\| &\leqslant \tau\|x_n - x_{n-1}\| + \|x_{n-1} - x_{n-2}\| + \ldots + \|x_1 - x_0\| \\ &\leqslant \tau(t_n - t_{n-1}) + (t_{n-1} - t_{n-2}) + \ldots (t_1 - t_0) = t - t_0.\end{aligned}$$

Hence, by condition 2)

$$\|S'(x)\| \leqslant \phi'(t).$$

Using the Remark in XVII.1.7, we see from this that

$$\|x_{n+1}-x_n\| = \left\| \int_{x_{n-1}}^{x_n} S'(x)\,dx \right\| \leqslant \int_{t_{n-1}}^{t_n} \phi'(t)\,dt = \phi(t_n)-\phi(t_{n-1}) = t_{n+1}-t_n.$$

Thus we have proved (13) for $k = n$. In addition, we have proved that $x_{n+1} \in \Omega_0$, since

$$\begin{aligned} \|x_{n+1}-x_0\| &\leqslant \|x_{n+1}-x_n\| + \|x_n - x_{n-1}\| + \ldots + \|x_1 - x_0\| \leqslant \\ &\leqslant (t_{n+1}-t_n)+(t_n-t_{n-1})+ \ldots + (t_1-t_0) = t_{n+1}-t_0 \leqslant \\ &\leqslant t'-t_0 = r. \end{aligned}$$

Hence, by induction, we have shown that $x_k \in \Omega_0$ and that (13) is true for $k = 0, 1, \ldots$. Furthermore, since (13) yields

$$\begin{aligned} \|x_{n+p}-x_n\| &\leqslant \|x_{n+p}-x_{n+p-1}\| + \ldots + \|x_{n+1}-x_n\| \leqslant \\ &\leqslant (t_{n+p}-t_{n+p-1}) + \ldots + (t_{n+1}-t_n) = t_{n+p}-t_n, \end{aligned} \tag{14}$$

it follows that $\{x_n\}$ is a Cauchy sequence, and therefore has a limit. Write $x^* = \lim_{n\to\infty} x_n$. Taking limits in (9) and bearing in mind that S is continuous, we obtain

$$x^* = S(x^*),$$

so that x^* is a root of (6).

We now obtain (10) from (14) by setting $n = 0$ and letting $p \to \infty$.

REMARK 1. The inequality

$$\|x^* - x_n\| \leqslant t^* - t_n \qquad (n = 0, 1, \ldots), \tag{15}$$

which we obtained by letting $p \to \infty$ in (14), gives a bound for the rate of convergence of the sequence $\{x_n\}$ to x^*.

REMARK 2. We did not make full use of condition 2) in the proof of the theorem. The proof required only that $\|S'(x)\| \leqslant \phi'(t)$ for corresponding points of the intervals $[x_{n-1}, x_n]$ and $[t_{n-1}, t_n]$ $(n = 1, 2, \ldots)$.

Equation (6) may have other roots as well as x^* in the ball Ω_0, even if the solution t^* of the majorant equation (7) is unique in $[t_0, t']$. However, we do have

THEOREM 2. *Suppose the conditions of the preceding theorem are satisfied, and suppose in addition that*

$$\phi(t') \leqslant t'.$$

If (7) *has a unique solution in* $[t_0, t']$, *then* (6) *has a unique root in* Ω_0 *and the process of successive approximations, starting from any* $\tilde{x}_0 \in \Omega_0$, *converges to this root.*

Proof. We apply the method of successive approximations to (7), taking $\tilde{t}_0 = t'$ as our initial approximation. A word-for-word repetition of the proof of Theorem 1 shows that the sequence

$$t_{n+1} = \phi(\tilde{t}_n) \qquad (n = 0, 1, \ldots)$$

is monotonically decreasing and bounded below ($\tilde{t}_n \geqslant t^*$), and so has a limit $\tilde{t}$; however, $\tilde{t}$ must be equal to t^*, as it is a root of (7).

Now we show that the process of successive approximations for equation (6), starting with any element $\tilde{x}_0 \in \Omega_0$, is convergent, and hence yields a root of (6).

The successive approximations have the form

$$\tilde{x}_{n+1} = S(\tilde{x}_n) \qquad (n = 0, 1, \dots).$$

We have

$$\tilde{x}_1 - x_1 = S(\tilde{x}_0) - S(x_0) = \int_{x_0}^{\tilde{x}_0} S'(x)\,dx$$

and so, using the Remark in XVII.1.7 again, we obtain

$$\|\tilde{x}_1 - x_1\| \leqslant \int_{t_0}^{\tilde{t}_0} \phi'(t)\,dt = \phi(\tilde{t}_0) - \phi(t_0) = \tilde{t}_1 - t_1.$$

It is therefore clear that

$$\|\tilde{x}_1 - x_0\| \leqslant \|\tilde{x}_1 - x_1\| + \|x_1 - x_0\| \leqslant (\tilde{t}_1 - t_1) + (t_1 - t_0) = \tilde{t}_1 - t_1 \leqslant r$$

and hence that $\tilde{x}_1 \in \Omega_0$.

Suppose we have already proved that

$$\tilde{x}_k \in \Omega_0, \qquad \|\tilde{x}_k - x_k\| \leqslant \tilde{t}_k - t_k \qquad (k = 1, 2, \dots, n). \tag{16}$$

Then

$$\tilde{x}_{n+1} - x_{n+1} = S(\tilde{x}_n) - S(x_n) = \int_{x_n}^{\tilde{x}_n} S'(x)\,dx.$$

However, if x and t are corresponding points of $[x_n, \tilde{x}_n]$ and $[t_n, \tilde{t}_n]$,

$$x = x_n + \tau(\tilde{x}_n - x_n), \qquad t = t_n + \tau(\tilde{t}_n - t_n) \qquad (0 \leqslant \tau \leqslant 1),$$

then

$$\begin{aligned}\|x - x_0\| &\leqslant \tau\|\tilde{x}_n - x_n\| + \|x_n - x_{n-1}\| + \dots + \|x_1 - x_0\| \leqslant \\ &\leqslant \tau(\tilde{t}_n - t_n) + (t_n - t_{n-1}) + \dots + (t_1 - t_0) = t - t_0,\end{aligned}$$

so for these x and t we have $\|S'(x)\| \leqslant \phi'(t)$ and

$$\|\tilde{x}_{n+1} - x_{n+1}\| \leqslant \int_{t_n}^{\tilde{t}_n} \phi'(t)\,dt = \phi(\tilde{t}_n) - \phi(t_n) = \tilde{t}_{n+1} - t_{n+1},$$

and thus

$$\begin{aligned}\|\tilde{x}_{n+1} - x_0\| &\leqslant \|\tilde{x}_{n+1} - x_{n+1}\| + \|x_{n+1} - x_0\| \leqslant \\ &\leqslant (\tilde{t}_{n+1} - t_{n+1}) + (t_{n+1} - t_0) = \tilde{t}_{n+1} - t_0 \leqslant r,\end{aligned}$$

and so $\tilde{x}_{n+1} \in \Omega_0$.

By induction, we conclude that (16) is true for all $k = 1, 2, \dots$.

Since the sequences $\{t_n\}$ and $\{t_n^*\}$ have a common limit (namely t^*), it follows from (16) that, if $\{x_n\}$ converges, then so does $\{\tilde{x}_n\}$, and we have

$$\lim_{n\to\infty} \tilde{x}_n = \lim_{n\to\infty} x_n = x^*. \tag{17}$$

We have thus proved that the process of successive approximations converges to x^*, whatever initial approximation $\tilde{x}_0 \in \Omega_0$ we use. The uniqueness of the solution to (6) is immediate from this. For if $\tilde{x} \in \Omega_0$ is a root of (6), then, taking $\tilde{x}_0 = \tilde{x}$, we obviously have, for $n = 1, 2, \ldots,$

$$\tilde{x}_n = \tilde{x}$$

and so, by (17), $\tilde{x} = x^*$, as we required to prove.

1.3. Let us turn to an investigation of Newton's method for equation (2).

In addition to (2) we consider the real equation

$$\psi(t) = 0. \tag{18}$$

We shall assume that P is defined in a ball Ω ($\|x - x_0\| < R$) in a space $\mathbf{X}$ and has a continuous second derivative in a closed ball Ω_0 ($\|x - x_0\| \leqslant r$). The function ψ is assumed twice continuously differentiable in the interval $[t_0, t']$ ($t' = t_0 + r$).

The following result on the convergence of the modified Newton's method is easily obtained with the aid of Theorem 1.

THEOREM 3. *Suppose the following conditions are satisfied*:

1) *there exists a continuous linear operator* $\Gamma_0 = [P'(x_0)]^{-1}$;

2) $c_0 = -\dfrac{1}{\psi'(t_0)} > 0$;

3) $\|\Gamma_0 P(x_0)\| \leqslant c_0 \psi(t_0)$;

4) $\|\Gamma_0 P''(x)\| \leqslant c_0 \psi''(t_0)$, *if* $\|x - x_0\| \leqslant t - t_0 \leqslant r$;

5) *equation* (18) *has a root* $\bar{t}$ *in* $[t_0, t']$.

Then the modified Newton's method for (2) *and* (18), *starting with* x_0 *and* t_0 *respectively, converges and yields solutions* x^* *and* t^* *of these equations, where*

$$\|x^* - x_0\| \leqslant t^* - t_0. \tag{19}$$

Proof. It was noted above that the modified Newton's method for (2) is equivalent to the method of successive approximations for (5)—that is, for the equation

$$x = S(x) \qquad (S(x) = x - \Gamma_0(P(x))). \tag{20}$$

In the same way, we can use in place of the modified Newton's method for (18) the process of successive approximations for the equation

$$t = \phi(t) \qquad (\phi(t) = t + c_0 \psi(t)). \tag{21}$$

Let us show that the equations (20) and (21) satisfy the conditions of Theorem 1.

We have

$$S(x_0) - x_0 = -\Gamma_0(P(x_0))$$

and similarly

$$\phi(t_0) - t_0 = c_0 \psi(t_0).$$

Hence, in view of condition 3), we have

$$\|S(x_0) - x_0\| \leqslant \phi(t_0) - t_0.$$

Also, using the differentiation rule (XVII.1.2), we find that

$$S'(x) = I - \Gamma_0 P'(x), \qquad S''(x) = -\Gamma_0 P''(x),$$

so that

$$S'(x) = S'(x) - S'(x_0) = \int_{x_0}^{x} S''(x)\,dx = -\int_{x_0}^{x} \Gamma_0 P''(x)\,dx,$$

and therefore, using the Remark in XVII.1.7 and condition 4) we find that

$$\|S'(x)\| \leqslant \int_{t_0}^{t} c_0\psi''(\tau)\,d\tau = c_0\psi'(t) - c_0\psi'(t_0) = 1 + c_0\psi'(t_0) = \phi'(t),$$

if x and t are related by $\|x - x_0\| \leqslant t - t_0$.

Hence (21) is a majorant for (20), so we can apply Theorem 1 to these equations, and this gives us the result we require.

Remark. As in Theorem 1, a bound for the rate of convergence of the sequence $\{x'_n\}$ (see(3)) to x^* is given by

$$\|x^* - x'_n\| \leqslant t^* - t'_n \qquad (n = 0, 1, \ldots; \quad x'_0 = x_0;\ t'_0 = t_0), \tag{22}$$

where t'_n are the successive approximations in the modified Newton's method for (18).

By applying Theorem 2 to (20) and (21) we can at once show that the solution of (2) is unique.

Theorem 4. *Suppose that the conditions of Theorem 3 are satisfied, and suppose in addition that*

$$\psi(t') \leqslant 0. \tag{23}$$

If (18) *has a unique root in* $[t_0, t']$, *then* (2) *has only one solution in* Ω_0.

For the proof we need only note that (23) implies

$$\phi(t') = t' + c_0\psi(t') \leqslant t',$$

so that the additional condition of Theorem 2 is satisfied.

Remark. As t^* is the smallest root of (18), the uniqueness theorem is applicable when we set $t' = t^*$. Hence we can always ensure the uniqueness of the solution x^* in the ball

$$\|x - x_0\| \leqslant t^* - t_0.$$

1.4. The convergence of the original Newton's method can be established using Theorem 3 on the convergence of the modified Newton's method.

Theorem 5. *Suppose the conditions of Theorem 3 are satisfied. Then the original Newton's method for* (2), *starting from* x_0, *is feasible, and leads to a sequence* $\{x_n\}$ *converging to a root* x^* *of* (2).

Proof. The first step in the modified Newton's method coincides with the first step in the original Newton's method, so x_1 is well defined and $x_1 \in \Omega_0$. We now verify that the conditions of Theorem 3 are not violated when we replace x_0 by x_1 and t_0 by t_1. Keeping the notation of Theorem 3, we consider the operator

$$I - \Gamma_0 P'(x_1) = -\Gamma_0(P'(x_1) - P'(x_0)) = -\int_{x_0}^{x_1} \Gamma_0 P''(x)\,dx.$$

Using the Remark in XVII.1.7, we find that

$$\|I - \Gamma_0 P'(x_1)\| \leqslant \int_{t_0}^{t_1} c_0\psi''(t)\,dt = 1 + c_0\psi'(t_1) = q.$$

Since $\psi''(t) \geqslant 0$ throughout $[t_0, t']$ and $\psi(t_0) \geqslant 0$, the function ψ cannot attain a minimum to the left of t^*. As $t_1 \leqslant t^*$ and $\psi'(t_0) < 0$, we have* $\psi'(t_1) < 0$, and so $q < 1$.

By Banach's Theorem (V.4.5), there exists a continuous linear operator

$$U = [\Gamma_0 P'(x_1)]^{-1} \in B(\mathbf{X}, \mathbf{X}),$$

and

$$\|U\| \leqslant \frac{1}{1-q} = \frac{\psi'(t_0)}{\psi'(t_1)} = \frac{c_1}{c_0} \qquad \left(c_1 = -\frac{1}{\psi'(t_1)}\right). \tag{24}$$

Hence there exists a continuous linear operator

$$\Gamma_1 = [P'(x_1)]^{-1} = U\Gamma_0.$$

Thus we have verified conditions 1) and 2) of Theorem 3.

We now prove that

$$\|\Gamma_0(P(x_1))\| \leqslant c_0\psi(t_1). \tag{25}$$

By Taylor's formula (XVII.2.5), we have

$$\Gamma_0(P(x_1)) = \Gamma_0(P(x_0)) + \Gamma_0 P'(x_0)(x_1 - x_0) + \int_{x_0}^{x_1} \Gamma_0 P''(x)(x_1 - x, \cdot)\,dx =$$

$$= (x_0 - x_1) + (x_1 - x_0) + \int_{x_0}^{x_1} \Gamma_0 P''(x)(x_1 - x, \cdot)\,dx =$$

$$= \int_{x_0}^{x_1} \Gamma_0 P''(x)(x_1 - x, \cdot)\,dx.$$

Similarly,

$$c_0\psi(t_1) = c_0 \int_{t_0}^{t_1} \psi''(t)(t_1 - t)\,dt.$$

Since, for corresponding points x and t in the intervals $[x_0, x_1]$ and $[t_0, t_1]$, we have

$$\|\Gamma_0 P''(x)(x_1 - x, \cdot)\| \leqslant \|\Gamma_0 P''(x)\| \, \|x_1 - x\| \leqslant c_0\psi''(t)(t_1 - t),$$

it follows that

$$\left\| \int_{x_0}^{x_1} \Gamma_0 P''(x)(x_1 - x, \cdot)\,dx \right\| \leqslant \int_{t_0}^{t_1} c_0\psi''(t)(t_1 - t)\,dt,$$

which yields (25).

Taking (25) and (24) into account, we obtain

$$\|\Gamma_1(P(x_1))\| = \|U\Gamma_0(P(x_1))\| \leqslant \|U\| \, \|\Gamma_0(P(x_1))\| \leqslant \frac{c_1}{c_0} c_0\psi(t_1) = c_1\psi(t_1),$$

so condition 3) is also satisfied.

---

* If $\psi'(t_1) = 0$, then $t_1 = t^*$ and $\psi(t_1) = 0$, and so

$$0 = \psi(t_1) - \psi(t_0) - \psi'(t_0)(t_1 - t_0) = \int_{t_0}^{t_1} \psi''(t)(t_1 - t)\,dt,$$

which is only possible if $\psi''(t) = 0$ in $[t_0, t_1]$. This in turn leads to the equation $\psi'(t_0) = \psi'(t_1) = 0$.

We verify condition 4) in a similar way. First note that, if $\|x-x_1\| \leqslant t-t_1$, then certainly $\|x-x_0\| \leqslant t-t_0$. Therefore

$$\|\Gamma_1 P''(x)\| = \|U\Gamma_0 P''(x)\| \leqslant \|U\| \, \|\Gamma_0 P''(x)\| \leqslant \frac{c_1}{c_0} c_0 \psi''(t) = c_1 \psi''(t).$$

Finally, condition 5) is not violated because the root $\bar{t}$ lies in $[t^*, t']$ and so also in the wider interval $[t_1, t']$.

Using similar arguments, one can show that the conditions of Theorem 3 are still not violated when we go from x_1 and t_1 to x_2 and t_2, and so on.

Thus all the x_n are well defined: that is, we have now shown the Newton's method is feasible.

As the sequence $\{t_n\}$ is obviously increasing and bounded, it has a limit $\tilde{t}$ and, in view of the inequality

$$\|x_{n+1}-x_n\| \leqslant t_{n+1}-t_n \qquad (n = 0, 1, \ldots)$$

the sequence $\{x_n\}$ also has a limit $\tilde{x} = \lim_{n\to\infty} x_n$.

From (1) we obtain

$$P(x_n) + P'(x_n)(x_{n+1}-x_n) = \mathbf{0} \qquad (n = 0, 1, \ldots). \tag{26}$$

For arbitrary $x \in \Omega_0$, the Mean Value Theorem yields

$$\Gamma_0[P'(x)-P'(x_0)] \leqslant \|x-x_0\| \sup_{0<\theta<1} \|\Gamma_0 P''(x_0+\theta(x-x_0))\| \leqslant r \max_{[t_0, t']} c_0\psi''(t),$$

and hence it follows that the $\|P'(x_n)\|$ are bounded in aggregate. Therefore, taking limits in (26), we obtain

$$P(\tilde{x}) = \mathbf{0},$$

that is, $\tilde{x}$ is a root of (2).

In the same way, $\tilde{t}$ is a root of (18), and, since $\tilde{t} \leqslant t^*$ and t^* is the smallest root, we have $\tilde{t} = t^*$.

Since it is obvious that $\|\tilde{x}-x_0\| \leqslant \tilde{t}-t_0 = t^*-t_0$, it follows from the Remark after Theorem 4 that $\tilde{x} = x^*$.

The proof of the theorem is therefore complete.

REMARK. The rate of convergence of the sequence $\{x_n\}$ to x^* can be estimated from the inequality

$$\|x^*-x_n\| \leqslant t^*-t_n \tag{27}$$

$$(n = 0, 1, \ldots).$$

This is a consequence of (22), as x_n and t_n can be regarded as first approximations in the modified Newton's method with initial approximations x_{n-1} and t_{n-1}, respectively.

1.5. It is often difficult to make direct use of Theorem 3 or Theorem 5 as they contain the undefined function ψ. In this connection, the following theorem acquires importance.

THEOREM 6. *Suppose, as above, that the operator P is defined in Ω and has a continuous second derivative in Ω_0. Suppose, in addition, that*

1) $\Gamma_0 = [P'(x_0)]^{-1}$ *exists and is a continuous linear operator*;

2) $\|\Gamma_0(P(x_0))\| \leqslant \eta$:

3) $\|\Gamma_0 p''(x)\| \leqslant K$ $(x \in \Omega_0)$.

Then, provided

$$h = K\eta \leqslant \tfrac{1}{2} \tag{28}$$

and

$$r \geqslant r_0 = \frac{1-\sqrt{1-2h}}{h}\eta, \tag{29}$$

equation (2) *has a solution* x^*, *and Newton's process* (*both the original and the modified version*) *converges to this solution. Furthermore,*

$$\|x^* - x_0\| \leqslant r_0. \tag{30}$$

Also, if for $h < \tfrac{1}{2}$ *we have*

$$r < r_1 = \frac{1+\sqrt{1-2h}}{h}\eta, \tag{31}$$

while for $h = \tfrac{1}{2}$ *we have*

$$r \leqslant r_1, \tag{32}$$

then x^* *is the unique solution in* Ω_0.

The rate of convergence is given, in the case of the original Newton's method, by

$$\|x^* - x_n\| \leqslant \frac{1}{2^n}(2h)^{2^n}\frac{\eta}{h} \qquad (n = 0, 1, \ldots), \tag{33}$$

and, in the case of the modified Newton's method, for $h < \tfrac{1}{2}$, *by*

$$\|x^* - x_n'\| \leqslant \frac{\eta}{h}(1-\sqrt{1-2h})^{n+1} \qquad (n = 0, 1, \ldots). \tag{34}$$

Proof. Consider the real-valued function in $[0, r]$ given by

$$\psi(t) = Kt^2 - 2t + 2\eta = Kt^2 - 2t + \frac{2h}{K} = \frac{h}{\eta}t^2 - 2t + 2\eta.$$

We shall verify that the operator P and the function ψ satisfy all the conditions of Theorem 3 (and hence of Theorem 5 as well). In fact, conditions 1)–4) are obviously satisfied; also since the roots of

$$\psi(t) = 0 \tag{35}$$

are

$$r_0 = \frac{1-\sqrt{1-2h}}{h}\eta, \qquad r_1 = \frac{1+\sqrt{1-2h}}{h}\eta,$$

(28) ensures that the roots are real, while (29) ensures that the least root r_0 lies in the interval $[0, r]$. Since $t^* = r_0$, equation (30) is simply the equation (19) of Theorem 3.

The claim about the uniqueness of solution is a consequence of Theorem 4, since for any $h < \tfrac{1}{2}$ we have $\psi(r) \leqslant 0$ under the conditions of the theorem and (35) has a unique root in $[0, r]$.

Let us now establish the inequalities (33) and (34) giving the rate of convergence of Newton's method. In view of the Remarks following Theorems 3 and 5, we need only consider the real sequences $\{t_n\}$ and $\{t_n'\}$ of successive approximations for the original Newton's method and the modified Newton's method, respectively, for equation (35).

First we consider the original method. We write

$$c_n = -\frac{1}{\psi'(t_n)}, \qquad \eta_n = c_n\psi(t_n),$$

$$K_n = c_n\psi''(t_n) = 2Kc_n, \qquad h_n = K_n\eta_n$$

and express η_n, K_n and h_n in terms of the corresponding terms with suffix $n-1$. To do this, we note that

$$t_{k+1} - t_k = -\frac{\psi(t_k)}{\psi'(t_k)} = \eta_k \qquad (k = 0, 1, \ldots). \tag{36}$$

Hence, by Taylor's formula for a polynomial of degree 2, we have

$$\eta_n = c_n\psi(t_n) = c_n\psi(t_{n-1} + \eta_{n-1}) =$$

$$= c_n\left[\frac{1}{2}\psi''(t_{n-1})\eta_{n-1}^2 + \psi'(t_{n-1})\eta_{n-1} + \psi(t_{n-1})\right] =$$

$$= c_n\left[K\eta_{n-1}^2 - \frac{\eta_{n-1}}{c_{n-1}} + \frac{\eta_{n-1}}{c_{n-1}}\right] = c_nK\eta_{n-1}^2 =$$

$$= \frac{1}{2}\frac{c_n}{c_{n-1}}2Kc_{n-1}\eta_{n-1}^2 = \frac{1}{2}\frac{c_n}{c_{n-1}}K_{n-1}\eta_{n-1}^2.$$

However,

$$\frac{c_{n-1}}{c_n} = \frac{\psi'(t_n)}{\psi'(t_{n-1})} = \frac{\psi'(t_{n-1}) + \psi''(t_{n-1})\eta_{n-1}}{\psi'(t_{n-1})} =$$

$$= 1 - K_{n-1}\eta_{n-1} = 1 - h_{n-1}, \tag{37}$$

so that

$$\eta_n = \frac{1}{2}\frac{K_n\eta_{n-1}^2}{1 - h_{n-1}} = \frac{\eta_{n-1}}{2}\frac{h_{n-1}}{1 - h_{n-1}}. \tag{38}$$

Similarly, using (37), we have

$$K_n = 2c_nK = 2Kc_{n-1}\frac{c_n}{c_{n-1}} = \frac{K_{n-1}}{1 - h_{n-1}}.$$

Hence

$$h_n = K_n\eta_n = \frac{1}{2}\frac{K_{n-1}\eta_{n-1}h_{n-1}}{(1 - h_{n-1})^2} = \frac{1}{2}\left[\frac{h_{n-1}}{1 - h_{n-1}}\right]^2. \tag{39}$$

Since $h_n \leqslant \frac{1}{2}$, we see from (38) and (39) that

$$\eta_n \leqslant h_{n-1}\eta_{n-1}, \qquad h_n \leqslant 2h_{n-1}^2 \qquad (n = 1, 2, \ldots). \tag{40}$$

Therefore $h_2 \leqslant \frac{1}{2}[2h_0]^{2^n} = \frac{1}{2}[2h]^{2^n}$ and

$$\eta_n \leqslant h_{n-1}\eta_{n-1} \leqslant h_{n-1}h_{n-2}\eta_{n-2} \leqslant \ldots \leqslant$$

$$\leqslant h_{n-1}h_{n-2}\ldots h_0\eta_0 \leqslant \frac{1}{2^n}[2h]^{2^n-1}\eta.$$

Finally, using (36), we deduce from this that

$$t^* - t_n = (t_{n+1} - t_n) + (t_{n+2} - t_{n+1}) + \ldots = \eta_n + \eta_{n+1} + \ldots \leqslant$$

$$\leqslant \frac{1}{2^n}[2h]^{2^n - 1}\eta\left\{1 + \frac{1}{2}[2h]^{2^n} + \frac{1}{2^2}[2h]^{2^{n+1}} + \ldots\right\} \leqslant$$

$$\leqslant \frac{2}{2^n}[2h]^{2^n - 1}\eta = \frac{1}{2^n}[2h]^{2^n}\frac{\eta}{h}.$$

In view of (27), this yields the inequality (33) we were seeking.

We now estimate the error of the modified method. Assume that $h < \frac{1}{2}$. Taking ϕ to be the same function as in the proof of Theorem 3, we can write

$$t^* - t'_n = \phi(t^*) - \phi(t'_{n-1}) = \phi'(\tilde{t}_n)(t^* - t'_{n-1})$$

$$\left(\tilde{t}_n = \frac{t^* + t'_{n-1}}{2}\right).$$

But

$$\phi'(t) = 1 + c_0\psi'(t) = Kt,$$

and so

$$\phi'(\tilde{t}_n) = K\tilde{t}_n \leqslant Kt^* = 1 - \sqrt{1 - 2h}.$$

Therefore

$$t^* - t'_n \leqslant [1 - \sqrt{1 - 2h}](t^* - t'_{n-1}).$$

In the same way, we find a bound for $t^* - t'_{n-1}$, and, finally, proceeding as before, we obtain

$$t^* - t'_n \leqslant [1 - \sqrt{1 - 2h}]^n(t^* - t'_0) = \frac{\eta}{h}[1 - \sqrt{1 - 2h}]^{n+1}.$$

Substituting this in (22), we find

$$\|x^* - x'_n\| \leqslant \frac{\eta}{h}[1 - \sqrt{1 - 2h}]^{n+1},$$

in other words, we have (34). This completes the proof.

Remark 1. Conditions 2) and 3) in the theorem can be replaced by

1′) $\|\Gamma_0\| \leqslant B'$;

2′) $\|P(x_0)\| \leqslant \eta'$;

3′) $\|P''(x)\| \leqslant K' \qquad (x \in \Omega_0)$.

Here h, r_0, r_1 are, respectively,

$$h = K'B'^2\eta',$$

$$r_0 = \frac{1 - \sqrt{1 - 2h}}{h}B'\eta',$$

$$r_1 = \frac{1 + \sqrt{1 - 2h}}{h}B'\eta'.$$

REMARK 2. Conditions (28), (29), (31) (or (32)) are exact, in the sense that if either of the first two fails to hold for the quadratic equation (35), then this equation has no solution, while if (31) or (32) fails to hold, then we do not have uniqueness.*

§ 2. Consequences of the convergence theorem for Newton's method

In this section we note some consequences of the convergence theorem for Newton's method (Theorem 1.6). As in the last section, we shall be considering the equation

$$P(x) = \mathbf{0}, \tag{1}$$

adhering to our previous assumptions about P (its domain of definition, differentiability properties, etc.) and to our previous notation.

2.1. First of all, we record the following generalization of Theorem 1.6, in which instead of requiring the existence of $\Gamma_0 = [P'(x_0)]^{-1}$ we require only that there exist an operator close to Γ_0.

THEOREM 1. *Suppose there exists a linear operator $\Gamma \in B(\mathbf{Y}, \mathbf{X})$ having a continuous inverse, and suppose the following conditions are satisfied:*

1) $\|\Gamma(P(x_0))\| \leqslant \bar{\eta}$;

2) $\|\Gamma P'(x_0) - I\| \leqslant \delta$;

3) $\|\Gamma P''(x)\| \leqslant \bar{K} \qquad (x \in \Omega_0)$.

Then, provided

$$\bar{h} = \frac{\bar{K}\bar{\eta}}{(1-\delta)^2} \leqslant \frac{1}{2}, \qquad \delta < 1 \tag{2}$$

and

$$r \geqslant \bar{r}_0 = \frac{1 - \sqrt{1-2\bar{h}}}{\bar{h}} \frac{\bar{\eta}}{1-\delta}, \tag{3}$$

equation (1) *has a solution $x^* \in \Omega_0$, which is unique if*

$$\begin{matrix} r < \bar{r}_1 \text{ for } \bar{h} < \frac{1}{2} \\ r \leqslant \bar{r}_1 \text{ for } \bar{h} = \frac{1}{2} \end{matrix} \quad \left(\bar{r}_1 = \frac{1 + \sqrt{1-2\bar{h}}}{\bar{h}} \frac{\bar{\eta}}{1-\delta} \right).$$

Moreover, if $\{x_n\}$ is the sequence of successive approximations for Newton's method, then we have

$$\|x^* - x_n\| \leqslant \frac{1}{2^n} [2\bar{h}]^{2^n} \frac{\bar{\eta}}{\bar{h}(1-\delta)} \qquad (n = 0, 1, \ldots).$$

Proof. We verify that the conditions of Theorem 1.6 are satisfied. The second condition implies the existence of the continuous linear operator

$$U = [\Gamma P'(x_0)]^{-1},$$

* In the case of condition (32), there is a unique solution for (30). However, the equation $\psi(t) - \varepsilon = 0$ ($\varepsilon > 0$) will then have two solutions in $[0, r]$.

and also implies that

$$\|U\| \leqslant \frac{1}{1-\delta}.$$

From this we deduce the existence of the continuous linear operator

$$\Gamma_0 = [P'(x_0)]^{-1} = U\Gamma.$$

Moreover,

$$\|\Gamma_0(P(x_0))\| \leqslant \|U\| \|\Gamma(P(x_0))\| \leqslant \frac{\eta}{1-\delta},$$

$$\|\Gamma_0 P''(x)\| \leqslant \|U\| \|\Gamma P''(x)\| \leqslant \frac{\overline{K}}{1-\delta} \qquad (x \in \Omega_0).$$

It remains only to replace η and K in Theorem 1.6 by $\dfrac{\overline{\eta}}{1-\delta}$ and $\dfrac{\overline{K}}{1-\delta}$.

2.2. Now we consider a process of successive approximations similar to the modified method but with the operator $\Gamma_0 = [P'(x_0)]^{-1}$ in the latter replaced by a close approximation Γ. In other words, we construct a sequence $\{\tilde{x}_n\}$, with

$$\tilde{x}_{n+1} = \tilde{x}_n - \Gamma(P(\tilde{x}_n)) \qquad (n = 0, 1, \dots; \tilde{x}_0 = x_0). \tag{4}$$

THEOREM 2. *With the hypotheses of Theorem* 1, *the process* (4) *is feasible: that is,* $\tilde{x}_n \in \Omega_0$ $(n = 0, 1, \dots)$ *and*

$$\tilde{x}_n \to x^*.$$

Proof. As in Theorem 3, we prove the convergence of the modified process by using Theorem 1.1, setting

$$S(x) = x - \Gamma(P(x)) \qquad (x \in \Omega_0)$$

and

$$\phi(t) = \frac{1}{2}\overline{K}t^2 + \delta t + \overline{\eta} = t + \psi(t)$$

$$\left(\psi(t) = \frac{1}{2}Kt^2 - (1-\delta)t + \overline{\eta}, \qquad t \in [t_0, t'], \qquad t_0 = 0, \qquad t' = r\right).$$

We have

$$\|S(x_0) - x_0\| = \|-\Gamma(P(x_0))\| \leqslant \overline{\eta} = \phi(t_0) - t_0.$$

Also, if $\|x - x_0\| \leqslant t \leqslant r$, then

$$\|S'(x)\| = \|I - \Gamma P'(x)\| \leqslant \|I - \Gamma P'(x_0)\| + \|\Gamma(P'(x_0) - P'(x))\| \leqslant$$

$$\leqslant \delta + \left\| \int_{x_0}^{x} \Gamma P''(y)\, dy \right\| \leqslant \delta + \int_0^t K\, d\tau = \delta + \overline{K}t = \phi'(t).$$

Hence the equation

$$x = S(x) \tag{5}$$

is majorized by the equation $t = \phi(t)$, which has the solution

$$t^* = \frac{1-\sqrt{1-2\bar{h}}}{\bar{h}} \frac{\bar{\eta}}{1-\delta} = \bar{r}_0 \leqslant r.$$

Hence (5) also has a solution, and this must coincide with x^* (since x^* is the unique solution of (1) in the ball $\|x-x_0\| \leqslant \bar{r}_0$).

REMARK. The rate of convergence of the sequence $\{\tilde{x}_n\}$ to x^* can be estimated exactly as for the modified process in 1.4. In fact, we leave it to the reader to prove that, when $h < \frac{1}{2}$,

$$\|x^* - \tilde{x}_n\| \leqslant \frac{1}{\bar{h}}[1-(1-\delta)\sqrt{1-2\bar{h}}]^{n+1} \frac{\bar{\eta}}{(1-\delta)^2}$$

$$(n = 0, 1, \dots).$$

2.3. Suppose the conditions of Theorem 1.6 are satisfied with $h < \frac{1}{2}$. Let us show that Newton's method is *stable*: that is, the convergence of the process is not destroyed when we replace x_0 by any initial approximation $x_0' \in \Omega_0$ sufficiently close to x_0.

THEOREM 3. *Suppose the conditions of Theorem* 1.6 *are satisfied with constants* η, K *and* $h = K\eta < \frac{1}{2}$. *Provided that*

$$\|x' - x_0\| \leqslant \varepsilon \qquad \left(\varepsilon = \frac{1-2h}{4K} = \frac{1-2K\eta}{4K}\right), \tag{6}$$

the original and modified versions of Newton's method both converge when x_0' *is taken as initial approximation.*

Proof. We apply Theorem 1, taking Γ to be $\Gamma_0 = [P'(x_0)]^{-1}$ and x_0 to be x_0'. Condition 1) is then

$$\|\Gamma_0(P(x_0'))\| = \left\|\Gamma_0\left[P(x_0) + P'(x_0)(x_0' - x_0) + \int_{x_0}^{x_0'} P''(x)(x_0' - x, \cdot)dx\right]\right\| \leqslant \eta + \varepsilon + K\frac{\varepsilon^2}{2},$$

so we need to take $\bar{\eta}$ to be $\eta + \varepsilon + K\frac{\varepsilon^2}{2}$. Now

$$\|\Gamma_0 P'(x_0') - I\| = \|\Gamma_0[P'(x_0') - P'(x_0)]\| = \left\|\int_{x_0}^{x_0'} \Gamma_0 P''(x)dx\right\|.$$

Hence $\delta = K\varepsilon$.

Finally, $\bar{K} = K$.

Now $\bar{h}$ will have the value

$$\bar{h} = \frac{\overline{K\eta}}{(1-\delta)^2} = \frac{K\left(\eta + \varepsilon + \frac{K}{2}\varepsilon^2\right)}{(1-K\varepsilon)^2} = \frac{1}{2}\frac{K^2\varepsilon^2 + 2K\varepsilon + 2K\eta}{K^2\varepsilon^2 - 2K\varepsilon + 1} = \frac{1}{2}.$$

Thus the theorem is proved.

REMARK. If $h \geqslant 4\sqrt{2} - 5.5 \approx 0.16$, then the ball $\|x - x_0\| \leqslant \varepsilon$ is contained in the ball $\|x - x_0\| \leqslant r_0$, and hence in Ω_0.

However, if $h < 4\sqrt{2} - 5.5$, then (6) no longer ensures, in general, that $x_0' \in \Omega_0$ and an additional condition is needed.

2.4. The operator Γ_0 occurs among the data used to estimate the closeness of the initial approximation x_0 to the solution x^*—that is, to estimate the value of r_0. When we know Γ_0, we can specify more precisely the region in which the solution lies. For if we know Γ_0, we can find the next approximation x_1 in Newton's method, and apply Theorem 1.

THEOREM 4. *Suppose* $\Gamma_0 = [P'(x_0)]^{-1}$ *exists and is a continuous linear operator, and that the following conditions are satisfied*:

1) $\|\Gamma_0(P(x_0))\| \leqslant \eta$;

2) $\|\Gamma_0(P(x_1))\| \leqslant \eta_1 \qquad (x_1 = x_0 - \Gamma_0(P(x_0)))$;

3) $\|\Gamma_0 P''(x)\| \leqslant K \qquad (x \in \Omega_0)$;

4) $h_1 = \dfrac{K\eta_1}{(1-K\eta)^2} \leqslant \dfrac{1}{2}$.

Then (1) *has a solution* x^*, *and*

$$\|x - x_1\| \leqslant \frac{1-\sqrt{1-2h_1}}{h_1}\,\frac{\eta_1}{1-K\eta}. \tag{7}$$

Proof. We verify the conditions of Theorem 1, taking $\Gamma = \Gamma_0$ and putting x_1 in place of x_0:

$$\|\Gamma_0(P(x_1))\| \leqslant \eta_1, \tag{8}$$

$$\|\Gamma_0 P'(x_1) - I\| \leqslant \|\Gamma_0[P'(x_1) - P'(x_0)]\| = \left\|\int_{x_0}^{x_1} \Gamma_0 P''(x)\,dx\right\| \leqslant K\|x_1 - x_0\| \leqslant K\eta.$$

Thus we need to take $\bar{\eta}$, δ and $\bar{h}$ to be η_1, $K\eta$ and h_1, respectively.

By Theorem 1, the solution x^* lies in the ball $\|x - x_1\| \leqslant \bar{r}_0$. Since we have

$$\bar{r}_0 = \frac{1-\sqrt{1-2\bar{h}}}{\bar{h}}\,\frac{\bar{\eta}}{1-\delta} = \frac{1-\sqrt{1-2h_1}}{h_1}\,\frac{\eta_1}{1-K\eta}$$

in the present case, this leads to (7).

REMARK 1. It can happen that $h_1 \leqslant \frac{1}{2}$, although $h = K\eta > \frac{1}{2}$. In this case, we can draw conclusions about the existence of solutions from the above theorem, although Theorem 1.6 is not applicable.

REMARK 2. Conditions 1)–4) do not guarantee that the ball $\|x - x_1\| < \bar{r}_0$ is contained in Ω_0. Hence we need to assume that the radius of Ω_0 is large enough for this to be so. For example, we can take

$$r \geqslant \eta + \bar{r}_0 = \eta + \frac{1-\sqrt{1-2h_1}}{h_1}\,\frac{\eta_1}{1-K\eta}.$$

2.5. If, in the conditions of Remark 1 following Theorem 1.6, we have a bound for the norm of $[P'(x)]^{-1}$ not only at x_0 but also throughout the whole of Ω_0, then we can relax the condition on h, requiring only $h < 2$ instead of $h < \frac{1}{2}$ (see Mysovskikh [1]).

THEOREM 5 (Mysovskikh). *Suppose the following conditions are satisfied*:

1) $\|P(x_0)\| \leqslant \eta$;

2) *for* $x \in \Omega_0$ *there exists a continuous linear operator* $\Gamma(x) = [P'(x)]^{-1}$, *with*

$$\|\Gamma(x)\| \leqslant B \qquad (x \in \Omega_0);$$

3) $\|P'(x)\| \leqslant K \quad (x \in \Omega_0)$.
Then, provided

$$h = B^2 K\eta < 2$$

and

$$r > r' = B\eta \sum_{k=0}^{\infty} \left(\frac{h}{2}\right)^{2^k - 1},$$

equation (1) *has a solution* $x^* \in \Omega_0$ *and the original Newton's method starting from* x_0 *converges to this solution. Moreover the rate of convergence is given by*

$$\|x^* - x_n\| \leqslant B\eta \frac{\left(\frac{h}{2}\right)^{2^n - 1}}{1 - \left(\frac{h}{2}\right)^{2^n}}$$

$$(x_{n+1} = x_n - \Gamma(x_n)\ (P(x_n)); \qquad n = 0, 1, \ldots). \tag{9}$$

Proof. First we show that Newton's method is feasible—that is, that $x_n \in \Omega_0$ ($n = 0, 1, \ldots$). We have

$$\|x_1 - x_0\| = \|-\Gamma(x_0)\ (P(x_0))\| \leqslant B\eta, \tag{10}$$

$$\|P(x_1)\| = \left\| P(x_0) + P'(x_0)(x_1 - x_0) + \int_{x_0}^{x_1} P''(x)(x_1 - x, \cdot)\, dx \right\| \leqslant$$

$$\leqslant \int_0^{B\eta} K(B\eta - t)\, dt = \frac{KB^2\eta^2}{2} = \frac{h\eta}{2} = \eta_1. \tag{11}$$

It follows from (10) that $x_1 \in \Omega_0$.

Let us calculate the value of h corresponding to the point x_1:

$$h_1 = B^2 K\eta_1 = \frac{B^2 K\eta h}{2} = \frac{h^2}{2}. \tag{12}$$

Suppose we have already proved that $x_n \in \Omega_0$. Writing h_k and η_k ($k = 0, 1, \ldots$) for the values of h and η corresponding to x_k, we have the following relations, analogous to (10), (11) and (12):

$$\|x_{n+1} - x_n\| \leqslant B\eta_n,$$

$$\eta_{k+1} = \frac{h_k \eta_k}{2} \qquad (k = 0, 1, \ldots), \tag{13}$$

$$h_k = 2\left(\frac{h_{k-1}}{2}\right)^2 = \ldots = 2\left(\frac{h_0}{2}\right)^{2^k} = 2\left(\frac{h}{2}\right)^{2^k} \qquad (k = 0, 1, \ldots, n).$$

From the last two of these we find that

$$\eta_n = \frac{h_{n-1}\eta_{n-1}}{2} = \frac{h_{n-1}h_{n-2}\eta_{n-2}}{4} = \ldots = \frac{h_{n-1}h_{n-2}\ldots h_0}{2^n}\eta_0 = \left(\frac{h}{2}\right)^{2^n-1}\eta.$$

Substituting this in (13), we obtain

$$\|x_{n+1} - x_n\| \leqslant B\eta\left(\frac{h}{2}\right)^{2^n-1}. \tag{14}$$

Hence, using similar estimates for $\|x_{k+1} - x_k\|$ $(k = 0, 1, \ldots, n-1)$, we have

$$\|x_{n+1} - x_0\| \leqslant \sum_{k=0}^{n} \|x_{k+1} - x_k\| \leqslant B\eta \sum_{k=0}^{n} \left(\frac{h}{2}\right)^{2^k-1} \leqslant r' \leqslant r$$

and therefore $x_{n+1} \in \Omega_0$.

From (14) we also obtain

$$\|x_{n+p} - x_n\| \leqslant \sum_{k=n}^{n+p-1} \|x_{k+1} - x_k\| \leqslant \sum_{k=n}^{n+p-1} B\eta\left(\frac{h}{2}\right)^{2^k-1} \xrightarrow[n\to\infty]{} 0, \tag{15}$$

so the sequence $\{x_n\}$ is a Cauchy sequence, and so $x^* = \lim_{n\to\infty} x_n$ exists. Clearly, x^* is a solution of (1).

Letting $p \to \infty$ in (15), we have

$$\|x^* - x_n\| \leqslant B\eta \sum_{k=n}^{\infty} \left(\frac{h}{2}\right)^{2^k-1} \leqslant$$

$$\leqslant B\eta\left(\frac{h}{2}\right)^{2^n-1} \sum_{k=0}^{\infty} \left(\frac{h}{2}\right)^{2^n k} = B\eta\frac{(h/2)^{2^n-1}}{1-(h/2)^{2^n}},$$

which gives the required bond (9) for the rate of convergence of the method.

2.6. It quite often turns out that we can replace the given equation (1) by a simpler equation that is close to (1), and is, in general, also non-linear. We now explain some conditions under which it is possible to judge the solubility of the given equation from a solution of the approximate equation.

Suppose (1) has the form

$$P(x) \equiv \pi(x) + R(x) = \mathbf{0}. \tag{16}$$

Assume that x_0 is a solution of the simplified equation:

$$\pi(x_0) = \mathbf{0}.$$

If the following conditions are satisfied:

1) $\|[\pi'(x_0)]^{-1}(R(x_0))\| \leqslant \eta$;

2) $\|[\pi'(x_0)]^{-1} R'(x_0)\| \leqslant \alpha < 1$;

3) $\|[\pi'(x_0)]^{-1}\pi''(x)\| \leqslant K, \quad \|[\pi'(x_0)]^{-1} R''(x)\| \leqslant L \quad (x \in \Omega_0)$

and if

$$h = \frac{\eta(K+L)}{(1-\alpha)^2} \leqslant \frac{1}{2} \tag{17}$$

and

$$r \geqslant r_0 = \frac{1-\sqrt{1-2h}}{h}\frac{\eta}{1-\alpha}, \tag{18}$$

then (16) has a solution x^* in Ω_0.

This proposition is obtained immediately from Theorem 1 by setting $\Gamma = [\pi'(x_0)]^{-1}$. Using Theorem 1, we can also establish the region in which x^* is the unique solution.

The situation we have just been considering is a special case of a more general situation, in which the left-hand side of (1) depends on a parameter, either numerical or of some other kind, and we know a solution of the equation for one value of the parameter and wish to establish the existence of solutions for neighbouring values. Here we restrict ourselves to the case where the parameter appears linearly in the equation. More precisely, we shall be considering the equation

$$P(x, \mu) \equiv \pi(x) + \mu R(x) = \mathbf{0}, \tag{19}$$

where μ is a linear operator on $\mathbf{Y}$ ($\mu \in B(\mathbf{Y}, \mathbf{Y})$). In particular, μ may be a numerical coefficient.

THEOREM 6. *Suppose the following conditions are satisfied:*

1) $P(x_0, \mathbf{0}) = \pi(x_0) = \mathbf{0}$;

2) $\Gamma_0 = [\pi'(x_0)]^{-1}$ *exists and is a continuous linear operator, with* $\|\Gamma_0\| \leqslant B$;

3) $\|R(x_0)\| \leqslant \eta$, $\|R'(x_0)\| \leqslant \alpha$;

4) $\|\pi''(x)\| \leqslant K$, $\|R''(x)\| \leqslant L \quad (x \in \Omega_0)$.

If μ *satisfies*

$$h_\mu = \frac{B^2\eta(K+L\|\mu\|)\|\mu\|}{(1-\alpha B\|\mu\|)^2} \leqslant \frac{1}{2}, \qquad \alpha B\|\mu\| < 1, \tag{20}$$

then, provided the radius r of the ball Ω_0 *is sufficiently large,* (19) *has a solution* x^* *in* Ω_0.

The proof is obtained immediately from what was said above, by means of equation (16).

Using Theorem 1, we can find the region in which the solution $x^*(\mu)$ is located and the region in which it is unique. The same theorem shows that both the original and the modified version of Newton's method, starting from x_0, converge to $x^*(\mu)$. Furthermore,

Theorem 2 shows that $x^*(\mu)$ is the limit of the sequence $\{\tilde{x}_n(\mu)\}$, defined by

$$\tilde{x}_{n+1}(\mu) = \tilde{x}_n(\mu) - [\pi'(x_0)]^{-1}(P(\tilde{x}_n(\mu), \mu))$$
$$(n = 0, 1, \ldots; \ \tilde{x}_0(\mu) = x_0). \tag{21}$$

This last process has the advantage that it makes use of the inverse operator corresponding to the initial point x_0 and the initial value $\mu = 0$ of the parameter.

REMARK 1. We draw attention to the fact that, for each μ, the real equation

$$\left(\frac{K}{2}t^2 - \frac{1}{B}t\right) + \|\mu\|\left(\frac{L}{2}t^2 + \alpha t + \eta\right) = 0$$

is a majorant for (19), and (20) is nothing more than the condition for this equation to have real roots.

REMARK 2. Using rather more complicated arguments, one can also deal with the general case, where the parameter μ appears non-linearly in the equation $P(x, \mu) = \mathbf{0}$.

We note that in certain cases it is possible to obtain a wider range of admissible values for the parameter, if Theorem 1 is applied, not to x_0 and Γ_0, but to the values x and Γ yielding a solution to the system of equations

$$P(x, \mu) \equiv \mathbf{0}, \qquad \Gamma P'(x, \mu) = I$$

to within "first" powers of the parameter μ.

We restrict ourselves to the case where the parameter has a numerical value and the approximate equation $\pi(x) = \mathbf{0}$ is linear—that is, equation (19) has the form

$$P(x, \mu) \equiv U(x - x_0) + \mu R(x) = \mathbf{0}, \tag{22}$$

where $U = (\pi'(x_0))^{-1}$ is a continuous linear operator.

In this case the solutions just referred to have the following form, to within first powers of μ:

$$x_1 = x_0 - \mu\Gamma_0(R(x_0)), \qquad \Gamma_1 = \Gamma_0 - \mu\Gamma_0 R'(x_0)\Gamma_0.$$

Hence we need to estimate the following quantities:

1) $\|\Gamma_1(P(x_1, \mu))\| = \|[\Gamma_0 - \mu\Gamma_0 R'(x_0)\Gamma_0][-\mu R(x_0) + \mu R(x_0 - \mu\Gamma_0 R(x_0))]\| \leqslant \eta_1$;

2) $\|\Gamma_1 P'(x_1, \mu) - I\| = \|[\Gamma_0 - \mu\Gamma_0 R'(x_0)\Gamma_0][U + \mu R'(x_0 - \mu\Gamma_0 R(x_0))] - I\| \leqslant \delta_1$;

3) $\|[\Gamma_0 - \mu\Gamma_0 R'(x_0)\Gamma_0]R''(x)\| \leqslant L_1 \qquad (x \in \Omega_0)$.

Then (22) is soluble, provided

$$\frac{L_1\eta_1|\mu|}{(1-\delta_1)^2} \leqslant \frac{1}{2}.$$

2.7. The existence of a solution and the convergence of the process determining it in a definite range of values of the parameter enable one to draw conclusions about the nature of the dependence of the solution $x^*(\mu)$ on the parameter. For example, since $\tilde{x}_0(\mu) = x_0$ depends continuously on μ and $\pi(x)$ and $R(x)$ are continuous functions of x, all the $\tilde{x}_n(\mu)$ in (21) are continuous functions of the parameter, and thus $x(\mu)$ also depends continuously on μ.

Under certain conditions, one can also draw conclusions about the analytic nature of the dependence of $x^*(\mu)$ on the parameter.

A function $x(\mu)$ with values in a space $\mathbf{X}$ is said to be *weakly analytic* if, for every linear functional $f \in \mathbf{X}^*$, the function $f(x(\mu))$ is analytic, in the usual sense.

The operator P is said to be *analytic* if it maps every weakly analytic function $x(\mu)$ into a weakly analytic function $P(x(\mu))$ (in the space $\mathbf{Y}$).

Taking μ to be a numerical parameter in (19), we now assume π and R are analytic operators. Then it is easily verified that all the approximations (21) will be analytic functions of μ. Let f be an arbitrary continuous linear functional on the space $\mathbf{X}$. Since the norms of the elements $\tilde{x}_n(\mu)$ are uniformly bounded (in μ), the analytic functions

$$\phi_n(\mu) = f(\tilde{x}_n(\mu)) \qquad (n = 0, 1, \ldots)$$

are bounded in aggregate, and so the function

$$\phi(\mu) = \lim_{n \to \infty} \phi_n(\mu) = f(x^*(\mu))$$

is also analytic. As f was an arbitrary functional, it follows that $x^*(\mu)$ is a weakly analytic function of μ.

If we define the expansion of $x^*(\mu)$ as a power series in μ in the natural way,

$$x^*(\mu) = x_0 + \mu x_1 + \ldots + \mu^n x_n + \ldots,$$

then it follows from what we have said that the series must be weakly convergent for all $|\mu| < \mu_0$.

There are a large number of works devoted to Newton's method and its applications and generalizations. In particular, see Krasnosel'skii *et al.*, Collatz and the literature there cited.

§ 3. Applications of Newton's method to specific functional equations

3.1. First we consider applications of Newton's method to a single real or complex equation

$$f(z) = 0. \tag{1}$$

Using Theorem 1.6 and bearing in mind that the η and K which occur there have the following meaning

$$\eta \geqslant \frac{|f(z_0)|}{|f'(z_0)|}, \qquad K \geqslant \max \frac{|f''(z)|}{|f'(z_0)|}$$

(z_0 being the initial approximation), we see that the existence of a root is guaranteed if

$$h = \eta K \leqslant \frac{1}{2} \qquad \text{or} \qquad \frac{|f(z_0)||f''(z_0)|}{|f'(z_0)|^2} \leqslant \frac{1}{2},$$

and that the root is located in the disc

$$|z - z_0| \leqslant r_0 = \frac{1 - \sqrt{1 - 2h}}{h}\eta \tag{2}$$

and is unique within the disc*

$$|z-z_0|<r_1=\frac{1+\sqrt{1-2h}}{h}\eta. \tag{3}$$

Newton's method can also be used to solve a system of algebraic equations

$$\phi_j(\xi_1, \xi_2, \ldots, \xi_m)=0 \qquad (j=1, 2, \ldots, m), \tag{4}$$

which we regard as a single equation

$$P(x)=\mathbf{0}$$

in an m-dimensional space $\mathbf{X}$:

$$y=P(x)$$

$$(y=(\phi_1(\xi_1, \ldots, \xi_m), \ldots, \phi_m(\xi_1, \ldots, \xi_m)),\ x=(\xi_1, \ldots, \xi_m)).$$

Let $x_0=(\xi_1^{(0)}, \xi_2^{(2)}, \ldots, \xi_m^{(0)})$ be the initial approximation. Substituting this in the equation for Newton's method,

$$P'(x_0)(x-x_0)+P(x_0)=\mathbf{0},$$

and using the expression for the derivative $P'(x_0)$ given in XVII.1.6, we obtain the following system of linear equations for the correction $\Delta x=x_1-x_0=(\Delta\xi_1, \Delta\xi_2, \ldots, \Delta\xi_m)$:

$$\sum_{k=1}^{m}\frac{\partial\phi_j(\xi_1^{(0)}, \xi_2^{(0)}, \ldots, \xi_m^{(0)})}{\partial\xi_k}\Delta\xi^k=-\phi_j(\xi_1^{(0)}, \xi_2^{(0)}, \ldots, \xi_m^{(0)}) \tag{5}$$

$$(j=1, 2, \ldots, m),$$

from which we can find Δx, and hence x_1. In the same way, we find x_2, and so on.

If we use the modified Newton's method, then the matrix of the system (5) remains unchanged at each stage—only the right-hand sides change.

The convergence conditions for the process depend on what norm is used for $\mathbf{X}$. We consider two cases: $\mathbf{X}=\ell_m^\infty$ and, in less detail, $\mathbf{X}=\ell_m^2$.

THEOREM 1. *Suppose the functions $\phi_1, \phi_2, \ldots, \phi_m$ and the initial approximation x_0 satisfy the following conditions:*

1) $|\phi_j(\xi_1^{(0)}, \xi_2^{(0)}, \ldots, \xi_m^{(0)})|\leqslant\eta'$ $(j=1, 2, \ldots, m)$;

2) *the Jacobian*

$$\left(\frac{\partial\phi_j}{\partial\xi_k}(\xi_1^{(0)}, \xi_2^{(0)}, \ldots, \xi_m^{(0)})\right) \qquad (j, k=1, 2, \ldots, m)$$

has determinant $D\neq 0$, and the following inequality holds, where $A_{j,k}$ denotes the cofactor

* Here we assume that the bound for the second derivative holds in the disc $|z-z_0|\leqslant r$, where $r\geqslant r_0$ in the case of (2) or $r\geqslant r_1$ in the case of (3).

of $\dfrac{\partial \phi_j}{\partial \xi_k}$,

$$\frac{1}{|D|} \sum_{j=1}^{m} |A_{j,k}| \leqslant B' \qquad (k = 1, 2, \ldots, m);$$

3) $\left| \dfrac{\partial^2 \phi_j}{\partial \xi_k \partial \xi_s} (\xi_1, \xi_2, \ldots, \xi_m) \right| \leqslant L (j, k, s = 1, 2, \ldots, m;\ |\xi_i - \xi_i^{(0)}| \leqslant r)$;

4) $h = B'^2 \eta' L m^2 \leqslant \frac{1}{2}$.

Then, provided

$$r \geqslant \frac{1 - \sqrt{1 - 2h}}{h} B'h', \tag{6}$$

the system (4) *has a solution* $x^* = (\xi_1^*, \xi_2^*, \ldots, \xi_m^*)$ *in a neighbourhood of* x_0, *and both the original and the modified version of Newton's method converge to this solution.*

The proof of this theorem amounts to verifying the conditions of Theorem 1.6—or, more precisely, of Remark 1 after that theorem—when we take $\mathbf{X} = \ell_m^\infty$. We leave this to the reader.*

REMARK 1. The rates of convergence of the original and modified versions of Newton's method can be estimated using inequalities (33) and (34) of Theorem 1.6.

REMARK 2. Since, for $0 \leqslant h \leqslant \frac{1}{2}$, we have

$$\frac{1 - \sqrt{1 - 2h}}{h} \leqslant 2,$$

inequality (6) will be satisfied if we take $r = 2B'\eta'$.

REMARK 3. Suppose the number of equations in the system (4) is $m = 2$. Then

$$A_{j,k} = \pm \frac{\partial \phi_{3-k}}{\partial \xi_{3-j}} (\xi_1^{(0)}, \xi_2^{(0)}) \qquad (j, k = 1, 2),$$

so, if we know a bound

$$\left| \frac{\partial \phi_j}{\partial \xi_k} (\xi_1^{(0)}, \xi_2^{(0)}) \right| \leqslant l \qquad (j, k = 1, 2),$$

then we can take B' to be

$$B' = \frac{2l}{|D|},$$

and condition 4) can be rewritten in the form

$$\frac{16 l^2 \eta' L}{|D|} \leqslant \frac{1}{2}$$

or

$$32 l^2 \eta' L \leqslant |D|.$$

---

* It is easy to check that $\|P'(x)\| \leqslant Lm^2$ (cf. XVII.2.2).

This condition was obtained by A. Ostrowski. It is strange that a direct proof of the convergence of Newton's method is extremely complicated, even in this very simple case.

If we use the norm of ℓ_m^2, then conditions 1)–4) of Theorem 1 must be replaced by

1′) $\|\Gamma_0(P(x_0))\| = \|\Delta x\| = \left[\sum_{k=1}^{m} |\Delta\xi_k|^2\right]^{1/2} \leqslant \eta$;

2′) $\|\Gamma_0\| = \sqrt{\Lambda} \leqslant B'$;

here Γ_0 means, as usual, $[P'(x_0)]^{-1}$, and Λ denotes the largest eigenvalue of the matrix $\Gamma_0\Gamma_0^*$; we can take B' to be, for example,

$$B' = \frac{1}{|D|}\left[\sum_{j=1}^{m}\sum_{k=1}^{m} |A_{j,k}|^2\right]^{1/2};$$

3′) $\|P''(x)\| \leqslant \left[\sum_{j=1}^{m}\sum_{k=1}^{m}\sum_{s=1}^{m} \left|\frac{\partial^2\phi_j}{\partial\xi_k\partial\xi_s}\right|^2\right]^{1/2} \leqslant K'$;

so we can take K' to be

$$K' = Lm\sqrt{m};$$

4′) $h = B'K'\eta \leqslant \frac{1}{2}$.

When these conditions are satisfied, the existence of a solution to the system (4) is assured, and so is the convergence of both the original and the modified version of Newton's method to this solution.

Let us illustrate what we have said with an example. Consider the system

$$3\xi_1^2\xi_2 + \xi_2^2 = 1,$$
$$\xi_1^4 + \xi_1\xi_2^3 = 1,$$

For the initial approximation $x_0 = (\xi_1^{(0)}, \xi_2^{(0)})$ we take

$$\xi_1^{(0)} = 0.98, \ \xi_2^{(0)} = 0.32.$$

The system used for determining the correction then has the form

$$1.880\,\Delta\xi_1 + 3.188\,\Delta\xi_2 = 0.045,$$
$$3.798\,\Delta\xi_1 + 0.301\,\Delta\xi_2 = 0.046,$$

from which we find that

$$\Delta\xi_1 = 0.0105, \ \Delta\xi_2 = 0.0075.$$

From this we obtain

$$-\Gamma_0 = \begin{pmatrix} -0.026 & 0.256 \\ 0.329 & -0.163 \end{pmatrix}.$$

Using the ℓ_2^∞ norm, we calculate the constants B', η' and L of Theorem 1. We have

$$\|\Gamma_0\| = 0.492 < 0.5 = B',$$
$$\|P(x_0)\| = \max[0.046, 0.045] < 0.05 = \eta'.$$

The second derivatives of ϕ_1 and ϕ_2 can be estimated for $\|x - x_0\| < 2B'\eta'$, that is, for $0.93 \leqslant \xi_1 \leqslant 1.03$, 0.27

$\leqslant \xi_2 \leqslant 0.37$. Since

$$\frac{\partial^2\phi_1}{\partial\xi_1^2} = 6\xi_2, \quad \frac{\partial^2\phi_1}{\phi\xi_1\partial\xi_2} = 6\xi_1, \quad \frac{\partial^2\phi_1}{\partial\xi_2^2} = 6\xi_2,$$

$$\frac{\partial^2\phi_2}{\partial\xi_1^2} = 12\xi_1^2, \quad \frac{\partial^2\phi_2}{\partial\xi_1\partial\xi_2} = 3\xi_2^2, \quad \frac{\partial\phi_2}{\partial\xi_2^2} = 6\xi_1\xi_2,$$

we can take L to be

$$L = 12.8 > \max\frac{\partial^2\phi_2}{\partial\xi_1^2} = 12(1.03)^2.$$

For these values of B', η' and L, the value of h will be

$$h = B'^2\eta' Lm^2 = 0.25 \times 0.05 \times 12.8 \times 4 = 0.64.$$

Hence we cannot deduce anything about the convergence of Newton's method from Theorem 1. This is because we have a very crude bound for $\|P''(x)\|$. In fact, we took

$$\|P''(x)\| \leqslant Lm^2 = 51.2,$$

whereas a more precise bound is actually

$$\|P''(x)\| \leqslant 16 = K'.$$

Using this, we obtain

$$h = B'^2\eta' K' = 0.25 \times 0.05 \times 16 = 0.2,$$

which does allow us to conclude that Newton's method converges.

Notice that if we use Theorem 1.6 directly, we can obtain an even smaller value for h. For

$$\|\Gamma_0(P(x_0))\| = \|\Delta x\| = \max[|\Delta\xi_1|, |\Delta\xi_2|] = 0.0105$$

and

$$\|\Gamma_0 P''(x)\| < 7.2.$$

Hence, if we take $\eta = 0.0105$, $K = 7.2$, we obtain

$$h = \eta K < 0.08. \tag{7}$$

Using the modified Newton's method, we obtain the successive approximations

$$\xi_1^{(1)} = 0.9905, \quad \xi_2^{(1)} = 0.3275,$$

$$\xi_1^{(2)} = 0.99117, \quad \xi_2^{(2)} = 0.32738,$$

$$\xi_1^{(3)} = 0.991189, \quad \xi_2^{(3)} = 0.327382.$$

An estimate for the error of the third approximation, calculated according to formula (34) of Theorem 1.6, yields the value 0.0008 (here h is taken to have the value (7)).

If we apply Theorem 2.4, we can determine an even more precise bound for the error; namely, the theorem gives

$$0.991173 \leqslant \xi_1^* \leqslant 0.991205, 0.327366 \leqslant \xi_2^* \leqslant 0.327398.$$

Thus the third approximation is correct to four decimal places.

If we use the ℓ_2^2 norm, then, starting with conditions 1′)–3′), we can obtain

$$\|\Gamma_0\| \leqslant 0.448 < 0.5 = B',$$

$$\|\Gamma_0(P(x_0))\| = \|\Delta x\| = 0.0133 < 0.015 = \eta,$$

$$\|P''(x)\| < 15.2 = K', h = B'K'\eta = 0.1026.$$

3.2. We now consider applications of Newton's method to non-linear integral equations. Suppose we are given the integral equation

$$x(s) = \int_0^1 K(s, t, x(t))\,dt, \tag{8}$$

where $K(s, t, u)$ is continuous in all its arguments and has continuous derivatives of all orders required. In the function space $\mathbf{X}$ we introduce the operator P:

$$y = P(x), y(s) = x(s) - \int_0^1 K(s, t, x(t))\,dt \tag{9}$$

and rewrite (8) as a functional equation

$$P(x) = \mathbf{0}.$$

We carry out Newton's method for this equation as follows: let x_0 be the initial approximation; assume that $\mathbf{X}$ and $K(s, t, u)$ are such that $P'(x_0)$ can be obtained by "differentiating under the integral sign": that is, that $z = P'(x_0)(x)$ means that

$$z(s) = x(s) - \int_0^1 K'_u(s, t, x_0(t))x(t)\,dt; \tag{10}$$

the correction $\Delta x = x_1 - x_0$ is determined from the equation

$$P'(x_0)(\Delta x) = -P(x_0),$$

which, in view of (10), has the form

$$\Delta x(s) - \int_0^1 K'_u(s, t, x_0(t))\Delta x(t)\,dt = \varepsilon_0(s), \tag{11}$$

where

$$\varepsilon_0(s) = \int_0^1 K(s, t, x_0(t))\,dt - x_0(s)$$

is the discrepancy in (8) corresponding to the initial approximation.

Hence to determine each of the subsequent approximations we need to solve a linear integral equation; moreover, if we used the modified Newton's method, then the kernel will be the same at each step.

On the basis of the general theorems we can state different results on the convergence of Newton's method, depending on what space we work with. For instance, if $\mathbf{X} = \mathbf{C}[0, 1]$, we obtain

THEOREM 2. *Suppose the following conditions are satisfied*:

1) *for the kernel* $K(s, t) = K'_u(s, t, x_0(t))$, *the integral equation* (11) *has resolvent* $G(s, t)$, *where*

$$\int_0^1 |G(s, t)|\,dt \leqslant B \qquad (s \in [0, 1]); \tag{12}$$

2) $|\varepsilon_0(s)| \leqslant \eta' \qquad (s \in [0, 1]);$

3) $\int_0^1 |K''_{u^2}(s, t, u)|\,dt \leqslant K'$

in the region determined by the values of s and u satisfying

$$0 \leqslant s \leqslant 1, \qquad |u - x_0(s)| \leqslant 2(1 + B)\eta';$$

4) $h = (1 + B)^2 K'\eta' \leqslant \frac{1}{2}$.

Then Newton's method (both the original and modified versions) converges to a solution of (8). *Moreover, the solution* $x^*(s)$ *is located in the region*

$$|x^*(s) - x_0(s)| \leqslant \frac{1 - \sqrt{1-2h}}{h}(1+B)\eta' \qquad (s \in [0, 1]),$$

and is unique in the region

$$0 \leqslant s \leqslant 1, \quad |u - x_0(s)| \leqslant \frac{1 + \sqrt{1-2h}}{h}(1+B)\eta'.$$

It can be verified without any difficulty that the conditions of Theorem 1.6 are satisfied, if one bears in mind that, by Theorem XVII.3.2, the derivative $P'(x_0)$ of P has the form (10) and, by the same theorem, $P''(x)$ is a bilinear integral operator with kernel $K'_{u^2}(s, t, x_0(t))$. As we showed in XIII.6.1, the operator $\Gamma_0 = [P'(x_0)]^{-1}$ has the form

$$z = \Gamma_0(y), \quad z(s) = y(s) + \int_0^1 G(s, t) y(t)\, dt$$

and so

$$\|\Gamma_0\| \leqslant 1 + B.$$

If we take $\mathbf{X}$ to be $\mathbf{L}^2(0, 1)$, then, using Theorem XVII.3.1, we can find other conditions for (8) to be soluble. Namely, these are the conditions:

1′) $\int_0^1 |\varepsilon_0(s)|^2\, ds \leqslant \eta'^2$;

2′) the following inequality is satisfied:

$$\left|\frac{\lambda_n}{1 - \lambda_n}\right| \leqslant B' \qquad (n = 1, 2, \ldots),$$

where λ_n are the characteristic values of the kernel $K(s, t) = K'_u(s, t, x_0(t))$, if this is symmetric (if $K(s, t)$ is not symmetric, this requirement becomes more complicated: cf. IV.2.6 and V.2.7);

3′) $|K''_{u^2}(s, t, u)| \leqslant K' \qquad (s, t \in [0, 1], \quad -\infty < u < \infty)$;

4′) $h = B'^2 K' \eta' \leqslant \frac{1}{2}$.

3.3. The conditions of Theorem 2 and conditions 1′)–4′) assume that we have an initial approximation x_0 sufficiently close to the exact solution. Now we consider a fairly general method of constructing such an initial approximation, which consists in replacing the given integral equation (8) by an equation of a simpler structure, whose solution reduces to the solution of an algebraic system.

In fact, assume that we can approximate the kernel $K(s, t, u)$ by a kernel of the simpler form

$$H(s, t, u) = \sum_{k=1}^{m} h_k(t, u)\omega_k(s),$$

where $\omega_k(s)$ $(k = 1, 2, \ldots, m)$ are certain functions, which we may assume, without losing any generality, are pairwise orthogonal and normalized. We could take H to be, for example, a truncated Fourier series for K with respect to the functions of a complete orthonormal system $\{\omega_k(s)\}$ $(k = 1, 2, \ldots)$.

It is natural to regard the equation

$$x(s) = \int_0^1 H(s, t, x(t))\,dt \tag{13}$$

as an approximation to (8). But a solution of (13) has the form

$$x_0(t) = \sum_{k=1}^{m} A_k\omega_k(t),$$

where the coefficients A_k $(k = 1, 2, \ldots)$ are defined by the non-linear algebraic system

$$A_j = \int_0^1 h_j\left(t, \sum_{k=1}^{m} A_k\omega_k(t)\right)dt \qquad (j = 1, 2, \ldots, m). \tag{14}$$

We shall take $x_0(t)$ as the initial approximation for (8).

Since in many cases the important problem is only to decide whether a solution exists and to find regions in which it is located or is unique, it is natural to use here the approach of Theorem 2.6. To do this we take $\mathbf{X} = \mathbf{C}[0, 1]$ and rewrite (8) in the form

$$P(x) \equiv \pi(x) + \mu R(x) = \mathbf{0},$$

$$\pi(x)(s) = x(s) - \int_0^1 H(s, t, x(t))\,dt,$$

$$R(x)(s) = \int_0^1 [K(s, t, x(t)) - H(s, t, x(t))]\,dt,$$

$$\mu = 1.$$

Then we can state the following theorem.

THEOREM 3. *Suppose the following conditions are satisfied:*

1) *the kernel* $H'_u(s, t, x_0(t))$ *has a resolvent* $G_0(s, t)$ *with*

$$\int_0^1 |G_0(s, t)|\,dt \leqslant B \qquad (s \in [0, 1]);$$

2) $$\left|\int_0^1 [K(s, t, x_0(t)) - H(s, t, x_0(t))]\,dt\right| = |\varepsilon_0(s)| \leqslant \eta \qquad (s \in [0, 1]);$$

3) $$\int_0^1 |K'_u(s, t, x_0(t)) - H'_u(s, t, x_0(t))|\,dt \leqslant \alpha \quad (s \in [0, 1]);$$

4) $$\int_0^1 |H''_{u^2}(s, t, u)|\,dt \leqslant K \qquad (s \in [0, 1],\ |u - x_0(s)| \leqslant r);$$

5) $$\int_0^1 |K''_{u^2}(s, t, u) - H''_{u^2}(s, t, u)|\,dt \leqslant L \qquad (s \in [0, 1],\ |u - x_0(s)| \leqslant r);$$

6) $$h = \frac{(1 + B)^2 (K + L)\eta}{[1 - \alpha(B + 1)]^2} \leqslant \frac{1}{2}, \qquad \alpha(B + 1) < 1.$$

Then, provided

$$r \geqslant r_0 = \frac{1-\sqrt{1-2h}}{h}\frac{\eta(1+B)}{1-\alpha(1+B)},$$

equation (8) *has a solution* x^*, *and*

$$|x^*(s) - x_0(s)| \leqslant r_0. \tag{15}$$

This theorem is convenient, in comparison with Theorem 2, since the kernel $H'_u(s, t, x_0(t))$ is degenerate, so that the resolvent $G_0(s, t)$ is easy to find.

The process described in 2.6 can be used to determine a solution of (8), although this will converge more slowly than in the case of the modified process.

To illustrate what has been said, let us consider an example. Suppose we are given the equation

$$x(s) = \frac{1}{2}\int_0^1 [x(t)]^2 \sin st\, dt + 1.$$

We take the approximate kernel to be

$$H(s, t, u) = \frac{1}{2}stu^2 + 1,$$

so that in the present case (13) will be

$$x(s) = \frac{1}{2}\int_0^1 [x(t)]^2 st\, dt + 1.$$

A solution of this has the form $x_0(t) = 1 + At$, where A is found from a quadratic equation: $A = 0.405887$. The constants appearing in the hypothesis of Theorem 3 may be calculated without difficulty. We obtain the following values:

$$B = 1.124,\ \eta = 0.0367,\ \alpha = 0.055,\ K = 0.5,\ L = 0.0417,$$

so that

$$0.1 < h < 0.103,\ \alpha(1+B) < 0.12,$$

and, by (15), we have

$$|x^*(s) - x_0(s)| < 0.096 \quad (s \in [0, 1]).$$

If we had taken

$$H(s, t, u) = \frac{1}{2}\left(st - \frac{s^3t^3}{6}\right)u^2 + 1,$$

then we would have obtained

$$x_0(s) = 1 + 0.38617s - 0.0345s^3$$

and the estimate (15) would have taken the form

$$|x^*(s) - x_0(s)| < 0.0119 \quad (s \in [0, 1]).$$

3.4. We now consider the problem of applying Newton's method to differential equations. Suppose we are given a differential equation

$$x'(t) - \phi(x(t), t) = 0 \qquad (x(0) = 0). \tag{16}$$

We introduce a norm in the space $\mathbf{C}^1$ of functions continuously differentiable in $[0, a]$ and vanishing at $t = 0$ by setting

$$\|x\| = \max_{t\in[0,a]} |x(t)| + \lambda \max_{t\in[0,a]} |x'(t)|,$$

where λ is a positive coefficient to be determined later. Assuming that $\phi(u, t)$ is continuous and has a continuous second derivative in u in the region

$$0 \leqslant t \leqslant a, \qquad |u - x_0(t)| \leqslant \delta \qquad (x_0 \in \mathbf{C}^1),$$

we consider the operator P defined by

$$y = P(x), \qquad y(t) = x'(t) - \phi(x(t), t).$$

It is not hard to check that P maps the ball $\|x - x_0\| \leqslant \delta$ (which we denote by Ω) into the space $\mathbf{C}[0, 1]$ and has continuous first and second derivatives in Ω. Furthermore,

$$P'(z)(x)(t) = x'(t) - \phi'_u(z(t), t)x(t), \tag{17}$$

$$P''(z)(x, \tilde{x})(t) = -\phi''_{u^2}(z(t), t)x(t)\tilde{x}(t). \tag{18}$$

Let us find $\Gamma_0 = [P'(x_0)]^{-1}$. By (17), the element $x = \Gamma_0(y)$ is a solution of the differential equation

$$x'(t) = \phi'_u(x_0(t), t)x(t) + y(t), \tag{19}$$

that is,

$$x(t) = \psi(t) \int_0^t \frac{y(s)}{\psi(s)} ds,$$

where we have written

$$\psi(t) = \exp\left(\int_0^t \phi'_u(x_0(s), s)\, ds \right).$$

Hence

$$\max_{t\in[0,a]} |x(t)| \leqslant a \frac{\max |\psi(t)|}{\min |\psi(t)|} \|y\|,$$

and we see from (19) that

$$\max_{t\in[0,a]} |x'(t)| \leqslant \max_{t\in[0,a]} |\phi'_u(x_0(t), t)| \max_{t\in[0,a]} |x(t)| + \|y\| \leqslant \theta \|y\|.$$

Thus

$$\|x\| \leqslant \left[a \frac{\max|\psi(t)|}{\min|\psi(t)|} + \lambda\theta \right] \|y\|,$$

so that

$$\|\Gamma_0\| \leqslant a \frac{\max|\psi(t)|}{\min|\psi(t)|} + \lambda\theta.$$

This bound enables us to apply Theorem 1.6, which leads to the following result.

THEOREM 4. *Suppose the following conditions are satisfied:*

1) $|x_0'(t) - \phi(x_0(t), t)| \leqslant \eta'(t \in [0, a])$;
2) $|\phi_u'(x_0(t), t)| \leqslant M_1 \quad (t \in [0, a])$;
3) $|\phi_{u^2}''(u, t)| \leqslant M_2 \quad (t \in [0, a],\ |u - x_0(t)| \leqslant \delta)$;
4) $h_0 = M_2 a^2 e^{4aM_1} \eta' < \dfrac{1}{2}$.

Then, provided

$$\delta > r_0 = \frac{1 - \sqrt{1 - 2h_0}}{h_0} e^{2aM_1} a\eta',$$

equation (16) *has a unique solution* $x^*(t)$ *in* $[0, a]$, *and*

$$|x^*(t) - x_0(t)| < r_0.$$

Proof. To verify the conditions of Theorem 1.6 we need only check that

$$\max |\psi(t)| \leqslant e^{aM_1}, \qquad \min |\psi(t)| \geqslant e^{-aM_1},$$

so we can take B' in Theorem 1.6 to be

$$B' = ae^{2aM_1} + \lambda\theta.$$

Furthermore, we can set $K' = M_2$. Hence the condition for (16) to be soluble can be written in the form

$$h = B'^2 K' \eta' = (ae^{2aM_1} + \lambda\theta)^2 M_2 \eta' \leqslant \frac{1}{2}. \tag{20}$$

Since $\lim_{\lambda \to 0} h = h_0$, we can ensure, by condition 4), that (20) is satisfied, by taking λ small enough.

REMARK 1. The case where the initial condition is non-zero can be reduced to the case considered above by an obvious change of variables.

REMARK 2. This method is also suitable for investigating systems of differential equations. The difficulty in carrying it out in this case is that, in general, it is not possible to find an analytic solution for a linear system, and hence a bound for the norm of Γ_0.

3.5. Now let us consider the differential equation

$$x''(t) + x(t) + \mu\phi(x(t), x'(t), t) = 0, \tag{21}$$

where $\phi(u, v, t)$ is a continuous function, having continuous second derivatives in u and v, and periodic in t, with period $\omega > 0$.

Using arguments similar to those of 3.4, we now find conditions under which (21) has a periodic solution with period ω—in other words, has a solution satisfying the boundary conditions

$$x(\omega) = x(0), \qquad x'(\omega) = x'(0). \tag{22}$$

This time we adopt the approach of Theorem 2.6. We introduce the space $\tilde{\mathbf{C}}^2$ of twice continuously differentiable periodic functions (with period ω). We define a norm in $\tilde{\mathbf{C}}^2$ by setting

$$\|x\| = \max_t |x(t)| + \max_t |x'(t)| + \lambda \max_t |x''(t)| \qquad (x \in \tilde{\mathbf{C}}^2),$$

where $\lambda > 0$ is to be determined later. We also introduce the operators π and R, defined by

$$y = \pi(x), \qquad y(t) = x''(t) + x(t), \qquad z = R(x), \qquad z(t) = \phi(x(t), x'(t), t).$$

Equation (21) can now be rewritten in the form

$$\pi(x) + \mu R(x) = \mathbf{0}. \tag{23}$$

Let us derive the bounds needed to apply Theorem 2.6.

The operator π is obviously continuous and linear, so $x_0 = 0$ is a solution of $\pi(x) = \mathbf{0}$. If $\cos\omega \neq 1$, then $\Gamma_0 = \pi^{-1}$ exists. We can find Γ_0 by solving the differential equation

$$x'' + x = y$$

subject to the conditions (22). We omit the tedious but essentially straightforward calculations, and state the following bound for Γ_0 straight away:

$$\|\Gamma_0\| \leqslant \omega\left[2 + \frac{1 + |\cos\omega| + |\sin\omega|}{1 - \cos\omega}\right] + \lambda\theta \leqslant \frac{4 + 1 + \sqrt{2}}{1 - \cos\omega}\omega + \lambda\theta < \frac{6.5\omega}{1 - \cos\omega} + \lambda\theta;$$

here θ is some expression containing ω.

Hence we may take B to be

$$B = \frac{6.5\omega}{1 - \cos\omega} + \lambda\theta = B_0 + \lambda\theta.$$

Bearing in mind that

$$R'(x_0)(x)(t) = \phi'_u(0, 0, t)x(t) + \phi'_v(0, 0, t)x'(t),$$

$$R''(z)(x, \tilde{x})(t) = \phi''_{u^2}(z(t), z'(t), t)x(t)\tilde{x}(t) + \phi''_{uv}(z(t), z'(t), t)[x(t)\tilde{x}'(t) +$$

$$+ x'(t)\tilde{x}(t)] + \phi''_{v^2}(z(t), z'(t), t)x'(t)\tilde{x}'(t),$$

we can now state the following theorem.

THEOREM 5. *Suppose the following conditions are satisfied:*

1) $\omega = 2n\pi \quad (n = 1, 2, \ldots)$;

2) $|\phi(0, 0, t)| \leqslant \eta$;

3) $|\phi'_u(0, 0, t)| \leqslant \alpha, \quad |\phi'_v(0, 0, t)| \leqslant \alpha$;

4) $|\phi''_{u^2}(u, v, t)| \leqslant L, \quad |\phi''_{uv}(u, v, t)| \leqslant L, \quad |\phi''_{v^2}(u, v, t)| \leqslant L.$

Then, provided

$$\mu < \frac{1 - \cos\omega}{6.5\omega(\alpha + \sqrt{2L\eta})}, \tag{24}$$

equation (21) *has a unique periodic solution of period* ω.

For the proof we need only verify that inequality (20) of Theorem 2.6, which in the present case takes the form

$$h_\mu = \frac{B^2 L\eta\mu^2}{(1 - \alpha\beta\mu)^2} \leqslant \frac{1}{2},$$

is satisfied when λ is sufficiently small and

$$\frac{B_0^2 L\eta\mu^2}{(1-\alpha B_0\mu)^2} < \frac{1}{2},$$

but this is equivalent to (24).

REMARK. If ϕ depends analytically on its arguments, then by what we said in 2.7 we can conclude that the periodic solution of (21) will also depend analytically on μ, for all μ satisfying (24).

Similar arguments can be used for equations of higher orders.

3.6. Now we consider an application of Newton's method to perturbation theory.

Let U and V be continuous linear operators mapping a B-space $\mathbf{X}$ into itself. Suppose we wish to find an eigenvalue and a corresponding eigenvector of the operator $U_t = U + tV$, assuming that the solution of this problem is known for $t = 0$—i.e. assuming we know an eigenvalue λ_0 of U and a corresponding eigenvector x_0.

Imposing certain additional requirements on U and V, we shall prove that, for sufficiently small t, the operator U_t has an eigenvalue λ_t and a corresponding eigenvector x_t, close to λ_0 and x_0, respectively. Namely, we have

THEOREM 6. *Suppose the following conditions are satisfied:*

1) $\overline{\lambda_0}$ *is an eigenvalue of* U^*:

$$U^*(f_0) = \overline{\lambda_0} f_0;$$

here f_0 *denotes an eigenvector normalized so that*

$$f_0(x_0) = 1.$$

2) *The operator* $U - \lambda_0 I$, *which takes values in the space* $\mathbf{X}_0 = \mathbf{N}(f_0)$ *of all* $y \in \mathbf{X}$ *for which* $f_0(y) = 0$ (*cf.* XII.2.1), *has a continuous inverse* T, *when regarded as an operator from* $\mathbf{X}_0$ *into* $\mathbf{X}$:

$$T(Ux - \lambda_0 x) = x, \qquad UTx - \lambda_0 Tx = x \qquad (x \in \mathbf{X}_0).$$

Then, provided

$$\|T\|\|V\||t| \leqslant \frac{1}{1 + 2\|f_0\|\|x_0\| + 2\sqrt{(1+\|f_0\|\|x_0\|)\|f_0\|\|x_0\|}}, \tag{25}$$

U_t *has an eigenvalue* λ_t *such that, if* x_t *is a corresponding eigenvector, normed so that*

$$f_0(x_t) = 1,$$

then

$$|\lambda_t - \lambda_0| \leqslant A|t|, \qquad \|x_t - x_0\| \leqslant B|t|,$$

where A *and* B *are constants determined by* U *and* V.

Proof. We need to define λ and x by the conditions

$$U(x) + tV(x) = \lambda x, \qquad f_0(x) = 1.$$

We introduce new unknowns z and δ, setting

$$x = x_0 + z, \qquad \lambda = \lambda_0 + \delta.$$

These unknowns must satisfy the conditions

$$U(z) + tV(x_0 + z) = \delta(x_0 + z) + \lambda_0 z, \qquad f_0(z) = 0,$$

so we need to find a solution in $\mathbf{X}_0$ for the equation

$$U(z) + tV(x_0 + z) = \delta(x_0 + z) + \lambda_0 z. \tag{26}$$

Now apply the functional f_0 to both sides of (26). Since

$$f_0(Uz - \lambda_0 z) = U^*(f_0)(z) - (\overline{\lambda_0} f_0)(z) = 0,$$

this yields the equation

$$\delta = tf_0(V(x_0 + z)), \tag{27}$$

expressing δ in terms of z. Substituting (27) in (26), we obtain an equation for z:

$$U(z) - \lambda_0 z + t[V(x_0 + z) - f_0(V(x_0 + z))(x_0 + z)] = \mathbf{0}. \tag{28}$$

Both terms in this equation belong to the subspace $\mathbf{X}_0$, so we can apply T to them; this yields

$$\pi(z) + tR(z) = \mathbf{0}, \tag{29}$$

where

$$\pi(z) = z,$$

$$R(z) = T[V(x_0 + z) - f_0(V(x_0 + z))(x_0 + z)].$$

We now apply Theorem 2.6 to (29). We have (with $z_0 = \mathbf{0}$)

$$\pi'(z_0) = I, \qquad \Gamma_0 = I, \qquad \pi''(z) = \mathbf{0},$$

$$R(z_0) = T[V(x_0) - f_0(V(x_0))x_0],$$

$$R'(z)(x) = T[V(x) - f_0(V(x))(x_0 + z) - f_0(V(x_0 + z))x],$$

$$R'(z)(x, \tilde{x}) = -T[f_0(V(x))\tilde{x} + f_0(V(x))\tilde{x}] = -f_0(V(x))T(\tilde{x}) - f_0(V(\tilde{x}))T(x).$$

This enables us to take

$$\eta = \|T\|\|V\|[1 + \|f_0\|\|x_0\|]\|x_0\|, \qquad B = 1,$$

$$\alpha = \|T\|\|V\|[1 + 2\|f_0\|\|x_0\|],$$

$$K = 0, \qquad L = 2\|T\|\|V\|\|f_0\|.$$

Hence the condition for (29) to be soluble is

$$h_t = \frac{L\eta|t|^2}{(1 - \alpha|t|)^2} \leqslant \frac{1}{2},$$

which is equivalent to (25).

Writing z_t for the solution of (29), we have

$$\|z_t\| = \|z_t - z_0\| \leqslant \frac{1 - \sqrt{1 - 2h_t}}{h_t}\eta|t| \leqslant 2\eta|t| = B|t|,$$

and we see from (27) that

$$|\lambda_t - \lambda_0| = |\delta| \leqslant |t|\|f_0\|\|V\|\|x_0 + z_t\| \leqslant A|t|.$$

The theorem is therefore proved.

REMARK. We note an important special case, where there is a simpler expression for η and α. Let $\mathbf{X}$ be a Hilbert space and U a self-adjoint operator. It is not difficult to see that in this case f_0 can be taken as the functional determined by x_0, that is,

$$f_0(y) = (y, x_0) \quad (y \in \mathbf{X}).$$

Using the fact that $\|x_0\|^2 = f_0(x_0) = 1$, we have

$$\|R(z_0)\| \leqslant \|T\| \| Vx_0 - (Vx_0, x_0)x_0\| \leqslant \|T\| \| V(x_0)\| \leqslant \|T\| \| V\|,$$

$$\begin{aligned}\|R'(z_0)(x)\| &\leqslant \|T\| \| Vx - (Vx, x_0)x_0 - (Vx_0, x_0)x\| \leqslant \\ &\leqslant \|T\| [\| Vx - (Vx, x_0)x_0\| + \|(Vx_0, x_0)x\|] \leqslant \\ &\leqslant \|T\| [\| Vx\| + |(Vx_0, x_0)| \|x\|] \leqslant 2\|T\| \| V\| \|x\|,\end{aligned}$$

so that this time we can take

$$\eta = \|T\| \| V\|, \qquad \alpha = 2\|T\| \| V\|,$$

and the condition for (29) to be soluble will then be

$$\frac{2(\|T\| \| V\| |t|)^2}{(1 - 2\|T\| \| V\| |t|)^2} \leqslant \frac{1}{2},$$

which enables us to replace (25) by the inequality

$$\|T\| \| V\| |t| \leqslant \frac{1}{4}.$$

A direct investigation of this case led M. K. Gavurin [3] to the condition

$$\|T\| \| V\| |t| \leqslant \frac{1}{2},$$

which in fact is best possible.

3.7. We now consider a quasi-linear differential equation of second order with two independent variables:

$$A(s, t, u, p, q)\frac{\partial^2 u}{\partial s^2} + B(s, t, u, p, q)\frac{\partial^2 u}{\partial s \partial t} + C(s, t, u, p, q)\frac{\partial^2 u}{\partial t^2} + D(s, t, u, p, q) = 0; \quad (30)$$

here p and q denote the first derivatives $p = \dfrac{\partial u}{\partial s}$, $q = \dfrac{\partial u}{\partial t}$.

We consider (30) in a region Q and seek a solution satisfying a boundary condition on a curve γ bounding Q.

By applying Newton's method to this problem one can determine conditions for the existence of a solution and the region in which a solution is located, if one knows an approximate solution $u_0 = u_0(s, t)$ (see Koshelev [1], [2]).

We shall assume that (30) is of elliptic type, or, more precisely, that all the values of the variables—or at least all those which we come across—satisfy the inequality

$$B^2 - 4AC < 0. \quad (31)$$

To apply Newton's method to (30), we rewrite it as a single equation

$$P(u) = \mathbf{0}, \quad (32)$$

where P is an operator mapping a certain function space $\mathbf{U}$ into another function space $\mathbf{V}$. We shall identify these spaces in more specific terms later on.

Let $F = F(s, t, u, p, q)$ be a function. We write F_0 for the function obtained by replacing u in F by u_0, and p and q by the partial derivatives $p_0 = \dfrac{\partial u_0}{\partial s}$ and $q_0 = \dfrac{\partial u_0}{\partial t}$.

If we do not concern ourselves with questions of rigorous justification, we find easily that

$$P'(u_0)(u) = A_0\frac{\partial^2 u}{\partial s^2} + \left[\left(\frac{\partial A}{\partial u}\right)_0 u + \left(\frac{\partial A}{\partial p}\right)_0 p + \left(\frac{\partial A}{\partial q}\right)_0 q\right]\frac{\partial^2 u_0}{\partial s^2} + B_0\frac{\partial^2 u}{\partial s\partial t} + \left[\left(\frac{\partial B}{\partial u}\right)_0 u + \left(\frac{\partial B}{\partial p}\right)_0 p + \left(\frac{\partial B}{\partial q}\right)_0 q\right]\frac{\partial^2 u_0}{\partial s\partial t} + C_0\frac{\partial^2 u}{\partial t^2} + \left[\left(\frac{\partial C}{\partial u}\right)_0 u + \left(\frac{\partial C}{\partial q}\right)_0 p + \left(\frac{\partial C}{\partial q}\right)_0 q\right]\frac{\partial^2 u_0}{\partial t^2} = \left[A_0\frac{\partial^2}{\partial s^2} + B_0\frac{\partial^2}{\partial s\partial t} + C_0\frac{\partial^2}{\partial t^2}\right]u + \dots . \quad (33)$$

Here the terms left out contain no second-order derivatives in u. Hence $P'(u_0)$ is an elliptic linear differential operator, and therefore each step of Newton's method for (32) reduces to solving a linear elliptic differential equation

$$P'(u_n)(u_{n+1}) = P'(u_n)(u_n) - P(u_n) \qquad (n = 0, 1, \dots). \quad (34)$$

When we use the modified Newton's method, we also obtain at each step an elliptic equation

$$P'(u_0)(u_{n+1}) = P'(u_0)(u_n) - P(u_n) \qquad (n = 0, 1, \dots), \quad (35)$$

but now with an unchanged left-hand side.

The second derivative of P, which is needed for studying the convergence of Newton's method, can also be found easily (omitting justifications, as above). In fact, $P''(u)(x, y)$ is clearly a bilinear form in x, y and their first and second derivatives, with coefficients depending on u. Here it is essential to note that there are no terms involving the product of a second derivative in x and a second derivative in y in this form.

Now we turn our attention to the spaces $\mathbf{U}$ and $\mathbf{V}$. We impose the following conditions on these:

1) P maps $\mathbf{U}$ (or some subset Ω of $\mathbf{U}$) into $\mathbf{V}$.

2) P must be twice differentiable, its derivatives being given by the expressions above.

3) For every $u \in \Omega$, the coefficients F_k of the bilinear form $P''(u)(x, y)$, regarded as operators of scalar multiplication, are continuous linear operators from $\mathbf{V}$ into $\mathbf{V}$, and satisfy

$$\|F_k\|_{\mathbf{V}}^{\mathbf{V}} \leqslant M \qquad (u \in \Omega). \quad (36)$$

Similarly, if $x \in \mathbf{U}$, then the functions $\dfrac{\partial^{\alpha+\beta} x}{\partial s^\alpha \partial t^\beta}$ $(\alpha, \beta = 0, 1;\ \alpha+\beta \leqslant 1)$, also regarded as operators of scalar multiplication, are continuous linear operators from $\mathbf{V}$ into $\mathbf{V}$, and

$$\left\|\frac{\partial^{\alpha+\beta} x}{\partial s^\alpha \partial t^\beta}\right\|_{\mathbf{V}}^{\mathbf{V}} \leqslant M_1 \|x\|_{\mathbf{U}}. \quad (37)$$

4) The operator associating with an element $x \in \mathbf{U}$ its derivative $\dfrac{\partial^{\alpha+\beta} x}{\partial s^\alpha \partial t^\beta}$ $(\alpha, \beta = 0, 1, 2;\ \alpha+\beta \leqslant 2)$ is a continuous linear operator from $\mathbf{U}$ into $\mathbf{V}$.

5) The operator $P'(u_0)$ has a continuous inverse $\Gamma_0 = [P'(u_0)]^{-1}$, mapping $\mathbf{V}$ into $\mathbf{U}$.

Spaces $\mathbf{U}$ and $\mathbf{V}$ satisfying these requirements will be called *corresponding* spaces. It is possible to apply Theorem 1.6 to such a pair of spaces, provided the initial approximation u_0 is appropriately chosen. In fact, conditions 3) and 4) ensure that $\|P''(u)\|$ is uniformly bounded, since for a term of the form $F\dfrac{\partial x}{\partial s}\dfrac{\partial^2 y}{\partial t^2}$, for example, we have, by (36) and (37),

$$\left\|F\frac{\partial x}{\partial s}\frac{\partial^2 y}{\partial t^2}\right\|_{\mathbf{V}} \leqslant \|F\|_{\mathbf{V}}^{\mathbf{V}}\left\|\frac{\partial x}{\partial s}\right\|_{\mathbf{V}}^{\mathbf{V}}\left\|\frac{\partial^2}{\partial t^2}\right\|\|y\|_{\mathbf{U}} \leqslant M \cdot M_1\left\|\frac{\partial^2}{\partial t^2}\right\|\|x\|\,\|y\|.$$

We give below some specific examples of pairs of corresponding spaces. However, we shall verify conditions 1)–5) only in the first, and simplest, case.

We consider the equation*

$$P(u) \equiv \pi(u) + R(u) \equiv \Delta u + \frac{\partial}{\partial s}(pH) + \frac{\partial}{\partial t}(qH) = \phi$$
$$\left(\pi(u) = \Delta u, \quad R(u) = \frac{\partial}{\partial s}(pH) + \frac{\partial}{\partial t}(qH)\right) \tag{38}$$

and solve the Neumann problem

$$\frac{\partial u}{\partial n} = 0 \tag{39}$$

on the boundary of the square $Q = [0, \pi; 0, \pi]$.

We shall assume that H is continuous, together with those of its derivatives that we require, and that it is bounded.

Using Green's formula; one can easily check that, if u satisfies the boundary condition, then

$$\int_Q\int [P(u)]\,ds\,dt = 0; \tag{40}$$

hence we must impose on ϕ the condition $\int_Q\int \phi\,ds dt = 0$.

We now consider, for our investigation of (38), the pair of spaces

$$\mathbf{U} = \mathring{\mathbf{W}}_2^{(3)}, \qquad \mathbf{V} = \mathring{\mathbf{W}}_2^{(1)},$$

where $\mathring{\mathbf{W}}_2^{(3)}$ consists of the functions $u \in \mathbf{W}_2^{(3)}$ satisfying the boundary condition (39) and the condition

$$\int_Q\int u(s, t)\,ds\,dt = 0. \tag{41}$$

* An example of this type of equation is the minimal surface equation

$$\frac{\partial}{\partial s}\frac{p}{\sqrt{1+p^2+q^2}} + \frac{\partial}{\partial t}\frac{q}{\sqrt{1+p^2+q^2}} = 0.$$

The same type of equation is also encountered in the problem of plastic torsion.

The space $\mathring{\mathbf{W}}_2^{(1)}$ is composed of the functions $v \in \mathbf{W}_2^{(1)}$ satisfying

$$\iint\limits_Q v(s, t)\, ds\, dt = 0. \tag{42}$$

We take the norms in these spaces to be

$$\|u\| = \left[\iint\limits_Q \sum_{\alpha+\beta=3} \left(\frac{\partial^3 u}{\partial s^\alpha \partial t^\beta}\right)^2 ds\, dt\right]^{1/2},$$

$$\|v\| = \left[\iint\limits_Q \left(\left(\frac{\partial v}{\partial s}\right)^2 + \left(\frac{\partial v}{\partial t}\right)^2\right) ds\, dt\right]^{1/2} \qquad (u \in \mathbf{U},\ v \in \mathbf{V}).$$

We now verify that the pair $\mathbf{U}$, $\mathbf{V}$ satisfies conditions 1)–5) (taking Ω to be a ball of sufficiently large radius).

By the embedding theorems, the second derivatives of the functions $u \in \mathbf{U}$ are summable to any power, so, in particular,

$$\left[\iint\limits_Q \left(\frac{\partial^2 u}{\partial s^2}\right)^4 ds\, dt\right]^{1/4} \leqslant M_2 \|u\| \tag{43}$$

and similarly for the other second derivatives. The function u itself and its first derivatives p and q are bounded.

Consider a typical term of the expression for $\|P(u)\|$:

$$\left\{\iint\limits_Q \left[\frac{\partial}{\partial s}\left(H \frac{\partial^2 u}{\partial s^2}\right)\right]^2 ds\, dt\right\}^{1/2} \leqslant \left\{\iint\limits_Q \left(H \frac{\partial^3 u}{\partial s^3}\right)^2 ds\, dt\right\}^{1/2} +$$

$$+ \left\{\iint\limits_Q \left[\left(\frac{\partial H}{\partial s} + \frac{\partial H}{\partial u} p + \frac{\partial H}{\partial p}\frac{\partial^2 u}{\partial s^2} + \frac{\partial H}{\partial q}\frac{\partial^2 u}{\partial s\, \partial t}\right)\frac{\partial^2 u}{\partial s^2}\right]^2 ds\, dt\right\}^{1/2} \leqslant$$

$$\leqslant \max|H|\, \|u\| + \max\left|\frac{\partial H}{\partial q}\right| \left\{\iint\limits_Q \left(\frac{\partial^2 u}{\partial s\, \partial t}\right)^4 ds\, dt\right\}^{1/4} \left\{\iint\limits_Q \left(\frac{\partial^2 u}{\partial s^2}\right)^4 ds\, dt\right\}^{1/4} + \dots.$$

The second term is finite, by (43); and the omitted terms are obviously also finite. Hence $P(u) \in \mathbf{V}$.

The same arguments show that P is differentiable, and that $P(u_0)$ has the form (33). First we establish that (33) is a weak derivative, and as P' is continuous, P is then differentiable. The same applies to the second derivative.

Now consider one of the coefficients of the bilinear form $P''(u)(x, y)$—this is a function, F, say. Obviously F is a partial derivative of H, so it is continuous and bounded. If $v \in \mathbf{V}$, then we have

$$\left\{\iint\limits_Q \left[\frac{\partial}{\partial s}(Fv)\right]^2 ds\, dt\right\}^{1/2} =$$

$$= \left\{ \iint_Q \left[F\frac{\partial v}{\partial s} + v\left(\frac{\partial F}{\partial s} + \frac{\partial F}{\partial u}\frac{\partial u}{\partial s} + \frac{\partial F}{\partial p}\frac{\partial^2 u}{\partial s^2} + \frac{\partial F}{\partial q}\frac{\partial^2 u}{\partial s\,\partial t} \right) \right]^2 ds\,dt \right\}^{1/2} \leqslant$$

$$\leqslant \max|F|\,\|v\| + \left\{ \iint_Q \left[v\left(\frac{\partial F}{\partial s} + \frac{\partial F}{\partial u}\frac{\partial u}{\partial s} + \frac{\partial F}{\partial p}\frac{\partial^2 u}{\partial s^2} + \frac{\partial F}{\partial q}\frac{\partial^2 u}{\partial s\,\partial t} \right) \right]^2 ds\,dt \right\}^{1/2}. \tag{44}$$

By the embedding theorems, v is summable to any power; in particular,

$$\left\{ \int_Q\int [v(s, t)]^4\, ds\,dt \right\}^{1/4} \leqslant M_3\,\|v\|.$$

We use this to find a bound for one of the terms in the second summand in (44). For instance,

$$\left\{ \iint_Q \left[v\frac{\partial F}{\partial p}\frac{\partial^2 u}{\partial s^2} \right]^2 ds\,dt \right\}^{1/2} \leqslant$$

$$\leqslant \max\left|\frac{\partial F}{\partial p}\right| \left\{ \iint_Q v^4\, ds\,dt \right\}^{1/4} \left\{ \iint_Q \left(\frac{\partial^2 u}{\partial s^2} \right)^4 ds\,dt \right\}^{1/4} \leqslant M_4\,\|v\|.$$

It is clear from what we have said that the operator we have been concerned with is a linear operator from $\mathbf{V}$ into $\mathbf{V}$, and that (36) is true. The second part of condition 3) follows from the fact that a function $u \in \mathbf{U}$ is bounded, as are its derivatives p and q.

Condition 4) is obviously satisfied.

Now we turn to condition 5). With a view to using the results of 2.6, we need only show that the Laplace operator, regarded as an operator from $\mathbf{U}$ into $\mathbf{V}$, has a continuous inverse. To do this, we consider the equation

$$\Delta u = v \qquad (v \in \mathbf{V}). \tag{45}$$

We expand the right-hand side v as a cosine series

$$v(s, t) = \sum_{k,\, m=0}^{\infty} a_{km} \cos ks \cos mt.$$

In view of (42), we have $a_{00} = 0$. Thus it is easy to see that the unique solution of (45) satisfying the boundary condition (39) is given by the series

$$u(s, t) = \sum_{k,\, m=0}^{\infty} \frac{a_{km}}{k^2 + m^2} \cos ks \cos mt.$$

Let us verify that $u \subset \mathbf{U}$. Since the derivatives of v are square summable, we have

$$\sum_{k,\, m=0}^{\infty} (k^2 + m^2) a_{km}^2 = M_4 \iint_Q \left[\left(\frac{\partial v}{\partial s} \right)^2 + \left(\frac{\partial v}{\partial t} \right)^2 \right] ds\,dt = M_4\,\|v\|^2.$$

But now it is obvious that the series defining the third derivatives of u, for example

$$\frac{\partial^3 u(s,t)}{\partial s^2\,\partial t} = \sum_{k,\,m=0}^{\infty} \frac{k^2 m}{k^2+m^2}\,a_{km}\cos ks\cos mt,$$

are convergent in mean. Moreover,

$$\iint_Q \left(\frac{\partial^3 u}{\partial s^2\,\partial t}\right)^2 ds\,dt = M_5 \sum_{k,\,m=0}^{\infty} \frac{k^4 m^2}{(k^2+m^2)^2}\,a_{km}^2 \leqslant M_6 \sum_{k,\,m=0}^{\infty} k^2 a_{km}^2 \leqslant M_7\,\|v\|^2.$$

Therefore $u \in \mathbf{U}$ and $\|u\| \leqslant M_8\|v\|$; hence we have proved the existence of Δ^{-1}.

Hence the convergence theorem for Newton's method is applicable in this case. Notice, in particular, that if instead of (38) we consider an equation with a parameter,

$$\Delta u + \mu\left[\frac{\partial}{\partial s}(pH) + \frac{\partial}{\partial t}(qH)\right] = \phi,$$

then, as the solution exists for $\mu = 0$, it must also exist for all sufficiently small μ, by Theorem 2.6.

We note that when we adopt the technique of Theorem 2.6 here we obtain a Poisson equation at each step of the process.

In conclusion, we present two more pairs of corresponding spaces, without going into any details. If we wish to solve the Dirichlet problem, we can take $\mathbf{U} = \mathring{\mathbf{W}}_p^{(2)}, \mathbf{V} = \mathbf{L}^p$. Here there is no essential difficulty in verifying conditions 1)–4). However, the same cannot be said of condition 5), whose verification is extremely complicated.

Finally, we can take $\mathbf{U}$ to be the space $\mathrm{Lip}^{(2)}\alpha$, consisting of all continuously differentiable functions whose second derivatives satisfy a Lipschitz condition of order α. In this case we must take $\mathbf{V}$ to be Lip α. As in the preceding case, the main difficulty is in verifying condition 5), for which we need to use deep theorems from the theory of partial differential equations.

In addition to the works mentioned above, those of Mysovskikh [2] and Nikolaeva [1] consider further applications of Newton's method.

§ 4. Newton's method in lattice-normed spaces

The idea of Newton's method, which we presented for normed spaces in § 1, carries over to a more general class of spaces, closely related to VLs—namely, the class of lattice-normed spaces. The results of 4.1 and 4.2 are due to L. V. Kantorovich [4], [10], [12], and those of 4.3 to A. N. Baluev [1]. In our presentation we shall follow the monograph Vulikh-I, where the reader can find detailed proofs of the theorems presented here, and also references to the literature.

4.1. A real vector space $\mathbf{X}$ is said to be *lattice-normed* by a VL $\mathbf{Z}$ if to each element $x \in \mathbf{X}$ there is associated a positive element $|x| \in \mathbf{Z}$, called its *abstract norm*, subject to the usual axioms:

1) $|x| > \mathbf{0}$ if $x \neq \mathbf{0}$;

2) $|\lambda x| = |\lambda|\,|x|$;

3) $|x_1 + x_2| \leqslant |x_1| + |x_2|$.

We shall call the VL $\mathbf{Z}$ the norming VL. A normed space is a special case of a lattice-normed space: in this case, the K-space of real numbers serves as the norming VL. More generally, every LCS can be represented as lattice-normed, by means of the K-space $\mathbf{s}(T)$, where the cardinality of T is equal to that of the defining system of semi-norms. Finally, if we take the abstract norm of an element in an arbitrary VL $\mathbf{X}$ to be the modulus of the element, then $\mathbf{X}$ becomes a lattice-normed space, whose norming VL is $\mathbf{X}$ itself.

Let $\mathbf{X}$ be a lattice-normed space, with norming space $\mathbf{Z}$. We say that a sequence $\{x_n\}$ is $(o\,\mathbf{Z})$-convergent to an element $x \in \mathbf{X}$ ($x_n, x \in \mathbf{X}$) if $|x_n - x| \overset{(o\sigma)}{\to} \mathbf{0}$ in $\mathbf{X}$ (we write $x_n \overset{(o\mathbf{Z})}{\to} x$). A sequence $\{x_n\} \subset \mathbf{X}$ is said to be $(o\,\mathbf{Z})$-*fundamental* if there exists a sequence $z_m \downarrow \mathbf{0}$ in $\mathbf{Z}$ such that

$$|x_n - x_m| \leqslant z_m \quad \text{for all } n > m.$$

If every $(o\mathbf{Z})$-fundamental sequence is $(o\mathbf{Z})$-convergent, then $\mathbf{X}$ is said to be $(o\mathbf{Z})$-*complete.* A B-space is $(o\mathbf{R}^1)$-complete; and a K-space $\mathbf{X}$ is $(o\mathbf{X})$-complete.

Important examples of lattice-normed spaces are furnished by the mixed norm spaces $\mathbf{X}[\mathbf{Y}]$ introduced in XI.1.3. We can regard $\mathbf{X}[\mathbf{Y}]$ as normed by the K-space $\mathbf{X}$, if an abstract norm is introduced according to the rule

$$|K|(t) = \|K(\cdot, t)\|_{\mathbf{Y}}.$$

We leave it to the reader to verify that $\mathbf{X}[\mathbf{Y}]$ is $(o\,\mathbf{X})$-complete.

We now present some auxiliary material, which we shall need for our investigation of Newton's method. Suppose, throughout 4.1 and 4.2, that $\mathbf{X}$ and $\mathbf{Y}$ are lattice-normed, by means of K-spaces $\mathbf{Z}$ and $\mathbf{W}$, respectively.

Let $U:\mathbf{X} \to \mathbf{Y}$ be a linear operator. A positive linear operator $U_0 : \mathbf{Z} \to \mathbf{W}$ is called a *modular majorant* of U if

$$|U(x)| \leqslant U_0(|x|)$$

for all $x \in \mathbf{X}$.

We now introduce the concept of the derivative of an operator, which will also be used for studying functional equations.

Let $P:\mathbf{X} \to \mathbf{Y}$ be an arbitrary operator. We take a fixed element $x_0 \in \mathbf{X}$ and assume that there exists a linear operator $U:\mathbf{X} \to \mathbf{Y}$ such that, for every $x \in \mathbf{X}$ and every sequence of numbers $t_k \to 0$ ($t_k \neq 0$), we have

$$(o\,\mathbf{W})\text{-lim}\,\frac{P(x_0 + t_k x) - P(x_0)}{t_k} = U(x).$$

In this situation the linear operator U is called the *derivative* of P at the point x_0. We then write

$$U = P'(x_0).$$

Now we consider the corresponding concept of an integral. Let $x_0, x \in \mathbf{X}$. We denote by $[[x_0, x]]$ the interval joining x_0 and x (see II.3.1; do not confuse this with the order interval!).

Suppose that a function $F(x)$ is defined on $[[x_0, x]]$, having as its values linear operators from $\mathbf{X}$ into $\mathbf{Y}$.

Suppose we are given a sequence $\{\sigma_n\}$ of partitions $0 = t_0 < t_1 < \ldots < t_n = 1$ of $[0, 1]$ such that $\lambda = \max(t_{k+1} - t_k) \to 0$, and fixed points $\tau_k \in [t_k, t_{k+1}]$ $(0 \leqslant k \leqslant n-1)$. Write $\xi_k = (1 - \tau_k)x_0 + \tau_k x$. If the sequence of integral sums

$$\sum_{k=0}^{n-1} F(\xi_k)(t_{k+1} - t_k)$$

has an $(o\,\mathbf{W})$-limit in $\mathbf{Y}$, and this limit is independent of the choice of $\{\sigma_n\}$ and $\{\tau_k\}$, then the limit is called the integral of F and is denoted by $\int_{x_0}^{x} F(x)\,dx$.

In particular, this definition is applicable to the derivative $P'(x)$ of an operator $P:\mathbf{X} \to \mathbf{Y}$. We shall say that the Newton–Leibniz formula holds for P on a set $\Omega \subset \mathbf{X}$ if $P'(x)$ exists for all $x \in \Omega$ and

$$P(x_0 + \Delta x) - P(x_0) = \int_{x_0}^{x_0 + \Delta x} P'(x)\,dx$$

for all Δx such that $x_0 + \Delta x \in \Omega$.

We shall not present conditions for the Newton–Leibniz formula to hold, leaving it to the reader to formulate a sufficient condition, starting from the proof in XVII.1.7 (cf. Vulikh-I, p. 352).

4.2. Now assume that $\mathbf{X}$ is $(o\,\mathbf{Z})$-complete. Consider the equation

$$P(x) = \mathbf{0}, \tag{1}$$

where P is an operator from $\mathbf{X}$ into $\mathbf{Y}$. Fixing $x_0 \in \mathbf{X}$, we can solve (1) by using both *Newton's method*

$$x'_{n+1} = x'_n - [P'(x_n)]^{-1}(P(x'_n)) \qquad (n = 0, 1, \ldots; \; x'_0 = x_0)$$

and the *modified Newton's method*

$$x_{n+1} = x_n - [P'(x_0)]^{-1}(P(x_n)) \qquad (n = 0, 1, \ldots). \tag{2}$$

For brevity, we shall restrict ourselves to the modified Newton's method. Set $\Gamma_0 = [P'(x_0)]^{-1}$ (here we require, for the present, only that Γ_0 be a linear operator defined on the whole of $\mathbf{Y}$). In addition to (1) we consider the equation

$$Q(z) = \mathbf{0}, \tag{3}$$

where Q is an operator from $\mathbf{Z}$ into $\mathbf{W}$, which will play the role of the *majorizing* equation. Suppose we are given an order interval $[z_0, z']$ in $\mathbf{Z}$; and let

$$D = \{x \in \mathbf{X}: |x - x_0| \leqslant z' - z_0\}.$$

THEOREM 1. *Suppose the operators P and Q satisfy the following conditions:*

1) *the Newton–Leibniz formula holds for P on the set D;*

2) *the Newton–Leibniz formula holds for Q on the interval* $[z_0, z']$;

3) *if* $\{z_n\}_{n=1}^{\infty} \subset [z_0, z']$ *and* $z_n \overset{(o)}{\to} z$, *then* $Q(z_n) \to Q(z)$;

4) *there exist inverse operators* $\Gamma_0 = [P'(x_0)]^{-1}$, *defined on the whole of* $\mathbf{Y}$, *and* $\Delta_0 = [Q'(z_0)]^{-1}$, *defined on the whole of* $\mathbf{W}$, *such that* $\Delta_0 \in \tilde{\mathbf{L}}_{n\sigma}(\mathbf{W}, \mathbf{Z})$ *and* Δ_0 *is a modular majorant for* Γ_0;

5) $|\Gamma_0 P(x_0)| \leqslant -\Delta_0 Q(z_0)$;

6) *if* $x \in D$ *and* $z \in [z_0, z']$ *satisfy* $|x - x_0| \leqslant z - z_0$, *then the operator* $I - \Delta_0 Q'(z)$ *is a modular majorant for* $I - \Gamma_0 P'(x)$;

7) *equation* (3) *has a solution* $\hat{z}$ *in the interval* $[z_0, z']$.

Then there exists a solution $\hat{x}$ *of* (1), *satisfying* $|\hat{x} - x_0| \leqslant \hat{z} - z_0$, *and the modified Newton's method, with initial point* x_0, *is* ($o\mathbf{Z}$)-*convergent to* $\hat{x}$.

If, under the conditions of Theorem 1, equation (3) has a unique solution in $[z_0, z']$, then (1) will also have a unique solution in D.

A proof of this may be found in the book by Vulikh-I (Theorems XII.5.1 and XII.5.2).

4.3. If a functional equation is solved in an ordered space, then it is natural to look for two monotonic sequences, one converging to the solution from either side. The results to be presented here can be regarded as an abstract generalization of *S. A. Chaplygin's method* in the theory of ordinary differential equations.

Consider the equation

$$U(x) = \mathbf{0}, \tag{4}$$

where U is an operator from a K-space $\mathbf{X}$ into a VL $\mathbf{Y}$.

LEMMA 1. *Suppose the following conditions are satisfied:*

1) *there exist* $x_0, x' \in \mathbf{X}$ *such that*

$$x_0 \leqslant x_0' \quad \textit{and} \quad U(x_0) \leqslant \mathbf{0} \leqslant U(x_0');$$

2) *there exists a linear operator T mapping* $\mathbf{X}$ *onto the whole of* $\mathbf{Y}$ *such that, if* $x_0 \leqslant x \leqslant x_0'$, *then*

$$U(x_0') + T(x - x_0') \leqslant U(x) \leqslant U(x_0) + T(x - x_0); \tag{5}$$

3) *T has a positive inverse* T^{-1}.

Then:

a) *the elements* x_1 *and* x_1', *defined by*

$$x_1 = x_0 - T^{-1}U(x_0), \qquad x_1' = x_0' - T^{-1}U(x_0'),$$

satisfy the inequalities

$$x_0 \leqslant x_1 \leqslant x_1' \leqslant x_0', \qquad U(x_1) \leqslant 0 \leqslant U(x_1'); \tag{6}$$

b) *if equation* (4) *has a solution* $\hat{x}$ *in the order interval* $[x_0, x_0']$, *then* $x_1 \leqslant \hat{x} \leqslant x_1'$.

Proof. Since $U(x_0) \leqslant \mathbf{0}$ and $T^{-1} \geqslant \mathbf{0}$, we have $T^{-1}U(x_0) \leqslant \mathbf{0}$, and so $x_1 \geqslant x_0$. Similarly, $x_1' \leqslant x_0'$.

Further,

$$x_0' - x_1 = x_0' - x_0 + T^{-1}U(x_0) = T^{-1}T(x_0' - x) + T^{-1}U(x_0) = T^{-1}[T(x_0' - x) + U(x_0)].$$

By the right-hand inequality of (5) and condition 1),

$$T(x_0' - x_0) + U(x_0) \geqslant U(x_0') \geqslant \mathbf{0},$$

so that $x_0' - x_1 \geqslant \mathbf{0}$. Therefore $x_0 \leqslant x_1 \leqslant x_0'$, and (5) is true with $x = x_1$. Hence

$$U(x_1) \leqslant U(x_0) + T(x_1 - x_0) = U(x_0) - TT^{-1}U(x_0) = \mathbf{0}$$

and

$$U(x_0') + T(x_1 - x_0') \leqslant U(x_1) \leqslant \mathbf{0}. \tag{7}$$

Further

$$\begin{aligned} x_1' - x_1 &= x_0' - T^{-1}U(x_0') - x_1 = T^{-1}T(x_0' - x_1) - T^{-1}U(x_0') \\ &= -T^{-1}[U(x_0') + T(x_1 - x_0')], \end{aligned}$$

and so, by (7), $x_1' - x_1 \geqslant \mathbf{0}$: that is, $x_1 \leqslant x_1'$.

Substituting $x = x_1'$ in the left-hand inequality of (5), we see that

$$U(x_1') \geqslant U(x_0') + T(x_1' - x_0') = U(x_0') - TT^{-1}U(x_0') = \mathbf{0}.$$

Hence inequality (6) is proved. It remains to verify that statement b) is true.

Suppose $x_0 \leqslant \tilde{x} \leqslant x'_0$, and $U(\tilde{x}) = \mathbf{0}$. Then, by (5),

$$\tilde{x} - x_1 = \tilde{x} - x_0 + T^{-1}U(x_0) = T^{-1}[U(x_0) + T(\tilde{x} - x_0)] \geqslant T^{-1}U(\tilde{x}) = \mathbf{0},$$

so that $\tilde{x} \geqslant x_1$. Similarly, $\tilde{x} \leqslant x'_1$. Thus the lemma is proved.

THEOREM 2. *Suppose the following conditions are satisfied:*

1) $x_0 \leqslant x'_0$ *and* $Ux_0 \leqslant \mathbf{0} \leqslant Ux'_0$;

2) *if* $\{\tilde{x}_n\}_{n=1}^{\infty} \subset [x_0, x'_0]$ *and* $\tilde{x}_n \overset{(o)}{\to} x$, *then* $U(\tilde{x}_n) \overset{(o\sigma)}{\to} U(x)$;

3) *the Newton–Leibniz formula holds for* U *in the interval* $[x_0, x'_0]$;

4) *there exists a linear operator* T, *mapping* $\mathbf{X}$ *onto the whole of* $\mathbf{Y}$, *such that, for all* $x \in [x_0, x'_0]$, *we have* $[U'(x)](h) \leqslant T(h)$ *for all* $h \in \mathbf{X}_+$;

5) T *has a positive* (o)*-continuous inverse* T^{-1}.

Then the set of solutions of (4) *belonging to* $[x_0, x'_0]$ *is non-empty and contains a smallest solution* x^* *and a largest one* x^{**}. *Furthermore, if elements* x_n *and* x'_n *are defined recursively by the formulae*

$$\begin{aligned} x_n &= x_{n-1} - T^{-1}U(x_{n-1}), \\ x'_n &= x'_{n-1} - T^{-1}U(x'_{n-1}) \qquad (n \in \mathbb{N}), \end{aligned} \tag{8}$$

then $x_n \uparrow x^*$ *and* $x'_n \downarrow x^{**}$.

Proof. We show that U and T satisfy the conditions of Lemma 1 on each interval $[x_n, x'_n]$.

Let us show that, for all $\bar{x}$, x and $\bar{x}'$ such that

$$x_0 \leqslant \bar{x} \leqslant x \leqslant \bar{x}' \leqslant x'_0,$$

the following analogue of (5) is true:

$$U(\bar{x}') + T(x - \bar{x}') \leqslant U(x) \leqslant U(\bar{x}) + T(x - \bar{x}). \tag{9}$$

In fact, using condition 3) and the definition of the integral, we obtain

$$U(x) - U(\bar{x}) = \int_{\bar{x}}^{x} U'(x)\,dx \leqslant \int_{\bar{x}}^{x} T\,dx = T(x - \bar{x}),$$

$$U(\bar{x}') - U(x) = \int_{x}^{\bar{x}'} U'(x)\,dx \leqslant \int_{x}^{\bar{x}'} T\,dx = T(\bar{x}' - x),$$

from which (9) follows.

Using (9) and condition 1), we see by induction that the conditions of Lemma 1 are satisfied for all x_n and x'_n, so that

$$x_0 \leqslant x_1 \leqslant \ldots \leqslant x_n \leqslant \ldots \leqslant x'_n \leqslant \ldots \leqslant x'_1 \leqslant x'_0.$$

Since $\mathbf{X}$ is a K-space, both $x^* = \sup x_n$ and $x^{**} = \inf x'_n$ exist, and $x_n \uparrow x^*$, $x'_n \downarrow x^{**}$. Taking (o)-limits in (8), and using 2) and 5), we obtain

$$T^{-1}U(x^*) = T^{-1}U(x^{**}) = \mathbf{0},$$

so that

$$U(x^*) = U(x^{**}) = \mathbf{0},$$

that is, x^* and x^{**} are solutions of (4). Moreover, by Lemma 1, any solution $\tilde{x}$ of (4) contained in $[x_0, x'_0]$ satisfies $x_n \leqslant \tilde{x} \leqslant x'_n$ for all n. Thus $x^* \leqslant \tilde{x}, \leqslant x^{**}$. This completes the proof of the theorem.

MONOGRAPHS ON FUNCTIONAL ANALYSIS AND RELATED TOPICS

AKHIEZER, N. I.; GLAZMAN, I. M. *Theory of linear operators in Hilbert space*. Transl. by M. Nestell. Ungar, N.Y., 1961.

AKILOV, G. P.; MAKAROV, B. M.; KHAVIN, V. P. *An elementary introduction to the theory of the integral* (Russian). Izdat. LGU, 1969.

ALEKSANDROFF, P. S. *Einführung in die Mengenlehre und die Theorie der reellen Funktionen*. VEB deutscher Verlag der Wiss., Berlin, 1956.

ANTONOVSKII, M. YA.; BOLTYANSKII, V. G.; SARYMSAKOV, T. A. *Topological Boolean algebras* (Russian). Izdat. Akad. Nauk UzSSR, 1963.

BAKEL'MAN, I. YA. *Geometric methods of solution of elliptic equations* (Russian). "Nauka", Moscow, 1965.

BANACH, S. *Théorie des opérations linéaires*. Warsaw, 1932.

BESOV, O. V.; IL'IN, V. P.; NIKOL'SKII, S. M. *Integral representations of functions and imbedding theorems*. Transl. ed. by M. H. Taibleson. N.Y., 1978.

BIRKHOFF, G. *Lattice theory*. Amer. Math. Soc. Colloquium Publications, Vol. 25, Providence, 1967.

BIRMAN, M. SH. *et al. Functional analysis* (Russian). In series *Spravochnaya matematicheskaya biblioteka*. "Nauka", Moscow, 1972.

BOURBAKI, N. *Elements of mathematics*. Hermann, Paris.

I. *Theory of sets*.
II. *General topology*.
III. *Topological vector spaces*.
IV. *Integration. Measures, integration of measures*.
V. *Integration. Vector integration, Haar measure, convolution and representation*.
VI. *Spectral theory*.

COLLATZ, L. *Functional analysis and numerical mathematics*. Transl. from German by H. Oser. Academic Press, N.Y. and London, 1966.

DAY, M. M. *Normed linear spaces*. Springer-Verlag, Berlin, 1958; 2nd ed. 1962.

DEM'YANOV, V. F.; MALOZEMOV, V. N. *Introduction to minimax* (Russian). "Nauka", Moscow, 1972.

DEM'YANOV, V. F.; RUBINOV, A. M. *Approximate methods of solution of extremal problems* (Russian). Izdat. LGU, Leningrad, 1968.

DIESTEL, J. *Geometry of Banach spaces. Selected topics*. Springer Lecture Notes in Math., 485, Berlin, 1975.

DIEUDONNÉ, J. *Foundations of modern analysis*. Academic Press, N.Y., 1960.

DINCULEANU, N. *Vector measures*. Pergamon Press, Oxford, 1967.

DUNFORD, N.; SCHWARTZ, J. T. *Linear operators*. Interscience, N.Y., 1958.

I. *General theory*.
II. *Spectral theory*.
III. *Spectral operators*.

EDWARDS, R. E. *Functional analysis: theory and applications*. Holt, Rinehart and Winston, N.Y., 1965.

FADDEEV, D. K.; FADDEEVA, V. N. *Computational methods of linear algebra*. Transl. by R. C. Williams. W. H. Freeman, San Francisco, 1963.

FIKHTENGOL'TS, G. M. *A course of differential and integral calculus* (Russian). 3 vols. "Nauka", Moscow, 1966.

FILIPPOV, A. F.; RYABEN'KII, V. S. *On the stability of difference equations* (Russian). Gostekhizdat, Moscow, 1956.

FREMLIN, D. H. *Topological Riesz spaces and measure theory*. Cambridge Univ. Press, 1974.

GAVURIN, M. K. *Lectures on methods of calculus* (Russian). "Nauka", Moscow, 1971.

GEL'FAND, I. M.; RAIKOV, D. A.; SHILOV, G. E. *Commutative normed rings*. (Transl.). Chelsea, Bronx, 1964.

GEL'FAND, I. M.; SHILOV, G. E. *Generalized functions*. Academic Press, N.Y. and London.

Vol. 1. *Properties and operations*. (Transl. by E. Salatan, 1964.)
Vol. 2. *Spaces of fundamental and generalized functions*. (Transl. by M. D. Friedman, A. Feinstein, C. P. Peltzer, 1968.)
Vol. 3. *Theory of differential equations*. (Transl. by M. E. Mayer, 1967.)

GEL'FAND, I. M.; VILENKIN, N. YA. *Generalized functions*. Academic Press, N.Y. and London.

Vol. 4 (of prec. ref.). *Applications of harmonic analysis*. (Transl. by A. Feinstein, 1964.)

GLAZMAN, I. M.; LJUBIC, JU. I. *Finite-dimensional linear analysis: a systematic presentation in problem form.* Transl. and ed. by G. P. Barker, G. Kuerti. MIT Press, Cambridge, Mass., 1974.
GOL'SHTEIN, E. G. *The theory of duality in mathematical programming and its applications* (Russian). "Nauka", Moscow, 1971.
HALMOS, P. R. *Measure theory.* Van Nostrand Reinhold, N.Y., 1950.
HARDY, G. H.; LITTLEWOOD, J. E.; PÓLYA, G. *Inequalities.* Cambridge Univ. Press, 1934; 2nd ed., 1952.
HAUSDORFF, F. *Set theory.* Transl. by J. R. Aumann *et al.* Chelsea, N.Y., 1957.
HILBERT, D. *Grundzüge einer allgemeine Theorie der Integralgleichungen.* Leipzig, 1912. Reprint Chelsea, 1953.
HILLE, E.; PHILLIPS, R. S. *Functional analysis and semigroups.* Amer. Math. Soc. Colloquium Publications. vol. 31, N.Y., 1948. Revised ed., Providence, 1957.
HOFFMAN, K. *Banach spaces of analytic functions.* Prentice-Hall, Englewood Cliffs, 1962.
HÖRMANDER, L. *Linear partial differential operators.* Springer-Verlag, Berlin, 1963.
IOFFE, A. D.; TIKHOMIROV, V. M. *The theory of extremal problems* (Russian). "Nauka", Moscow, 1974.
KADETS, M. I. *The geometry of normed spaces* (Russian). In *Matematicheskii Analiz, 1975.* Itogi Nauki, VINITI AN SSSR, "Nauka", Moscow, 1975.
KANTOROVICH, L. V.; KRYLOV, V. I. *Approximate methods of higher analysis.* Transl. from 3rd Russian ed. by C. D. Benster. Noordhoff, Groningen, 1958.
KANTOROVICH, L. V.; VULIKH, B. Z.; PINSKER, A. G. *Functional analysis in partially ordered spaces.* Gostekhizdat, Moscow–Leningrad, 1950.
KARLIN, S. *Mathematical methods and theory in games, programming and economics,* 2 vols. Addison-Wesley, 1959.
KELLEY, J. L. *General topology.* Van Nostrand, N.Y., 1955.
KHAVIN, V. P. *Spaces of analytic functions* (Russian). In *Matematicheskii Analiz, 1964.* Itogi Nauki, VINITI AN SSSR, Moscow, 1966.
KOLMOGOROV, A. N.; FOMIN, S. V. *Introductory real analysis.* Transl. and ed. by R. A. Silverman. Dover, N.Y., 1975.
KOROTKOV, V. B. *Integral operators* (Russian). Izdat. NGU, Novosibirsk, 1977.
KRASNOSEL'SKII, M. A. I. *Topological methods in the theory of nonlinear integral equations.* Transl. by A. H. Armstrong. Pergamon Press, Oxford, 1964.
II. *Positive solutions of operator equations.* Transl. by R. E. Flaherty. Noordhoff, Groningen, 1964.
KRASNOSEL'SKII, M. A. *et al. Approximate solution of operator equations* Transl. by D. Louvish. Wolters-Noordhoff, Groningen, 1972.
KRASNOSEL'SKII, M. A. *et al. Integral operators on spaces of summable functions* (Russian). "Nauka", Moscow, 1966.
KRASNOSEL'SKII, M. A.; RUTICKII, Y. B. *Convex functions and Orlicz spaces.* Transl. by L. F. Boron. Noordhoff, Groningen, 1961.
KRYLOV, N. M. *Les méthodes de solution approchée des problèmes de la physique mathématique.* Mém. des sci. math., 49, Paris, 1931.
KUTATELADZE, S. S.; RUBINOV, A. M. *Minkowski duality and its applications* (Russian). "Nauka", Novosibirsk, 1976.
LADYZHENSKAYA, O. A. *Boundary value problems of mathematical physics and related aspects of function theory.* (Transl.) Ed. by V. P. Il'in. Sem. in Math., V. A. Steklov Inst., Vol. 5, Consultants Bureau, N. Y., 1969.
LINDENSTRAUSS, J.; TZAFRIRI, L. *Classical Banach spaces.* Springer Lecture Notes in Math., 338, Berlin, 1973.
LUSTERNIK, L. A.; SOBOLEV, V. I. *Elements of functional analysis.* Transl. by A. E. Labarre, Jr., H. Izbicki, H. W. Crowley. Gordon and Breach, N. Y., 1961.
LUXEMBURG, W. A. J.; ZAANEN, A. C. *Riesz spaces,* Vol. 1. North-Holland, Amsterdam, 1971.
MIKHLIN, S. G. I. *The numerical performance of variational methods.* Transl. by R. S. Anderssen. Noordhoff, Groningen, 1971.
II. *Variational methods in mathematical physics.* Transl. by T. Boddington. Pergamon Press, Oxford, 1964.
NAIMARK, M. A. *Normed rings.* Transl. by L. F. Boron. 3rd American ed., entitled *Normed algebras.* Wolters-Noordhoff, Groningen, 1972.
NAKANO, H. *Modulared semi-ordered linear spaces.* Maruzen. Tokyo, 1950. 2nd ed., entitled *Linear lattices.* Savoyard Books, Detroit, 1966.
NATANSON, I. P. I. *Constructive function theory.* Transl. by A. N. Obolensky. Ungar, N.Y., 1964.
II. *Theory of functions of a real variable.* 2 vols. Transl. by L. F. Boron. Ungar, N.Y., 1955.
NIKAIDO, H. *Convex structures and economic theory.* Academic Press, N.Y., 1968.
NIKOL'SKII, N. K. *Selected problems in weight approximation and spectral analysis* (Russian). Trudy Mat. Inst. im. V. A. Steklova, 120. "Nauka", Leningrad, 1974.
PETROVSKII, I. G. *Lectures on the theory of integral equations* (Russian). "Nauka", Moscow, 1965.
PHELPS, R. R. *Lectures on Choquet's Theorem.* Van Nostrand, Princeton, 1966.
PLESNER, A. I. *Spectral theory of linear operators* (Russian). "Nauka", Moscow, 1965.

PSHENICHNII, B. N. *Necessary conditions for an extremum.* Transl. by K. Makowski. M. Dekker, N.Y., 1971.
RIESZ, F.; SZ-NAGY, B. *Functional analysis.* Transl. from 2nd French ed. by L. F. Boron. Ungar, N.Y., 1955.
ROBERTSON, A. P.; ROBERTSON, W. *Topological vector spaces.* Cambridge Univ. Press, 1973.
ROCKAFELLAR, R. T. *Convex analysis.* Princeton Math. Ser. 28, Princeton, 1970.
RUBINSHTEIN, G. SH. *Finite-dimensional optimization models* (Russian). "Nauka", Novosibirsk, 1970.
RUDIN, W. *Functional analysis.* McGraw-Hill, N.Y., 1973.
SCHAEFER, H. H. I. *Topological vector spaces.* Macmillan, N.Y., 1966. Reprinted, Springer-Verlag, N.Y., 1971.
II. *Banach lattices and positive operators.* Springer-Verlag, Berlin, 1974.
SCHWARTZ, L. *Théorie des distributions.* Actualités sci. et ind., 1091, 1122, Paris, 1950, 1951, Nouv. éd., 1966.
SHILOV, G. E. *Mathematical analysis.* Transl. ed. by D. A. R. Wallace. Pergamon Press, Oxford, 1965, 1968. Revised English ed. transl. and ed. by R. A. Silverman. Cambridge, Mass., 1973.
SHILOV, G. E.; GUREVICH, B. L. *Integral, measure and derivative: a unified approach.* Transl. by R. A. Silverman. Prentice-Hall, Englewood Cliffs, 1966.
SINGER, I. *Bases in Banach spaces.* Vol. 1. Springer-Verlag, Berlin–N.Y., 1970.
SMIRNOV, V. I. *A course of higher mathematics.* Transl. by D. E. Brown. 5 vols. Pergamon Press, Oxford, 1964, 1968.
SOBOLEV, S. L. I. *Applications of functional analysis in mathematical physics.* Transl. by F. E. Browder. Amer. Math. Soc. Translations of math. monographs, vol. 7. Providence, 1963.
II. *Introduction to the theory of cubic formulae* (Russian). "Nauka", Moscow, 1974.
SZEGÖ, G. *Orthogonal polynomials.* Amer. Math. Soc. Colloquium Publications, vol. 23, N.Y., 1939. Revised ed., N.Y. 1959.
TIKHOMIROV, V. M. *Some questions in the theory of approximation* (Russian). Izdat. MGU, Moscow, 1976.
VAINIKKO, G. M. *Compact approximation of operators and approximate solution of equations* (Russian). Izd. Tartuskogo Univ., Tartu, 1970.
VLADIMIROV, D. A. *Boolean algebras* (Russian). "Nauka", Moscow, 1967. (German transl.: *Boolesche Algebren*, transl. by G. Eisenreich. Akademie-Verlag, 1972.)
VOROBEV, YU. V. *Method of moments in applied mathematics.* Transl. by B. Seckler. Gordon and Breach, N.Y., 1965.
VULIKH, B. Z. I. *Introduction to the theory of partially ordered spaces.* Transl. by L. F. Boron. Noordhoff, Groningen, 1967.
II. *Introduction to functional analysis for scientists and technologists.* Transl. by I. N. Sneddon. Pergamon Press/Addison-Wesley, 1963.
III. *A short course on the theory of functions of a real variable* (Russian). "Nauka", Moscow, 1973.
IV. *An introduction to the theory of cones in normed spaces* (Russian). Izd. KGU. Kalinin, 1977.
YOSIDA, K. *Functional analysis.* Springer-Verlag, 1965.
ZAANEN, A. C. I. *Linear analysis.* North-Holland, Amsterdam, 1953.
II. *Integration.* North-Holland, Amsterdam, 1967.
ZYGMUND, A. *Trigonometric series.* 2nd ed., Cambridge Univ. Press, 1959.

REFERENCES

Titles marked with an asterisk are of papers in Russian.

ABRAMOVICH, YU. A.

[1] Theorems on normed lattices.* *Vestnik LGU*, No. 13 (1971), 5–11.

AGAKHANOV, S. A.; NATANSON, G. I.

[1] The Lebesgue function of Fourier–Jacobi sums.* *Vestnik LGU*, No. 1 (1968), 11–23.

AKILOV, G. P.

[1] On the extension of linear operators.* *Dokl. Akad. Nauk SSSR*, **57**, No. 7 (1947), 643–646.

[2] Necessary conditions for the extension of linear operators.* *Dokl. Akad. Nauk SSSR*, **59**, No. 3 (1948), 417–418.

[3] On the application of a method of solution for non-linear differential equations to the study of systems of differential equations.* *Dokl. Akad. Nauk SSSR*, **68**, No. 4 (1949), 645.

ALSYNBAEV, K.; IMOMNAZAROV, B.; RUBINSHTEIN, G. SH.

[1] The problem of translocation of mass on a compactum with asymmetric metric.* In *Optimizatsiya*, 17 (34), Novosibirsk, 1975, 94–107.

ANDO, T.

[1] Linear functionals on Orlicz spaces. *Nieuw Arch. Wiskunde*, **8**, No. 1 (1960), 4–16.

ARONSZAIN, N.; SMITH, K.

[1] Invariant subspaces of completely continuous operators. *Ann. Math.*, **60** (1954), 345–350.

BALUEV, A. N.

[1] On S. A. Chaplygin's method.* *Vestnik LGU*, No. 13 (1956), 27–42.

BANACH, S.

[1] Sur les opérations dans les ensembles abstraits et leur applications aux équations intégrales. *Fund. Math.*, **3** (1922), 131–181.

[2] Sur les fonctionelles linéaires II. *Studia Math.*, **1** (1929), 223–240.

BANACH, S.; STEINHAUS, H.

[1] Sur le principe de la condensation de singularités. *Fund. Math.*, **9** (1927), 50–61.

BERKOLAIKO, M. Z.; RUTITSKII, YA. B.

[1] On operators in generalized Hölder spaces.* *Sibirsk. Mat. Zh.*, **12**, No. 5 (1971), 1015–1025.

BERMAN, D. L.

[1] On a class of linear operators.* *Dokl. Akad. Nauk SSSR*, **85**, No. 1 (1952), 13–16.

BIRMAN, M. SH.

[1] Some bounds for the method of steepest descent.* *Uspekhi Mat. Nauk*, **5**, No. 3 (1950), 152–155.

[2] On the calculation of eigenvalues by the method of steepest descent.* *Zap. Leningradsk. Gornogo Inst.*, **27**, No. 1 (1952), 209–216.

BIRMAN, M. SH.; SOLOMYAK, M. Z.

[1] Bounds for the singular values of integral operators.* *Uspekhi Mat. Nauk*, **32**, No. 1 (1977), 17–84.

BROUWER, L. E. F.

[1] Über Abbindung von Mannigfaltigkeiten. *Math. Ann.*, **71** (1911), 97–115.

BUKHVALOV, A. V.

[1] On locally convex spaces generated by weakly compact sets.* *Vestnik LGU*, No. 7 (1973), 11–17.

[2] On mixed norm spaces.* *Vestnik LGU*, No. 19 (1973), 5–12.

[3] On integral representations of linear operators.* *Zap. Nauchn. Seminarov Leningradsk. Otdel. Mat. Inst. Akad. Nauk SSSR*, **47** (1974), 5–14.

[4] Integral operators and representations of completely linear functionals on mixed norm spaces.* *Sibirsk. Mat. Zh.*, **16**, No. 3 (1975), 483–493.

[5] On analytic representations of operators with abstract norm.* *Izvestiya VUZ, Matem.*, No. 11 (1975), 21–32.

BUKHVALOV, A. V.; LOZANOVSKII, G. YA.

[1] On sets closed in measure in spaces of measurable functions.* *Trudy Mosk. Mat. Ob-va.*, **34** (1976), 128–149.

CURRY, H. B.

[1] The method of steepest descent for nonlinear minimization problems. *Quart. J. Appl. Math.* **2**, No. 3 (1944), 268–271.

DAUGAVET, I. K.

[1] An application of the general theory of approximation methods to the investigation of the convergence of

Galerkin's method for certain boundary-value problems of mathematical physics.* *Dokl. Akad. Nauk SSSR*, **98**, No. 6 (1954).
[2] On the method of moments for ordinary differential equations.* *Sibirsk. Mat. Zh.*, **6**, No. 1 (1965), 70–85.
DUBOVITSKII, A. YA.; MILYUTIN, A. A.
[1] Extremum problems in the presence of constraints.* *Zh. Vychisl. Mat. i Mat. Fiz.*, **5**, No. 3 (1965), 395–453.
[2] Necessary conditions for a weak extremum in problems of optimal control with mixed constraints in the form of inequalities.* *Zh. Vychisl. Mat. i Mat. Fiz.*, **8**, No. 4 (1968), 725–779.
DUNFORD, N.; PETTIS, B. J.
[1] Linear operators on summable functions. *Trans. Amer. Math. Soc.*, **47**, No. 3 (1940), 323–392.
EKHLAKOV, N. P.
[1] The existence of optimal transports of Lebesgue measure with densities.* *In Optimizatsiya*, **15** (32), Novosibirsk, 1974, 90–114.
ENFLO, P.
[1] A counterexample to the approximation problem in Banach spaces. *Acta Math.*, **130**, No. 3–4 (1973), 309–317.
ERMOL'EV, YU. M.; SHOR, N. Z.
[1] On the minimization of non-differentiable functions.* *Kibernetika*, **1** (1967), 101, 102.
FRECHET, M.
[1] Sur quelques points du calcul fonctionel. *Rend. Mat. di Palermo*, **22** (1906), 1–74.
[2] Sur les ensembles des fonctions et les opérations linéaires. *C. R. Acad. Sci. (Paris)*, **144** (1907), 1414–1416.
[3] Sur les fonctionelles continues. *Ann. Ecole Norm.*, **27**, No. 3 (1910), 193–216.
FREUDENTHAL, H.
[1] Teilweise geordnete Moduln. *Nederl. Akad. Wetensch. Proc.*, **39** (1936), 641–651.
FRIDMAN, V. M.
[1] On the convergence of methods of the steepest descent type, *Uspekhi Mat. Nauk*, **17**, No. 3 (1962), 201–204.
GATEAUX, R.
[1] Sur les fonctionelles continues et les fonctionelles analytiques. *C. R. Acad. Sci. (Paris)*, **157** (1913), 325–327.
GAVURIN, M. K.
[1] Über die Stieltjessche Integration abstrakter Funktionen. *Fund. Math.*, **27** (1936), 254–268.
[2] Analytical methods of studying nonlinear functional mappings.* *Uch. zap. LGU, ser. mat.*, **19** (1950), 59–154.
[3] On bounds for the eigenvalues and eigenvectors of a perturbed operator.* *Dokl. Akad. Nauk SSSR*, **96**, No. 6 (1954), 1093–1095.
[4] Non-linear functional equations and continuous analogues of iterative methods.* *Izv. VUZ, Matem.*, No. 5 (1958), 18–31.
GEL'FAND, I. M.
[1] Abstrakte Funktionen und lineare Operatoren. *Mat. Sb.*, **4** (46) (1938), 235–286.
GONCHAROV, V. L.
[1] *The theory of interpolation and approximation of functions.** 2nd ed. Gostekhizdat, Moscow–Leningrad, 1954.
GRIBANOV, YU. I.
[1] On the measurability of a function.* *Izv. VUZ, Matem.*, No. 3 (1970), 22–26.
[2] On the measurability of kernels of integral operators.* *Izv. VUZ, Matem.*, No. 7 (1972), 31–34.
GROTHENDIECK, A.
[1] Sur les applications linéaires faiblement compact d'espaces du type C(K). *Canad. J. Math.*, **5** (1953), 129–173.
[2] Produits tensoriels topologiques et espaces nucléaires. *Mem. Amer. Math. Soc.*, **16** (1955).
GYUNTER, N. M.
[1] *Potential theory.** Gostekhizdat, Moscow–Leningrad, 1952.
HELLINGER, E.; TOEPLITZ, O.
[1] Grundlagen für eine Theorie der unendlichen Matrizen. *Math. Ann.*, **69** (1910), 289–330.
[2] Integralgleichungen und Gleichungen mit unendlich vielen Unbekannten. *Enciclopädie der matematische Wissenschaften*, Bd II, C 13 (1927), 1335–1616.
HILDEBRANDT, T. H.
[1] Über vollstetige lineare Transformationen. *Acta Math.*, **51** (1928), 311–318.
JAMES, R.
[1] A non-reflexive Banach space isometric with its second conjugate space. *Proc. Nat. Acad. USA*, **37**, No. 3 (1951), 174–177.
KADETS, M. I.; MITYAGIN, B. S.
[1] Complemented subspaces in Banach spaces.* *Uspekhi Mat. Nauk*, **28**, No. 6 (1973), 77–94.

KAKUTANI, S.
[1] Some characterizations of Euclidean space. *Jap. J. Math.*, **16**, No. 2 (1939), 93–98.
[2] A generalization of Brouwer's fixed point theorem. *Duke Math. J.*, **8**, No. 3 (1941).
KALANDIYA, A. I.
[1] On a direct method of solution of an equation in the theory of the aerofoil and its application to the theory of elasticity.* *Mat. Sb.*, **42** (1957).
KANTOROVICH, L. V.
[1] On a new method of approximate solution of partial differential equations.* *Dokl. Akad. Nauk SSSR*, **2**, No. 8–9 (1934), 532–536.
[2] On partially ordered linear spaces and their applications in the theory of linear operations.* *Dokl. Akad. Nauk SSSR*, **4** (1935), 11–14.
[3] Partially ordered linear spaces.* *Mat. Sb.*, **2** (44) (1937), 121–168.
[4] On functional equations.* *Uch. zap. LGU*, **3** (17) (1937), 24–50.
[4a] Linear operations on partially ordered spaces.* *Mat. Sb.*, **7** (1940), 209–284.
[5] *Mathematical methods of organization and planning of production.** Izd. LGU, Leningrad, 1939.
[6] On the translocation of masses.* *Dokl. Akad. Nauk, SSSR*, **37**, No. 7–8 (1942), 227–229. Transl.: *Management Sci.*, **5** (1958), 1–4.
[7] On an effective method of solution of extremal problems for quadratic functionals.* *Dokl. Akad. Nauk SSSR*, **48**, No. 7 (1945), 483–487.
[8] On a problem of Monge.* *Uspekhi Mat. Nauk*, **3**, No. 2 (1948), 225–226.
[9] Functional analysis and applied mathematics.* *Uspekhi Mat. Nauk*, **3**, No. 6 (1948), 89–185. Transl. by C. D. Benster, U.S. Department of Commerce, Nat. Bureau of Standards, L.A., 1952.
[10] The principle of majorants and Newton's method.* *Dokl. Akad. Nauk SSSR*, **76** (1951), 17–20.
[11] On integral operators.* *Uspekhi Mat. Nauk*, **11**, No. 2 (1956), 3–29.
[12] Further applications of Newton's method.* *Vestnik Leningradsk. Univ.*, No. 7 (1957), 68–103.
KANTOROVICH, L. V.; AKILOV, G. P.; RUBINSHTEIN, G. SH.
[1] Extremal states and extremal controls.* *Vestnik LGU*, No. 7 (1957), 30–37.
KANTOROVICH, L. V.; GAVURIN, M. K.
[1] The application of mathematical methods to problems in the analysis of freight transport.* In *Problems of increasing transportation effectiveness* (Russian). Izdat. Akad. Nauk SSSR, 1949, pp. 110–138.
KANTOROVICH, L. V.; RUBINSHTEIN, G. SH.
[1] On a function space and certain extremal problems.* *Dokl. Akad. Nauk SSSR*, **115**, No. 6 (1957), 1058–1061.
[2] On a space of completely additive functions. *Vestnik LGU*, No. 7 (1958), 52–59.
KANTOROVICH, L. V.; VULIKH, B. Z.
[1] Sur la représentation des opérations linéaires. *Comp. Math.*, **5** (1937), 119–165.
KARPILOVSKAYA, E. B.
[1] On the convergence of interpolational methods for ordinary differential equations.* *Uspekhi Mat. Nauk*, **8**, No. 3 (1953), 111–118.
[2] On the convergence of the method of subregions for ordinary integro-differential equations.* *Sibirsk. Mat. Zh.*, **4**, No. 3 (1963), 632–640.
KELDYSH, M. V.
[1] On B. G. Galerkin's method for the solution of boundary problems.* *Izv. Akad. Nauk SSSR, Ser. Mat.*, **6** (1942), 309–330.
KELLEY, J.
[1] Banach spaces with the extension property. *Trans. Amer. Math. Soc.*, **72**, No. 2 (1952), 323–326.
KHARRIK, I. YU.
[1] On an analogue of Markov's inequality.* *Dokl. Akad. Nauk SSSR*, **106**, No. 2 (1956), 203–206.
[2] On the approximation of functions vanishing on the boundary of a region, together with their partial derivatives, by means of functions of a special type.* *Sibirsk. Mat. Zh.*, **4**, No. 2 (1963), 408–425.
KIS, O.
[1] On the convergence of the method of coincidence. *Acta Math., Acad, Sci. Hung.*, **17** (1966), 3–4.
KOLMOGOROV, A. N.
[1] Zur Normierbarkeit eines allgemeinen topologischen linearen Raumes, *Studia Math.*, **5** (1934), 29–33.
KOROTKOV, V. B.
[1] Integral operators with kernels satisfying the conditions of Carleman and Akhiezer, I.* *Sibirsk. Mat. Zh.*, **12**, No. 5 (1971), 1041–1055.
[2] Integral representations of linear operators.* *Sibirsk. Mat. Zh.*, **15**, No. 3 (1974), 529–545.
KOSHELEV, A. I.
[1] Newton's method and generalized solutions of non-linear equations of elliptic type.* *Dokl. Akad. Nauk SSSR*, **91**, No. 6 (1953), 1263–1266.

[2] The existence of a generalized solution of the elastico-plastic torsion problem.* *Dokl. Akad. Nauk SSSR*, **99**, No. 3 (1954), 357–360.

KRASNOSEL'SKII, M. A.

[1] Some problems of non-linear analysis.* *Uspekhi Mat. Nauk*, **9**, No. 3 (1954), 57–114.

[2] On some new fixed point principles.* *Dokl. Akad. Nauk SSSR*, **208**, No. 6 (1973), 1280–1281.

[3] On quasi-linear operator equations.* *Dokl. Akad. Nauk SSSR*, **214**, No. 4 (1974), 761–764.

KRASNOSEL'SKII, M. A.; CHECHIK, V. A.

[1] On a theorem of L. V. Kantorovich.* *Proc. seminar on functional anal.*, vols 3–4, Rostovsk. n/D. Univ., VGU (1960), 50–53.

KRASNOSEL'SKII, M. A.; KREIN, S. G.

[1] An iterative process with minimal discrepancies.* *Mat. Sb.*, **31** (1952), 315–334.

KREIN, S. G.; PETUNIN, YU. I.

[1] Scales in Banach spaces.* *Uspekhi Mat. Nauk*, **21**, No. 2 (1966), 89–168.

KREIN, S. G.; RUTMAN, M. A.

[1] Linear operators leaving invariant a cone in a Banach space.* *Uspekhi Mat. Nauk (N.S)*, **3**, No. 1 (1948), 3–95. *Amer. Math. Soc. Transl.*, No. 26 (1950).

LERAY, J.; SCHAUDER, YU.

[1] Topology and functional equations.* *Uspekhi Mat. Nauk*, **1**, No. 3–4 (1946), 71–95.

LEVIN, A. YU.; STRYGIN, V. V.

[1] On the rate of convergence of the Newton–Kantorovich method.* *Uspekhi Mat. Nauk*, **17**, No. 3 (1962), 185–187.

LEVIN, V. L.

[1] The problem of translocation of mass.* *Dokl. Akad. Nauk SSSR*, **224**, No. 5 (1975), 1016–1019.

[2] Extremal problems with convex functionals semi-continuous from below with respect to convergence in measure.* *Dokl. Akad. Nauk SSSR*, **224**, No. 6 (1975), 1256–1259.

[3] Duality theorems for the Monge–Kantorovich problem.* *Uspekhi Mat. Nauk*, **32** (1977).

[4] The Monge–Kantorovich problem on translocation of masses.* In *Methods of functional analysis in mathematical economics* (Russian). "Nauka", Moscow, 1977.

LOMONOSOV, V. I.

[1] On invariant subspaces of a family of operators commuting with the completely continuous operators.* *Funkts. Analiz i ego Pril.* **7**, No. 3 (1973), 55–56.

LOZANOVSKII, G. YA.

[1] On isomorphic Banach lattices.* *Sibirsk. Mat. Zh.*, **10**, No. 1 (1969), 93–98.

[2] On certain Banach lattices.* *Sibirsk. Mat. Zh.*, **10**, No. 3 (1969), 584–599.

[3] On localized functionals in vector lattices.* In *Theory of functions, functional analysis and their applications* (Russian). Khar'koy, vol. 19 (1974), 66–80.

LOZINSKII, S. M.

[1] The spaces $\tilde{C}_\omega$ and $\tilde{C}^*_\omega$ and the convergence of interpolational processes in them.* *Dokl. Akad. Nauk SSSR*, **59**, No. 8 (1948), 1389–1392.

[2] Inverse functions, implicit functions and solutions of equations.* *Vestnik LGU*, No. 7 (1957), 131–142.

LUXEMBURG, W. A. J.

[1] Notes on Banach function spaces. *Proc. Acad. Sci. Amsterdam*, A **68** (1965), 229–248, 415–446, 646–667.

LUXEMBURG, W. A. J.; ZAANEN, A. C.

[1] Notes on Banach function spaces. *Proc. Acad. Sci. Amsterdam*, A **66** (1963), 135–153, 239–263, 496–504, 655–681; A **67** (1964), 104–119, 360–376, 493–543.

LYUBICH, YU. I.; MAISTROVSKII, G. D.

[1] A general theory of relaxation processes for convex functionals.* *Uspekhi Mat. Nauk*, **25**, No. 1 (1970), 57–112.

MYSOVSKIKH, I. P.

[1] On the convergence of L. V. Kantorovich's method of solution of functional equations and its application.* *Dokl. Akad. Nauk SSSR*, **70**, No. 4 (1950), 565–568.

[2] On the boundary-value problem for the equation $\Delta u = k(x, y)u^2$.* *Dokl. Akad. Nauk SSSR*, **94**, No. 6 (1954), 995–998.

NACHBIN, L.

[1] A theorem of the Hahn–Banach type for linear transformations. *Trans. Amer. Math. Soc.*, **68**, No. 1 (1950), 28–46.

NATANSON, I. P.

[1] Some non-local theorems on singular integrals.* *Dokl. Akad. Nauk SSSR*, **19**, No. 5 (1938), 357–360.

VON NEUMANN, J.

[1] Allgemeine Eigenwerttheorie Hermitescher Funktionaloperatoren. *Math. Ann.*, **102** (1929), 49–131.

NIKOLAEV, V. F.
[1] On the problem of approximating continuous functions by polynomials.* *Dokl. Akad. Nauk SSSR*, **61**, No. 2 (1948), 201–204.

NIKOLAEVA, G. A.
[1] On the approximate construction of a conformal transformation by the method of conjugate trigonometric series.* *Trudy Mat. Inst. im. V. A. Steklova*, **53** (1959), 236–266.

NIKOL'SKII, S. M.
[1] Linear equations in normed linear spaces.* *Izv. Akad. Nauk SSSR, Ser. Mat.*, **7**, No. 3 (1943), 147–163.

PELCHINSKII, A.; FIGEL', T.
[1] On Enflo's method for constructing Banach spaces without the approximation property.* *Uspekhi Mat. Nauk*, **28**, No. 6 (1973), 95–108.

PHILLIPS, R. S.
[1] On weakly compact subsets of a Banach space. *Amer. J. Math.*, **65** (1943), 108–136.

POLYAK, B. T.
[1] Gradient methods for solving equations and inequalities.* *Zh. Vychisl. Mat. i Mat. Fiz.*, **4**, No. 6 (1964), 995–1005.
[2] A general method for solving extremal problems.* *Dokl. Akad. Nauk SSSR*, **174**, No. 1 (1967), 33–36.

PUGACHEV, B. P.
[1] On the convergence of methods locally close to the Newton–Kantorovich method.* *Proc. seminar on functional anal.*, vol. 7, *Voronezh, Izdat. VGU* (1963), 130–136.

RELLICH, F.
[1] Spektraltheorie in nichtseparablen Räumen. *Math. Ann.*, **110** (1934), 342–356.

RIESZ, F.
[1] Sur les opérations fonctionelles linéaires. *C. R. Acad. Sci. (Paris)*, **149** (1909), 974–977.
[2] Untersuchung über Systeme integrierbarer Funktionen. *Math. Ann.*, **69** (1910), 449–497.
[3] *Leçons sur les systèmes d'équations linéaires a une infinité d'inconnues.* Gauthier-Villars, Paris, 1913.
[4] Sur la décomposition des opérations fonctionelles. *Atti Congresso Bologna*, **3** (1928), 143–148.
[5] Sur quelques notions fundamentales dans la théorie générale des opérations linéaires. *Ann. of Math.*, **41** (1940), 174–206.

RUBINSHTEIN, G. SH.
[1] Some examples of dual extremal problems.* In *Mathematical programming* (Russian), "Nauka", Moscow, 1966.
[2] Duality in mathematical programming and some problems in convex analysis.* *Uspekhi Mat. Nauk* **25**, No. 5 (155) (1970).
[3] On translocation of mass in a compactum. *Dokl. Akad. Nauk SSSR* **223**, No. 3 (1975), 572–575.

RVACHEV, M. A.
[1] On the problem of translocation of mass.* In *Optimizatsiya*, vol. 9 (26), Novosibirsk, 1974, 203–208.

SAMOKISH, B. A.
[1] An investigation of the rate of convergence of the method of steepest descent.* *Uspekhi Mat. Nauk*, **12**, No. 1 (1957), 238–240.
[2] The method of steepest descent in the problem of eigenvalues for semibounded operators.* *Izv. VUZ, Mat.*, No. 5 (1958), 105–114.

SCHAUDER, J.
[1] Zur Theorie stetiger Abbildungen in Funktionalräumen. *Math. Zeit.*, **26**, No. 1 (1927), 47–65.
[2] Über lineare vollstetige Operatoren. *Studia Math.*, **2** (1930), 183–196.
[3] Über den Zusammenhang zwischen der Eindeutigkeit und Lösbarkeit partieller Differentialgleichungen zweiter Ordnung vom elliptischen Typus. *Math. Ann.*, **106** (1932), 611–721.
[4] Das Anfangswertproblem einer quasilinearen hyperbolischen Differentialgleichung zweiter Ordnung in beliebiger Anzahl von unabhängigen Veränderlichen. *Fund. Math.*, **24** (1935), 213–246.

SEDAEV, A. A.
[1] On a problem of G. Ya. Lozanovskii.* *Trudy NIIM VGU, Voronezh*, vol. 14 (1974), 63–67.

SEMENOV, E. M.
[1] Embedding theorems for Banach spaces of measurable functions.* *Dokl. Akad. Nauk SSSR*, **156**, No. 6 (1964), 1292–1295.

SIRVINT, YU. F.
[1] Convex sets and linear functionals in abstract spaces.* *Izv. Akad. Nauk SSSR, Ser. Mat.*, **6** (1942), 143–170, 189–211.

SMOLITSKII, KH. L.
[1] On the summability of potentials. *Uspekhi Mat. Nauk*, **12**, No. 4 (1957), 349–356.

STEINHAUS, H.
[1] Additive und stetige Funktionaloperatoren. *Math. Zeit.*, **5** (1919), 186–221.

STEPANOV, V. V.
[1] *A course on differential equations.** 8th ed., Fizmatgiz, Moscow, 1959.
SUDAKOV, V. N.
[1] Geometrical problems in the theory of infinite-dimensional probability distributions.* *Trudy MIAN*, **141**, "Nauka", Leningrad, 1976.
TROITSKAYA, E. A.
[1] On eigenvalues and eigenvectors of completely continuous operators.* *Izv. VUZ, Mat.*, No. 3 (1961), 148–156.
TULAIKOV, A. N.
[1] Zur Kompaktheit im Raum L_p für $p = 1$. *Nachr. Ges. Wiss. Göttingen*, **1**, No. 39 (1933), 167–170.
VEKSLER, A. I.
[1] On the localness of vector lattices of functions.* *Sibirsk. Mat. Zh.* **12**, No. 1 (1971), 54–64. Transl.: *Siberian Math. J.*, **12** (1971), 39–46.
VERSHIK, A. M.
[1] Some remarks on infinite-dimensional problems of linear programming.* *Uspekhi Mat. Nauk*, **25**, No. 5 (1970), 117–124.
[2] On D. Ornstein's papers, weak dependence conditions and classes of stationary measures.* "*Teoriya veroyatnostei i ee primeneniya*", **21**, No. 3 (1976). Transl.: *Theory Probab. Appl.*, **21** (1976), 655–657 (1977).
VERTGEIM, B. A.
[1] On some methods of solution of non-linear operator equations in Banach spaces. *Uspekhi Mat. Nauk*, **12**, No. 1 (1957), 166–169.
VLADIMIROV, V. S.
[1] *Mathematical problems of the one-velocity theory of particle transport.** "Nauka", Moscow, 1961. Transl.: Atomic Energy of Canada Ltd., Chalk River, Ont., 1963.
ZABREIKO, P. P.
[1] Ideal spaces of functions.* *Vestnik Yarosl. Univ.*, vol. 8 (1974), 12–52.

SUBJECT INDEX

INDEX OF NOTATION

Spaces

Other notation

INDEX OF ABBREVIATIONS